1st International School and Conference "Saint-Petersburg OPEN 2014" on Optoelectronics, Photonics, Engineering and Nanostructures

Journal of Physics: Conference Series Volume 541

St. Petersburg, Russia
25-27 March 2014

ISBN: 978-1-63439-593-9
ISSN: 1742-6588

Printed from e-media with permission by:

Curran Associates, Inc.
57 Morehouse Lane
Red Hook, NY 12571

Some format issues inherent in the e-media version may also appear in this print version.

Printed by Curran Associates, Inc. (2015)

For permission requests, please contact the Institute of Physics at the address below.

Institute of Physics
Dirac House, Temple Back
Bristol BS1 6BE UK

Phone: 44 1 17 929 7481
Fax: 44 1 17 920 0979

techtracking@iop.org

Additional copies of this publication are available from:

Curran Associates, Inc.
57 Morehouse Lane
Red Hook, NY 12571 USA
Phone: 845-758-0400
Fax: 845-758-2634
Email: curran@proceedings.com
Web: www.proceedings.com

Table of contents

Accepted papers received: 4 September 2014
Published online: 27 October 2014

Preface

011001

1st International School and Conference "Saint–Petersburg OPEN 2014" on Optoelectronics, Photonics, Engineering and Nanostructures OPEN ACCESS

011002

Peer review statement OPEN ACCESS

Papers

Invited Speakers

012001

Recipes for crystal phase design in Au-catalyzed III-V nanowires OPEN ACCESS V G Dubrovskii and J Grecenkov 1

012002

Semiconductors and nanostructures-based sources of terahertz radiation OPEN ACCESS D Firsov 10

012003

A new method of replacement atoms for the synthesis of epitaxial layers of SiC on Si: From theory to practice OPEN ACCESS S A Kukushkin and A V Osipov 15

Crystal Growth

012004

Pulsed laser deposition and investigation of antimony-doped ZnO films

A S Puzikov, N V Lyanguzov and E M Kaidashev 24

012005

Self-arrangement of periodic layers of silver nanoparticles in silicate glass

A V Red'kov, A A Lipovskii and V V Zhurikhina 28

012006

Express–control of biological solutions by portable nuclear – magnetic spectrometer
OPEN ACCESS V Davydov, A Cheremiskina, E Velichko and A Karseev 34

012007

Initial stages of misfit stress relaxation by rectangular prismatic dislocation loops in composite nanostructures OPEN ACCESS M Yu Gutkin and A M Smirnov 38

012008

A finite-length capped single-walled carbon nanotube (5, 5) under an applied electric field OPEN ACCESS A V Tuchin, A V Nestrugina, L A Bityutskaya and E N Bormontov 43

012009

Studying field emission characteristics of point and wedge–shaped surface defects OPEN ACCESS K A Nikiforov and A N Zartdinov 48

012010

Influence of process pressure on β-SiC growth by CVD OPEN ACCESS A A Andreev, A O Sultanov, A S Gusev, N I Kargin and E P Pavlova 52

012011

Oxide nanofilms on copper and silver: anodic synthesis and semiconducting properties
OPEN ACCESS D S Eliseev, S N Grushevskaya, A V Vvedenskii and D A Kudrayshov 57

012012

Automated method of measuring the temperature dependences of the dielectric parameters of ferroelectrics with second kind phase transition OPEN ACCESS D V Ryabov, E A Pecherskaya, J V Shepeleva and R M Pecherskaya 61

012013

Two-scale model of hydrothermal synthesis of nanotubes OPEN ACCESS D A Kucher, S A Chivilikhin, I Yu Popov and V V Gusarov 65

012014

Modelling the interaction of nanotube with particles of the medium by the Green's function method OPEN ACCESS A V Tulub and D A Nevsky 70

012015

Synthesis of core-shell Ag/SiO_2 nanoparticles for SPASER structures OPEN ACCESS D V Viazmitinov, L B Matyushkin and A I Maximov 74

012016

Model of porous aluminium oxide growth during initial stage of anodization OPEN ACCESS E M Aryslanova, A V Alfimov and S A Chivilikhin 78

012017

Self-formation of lead telluride nanostructures during argon plasma etching of single-crystal wafers OPEN ACCESS G A Dubov, S P Zimin, E S Gorlachev, I I Amirov, V V Naumov and G Z Bagiyeva 83

012018

Resonance properties of metallic nanocantilevers OPEN ACCESS I V Uvarov, V V Naumov and I I Amirov 87

012019

Transverse optical phonon dispersion for multi-layer graphene OPEN ACCESS I Kurkina and S Smagulova 92

012020

Studying fringe field effect of a field emitter array OPEN ACCESS M F Sayfullin and K A Nikiforov 96

012021

Optical spectroscopy of organic materials based on C_{60}<A_2B_6> OPEN ACCESS M Elistratova, N Romanov, I Zakharova and O Kvyatkovskii 100

012022

Structural and optical properties of the bismuth-containing quartz-like glasses OPEN ACCESS M A Girsova, S V Firstov and T V Antropova 106

012023

Permittivity of anisotropic dielectric near surface with local field effects OPEN ACCESS M N Anokhin, A A Tishchenko, M I Ryazanov and M N Strikhanov 119

012024

Local field effects in anisotropic metamaterials OPEN ACCESS O V Porvatkina, A A Tishchenko and M N Strikhanov 123

012025

Modelling of ion convection during electrochemical oxidation of aluminium OPEN ACCESS P Zun and A Svitenkov 127

012026

Low temperature growth of the epitaxial Ge layers on Si(100) by Hot Wire Chemical Vapor Deposition OPEN ACCESS S A Matveev, S A Denisov, D V Guseinov, V N Trushin, A V Nezhdanov, D O Filatov and V G Shengurov 131

012027

Transparent conductor based on aluminum nanomesh OPEN ACCESS B Kazarkin, A S Mohammed, A Stsiapanau, S Zhuk, Y Satskevich and A Smirnov 136

012028

Drift mobility in thin organic layers: joint Monte-Carlo and analytic modeling OPEN ACCESS V R Nikitenko, V M Sukharev and N A Sannikova 142

012029

Photoelectrical characteristics of Ga_2O_3-GaAs structures OPEN ACCESS V V Vishnikina, V M Kalygina, Yu S Petrova and I A Prudaev 146

012030

Features and benefits of III-N growth by ammonia-MBE and plasma assisted MBE OPEN ACCESS A N Alexeev, V P Chaly, D M Krasovitsky, V V Mamaev, S I Petrov and V G Sidorov 150

012031

Synthesis and characterization of ZnO nanorods obtained by catalyst-free thermal technique OPEN ACCESS V I Pushkariov, A L Nikolaev and E M Kaidashev 154

012032

Investigation of defects formation on different stages of sapphire crystals growth OPEN ACCESS S P Malyukov and Y V Klunnikova 158

012033

Photoelectrical properties of TiO_2-Si structures OPEN ACCESS Yu S Petrova, A N Zarubin, V M Kalygina and S Yu Zupiy 163

012034

The application of the modified Blumlein scheme for electrodischarge obtainment of nanopowders in liquids OPEN ACCESS Y Izotov and A Tikhomirov 168

012035

Model of electron transport in cell of a thin-film vacuum nanotriode OPEN ACCESS Z R Gallyamov and K A Nikiforov 172

Nanobiotechnology

012036

Localization microscopy study of FtsZ structures in *E. coli* cells during SOS-response OPEN ACCESS A D Vedyaykin, A V Sabantsev, I E Vishnyakov, S N Borchsenius, Y V Fedorova, A S Melnikov, P Yu Serdobintsev and M A Khodorkovskii 177

012037

Complexes of carbon nanotubes with oligonucleotides in thin Langmuir-Blodgett films to detect electrochemically hybridization OPEN ACCESS A S Egorov, V P Egorova, H V Krylova, I V Lipnevich, T I Orekhovskaya, A A Veligura, M I Govorov and B G Shulitsky 182

012038

Photoelectric properties of MSM structure based on ZnO nanorods, received by thermal evaporation and carbothermal syntesis OPEN ACCESS D A Zhilin, N V Lyanguzov, L A Nikolaev, V I Pushkariov and E M Kaidashev 186

012039

Optoelectronic method for analysis of biomolecular interaction dynamics OPEN ACCESS E Nepomnyashchaya, E Velichko, E Aksenov and T Bogomaz 190

012040

Different photodynamic effect between continuous wave and pulsed laser irradiation modes in k562 cells in vitro OPEN ACCESS V V Klimenko, A A Bogdanov, N A Knyazev, A A Rusanov and M V Dubina 195

012041

Angular measurements of light scattered by the glucose containing biological tissues and their phantoms OPEN ACCESS K G Domnin, G A Cherevatenko and E T Aksenov 199

012042

Specialized probes with nanowhisker structures for scanning probe microscopy OPEN ACCESS M V Zhukov, I V Kukhtevich, V V Levichev, I S Mukhin and A O Golubok 205

012043

Investigation of blood cell properties using laser diffractometry and kinetic nephelometry OPEN ACCESS S A Kavalenka, A I Kavalenka, V V Popov, T D L Nguyen and V A Loban 211

012044

Influence of a diffusion mass flux on instability development of surface shape of high-viscosity liquid OPEN ACCESS T Pogosian and S Chivilikhin 217

012045

Photoinduced modification of quantum dot optical properties affects bacteriorhodopsin photocycle in a (quantum dot)- bacteriorhodopsin hybrid material OPEN ACCESS V A Krivenkov, D O Solovyeva, P S Samokhvalov, K I Brazhnik, G E Kotkovskiy, A A Chistyakov, E P Lukashev and I R Nabiev 221

Optoelectronics and Lasers

012046

Investigation the bit rate of quantum key using Si single photon detectors OPEN ACCESS A V Kolyako, I G Neizvestny and V L Kurochkin 226

012047

Influence of external electrooptical modulator biasing on gain and nonlinear distortions in analog fiber-optic links OPEN ACCESS A Petrov, I Ilichev, P Agruzov, V Lebedev, E Velichko and A Shamray 231

012048

Analysis of efficiency of solar energy conversion by tandem $Cd_xZn_{1-x}Te/Si$ solar cell OPEN ACCESS D V Grigoryev, K A Lozovoy and A A Pishchagin 235

012049

Lasing in microdisk resonators with InAs/InGaAs quantum dots transferred on a silicon substrate OPEN ACCESS E I Moiseev, N V Kryzhanovskaya, A M Nadtochiy, M V Maximov, A E Zhukov, I I Shostak, M M Kulagina, A V Savelyev, K A Vashanova and A A Lipovskii 239

012050

Noise reduction methods of single photon detector based on InGaAs/InP avalanche photodiodes OPEN ACCESS G K Krivyakin, A S Pleshkov, A V Zverev, I I Ryabtsev and V L Kurochkin 244

012051

Application of the time-space distributed model for the description of Q-switched regime in solid-state lasers with a saturable absorber OPEN ACCESS M Belov, L Burov and L Krylova 250

012052

Thermal lensing in diode pumped $Nd^{3+}:YVO_4/LiNbO_3$ green microchip laser OPEN ACCESS M Kerobyan, R Kostanyan, V Nersesyan and S Soghomonyan 254

012053

Injection currents in thin disordered organic films OPEN ACCESS V R Nikitenko, N A Sannikova and M N Strikhanov 258

012054

Study of the characteristics of UVA LEDs grown by HVPE: active region thickness-dependent performance OPEN ACCESS E A Menkovich, S A Tarasov, I A Lamkin, A V Solomonov, S Yu Kurin, A A Antipov, I S Barash, A D Roenkov, A S Usikov, H I Helava and Yu N Makarov 262

012055

Comparative analysis of efficiency droop in InGaN/GaN light- emitting diodes for electrical and optical pumping conditions OPEN ACCESS V V Kopyev, I A Prudaev and I S Romanov 267

012056

The analytical approach to optimization of active region structure of quantum dot laser OPEN ACCESS V V Korenev, A V Savelyev, A E Zhukov, A V Omelchenko and M V Maximov 271

012057

Study of electric field distribution in AlGalnP light-emitting diode by Kelvin probe force microscopy OPEN ACCESS V L Oleynik, I A Prudaev, V A Novikov, Yu L Ryaboshtan and P V Gorlachuk 276

Photonics

012058

Influence of magnetic field on the surface waves properties in the photonic crystal / graphene structure for terahertz frequency range OPEN ACCESS A Denisultanov, S Azbite and M Khodzitsky 280

012059

Investigation of device imperfection influence on measurement results in beamsplitter-based quantum random number generation schemes OPEN ACCESS A E Ivanova, S A Chivilikhin, V I Egorov and A V Gleim 286

012060

Optimization of the structure nanoporous TiO_2 film in a dye- sensitized solar cell OPEN ACCESS S P Malyukov, I V Kulikova, A V Sayenko and Yu V Klunnikova 291

012061

Bent and branched chains of nanoresonators OPEN ACCESS A S Melikhova and I Yu Popov 297

012062

Detection of beamsplitting attack in a quantum cryptographic channel based on photon number statistics monitoring OPEN ACCESS A A Gaidash, V I Egorov and A V Gleim 302

012063

An analytical method for determining the interaction energy between multiple identical spherical colloidal zinc oxide nanoparticles OPEN ACCESS A V Alfimov, E M Aryslanova and S A Chivilikhin 306

012064

Plasmon-Exciton Interaction in AuNP-Phtalocyanine Core/Shell Nanostructures OPEN ACCESS A Zasedatelev, A Karpo, I Feofanov, V Krasovskii and V Pushkarev 312

012065

Collimation and focusing of initially single-cycle paraxial optical beams OPEN ACCESS D N Puzyrev and A A Drozdov 318

012066

Determining influence of four-wave mixing effect on quantum key distribution OPEN ACCESS D N Vavulin, V I Egorov, A V Gleim and S A Chivilikhin 324

012067

Simulation of beam-splitter made of metamaterials with angle spatial distribution of constitutive parameters based on transformation optics for THz frequency range OPEN ACCESS E A Gurvitz, A V Vozianova and M K Khodzitsky 330

012068

Transmission of few-cycle pulses by Fabry-Perot interferometer OPEN ACCESS E Buyanovskaya and E Kryshkovets 335

012069

Band-gap structure of photonic crystal with metasurface-teflon layers OPEN ACCESS E A Sedykh, A K Denisultanov and M K Khodzitsky 339

012070

Raman scattering in conducting metal-organic films deposited on nanoporous anodic alumina OPEN ACCESS E N Golubeva, H V Grushevskaya, N G Krylova, I V Lipnevich and T I Orekhovskaya 345

012071

Light bullets from Mid-IR femtosecond filament in air OPEN ACCESS E O Smetanina, V Yu Fedorov, A E Dormidonov and V P Kandidov 350

012072

Enhanced third harmonic generation using the surface states of light in periodic photonic structures OPEN ACCESS K I Zaytsev, V S Gorelik, G M Katyba and S O Yurchenko 355

012073

SERS-applicable silver nanoisland film grown under protective coating OPEN ACCESS I Reduto, S Chervinskii, A Matikainen, A Baklanov, A Kamenskii and A Lipovskii 362

012074

Fabrication and optical properties of silicon nanopillars OPEN ACCESS L S Golobokova, Yu V Nastaushev, F N Dultsev, D V Gulyaev, A B Talochkin and A V Latyshev 368

012075

Spectral properties of random LC networks with uniformly distributed entries OPEN ACCESS N A Olekhno, Y M Beltukov and D A Parshin 373

012076

Plasmonic resonance of colloidal silver in nanoporous matrix OPEN ACCESS O V Andreeva, S V Saitov, N V Andreeva and A I Sidorov 379

012077

Electric properties of hemispherical metal nanoparticles: influence of the dielectric substrate OPEN ACCESS S A Scherbak, O V Shustova, V V Zhurikhina and A A Lipovskii 384

012078

Implementation of multiplexing in a subcarrier-wave quantum cryptography system OPEN ACCESS V V Chistyakov, A V Gleim, V I Egorov and Yu V Nazarov 390

Quantum Dots and Wells

012079

Edge modes splitting by nanostructured surface OPEN ACCESS A A Zalutskaya and A V Prokaznikov 395

012080

Investigation of $In_{0.7}Ga_{0.3}As/In_{0.7}Al_{0.3}As$ metamorphic HEMT- heterostructures by photoluminescence spectroscopy OPEN ACCESS D S Romanovskiy, S A Tarasov, G B Galiev and S S Pushkarev 399

012081

Intersubband light absorption in tunnel-coupled GaAs/AlGaAs quantum wells for electrooptic studies OPEN ACCESS D A Rybalko, M Ya Vinnichenko, L E Vorobjev, D A Firsov, R M Balagula, V Yu Panevin, M M Kulagina and A P Vasil'iev 406

012082

Far- and near-infrared photoluminescence from n-GaAs/AlGaAs multiple quantum wells OPEN ACCESS L E Vorobjev, D A Firsov, V Yu Panevin, A N Sofronov, R M Balagula, I S Makhov and A P Vasil'ev 410

012083

Effect of growth temperature of GaN:Mg layer on internal quantum efficiency of LED structures with InGaN/GaN quantum wells OPEN ACCESS I S Romanov, I A Prudaev, V V Kopyev, A A Marmalyuk, V A Kureshov, D R Sabitov and A V Mazalov 415

012084

Elongated quantum dots of Ge on Si formation modelling OPEN ACCESS K A Lozovoy, A P Kokhanenko and A V Voitsekhovskiy 419

012085

Determination of InSb/AlInSb quantum well energy spectrum OPEN ACCESS M S Mironova, O S Komkov, D D Firsov and G F Glinskii 423

012086

Cyclometalated complexes of platinum metals – the new luminescent activators of OLED OPEN ACCESS M V Nikolaeva, E A Katlenok, M S Khakhalina, M V Puzyk and K P Balashev 428

012087

Photoinduced absorption and photoconductivity of Ge/Si quantum dots in mid-infrared range under interband excitation OPEN ACCESS R M Balagula, A N Sofronov, V Yu Panevin, D A Firsov, L E Vorobjev, A A Tonkikh and P Werner 434

012088

Electronic energy transfer from the semiconductor quantum wire excitons to an organic media OPEN ACCESS Y A Strokova and M G Kucherenko 439

012089

Size distributions and scaling relations for heterogeneous nucleation and growth of atomic chains OPEN ACCESS Y Berdnikov and V G Dubrovskii 444

Other Aspects of Nanotechnology

012090

Peculiarities of optical and ODMR Spectra of Nitrogen–Vacancy Color Centers in Diamond Crystals OPEN ACCESS A K Dmitriev and A K Vershovskii 450

012091

Investigation of lead-free thin films based on barium titanate for electrocaloric devices OPEN ACCESS E V Abrashova and A K Fominykh 455

012092

Chain of point-like potentials in $\mathcal{R}^3$ and infiniteness of the number of bound states OPEN ACCESS A A Boitsev, I Yu Popov and O V Sokolov 461

012093

Optical phenomena associated with surface plasmon polaritons in GaN-based microstructures OPEN ACCESS D S Mostovoy, G A Melentyev, V A Shalygin, M D Moldavskaya and L E Vorobjev 467

012094

Algorithm of molecular computing on the base of membranes OPEN ACCESS D S Nikiforov and I Yu Popov 471

012095

A change in the electro-physical parameters of narrow-gap CdHgTe solid solutions acted upon by a diffusion discharge in the air, argon and nitrogen at atmospheric pressure OPEN ACCESS D V Grigoryev and E A Stavkov 475

012096

Dielectric spectroscopy of polymer nanocomposites based on tetrazol and KNO_3 OPEN ACCESS R A Castro and E N Lushin 479

012097

Prediction of phase equilibria in the W-Si system in the generalized lattice model OPEN ACCESS G A Panov and M A Zakharov 483

012098

Diffuse reflectance measurements using lensless CMOS imaging chip OPEN ACCESS I Schelkanova, A Pandya, D Shah, L Lilge and A Douplik 487

012099

The discrete spectrum of the multiparticle Hamiltonian in the framework of the Hartree-Fock approximation OPEN ACCESS I F Melikhov and I Yu Popov 495

012100

Assessment of surface roughness of substrates subjected to plasma-chemical etching OPEN ACCESS M S Glyanko, A V Volkov and S A Fomchenkov 499

012101

Quantum state tomography of the microwave field OPEN ACCESS G P Miroshnichenko and O M Korneeva 503

012102

Analysis of applicability of the transport level concept in organic materials, providing correlated disorder OPEN ACCESS O V Porvatkina, V R Nikitenko and M N Strikhanov 509

012103

Dependence of the oscillations amplitude on the thickness of magnetostrictive-piezoelectric bilayer structure in the theory of magnetoelectric effect OPEN ACCESS T A Galichyan and D A Filippov 513

012104

Relaxation of conductivity in amorphous multi-component chalcogenides of copper and silver OPEN ACCESS V E Zaikova, N V Melnikova, K V Kurochka and O L Kheifets 519

012105

Conditional evolution of vacuum state in dynamical Casimir effect OPEN ACCESS Sheremetyev V O, Trifanova E S and Trifanov A I 523

Editorial Preface

Dear Colleagues,

1st International School and Conference "Saint–Petersburg OPEN 2014" on Optoelectronics, Photonics, Engineering and Nanostructures was held on March 25 – 27, 2014 at St. Petersburg Academic University – Nanotechnology Research and Education Centre of the Russian Academy of Sciences. The School and Conference included a series of invited talks given by leading professors with the aim to introduce young scientists with actual problems and major advances in physics and technology. The keynote speakers were

Mikhail Glazov (*Ioffe Physico–Technical Institute RAS, Russia*)

Vladimir Dubrovskii (*Saint–Petersburg Academic University RAS, Russia*)

Alexey Kavokin (*University of Southampton, United Kingdom* and *St. Petersburg State University*)

Vladimir Korenev (*Ioffe Physico–Technical Institute RAS*)

Sergey Kukushkin (*Institute of Problems of Mechanical Engineering RAS, Russia*)

Nikita Pikhtin (*Ioffe Physico–Technical Institute RAS, Russia* and *"Elfolum" Ltd., Russia*)

Dmitry Firsov (*Saint–Petersburg State Polytechnical University, Russia*)

During the poster session all undergraduate and graduate students attending the conference presented their works. Sufficiently large number of participants with more than 160 student attendees from all over the world allowed the Conference to provide a fertile ground for the fruitful discussions between the young scientists as well as to become a perfect platform for the valuable discussions between student authors and highly experienced scientists. The best student papers, which were selected by the Program Committee and by the invited speakers basing on the theses and their poster presentation, were awarded with diplomas of the conference – see the photos.

This year School and Conference is supported by SPIE (The International Society for Optics and Photonics), OSA (The Optical Society), St. Petersburg State Polytechnical University and by Skolkovo Foundation. It is a continuation of the annual schools and seminars for youth on topical problems of physics and technology that is organized by the Academic University since 2009. **We invite all the students and young scientists to attend "Saint-Petersburg OPEN" in 2015!**
Please, find details at *http://spbopen2015.spbau.com/*

With best wishes,
Editorial Board, Program and Organizing Committees

Journal of Physics: Conference Series **541** (2014) 012001 doi:10.1088/1742-6596/541/1/012001

1st International School and Conference "Saint–Petersburg OPEN 2014" on Optoelectronics, Photonics, Engineering and Nanostructures. Conference Group Photo

Conference Organizers

St. Petersburg Academic University – Nanotechnology Research and Education Centre of the Russian Academy of Sciences

SPIE & OSA Student Chapters at St. Petersburg Academic University

St. Petersburg State Polytechnical University

Foundation for support of education and science (Alferov's foundation)

Acknowledgements

The Optical Society

The International Society for Optics and Photonics

Ioffe Physical–Technical Institute of the Russian Academy of Sciences

Skolkovo Foundation

Institute of Physics

St. Petersburg Scientific Center of RAS

Conference Photos

The head of Organizing Committee, Professor Alexey Zhukov, is making opening remarks

Journal of Physics: Conference Series **541** (2014) 012001 doi:10.1088/1742-6596/541/1/012001

The Conference Stage

The talk of Prof. Mikhail Glazov on graphene and its transport properties opens the conference

Journal of Physics: Conference Series **541** (2014) 012001 doi:10.1088/1742-6596/541/1/012001

The talk of Prof. Vladimir Dubrovskii on nanowires

Journal of Physics: Conference Series **541** (2014) 012001
doi:10.1088/1742-6596/541/1/012001

During the lectures

Journal of Physics: Conference Series **541** (2014) 012001 doi:10.1088/1742-6596/541/1/012001

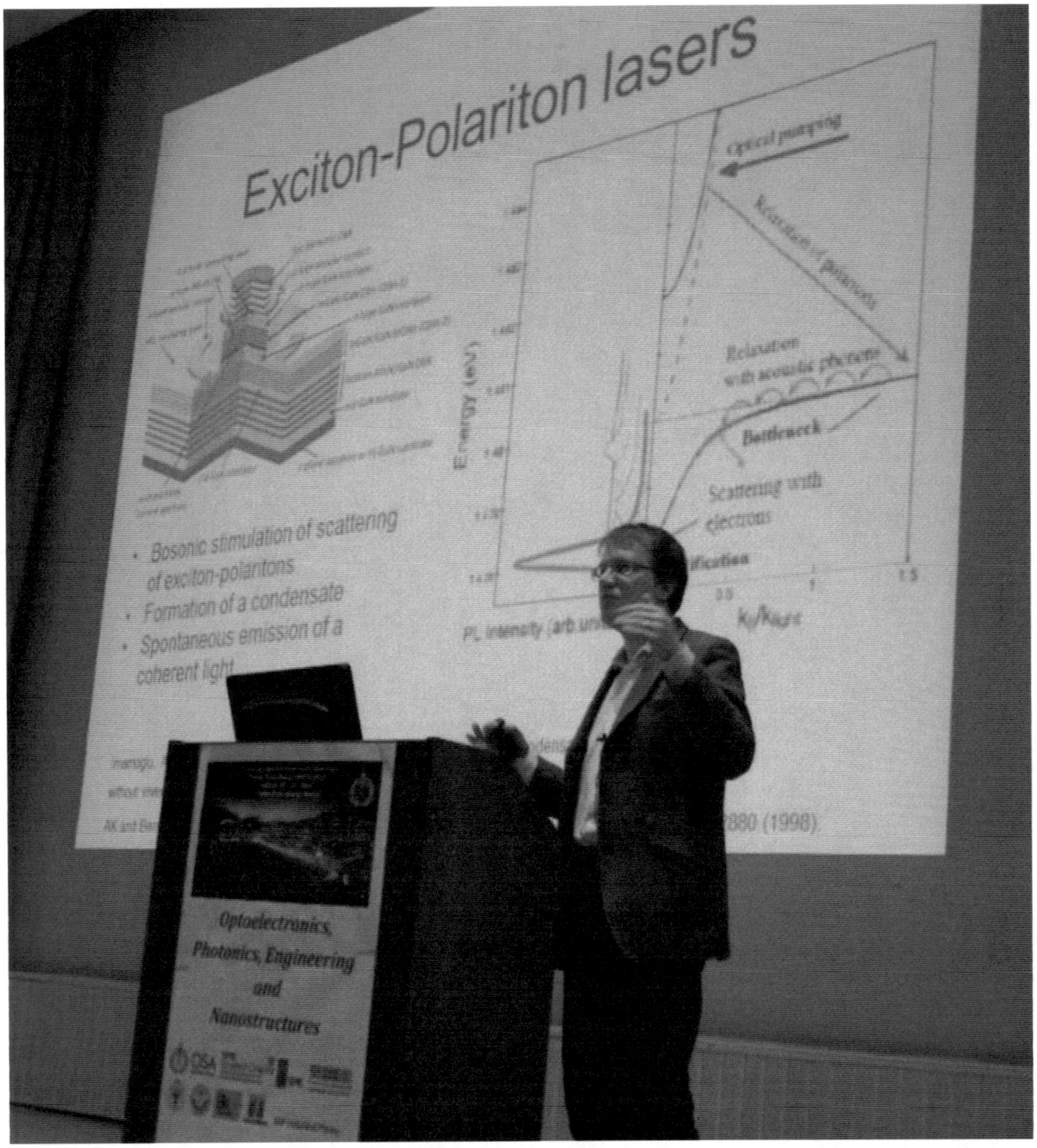

Prof. Kavokin from the University of Southampton delivers his talk on exciton-polariton lasers

Prof. Korenev explains basics of ferromagnetic-based memory

Students and guests of the conference after the break

Journal of Physics: Conference Series **541** (2014) 012001 doi:10.1088/1742-6596/541/1/012001

Prof. Sergey Kukushkin explains the features of production of SiC

Journal of Physics: Conference Series **541** (2014) 012001 doi:10.1088/1742-6596/541/1/012001

Prof. Nikita Pikhtin is talking on semiconductor lasers

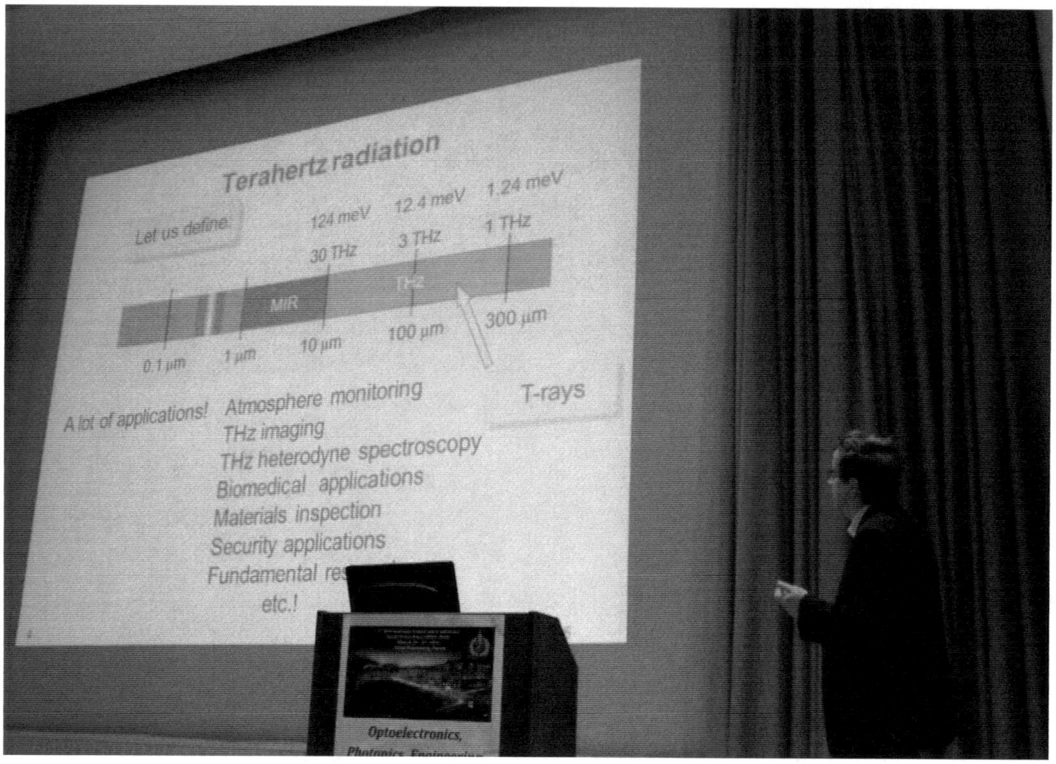

Prof. Dmitry Firsov explains the objectives of studying terahertz lasers

Journal of Physics: Conference Series **541** (2014) 012001 doi:10.1088/1742-6596/541/1/012001

Award ceremony. Prof. Vladimir Dubrovskii and students

Journal of Physics: Conference Series **541** (2014) 012001 doi:10.1088/1742-6596/541/1/012001

Award ceremony. Prof. Vladimir Dubrovskii and students

Journal of Physics: Conference Series **541** (2014) 012001 doi:10.1088/1742-6596/541/1/012001

Award ceremony. Prof. Vladimir Dubrovskii and students

Journal of Physics: Conference Series **541** (2014) 012001 doi:10.1088/1742-6596/541/1/012001

Award ceremony. Prof. Vladimir Dubrovskii and students

Closing remarks from Prof. Dubrovskii and Prof. Maximov

Journal of Physics: Conference Series **541** (2014) 011002 doi:10.1088/1742-6596/541/1/011002

Peer review statement

All papers published in this volume of *Journal of Physics: Conference Series* have been peer reviewed through processes administered by the proceedings Editors. Reviews were conducted by expert referees to the professional and scientific standards expected of a proceedings journal published by IOP Publishing.

Journal of Physics: Conference Series **541** (2014) 012001 doi:10.1088/1742-6596/541/1/012001

Recipes for crystal phase design in Au-catalyzed III-V nanowires

Vladimir G. Dubrovskii[1,2], **and Jurij Grecenkov**[1]

1 - St.Petersburg Academic University, Khlopina 8/3, 194021 St. Petersburg, Russia,
2 - Ioffe Physical Technical Institute RAS, Politekhnicheskaya 26, 194021, St. Petersburg, Russia
3 - ITMO University, Kronverkskiy pr. 49, 197101 St. Petersburg, Russia

E-mail: dubrovskii@mail.ioffe.ru

Abstract. We develop a kinetic model for the crystal structure of Au-catalyzed III-V nanowires which is capable of describing the wurtzite content as a function of the growth temperature, group V flux, and the nanowire elongation rate. The self-consistency condition with the correct pre-exponent in the Zeldovich nucleation rate allows us to estimate the actual group III concentration in the droplet which corresponds to a given elongation rate. Our model predicts a non-monotonic dependence of the wurtzite percentage on the group III concentration. We analyze relevant experimental data on the preferred crystal structure of Au-catalyzed GaAs nanowires obtained by three different epitaxy techniques and explain why the MBE-grown GaAs nanowires at near 550°C are predominantly wurtzite, while the MOCVD and HVPE-grown GaAs nanowires obtained with very high As fluxes and at very different temperatures (464°C and 715°C) are zincblende. Overall, the model provides some simple recipes for structural design of III-V nanowires by tuning the technologically controlled growth conditions such as temperature and fluxes.

1. Introduction

III-V semiconductor nanowires (NWs) offer a unique possibility of sophisticated design of their crystal structure. Stabilization of hexagonal wurtzite (WZ) crystal phase of III-V materials has only been observed in III-V NWs [1]. In the case of III-V NWs fabricated by the Au-catalyzed vapor-liquid-solid (VLS) method, the preferred crystal structure strongly depends on the growth conditions which influence the chemical potential of a liquid Au-III-V alloy in the droplet [2-6]. Recently, the synthesis of NWs containing heterostructures of the zincblende (ZB) and WZ crystal phases was demonstrated within a single nanowire [6-14]. The WZ/ZB superlattices in III-V NWs show very interesting optical properties [8]. In particular, insertions of a smaller bandgap ZB segments into the matrix of WZ III-V NWs yield a principally new type of quantum dots based on different crystal phases of same material rather than a combination of different materials [7]. These crystal phase heterostructures are achieved by tuning the growth parameters such as the group III or group V fluxes and the growth temperature. Such structural design requires advanced modeling of NW growth and the resulting crystal phase. VLS III-V NWs can be obtained not only by the Au-catalyzed VLS procedures but also with a group III metal (Ga) in the self-catalyzed approach [15-19]. For Ga-catalyzed VLS growth of GaAs NWs, it is known that the NW elongation rate is limited by the As flux [17-19], while the crystal structure is predominantly ZB due to the low surface energy of liquid Ga relative to Au [16]. Recently, Glas *et al.* presented a predictive model of Ga-catalyzed, molecular beam epitaxy (MBE) growth of GaAs NWs based entirely on the kinetics of As [19]. In this work, we try to extend this approach to a more complex case of Au-catalyzed III-V NWs. In Au-catalyzed VLS growth, the liquid Au-III-V alloy is ternary rather than binary and the growth rate is commonly believed to be limited by the kinetic of group III atoms, including those collected by the NW sidewalls [6,20]. We use the model of Ref. [21] to circumvent the uncertainty in the As chemical potential in the liquid phase [22] by expressing it through the As flux and the NW elongation rate in the steady state. In the self-consistent approach where the transport-limited NW elongation rate is equated to the nucleation-mediated growth rate in the mononuclear mode [19,21,23], we include the Kashchievs renormalization [24] for the pre-exponential factor in the Zeldovich nucleation rate. This allows us to estimate more precisely the actual Ga concentration in the droplet during growth which corresponds to a particular elongation rate. Finally, we consider the experimental data demonstrating predominantly WZ phase of Au-catalyzed GaAs NWs grown by molecular beam epitaxy (MBE) at 550°C [2,6], pure ZB phase of Au-catalyzed GaAs NWs grown by metal organic chemical vapor deposition (MOCVD) at 464°C [25,26] and Au-catalyzed GaAs NWs grown by hydride vapor phase epitaxy (HVPE) at 715°C [27]. These different crystal structures are explained very well within our model. We speculate that the pure ZB phase of VLS GaAs NWs obtained by chemical vapor epitaxies is due to a very high Ga content in the Au-Ga-As droplets where the necessary condition for the WZ phase formation the so-called triple phase line (TPL) nucleation [3] is suppressed on surface energetic grounds, as in the case of Ga-catalyzed GaAs NWs [16]. This large Ga concentration is required to ensure a high elongation rates (10 nm/s or more) and to compensate for a high As content which originates from a huge group V input in chemical vapor deposition methods compared to the MBE case.

2. Experimental data

The Au-catalyzed GaAs NWs described in detail in Ref. [26] were grown by MOCVD on GaAs(111)B substrates from trimethylgallium (TMGa) and arsine (AsH_3) at a system pressure of 100 Torr, constant surface temperature of 464°C and with V/III ratio of 70. The measured NW elongation rate was found to be radius-independent and equaled approximately 10 nm/s. Transmission electron microscopy (TEM) studies revealed the pure ZB crystal structure of these GaAs NWs regardless of their radius from 80 nm down to 12 nm, as shown in Figs. 1 (a1) and (a2). The NWs were oriented along <111> crystallographic direction, their shape was regular hexahedral, with the side facets defined by six equivalent (211) planes.

Figure 1 (b) shows the TEM image of the body of a GaAs NW obtained by the Au-assisted MBE [2]. These NWs (usually 20-60 nm in radius) show the predominant WZ structure (although with some twins and stacking faults) in the temperature window 530 – 600°C, V/III fluxes ratios of about 3 and typical equivalent planar growth rates of the order of 0.2-0.3 nm/s. The NW shown in Fig. 1 (b) was obtained at 550°C and was found to have WZ side facets. In contrast to the MOCVD case, the NW elongation rate in MBE strongly depends on the NW radius: the typical length-radius correlation is decreasing as in the diffusion-induced growth mode [20]. Collection of sidewall adatoms increases the elongation rate of thinnest NWs up to 1 nm/s. However, this does not change the prevalence of the WZ structure. Similar findings for Au-catalyzed, MBE-grown GaAs NWs are published, e.g., in Refs. [3,20,21] and show the prevalence of the WZ structure. In accordance with experiments [2,3] the model of Ref. [3] shows that the ZB phase can be present in MBE-grown, Au-catalyzed GaAs NWs only at the transient growth steps at the beginning and at the end of growth where the Ga concentration in the droplet is lower than in the steady state. Figures 1 (c1) and (c2) show the defect-free ZB structure of a 60 nm radius GaAs NWs obtained by the Au-catalyzed HVPE from GaCl and As_4 vapors on Au-coated 4-misoriented GaAs(100) substrates in a hot wall

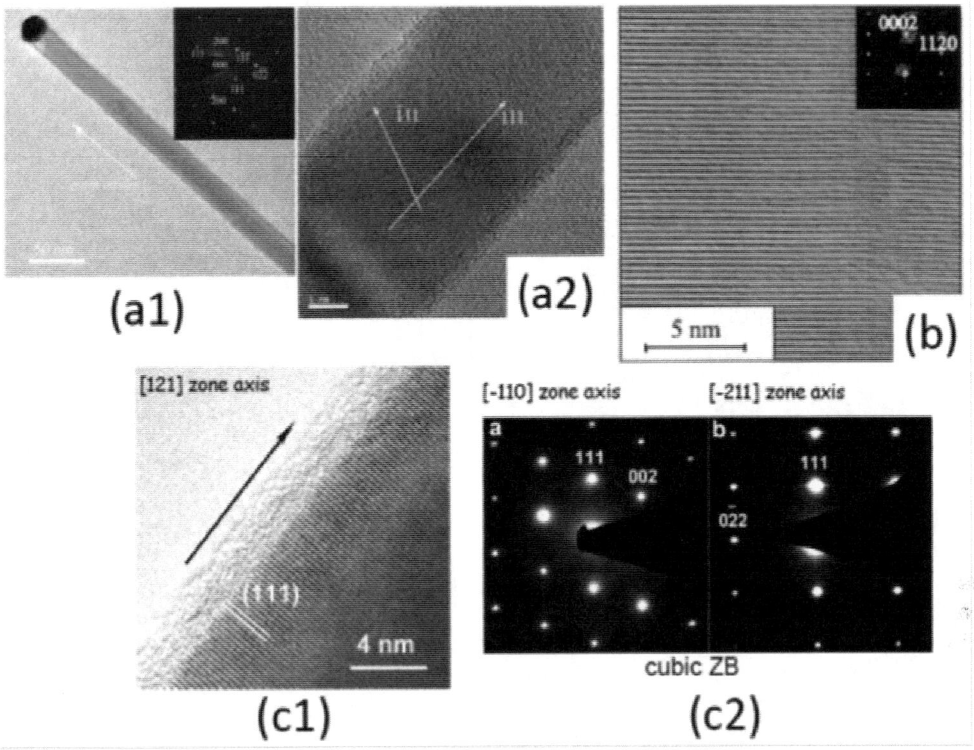

Figure 1. TEM (a1) and high resolution TEM (HRTEM) (a2) images of a <111>-oriented, 12 nm radius GaAs NW grown by the Au-catalyzed MOCVD at 464°C, with the corresponding selected area electron diffraction (SAED) pattern in the insert of (a1) showing its ZB phase [26]. HRTEM image of a WZ GaAs NW grown by MBE at 550°C, with the insert showing the SAED pattern (b) [2]. HRTEM image of a <111>-oriented, 60 nm radius GaAs NW obtained by HVPE at 715°C (c1), SAED patterns viewed in different zone axes (c2) demonstrate the defect-free ZB structure [27].

horizontal reactor kept at atmospheric pressure [27]. These NWs are typically obtained at 715C with a partial pressure GaCl of 300 Pa, a partial pressure of As_4 of 70 Pa and with a H2 vector gas of 3000 SCCM including additional HCl flow used to tune the supersaturation of the vapor phase. These NWs grow in <111> direction and are defined predominantly by (211) side facets. Very importantly, the radius-independent elongation rate of GaAs NWs obtained by Au-catalyzed VLS HVPE reaches the record value of the order of 30 nm/min. Therefore, the Au-catalyzed GaAs NWs grown by MOCVD and HVPE at rather low (464°C) and high (715°C) surface temperatures have the ZB crystal structure, while the MBE-grown GaAs NWs obtained at intermediate temperatures (550°C) are predominantly WZ. Therefore, the growth temperature is not the sole key factor determining the phase, as commonly assumed [6]. The NW radii in these experiments are also close, while the structure is different, so size-dependent effects on the preferred crystal structure [4] should not be important. Hence, the observed difference in the resulting crystal structure must originate from the main features of chemical vapor epitaxies versus MBE. Obviously, MOCVD and HVPE involve much higher amounts of materials deposited per unit time (yielding much larger elongation rates), and huge values of V/III ratios compared to MBE. It is the aim of the foregoing theoretical analysis to attribute the

WZ phase of Au-catalyzed MBE-grown GaAs NWs and ZB phase of MOCVD and HVPE-grown GaAs NWs to these differences in the growth techniques.

3. Chemical potentials

As shown in Ref. [21], if the group V (As) concentration in the liquid Au-Ga-As droplet remains constant during growth, the chemical potential of GaAs in the ternary liquid alloy with respect to the stoichiometric ZB solid state (as defined in Ref. [22]) is given by

$$\Delta\mu = \mu_3^L + \frac{k_B T}{n} ln\left(\frac{\omega_5 v_5 (1+\varepsilon) - \omega_3 dL/dt}{v_5^0}\right) \tag{1}$$

with

$$v_5^0 = \frac{2\omega_3}{1+cos(\beta)} \frac{n\Omega_{35} k_5 P_{5(n)}^0}{\sqrt{2\pi m_{5(n)} k_B T}} \tag{2}$$

Here, $\mu_3^L(c_3, T)$ is the chemical potential of Ga atoms in the alloy, which is determined primarily by the Ga concentration in the droplet and the surface temperature and is not expected to be strongly dependent of the As concentration c_5 in view of $c_5 << c_3$, k_B is the Boltzmann constant, n is the number of As atoms in an $mathrmAs_n(X)$ molecule in vapor, $v_5 = I_5\Omega_{35}$, I_5 is the direct atomic As flux, $\Omega_{35} = 0.0452\,\mathrm{nm}^3$ is the elementary volume per a GaAs pair in the solid state, ε is the As re-emission coefficient [17,19], $\omega_5 = 1/sin^2(\beta)$ in MBE and 1 for chemical vapor deposition, $\omega_3 = 1$ for MBE and $(1+cos(\beta))/2$ for chemical vapor deposition (with β being the contact angle of the droplet at the NW top), dL/dt is the nanowire elongation rate, and μ_{35}^s is the chemical potential of a GaAs pair in the ZB solid state. In Eq. (2), k_5 is the Knudsen adsorption coefficient for As ($k_5 \cong 1$) and $m_{5(n)}$ is the mass of an $\mathrm{As_n(X)}$ molecule in vapor, while the temperature-dependent $P_{5(n)}^0$ is defined as

$$\mu_{5(n)}^v = k_B T ln\left(P_{5(n)}/P_{5(n)}^0\right) \tag{3}$$

where $\mu_{5(n)}^v$ is the chemical potential of an $mathrmAs_n(X)$ perfect gas at the pressure $P_{5(n)}$. According to Refs. [3,4], the prevalence of the WZ phase over ZB in VLS NWs occurs when the $\Delta\mu$ is larger than the two critical chemical potentials. The first one corresponds to the equality of the nucleation barriers for the WZ and ZB nuclei at the TPL and the second to the equality of the nucleation barrier of the WZ nucleus at the TPL and the ZB nucleus in the center (C) of the liquid-solid interface under the droplet. Consequently, the WZ structure forms when

$$\Delta\mu > max\{\Delta\mu_{TPL}; \Delta\mu_C\}$$

$$\Delta\mu_{TPL} = \frac{\Psi_{WZ}}{1-(\Gamma_{WZ}/\Gamma_{ZB})^2}; \quad \Delta\mu_C = \frac{\Psi_{WZ}}{1-(\Gamma_{WZ}/\gamma_{SL})^2} \tag{4}$$

with Ψ_{WZ} being the bulk energy difference between WZ and ZB structures. The effective surface energies of nuclei at the TPL positions are given by [3]: $\Gamma_{ZB} = (\gamma_{ZB} - \gamma_{LV} sin(\beta))x + \gamma_{SL}(1-x)$; $\Gamma_{WZ} = (\gamma_{WZ} - \gamma_{LV} sin(\beta))x + \gamma_{SL}(1-x)$. The γ_{ZB}, γ_{WZ} are the surface energies of relevant ZB and WZ NW sidewall facets, γ_{LV} is the liquid-vapor surface energy of the droplet, γ_{SL} is the surface energy of the lateral solid-liquid interface of a two-dimensional island, and x is the fraction of the island perimeter at the TPL. In the stationary model, we do not consider possible variations of the contact angle β during growth [28]. The critical chemical potentials obtained from Eqs. (4) are shown in Fig. 2 for the following parameters of Au-catalyzed GaAs NWs: $\Psi = 24$ meV/pair [29], $\gamma_{WZ} = 1.30\,\mathrm{J/m^2}$ for the lowest energy$(1\bar{1}00)$ WZ side facets and $\gamma_{ZB} = 1.543\,\mathrm{J/m^2}$ for the lowest energy (110) ZB side facets and $\gamma_{ZB} = 1.79\,\mathrm{J/m^2}$ for the

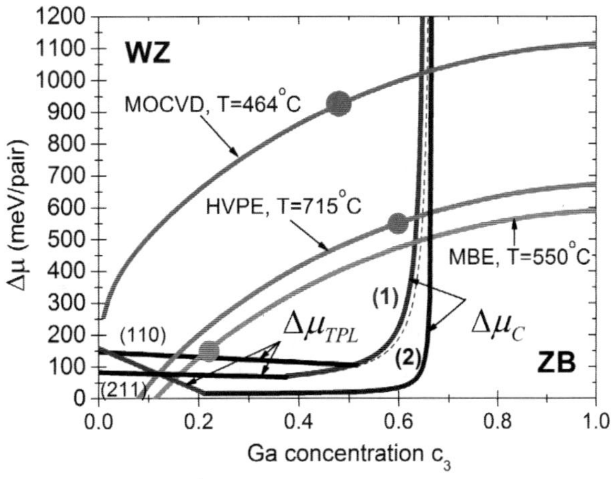

Figure 2. Critical chemical potentials, obtained from Eqs. (4): black curves - $\Delta\mu_{TPL}$ for (100) and (211) ZB side facets, light blue curves - $\Delta\mu_C$ at 715°C (solid line) and 550°C (dotted line), these critical curves show a negligible temperature dependence (1); violet curve - with the Kashchiev renormalization, dark blue curve - $\Delta\mu_{TPL}$ with the Kashchiev renormalization (2). The WZ phase is prevalent in the top-left corner, while the ZB structure forms below the critical curves. Red, magenta and green lines correspond to the chemical potentials in Au-Ga-As droplets for GaAs NWs by HVPE at 715°C, MBE at 550°C, and MOCVD at 464°C, calculated from Eq. (1) for the parameters listed in Table 1. Dots show the actual Ga concentrations during growth, obtained from the self-consistency condition. GaAs NWs obtained by HVPE and MOCVD form at high Ga concentrations larger than 0.67, where the WZ phase is suppressed on surface energetic grounds (the TPL nucleation becomes unfavorable). MBE-grown GaAs NWs are formed at a much lower Ga concentrations and hence their preferred crystal structure is WZ.

more energetic (211) side facets [30], $\beta = 110^o$, and a triangle island ($x=$ 0 or 1/3). The dependence $\gamma_{LV}(c_3)$ is taken as the linear approximation between pure liquids ($\gamma_{Ga} = 0.663\,\mathrm{J/m^2}$, $\gamma_{Au} = 1.215\,\mathrm{J/m^2}$at715°C and $\gamma_{Au} = 1.248\,\mathrm{J/m^2}$at550°C (Ref. [31])). The is the linear approximation between 1.0 J/m2 at $c_3 = 0.2$ (Ref. [3]) and 0.123 J/m2 for a pure Ga droplet (at $c_3 = 1$) [19]. It is seen that the critical chemical potentials are almost temperature-independent (due to a slow temperature dependence of the liquid and solid surface energies). Consequently, we have a temperature-independent WZ corner in the top-left part of the plane, and the WZ phase forms when the liquid chemical potential during growth enters this corner. Very importantly, the ZB phase should form for all $c_3 > 0.67$ actually regardless of the growth conditions such as the arsenic flux and the NW elongation rate. We also note that the first principle calculations by Rosini and Magri [32] and Pankoke *et al.* [33] give much lower values for the sidewall surface energies and compared to those used in Fig. 2, which could lead to a very different predictions regarding the preferred crystal structure. This will be studied in detail elsewhere. The increasing curves in Fig. 2 show the dependences of the chemical potentials on the Ga concentration, calculated for the growth parameters listed in Table 1, with the Ga chemical potential obtained from the model of Glas [22]. It is seen that the first ZB-to-WZ transition occurs at low Ga concentration within the range of approximately 0.1-0.2 depending on the facet type, growth temperature and the model used for $\Delta\mu_{TPL}$. This transition takes place due to a lower surface energy of WZ nuclei relative to ZB nuclei at the TPL, and is explained within the original model of Glas *et al.* [3]. The reverse WZ-to-ZB transition, discussed earlier by Dubrovskii [21], occurs abruptly when the Ga concentration in the Au-Ga-As droplets exceeds the critical value of 0.67, and is due to the energetic suppression of the TPL nucleation in droplets having low surface energy. Therefore, the preferred crystal structure now depends on the actual Ga concentration during the VLS growth, which will be studied in the foregoing sections.

Table 1. Growth parameters of Au-catalyzed GaAs

Growth method	T(°C)	As vapor	dL/dt (nm/s)	v_5 (nm/s)	ε	v_5^0 (nm/s)	μ_{35}^s (meV)	$P_{5(n)}^0$ (Pa)
MOCVD	464	AsH_3	10	2.1×10^3	3.25	367	-1520	0.51
MBE	550	As_2	1	9.0	3.25	1.1×10^4	-1615	16
HVPE	715	As_4	28	6.2×10^4	3.25	8.5×10^14	-1820	2800

4. Zeldovich nucleation rate and Kashchiev renormalization

In their modeling of MBE-grown, Ga-catalyzed GaAs NWs, Glas *et al.* [19] used the following classical expression [34] for the Zeldovich nucleation rate:

$$J = Ac_5 \Delta\bar{\mu}^{1/2} exp\left(-\frac{a^2}{4\Delta\bar{\mu}}\right) \tag{5}$$

Here, c_5 is the As atomic concentration in the liquid Au-Ga-As alloy, $\Delta\bar{\mu} = \Delta\mu/(k_B T)$ is the dimensionless chemical potential of GaAs in the droplet in thermal units,

$$a = 2 \times 3^{3/4}(\Omega_{35}h)^{1/2}\frac{\Gamma}{k_B T} \tag{6}$$

is the surface energy constant for a regular triangle island (with Γ as the effective surface energy of the island lateral surface for a given nucleation scenario, $\Gamma = \gamma_{SL}$ in the case of Ga-catalyzed GaAs NWs), h=0.326 nm is the height of a GaAs monolayer and A is a certain pre-factor which may depend on temperature but neither on c_5 nor on $\Delta\bar{\mu}$. It was found that the experimental data on the NW growth rate dependence on temperature and As flux are best fitted at a temperature-independent $A = 7\,\text{nm}^{-2}\text{s}^{-1}$ and $\gamma_{SL} = 0.123\text{J/m}^2$ (within the temperature window $580 - 640$°C) [19]. Our analysis involves the so-called Kashchiev re-normalization of the nucleus formation energy [24] which has been used extensively in general studies [35,36] but overlooked in modeling of VLS NWs. In brief, Kaschiev proposed to change the equation for the macroscopic formation energy of a nucleus consisting of i monomers (GaAs pairs in our case) $F(i)$ to $F(i) - F(1)$ in order to ensure that the quasi-equilibrium distribution $n(i) = n_1 exp[-F(i)]$ gives correctly the monomer concentration at $i = 1$. Under the assumption of the ballistic growth mode for near-critical nuclei, this yields the re-normalized expression for the pre-factor A in Eq. (5) of the form (a more detailed analysis will be presented elsewhere):

$$A = \frac{3^{3/4}}{\sqrt{pi}}\left(\frac{h}{\Omega_{35}}\right)^2 D_5 exp(\bar{\mu}_{35}^s + a) \tag{7}$$

for a triangle island. Here, D_5 is the diffusion coefficient of As in liquid. This shows the validity of the Glas assumption [19]: for Ga-catalyzed VLS growth, $A = A(T)$ indeed depends only on temperature. However, in the more general case of ternary Au-Ga-As alloy in the droplet, the value would depend on the Ga concentration c_3 via the surface energy Γ in the constant a[Eq. (6)] and also through the As diffusion coefficient $D_5 = D_5(T, c_3)$. When the Kashchiev renormalization of the formation energy and the corresponding modification of the nucleation barrier are included, the elongation rates dL/dt for NWs having different crystal structures contain the exponential terms $exp[-a^2/(4\bar{\mu}) - a]$ rather than just $exp[-a^2/(4\bar{\mu})]$ in the Glas model [3]. This modifies the Glas conditions given by Eqs. (4) to

$$\frac{a_{WZ}^2}{4(\bar{\mu}-\bar{\Psi})} - a_{WZ} < \frac{a_{ZB}^2}{4(\bar{\mu}-\bar{\Psi})} - a_{ZB} \quad (8)$$

with $\bar{\Psi} = \Psi/(k_B T)$, for the two possible ZB scenarios [(ZB,TPL) and (ZB,C)]. This gives a rather different behavior of the critical chemical potentials compared to $\Delta\mu_C = \Psi/[1-(a_{WZ}/a_{ZB})^2]$ in the case of Au-catalyzed VLS growth, because both a_{WZ} and a_{ZB} depend on c_3 via the surface energies Γ_{WZ}, Γ_{ZB} and γ_{SL} entering Eq. (6). The modified liquid chemical potentials are shown in Fig. 2 and give a wider WZ region, where the WZ phase formation at intermediate Ga concentrations becomes much easier (since it requires a much lower chemical potentials during growth). A more detailed analysis of these new features will be presented elsewhere. We note, however, that the Kashchiev renormalization does not change significantly the abrupt transition to the ZB structure at high Ga concentrations ($c_3 > 0.67$ for our model parameters).

5. Self-consistent determination of Ga concentrations

In order to find the actual Ga concentration during the VLS NW growth at a given set of the growth conditions, we write the self-consistency equation in the form

$$\frac{dL}{dt} = \frac{3\sqrt{3}}{2}R^2 h \sum_{k=1}^{3} \eta_k J_k; \quad J_k = A_k c_5 (\Delta\bar{mu} - \bar{\Psi}_k)^{1/2} exp\left(-\frac{a_k^2}{4(\Delta\bar{\mu} - \bar{\Psi}_k)}\right) \quad (9)$$

The left-hand side is the transport-limited NW elongation rate which was directly measured for each set of the experimental data discussed later on. The right-hand side gives the nucleation-mediated elongation rate of a hexahedral NW with side R in the mononuclear mode [3,4,19,23] in three possible nucleation scenarios k=(WZ,TPL), (ZB,TPL) and (ZB,C) relating to different Γ_k in Eq. (6) and different $\bar{\Psi}_k = \Psi/(k_B T)$ for the WZ structure and 0 for ZB structure. The $\eta_k = 2r_c^{(k)}/R$ for the TPL nucleation position, with $r_c^{(k)} = (1/3^{1/4})(\Omega_{35}/h)^{1/2}[a/(\Delta\bar{\mu} - \bar{\Psi}_k)]$ as the critical radii, are the geometrical factors in the TPL nucleation scenarios which account for the reduction of the available nucleation area to the TPL ring of the critical radius [4]. We ignore the pulsed fluctuations of the liquid chemical potential in mononuclear growth [36-39] by averaging the NW elongation rate over a sufficiently long time interval consisting of many monolayer pulses.

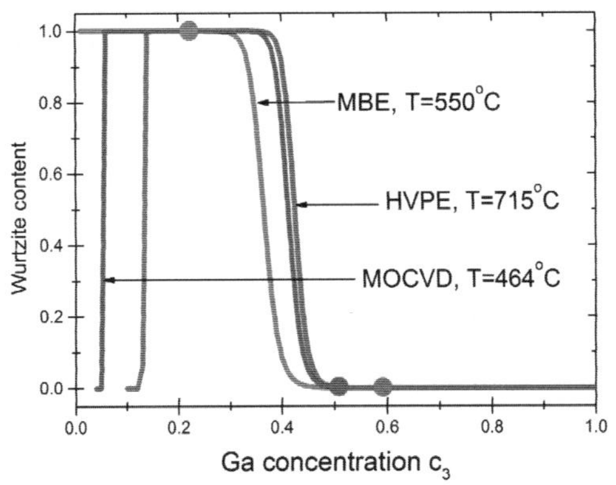

Figure 3. WZ contents versus Ga concentration for the three growth methods; dots correspond to the actual Ga concentrations obtained from the self-consistent model with the Kashchiev renormalization of the nucleation barriers.

We then use Eq. (1) for the chemical potentials in which the Ga contribution is computed by means of the Glas model [22] for each temperature and the growth conditions listed in Table 1, as in Fig. 2. After that, Eq. (9) contains only one unknown (in both sides): the Ga concentration , which is obtained by numerically solving Eq. (9) for each set of the growth parameters. The corresponding Ga concentrations during growths are shown by dots in Fig. 2. Figure 3 shows the WZ content versus Ga concentration for each set of the growth conditions, obtained simply by

$$p_{WZ} = \frac{(dL/dt)_{WZ,TPL}}{dL/dt} \tag{10}$$

where $(dL/dt)_{WZ,TPL}$ is the term in the right-hand side of Eq. (9) corresponding to the WZ phase formation via the (WZ,TPL) nucleation scenario. These graphs were obtained with the parameters of Table 1 and NWs with 20 nm radius. In calculations, we assumed the same diffusion coefficients at 550 and 715°C (based on the data of Ref. [19] which reveal a temperature-independent coefficient $A = 7$ nm-2s-1 in Eq. (5) for temperatures $580 - 640$°C), and the ten times lower diffusivity of As at 464°C. It is seen that the HVPE and MOCVD-grown GaAs NWs should be pure ZB regardless of their radius, although the growth temperatures are very different. Prevalence of the ZB structure in these NWs is due to the high Ga concentrations in the droplets, the effect which decreases the liquid-vapor surface energy and suppresses the TPL nucleation on surface energetic grounds, as in the case of Ga-catalyzed GaAs NWs [16,40]. High Ga content originates from very large input amounts of both growth species in HVPE and MOCVD, yielding large values of the NW elongation rates (10-30 nm/s). This benefit of high growth rates for the phase purity has been discussed earlier by Ren *et al.* [26] and Joyce *et al.* [41]. For Au-catalyzed GaAs NWs grown by MBE with modest fluxes of Ga and As, the elongation rate is much lower, which yields a low Ga content corresponding to the WZ corner. This prevalence of WZ structure in Au-catalyzed, MBE-grown GaAs NWs is quite stable and cannot be changed by the achievable variations in the fluxes or growth temperatures within the growth window (usually $500 - 600$°C [36]). Hence, the ZB phase can be present in these NWs only at the transient growth steps at the beginning and at the end of MBE growth where the Ga concentration in the droplet is lower than in the steady state [2,3], or when the upper parts of the NWs are grown by MBE at very high temperatures about 630°C [6]. All these conclusions of the model are consistent with the experimental data [2,3,25-27]. Summarizing, we have developed a model for the VLS growth and crystal structure of Au-catalyzed GaAs and other III-V NWs obtained by different epitaxy techniques. Our model accounts for the complete set of the growth parameters: temperature, As flux, V/III flux ratio and NW elongation rate. Some of the key elements of the model (such as the pre-exponent in the Zeldovich nucleation rate, the modified critical chemical potentials for the WZ-ZB phase transitions, the dependence of As diffusivity in the liquid phase on temperature and Ga concentration) still need to be developed. However, the model represents very well the main structural trends of Au-catalyzed GaAs NWs obtained by three epitaxy techniques and in very different conditions. Our results can be used for tuning the crystal structure of Au-catalyzed III-V NWs by changing the technologically controlled growth parameters, e.g., by alternating the flux of group V element. This brings about some new and useful ways for fabrication of the crystal phase heterostructures and quantum dots in III-V NWs. This work was supported by scientific program 21 of the Russian Academy of Sciences, grants 13-02-12405 and 13-02-00662 of the Russian Foundation for Basic Research, and the FP7 projects NANOEMBRACE (Grant agreement 316751) and FUNPROB (Grant agreement 269169).

References

[1] A.I. Persson, M.W. Larsson, S. Stengstrom, B.J. Ohlsson, L. Samuelson, and L.R. Wallenberg. Nature Mater. **3**, 677 (2004)

[2] I.P. Soshnikov, G.E. Cirlin, A.A. Tonkikh, Y.B. Samsonenko, V.G. Dubrovskii, V.M. Ustinov, O.M. Gorbenko, D. Litvinov, and D. Gerthsen, Phys. Sol. State **47**, 2213 (2005)
[3] F. Glas, J.C. Harmand, and G. Patriarche, Phys. Rev. Lett. **99**, 146101 (2007)
[4] V.G. Dubrovskii, N.V. Sibirev, J.C. Harmand, and F. Glas, Phys. Rev. B **78**, 235301 (2008)
[5] H. Shtrikman, R. Popovitz-Biro, A. Kretinin, and M. Heiblum, Nano Lett.**9**, 215 (2009)
[6] V.G. Dubrovskii, N.V. Sibirev, G.E. Cirlin, A. D. Bouravleuv, Yu.B. Samsonenko, D.L. Dheeraj, H.L. Zhou, C. Sartel, J.C. Harmand, G. Patriarche, and F. Glas, Phys. Rev. B **80**, 205305 (2009)
[7] N. Akopian, G. Patriarche, L. Liu, J.C. Harmand, and V. Zwiller, Nano Lett. Nano Lett. **10**, 1198 (2010)
[8] D. Spirkoska, J. Arbiol, A. Gustafsson, S. Conesa-Boj, F. Glas, I. Zardo, M. Heigoldt, M.H. Gass, A.L. Bleloch, S. Estrade, M. Kaniber, J. Rossler, F. Peiro, J.R. Morante, G. Abstreiter, L. Samuelson, and A. Fontcuberta i Morral, Phys. Rev. B **80**, 245325 (2009)
[9] D.L. Dheeraj, A.M. Munshi, M. Scheffler, A.T. J van Helvoort, H. Weman, and B.O. Fimland, Nanotechnology **24**, 015601 (2013)
[10] K.A. Dick, C. Thelander, L. Samuelson, and P. Caroff, Nano Lett. **10**, 3494 (2010)
[11] S. Lehmann, D. Jacobsson, K. Deppert, and K. Dick, Nano Research **5**, 470 (2012)
[12] S. Lehmann, J. Wallentin, D. Jacobsson, K. Deppert, and K.A. Dick, Nano Lett. **13**, 4099 (2013)
[13] P. Krogstrup, M.H. Madsen, H. Wen, K. Miwa, J. Nygrd, T. Masamitu, and R.K. Feidenhans, Appl. Phys. Lett. **100**, 093103 (2012)
[14] A. Fakhr, Y.M. Haddara, and R.R. LaPierre, Nanotechnology 21, 165601 (2010)
[15] C. Colombo, D. Spirkoska, M. Frimmer, G. Abstreiter, and A. Fontcuberta i Morral, Phys. Rev. B **77**, 155326 (2008)
[16] V.G. Dubrovskii, G.E. Cirlin, N.V. Sibirev, F. Jabeen, J.C. Harmand, and P. Werner, Nano Lett. **11**, 1247 (2011)
[17] M.R. Ramdani, J.C. Harmand, F. Glas, G. Patriarche, and L. Travers, Cryst. Growth Des.**13**, 91 (2013)
[18] G. Priante, S. Ambrosini, V.G. Dubrovskii, A. Franciosi, and S. Rubini, Cryst. Growth Des.**13**, 3976 (2013)
[19] F. Glas, M.R. Ramdani, G. Patriarche, and J.C. Harmand, Phys. Rev. B **88**, 195304 (2013)
[20] G.E. Cirlin, V.G. Dubrovskii, N.V. Sibirev, I.P. Soshnikov, Yu.B. Samsonenko, A.A. Tonkikh, and V.M. Ustinov, Semiconductors **39**, 557 (2005)
[21] V.G. Dubrovskii, Appl. Phys. Lett. **104**, 053110 (2014)
[22] F. Glas, J. Appl. Phys. **108**, 073506 (2010)
[23] V.G.Dubrovskii and N.V.Sibirev, Phys. Rev. E **70**, 031604 (2004)
[24] D. Kashchiev, Nucleation: Basic Theory with Applications. Butterworth-Heinemann, Oxford, 2000
[25] H. Huang, X. Ren, X. Ye, J. Guo, Q. Wang, Y. Yang, S. Cai, and Y. Huang, Nano Lett. **10** 64 (2010).
[26] X. Ren, H. Huang, V.G. Dubrovskii, N.V. Sibirev, M.V. Nazarenko, A.D. Bolshakov, X. Ye, Q. Wang, Y. Huang, X. Zhang, J. Guo, and X. Liu, Semicond. Sci. Technol. **26**, 014034 (2011)
[27] M.R. Ramdani, E. Gil, Ch. Leroux, Y. Andre, A. Trassoudaine, D. Castelluci, L. Bideux, G. Monier, C. Robert-Goumet, and R. Kupka, Nano Lett. **10**, 1836 (2010)
[28] V.G.Dubrovskii, I.P.Soshnikov, G.E.Cirlin, A.A.Tonkikh, Yu.B.Samsonenko, N.V.Sibirev, and V.M.Ustinov, Phys. Stat. Sol. (b) **241**, R30 (2004)
[29] C.-Y.Yeh, Z. W. Lu, S. Froyen, and A. Zunger, Phys. Rev. B **46**, 10086 (1992)
[30] N.V. Sibirev, M.A. Timofeeva, A.D. Bolshakov, M.V. Nazarenko, and V.G. Dubrovskii, Phys. Solid State **52**, 1531 (2010)
[31] J.J. Jasper, J. Phys. Chem. Ref. Data **1**, 841 (1972)
[32] M. Rosini and R. Magri, Nano Letters **10**, 6021 (2010)
[33] V. Pankoke, P. Kratzer, and S. Sakong, Phys. Rev. B **84**, 075455 (2011)
[34] I. Markov, Crystal growth for beginners. 2nd Edition. World Scientific, New Jersey London Singapore Hong Kong (2003)
[35] V.G. Dubrovskii and M.V. Nazarenko, J. Chem. Phys.**132**, 114507 (2010)
[36] V.G. Dubrovskii, Nucleation theory and growth of nanostructures. Springer, Heidelberg New York Dordrecht London, 2014
[37] C.-Y. Wen, J. Tersoff, K. Hillerich, M.C. Reuter, J.H. Park, S. Kodambaka, E.A. Stach,and F. M. Ross, Phys. Rev. Lett. 107, 025503 (2011)
[38] F. Glas, J.C. Harmand, and G. Patriarche, Phys. Rev. Lett. 104, 135501 (2010)
[39] V.G. Dubrovskii, Phys. Rev. B. **87**, 195426 (2013)
[40] G.E. Cirlin, V.G. Dubrovskii, Yu.B. Samsonenko, A.D. Bouravleuv, K. Durose, Y.Y. Proskuryakov, Budhikar Mendis, L. Bowen, M. A. Kaliteevski, R.A. Abram, and Dagou Zeze, Phys. Rev. B **82**, 035302 (2010)
[41] H.J. Joyce, Q. Gao, H.H. Tan, C. Jagadish, Y. Kim, M.A. Fickenscher, S. Perera, T.B. Hoang, L.M. Smith, H.E. Jackson, J.M. Yarrison-Rice, X. Zhang, and J. Zou, Nano Lett. **9**, 695 (2009)

Semiconductors and nanostructures-based sources of terahertz radiation

D Firsov[1]

Saint Petersburg State Polytechnical University, Polytechnicheskaya str. 29, St. Petersburg 195251, Russia

E-mail: dmfir@rphf.spbstu.ru

Abstract. A review of several approaches to obtaining the emission of terahertz spectral range based on the use of semiconductor micro- and nanostructures is presented. Physical principles of terahertz sources based on impurity related carrier transitions including transitions involving resonant impurity states and on carrier heating in strong electric field are discussed. Emitter of millimeter radiation based on transit time resonance in n-InP is described and the way to rise the frequency up to several THz is suggested

1. Introduction

Considerable recent attention has been focused on practical application of the radiation in terahertz (or far-infrared) spectral range. This spectral range ($\lambda \approx 30 - 300$ μm) lies between optical and radio frequencies. Special features of the interaction of terahertz radiation with matter determine the wide range of possible application of terahertz-based devices in non-destructive testing, environmental monitoring, biology, medicine, space communication, spectroscopy etc.

The main problem facing researchers in this field is the lack of low-cost effective and convenient in use sources and detectors of radiation for terahertz spectral range. It is difficult to take full advantage of both optical and radio frequency methods. At the same time, terahertz radiation sources with the specified properties can be realized using semiconductors or semiconductor nanostructures. Semiconductor quantum cascade laser [1] is the most promising solid state emitter of terahertz radiation but a wide application of this perfect device is restricted with very complicated technology of its production.

The present paper is devoted to the brief review of several methods of obtaining the terahertz radiation based on the use of semiconductor micro- and nanostructures. We consider mainly the sources of spontaneous terahertz radiation.

2. Terahertz sources based on impurity related carrier transitions

The energies of intracenter carrier transitions in doped semiconductors and semiconductor nanostructures correspond to terahertz spectral ranges, so the use of impurity related carrier transitions is one of the ways to make THz emitter. The common feature of such emitters is the need to use low temperatures.

[1] To whom any correspondence should be addressed.

The methods of obtaining the radiation differ in the way to depopulate final states for carrier transition. Usually, impurity ionization in electric field or optical excitation are used.

2.1. Stressed GaAsN/GaAs microstructures

The first THz laser based on carrier transitions between impurity states was realized on stressed p-Ge [2]. External mechanical strain in bulk p-Ge splits not only light hole and heavy hole subbands of valence band but also acceptor states. At a certain value of strain, the acceptor states related to the heavy hole subband become resonant with the continuum of light hole subband. At low temperature, applied electric field ionizes impurity; free light holes are heated in electric field and can be captured with resonant impurity states. As a result, the population inversion can appear between some resonant states and localized ones resulting in THz emission. A significant drawback of this approach is the use of external stress.

In frame of the approach described in [2], we suggested to use internal strain appearing in layered structure with different lattice constants of the layers [3]. Diagram of localized and resonant impurity states and possible hole transitions is shown in figure 1 (a). Terahertz luminescence spectra under strong electric fields were studied in strained GaAsN layers grown on GaAs substrate and doped with Be. Results are shown in figure 1. Position of the emission spectrum correlates well with the calculations.

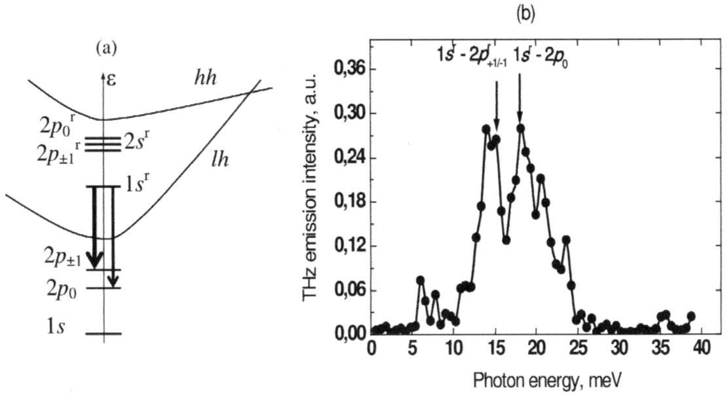

Figure 1. Diagram of optical hole transitions in $GaAs_{0.982}N_{0.018}$/GaAs structure in strong electric field (a); emission spectrum at T = 10 K and E = 1600 V/cm, arrows show calculated energies of marked hole transitions (b).

2.2. Nanostructures with GaAs/AlGaAs quantum wells

Resonant impurity states can appear not only owing to mechanical strain but also due to size quantization. In doped quantum wells, the formation of subbands of size quantization is accompanied also by quantization of impurity levels and formation of series of impurity levels below each subband (it is correct if the impurity bound energy is less than energy distance between subbands). Some of these states may be resonant with the states of underlying subband. At low temperature, longitudinal electric field applied in the plane of the structure ionized the impurity. Free electrons are heated with electric field and can be captured with upper resonant states. Electron accumulation on the resonant state makes possible radiative electron transitions from the resonant states to localized ones.

Terahertz radiation of this type was observed in 30 nm thick $GaAs/Al_{0.3}Ga_{0.7}As$ quantum wells [4]. Diagram of the possible electron transitions and emission spectra are shown in figure 2. Impurity-related nature of the observed radiation is confirmed by good agreement of the experimental spectra with calculated energies of impurity-related electron transitions.

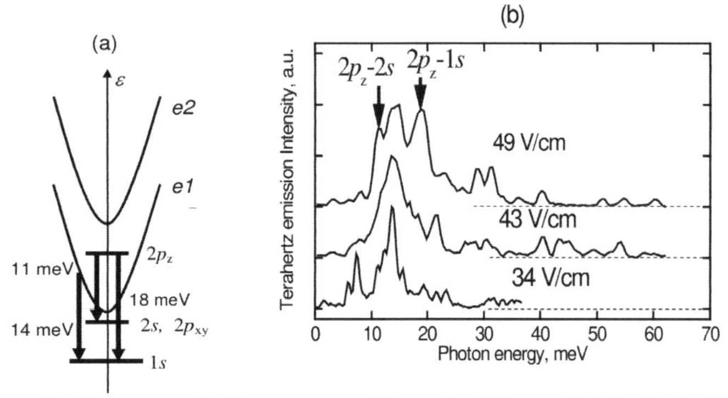

Figure 2. Diagram of electron transitions in 30 nm thick GaAs/AlGaAs quantum well (a); emission spectrum at T = 6 K at different longitudinal electric fields. Arrows show calculated energies of electron transitions (b).

2.3. THz emission under interband optical excitation

Low-temperature interband photoluminescence in doped semiconductor nanostructures can be used for the depopulation of the ground impurity states. Under interband optical excitation, the recombination of nonequilibrium holes and electrons initially localized at non ionized donors can exist along with exciton or electron-hole recombination. After interband optical excitation the electrons on ground donor state can radiatively recombine with nonequilibrium holes, so ground donor states are depopulated. Nonequilibrium electrons in the first subband of size quantization can be captured on the ground states of positively charged donors with emission of terahertz photons. Diagram explaining these processes in GaAs/AlGaAs quantum well is shown in figure 3 (a).

Terahertz emission under interband optical excitation in bulk n-GaAs and p-Ge was reported in [5]. We present the results obtained in n-GaAs/AlGaAs quantum wells. Quantum well structures are of considerable interest because the energy distances between impurity levels can be easily changed with changing the structure parameters.

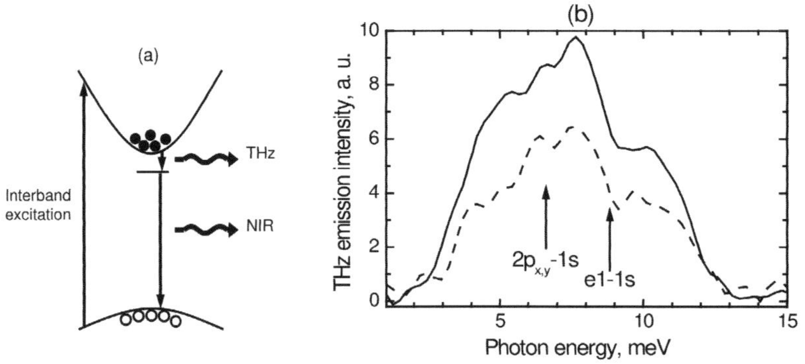

Figure 3. Schematic diagram of optical electron transitions in doped GaAs/AlGaAs quantum well in terahertz (THz) and near-infrared (NIR spectral ranges (a); emission spectra at T = 4.4 K (solid line) and T = 10 K (dashed line) measured in 30 nm thick GaAs/$Al_{0.3}Ga_{0.7}$As quantum well under interband optical excitation. Arrows show calculated energies of allowed electron transitions (b).

Emission spectra measured in GaAs/AlGaAs quantum well structure are shown in figure 3 (b). In this case terahertz emission was associated not only with e1→1s electron transitions from the first subband e1 but also with $2p_{x,y}$→1s transitions between excited and ground impurity states (energy level diagram is similar to one shown in figure 2(a)). This conclusion is confirmed by calculations of the energies of the corresponding transitions.

3. Terahertz sources on hot carriers

Free carriers in semiconductor and semiconductor nanostructures can be heated with strong electric field. Under conditions of heating, the electron temperature rises and exceeds the temperature of lattice. Black-body like spectrum of radiation emitted by electron gas contains the part belonging to terahertz spectral range. Because electrons can be heated to high temperatures without significant heating of lattice, such approach is perspective for simple obtaining terahertz emission with relatively high intensity.

High electron temperatures in electric field can be achieved in structures with high electron mobility such as modulation-doped GaN/AlGaN heterojunctions [6]. Heterostructure under investigations represented a standard HEMT (high electron mobility transistor) structure consisting of a set of GaN/$Al_{0.3}Ga_{0.7}N$ layers. At high electric field and lattice temperature $T = 4.2$ K, the donors are ionized and temperature of free electrons can reach the values of 400 K (see figure 4). The experimental and calculated dependences of terahertz emission intensity on electric field in this structure are also shown in figure 4. Calculations used blackbody model for the emission from hot 2D electrons. The experimental results are in a good qualitative agreement with the considered theory. The noticeable quantitative discrepancy rising at high electric fields can be attributed to the optical phonon accumulation.

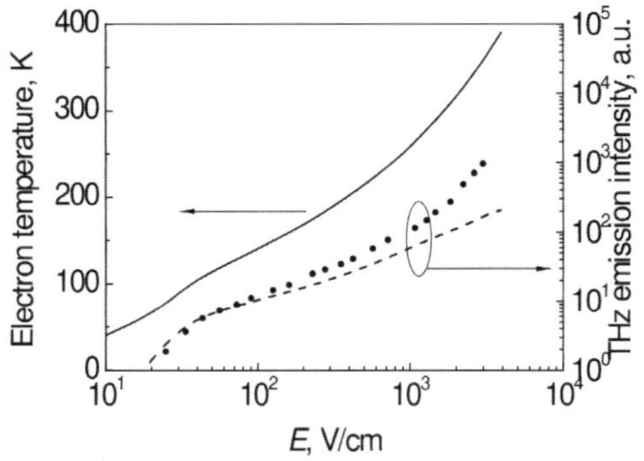

Figure 4. Field dependence of the integral terahertz emission (dashed line – experiment, circles – calculation) and electron temperature (solid line, calculation) in GaN/AlGaN heterostructure. Lattice temperature $T = 4.2$ K.

4. Electron transit time resonance

Last part is devoted to optical phonon transit time resonance. This phenomenon can be observed in semiconductors with sharp onset of optical phonon scattering. For the first time, this mechanism was considered in [7]. It relies on the quasi-periodic dynamics of carriers in strong electric fields inside the optical phonon sphere in the momentum space. Such an electron transport gives rise to the electrical instability and as a result to amplification and generation in the frequency range corresponding to the period of the electron motion in the momentum space. For the first time short millimeter wave generation caused by this effect has been observed experimentally in bulk n-InP (spectra of stimulated emission are shown in figure 5) [8].

Spectral range of the radiation observed in [8] (0.06-0.2 THz) is not very convenient and interesting for practical use. The theory tells us that the stronger is the carrier-phonon interaction the higher will be the upper generation frequency. On this basis, wide band gap semiconductors (nitrides) are very promising materials to realize a terahertz maser due to a high value of the optical phonon energy and the strong interaction of electrons with polar optical phonons which are expected to increase the frequency of the resonance up to a few THz. It was also theoretically shown that for 2D electrons confined in a quantum well the transit time resonance is more pronounced due to abrupt turn on of the optical phonon scattering. From this point of view, the use of GaN/AlGaN HEMT (high electron mobility transistor) heterostructures is the most perspective way to realize emitter of stimulated terahertz radiation due to high electron mobility in these structures and increasing the quality of GaN-based structures.

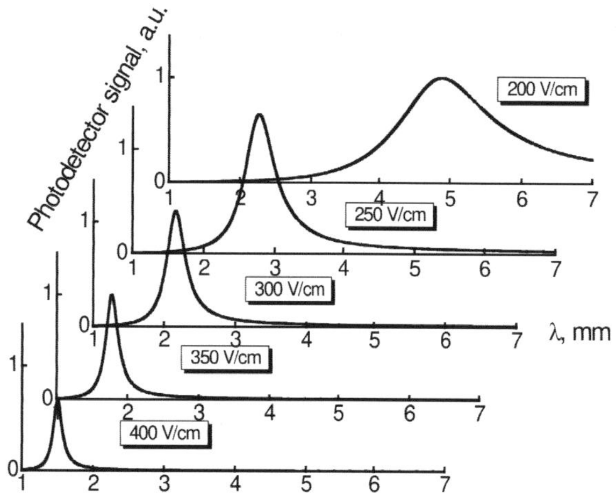

Figure 5. Stimulated emission spectra of bulk n-InP in different electric field. Lattice temperature $T = 4.2$ K. Resonator on total internal reflection was used.

5. Summary

In this paper, several methods to produce terahertz radiation are presented. To date, spontaneous terahertz emission only with the use of these methods was obtained. Nevertheless, the experimental results are in a good qualitative agreement with the theory and may be considered as a first steps to the realization of terahertz emitter having a practical application.

Acknowledgments

This work is supported by RFBR (Grant 12-02-01155), Ministry of Education and Science RF and German-Russian Program of the Federal Ministry of Education and Research (BMBF).

References

[1] Kohler R, Tredicucci A, Beltram F, Beere H E, Linfield E H, Davies A G, Ritchie D A, Iotti R C and Rossi F 2002 *Nature* **417** 156.

[2] Altukhov I V, Kagan M S, Korolev K A and Sinis V P 1994 *JETP Lett.* **59** 476

[3] Shalygin V A, Vorobjev L E, Firsov D A, Panevin V Yu, Sofronov A N, Andrianov A V, Zakhar'in A O, Egorov A Yu, Gladyshev A G, Bondarenko O V, Ustinov V M, Zinov'ev N N and Kozlov D V 2007 *Appl. Phys. Lett.* **90** 161128.

[4] Firsov D A, ShalyginV A, Panevin V Yu, Melentyev G A, Sofronov A N, Vorobjev L E, Andrianov A V, Zakhar'in A O, Mikhrin V S, Vasil'ev A P, Zhukov A E, Gavrilenko L V, Gavrilenko V I, Antonov A V and Aleshkin V Ya 2010 *Semiconductors* **44** 1394

[5] Andrianov A V, Zakhar'in A O, Ivanov Yu L and Kipa M S. 2010 *JETP Letters* **91** 96

[6] Shalygin V A, Vorobjev L E, Firsov D A, Sofronov A N, Melentyev G A, Lundin W V, Nikolaev A E, Sakharov A V and Tsatsulnikov A F 2011 *Journ. of Appl. Phys.* **109** 073108

[7] Andronov A A and Kozlov V A 1973 *JETP Letters* **17** 87

[8] Vorob'ev L E, Danilov S N, Tulupenko V N and Firsov D A 2001 *JETP Letters* **73** 219

A new method of replacement atoms for the synthesis of epitaxial layers of SiC on Si: From theory to practice

S.A. Kukushkin[1,2], A.V. Osipov[1]

[1]Institute of Problems of Mechanical Engineering RAS, V.O., Bolshoj pr. 61, St. Petersburg, 199178 Russia
[2]St. Petersburg State Polytechnical University, Polytechnicheskaya 29, 195251 St.Petersburg, Russia

E-mail: sergey.a.kukushkin@gmail.com

Abstract. The new method of epitaxy of low-defect unstressed nanoscaled silicon carbide (SiC) films on silicon (Si) substrates is theoretically developed and experimentally realized in the case of the big mismatch of lattices of a film a substrate (~20 %). The method consists in the replacement of a part of atoms of the Si matrix by atoms of C with the formation of SiC molecules. The laboratory model of the operating light-emitting diode is originally obtained based on these structures.

1. Introduction

The tremendous elastic stress appearing during the growth of single crystal films with the lattice parameter strongly differing from the lattice parameter of the substrate do not allow us to obtain high quality layers of new wide-bandgap semiconductors on silicon (Si) without misfit dislocations. It is assumed that the most promising of these are silicon carbide (SiC), gallium nitride (GaN), aluminum nitride (AlN), zinc oxide (ZnO), and some others. Integration of these materials into the silicon electronics plays a key role for the development of industrial technologies; therefore, it is very important to deposit epitaxial films of these materials just on silicon. However, due to the large difference in lattice parameters of Si and all the mentioned wide-bandgap semiconductors (~20%), during the ordered growth of these materials the stress energy rises. This leads to the appearance of a tremendous number of misfit dislocations in growing films and even to complete film cracking.

2. Theory

In this work, we suggest new relaxation mechanism of the elastic energy for growing dislocation-free heteroepitaxial films. The essence of this approach, which differs from all the existing methods of film growth, is based on the idea of preliminary incorporation of point defects into the crystal lattice of the silicon host. When growing the SiC film on the Si substrate, such defects are the carbon atom C placed in the Si interstitial position and the vacancy formed as a result of removal of one of the Si atoms. If these defects are attracted to each other by the elastic interaction in the Si matrix, the resulting elastic energy caused by their incorporation into the substrate host is considerably lower than the energy of no interacting defects. It is well known that the spherically symmetric dilatation centers do not interact with each other at all in an isotropic medium of infinite size. We show [1,2] that the dilatation centers can attract to each other in substantially anisotropic media, such as the crystal with the cubic lattice symmetry, thereby considerably decreasing the total elastic energy. Such attractive centers form stable objects of a new type, namely, the elastic dilatation dipoles. We calculated the elastic energy of the system using heteroepitaxy of the SiC film on the Si substrate (Si has a cubic lattice) as the example and showed that the elastic energy can relax completely only due to the ensemble of dilatation dipoles. It has been proven in [1,2] that the total elastic energy of a crystal during the formation of interacting defects inside its volume is

$$\mathrm{W}\big(cos\varphi(x,y,z)\big) = \mathrm{E}_C + \mathrm{E}_v + E_{int}\big(cos\varphi(x,y,z)\big). \tag{1}$$

Here E_C and E_v are the elastic energies of the lattice distortion by the carbon atom and the vacancy

$$\mathrm{E}_{C,v} = \frac{2\mu^{Si}K^d}{3K^d + 4\mu^{Si}}\,\frac{\left(V^d - V^{Si,m}\right)^2}{3V}, \tag{2}$$

E_{int} denotes the energy of defects interaction

$$E_{int}(cos\varphi(x,y,z)) = -E_0\left(\cos^4\varphi_x + \cos^4\varphi_y + \cos^4\varphi_z - \frac{3}{5}\right), \tag{3}$$

$$E_0 = \frac{15K(3c_{11} - 4c_{44})\Omega^I\Omega^{II}\eta}{8\pi(c_{12} + 2c_{44})r^3}, \tag{4}$$

$\cos\varphi_i = x_i/r$ are the cosines between the axes x, y, z and a direction of the line connecting the centers of interacting defects. The expression $\cos^4\phi_x + \cos^4\phi_y + \cos^4\phi_z - 3/5$ reaches its minimum, equaled to -0.27, in the direction <111> corresponding to an attraction of similar defects. Its maximum, equaled to 0.4, is reached in the direction <100> corresponding to a repulsion of similar defects. Also it has a saddle point -0.1 in the direction <110>. Therefore, the direction <111> is the most energetically favorable for similar point defects. It corresponds to a situation when C atom is located along $(\bar{1}10)$ plane and under (111) planes in an interstitial position, whereas a vacancy is located along <111> direction with respect to it. In this case, the attraction between the silicon vacancy and the carbon atom (both of them are compressing dilation centers) reaches its maximum. That is the considered dilatation dipole provides the maximal relaxation of the elastic energy [1,2]. Taking into account that $n_1^2 + n_2^2 + n_3^2 = 1$ $(n_1 = \cos\phi_x, n_2 = \cos\phi_y, n_3 = \cos\phi_z)$, Eq. (3) yields

$$E_{int} = -E_0\left[{}^2/_5 - 2(n_1^2n_2^2 + n_2^2n_3^2 + n_1^2n_3^2)\right] \approx 6\cdot 10^{-19}\, J/atom. \tag{5}$$

This mechanism of elastic energy relaxation is extreme efficient. Indeed, if dilatation centers do not attract to each other, the total elastic energy per SiC molecule is $W_{dil} = 0.22\ eV$. But if one part of the dilatation dipole is the carbon atom, the elastic energy per SiC molecule is $W_{dip} = 0.13\ eV$. Every 8 atoms of silicon in the original matrix could be replaced by no more than 4 dilatation dipoles, therefore, the maximum relaxation of the elastic energy owing to dipoles is $\Delta W = (W_{dil} - W_{dip})/2 = 0.2\ eV$ per one molecule of SiC. In other words, as shown in [35-37], the elastic energy can relax completely by dilatation dipoles alone, because $W_{dil} - \Delta W \approx 0$. And the lifetime of dilatation dipoles at the temperatures 1100-1250°C is about $\sim 10^{-1} \div 1\ sec$, that is three orders of magnitude greater than the formation time of the first SiC monolayer. So, the elastic dipoles play a role of ordering epitaxy centers. In the opposite case (when the point defects repel being located along

<100> direction), the dipoles will not arise and increase the elastic energy, that leads to layers disorientation.

3. The formation of transition state "elastic dipoles – silicon matrix" as a necessary condition for the synthesis of low-defective SiC layer on Si

In order to provide the effective usage of a new relaxation mechanism of elastic energy due to the interaction of point defects, we suggest the deposition process of SiC not from the vapor phase but immediately from the matrix of the single crystalline Si substrate due to the chemical reaction between the crystalline Si and gaseous carbon monoxide CO [3]

$$2\text{Si (solid)} + \text{CO (gas)} = \text{SiC (solid)} + \text{SiO} \uparrow \text{(gas)} \; . \quad (6)$$

We selected this reaction because of the fact that the forming gaseous silicon monoxide SiO partially carries the atoms from the Si matrix inducing vacancies in it, while gaseous carbon monoxide CO is the source of carbon atoms C arranged in atomic voids of the silicon lattice. Both Si vacancies and incorporated C atoms are the compressing dilatation centers in the cubic Si lattice and interact with each other [1] (see Fig.1).

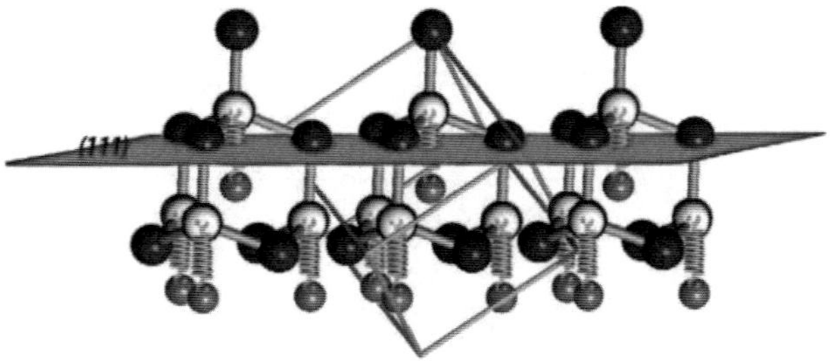

Fig. 1. Schematic illustration of elastic dipoles in the cubic lattice of Si. Large blue balls are Si atoms, small green ones are C atoms in the interstitial positions. The strings indicate the elastic interaction between dilatation centers (vacancies and C atoms) with dipole formation.

Figure 2 represents the difference between the classical mechanism of thin film growth and the mechanism of epitaxy suggested here. In the first case the film grows on the substrate surface which results in enormous stress energy, whereas in the last case the film grows inside the substrate, and the attraction between point defects provide the complete relaxation of the stress energy. The merging process of vacancies after film nucleation results in the pore formation under the film [1,2].

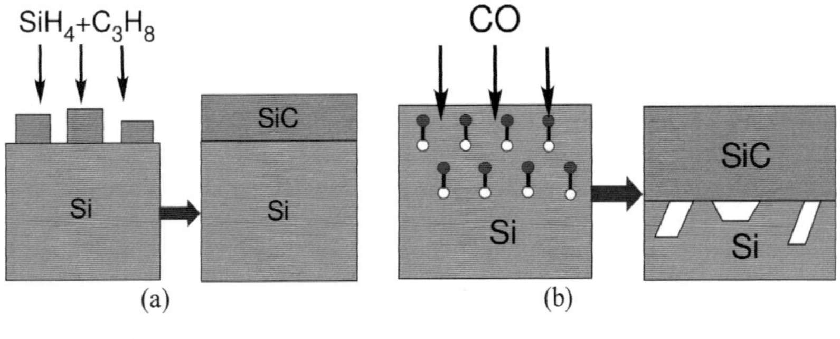

Fig. 2. Schematic pattern of regular thin film growth (a) and the suggested one (b). In the second case dumbbells represent the attraction between different point defects (dilatation dipoles).

The chemical reaction (6) proceeds in two stages. On the first stage the intermediate compound consisting of activating complex, namely, Si vacancy – C atom – Si matrix is formed. C atoms are placed in interstitial positions of Si matrix. On the second one the activating complex transforms into SiC, and released Si vacancies merge into pores in Si under the layer of SiC. It is found that the activating complex consists of the ensemble of dilatation dipoles ordered by the original Si matrix. The transition of the activating complex into SiC and pores is the structural phase transition of shifting type. In this transition the ordering of SiC molecules is followed by the original crystalline Si matrix. The mechanism of heterogeneous chemical reaction between gas and solid is investigated on the example of the reaction (6).

New type of phase transitions, namely, the transition through the intermediate stage, when one of the phase stimulates the nucleation of another one, is described theoretically. Formation work and all the main thermodynamic and kinetic characteristics of two-phase phase transitions are calculated. Conditions are found at which the dislocation-free growth of SiC films on Si takes place [4].

Assembly of nano-layers of SiC on Si by the synthesis of dilatation dipoles is realised on practice for the first time and high quality templates SiC/Si is obtained [5]

The process of phase transformation suggested in this work is very similar to the process of full replacement of living cells of a tree by inorganic. Such processes happened to the trees which originally got to sea hot water and have lain under a layer of earth for one hundred millions years. The photo of such mineral tree of 250 million years is given in Fig. 3.

Fig.3. The hardened tree which has lain in hot water saturated with mineral salts. Living cells of this tree were completely replaced with inorganic minerals as a result of slow exchange process

4. Experimental and Results

To confirm experimentally the suggested relaxation mechanism of elastic energy, Si (111) substrates 35 mm in diameter were held in a vacuum furnace at T = 1100–1300°C in the atmosphere of CO at p = 10–300 Pa for 5–60 min [3]. A SiC film 50–200 nm thick grew inside the Si substrate during this time. The average value of tensile elastic stresses in SiC films measured by an FLX-2320-S thin film stress measurment system is 0.5 GPa in the absence of lattice misfit and cracks [1-4]. Microscopic analysis revealed an almost ideal conjugation of the lattices of silicon and silicon carbide [5]. Such a low measured elastic energy with the ideal conjugation of lattices in the absence of misfit dislocations and cracks can be interpreted by the relaxation of elastic stresses due to the ensemble of dilatation dipoles. Monocrystal nanoscaled layers of SiC of hexagonal polytypes 4H and 6H are obtained on Si for the first time. Best of them have $\mathbf{FWHM}_{\omega-2\theta}$ in the range $\mathbf{0.4^0} - \mathbf{0.65^0}$ at the SiC thickness of 100 nm, and $\mathbf{FWHM}_{\omega-2\theta}$ in the range $\mathbf{0.1^0} - \mathbf{0.15^0}$. At the present time SiC layers on Si substrates of 6 inch are available [6].

5. The growth of semiconductor films AlN, GaN and ZnO on SiC/Si(111) substrates

Nowadays III-nitride heterostructures are grown mainly on the substrates of sapphire, Al_2O_3 or SiC. However, the growth of such structures on the silicon substrates provokes the greatest interest due to the prospect of integration of the gallium nitride and silicon electronics, the possibility of exploiting of a large-diameter substrates, their low cost and good electrical and thermal conductivity.

The large difference of lattice constants (17%) and thermal expansion coefficients (33%) of GaN and Si is the main reason of deformation of GaN/Si heterostructures leading to a high defect density in GaN epitaxial layer directly grown on Si [7,8]. The essential difference in the lattice parameters of SiC and Si (20%) resulting in the formation of misfit dislocation arrays on their interface also affects GaN films grown on Si/SiC substrates. Even the best GaN/SiC/Si samples obtained by MOVPE on 1μm SiC layers contain ~10^9/cm^2 of misfit dislocations. In [9] has been made an attempt to improve the layer structural quality by annealing and formation of helium bubbles in the silicon bulk. According to the authors the stress field around the bubbles redirects the misfit dislocations in the substrate, thereby improving the structural quality of the epitaxial layer. Most researchers try to reduce the defect density in GaN layer by adjusting the growth conditions and composing the buffer layers of different composition and thickness ($Al_xGa_{1-x}N$, AlN, SiC or theirs combination). GaN epitaxial layers with thickness up to 3μm have been successfully synthesized on Si by MOCVD using AlN intermediate layers [10], but the growth of thicker layers (over 10μm) of AlN or GaN by HVPE became very difficult due to the deformation of the structure. To suppress the process of defect formation and to reduce the deformation of GaN epitaxial layer the different intermediate layers are used (TiN, HfN, ZnO, SiN_x, etc.). Recently, the attempts are made to use the intermediate layers of silicon carbide of cubic modification, β-SiC [11].

To answer the questions put in the beginning of this section we have performed a comprehensive analysis of structural, crystallographic, physicochemical, electro-optical and spectroscopic studies of AlN, GaN and ZnO films grown on the SiC/Si substrates [12-17]. Two methods, MOCVD and HVPE, are commonly used for the deposition of the films of AlN, GaN or $Al_xGa_{1-x}N$ solid solution. ZnO films are grown by the magnetron sputtering.

A distinctive feature of SiC films synthesized by the solid-phase epitaxy [1-4] is the ability to obtain in a single process the heterolayers of silicon carbide of both cubic β-SiC and hexagonal α-SiC polytypes [1-4]. Moreover, the inverted layers could be grown, i.e. β-SiC bottom layer (lying on the silicon substrate) and α-SiC upper one, and vice versa. Or even the layer containing the polytype periodically distributed over the substrate. And grown single-crystal defectless SiC layers ensure a substantial relaxation of the elastic thermal stresses covering the pores between the silicon carbide and the silicon, even if β-SiC films are used as buffer layers.

In Ref.[12] the results of electron microscopic analysis of GaN/AlN/nanoSiC/Si(111) have been presented. The layers of gallium nitride and aluminum chloride were grown using the hydrogen chloride gas epitaxy (HVPE) and the layers of nano silicon carbide – by the original method of the solid-phase epitaxy [1-4]. The grown GaN layer (of 20 μm thickness) had a good structural quality with no misfit dislocations parallel to the growth axis [0001] "c" (the overall dislocation density in GaN layer was less than 10^7cm^{-2}). In several samples cut from the different portions of the 3-inch diameter substrate studied in [4] these dislocations were absolutely absent. However, we do not insist that they do not present on the whole 3-inch surface without exceptions. Number of screw dislocations in the layer was about $4 \cdot 10^8$cm^{-2}and the total dislocation density of mixed type was $8 \cdot 10^8$cm^{-2}. This value is significantly less than the density of dislocations in GaN layers grown on Si with an intermediate layer of AlN by MBE and it competes with the dislocation density in gallium nitride layers grown on the sapphire by MOCVD. According [18] the density of dislocation oriented parallel to the growth in GaN films grown on the sapphire by MOCVD is (3.4±0.5) $10^7 cm^{-2}$, of the perpendicularly oriented – (4.5±0.5) $10^7 cm^{-2}$ and the density of inclined dislocations is (1.1±0.3) $10^9 cm^{-2}$. It should be noted that the dislocations oriented parallel to the growth axis and inclined dislocation (to a lesser extent) have a significant effect on the electrical characteristics of the structure. These dislocations decrease the electrical conductivity of the layer and thus increase the heat generation reducing the efficiency of the LEDs fabricated on such structures. The samples of GaN obtained on nano-SiC/Si substrates these dislocations are practically not observed [12]. It should be noted that GaN films has been grown using a thin (<100nm) AlN nucleation layer on the pre-synthesized SiC/Si substrate. The combination of thin AlN layer and of the layer of SiC on the surface of the pores (Fig.4) results in substantially complete elastic relaxation in thick (over 2 μm) GaN layers.

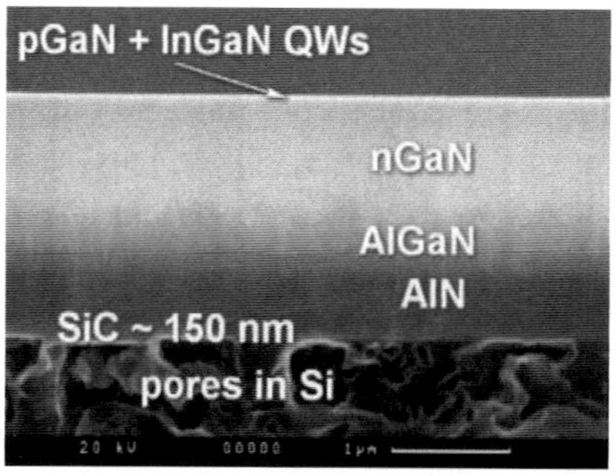

Fig. 4. SEM image of the light-emitting diode structure on SiC/Si template received by the method of solid state epitaxy.

In Ref. [15] the LED structure grown by MOCVD on the basis of III-nitrides on SiC nano-layer (50-200nm) formed by the solid-phase epitaxy on Si(111) substrate has been reported. Firstly, the nucleating layer of AlN was deposited on SiC/Si template. Then the transition layer of $Al_xGa_{1-x}N$ solid solution was grown with Al content varying from x=1 to x=0 along the growth direction. After that the layer of GaN was formed in order to provide the planarity of structure surface. Finally, standard LED structure was grown containing n- and p-type GaN layers and the active region with five InGaN quantum wells separated by barrier GaN layers. To prevent cracks its total thickness was limited by 2.5μm. The density of misfit dislocations of the structure was less than $10^8 cm^{-2}$ and overall dislocation density ~$8\cdot10^8$ cm^{-2}. In this case the thickness of continuous AlN layer was about 200 nm. The increase of AlN layer thickness results in the increase of the dislocation density in grown GaN film as compared to the dislocation density in the films of GaN grown on the nucleation layer of AlN. The number of dislocations in GaN films grown by MOCVD on 400 nm AlN layer was ~$5\cdot10^9$ cm^{-2} with no further increase with AlN layer thickness [6] Data on the growth of GaN films on SiC/Si by MBE are not accessible at the moment. But in this case one should expect a decrease in dislocation density till <$10^7 cm^{-2}$. Thus, a combination of SiC layer on Si (grown by the substitution of Si atoms in the substrate) and of a thin nucleation layer of AlN (deposited on SiC surface) can dramatically reduce the dislocation density in the grown layer of GaN. This method allows growing the structures with the dislocation density comparable with these grown by MOCVD and HVPE on sapphire and bulk crystals of SiC (<$10^8 cm^{-2}$) [18]. SEM photomicrographs of the working LED heterostructure and its ellipsometric spectrum are presented at Fig.5. The analysis of ellipsometric spectrum reveals that the basis of this structure is a single-crystal GaN layer with a band gap of about 3.4eV, the electroluminescence spectra has a maximum in the wavelength range 430-460nm.

Fig. 5. A GaN-based operating light-emitting diode structure grown on the SiC/Si template with dilatation dipoles.

Besides, in Ref. [14] the utilization of SiC layers (grown by the new method) as a buffer foundation for the growth of not strained ZnO films by the magnetron sputtering has been discussed.

Obviously, SiC nano-layers also can be used as a buffer layer for the growth of thick (over 1 μm) SiC films. It will get a possibility to obtain the hexagonal polytypes of the silicon carbide on the silicon. However, this problem is not yet investigated.

6. The growth of semi-polar layers of GaN($20\bar{2}3$) on SiC/Si(100) substrates

One of the most important properties of nanoSiC synthesized by the solid-phase epitaxy method is a unique opportunity to obtain the semi-polar layers of hexagonal crystals and, in particular, of the semi-polar epitaxial layers of GaN. Wurtzite gallium nitride crystals possess a significant internal electric field in the direction of "C" axis due to the spontaneous polarization and associated piezoelectric effect. It plays a negative role in the utilization of such layers in light-emitting devices. However, as has been theoretically predicted in Ref. [19], in InGaN/GaN structures there are some directions where the piezoelectric polarization becomes zero. And this property is inherent not only for the nonpolar but also for the semi-polar orientation if the crystal plane is inclined to "C" plane on about 45°. It has stimulated a large number of studies dealing with the GaN growth in so-called nonpolar or semipolar directions.

In Ref. [20] the semipolar GaN has been grown on nanoSiC surface synthesized on Si (210). However, the structural perfection of GaN crystal films remains poor [20]. More perfect semipolar GaN films, e.g. GaN ($\mathbf{1\bar{1}01}$) [20], could be grown by masking the surface of Si(100) and etching it by the etchant providing the different etch rates for the different crystallographic axis. In this case, it is possible to obtain Si(111) planes with an inclination of 55° to Si(100) and to grow GaN ($\mathbf{1\bar{1}01}$) on these planes.

But a longitudinal wedge-shaped hump of silicon carbide can be formed by the method of solid-phase epitaxy [1-4] without any surface masking devices or chemical etching substances. As described in section 4, the method is based on a preliminary introduction of the carbide atoms (from carbon monoxide) into interstitial positions of silicon matrix crystal lattice and on a following transformation of the intermediate complex "dilatation diploes – silicon matrix" into the silicon carbide. According to the thermodynamics, if Si(100) plane will be inclined on 1°-10° from <100> towards <011> direction and then heated above 600°C then Si(100) plane will be covered by steps. Upper and lower planes will be (100) and the steps will be restricted by (011) planes. Due to the silicon lattice "looseness" along

<011> direction (determined by the specific crystallographic lattice structure of the silicon) the molecules of CO will rush along this direction deep into the silicon substrate (perpendicular to the steps). The silicon surface will become saturated with CO and the dipoles "silicon vacancy – carbon atom – silicon matrix" will be formed owing to Si–CO reaction. Since the attraction between the silicon vacancy and the carbon atom in the silicon matrix is maximum along the direction <011>, then Si(011) step becomes SiC(112). The angle between (112) and (100) planes in a cubic crystal with diamond lattice is about 35°. Therefore, a longitudinal wedge-shaped hump of silicon carbide will be formed having a top above the upper step and an inclined face (with 35° slop) extending to the terrace of the underlined step. In general, Si(100) surface will be composed by an array of the parallel steps having a shape of a triangular prism. The prismatic symmetry is typical for both cubic and hexagonal($\mathbf{20\bar{2}3}$) crystals. That is these surfaces can be used to grow both type of crystals depending on the growth conditions (not on the substrate orientation). It provides completely new opportunities for growth of hexagonal crystals of semipolar orientation. This method has been firstly implemented in Ref. [21] to obtain the semipolar 10μm GaN($\mathbf{20\bar{2}3}$) and ($\mathbf{11\bar{2}0}$) layers with $FWHM_{,\omega-\theta}$ =24 arcmin.

7. Summary

Thus, a new relaxation mechanism of elastic energy due to the formation of dilatation dipoles is put forward. Usage of this mechanism allowed us to originally grow the light-emitting diode epitaxial structure on the silicon substrate and fabricate the operating laboratory model of the light-emitting diode. The employment of this relaxation mechanism also permits us to obtain heteroepitaxial films of wide-bandgap semiconductors (such as SiC, AlN, GaN, AlGaN) on silicon without misfit dislocations and cracks having sufficient quality to fabricate many devices of microelectronics and optoelectronics.

The authors gratefully acknowledge the partial support of this work by the Russian Foundation for Basic Research (grants 11-02-12154-ofi-m and 12-02-00935), and by Program of the Presidium of the Russian Academy of Sciences (No 24 "Fundamentals of Technology of Nanostructures and Nanomaterials".

References

[1] Kukushkin S A, Osipov A V, 2012 *Doklady Physics* **57**, 217
[2] Kukushkin S A, Osipov A V, 2013 J. *Appl. Phys.* **113**, 4909
[3] Kukushkin S A, Osipov A V, 2008 *Physics of the Solid State* **50** 1238
[4] Kukushkin S A, Osipov A V, 2014 *Physics of the Solid State*, **56**, 792
[5] Sorokin L M, Veselov N V, Shcheglov M P, Kalmykov A E, Sitnikova A A, Feoktistov N A, Osipov A V, and Kukushkin S A 2008 *Technical Physics Letters* **34** 992
[6] Kukushkin S A, Lukyanov A V, Osipov A V, and Feoktistov N A, 2014 *Tech. Phys. Lett.* **40**, 36
[7] Kumagai Y, Nagashima T, Murakami H, Takada K, and Koukitu A 2008 *Phys. Stat. Sol.(c)* **5** 1512
[8] Zhao Y, Kita K, Kyuno K, Toriumi A 2009 *Appl. Phys. Lett.* **94** 042901
[9] Liliental-Weber Z, Maltez R.L., Xie J., Morkoc H 2008 *J. Cryst. Growth* **310** 3917
[10] Dadgar A, Hums C, Diez A, Blasing J, Krost A 2006 *J. Cryst. Growth* **297** 279
[11] Severinoa A, Lockeb C, Anzalonea R, Camardaa M, Pilusoa N, Magnaa A La, Saddowb S E, Abbondanzac G, D'Arrigoa G, Viaa F La 2011 *ECS Transactions* **35** 99
[12] Sorokin L M , Kalmykov A E, Bessolov V N, Feoktistov N A , Osipov A V, Kukushkin S A, and Veselov N V. 2011 *Technical Physics Letters* **37** 326
[13] Kukushkin S A, Osipov A V, Osipova E V, Razumov S V, Kandakov, A V 2011 *J. Opt. Technol.* **78** 440

[14] Osipov A V, Kukushkin S A, Feoktistov N A, Osipova E V, Venugopalb N, Vermab G D, Gupta Bipin Kumar, Mitra Anirban, 2012 *Thin Solid Films* **520** 6836.
[15] Kukushkin S A, Osipov A V,. Zhukov S. G,. Zavarin E. E and. Lundin W. V, et al. 2012 *Tech.Phys. Lett.* **38** 297
[16] Sorokin L M, Kalmykov A E, Bessolov V N, Feoktistov N A and Osipov A V, et al, 2011 *Tech. Phys. Lett.* **37**, 326
[17] Ratnikov V V, Kalmykov A E, Mysoedov A V, Kukushkin S A, Osipov A V, and Sorokin L M, 2013 *Technical Physics Letters* **39** 994
[18] Kyutt R N 2010 *Technical Physics Letters* **36** 690
[19] Romanov A E, Baker T J, Nakamura S, and Speck J S 2006 *J. Appl. Phys.* **100**, 023522
[20] Bessolov V, Konenkova E, Shcheglov M, Sharofidinov Sh, Kukushkin S, Osipov A, and Nikolaev V 2013 *Phys. Status Solidi C* **10** 433
[21] Bessolov V, Konenkova E, Kukushkin S A, Mysoedov A V, Osipov A V, Rodin S N, Shcheglov M P, Feoktistov N A, 2014 *Technical Physics Letters* **40**, 802

Pulsed laser deposition and investigation of antimony-doped ZnO films

A S Puzikov[1], N.V. Lyanguzov[2], E M Kaidashev[3]

[1,2,3] Vorovich Mechanics and Applied Mathematics Research Institute, Southern Federal University,Stachki 200/1, Rostov-on-Don, Russia

E-mail: [1]alexander@puzikov.ru, [3]kaidashev@mail.ru

Abstract. We have investigated the influence of oxygen partial pressure, temperature of synthesis and annealing conditions on nanocsrystallineSb-doped thin films, grown by pulsed laser deposition. It is shown that the minimum resistivity (~$8{\bullet}10^{-3}\Omega{\bullet}$cm) and the maximum carriers density (~ $2{\bullet}10^{19}$ cm-3) corresponds to the pressure range $5{\bullet}10^{-3}$–$7{\bullet}10^{-3}$ mbar, to the temperature 550 ° C and *in situ* annealing at 700 °C.Also we show the features of the crystal lattice's dynamics, which are found in the Raman research.

1. Introduction

Owing to a wide band gap of ~3.3 eV at room temperature and a large exciton binding energy of 60 meV, ZnO holds great promise for various applications in electronic and optoelectronic devices. The absence of reliable p-type doping is the major challenge limiting the development of various types of bipolar ZnO-based devices, including light emitting and laser diodes. Antimony-doped p-type ZnO thin films grown by PLD, MBE, and thermal oxidation have been investigated by different groups however, films exhibits diverse characteristics and properties with poor stability [1, 2, 3]. In this work, we undertook an experimental investigation of the structural, electrical and optical properties of ZnO:Sb films.

2. Objects and methods of research

The samples were obtained on the sapphire (Al_2O_3) substrates by means of pulsed laser deposition. KrF laser (λ = 248 nm, E = 300 mJ) was used for ablation of ZnO and ZnO:Sb_2O_3 ceramic targets with Sb concentration 1 – 5 at. %. The scheme of the synthesis chamber is presented in figure 1. The distance between targets and substrates was 50 mm. The synthesis was carried out in oxygen atmosphere.

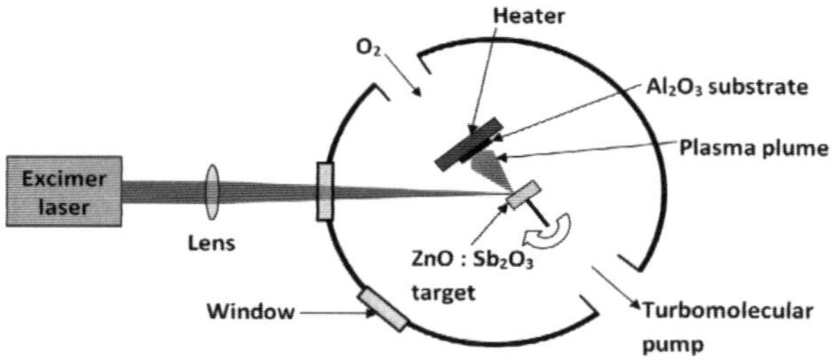

Figure 1. The scheme of the synthesis chamber for pulsed laser deposition.

The ZnO sublayers were deposited before ZnO : Sb. ZnO:Sb films have been grown at the range of pressure $2x10^{-3} - 5x10^{-2}$ mbar; the growth temperature was varied from 500 to 650 °C. The influence of the growth temperature, oxygen pressure and annealing conditions on the resistivity, carriers concentration and mobility was investigated.

An X-ray diffraction apparatus (XRD DRON-7) with Co Kα, $K\alpha_1$, Kβ incident radiation was used to identify the phase structure of the films. The surface morphology and roughness were investigated by a scanning electron microscope (SEM; Zeiss SUPRA 25). Raman backscattering spectra were measured by Renishaw inVia spectrograph at room temperature using the 514.5 nm line of an Ar+ ion laser for the excitation. The Hall effect and the resistivity measurements at room temperature were performed under Van der Pauw configuration.

3. Experimental results and discussion

It was found by using scanning electron microscopy that the uniformity of the surface is improved by decreasing the concentration of antimony in samples.

One of X-ray diffraction spectra is shown in figure 2.

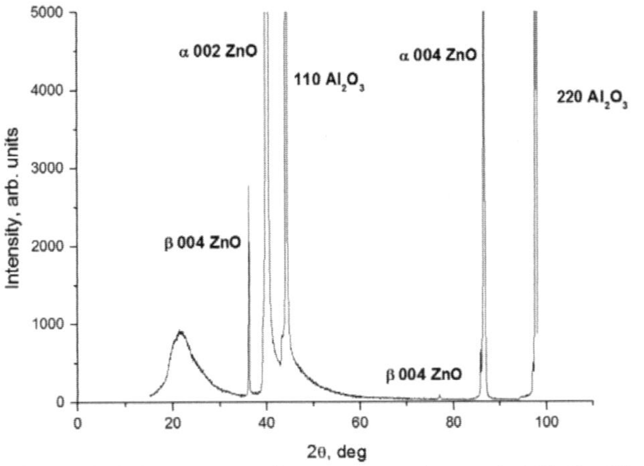

Figure 2. XRD pattern of as grown and annealed ZnO : Sb thin film deposited at $2\cdot10^{-3}$ mbar and 550 ° C.

Diffraction patterns contain intensive reflections corresponding to the family of {001} planes of ZnO. Besides ZnO peaks, reflection of the Al_2O_3 (110) were found, because films' thickness (about 400 nm) is not sufficient to fully absorb the incident X-ray beam. The presence of such a set of peaks leads to the conclusion that in the normal direction mutual crystallographic orientation $(001)_{ZnO} \parallel (110)_{Al2O3}$ is observed. This fact suggests the possibility of an epitaxial film growth, or at least the presence of an axial texture, but is not blocked rigorous proof. However, Sb-related secondary phase was detected in the films (wide band around 22°).

In Raman spectra of the films the broad band between 500 and 600 cm^{-1} was detected (see figures 3a, 3b), which is in good agreement with reports of other authors. Most likely, this band appears due to Sb-related in-band modes which coupling to the LO phonon modes of the host lattice [4]. Also the additional phonon mode was detected around 277 cm^{-1} (marked '*' in the figures). It is due to the fact of doping ZnO with antimony, such mode was detected by different researchers in the films of ZnO doped with gallium, aluminum and iron [5]. The intensity of the E_2^{high} phonon mode decreases with increasing of antimony concentration. It indicates an increase in the compressive strain in the ZnO:Sb layers. In addition, a broad phonon mode appears at ~700 cm^{-1} that is attributed to Sb–O–Sb vibrations [6, 7, 8].

The investigation of electrical properties of films showed (see figures 4a, 4b, table 1) that the minimum resistivity (~8x10^{-3} Ωxcm) and the maximum carriers density (~ 2x10^{19} cm^{-3}) corresponds to the pressure range 5x10^{-3}–7x10^{-3} mbar.

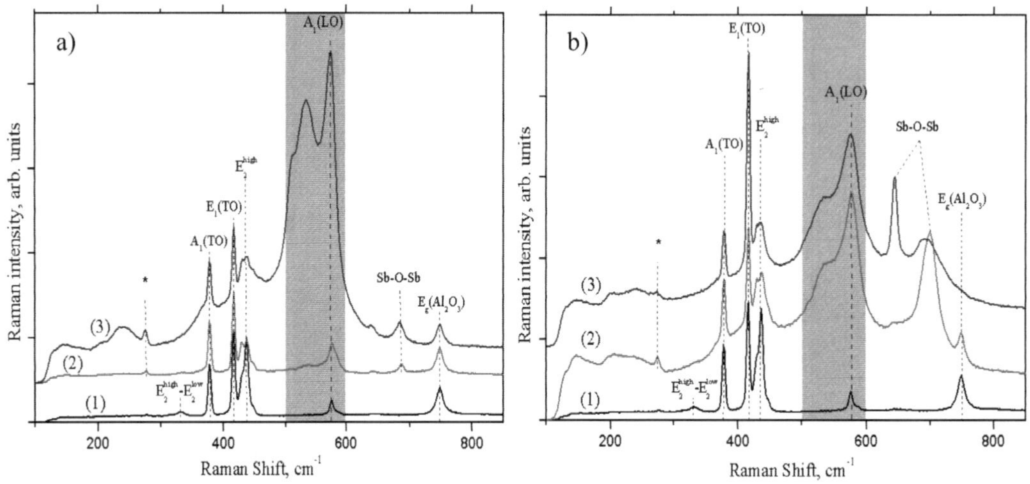

Figure 3. Raman spectra of Sb-doped ZnO films: a.1, b.1 – spectrum of the pure ZnO film;
a.2 – 1 at. % of Sb, synthesis temperature 550°C, annealing at 750 °C, O_2 pressure 5·10^{-3} mbar;
a.3 – 1 at. % of Sb, synthesis temperature 550°C, annealing at 650 °C, O_2 pressure 5·10^{-3} mbar;
b.2 – 2 at. % of Sb, synthesis temperature 550°C, annealing at 750 °C, O_2 pressure 5·10^{-3} mbar;
b.3 – 2 at. % of Sb, synthesis temperature 550°C, annealing at 650 °C, O_2 pressure 5·10^{-3} mbar.

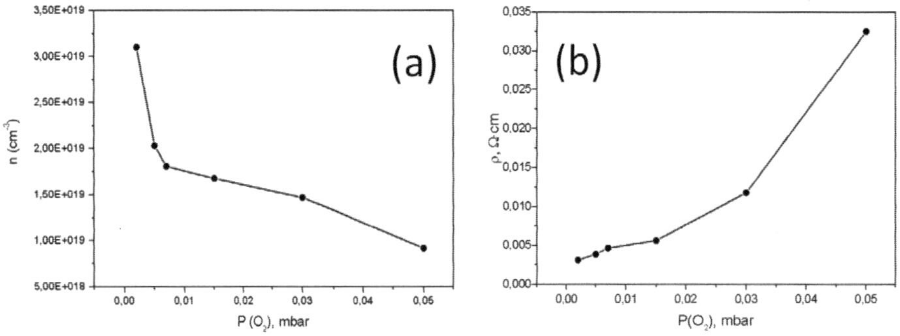

Figure 4. Electrical properties of Sb-doped ZnO films.
(a) – electron concentration versus oxygen pressure, (b) – resistivity versus oxygen pressure.
The figure shows graphs for the films with 1 at. % antimony concentration.

Significant differences in the photoluminescence spectra of samples obtained under different conditions were demonstrated (figure 5). The conditions under which there is no band in the region 500 - 600 nm the PL spectra were found: 1 at. % of Sb, synthesis temperature 550°C, annealing at 650 °C, O_2 pressure 2·10^{-3} mbar. The absence of this band can indicate low concentrations of point defects related with oxygen vacancies in films. It was also found that the peak corresponding to exciton transitions is red-shifted compared with such peak for pure zinc oxide.

Table 1. Electrical properties at room temperature of ZnO : Sb films. All obtained samples show n-type of conductivity.

Concentration of Sb, at. %	O_2 pressure, mbar	Resistivity, Ω·cm	Mobility, $cm^2/V·s$	Electron concentration, cm^{-3}
1	$5·10^{-3}$	$8.072·10^{-3}$	38.08	$2.031·10^{19}$
	$7·10^{-3}$	$8.859·10^{-3}$	38.98	$1.808·10^{19}$
	$1.5·10^{-2}$	$5.609·10^{-3}$	49.04	$2.269·10^{19}$
	$3·10^{-2}$	$1.180·10^{-2}$	35.97	$1.471·10^{19}$
	$5·10^{-2}$	$3.253·10^{-2}$	20.94	$9.162·10^{18}$
2	$5·10^{-3}$	$2.925·10^{-1}$	12.48	$1.709·10^{18}$
	$7·10^{-3}$	$5.815·10^{-2}$	42.30	$2.538·10^{18}$
	10^{-2}	$9.955·10^{-1}$	13.08	$4.792·10^{17}$
	$3·10^{-2}$	1.295	5.881	$8.199·10^{17}$

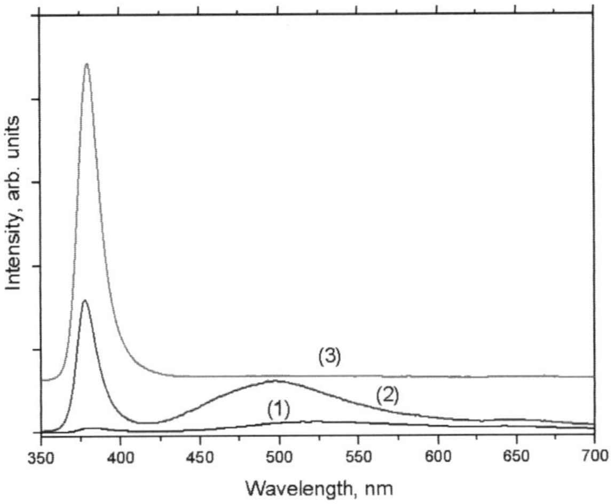

Figure 5. Room-temperature PL spectra of ZnO:Sb films (1 at. %):
(1) – synthesis temperature 550°C, annealing at 750 °C, O_2 pressure $2·10^{-3}$ mbar;
(2) – synthesis temperature 550°C, not-annealed, O_2 pressure $2·10^{-3}$ mbar;
(3) – synthesis temperature 550°C, annealing at 650 °C, O_2 pressure $2·10^{-3}$ mbar.

4. References

[1] Aoki T, Shimizu Y, Miyake A, Nakamura A, Nakanishi Y, Hatanaka Y 2002 *Phys. Status Solidi B* **229** 911
[2] Chu S, Lim J H, Mandalapu L J, Yang Z, Li L, Liu J L 2008 *Appl.Phys. Lett.* **92** 152103
[3] Pan X H, Guo W, Ye Z Z, Liu B, Che Y, He H P, Pan X Q 2009 *J.Appl. Phys.* **105** 113516
[4] Friedrich F, Sieber I, Klimm C, Klaus M, Genzel Ch 2011 Appl. Phys. Lett. **98** 131902
[5] Bundesmann C, Ashkenov N, Schubert M, Spemann D, Butz T, Kaidashev E M, Lorenz M, Grundmann M 2003 *Appl. Phys. Lett.* **81** 10
[6] Cody C A, DiCarlo L, Darlington R K 1979 *Inorg. Chem.* **18**, 1572
[7] B. Naidu, M. Pandey, V. Sudarsan, R. Vatsa, and R. Tewari 2009 *Chem. Phys.Lett.* **474** 180
[8] Gilliam S J, Jensen J O, Banerjee A, Zeroka D, Kirkby S J, Merrow C N 2004 *Spectrochim. Acta, Part A* **60** 425

Self-arrangement of periodic layers of silver nanoparticles in silicate glass

A V Red'kov[1], A A Lipovskii[1,2] and V V Zhurikhina[2]

[1] St.Petersburg Academic University, 8/3 Khlopina str., St.Petersburg, 194021, Russia
[2] St.Petersburg State Polytechnic University, 29, Politechnicheskaya str., St.Petersburg, 195251, Russia

E-mail: red-alex@mail.ru

Abstract. We studied the influence of hydrogen-reducible metal ions concentration profile under the surface of ion exchanged glass on the behavior of the layered glass-metal nanocomposite which can be formed via self-arranging in the course of thermal processing of the glass in hydrogen atmosphere. The modeling shows that depending of the ascending or descending type of the concentration profile the layered nanocomposite with decreasing or increasing gaps between the layers of nanoparticles can be formed. In certain cases the mean radius and the particle size-distribution function may be the same among the subsequent layers. A proper choice of the ascending concentration profile can provide practically equidistant layers of the nanocomposite applicable for photonic crystals and Bragg gratings.

1. Introduction

Since famous R. Feymann's talk "There's Plenty of Room at the Bottom" at the annual meeting of the American Physical Society at the California Institute of Technology in 1959 a lot of regular nanostructures were formed using the bottom-up approach. Firstly presented in 1981 [1] the self-arrangement, both spontaneous and externally stimulated, has been demonstrated in the systems of different nature. Below we concentrate on metal nanoparticles whose plasmonic properties [2] have attracted a lot of attention because of application potential [3]. Additionally to regular arrangements of metal nanoparticles under electric charging [4], with electron beam [5] or laser [6] irradiation, and with the use of organic templates [7], the self-arranging of multiple layers of silver [8] and copper [9] nanoparticles within glasses have been reported. The latter experiments consisted in hydrogen processing of glasses containing metal ions introduced in the glasses using ion exchange techniques. This kind of processing, being the conventional one for the formation of glass-metal nanocomposites (GMN), has resulted [8,9] in an irregular layered spatial distributions of the nanoparticles contrary to usually observed monotonic decrease [10,11] in nanoparticles concentration with the distance from the glass substrate surface. Modification of the existing theory of the nanoparticles formation [12] have allowed us to present the first model of the layering [13], which is capable of description of available experimental data [8,9], and the mechanism to control the gap between the GMN layers via varying the pressure of hydrogen in the course of the processing. Here we present another approach to govern the GMN layering, which is based on the formation of prescribed metal ions concentration profile in glasses to be processed in hydrogen.

2. Formation of layers

The problem under consideration is illustrated with figure 1. Hydrogen penetrates into heated glass containing positive ions capable of hydrogen reduction, and the probability of the reduction is proportional to the product of the ions and hydrogen concentrations. The glass contains one more type of positive ions, usually sodium, which possesses mobility comparable with the mobility of reducible ions. The mobilities of neutral hydrogen and reduced metal are also accounted for, while the mobility of oxidized hydrogen is low [14]. The set of equations describing the problem is presented in Appendix.

It is worth to note that proper relation of ionic mobilities and concentrations of species participating in the formation of metal nanoparticles (ionic and molecular hydrogen, ionic and atomic metal, e.g. silver, and ionic sodium initially presenting in glass) can provide non-monotonous profile of nanoparticles concentration in the glass due to the concurrence of two processes occurring in the glass enriched with reducible ions [13]. The first one is the formation of metal nuclei in the oversaturated solid solution of hydrogen-reduced metal in the glass, and the second is the growth of metal nanoparticles after the nucleation stage. This concurrence is because growing nanoparticles "consume" metal atoms via diffusion flow from the surrounding glass and, thus, decrease the concentration of the solid solution, while the size of critical nuclei increases with the solution drop [15]. That is why the probability of nucleation and, respectively, formation of new nanoparticles is hardly possible in the vicinity of existing ones.

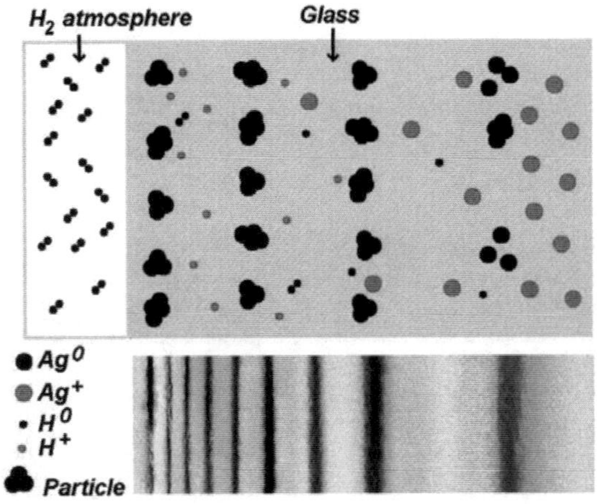

Figure 1. Formation of metal nanoparticles in glass: the scheme and a fragment of cross-sectional view of the glass sample from paper [8], darker regions correspond to GMN layers. The cross sectional view reproduced with the permission of the IOP Publishing Ltd from Ref. 8 (doi:10.1088/0953-8984/13/3/312).

In the course of hydrogen processing the gas diffuses into the glass through the surface and reduces the metal. This reactive diffusion results in hydrogen and reduced metal concentrations decaying with the distance, the same way decays the oversaturation of the glass-metal solid solution. With the account for the depletion of glass with metal atoms under the region where the nanoparticles are already formed, the distance to the next region where the nuclei can arise increases with moving away from the glass surface. This results in the increase in the gap between the layers of GMN for deeper layers. It was both experimentally demonstrated by Mohr [8] and Estournes [9] and theoretically described in our previous work [13]. The probability of metal reduction is, however, proportional to both metal ions and hydrogen concentration, and a non-uniformity of the ionic concentration can strongly influence mutual position of the GMN layers. This influence is essential if the distance where the concentration of the reducible ions in the glass changes is comparable with the size of the region depleted with the ions in the vicinity of growing metal nanoparticle. Thus relatively sharp concentration profiles are necessary to modify GMN layering and, if the reducible ions are introduced into the glass using ion exchange technique, only short processing duration should be used. According to the measured gap between the GMN layers [8], to modify the layering the concentration of the reducible ions should essentially differ at easily achievable distance of the order of a micron. Descending concentration profile can provide the gap between GMN layers growing with the distance from the glass surface and vice versa.

3. Results

To illustrate the influence of the concentration profile on the formation of the GMN layers we simulated the distribution of nanoparticles in soda-lime glass subjected to silver-sodium ion exchange and subsequent burying of the ion-exchanged layer using reverse sodium-silver ion exchange. Silver ions diffusion coefficient and concentration data were taken from [16] and other parameters from our paper [13]. Calculated for different durations of direct and reverse ion exchange ascending and descending profiles of silver ions in glass are presented in figure 2a.

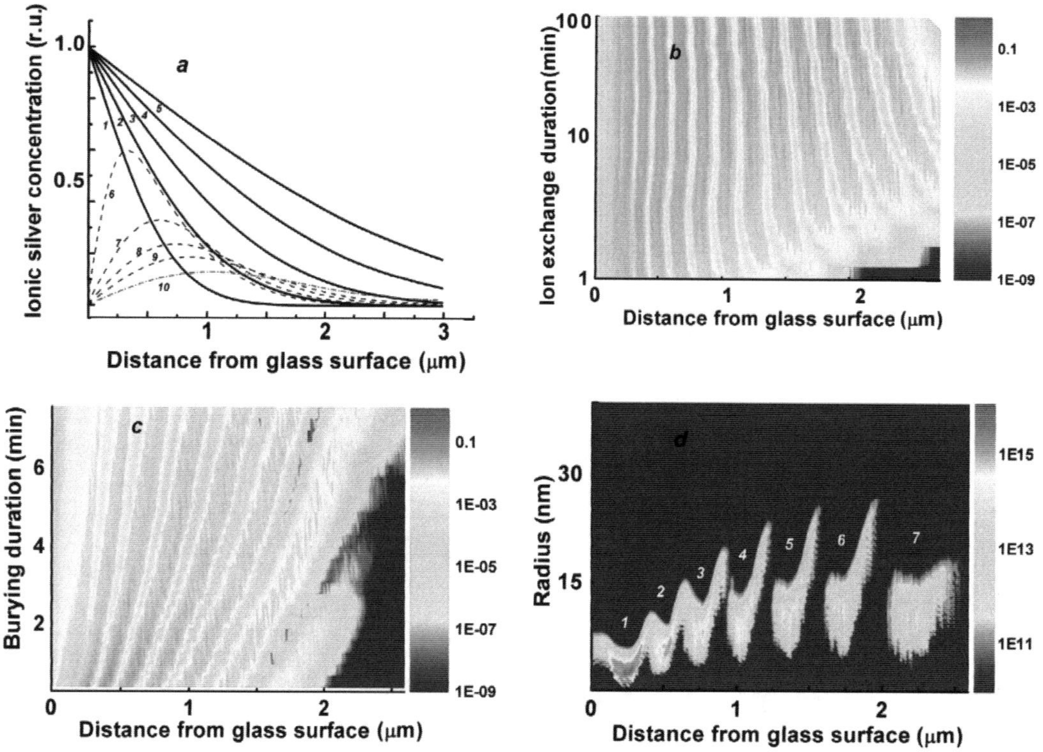

Figure 2. Concentration profiles of silver ions in glass formed with one-step (solid lines) and two-step (dashed lines) ion exchange (a), 1-5: 0.5 - 8.5 min of ion exchange; 6-10: 20s - 7 min of out-diffusion after 5 min ion exchange; depth distribution of volume fraction of silver nanoparticles calculated for descending silver concentration profile for different ion-exchange duration (b) and for ascending silver concentration profile for different durations of silver ions burying at the second step of ion exchange processing (c); nanoparticles size distribution function in cm, corresponding to profile 10 in (c) (5 min of silver to sodium and then 7 min of sodium to silver ion exchange). Color denotes the concentration of silver nanoparticles in cm^{-3} (d).

We used these silver ions concentration data to compute spatial distribution of silver nanoparticles resulted from hydrogen processing. In figure 2b,c one can see that shorter ion exchange time provides sharper nanoparticles concentration profiles that are narrower layers of the glass-metal nanocomposite.

For descending concentration of silver ions (solid lines in figure 2a) this corresponds to longer distances between the layers (figure 2b). Contrary, ascending concentration of silver (dashed lines in figure 2a), which can be obtained by two stage ion exchange, short sodium to silver after long silver to sodium, results in decreasing the gaps between the nanocomposite layers with the distance from the glass surface (figure 2c,2d). This phenomenon can be used to form equidistant layers of GMN in glasses. Also the mean radius and the particle size-distribution function are almost the same among the equidistant layers (figure 2d, layers 3-6).

4. Conclusions

Finally we have shown that varying concentration profile of hydrogen-reducible ions in glass allows formation within the glass the layers of GMN which differ in the behavior of spacing between the layers from increasing to decreasing with the distance from the glass surface in the cases of descending and ascending profiles, respectively. Our modeling has demonstrated that bell-shaped profile resulting from silver to sodium and then sodium to silver ion exchange processing of soda-lime glass can provide practically equidistant layers of GMN in the glass that is of interest for variety of optical and plasmonics applications.

Acknowledgements
Authors are grateful to S.A. Kukushkin for his valuable advices. The study was supported by RFBR project #14-02-00127

Appendix: Formulation of the problem

Here we formulate the equations describing the whole process which involves diffusion and ion exchange in the presence of sinks and sources, nucleation, growth and repining of nanoparticles. There are six entities involved into the process: ionic and atomic silver and hydrogen, ionic sodium, which diffuses itself and, thus, influences diffusion of ionic silver and ionic hydrogen, and silver nanoparticles. The equation (1) for hydrogen is

$$\frac{\partial C_{\mathrm{H}^0}}{\partial t} = D_{\mathrm{H}^0} \frac{\partial^2 C_{\mathrm{H}^0}}{\partial x^2} - k_1 C_{\mathrm{H}^0} C_{\mathrm{Ag}^+}, \tag{1}$$

where C_{H^0}, C_{Ag^+} are the concentrations of neutral hydrogen and ionic silver, D_{H^0} is hydrogen diffusion coefficient, k_1 is the rate of reduction reaction. The equation for ionic silver is

$$\frac{\partial C_{\mathrm{Ag}^+}}{\partial t} = D_{\mathrm{Ag}^+} \frac{\partial^2 C_{\mathrm{Ag}^+}}{\partial x^2} - k_1 C_{\mathrm{H}^0} C_{\mathrm{Ag}^+} + D_{\mathrm{Ag}^+} \frac{\partial \left(C_{\mathrm{Ag}^+} \dfrac{\alpha \nabla C_i}{1-\alpha C_i} \right)}{\partial x}, \tag{2}$$

where C_i is total concentration of ionic silver and sodium, D_{Ag^+}, D_{H^+} are diffusion coefficients of ionic silver and hydrogen, respectively, $\alpha = 1 - D_{Ag^+} \cdot (D_{H^+})^{-1}$, and we assume that silver and sodium diffusion coefficients are equal. The third term in the right part of equation (2) is responsible for self-consistent process of hydrogen, sodium and silver ions diffusion.

Sodium does not participate in silver reduction but affects the distributions of other charged species, ionic silver and hydrogen. The equation (3) for ionic sodium is

$$\frac{\partial C_{Na^+}}{\partial t} = D_{Na^+} \frac{\partial^2 C_{Na^+}}{\partial x^2} + D_{Na^+} \frac{\partial \left(C_{Na^+} \frac{\alpha \nabla C_i}{1 - \alpha C_i} \right)}{\partial x}, \tag{3}$$

where C_{Na^+} is the concentration of sodium, D_{Na^+} is sodium diffusion coefficient.

The particle critical radius, at which it becomes stable, can be written as [15]

$$R_{cr}(\xi) = \frac{2\gamma\omega}{k_B T \ln \xi}, \tag{4}$$

where γ is the coefficient of surface tension, ω is the atomic volume of silver, and ξ is the oversaturation of the solid solution of neutral silver in glass.

Equation (5) for atomic silver accounts for its diffusion, reduction of silver ions, formation and growth of nanoparticles:

$$\frac{\partial C_{Ag^0}}{\partial t} = D_{Ag^0} \frac{\partial^2 C_{Ag^0}}{\partial x^2} + k_1 C_{H^0} C_{Ag^+} - k_2 C_{Ag^0} \ln \xi \frac{4\pi}{3\omega} R_{cr}^3 \exp\left(- \frac{k_3}{\ln^2(\xi)} \right) - \int_0^\infty 4\pi R^2 V_r N dR \tag{5}$$

Here the growth rate V_r can be described [17] as

$$V_r(R) = \partial R / \partial t = D_{Ag^0} (C_{Ag^0} - S(1 + 2\gamma\omega / k_B TR)) / R,$$

and k_2 and k_3 are the nucleation rate and a kind of threshold dimensionless nucleation energy, respectively, $k_3 = 16\pi\gamma \cdot (3k_B T)^{-1}$. The expression under the exponent equals to the ratio of the energy to create a nanoparticle of critical radius and $k_B T$.[18]

The last equation is the equation of continuity for particle size-distribution function *N*. It describes nanoparticle growth and formation [18]

.

$$\frac{\partial N}{\partial t} + \frac{\partial}{\partial R}(N V_r) = k_2 C_{Ag^0} \ln \xi \exp\left(- \frac{k_3}{\ln^2(\xi)} \right) \delta(R - R_{cr}) \tag{6}$$

Here the right part describes the nucleation of particles of critical radius R_{cr}. We solved one-dimensional problem using the boundary conditions of constant hydrogen concentration at the glass surface, the absence of fluxes of other species through the surface, and taking into account that the sample thickness is much larger than the diffusion depths.

References

[1] K. E. Drexler, Proc. Natl. Acad. Sci. USA 78, 5275, (1981).
[2] U. Kreibig, M. Vollmer, Optical properties of metal clusters (Springer, N.Y., 1995)
[3] M. Pelton, J. Aizpurua, G. Bryant, Laser & Photon. Rev. 2, 136 (2008).
[4] S. Pyrpassopoulos, D. Niarchos, G. Nounesis, N. Boukos, I. Zafiropoulou, V. Tzitzios, Nanotechnology 18, 485604 (2007).
[5] A. I. Sidorov, A. V. Nashchekin, V. N. Nevedomskiy, O. A. Usov, O. A. Podsvirov, Int. J. Nanosci. 10, 1265 (2011).
[6] F. Goutaland, J.-P. Colombier, M. C. Sow, N. Ollier, F. Vocanson, Opt. Express 21, 31789 (2013).
[7] B. W. Jacobs, R. J. T. Houk, M. R. Anstey, S. D. House, I. M. Robertson, A. A. Talin, M. D. Allendorf, Chem. Sci. 2, 411 (2011).
[8] C. Mohr, M. Dubiel, H. Hofmeister, J.Phys-Condens Mat. 13, 525 (2001).
[9] C. Estournes, N. Cornu, J.L. Guille, J. Non-Cryst.Solids, 170, 287 (1994).
[10] G. De Marchi, F. Caccavale, F. Gonella, G. Mattei, G. Battaglin, A. Quaranta, Appl. Phys. A 63, 403 (1996).
[11] V.V. Afrosimov, B.Ya. Ber, V.V. Zhurikhina, M. V. Zamoryanskaya, D. Yu. Kazantsev, E. V. Kolesnikova, A. A. Lipovskii, M. I. Petrov, Tech. Phys.+ 55, 1601 (2010).
[12] Yu. Kaganovskii, A. Lipovskii, M. Rosenbluh, V. Zhurikhina, J. Non-Cryst.Solids 353, 2263 (2007).
[13] A.V. Redkov, A.A. Lipovskii, V.V. Zhurikhina, J. Non-Cryst.Solids 376, 152 (2013).
[14] R. H. Doremus, Appl. Phys. Lett. 87, 232904 (2005).
[15] E.M. Lifshits, L.P. Pitaevski, Physical Kinetics (Pergamon, Oxford, 1981).
[16] V. V. Zhurikhina, M. I. Petrov, K. S. Sokolov, O. V. Shustova, Tech. Phys.+ 55, 1447 (2010).
[17] S.A. Kukushkin, A.V. Osipov, Usp. Fiz. Nauk+, 168, 1083 (1998).
[18] A.V. Redkov, Phys. Solid State+ 54, 1875 (2012).

SPbOPEN2014 IOP Publishing
Journal of Physics: Conference Series **541** (2014) 012006 doi:10.1088/1742-6596/541/1/012006

Express – control of biological solutions by portable nuclear – magnetic spectrometer

V. Davydov, A. Cheremiskina, E. Velichko, A. Karseev

Saint Petersburg State Polytechnical University, Saint Petersburg, Russia

E-mail: davydov_vadim66@mail.ru

Abstract. The portable nuclear–magnetic spectrometer for express control of biological solutions and ecological monitoring is presented. The designed construction of nuclear–magnetic spectrometer allows measuring longitudinal relaxation time T1 and transverse time T2 of condensed media in low magnetic field B_0 = 62 mTl. The presented nuclear magnetic resonance signal accumulation scheme can measure the relaxation times of small amounts of liquid media (1 ml).

1. Introduction

Express-control of biological solutions is actual task for ecological monitoring of liquid media. The longitudinal relaxation time (T1) and transverse time (T2) of condensed media give information about impurities presence, changes of chemical composition and other changes of medium state. Multifunctional X-ray, optical and magnetic resonance spectrometers provide precise analysis of liquid media but they are expensive, power consuming and not portable.

It is known that state of liquid medium can be evaluated by analysis of shifts in temperature, T1 and T2 relaxation times in comparison with initial values [1-4]. The valid method of analysis should include simultaneous measurement of T1 and T2 [2-6]. But commercially-available portable nuclear – magnetic spectrometers usually measure only T2 time, and the weight of the spectrometers for simultaneous measurement of T1 and T2 is usually higher than 40 kg.

Development of small portable spectrometer for preliminary express analysis of biological media or water in field conditions is actual task for timely ecological monitoring. We report about a new construction of portable nuclear – magnetic spectrometer for analysis of longitudinal relaxation time and transverse time of small amount of condensed liquid media (1 ml) in low magnetic field (B_0 = 62 mTl).

2. The construction of portable nuclear – magnetic spectrometer

Based on our previous work a new construction of compact spectrometer on the nuclear-magnetic resonance (NMR-spectrometer) was developed.

The structural scheme of the designed NMR-spectrometer is presented in figure 1.

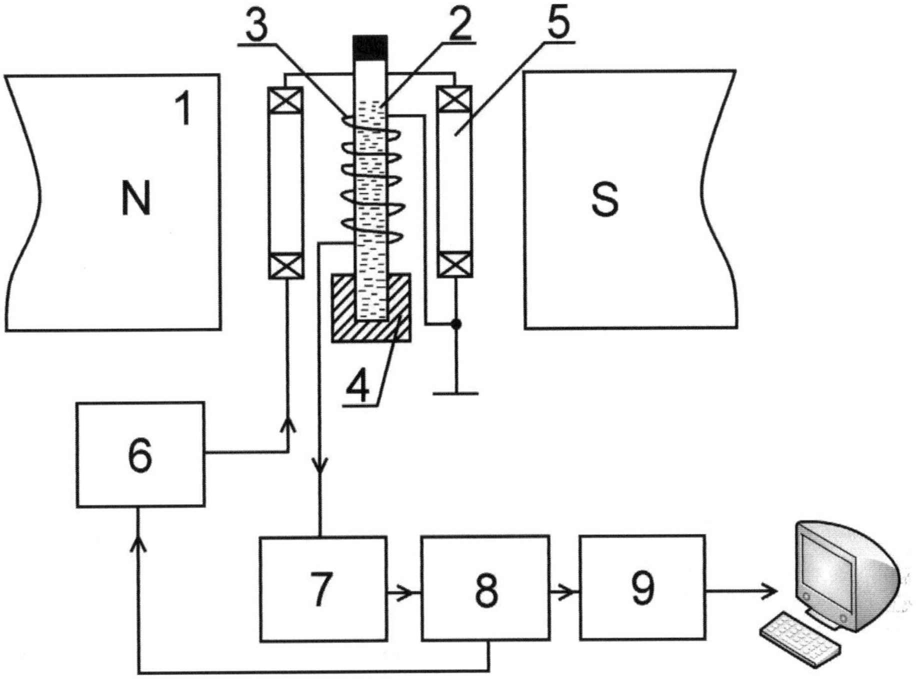

Figure 1. Block scheme of NMR spectrometer: 1 –permanent magnet, 2 – container with test media, 3 – coil for NMR signal registration, 4 – fixing gripper for container, 5 – modulation coil, 6 – radiofrequency generator, 7 – weak fluctuations generator (autodyne receiver), 8 – processing and control equipment, 9 – conversion diagram.

It is known that magnet systems make the main contribution to weight of the devices based on NMR effect [1,5-6]. So, the miniaturization of the spectrometer is related to reduction of magnets size. In this situation the modern high-energy magnetic materials with high level of magnetic field induction and homogeneity should be used.

In the presented construction of the device the distance between the magnetic poles is 12 mm, the magnetic field induction is $B_0 = 62$ mTl, inhomogeneity level is $0.5{\cdot}10^{-3}$ cm^{-1} and frequency of NMR signal registration is $f_{NMR} = 2614$ kHz.

The NMR signal in weak field B_0 was registered by coil for NMR signal registration (3) and the developed weak fluctuations generator (7). The special models of the coil and autodyne receiver allowed us to measure signal from minute volumes of liquid media (1 –1.5 ml). The circuit of weak fluctuations generator (autodyne) was based on the differential stage with planar field-effect transistor with drain detection and amplification of NMR signal. The autodyne receiver based on suggested scheme is characterized by steady operation in broad band and has high level of common noise suppression.

3. Results and discussion

In figure 2 the NMR signal on generator output is presented. The signal was registered in wide bandwidth (10 kHz) with modulation frequency $f_{mod} = 50$ Hz.

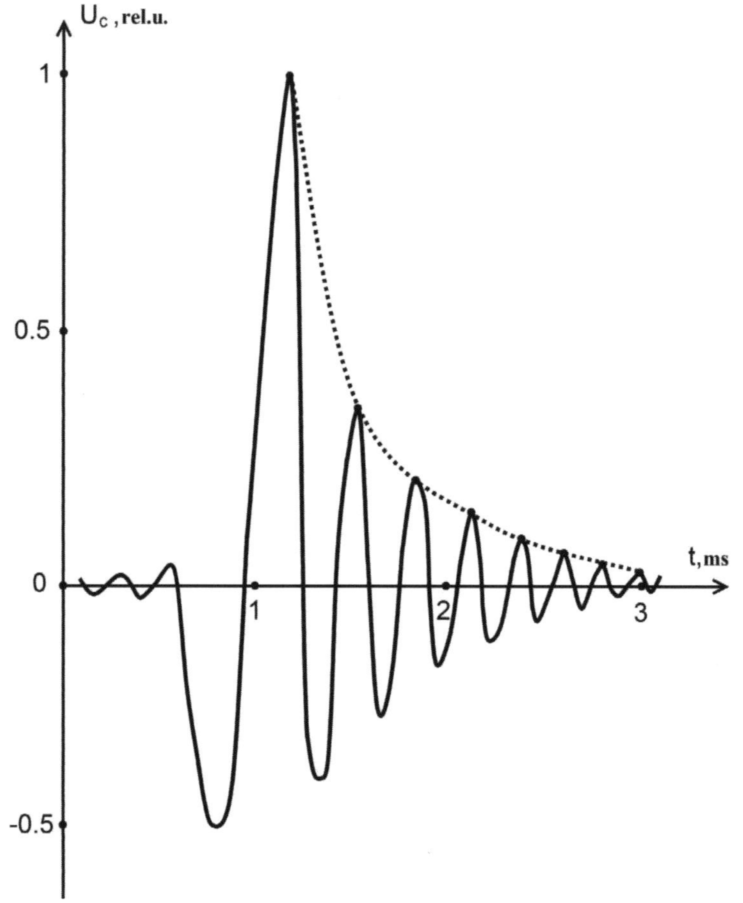

Figure 2. Signal curve of NMR with the envelope line for T_2^* measurements.

The shape of NMR signal line is approximated by function [7,8]:

$$U_c(t) = U_0 e^{-t/T_2^*} \cdot cos\left(at^2/2\right),$$

where U_0 is the maximal value of signal amplitude, T_2^* is the effective transverse relaxation time, a is the rate of magnetic field detuning change. The rate of magnetic field detuning change is defined by:

$$a = \gamma \frac{\partial H_z}{\partial t} = \partial\left(\frac{\Delta\omega}{\partial t}\right),$$

where γ is the nuclear gyromagnetic ratio, $\Delta\omega$ is the resonance frequency detuning.

The effective transverse relaxation time T_2^* is calculated from the envelope curve decay (dashed line in figure 2). The transverse relaxation time T_2 is defined by:

$$\frac{1}{T_2^*} = \frac{1}{T_2} + \frac{\gamma \Delta H}{\pi},$$

where ΔH is the magnetic field nonuniformity in the area of NMR signal registration coil.

The Julotto method for registration of NMR signal in weak magnetic field was implemented for the first time. The method was used for registration of T_1.

Let us consider the periodic modulation which is symmetric with respect to:

$$B_0 = {f_{NMR}}/{\gamma}.$$

The maximal amplitude of the signal is registered if time between two resonance crossings is much more than T_1. Otherwise the registered NMR signals decrease according to:

$$M = M_0 \frac{1-exp\left(-\tau/T_1\right)}{1+exp\left(-\tau/T_1\right)}, \quad (1)$$

where M is the value of magnetization (which is directly proportional to amplitude of the signal), τ is modulation period, M_0 is the magnetization of liquid medium in the absence of magnetic field modulation:

$$M_0 = \chi_0 B_0,$$

where χ_0 is the static nuclear magnetic susceptibility.

From equation (1) it follows that for computation of T_1 it is necessary to measure NMR signal amplitude at two different modulation frequencies (which correspond to different modulation periods τ). T_1 is usually defined graphically.

Size reduction of registration coil allowed us to decrease ΔH, improve accuracy of T_2 measurements and decrease the sample volume to 1 –1.5 ml. But it caused the signal to noise ratio degradation. To solve this problem a new process and control unit (based on microcontroller STM32) was developed. The process and control unit use the effective signal-summing scheme and can regulate the parameters for optimal registration of the signal. The microcontroller stores the signal to suppress the noise, and at necessary level of signal to noise ratio the process and control unit calculates the relaxation times. The advantage of suggested approach is the flexible system of signal-summing, which provides the optimal choice of accumulation cycles number and save measuring time.

The experimental results showed that the spectrometer can measure the longitudinal relaxation time T1 from 1 ms to 20 s, the transverse time – from 0.5 ms to 4 s. The relative error of measured parameters was less than 1 %.The operating range of temperatures of the spectrometer is 4-45^{0}C.

The obtained experimental results confirmed effectiveness of developed spectrometer for measurement of T_1 and T_2 relaxation times for biological liquids and showed possibilities of detection impurities in low concentration.

4. Conclusion

The portable nuclear–magnetic spectrometer for express control of biological solutions and ecological monitoring is presented. We developed the scheme of autodyne receiver based on differential amplification stage for signal detection, which increase signal to noise ratio and save measuring time.

The main advantages of the device are the possibility to analyze small amounts of liquid media (volume is about 1 ml), short time of measurements, transportability and capability of field analysis.

References

[1] Davydov V V and Karseev A U 2013 *Proc. of XIV Int. Environmental Forum Baltic Sea Day "BSD'13"* **13** 34
[2] Davydov V V 1997 *J. Phys. B: Atomic, Molecular and Optical Physics* **30** 3993
[3] Chen B, Ivanov I, Klein M L and Parrinello M 2003 *Phys. Rev. Lett* **91(21)** 2155
[4] Davydov V V, Dudkin V I and Karseev A U 2013 *Opt. Mem. & Neural Networks (Information Optics)* **22** 112
[5] Lopez E, Ortiz W and Quintana I M 1998 *Chemical Physics Letters* **287** 429
[6] Davydov V V 1998 *Int. J. of Modern Phys.* **7** 798
[7] Abragam A 1963 *Nuclear magnetism* 646
[8] Leshe A 1963 *Nuclear induction* 684

Initial stages of misfit stress relaxation by rectangular prismatic dislocation loops in composite nanostructures

M. Yu. Gutkin,[1-3] **A.M. Smirnov**[2,3]

[1]Institute of Problems in Mechanical Engineering, Russian Academy of Sciences, Bolshoi pr. 61, St. Petersburg, 199178 Russia
[2]St. Petersburg State Polytechnical University, ul. Polytekhnicheskaya 29, St. Petersburg, 195251 Russia
[3]ITMO University, Kronverkskii pr. 49, St. Petersburg, 197101 Russia

Email: m.y.gutkin@gmail.com, smirnov.mech@gmail.com

Abstract. Theoretical models are proposed for describing the initial stages of misfit stress relaxation in bulk and hollow core-shell nanoparticles and nanowires, and in planar bi- and tri-nanolayers through generation of rectangular prismatic dislocation loops at the inner and outer interfaces. We obtain the changes in the total energy of the system due to the appearance of dislocation loops and calculate the critical conditions of their formation. We also determine the most profitable shape of the loops and the preferred places of their generation. Finally, we choose the nanostructures which are the most stable against dislocation loop generation.

1. Introduction

To date production and applications of nanoparticles, nanowires and nanolayers is one of the most promising areas of nanotechnology. Electronic, magnetic and optical properties of such nanostructures depend on their shape, size, chemical composition, crystal lattice type and the presence of various defects in them. Composite nanoparticles, which represent a significant fraction among the nanoparticles synthesized and studied today, consist of different materials and have many applications in modern optoelectronics, photonics, spintronics, solar cells, sensors, storage devices, catalysis, medicine, etc. [1-3]. Composite nanowires also have excellent electronic and optical properties, so they are used in various devices of optoelectronics, in nanoscale field effect transistors, logic devices, etc. [4,5]. Use of composite nanolayers in engineering systems can give them new properties and functions. One can mention opto- and microelectronic devices, NEMS and MEMS, and various coatings designed to improve the mechanical, tribological, environmental, optical, electrical, magnetic and biological properties of surfaces [6].

Fabrication of composite nanostructures is accompanied by the appearance of misfit stresses caused by the mismatch between crystal lattices and material properties of their components, which leads to substantial changes in their properties. Under some critical conditions, these stresses can relax through development of various defect structures [6-10] or through the destruction of the nanostructures, which explains a special interest to this problem. One of possible relaxation mechanisms is generation and expansion of prismatic dislocation loops (PDLs). The purpose of this work is an approximate calculation and a comparison of critical conditions for the generation of PDLs at the outer and inner boundaries in spherical bulk and hollow core-shell nanoparticles, cylindrical bulk and hollow core-shell nanowires and planar bi- and tri-nanolayers.

2. Models

Consider some typical composite nanostructures which are bulk core-shell nanoparticles and nanowires (Figure 1a,c), hollow core-shell nanoparticles and nanowires (Figure 1b,d), and bi- and tri-nanolayers (Figure 1e,f).

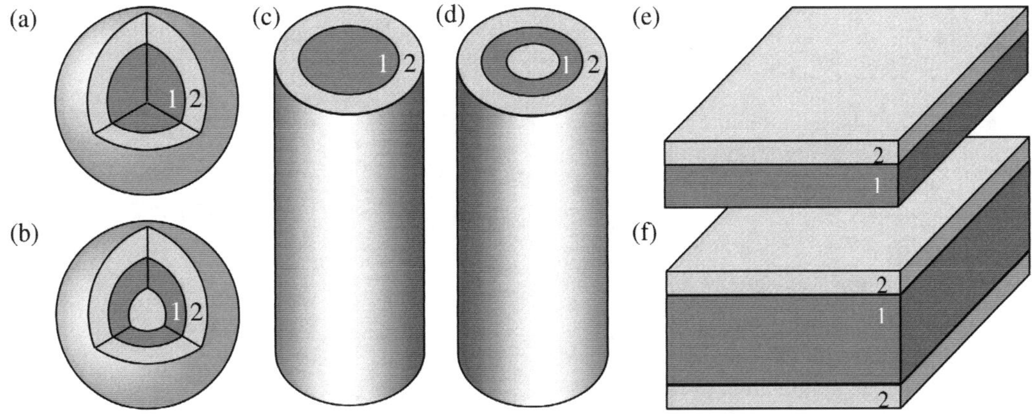

Figure 1. Sketch of composite nanostructures in the form of thin shells (films) 2 on different cores (substrates) 1: (a) bulk core-shell nanoparticle, (b) hollow core-shell nanoparticle, (c) bulk core-shell nanowire, (d) hollow core-shell nanowire, (e) bi-nanolayer, and (f) tri-nanolayer.

The nanostructures are assumed to be elastically isotropic and homogeneous. Further, we suppose that in the nanoparticles and nanowires, the shell thickness is much smaller than the core radius, in which case one can use a planar model of a thin layer on a semi-infinite substrate for correct calculation of strain energies of rectangular PDLs generated in the nanostructures. With these approximations, we find strict analytical formulas for the energy changes accompanied the PDL generation on either core-shell (film-substrate) interface or outer shell (film) surface, and compare these cases in terms of critical misfit parameters with account for the PDL shape. We consider the following three characteristic shapes for the PDLs: (i) extended normally to the interface (NI-loop, Figure 2a), (ii) square (S-loop, Figure 2b), and (iii) extended along the interface (AI-loop, Figure 2c).

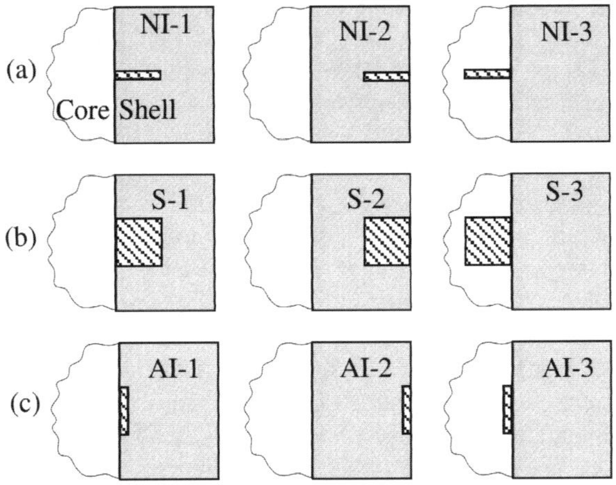

Figure 2. Schematics of prismatic dislocation loops of types (a) NI-(1, 2, 3), (b) S-(1, 2, 3), and (c) AI-(1, 2, 3).

3. Results

Our calculations have shown that the AI-loops are the most preferable for all nanostructures under consideration and that the free surface is the most favorable place for generation of these loops. On the other hand, the core-shell (substrate-film) interface is more profitable than the shell (film) free surface for generation of the NI- and S-loops. The critical misfit for PDLs, which are generated in the core/substrate (shell/film), decreases (increases) with an increase in the shell/film thickness h and with a decrease in the nanoparticle/nanowire radius R or in the bi-nanolayer thickness H.

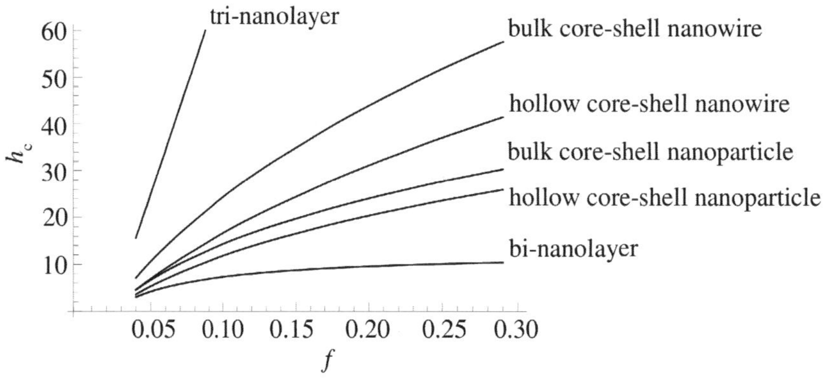

Figure 3. Dependence of the critical shell/film thickness h_c on the misfit parameter f for the case when $r_0 = h_s = 24$ nm, $H_s = 48$ nm, $r_p = 15$ nm and PDLs are generated from the shell/film free surface. Here r_p is the inner radius of hollow core-shell nanoparticles/nanowires, r is the inner radius of bulk core-shell nanoparticles/nanowires, h_s – thickness of bi-nanolayer substrate, H_s – thickness of tri-nanolayer substrate.

Figure 3 shows the dependences of the critical shell/film thickness h_c on the misfit parameter f for the nanostructures under consideration in the most favourable case when PDLs are generated from the shell/film free surface and extend along the interface (AI-2-loops, see Figure 2c). The generation of PDLs in a nanostructure is energetically favorable when the values of h_c and f fall into the area under the curve corresponding to the nanostructure. Therefore, the most stable nanostructure in this sense is the bi-nanolayer, as it is characterized by the smallest area under its curve. The reason is its bending which is an effective channel of misfit stress relaxation. In contrast, the tri-nanolayer (two thin films deposited on both the sides of a planar substrate, see Figure 1f) is the least resistant to the PDL formation. Indeed, when the nanoparticle/nanowire core diameter $2r_0$ is equal to the substrate thickness H_s and the nanoparticle/nanowire shell thickness is equal to the film thickness, the misfit stress in the nanoparticle shell is lower than that in the nanowire shell and lower than that in the film. Further reduction of the misfit stress in the nanoparticle/nanowire shell can be achieved through the formation of a void in its core (Figures 1b,d) [11]. The larger the void radius, the lower the misfit stress in the nanoparticle/nanowire shell and the harder the PDL formation.

To determine the preference of the PDL shape and its generation site in the nanostructures shown in Figure 1, one can use Table 1. The less grade in the table cell the easier formation of the PDL in the corresponding nanostructure. As is seen, PDL generation is easier in the shell of nanowire than in the shell of nanoparticle or in the film of bi-nanolayer. On the other hand, PDL generation is easier in the substrate of bi-nanolayer than in the core of nanowire or nanoparticle. PDL generation in a longitudinal section of nanowire shell is more profitable than that in its cross section, however, for the nanowire core it is vice versa. It is also shown, that for the PDLs generated from free surfaces and extended along interfaces (AI-2-loops), the most stable nanostructure is the bi-nanolayer (grade 11), while the most unstable is the tri-nanolayer (grade 1). The most disadvantageous case for PDL generation is when square PDLs (S-3-loops) are generated from the interface to the nanowire core in its longitudinal section.

Table 1. Preference of loop shapes and generation sites in different nanostructures

	Loop generation from the interface to the shell/film			Loop generation from the free surface to the shell/film			Loop generation from the interface to the core/substrate		
	AI-1	S-1	NI-1	AI-2	S-2	NI-2	AI-3	S-3	NI-3
Bulk core-shell nanoparticle	13	17	16	6	35	26	28	34	29
Hollow core-shell nanoparticle	-	-	-	7	-	-	-	-	-
Bulk core-shell nanowire (loop in cross section)	12	15	10	3	33	23	25	31	27
Bulk core-shell nanowire (loop in longitudinal section)	9	14	8	2	32	22	38	**39**	36
Hollow core-shell nanowire (loop in cross section)	-	-	-	5	-	-	-	-	-
Hollow core-shell nanowire (loop in longitudinal section)	-	-	-	4	-	-	-	-	-
Bi-nanolayer	18	20	19	11	37	30	21	25	24
Tri-nanolayer	-	-	-	**1**	-	-	-	-	-

Evaluated for $R = H = 60$ nm, $h = 10$ nm, $r_p = 15$ nm.

References

[1] S. Behrens, Nanoscale **3** (2011) 877.
[2] C. De Mello Donega, Chem. Soc. Rev. **40** (2011) 1512.
[3] D. Shi, N.M. Bedford, H.S. Cho, Small **7** (2011) 2549.
[4] V.G. Dubrovskii, G.E. Cirlin, V.M. Ustinov, Semiconductors **43** (2009) 1539.

[5] S. Barth, F. Hernandez-Ramirez, J.D. Holmes, et al., Prog. Mater. Sci. **55** (2010) 563.
[6] L.B. Freund, S. Suresh. Thin film materials: stress, defect formation and surface evolution. Cambridge University Press, 2004.
[7] L.I. Trusov, M.Yu. Tanakov, V.G. Gryaznov, et al., J. Cryst. Growth **114** (1991) 133.
[8] M.Yu. Gutkin, Int. J. Engng. Sci. **61** (2012) Special Issue, 59.
[9] M.Yu. Gutkin, A.L. Kolesnikova, S.A. Krasnitckii, A.E. Romanov, Phys. Solid State **56** (2014) 723.
[10] M.Yu. Gutkin, A.M. Smirnov, Phys. Solid State **56** (2014) 731.
[11] M.Yu. Gutkin, A.L. Kolesnikova, S.A. Krasnitckii, A.E. Romanov, A.G. Shalkovskii, Scripta Mater., 2014, in press.

A finite-length capped single-walled carbon nanotube (5, 5) under an applied electric field

A V Tuchin, A V Nestrugina, L A Bityutskaya, E N Bormontov

Voronezh State University, 394006, Universitetskaya pl.1, Voronezh, Russia

E-mail: a.tuchin@bk.ru

Abstract. The paper describes results of a theoretical study of the electronic structure of the capped finite-length single-walled carbon nanotube (5, 5) under an applied electric field in the range from 0 to 0.5 V/Å. Clear oscillations of an ionization potential, electron affinity, gap energy and an effective work function on the length of the nanotube are presented. An applied electric field leads to the energy gap decrease and the work function increase. An estimation of the electron emission current density changes is revealed taking into account the work function dependence both on the nanotube length and the strength of the electric field.

1. Introduction

An electric field affects fullerenes and carbon nanotubes (CNT) in many cases, such as, the arc discharge syntheses, scanning microscopy investigations, in the polymer solar cells, field emission and field effect devises e.t.c. [1-4]. A high curvature and small sizes lead to the high local electric field even if a low voltage is applied [5-6]. The electric field can significantly change the electronic structure of nanotubes [6-8]. Klinovaja at al. [6] showed that electric fields can be used for spin manipulation through spin-orbit interaction in the infinite-length defect-free CNTs. Kim et al. [7] revealed that the linear effective work function of the capped armchair (5, 5) nanotube decreases with the field strength increase. There are many external and internal degrees of freedom, such as field strength, CNT radius, chirality, electronegativity allow an extensive tunability of the electronic structure [6, 8]. From this point of view finite-length single-walled carbon nanotubes (SWCNTs with the length less than 10 nm) are very interesting object with the length as additional internal parameter.

Finite-length SWCNTs are the special class of objects which occupy intermediate position between fullerenes and nanotubes [9, 10]. Buonocore et al. [8], Rocherfort et al. [11] and Wang et al. [12] showed that finite-length open-ended armchair SWCNTs have a non-zero band gap. The energy gap, ionization potential and electron affinity show clear oscillations with the length of tubes. The capped finite-length tube (5, 5) shows oscillations of fundamental parameters and standard enthalpy of the formation with the length [9, 10]. It should be noted, that infinite-length armchair nanotubes must have a zero band gap because their chiral indexes n and m satisfy condition $n–m=3k$, $k= 0, 1, 2...$ [13-15]. The influence of the electric field on the electronic structure of the capped finite-length SWCNTs is poorly understood. Tuchin et al. [10] showed quadratic decreases of the energy gap of the capped finite-length SWCNT (5, 5) with the field strength increases. The dependences of other fundamental parameters were not discussed.

The aim of the investigation is to reveal the impact of the length and the electric field on the electronic structure of the capped finite-length SWCNT (5, 5).

2. Computational details

All electron simulations of the electronic structure of the capped finite-length SWCNTs were performed using the density functional theory (DFT) computational method with a local spin density approximation (LSDA) and 3-21*G basis set with the help of the Gaussian software in the Supercomputing center of Voronezh State University [16-19]. A stoichiometric formula of the capped SWCNT (5, 5) is C_{60+10i}, where i denotes the number of segments which form the body of the tube while caps essentially are halves of the fullerene molecule dissected perpendicularly to the C_5 axis [9, 10]. The SWCNT has D_{5h} and D_{5d} symmetry at odd and even i. A diameter of the tube equals to the 7Å. To take an orientation deformation under the influence of ponderomotive force into account, a geometry optimization of the molecule in the range of electric field ε=0–0.5 V/Å was performed without symmetry retention. The electric field was directed parallel to the tube axis.

There are two approach to calculate the ionization potential (IP) and electron affinity (EA) as IP= $E_{vac}-E_{LUMO}$, EA= $E_{vac}-E_{HOMO}$ and IP= $E_{tot}(q=-1)-E_{tot}(q=0)$, EA= $E_{tot}(q=0)-E_{tot}(q=+1)$, where $E_{vac}=0$ the electrostatic potential energy in the vacuum, far away from system, E_{LUMO}, E_{HOMO} energy of the lowest unoccupied (LUMO) and highest occupied (HOMO) molecular orbitals, $E_{tot}(q=0)$, $E_{tot}(q=-1)$, $E_{tot}(q=+1)$ total ground-state energies in the neutral and single charged configuration. Buonocore et al. [8] showed that work function of the finite-length SWNTs defined as φ= (EA+IP)/2 not depend on the used approach. In the current work we defined the IP, EA and hence φ out of LUMO and HOMO energies because this method requires fewer calculations. The energy gap and an effective work function assigned as $E_{LUMO\text{-}HOMO}= E_{LUMO}-E_{HOMO}$ and $\varphi_{eff}= (E_{LUMO}-E_{HOMO})/2$ respectively.

3. Results and discussion

3.1. The fundamental parameters oscillation of the finite-length capped SWCNT (5, 5)

Results of the fundamental parameters calculation of the capped finite-length armchair SWCNT (5, 5) are presented in the figure 1. The ionization potential, electron affinity (figure 1a), energy gap, effective work function (figure 1b) show clear oscillations with the tube length. The nature of oscillations discussed in several works for open-ended [8, 12] and capped [9, 10] armchair SWCNTs and can were explained as a quantum confinement effect and an electron density redistribution of the LUMO and HOMO. Rocherfort et al. [11] introduced a division on the number of segments i of the finite–length open-ended armchair SWCNTs: i=3p+1, 3p and 3p+2 with the maximum, average and minimal values. That was further confirmed for capped tubes in the region of ultra-small lengths [9, 10].

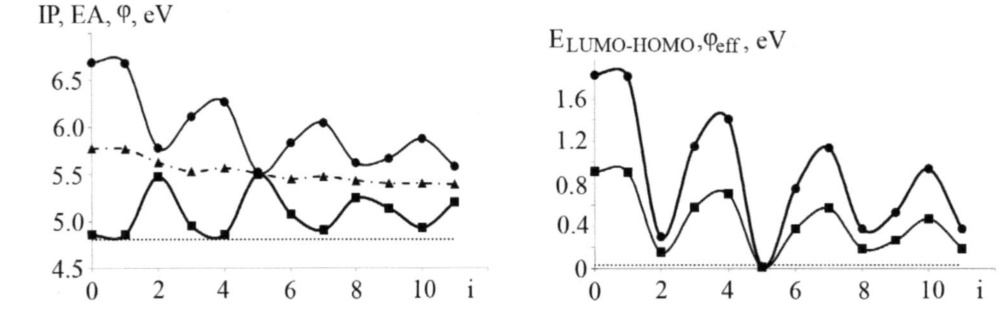

a b

Figure 1. The ground-state electronic structure of the finite-length capped SWCNT (5, 5) versus the number of segments i. a – the ionization potential IP (circles), electron affinity EA (squares) and work function φ (triangles), the dot line corresponds to the calculated φ=4.81 eV of the infinite-length open-ended SWCNTS (5, 5). b – the energy gap $E_{LUMO\text{-}HOMO}$ (circles) and effective work function φ_{eff} (squares) the dot line corresponds to the calculated gap 0.03 eV of the infinite-length open-ended SWCNTS (5, 5)

Our calculations have revealed that introduced division is true in the studied range i from 0 to 11 (figure 1b). Therefore, properties of finite-length armchair SWCNTs of the ultra-small diameters are primarily determined by the tube length. It is important to consider the influence of curvature effects on the energy gap of the SWCNTs with diameters less than 2 nm [9, 10, 20-23]. Our calculated value for infinite-length open-ended SWCNT (5, 5) is 30 meV. An additional contribution to the nonzero $E_{LUMO\text{-}HOMO}$ is the caps influence.

The work function shows low amplitude oscillations because of antiphase dependencies on the length with close amplitudes of the ionization potential and electron affinity (fig.1a). In the range i from 0 to 11 φ decreases from 5.77 eV to 5.39 eV. The electron field emission current density j is very sensitive to the magnitude of the work function [24]. According to the Fowler-Nordheim model [25] the current density can be expressed as follows:

$$j = \frac{A\beta^2\varepsilon^2}{\varphi}\exp\left(-\frac{B\varphi^{\frac{3}{2}}}{\beta\varepsilon}\right) \tag{1}$$

, where ε is the applied field (V/nm), $A=1.56\cdot10^{-6}$ ($A\cdot eV/V^2$), $B=6.83\cdot10^{9}$ ($V\cdot eV^{-3/2}/m$), β is the field enhancement factor. The expression (1), called the Fowler-Nordheim formula, is convenient for treatment of experiments on electron field emission [26-27]. Using (1) we evaluated the relative current density change $j(i=11)/j(i=0)\approx 6$ at $\varepsilon=0.5$ V/Å.

3.2. The finite-length capped SWCNT (5, 5) excited by the electric field

An applied electric field leads to the charge redistribution and polarization of the SWCNT. A dipole moment of capped SWCNT (5, 5) is proportional to the field strength (figure 2a). The absolute change of the dipole moment, relative to the unexcited state, is proportional to the length square. The dependence of a module of a maximum effective charge $|Q_{effmax}|$ on the field strength is close to linear for tubes with i>5 (figure 2b). There is a pronounced threshold field strength $\varepsilon\approx 0.25$ V/Å under which the field-induced charge is less than caps-induced for tubes with i=1, 2. It is known that the $|Q_{effmax}|$ of the unexcited capped SWCNT (5, 5) rapidly converges with the length to the constant value [28] that also markedly in the figure 2b at $\varepsilon=0$. The influence of caps decreases with the tube length. A first low-charged atom ring layer appears at i=3. The almost linear dependence of the $|Q_{effmax}|$ on the field strength can be attributed to the wall polarization of the SWCNTs body. Hence, the rise of the electric field influence on the electronic structure of capped finite-length SWCNT (5, 5) expects with the tube length increase.

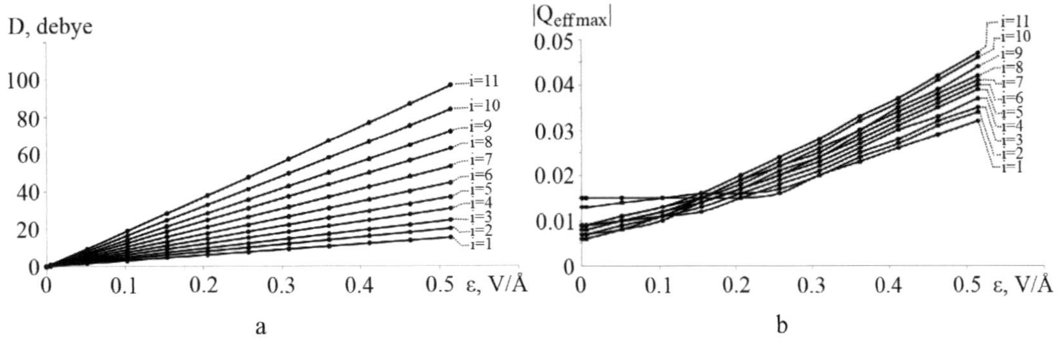

Figure 2. The dipole moment D (a) and the module of the maximum effective charge $|Q_{effmax}|$ (b) of the capped finite-length SWCNT (5, 5) in the range of the number of segments i=0–11 under an applied electric field ε=0–0.5 V/Å.

Let us present results of the calculation of the energy gap and work function of the capped finite-length SWCNT (5, 5) under an applied electric field. The amplitude of the energy gap oscillation decreases with the increase of the field strength (figure 3a). The quadratic $E_{LUMO-HOMO}$ decline was previously observed in the fullerene C_{60} as a result of a Stark shift of frontier orbitals [29, 30]. In the case of the capped SWCNT (5, 5) the quadratic energy gap decrease refers to tubes with the number of segments i=3p. While tubes of the 3p+2 group show weak linear dependence $E_{LUMO-HOMO}(\varepsilon)$. The energy gap of SWCNTs of the 3p+1 group decreases as a straight line with a low slope coefficient ~0.1 eVÅ/V at the field strength less than 3.5, 3.0 and 2.5 V/Å for tubes with i=4, 7 and 10 respectively. At the higher field the slope coefficient dramatically increases to 1.0 eVÅ/V (i=4) and 1.8 eVÅ/V (i=10). Thus, the field energy gap tuning $\Delta E_{LUMO-HOMO} = |E_{LUMO-HOMO}(\varepsilon=0) - E_{LUMO-HOMO}(\varepsilon)|$ strongly depends on the tube type and its length (figure 3b).

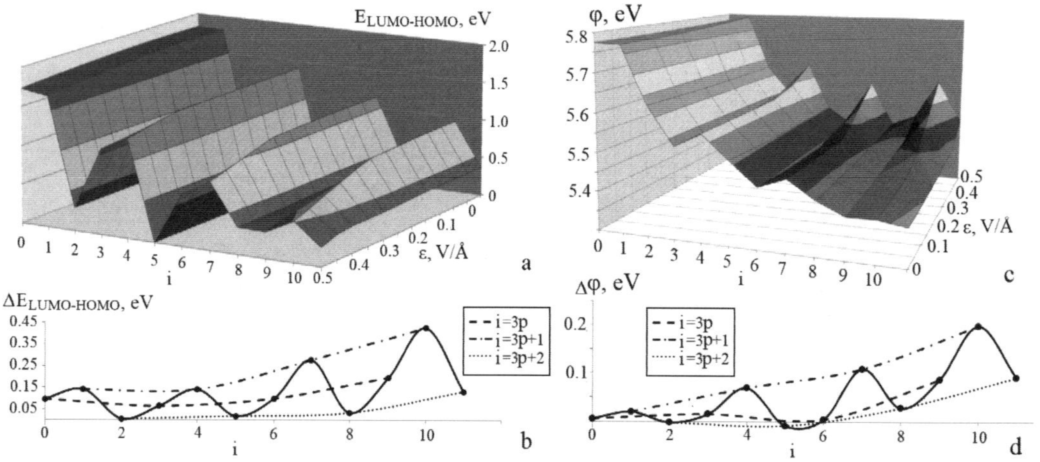

Figure 3. The energy gap LUMO-HOMO (a), energy gap tuning $\Delta E_{LUMO-HOMO}$ (ε=0.5 V/Å) (b), work function φ (c) and work function tuning $\Delta\varphi$ (d) of the capped finite-length SWCNTs (5,5) under an applied electric field in the range from 0 to 0.5 V/Å

The field-induced energy gap decrease in the fullerene C_{60} occurs due to the E_{LUMO} drop and the rise of the E_{HOMO} [29, 30]. This is equivalent to the ionization potential decrease and the electron affinity increase. The response of the EA prevails over IP that is why the work function increases with the strength of the electric field (figure 3c). Taking into account the field-induced work function decrease, the current density reduced in 1.0-1.5 and 1.1-2.6 times at ε=0.5 V/Å for tubes of 3p, 3p+2 and 3p+1 groups respectively. The field effect increases with the length of the SWCNTs especially for tubes of the 3p+1 group that clearly follows from the dependence of the work function tuning $\Delta\varphi=|\varphi(\varepsilon=0) - \varphi(\varepsilon)|$ on the field strength. Their work function converges to the value 5.6 eV at i=7, 10 (figure 3c). Thus, the electron emission current density also saturates.

4. Conclusion

Fundamental parameters and properties of capped finite-length SWCNTs are defined by the quantum confinement effect, curvature effect and influence of capes. There are three groups of the armchair SWCNT (5, 5) with the number of segments i=3p, 3p+1 and 3p+2 (p=0, 1, 2,...) in the region of ultra-small length. These groups behave not only the different set parameters but also response on the electric field. The quantum confinement effect tends to decrease both the energy gap and work function with the tube length increase. The strong electric field leads to the additional drop of the energy gap but the rise of the work function that is more sufficient for longer tubes of the 3p+1 group.

The evaluation of the electron current density for tubes of these group shows the saturation at the field strength 0.5 V/Å.

5. References

[1] Huang C-P, Su W-S, Su C-C and Ho M-S 2013 *RSC Adv.* **3** 9234
[2] Gearba R I, Mills T, Morris J, Pindak R, Black C T and Zhu X 2011 *Adv. Funct. Mater.* **21** 2666
[3] Liu L, Liu S, Chen X, Li C, Ling J, Liu X, Cai Y, Wang L 2013 *Scientific Reports* **3** 3062
[4] Lou L, Nordlander P and Smalley R E 1995 *Phys. Rev.* B **52** 1429
[5] Lin M E, Andres R P, Reifenberger R, Huffman D R 1993 *Phys. Rev.* B. **47** 7546
[6] Klinovaja J, Schmidt M J, Braunecker B and Loss D 2011 *Phys. Rev.* B **84** 085452
[7] Kim C, Kim B, Lee S M, Jo C and Lee Y H 2002 *Phys. Rev.* B **65** 165418
[8] Buonocore F, Trani F, Ninno D, Matteo A Di, Cantele G and Iadonisi G 2008 *Nanotech.* **19** 025711
[9] Cioslowski J, Rao N and Moncrieff D 2002 *J. Am. Chem. Soc.* **124** 8485
[10] Tuchin A V, Ganin A A, Zhukalin D A, Bitytskaya L A and Bormontov E N 2014 *Recent Adv. In Biomedical & Chem. Eng. and Mat. Sc.* (Venice, Italy) 40
[11] Rocherfort A, Salahub D R and Avouris 1999 *J. Phys. Chem.* B **103** 641
[12] Wang B-C, Wang H-W, Lin I-C, Lin Y-S, Chou Y-M and Chiu H-L 2002 *J. Chin. Chem. Soc.* **50** 939
[13] Saito R, Fujita M, Dresselhaus G and Dresselhaus M 1992 *Appl. Phys. Lett.* **60** 2204
[14] Dresselhaus M S, Dresselhaus G and Saito R 1995 *Carbon* **33** 883
[15] Odom T W, Huang J-L, Kim P and Lieber C M 1998 *Nature* **391** 62
[16] Hohenberg P and Kohn W 1964 *Phys. Rev.* **136** 864
[17] Kohn W and Sham L J 1965 *Phys. Rev.* **140** 1133
[18] Vosko S H, Wilket L, Nusair M 1980 *Can. J. Phys* **58** 1200
[19] http://www.gaussian.com
[20] Ouyang M, Huang J-L and Lieber C M 2002 *Acc. Chem. Res.* **35** 1018
[21] Bormontov E N, Ganin A A and Bitytskaya L A 2012 *Semiconductors* **13** 1573
[22] Bormontov E N, Ganin A A and Bitytskaya L A 2012 *Proc. of SPIE* **8700** 870011
[23] Zhao X, Liu Y, Suzuki T, Jones R O, and Ando Y 2004 *Phys. Rev. Lett.* **92** 125502
[24] Eletskiy A V 2002 *Phys.-Uspekhi* **45** 369
[25] Fowler R H, Nordheim L 1928 *Proc. R. Soc. Lond. A* **119** 173
[26] Gautier L-A, Borgne V L, Moussalami S A and Khakani M A E 2014 *Nanoscale Res. Lett.* **9**:55
[27] Musatov A L, Izrael'yants K R, Chirkova E G, Krestinin A V 2011 *Phys. of the Sol. St.* **53** 1504
[28] Zhukalin D A, Tuchin A V, Avilov S V, Bitytskaya L A and Bormontov E N 2014 *Recent Adv. In Biomedical & Chem. Eng. and Mat. Sc.* (Venice, Italy) 79
[29] Wehrli S, Koch E and Sigrist M 2003 *Phys. Rev B* **68** 115412
[30] Tuchin A, Bityutskaya L, Bormontov E 2013 *19th International Vacuum Congress* (Paris, France) 1168

Acknowledgments

This work was supported by RFBR, research project No. 14-02-31315 mol_a and by the project FP7-IRSES-295260 “ECONANOSORB“ under Marie Curie Actions of the 7th Framework Program of European Union. Special thanks to Galina Tuchina for the help and support.

Journal of Physics: Conference Series **541** (2014) 012009 doi:10.1088/1742-6596/541/1/012009

Studying field emission characteristics of point and wedge-shaped surface defects

K A Nikiforov and A N Zartdinov

Saint Petersburg State University, 7-9, Universitetskaya nab., St. Petersburg, 199034, Russia

E-mail: knikiforov@apmath.spbu.ru

Abstract. Electric field distribution and field emission current over the surface of point (sphere-on-cone) and wedge-shaped defects on a plane substrate are studied. It is found emission current due to wedge-shaped defects is more sensitive to emitter-anode distance than in case of point conical defects. This property is consequence of greater emission area of wedge-shaped defects and has important practical applications. The computer model in Matlab of field electron emission from emitter-defects arrays is used. Electrostatic simulation of electron transport processes with PDE Toolbox finite element solutions is implemented. The effects of the variations in defects geometrical structure and parameters on its potential distribution, electric field, and emission current are discussed. It is found that with increase of distance from the tip or the edge of defect the current density falls very fast and greater part of emission current is in fact produced by a very small area near the emitter tip. The closer the emitter and collector are to each other, the less significant the inter-defect distance becomes.

Field electron emitters are generally characterized by surface non-uniformity, because field emission is process in intensive electric field and surface protrusions or defects are necessary for field enhancement. For example classical field emitter tip from metal or semiconductor wire have different crystal faces on the surface with steps, kinks and edges. Its shape can be very different: ball-models for atomistic surface structure of monocrystal tip and analysis of crystal faces forms and sizes were presented in [1]. Field emitters array can also be imaged as regular grid of surface defects or protrusions on substrate plane. Emitter in unit cell of the array might have sphere-on-cone shape (like Spindt-cathodes [2]) or edged straight line shape (or circular/cylinder edge shape [3]) and etc.

In this paper we study two types of multiemitter systems – models for point and wedge-shaped defects on the surface: ones with conic and straight-edge emitters. Employing the methods of mathematical and computer modeling, we analyze interactions between defects in the array and the influence its geometrical features have on the emission current value. We take in account such geometrical properties as distance between emitter and collector d, curvature radius of emitter surface r, distance between the emitters b, as shown on figure 1 and figure 2.

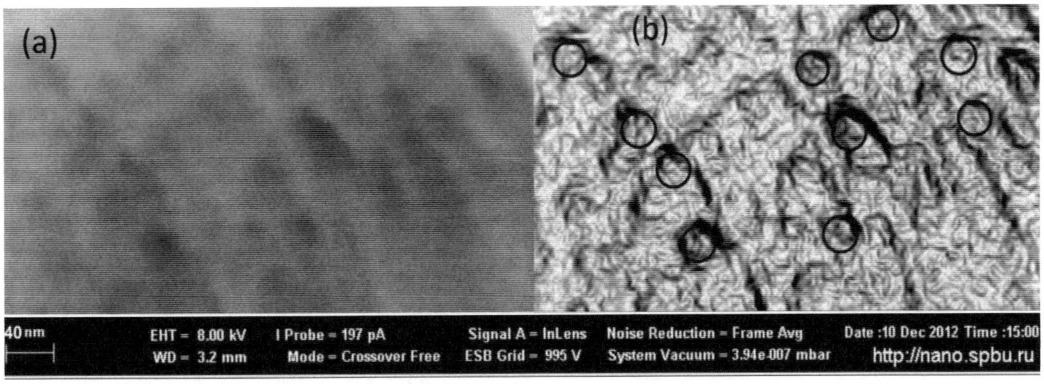

Figure 1. (a) SEM image of nanostructure of emitter surface relief [3]. **(b)** Results of Sobel-mask filtering, defects with spherical tops are shown.

The relation between field emission current density and strength of the field that generates the current is calculated within the framework of Fawler-Nordheim theory. Multiemitter system is regarded as a periodically structured emitter array. The tip of conic emitter is approximated with a spherical surface, and the edge of a straight-blade one with a orthogonal cylinder surface (figure 2).

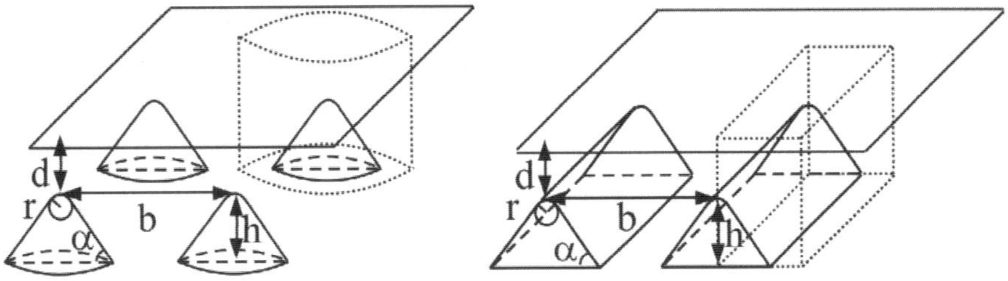

Figure 2. Scheme of conical and edge-shaped emitters. The cells for modelling with conditions of symmetry (periodicity) on the side surfaces are shown.

Program complex [4] is used for computer simulation. Non-uniformity of physical conditions on defects surface (variance of work function etc.) is accounted for with a local field enhancing coefficient [1]. The influence of the space charge is neglected and electrostatic approximation is used.

The results of electrostatic field calculation are shown on figure 3.

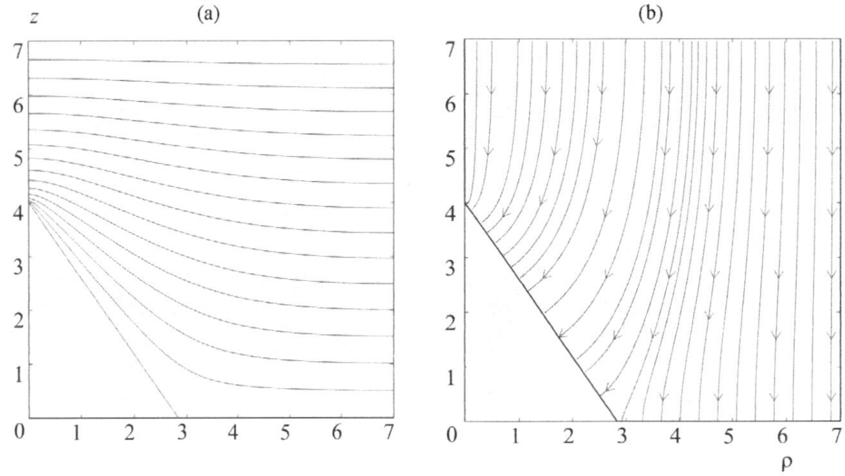

Figure 3. (a) The equipotential lines of electric field (in cylindrical coordinate system) near conical emitter (voltage of 30 V, augment of equipotentials of 2 V); **(b)** electric field lines of force.

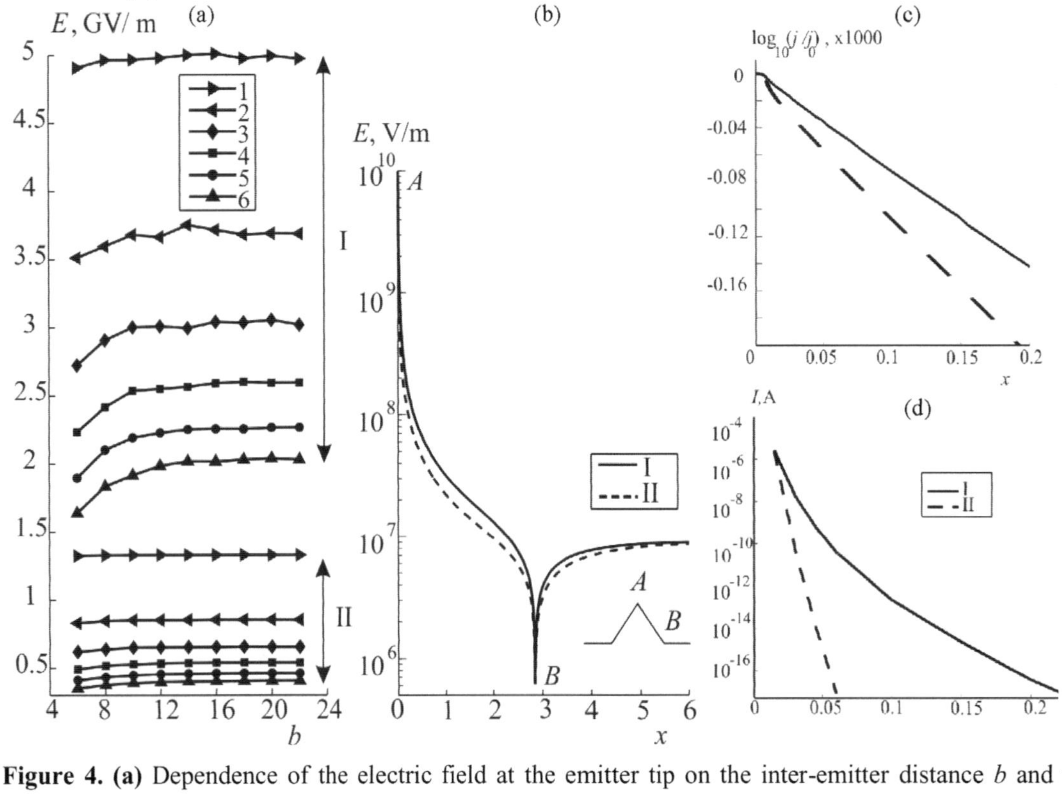

Figure 4. (a) Dependence of the electric field at the emitter tip on the inter-emitter distance b and the emitter-collector distance d with constant radius r=10 nm of top curvature, the voltage of 30 V, the inset shows the values of d in mkm; the curves I correspond to conic emitters and II to the edged ones. **(b)** Plot of the electric field strength on the surface between the emitter tip and middle of a segment between two emitters with constant voltage of 5 V; the X-axis corresponds to horizontal coordinate on the surface points. **(c)** Emission current density on the emitter surface from its tip up to the base. **(d)** Dependence of integral emission current from the emitter-collector distance with r=10 nm, voltage 5 V.

Distance between adjacent emitters-defects is one of the parameters that define interactions between defects and their electric field. As shown on figure 4 (a) strength of the electric field on the tips of blade edges decreases when emitters are moved closer to each other, as the interaction between emitters gets stronger.

Influence of distance *d* on the electric field is also illustrated on figure 4 (a). The closer the emitter and collector are to each other, the less significant the inter-emitter distance becomes. This occurs because the mutual shielding effect disappears and the emitters start working as individual tips or blades. Changes in *d* has greater effect for conic emitters than for blade ones.

Figure 4 (b) shows the emitter electric field plotted against the surface points from the tip and up to the middle of a segment between two adjacent tips or blades. On the plot one can see rapid growth of the field near the emitter tip due to appropriately named "tip effect" and the minimum of field strength at its base because of a cavity formed by emitter side surface and the base plane.

Figure 4 (c) shows the distribution of normalized current density along the emitter surface.

It's easy to see that with increase of distance from the tip or the edge of the blade the current density falls very fast and greater part of emission current is in fact produced by a very small area near the emitter tip. Thus, emission current is strongly localized near the tip.

The other important feature is shown on figure 4 (d): growth of emitter current with decrease of emitter-collector distance *d*. When *d* is decreased to the value of the same order as the tip curvature radius, the blade emitter current gets bigger than the conic emitter one (figure 4 (d)). Such behaviour can be explained by the fact that with decrease of *d* the current density becomes similar for both cases and the current value gets larger for a blade emitter because of larger emission surface. An important feature that can be seen on figure 4 (d) is that speed of change of emission current dependent on *d* for edge emitters is higher than for conic ones, the fact that is again connected with larger emission surface. Thus edge-emitter defects are more sensitive to inter-electrode distance than conic ones. Obviously, this is right to edges of any shapes, not necessarily straight-line ones, which is practically important for several kinds of applications.

In this paper we present modelling results for potential, electric field strength and emission current of multiemitter systems of surface defects with point- and wedge-shapes. Results were obtained in a software complex created with Matlab and Matlab PDE Toolbox [4].

Acknowledgments

Financial support has been obtained from RFBR (13-01-00150) and partially from SPbSU (9.38.673.2013). Research was carried out using computational resources provided by Resource Center "Computer Center of SPbU" (http://cc.spbu.ru/en) and using experimental equipment of the Interdisciplinary Resource Center for Nanotechnology of St. Petersburg State University.

[1] Nikiforov K A, Egorov N V and Shen C-C 2009 Surface reconstruction of a field electron emitter *Surface Investigation X-Ray, Synchrotron and Neutron Techniques* **5** 833–39

[2] Spindt C A 1968 *J. Appl. Phys.* **39** 3504

[3] Nikiforov K A, Antonova L I, Egorov N V, Trofimov V V, Makeev V V and Ogurtsov O F 2012 Non-gated field emission array as low-energy electron source: Experiment and simulation *RuPAC 2012 Contributions to the Proc. of 23 Russian Particle Accelerator Conf. (St. Petersburg)*, (Cern: JACoW) 218–20

[4] Nikiforov K A and Egorov N V 2012 Program complex for vacuum nanoelectronics finite element simulations *RuPAC 2012 Contributions to the Proc. of 23 Russian Particle Accelerator Conf. (St. Petersburg)*, (Cern: JACoW) 409–11

Influence of process pressure on β-SiC growth by CVD

A A Andreev, A O Sultanov, A S Gusev, N I Kargin, E P Pavlova

National Research Nuclear University «MEPHI», Moscow, Russia

E-mail: AOSultanov@mephi.ru

Abstract. 3C-SiC films grown on Si (100) substrates by CVD method using silane-propane-hydrogen system were analyzed for crystallinity at various process pressures. The deposition experiments were carried out in a shower-head type cold-wall CVD reactor. The influence of growth conditions on a structural modification of experimental samples was studied by X-ray diffraction (XRD) measurements, Fourier transform infrared spectroscopy (FTIR) and spectroscopic ellipsometry (SE).

1. Introduction

Silicon carbide is a wide bandgap semiconductor with the combination of unique properties such as high breakdown field, high electron drift velocity, high thermal stability and conductivity. It is an attractive material for wide variety of applications ranging from microelectronic and optoelectronic devices to protective and tribological coatings. Silicon carbide crystallizes in numerous (approximately 200) different polytypes. The most widespread modifications are hexagonal (2H-, 4H-, 6H-SiC) and cubic (3C-SiC). 3C-SiC is the only cubic SiC polytype with the zincblende structure which leads to isotropic physical properties and absence of spontaneous polarization in the material. 3C–SiC is more advantageous for MOSFETs with higher channel mobility since near interface traps are not active in the SiO_2/3C-SiC system [1]. However, the cubic polytype is a metastable phase, and it cannot be grown in large size by spontaneous nucleation with the classical sublimation method [2]. For this reason single-crystal β-SiC substrates are not commercially available, while the α-SiC wafers have been on the market for about 20 years. At the same time unlike the hexagonal polytypes β-SiC has the ability to be epitaxially grown on large-area (up to 12 inch in diameter) and relatively cheap Si wafers. β-SiC epilayers on Si are promising for microsystem devices manufacturing as well as for adverse environment optical systems. High quality β-SiC thin film might be one of a suitable buffer layers for the cubic or even hexagonal GaN growth (atomic arrangement of the (111) plane of 3C-SiC is the same as that of the (0001) plane of 6H-SiC) on Si substrate [3, 4].

Chemical vapor deposition (CVD) using mixtures of silane with hydrocarbons or various organosilicon compounds in hydrogen flow is a conventional way to fabricate the silicon carbide epilayers [5]. This technology provides an opportunity to produce high quality epitaxial films, but the main disadvantage of CVD-method is a rather high operational temperature (1400°C and higher). The development of a low temperature techniques for growth of 3C-SiC films is necessary for increasing the feasibility of device fabrication. Therefore, we have focused on the physical and technological features of low-temperature synthesis of 3C-SiC thin films on Si by the CVD method and, in particular, on the influence of process pressure on structure and properties of experimental samples.

2. Experimental setup

The deposition experiments were carried out in a shower-head type cold-wall CVD reactor. As substrates, we used 3-inch silicon wafers with (100) crystalline orientation. The substrates were cleaned in acetone and propanol mixture in ultrasonic bath. Before being transferred to the reactor, the Si wafers were placed in 5% HF solution (to remove the native oxide layer) and then rinsed in distilled water. After that the substrate was placed to the substrate holder and transported through the load lock into the growth chamber using the manipulator. In turn the substrate holder was mounted on the rotator, the rotation rate which can reach more than 1000 rpm. In situ cleaning was carried out at 1100°C in a hydrogen atmosphere. Then the temperature was slowly raised to the growth value in the presence of hydrogen flow and the mixture of SiH_4 (99,999%) and C_2H_4 (99,9%) was carried by H_2 into the CVD chamber.

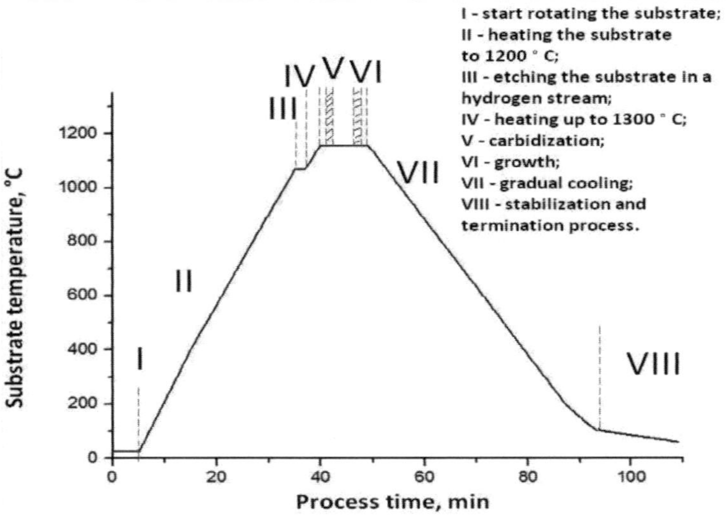

Figure 1. Process schedule for the growth of 3C-SiC on Si

It is well known [6] that the epitaxy of 3C-SiC on Si is limited by a 20% lattice mismatch and 8% CTE mismatch between silicon and silicon carbide. For this reason the deposition experiments were conducted applying a two-step technique: carbonization and growth of 3C-SiC on the initial carbonized layer. Figure 1,a shows a time–temperature diagram of the typical growth process. During carbonization, the surface of the Si wafer is converted to SiC, which helps in minimizing the stress in growing layers. The growth pressure was varied by tuning the throttle valve between the growth chamber and the rotary vane vacuum pump. The growth temperature was 1150°C for all experiments. It was ramped up at 36,8°C/min. Experimental parameters for each sample are listed in Table 1.

Table 1. The parameters of SiC deposition process

№	Growth duration, min	Chamber pressure, Torr	SiH_4 flow rate, sccm	C_2H_4 flow rate, sccm	Growth temp., °C	Substrate
1	2	300	5	5	1150	Si (100)
2	2	150	5	5	1150	Si (100)
3	2	100	5	5	1150	Si (100)
4	2	50	5	5	1150	Si (100)
5	2	5	5	5	1150	Si (100)

3. Results

The influence of growth conditions on structural modifications of experimental samples was studied by X-ray diffraction (XRD) measurements and Fourier transform infrared spectroscopy (FTIR). XRD patterns of the samples were collected using CuKα radiation (λ_{Cu} = 0.154056 nm) with «Ultima IV» (Rigaku) diffractometer in thin film configurations. It was found that the spectra of the samples №1 – 4 (P_{growth} = 300 ÷ 50 Torr) contain two well pronounced peaks typical for β-SiC at 2θ = 35,54° and 41,22° corresponding to the (111) and (200) reflection planes, respectively. The decrease of SiC (200) and (111) peaks FWHM with chamber pressure (Figure 1,a) indicates an improvement in crystallinity of thin films. The XRD pattern for the film grown at the growth pressure of 5 Torr is shown in Figure 1,b. No XRD peaks corresponding to the (111) or (110) planes of 3C-SiC were observed for the sample №5, indicating that the film is highly oriented.

Additionally, FTIR measurements were performed in order to study the bonding configurations. Fourier transform infrared spectrometer «FTIR-8400S» (Shimadzu) was used to measure the infrared absorption of the prepared films. The measurements of IR spectra were performed by means of a standard technique with a resolution of 2 cm^{-1}. The FTIR spectra of all samples show a strong band at 780 cm^{-1} assigned to Si–C stretching mode [7]. No absorption bands corresponding to Si-H or C-H chemical bonds were detected. As an example, Figure 3,a illustrates the evolution of a FTIR spectrum during the deposition process: *I* – after the carbonization; *II* - during the deposition process; *III* - after the growth process. The peak at 2400 cm^{-1} is due to the small amount of carbon dioxide in air. FWHM of the Si–C FTIR peak for the sample №5 was 116 cm^{-1}.

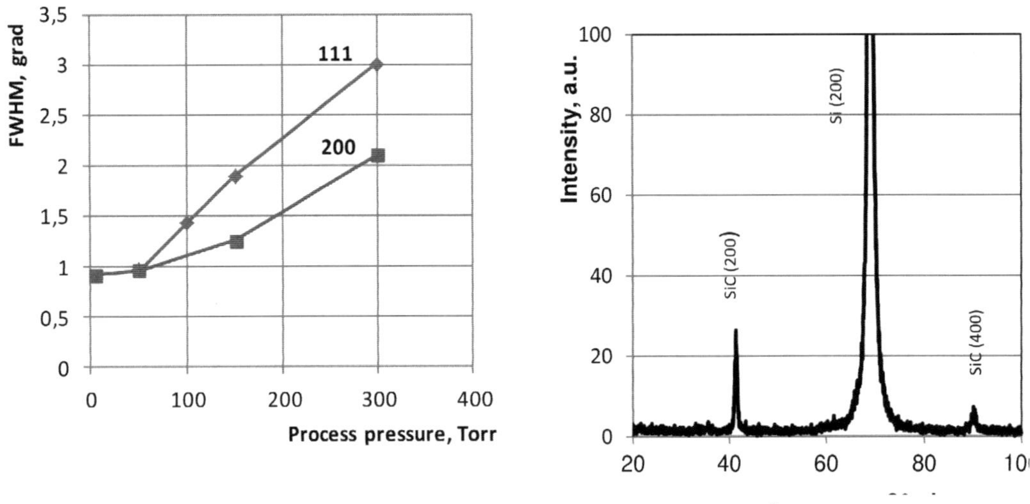

a b

Figure 2. a) FWHM of SiC (200) and (111) XRD peaks vs. process pressure; b) XRD spectrum of experimental sample №5.

The thickness and refractive index of SiC thin films were determined using a spectroscopic ellipsometry with SE-850 («Sentech») ellipsometer. It was found that the average layer thickness weakly depends on the pressure in the growth chamber. The growth rate of SiC films was approximately 13,74 μm/h. However, a significant improvement in uniformity of thickness at low process pressure was observed. The obtained thickness map of sample №5 (P_{growth} = 5 Torr) showed a good uniformity unlike the sample №3 (P_{growth} = 100 Torr) as illustrated in Figures 4,a and b. Figure 3,b contains a dependence of refractive index value on reactor pressure. The decrease in refractive index of SiC thin films and a corresponding decrease in XRD peaks FWHM (Fig. 1,a) with reactor pressure suggest the obtaining of good-quality crystalline SiC films at growth temperature of 1150°C.

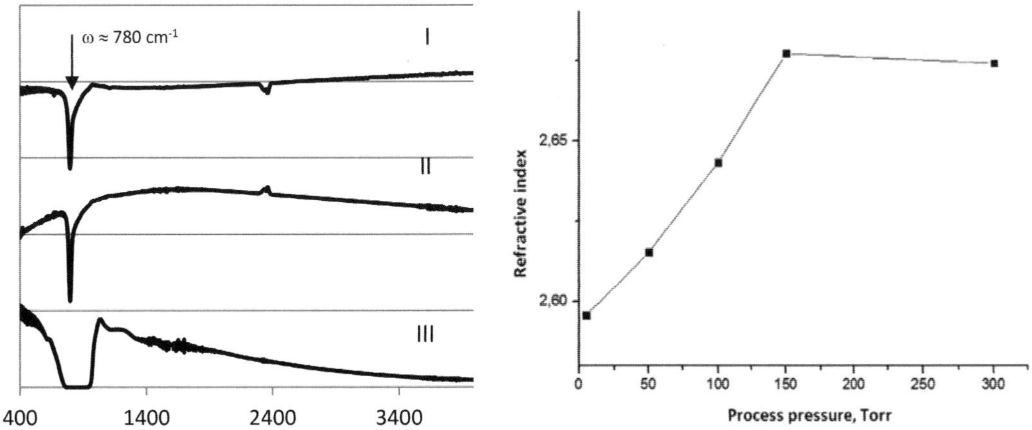

a b

Figure 3. a) Evolution of FTIR spectrum during the deposition process (*I* –after the carbonization process; *II* - during the growth process; *III* - after the growth process; b) Refractive index as a function of process pressure

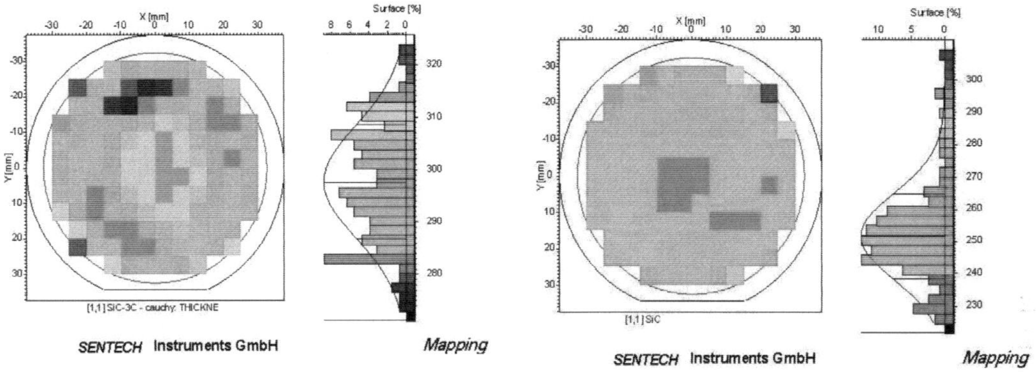

a b

Figure 4. Spectroscopic ellipsometry maps of the thickness of experimental samples №3 (a) and №5 (b)

4. Conclusion

3C-SiC films grown on Si (100) substrates by CVD method using silane-propane-hydrogen system were analyzed for crystallinity at various process pressures. The deposition experiments were carried out in a shower-head type cold-wall CVD reactor. The influence of growth pressure on a structural modification of experimental samples was studied by X-ray diffraction (XRD) measurements, Fourier transform infrared spectroscopy (FTIR) and spectroscopic ellipsometry (SE). The following results were obtained:

(1) The crystallinity of β-SiC thin films on Si strongly depends on growth pressure that is confirmed by X-ray analysis data;

(2) The epitaxial growth of 3C-SiC film on Si (100) was carried out at T_{sub} = 1150°C and P_{growth} = 5 Torr;

(3) It was found that the average layer thickness weakly depends on the pressure in the growth chamber. However, a significant improvement in uniformity of thickness at low process pressure was observed.

Acknowledgements

This work was supported by the Ministry of Education and Science of the Russian Federation. The work was carried out using the scientific equipment of the NRNU MEPhI Common Use Center.

References

[1] Schoner A. et al 2006 *Chem. Vapor Deposition.* **12** pp 523 – 526
[2] Tairov Yu M et al. 1978 *J. Crystal Growth.* **43** pp 209
[3] Kukushkin S A et al. 2012 *Technical Physics Letter.* **38,** pp 298 – 299
[4] Jin-Hyo Boo et al. 1998 *J. Crystal Growth* **189/190,** pp 183-188
[5] Nishino S et al. 1983 *Appl. Phys. Lett.* **42,** pp 460 – 462
[6] Nishino S et al. 1980 *J. Electrochem. Soc.* **127**, 2674
[7] Tolstoy V P et al 2003 *Handbook of infrared spectroscopy of ultrathin films* (Hoboken N.J.: Wiley-Interscience).

Oxide nanofilms on copper and silver: anodic synthesis and semiconducting properties

D S Eliseev[1], S N Grushevskaya[1, 3], A V Vvedenskii[1], D A Kudrayshov[2]

[1] Voronezh State University, 1 Universitetskaya pl, 394006, Voronezh, Russia

[2] St Petersburg Academic University, 8/3 Khlopina Str., St Petersburg, 194021, Russia

[3] E-mail: sg@chem.vsu.ru

Abstract. Structure-sensitive properties of oxide nanofilms anodically formed on single crystals and polycrystalline silver and copper were determined by photoelectrochemical methods and scanning electron microscopy. At transition from polycrystalline substrate to single crystals the structural disordering decreases for Ag(I) oxide and increases for Cu(I) oxide in accordance with morphological modification.

1. Introduction

Semiconducting metal oxides are perspective materials for photolytic hydrogen evolution, electrocatalysis, current-generating devices, UV- and IR-sensors. The field of application of semiconducting oxides is determined by their structure-sensitive parameters. These parameters depend on the conditions of oxide formation that involve chemical composition of solution and substrate, the structural and crystallographic orientation of metal surface. The list of structure-sensitive *in situ* methods of investigation of oxide films is short; among them the photocurrent measurements occupy an important place [1-3].

The aim of the paper was to reveal the influence of the crystallographic orientation of metal surface on the structural parameters of anodically formed oxide nanofilms. We carried out photoelectrochemical *in situ* and physical *ex situ* measurements.

2. Experimental

The potentiostatic formation of Ag(I) and Cu(I) oxides was carried out in deoxygenated aqueous solution of 0.1 M KOH (analytical grade) on polycrystalline silver or copper and their single crystals 100, 110 and 111 (0.9999). Copper single crystals with a working surface of 0.2 cm^2 were produced by “Materials Technology & Crystals for Research, Development and Production” (Juelich, Germany) with an accuracy of 2^0. Silver (100), (110) and (111) single crystals with a working surface of 0.38; 0.52 and 0.60 cm^2 were grown in a horizontally moving furnace at temperatures from 1273 to 673 K with the subsequent cooling and orientation with an accuracy of 3^0.

Before the electrochemical measurements the electrode surface was mechanically ground on grain paper, polished with aqua-MgO suspension, chemically polished, rinsed with double-distilled water and polarized in cathodic potential range during 10 minutes.

The morphology of the surface was monitored before and after electrochemical measurements was carried out by the scanning electron microscope JEOL 6380LV.

The concentration of prevailing structure defects, the coefficient of optical transition and the length of the space charge region (SC) for oxides were determined from photocurrent i_{ph} measurements in accordance with the equation obtained in [4] for thin oxide films:

$$i_{ph} = e\eta f \Phi_0 \left(1 - R^{out}\right)\left(1 - e^{-2\alpha L}\right) \approx i_{ph}^{max}\left(1 - e^{-2\alpha L}\right). \quad (1)$$

Here η is the quantum yield; f is the coefficient, that takes into account the difference in the rate constants of electron-hole recombination and hole consumption in an electrochemical reaction on the oxide/solution interface, R^{out} is the reflection coefficient of light flux from the oxide/solution interface; α is the light absorption coefficient and L is the film thickness. If the film thickness reaches the length of SC then the photocurrent reaches maximal values of i_{ph}^{max}.

Photocurrent was measured using original equipment with a low-noise amplifier under potentiostatic polarization conditions with accuracy of 10 nA. Emitting diodes with wave length of 400 or 470 nm were used as a quasi-monochromatic light source. The density of light flux Φ_0 was equal to $3.6 \cdot 10^{15}$ photon s^{-1} cm^{-2}. The details of the environment were described in [4].

3. Results and discussion

3.1. Morphology of oxide films

The Ag_2O film formed on polycrystalline silver and its single crystals consists of well-resolved individual clusters of the oxide phase (figures 1). At transition from polycrystalline silver to single crystals the morphology of the Ag(I) oxide film slightly changes. The size of separate agglomerates reaches 400 nm although an average thickness of the oxide film does exceed 120 nm. The size of agglomerates and the distance between them increase in a series: $Ag_{poly} < Ag_{110} < Ag_{100} < Ag_{111}$.

Figure 1. SEM-images of Ag(I) oxide formed in 0.1 M KOH at $E = 0.56$ V on Ag_{poly} (a), Ag_{110} (b), Ag_{100} (c) and Ag_{111} (d).

On polycrystalline copper (figure 2a) the size of agglomerates is lower as compared with polycrystalline silver. The film thickness reaches 20 nm. The crystallinity of Cu(I) oxide phase decreases in the following series: Cu_{poly} > Cu_{100} > Cu_{110} > Cu_{111}. It is hard to observe separate agglomerates of the oxide phase on single crystals of copper. The most patterned film is formed on Cu_{100} (figure 2b). Most probably, on Cu_{poly} and Cu_{100} 3D-nucleation is realized while on Cu_{110} and Cu_{111} 2D-nucleation is possible [5]. The change of nucleation mechanism can be the reason for the difference in the Cu_2O properties formed on different substrates.

Figure 2. SEM-images of Cu(I) oxide formed in 0.1 M KOH at E = -0.20 V on Cu_{poly} (a), Cu_{100} (b), Cu_{110} (c) and Cu_{111} (d).

3.2. Photocurrent measurements in oxide films

The photocurrent in silver oxide was measured at the potentials from 0.51 to 0.56 V/SHE. The first 5-7 seconds of potentiostatic polarization the photocurrent is not registered or does not exceed the noise level. When the oxide film with a significant thickness is formed the anodic photocurrent will arise. At the constant values of L and E the photocurrent in Ag_2O on single Ag crystals is always less as compared with Ag_{poly} substrate. Positive photocurrent proves n-type conductivity and prevailing of donor defects in Ag_2O on all substrates.

The growth of the amplitude with the film thickness proves that the film is thin. It does not exceed the length of SC and the equation (1) is valuable. The numerical processing of experimental data in accordance with this equation makes it possible to obtain the coefficient of optical transition, the length of SC for oxides and the concentration of prevailing structure defects. On single crystals the length of SC region of Ag_2O is higher by an order and α is noticeably lower as compared with Ag_{poly}-substrate (table). The concentration of donor defects in Ag(I) oxide formed on single silver crystals is almost 10 times lower than in Ag(I) oxide formed on polycrystalline silver. Hence, the degree of deviation from the stoichiometric composition Ag_2O decreases. For Ag_2O formed on single crystals the length of SC increases in the following series: Ag_{poly} < Ag_{100} < Ag_{111} < Ag_{110}. The values of α and N_D decrease in the same manner. This series does not coincide accurately with the series of morphology changing.

On polycrystalline copper and its single crystals the cathodic photocurrent arises in Cu_2O after 20-30 seconds of the polarization at the potentials between -0.22 and -0.05 V/SHE. The cathodic photocurrent reflects the p-type of conductivity and the prevailing of acceptor defects in Cu_2O formed on all substrates. The amplitude of photocurrent on single crystals is significantly lower than on Cu_{poly} which is caused most probably by the structural and electronic state of Cu_2O film. The numerical processing was carried out for Cu_{poly}, Cu_{100} and Cu_{111} substrates where the amplitude of photocurrent in Cu_2O reached sufficient values. In Cu_2O on Cu_{110} we did not observe photocurrent hence we could not calculate the structural parameters.

The amplitude of photocurrent as well as the length of SC decrease at transition from Cu_{poly} to Cu_{100} and then to Cu_{111} (table). The optical density and the concentration of acceptor defects increase in the same manner. Note that this series in general coincide with the series of decreasing crystallinity in Cu_2O formed on different substrates.

Table. The coefficient of optical transition α, the length of SC W and the concentration of donor or acceptor defects N_D or N_A in Ag(I) and Cu(I) oxides anodically formed on silver and copper.

Potential	0.56 V				-0.20 V		
Substrate	Ag_{poly}	Ag_{100}	Ag_{111}	Ag_{110}	Cu_{poly}	Cu_{100}	Cu_{111}
$\alpha \cdot 10^{-5}$ / cm^{-1}	2.3	1.4	1.0	0.7	7.6	11.1	19.2
W / nm	100	164	230	329	40.7	28.9	21.5
N_D*10^{-15} / cm^{-3}	22.4	8.3	4.2	2.1			
N_A*10^{-15} / cm^{-3}					48.0	113.0	204.0

4. Conclusion

The polarity of photocurrent points to n-type conductivity in Ag(I) oxides and p-type conductivity in Cu(I) oxides anodically formed in alkali on polycrystalline silver or copper and their single crystals. A decrease of photosensitivity of oxide nanofilms grown on single crystals is determined by their morphology, structure and electronic state. The transition from polycrystalline silver to its single crystals results in a decrease of optical density and donor defects. The transition from polycrystalline copper to its single crystals results in a growth of optical density and acceptor defects. Hence, the degree of deviation from the stoichiometric composition or the structural disordering decreases for Ag(I) oxide and increases for Cu(I) oxide.

Acknowledgements

The investigations were carried out under financial supporting of the Ministry of Education and Science of the Russian Federation (project 675).

References

[1] Gurevich YV and Pleskov YY 1986 *Semiconductor photoelectrochemistry* (New York: Consultants Bureau)
[2] Bard AJ and Faulkner LR 2001 *Electrochemical methods: Fundamentals and Applications* (Hamilton: John Willey)
[3] Bockris JO'H and Khan SUM 1993 *Surface Electrochemistry: A Molecular Level Approach* (New-York, London: Plenum Press)
[4] Kudryashov DA, Grushevskaya SN, Ganzha SV and Vvedenskii AV 2009 *Prot Met* **45** 501
[5] Gamburg YD and Zangari G 2011 *Theory and Practice of Metal Electrodeposition* (Boston, New York, Heidelberg: Springer)

Journal of Physics: Conference Series **541** (2014) 012012 doi:10.1088/1742-6596/541/1/012012

Automated method of measuring the temperature dependences of the dielectric parameters of ferroelectrics with second kind phase transition

D V Ryabov[1], E A Pecherskaya[2], J V Shepeleva[3] and R M Pecherskaya[4]
[1] Postgraduate, Penza State University, 40 Krasnaya Street, 440026, Penza, Russia
[2] Professor, Penza State University, 40 Krasnaya Street, 440026, Penza, Russia
[3] Teacher, Penza State University, 40 Krasnaya Street, 440026, Penza, Russia
[4] Professor, Penza State University, 40 Krasnaya Street, 440026, Penza, Russia

E-mail: pea1@list.ru

Abstract. An automated method of indirect measurements of the temperature dependences of the dielectrics parameters of ferroelectrics is proposed. The basic of its principle – linear dependences of the inverse dielectric susceptibility of the temperature in the ferroelectric and paraelectric phases. The method allows to optimize the research process by reducing the number of measurement procedures, measuring in a limited temperature range (not exceeding the phase transition temperature) and dependences modeling in the range which is characterized by the greatest instrumental components of the measurements errors.

The dielectric parameters changes of ferroelectrics in the ferroelectric phase under the influence of influencing factors make their use in small capacitors, various converters, including the resistors for temperature measurement and control. Study of the dependence of relative permittivity *E*, polarized *P*, dielectric loss tangent $\mathrm{tg}\delta$ on temperature *T* allows to set the Curie temperature, and consequently, to identify operating temperature range of functional elements of electronics based on these materials. Getting the listed functional dependences is possible either by indirect measurements or the basic of empirical and theoretical modeling. The phenomenological theory of Landau – Ginzburg – Devon shire explains the difference in the behavior of the temperature dependences of the dielectric parameters of ferroelectric with the phase transition of the first and the second kind indicating need of the development of various techniques and methods of their measurements.

This working paper proposes an automated method of the empirical and theoretical modeling of the dielectric parameters of ferroelectrics with a second kind phase transition, for which according to the phenomenological theory the Curie – Weiss, the following relations are made:

$$\begin{cases} \dfrac{\partial E}{\partial P} = -\dfrac{2\chi(T)}{\varepsilon_0}, \text{ при } T < T_c, \\ \dfrac{\partial E}{\partial P} = \dfrac{\chi(T)}{\varepsilon_0}, \text{ при } T > T_c, \end{cases} \tag{1}$$

where E – electric field tension, T_c – Curie temperature, ε_0 – electric constant.

Accordance with the expressions (1), dependences $\chi(T)$ in ferroelectric and paraelectric phases can be approximated by straight lines, while the slope in the ferroelectric phase is twice higher than in the paraelectric phase (figure 1). The essence of the proposed method consists of the sufficiently of the inverse dielectric susceptibility χ measuring at least for these values of temperature T and then calculating the inverse dielectric susceptibility χ in the whole temperature range.

Thus, in the dielectric phase the function $\chi_s(T)$ is described by

$$\chi_s(T) = a_s + b_s T,$$

where a_s has a sense of the back dielectric susceptibility at $T = 0^\circ C$, it always takes positive values; the coefficient b_s determines the slope of the line. As in the paraelectric phase the function $\chi_s(T)$ is decreasing, b_s always takes negative values, its measure unit is $\frac{1}{^\circ C}$.

In the paraelectric phase the slope $\chi_p(T)$ of the line is twice less than the slope $\chi_s(T)$ in the ferroelectric phase, which is described by

$$\chi_p(T) = a_p + b_p T = a_p - 0{,}5 \cdot b_s T,$$

where the coefficient $a_p < 0$ is numerically defined as the ordinate of the point of intersection of the line continuation $\chi_p(T)$ with the vertical axis. Similarly, the coefficient b_p represents the slope of the line $\chi_p(T)$. Both direct lines $\chi_s(T)$ and $\chi_p(T)$ cross the X - axis at point (T_c , 0).

The basic steps of the methodology of function $\chi(T)$ measurement in both phases are below:

1. Values of temperature are set up at which the measurement of reciprocal dielectric susceptibility $\chi(T)$ is conducted.
2. Measurements of $\chi(T)$ at temperatures are specified in paragraph 1. To avoid accidental errors it is advisable to conduct multiple measurements with its subsequent processing (averaging of measurements).
3. Analysis of the measurements, i.e. their assignment to a particular phase. Thus the measurement results can satisfy one of the three situations:

a) all measurement results correspond to the ferroelectric phase. This decision is made, if the function $\chi(T)$ is decreasing at a given temperature range ($T_{i+1} > T_i$), i. e. with increasing temperature T the reciprocal dielectric susceptibility χ values are decreasing ($\chi_{i+1} < \chi_i$);

b) all measurement results correspond to the paraelectric phase. It means that the function $\chi(T)$ at the determined temperature range is increasing, i.e. with increasing temperature ($T_{i+1} > T_i$) the reciprocal dielectric susceptibility χ values are increasing ($\chi_{i+1} > \chi_i$);

c) all measurement results correspond to both phases and the Curie temperature T_c belongs to the determined temperature range.

In the beginning of the temperature range the function $\chi(T)$ is decreasing, and then it is increasing. If dependence $\chi(T)$ changes the values from decreasing to increasing more than once, it means there are errors at measurement results that must be eliminated by carrying out repeated multiple measurements or it indicates that the monitoring ferroelectric has blurred phase transition and, as a result, the method of indirect measurement of dependences of dielectric parameters of ferroelectrics is not applied to it.

The situation described above (after errors eliminating and in the condition that ferroelectric parameters obeys to the Curie-Weiss law) are formalized by the following system:

$$\begin{cases} \text{if } \chi_{i+1} < \chi_i,\ T_{i+1} > T_i, \text{the phase is ferroelectric,} \\ \text{if } \chi_{i+1} > \chi_i, T_{i+1} > T_i, \text{the phase is paraelectric,} \\ \text{if } \chi_{i+1} < \chi_i, T_{i+1} > T_i,\ T_i \le T_c \text{ and } \chi_{i+1} > \chi_i, T_{i+1} > T_i, T_i > T_c, \\ \text{at both phases,} \end{cases}$$

where the temperature corresponds to the minimum reciprocal dielectric susceptibility (respectively to maximum relative permittivity).

4. Approximation of the results of dependence measurements $\chi(T)$. Approximation of given results by a straight line can be accomplished by any known method. In this paper, the method of least squares was used, which is applicable due to its accuracy in terms of the problem being solved. Approximation algorithm is different for the situations described in paragraphs 3a), 3b) and for 3c).

For example, in cases where all measurement results correspond to the same phase by the number of measured points approximating line parameters a_s (or a_p), as and b_s (or b_p) are determined by formulas:

$$b = \frac{\left[n\sum_{i=1}^{n} T_i \cdot \chi_i - \sum_{i=1}^{n} T_i \sum_{i=1}^{n} \chi_i \right]}{n\sum_{i=1}^{n} T_i^2 - \left(\sum_{i=1}^{n} T_i \right)^2}, \quad a = \frac{\sum_{i=1}^{n} \chi_i - b\sum_{i=1}^{n} T_i}{n},$$

where n - number of measurements.

5. The Curie temperature T_c determination for ferroelectrics with the second phase transition is possible by the following ways: abscissa of reciprocal dielectric susceptibility χ function of the temperature *T*, corresponding to the minimum χ (see figure 1); as abscissa of lines intersection point, approximating the function $\chi(T)$ in the ferroelectric and paraelectric phases.

Algorithm in the base of this methodology allows to automate the measuring and data processing. The measurements were conducted using an automated measuring system, its structure and main metrological characteristics were presented in [1]. The results of research temporal instability of the dielectric parameters of ferroelectrics were shown in [2]. In the figure

1(shown by the continuous line) the dependence for triglitcinsulfat made with the developed program in Delphi is presented.

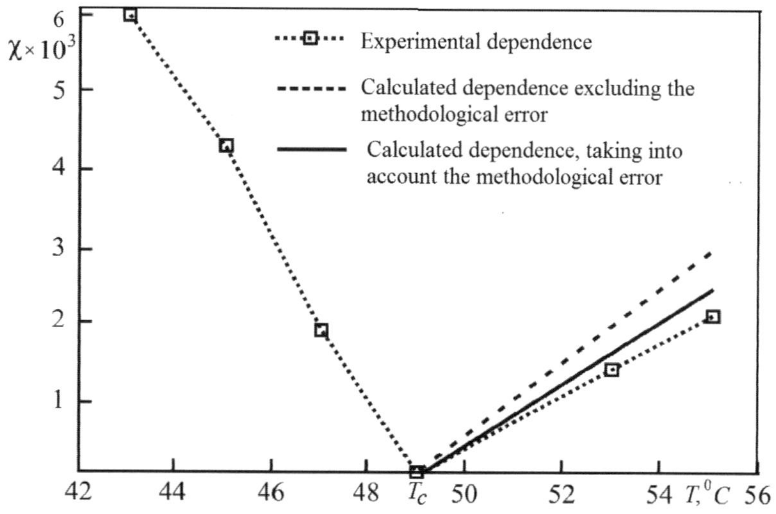

Figure 1. Temperature dependences of inverse dielectric susceptibility for trigletcinsulfat.

In this example the measurements were performed in the ferroelectric phase and modeling was carried out in the paraelectric phase. In addition to instrumental error, there is a methodological error. It is due to the difference of adiabatic conditions of experimental determination of the temperature dependences of dielectric parameters on the sinusoidal signal from isothermal condition in which obtained formulas are made according to the thermodynamic theory.

In relative unities the methodical error $\delta\chi$ is given by:

$$\delta\chi = \frac{16\pi^2 T_c}{C^2 C_P \varepsilon_0 \chi} P^2,$$

where C_P – the heat capacity of volume unit at constant polarization P; C – Curie constant.

Based on known interactions with inverse dielectric susceptibility dielectric material parameters can be calculated. It was found that in ferroelectrics with heterogeneous structure, for example, in films type $Ba_x St_{1-x} TiO_3$ (BSTO) diffuse phase transition is observed and consequently, the Curie – Weiss law is not applied to them. This fact limits the application of the proposed measurement method. In turn, for modeling of the dielectric response of these ferroelectrics the Monte Carlo method was used in [3]. The proposed simulation method of the temperature dependences of dielectric parameters of ferroelectrics with a second kind phase transition allows to improve the research efficiency in 1,4 times by reducing measuring time and also it does not require temperature measurements in a wide range.

References

[1] Pecherskaya E A 2007 *Measurement Techniques*. **50** (10) 1101 - 1107
[2] Pecherskaya E A, Solov'ev V A, Metal'nikov A M, Varenik Yu V, Gladkov I M and Ryabov D V 2013 *Semiconductors*. **47** (13) 1720 - 1722
[3]Yudin P N, Nikol'sky M A and Zubko S P 2003 *Technical Physics*. **48** (8) 995 - 1000

Journal of Physics: Conference Series **541** (2014) 012013 doi:10.1088/1742-6596/541/1/012013

Two-scale model of hydrothermal synthesis of nanotubes

D A Kucher[1], S A Chivilikhin[1], I Yu Popov[1] and V V Gusarov[2]

[1]St. Petersburg National Research University of Information Technologies, Mechanics and Optics, Kronverkskiy, 49, St. Petersburg, 197101, Russia
[2]Ioffe Physical-Technical Institute RAS, Polytehnicheskaya, 26, St. Petersburg, Russia

dmitry.a.kucher@gmail.com

Abstract. The hydrodynamic two-scale model of hydrothermal synthesis inorganic nanotubes is proposed. The model, which proposed at this paper is consist from two scale of parameters: micro and macro. The macro parameters are: distribution of temperature, velocity and concentrations. The distribution of concentration is binds two scales. Boundary conditions for micro-scale is concentration, which calculated from macro-scale parameters.

1. Introduction

Hydrothermal synthesis is one of the methods for synthesis nanotubes. The main characteristics of hydrothermal synthesis are distribution of velocity, temperature and concentrations. Different configuration of these characteristics determines the efficiency of synthesis. Authors have autoclave in his laboratory. This experimental setup allows investigating early step of creation nanotubes. Understanding of these processes can help create configuration of characteristics for more efficient synthesis of nanotubes. In this work a model of the formation and evolution of the ensemble nanotubes formed from particles of compounds with a layered structure is described. The nanotube source is a system consisting of crystalline nanoplates which have a layered structure having and distributed in the liquid. Crystals consist of double strained layers. Due to intercalation of the liquid occurs delamination between the layers of the outer layers. Delaminated layer is twisted in nanotube for intervals that are small in comparison with the time of diffusive growth. Internal elastic stresses in nanotube less than in the layer, which determines the energy gain nanotube rolling process.

2. Hydrodynamic model of nanotubes hydrothermal synthesis

Two-scale model of hydrothermal synthesis of nanotubes in autoclave is proposed for estimation characteristics of process. The macro-scale process includes the thermal convection of liquid, due to heating of autoclave wall, heat transfer and diffusion of chemical components dissolved in the liquid. Distribution of velocity, temperature and concentrations obtained in macro-scale model let us to describe the nano-scale processes - diffusion evolution of nanotube ensemble [1-6]. The concentrations, calculated in macro-model use in nano-model as boundary conditions on infinity. In the other hand, the summary mass flow on the nanotubes in some small volume is the mass source in diffusion equation of macro-scale model.

3. Evolution of nanotubes ensemble

Let us characterize the individual nanoplates its characteristic diameter L_p and thickness H. Distribution of nanoplates according to the diameter and thickness of each time moment is characterized by function $f_p(L_p, H, t)$. We characterize the individual nanotube by its length L_T and cross-sectional area S (excluding the area of the inner hole). Distribution of nanotubes at lengths and cross-sectional area at any given time has a density $f_T(L_T, S, t)$.

Diffusive growth (dissolution) of nanoplates and nanotubes determined to transport of substances to their surfaces through the liquid phase. Density solutes ρ at each time can be determined from the macroscopic model. We assume that the equilibrium concentration of solute on the solid phase density ρ_e is a function of the surface energy of the strained layer. Elastic energy density in the plates exceeds the energy density of the mechanical stresses in the nanotubes. Reducing the strain and, accordingly, mechanical energy, is driving the twisting of nanotubes. Note that in this discussion we will assume that the mass transfer between the solid and liquid phase occurs in the diffusion mode, i.e. neglect kinetic effects of dissolution and precipitation of substances on the surface.

Rate of change of plate and nanotubes sizes by diffusive mass transfer is determined by the density gradient in the vicinity of the boundary, which is as a boundary value of the normal derivative of the solution corresponding to the external (for nanotubes) of the Dirichlet's problem.

The calculated value of the normal derivative boundary density let us to determine the mass flow density j on the boundary and the normal velocity of the interface V:

$$j = -D\frac{\partial \rho}{\partial N} \sim \rho_e - \rho, \quad V = \frac{j}{\rho_m}, \tag{1}$$

where D is the diffusion coefficient of the solute, ρ_e is the equilibrium solute concentration on the solid surface, ρ_m is the density of nanoplates and nanotubes matter. We assume that this is the equilibrium concentration of the energy density of the strained layer:

$$\rho_e = \rho_e^0\left(1 + \gamma e^2\right), \tag{2}$$

where ρ_e^0 is the equilibrium density on the surface of the unstrained layer, e is the density of elastic energy of strained layer, γ is a constant.

Elastic energy density for nanoplates is greater than the corresponding energy for nanotubes. Reducing the strain and, accordingly, the elastic energy is the driving force for twisting of the flat layer and forming of nanotube. We assume that the mass transfer between the solid and the liquid phase takes place in the diffusion regime (we do not take into account the kinetic effects on the surface of nanoplates and nanotubes).

Density of elastic energy for double layer with the thickness δ and radius of curvature r may be expressed as:

$$e = \frac{E\delta^3}{24}\left(\frac{1}{r} - \frac{1}{R_0}\right)^2, \tag{3}$$

where R_0 is the equilibrium radius of curvature of the layer, E is theYoung's modulus. Nanoplatelets energy density is

$$e_p = \frac{E\delta^3}{24R_0^2}. \tag{4}$$

The energy density on the side surface of nanotube is

$$e_R = \frac{E\delta^3}{24}\left(\frac{1}{R} - \frac{1}{R_0}\right) \approx \frac{E\delta^3 h^2}{24R_0^4}, \tag{5}$$

where h is the wall thickness of the nanotube. The average value of the elastic energy density at the end of the nanotube may be calculated as

$$\bar{e} = \frac{1}{h}\int_{R-h}^{R} e(r)dr \approx \frac{E\delta^3}{192R_0^4}. \tag{6}$$

We can see, that $\bar{e} < e_R << e_p$.

The evolution of the density distribution of plates and nanotubes size is described by the kinetic equations:

$$\begin{aligned} &\frac{\partial f_p}{\partial t} + \frac{\partial}{\partial L_p}\left(V_{L_p} f_p\right) + \frac{\partial}{\partial H}\left(V_H f_p\right) = \nu\left[f_p\left(L_p, H+a, t\right) - f_p\left(L_p, H, t\right)\right], \\ &\frac{\partial f_T}{\partial t} + \frac{\partial}{\partial L_T}\left(V_{L_T} f_T\right) + \frac{\partial}{\partial S}\left(V_S f_T\right) = \nu\delta\left(S - aL_T\right)\int_0^\infty f_p\left(L_T, H, t\right)dH, \end{aligned} \tag{7}$$

where a is the thickness of the double layer, ν is the frequency of double layers of peeling from the plate, $\delta(x)$ is the Dirac's delta-function, V_Z is the rate of change of the parameter Z due to the diffusion associated with the above-mentioned normal derivative of the density. Source the right side of equation for f_T describes the occurrence of nanotubes delamination by double layers of nanoplates and subsequent twisting of nanotube. By specifying the initial distribution of the size of the plates, and assuming that the nanotubes absent at the initial moment, we obtain the temporal evolution of the size distribution of nanoplates and nanotubes. Calculations show that over time the nanotubes receive approximately the same radius R with approaching to the the equilibrium radius R_0. Calculation results correlated with the available experimental data on hydrothermal synthesis hydrosilicate nanotubes. Namely, there is an increase in the proportion of nanotubes having a diameter close to the equilibrium that is obtained in this model.

4. Model of macro-scale process

To describe the macro-scale hydrothermal synthesis is convenient to use the Navier-Stokes equations supplemented by the heat equation. The process which described by this system is named free convection. Those equations are allowed to evaluate distribution of temperature and velocity from initial condition.

$$\begin{cases} \rho\left[\frac{\partial V}{\partial t}+(V\cdot\nabla)V\right]=-\nabla p+\eta\nabla^2V+\frac{1}{3}\eta\nabla(\nabla\cdot V)+\rho g, \\ \frac{\partial\rho}{\partial t}+\nabla\cdot(\rho V)=0, \\ c\rho\left(\frac{\partial T}{\partial t}+V\cdot\nabla T\right)=\lambda\nabla^2T; \end{cases} \tag{8}$$

where V – velocity vector; p – pressure; T – absolute temperature of the fluid; ρ – density; η – dynamic viscosity; λ – thermal conductivity; c – specific heat; t – time; g – acceleration of free fall.
The first equation of this system at general case may be rewrite as:

$$\frac{\partial V}{\partial t}=\eta\nabla^2V-(V\cdot\nabla)V-\nabla p+f \tag{9}$$

where f - is the body force such as gravity or buoyancy.
This equation can be split into two parts (assuming a uniform distribution of pressure):

$$\frac{\partial V_1}{\partial t}=\eta\nabla^2V+f \tag{10}$$

$$\frac{\partial V_2}{\partial t}=-(V_1\cdot\nabla)V_2 \tag{11}$$

The first part is diffusion and describing by eq(10). This equation can be solved by relaxation method. The second part is advection and describing by eq(11) can be solved by MacCormack method, which consist of two steps the first one is predictor and the second one is corrector.
In our case the second equation of system can be rewrite as:

$$\nabla\cdot V=0 \tag{12}$$

Using the Helmholtz decomposition this equation can be split into the following form:

$$V=V'+\nabla\varphi \tag{13}$$

where φ is a scalar field.
If we apply the gradient operator to both sides we obtain:

$$\nabla\cdot V=\nabla^2\varphi \tag{14}$$

This is a Poisson equation and it can be solved by multigrid method.
The implementations of these methods allow creating a good instrument to get distribution of temperature and velocity. At fig. 1 is presented visualization of the fields of velocity vectors. The data also can be saved at a text document for following calculated. At fig. 2 is presented visualization of the isotherms for natural convection.
The future step to developing this model is an implementation of equation for concentration which will allow binding macro- and micro-scales.

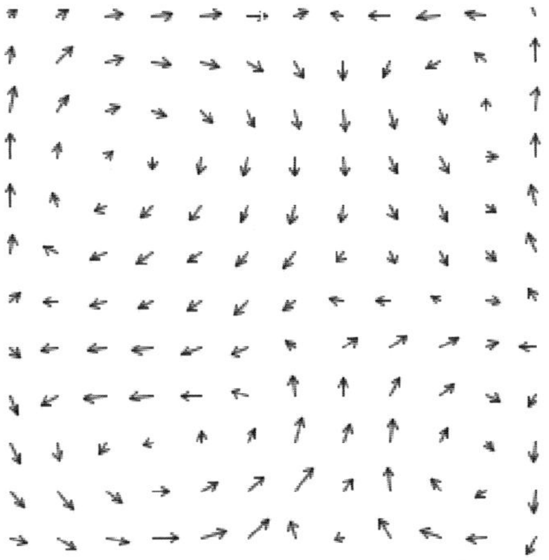

Figure 1. Fields of velocity vectors

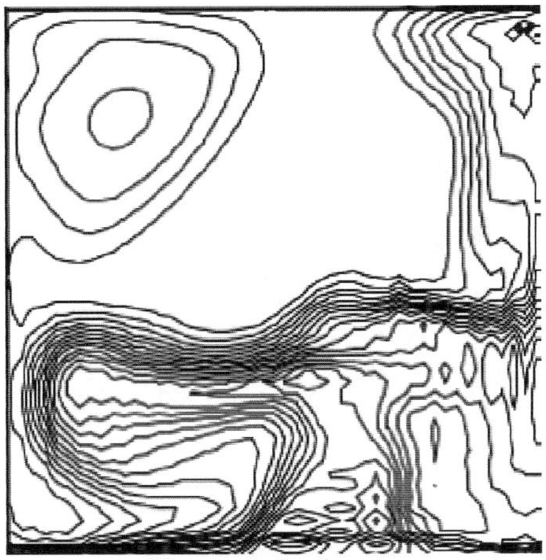

Figure 2. Isotherms

5. Conclusion

At the paper described the principles of modelling hydrothermal synthesis of nanotubes by two scale model. This way of modelling can allow to estimate many important parameters for hydrothermal synthesis in practice. Also it can help to find the best parameters for synthesis nanotubes with defined configuration.

Acknowledgments

This work was financially supported by Government of Russian Federation, Grant 074-U01.

References

[1] Schmidt O G and Eberl K 2001 *Nature* **410** 168
[2] Pauling L 1930 *Proc. Nat. Acad. Sci. U.S.A.* **16** 578
[3] Chivilikhin S A, I. Yu. Popov and V. V. Gusarov 2007 *Dokl. Phys.* **412** (2) 201
[4] Chivilikhin S A, Popov I Yu, Svitenkov A I, Chivilikhin D S and Gusarov V V 2009 *Dokl. Phys.* **54** (11) 491
[5] Deneke Ch and Schmidt O G 2004 *Appl. Phys. Lett.* **85** (14) 2914
[6] Braga S F, Coluci V R, et al. 2004 *Nano Lett.* **4**(5) 881

Modelling the interaction of nanotube with particles of the medium by the Green's function method

A V Tulub[1] and D A Nevsky[1]
[1]Institute of Chemistry, St. Petersburg State University, 26 University Ave, Petrodvorets, 198504, Russia

E-mail: nevsky14@mail.ru

Abstract. The end surface of a cylindrical nanotube is considered as an annular and is replaced by a torus by solving the electrostatic problems, they are 1) the interaction of a charged particle with a torus and 2) the interaction of a charged torus with a neutral particle. Solution of the Poisson equation is given in the toroidal coordinate system and expressed by a half-order Legendre function of the first and second kind. The convergence of series expansion is improved by a given transformation of corresponding hypergeometric functions. The calculations are represented for the axially symmetric case.

In recent years interest in such nanostructures in which the particle, or fullerene molecule with a great polarizability is situated inside a carbon nanotube is sharply increased. These objects can be widely used in microelectronics, optics and biophysics. Materials based on them are promising nanocomposites [1-2]. Also annular and coiled nanostructure are of special interest [3].

An approach and a retraction of a particle to a nanotube is mainly due to non-covalent interactions having a classical electrostatic nature.

The open end of a nanotube is a ring of finite thickness and it is approximated by a torus of radius a by the solution of electrostatic problems. The toroidal coordinates (η, θ, φ) are expressed through the Cartesian by

$$x = a\frac{\sinh\eta\cos\varphi}{\cosh\eta-\cos\theta}, \quad y = a\frac{\sinh\eta\sin\varphi}{\cosh\eta-\cos\theta}, \quad z = a\frac{\sin\theta}{\cosh\eta-\cos\theta} \tag{1}$$

and are used by the solution of the Poisson's equation in terms of a Green function. Radius a, a scale parameter of torus, is

$$a=\sqrt{R_0^2-R_t^2}, \tag{2}$$

where R_0 is a exterior radius, or in our case it is a radius of a nanotube, R_t is a thickness of torus. Thickness is considered to be equal to the van der Waals radius of carbon, 0.17 nm [4].

The essence of the method used for solving such problems is to find the Green's function which satisfies Poisson's equation in chosen coordinate system

$$\Delta G(\mathbf{r},\mathbf{r}_0) = -4\pi\varepsilon\delta(\mathbf{r}-\mathbf{r}_0), \tag{3}$$

where $\mathbf{r}_0$ is radius vector of a point charge, $\delta(\mathbf{r}-\mathbf{r}_0)$ is Dirac delta function and ε is dielectric permittivity, and the bounary conditions at the surface, in this case on the toroidal surface.

$$G(\mathbf{r},\mathbf{r}_0)|_{\partial S}=0 \tag{4}$$

Firstly, the problem of a point unit charge above the torus is considered. Green's function consists of two components – the Coloumb potential part $G(\mathbf{r}, \mathbf{r}_0)$ of a point charge in an unbounded region

$$G_0(\mathbf{r},\mathbf{r}_0)=\frac{q}{4\pi\varepsilon|\mathbf{r}-\mathbf{r}_0|} \tag{5}$$

and reactive part $R(\mathbf{r}, \mathbf{r}_0)$, describing a reactive field, which arises as a result of the occurence of the charge induced on the end of nanotube:

$$G(\mathbf{r},\mathbf{r}_0)=G_0(\mathbf{r},\mathbf{r}_0)+R(\mathbf{r},\mathbf{r}_0) \tag{6}$$

After the separation of variables in the equation (2), see for instance [4], the expression for the Coulomb potential can be written as

$$G_0(\mathbf{r},\mathbf{r}_0)=\frac{q}{4\pi^2\varepsilon a}\sqrt{(\cosh\eta-\cos\theta)}\sqrt{(\cosh\eta_0-\cos\theta_0)}\sum_{m=0}^{\infty}\xi_m\cos m(\varphi-\varphi_0)\sum_{n=0}^{\infty}\xi_n\cos n(\theta-\theta_0)\times$$
$$\times(-1)^m\frac{\Gamma(n-m+1/2)}{\Gamma(n+m+1/2)}[Q^m_{n-1/2}(\cosh\eta)P^m_{n-1/2}(\cosh\eta_0)\Theta(\eta-\eta_0)+P^m_{n-1/2}(\cosh\eta)Q^m_{n-1/2}(\cosh\eta_0)\Theta(\eta_0-\eta)]$$

where $\xi_{n,\,m}$ takes value 1 for n, m = 0 and 2 for n, m > 0, $P^m_{n\text{-}1/2}$ and $Q^m_{n\text{-}1/2}$ are associated Legendre functions of first and second kind respectively, $\Theta(\eta - \eta_0)$ is a unit step function, q is a unit charge. The expression for the reactive fields takes correspondingly the form

$$R(\mathbf{r},\mathbf{r}_0)=-\frac{q}{4\pi^2\varepsilon a}\sqrt{(\cosh\eta-\cos\theta)}\sqrt{(\cosh\eta_0-\cos\theta_0)}\sum_{m=0}^{\infty}\xi_m\cos m(\varphi-\varphi_0)\sum_{n=0}^{\infty}\xi_{\text{nital}}\cos n(\theta-\theta_0)\times$$
$$\times(-1)^m\frac{\Gamma(n-m+1/2)}{\Gamma(n+m+1/2)}P^m_{n-1/2}(\cosh\eta_0)P^m_{n-1/2}(\cosh\eta)[Q^m_{n-1/2}(\cosh s)/P^m_{n-1/2}(\cosh s)] \tag{7}$$

s is a constant which specifies toroidal surface and $\cosh s$ is equal to the ratio of radius of the nanotube R_0 and nanotube thickness R_t, see (2).

The resulting formulas are considerably simplified in the axially symmetric case, when the charge is located on the axis z and accordingly η_0=0, then the dependence of angle φ disappears and we obtain

$$G_0(\mathbf{r},\mathbf{r}_0)=\frac{q}{4\pi^2\varepsilon a}\sqrt{(\cosh\eta-\cos\theta)}\sqrt{(1-\cos\theta_0)}\sum_{n=0}^{\infty}\xi_n\cos n(\theta-\theta_0)Q_{n-1/2}(\cosh\eta) \tag{8}$$

$$R(\mathbf{r},\mathbf{r}_0)=-\frac{q}{4\pi^2\varepsilon a}\sqrt{(\cosh\eta-\cos\theta)}\sqrt{(1-\cos\theta_0)}\sum_{n=0}^{\infty}\xi_n\cos n(\theta-\theta_0)P_{n-1/2}(\cosh\eta)\frac{Q_{n-1/2}(\cosh s)}{P_{n-1/2}(\cosh s)} \tag{9}$$

Standard expression of Legendre functions with half-integer index through hypergeometric series in the case of m = 0 has the form [6]:

$$P_{n-1/2}(\cosh\eta)={}_2F_1\left(\frac{1}{2}-n,\frac{1}{2}+n;1;\frac{1-\cosh\eta}{2}\right) \tag{10}$$

$$Q_{n-1/2}(\cosh\eta)=\left(\Gamma\left(n+\frac{1}{2}\right)/\Gamma(n+1)\right)(2\cosh\eta)^{-n-1/2}{}_2F_1\left(\frac{n}{2}+\frac{3}{4},\frac{n}{2}+\frac{1}{4};n+1;\cosh^{-2}\eta\right) \tag{11}$$

However, for practical purposes, these expressions for the function of the first kind and the functions of the second kind are inconvenient due to the slow convergence hypergeometric series and need accommodate a large number of terms in the partial sum.

To improve the convergence of the hypergeometric function transform using known relationships by Kummer. In the case of functions of the first kind of the optimal transformation is done as follows:

$${}_2F_1(a,b;c;z)=(1-z)^{-a}{}_2F_1\left(a,c-b;c;\frac{z}{z-1}\right) \tag{12}$$

Applying this transformation to (10) is easy to obtain

$$P_{n-1/2}(\cosh\eta)=\left(\frac{\cosh\eta+1}{2}\right)^{n-1/2}{}_2F_1\left(\frac{1}{2}-n,\frac{1}{2}-n;1;\tanh^2\frac{\eta}{2}\right) \tag{13}$$

Thus we get series in powers of the hyperbolic tangent, which converges quickly enough. Hypergeometric series which expresses the Legendre function of the second kind is conveniently subjected to the following sequence of transformations: first, to make the quadratic transformation:

$${}_2F_1\left(a,a+\frac{1}{2};c;z\right)=(1-z)^{-a}{}_2F_1\left(2a,2c-2a-1;c;\frac{1}{2}\left(1-\frac{1}{\sqrt{1-z}}\right)\right) \tag{14}$$

and then use the transform (12). The result is the following rapidly converging series in powers of exponent:

$$Q_{n-1/2}(\cosh\eta)=\sqrt{\pi}(\Gamma(n+\frac{1}{2})/\Gamma(n+1))e^{-(n-1/2)\eta}{}_2F_1(\frac{1}{2},n+\frac{1}{2};n+1;e^{-2\eta}) \tag{15}$$

Thus the series obtained using (13) and (15) can be used for further quantitative assessment of the effect. Table 1 shows a comparison of values of Legendre functions with an integer index, which can be calculated with high accuracy, given in [7] and obtained by the proposed method. Table 2 shows a comparison of Legendre functions with half-integer index calculated through elliptic integrals and the proposed method.

Table 1. Comparison of Legendre functions values with integer index n = 0(1)4 obtained in [7] and by the proposed method.

n	$P_n(z)$, z = 2.6000		$Q_n(z)$, z = 2.6000	
	[7]	(13)	[7]	(15)
0	1.0000	1.0000	0.405460	0.405460
1	2.6000	2.6001	0.054209	0.054209
2	9.6400	9.6400	0.008684	0.008684
3	40.0400	40.0395	0.001489	0.001490
4	174.9520	174.9500	0.000262	0.000262

Table 2. Comparison of Legendre functions values with half-integer index $n = \pm$ ½ obtained by using full elliptic intergal representation [7] and by the proposed method.

n	$P_{n-1/2}(z)$, z = 8.8235		$Q_{n-1/2}(z)$, z = 8.8235	
	[7]	(13)	[7]	(15)
- ½	0.6461	0.6420	0.74966	0.74968
½	2.6862	2.6864	0.02131	0.02130

Polarization energy in the case of the set of point charges consists of summ of single-particle contributions U_i and two-particle contributions U_{ij} which are due to the interaction of charge q_i with the reactive field of q_j.

$$U_i=\frac{q_i}{2}\int\delta(\mathbf{r}-\mathbf{r}_i)R(\mathbf{r},\mathbf{r}_i)d^3\mathbf{r},\quad U_{ij}=\frac{q_iq_j}{2}\int d^3\mathbf{r}'\int\delta(\mathbf{r}'-\mathbf{r}_i)R(\mathbf{r},\mathbf{r}')\delta(\mathbf{r}-\mathbf{r}_j)d^3\mathbf{r} \tag{16}$$

Polarization energy of unit point charge is given by the first term in (16). After substitution (9) with the assumption that the charge is on the z-axis in (16) and simplifying we get

$$U=-\frac{a\Lambda(s)}{4\pi^2\varepsilon}\frac{q^2}{a^2+z_0^2}, \tag{17}$$

where by $\Lambda(s)$ denotes the series, which, as can be shown by using the asymptotic formulas for the Legendre functions converges uniformly:

$$\Lambda(s)=\sum_{n=0}^{\infty}\xi_n(Q_{n-1/2}(\cosh s)/P_{n-1/2}(\cosh s)) \tag{18}$$

Values of constant Λ(s) calculated for different radii R_0 of nanotube and respectively for different s are given in table 3

Table 3. Values of $\Lambda(s)$.

R_0, nm	cosh s	$\Lambda(s)$
1.0	5.8824	1.3332
1.5	8.8235	1.1837
2.0	11.7647	1.1100
5.0	29.4118	1.1079

Force component acting upon a unit charge can be calculated as a derivative of the reactive field

$$F_\theta=-q\frac{1}{H_\theta}\frac{\partial R}{\partial\theta}\Big|_{r=r_0}, \tag{19}$$

where H_θ is a Lame coefficient which is equal to $a/(\cosh\eta - \cos\theta)$. Also force can be calculated by the gradient energy. Of course, both approaches should give the same result. So component of the force acting in the z direction is given by

$$F_z = \frac{q^2 a \Lambda(s)}{2\pi^2 \varepsilon} \frac{z_0}{(a^2 + z_0^2)^2} \tag{20}$$

In the axial symmetric case the computation of force, acting the charge, and polarization energy was done. Thickness of torus is considered to be 0.17 nm as was told earlier, radius of nanotube is changed. Plots for calculated force (figure 1) and energy (figure 2) are given for the case of unit point charge located on z-axis.

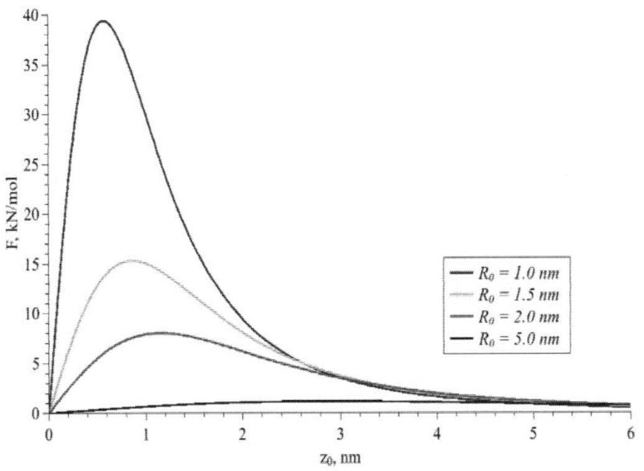

Figure 1. Force F, N/mol acting upon a unit point charge as functions of distance z_0, nm to the centre of ring.

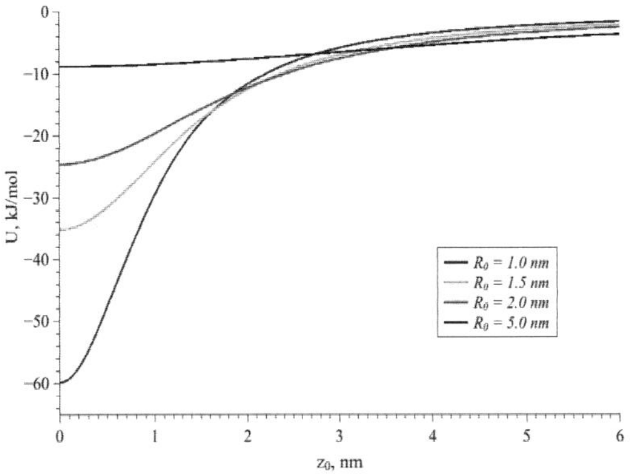

Figure 2. Polarization enegry U, kJ/mol as functions of distance z_0, nm to the centre of ring.

Summing up, solution of the Poisson's equation was presented in the toroidal coordinate system and expressed by a half-order Legendre function of the first and second kind. Transformation of hypergeometric functions was done. Numerical calculations, plots of energy and force were represented for the axially symmetric case.

References

[1] Smith B W, Monthioux M and Luzzi D E 1998 Nature **396** 323-324 1197
[2] Khlobystov A N, Britz D A and Briggs G A D 2005 *Acc. Chem. Res.* **38** 901
[3] Dappe Y J 2014 *J. Phys.* D **47** 083001
[4] Rowland R S and Taylor R 1996 *J. Phys. Chem.* **100** 7384-7391
[5] Lebedev N N 1965 *Special functions and their Applications* (Prentice-Hall INC New Jersey: Englewood Cliffs) pp 221-226
[6] Bateman H and Erdelyi A 1953 *Higher transcendental functions.* (New-York Toronto London: McGraw-Hall Book INC) Vol. 1 pp 125-174
[7] Abramowitz M and Stegun I A 1972 *Handbook of Mathematical Functions with Formulas, Graphs and Mathematical Tables* (New York: Dover Publications) pp 153-174

Synthesis of core-shell Ag/SiO$_2$ nanoparticles for SPASER structures

D V Viazmitinov, L B Matyushkin, A I Maximov
St Petersburg Electrotechnical University "LETI", 197376 St Petersburg, Russia

E-mail: dviazmitinov@gmail.com

Abstract. Aqueous solution of colloidal silver was prepared using a method based on a chemical reaction of silver nitrate water-soluble salts and sodium borohydride. Synthesized silver nanoparticles were analyzed by transmission optical spectroscopy. Silica shells were formed by the modified Stöber process. During shell formation nanoparticles were also doped by organic dye (Rhodamine 6G). The resulting structures were deposited on silicon substrates. Half of the samples were annealed at 300°C, and the other half was just dried. Processed samples were investigated by atomic force microscopy.

1. Introduction

Nowadays nanoplasmonics is one of the most rapidly growing areas of physics and nanotechnology. It has many possible applications ranging from sensing and biomedicine to imaging and information technology. One of the most significant problem in this field of science is development of devices that can generate coherent plasmonic fields – SPASER's (Surface Plasmon Amplification by Stimulated Emission of Radiation) [1]. There are two main types of such devices – planar structures, consist of metal nanostructure separated from a two-level energy system by a dielectric layer [1] and core-shell structures in which the dielectric layer surrounds the metallic core, separating it from the two-level system [2]. Core-shell structures can be synthesized by methods of colloid chemistry, which allows to produce them cheaper and more massively than planar structures requiring quite expensive and inefficient process technologies. Different methods and materials are used to produce core-shell structures for nanoplasmonics. The most suitable material for the core are considered noble metals such as gold [2] or silver [3]. Stable and monodisperse silver nanoparticles can be simple synthesized by method, described below. Technology of producing shells with predetermined thickness and composition still needs to be elaborated.

2. Experimental

Aqueous solution of colloidal silver was prepared using a method based on a chemical reaction of water-soluble salts of silver nitrate and sodium borohydride:

$$AgNO_3 + NaBH_4 \rightarrow Ag + \tfrac{1}{2} H_2 + \tfrac{1}{2} B_2H_6 + NaNO_3$$

A volume fraction of 10^{-3} M aqueous silver nitrate solution was slowly added dropwise to three volumetric fractions cooled in an ice bath $2 \cdot 10^{-3}$ M aqueous solution of sodium borohydride. Use of an excess of sodium borohydride is because it acts both as reducing and stabilizing agent. As a supplement to sodium borohydride, polyvinylpyrrolidone was added for greater stabilization of the particles size. The resulting silver colloidal solution is due to a plasmon resonance absorption spectrum with a peak near 400 nm λ_{pl} has a transparent yellow color. Full width at half maximum (FWHM) is 50-70 nm. The peak position essentially depends on particle size (Table 1).

Table 1. Relation between particle diameter and form of the optical spectrum

Particle diameter *d*, nm	λ_{pl}, nm	FWHM, nm
10-14	395-405	50-70
35-50	420	100-110
60-80	438	140-150

Dynamics of particle stability is shown at fig.1.

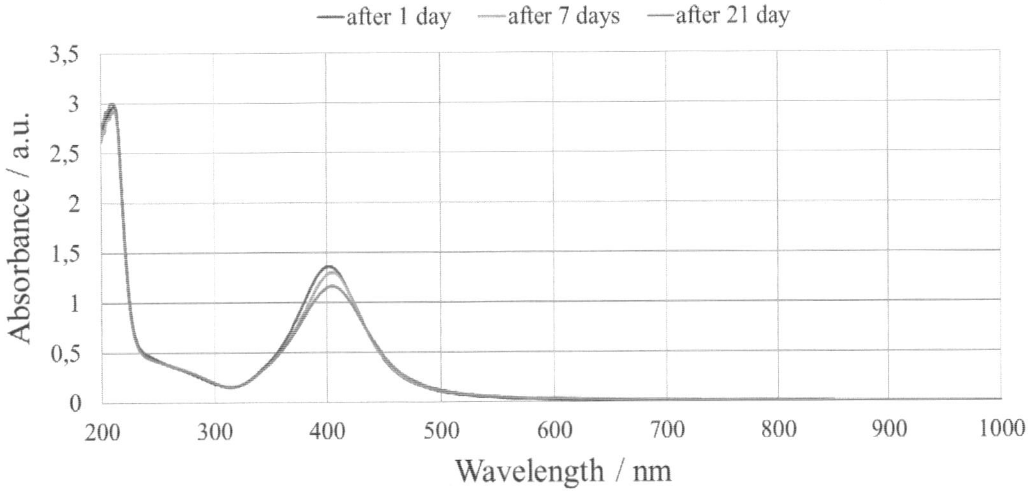

Fig. 1. Ag nanoparticles stability dynamics

Transmission optical spectroscopy showed that the particle size in the result solution of colloidal silver is 12 nm.

Silica shells were formed by the modified Stöber process:

1. Silver nanoparticle sol was diluted with isopropyl alcohol in a volume ratio of 1/200 in an ultrasonic bath.
2. The resulting solution was stirred in an ultrasonic bath at 40°C for 1 hour.
3. Then, aqueous ammonia solution was added into the solution.
4. After temperature stabilization, tetraethoxysilane (TEOS) was injected in an amount that depends on the total surface of the silver particles and the desired thickness of the shell (it was assumed that all TEOS transformed into silica).

The final ratio of the reactants is shown in table 2.

Table 2. The final ratio of the reactants

Reactant	Value, ml
NH_4OH	10,50
Distilled water	18,00
TEOS	0,13
Isopropyl alcohol	50,00

During shell formation nanoparticles were also doped by organic dye (Rhodamine 6G). It was made by adding aqueous solution of rhodamine 6G (10 ml, 25%) before injection of TEOS in the above-described silica shells formation technology. The resulting structures were cleared by sedimentation process and then investigated using optical transmission spectroscopy (fig. 2) and photoluminescence test (fig. 3).

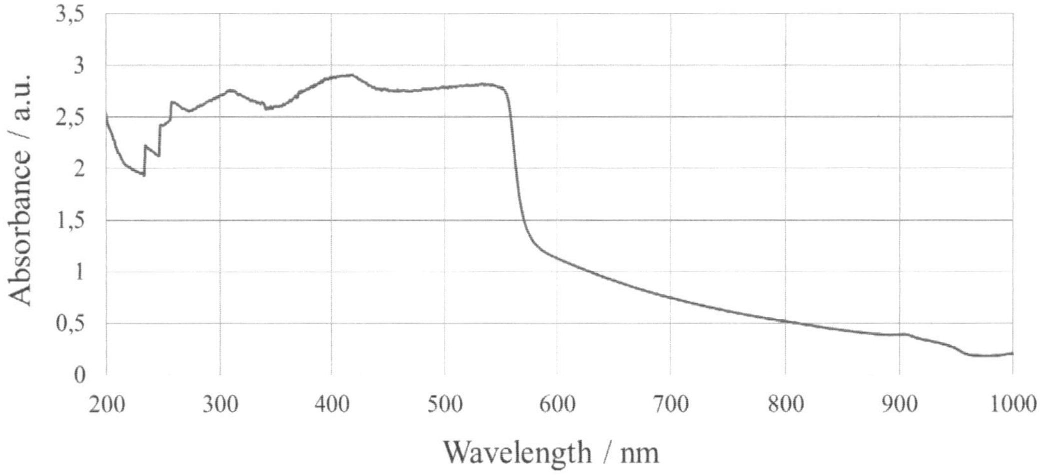

Fig. 2. Absorption spectrum of the resulting core-shell nanostructures

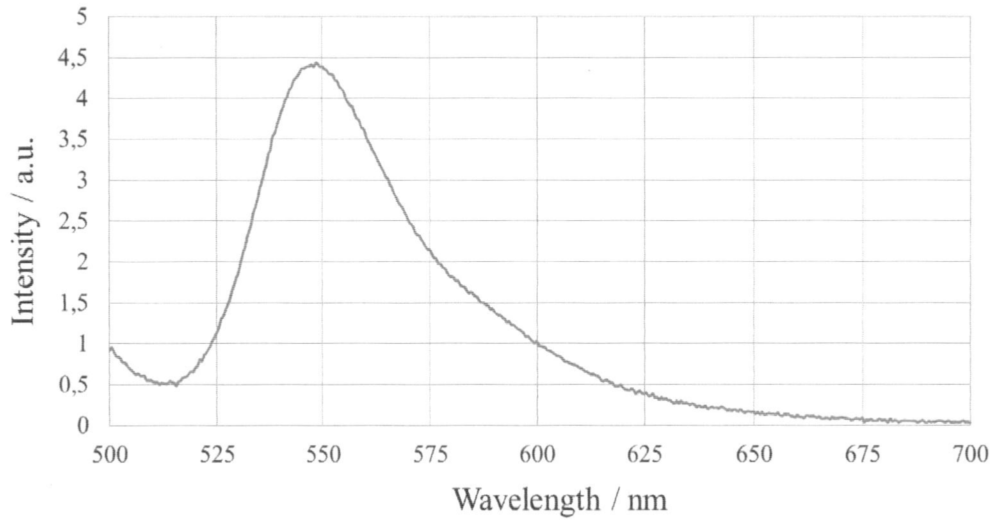

Fig. 3. Photoluminescence spectrum of the resulting core-shell nanostructures

The resulting structures were deposited on silicon substrates. Half of the samples were annealed at 300°C for 15 minutes, and the other half was just dried for 1 hour. Then all of the samples were analyzed using the atomic force microscopy in tapping mode (fig. 4).

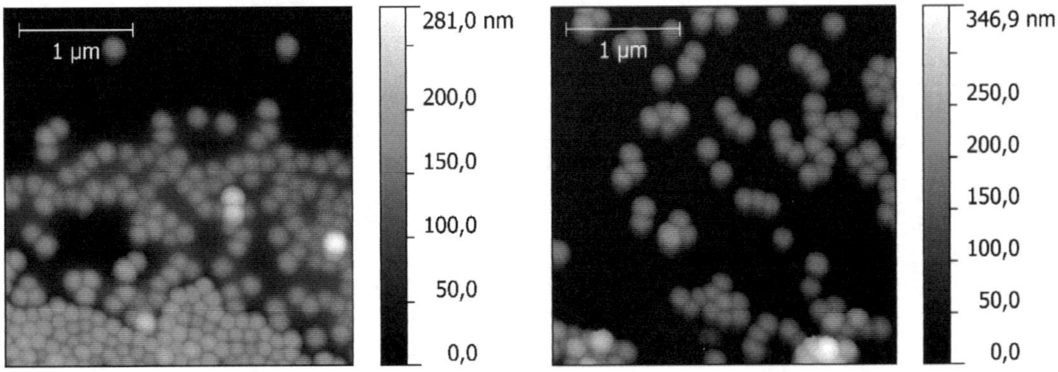

Fig. 4. Atomic force microscopy images for dried (left) and annealed (right) samples

3. Results

Atomic force microscopy showed that the synthesized core-shell structures have a size approximately equal to 150 nm and variation in size is approximately 5 nm. The images obtained allow us to conclude that the calculation of the thickness of the silica shell is quite exactly the same as the actual thickness of the shells of the synthesized nanoparticles. By comparing the two groups of samples (dried and annealed), it can be concluded that annealing increases the uniformity of particle distribution over the substrate, and also reduces the variation in size. Analysis of the absorption and fluorescence spectra indicates that the resulting nanoparticles are able to absorb a wide wavelength range, while having a relatively narrow photoluminescence peak. Considering that source for photoluminescence excitation radiation is a laser with a wavelength of 455 nm, and the photoluminescence peak falls on 540 nm, we can assume that the rhodamine molecules are excited by the energy generated by silver nanoparticles.

References

[1] David J Bergman and Mark I Stockman 2003 *Physical Review Letters* **90(2)** 027402-1.

[2] M A Noginov, G Zhu, A M Belgrave, R Bakker, V M Shalaev, E E Narimanov, S Stout, E Herz, T Suteewong & U Wiesner 2009 *Nature* **460(27)** 1110.

[3] V V Hardikar, E Matijevic 2000 *Journal of Colloid and Interface Science* **221** 133.

[4] A I Maximov, V A Moshnikov et al, *Fundamentals of sol-gel technology of nanocomposites*, St Petersburg: Technomedia, Elmor, 2007.

Model of porous aluminium oxide growth during initial stage of anodization

E M Aryslanova[1], A V Alfimov[1], S A Chivilikhin[1]

[1]St. Petersburg National Research University of Information Technologies, Mechanics and Optics, Kronverkskiy, 49, St. Petersburg, 197101, Russia

E-mail: elizabeth.aryslanova@gmail.com

Abstract. Currently, the development of nanotechnology and metamaterials requires the ability to obtain regular self-assembled structures with different parameters. One such structure is porous alumina in which the pores grow perpendicular to the substrate and are hexagonally packed. Pore size and the distance between them can be varied depending on the anodization voltage, the electrolyte and the anodization time (pore diameter - from 2 to 350 nm, the distance between the pores - from 5 to 50 nm). At the moment, there are different models describing the process of anodizing aluminum, in this paper we propose a model that takes into account the effect of layers of aluminum, aluminum oxide, and the electrolyte, as well as the influence of the effect of surface diffusion.

1. Introduction

Currently the scope of anodic aluminum oxide (AAO) use expanded from protection against corrosion, electrical protection, and thermal protection to a development of a template for the synthesis of various nanocomposites. Such as: synthesizing nanotubes with matrix method [1-2]: 50-60 nm thick films, with an ordered system of nanopores (diameter of 40-100 nm) is used for synthesizing oriented carbon nanotubes [3]; the ability to control parameters of the porous structure of Al_2O_3 can be used as filters, carriers for catalysts [2]; films with high regularity of the porous structure are increasingly used for creating nanoscale structures in electronic, magnetic, and photonic devices [4]; with sorption of silver ions in the matrix of porous alumina, followed by chemical deposition nanocomposites are synthesized with biochemical activity properties [5].

Porous anodic alumina films are self-organizing structures, which can be represented in a hexagonal packing of cylindrical pores normal to the plane of the aluminum film and used as a template for synthesis of various nanocomposites (Figure 1).

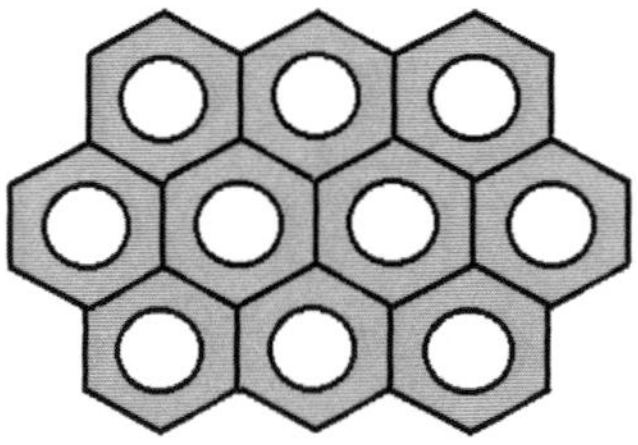

Figure 1. Hexagonal packing of the porous AAO

2. Model

In this paper we consider the motion of the interfaces between the electrolyte-Al_2O_3 (dissolution), and between Al_2O_3-aluminum (oxidation), as well as consider the dynamics of moving boundaries and the change of small perturbations of these boundaries. In each area under Laplace's equation is solved for the potential of the electric field. The growth process of the porous alumina is described by the theory of small perturbations. In zero approximation boundaries are considered flat and the speed of their movements is proportional to the current density at these boundaries. In first approximation small perturbations of the interface are considered, which lead to small changes in potential and current on these boundaries. The evolution of small perturbations of the interface is defined as a disturbance of the current density at the borders, and the process of surface diffusion.

2.1. The evolution of perturbations of the film boundaries of Al_2O_3-layer model:
Figure 2 shows the geometry of the area under consideration. Here h_1 and h_2 are small perturbations of the boundaries of Al-Al_2O_3 and Al_2O_3-electrolyte, respectively.

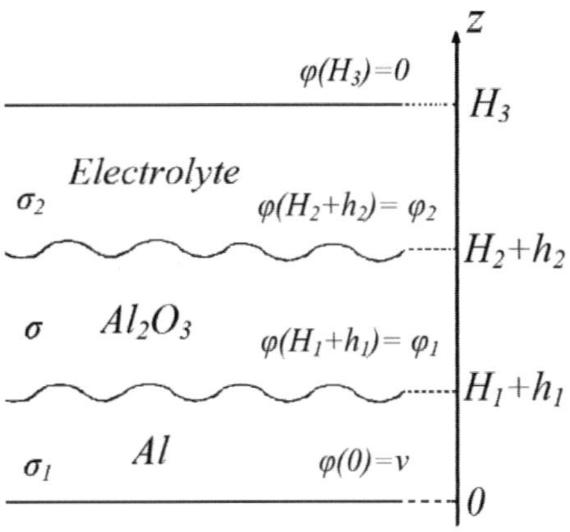

Figure 2. Inhomogeneous film with rough boundaries.

Using conditions of continuity of the current density at both interfaces (aluminum-alumina, alumina-electrolyte) we obtain the system of equations relating the potential disturbance on the interfaces:

$$\begin{cases} \left(-\sigma_2 sh(kH) - \sigma ch(kH)\right)\hat{\varphi}_{2\bar{k}} + \sigma\hat{\varphi}_{1\bar{k}} = \beta e^{kH} - \alpha \\ -\sigma\hat{\varphi}_{2\bar{k}} + \left(\sigma_1 sh(kH) + \sigma ch(kH)\right)\hat{\varphi}_{1\bar{k}} = \beta - \alpha e^{kH} \end{cases} \quad (1)$$

where $\alpha = \frac{v}{\vartheta} h_{1\bar{k}}$, $\beta = \frac{v}{\vartheta} h_{2\bar{k}}$, $\hat{\varphi}_{\bar{k}}$ - the Fourier transform of the potential disturbance to the coordinates x, y: $\hat{\varphi}_{\bar{k}} = \int_{-\infty}^{\infty} \exp(-ik_y y) \int_{-\infty}^{\infty} \hat{\varphi} \exp(-ik_x x) dx dy$, $k^2 = k_x^2 + k_y^2$, $\varphi = \varphi^0_{(z)} + \hat{\varphi}$, $\varphi^0_{(z)}$- stationary potential for the homogeneous problem, $\hat{\varphi}$- potential oscillations, φ_1 – electrostatic potential of the aluminum layer on Al-Al_2O_3 interface, φ_2 – electrostatic potential of the aluminum layer on electrolyte-Al_2O_3 interface, σ_1 – conductivity of the aluminum, σ – conductivity of aluminum oxide, σ_2 – conductivity of the electrolyte, $\vartheta = \left(\frac{H}{\sigma} + \frac{H_1}{\sigma_1} + \frac{H_3 - H_2}{\sigma_2} \right)$, v – anodization voltage, H–

thickness of aluminum oxide, H_1+h_1 – thickness of the aluminum layer, $H_3-(H_2+h_2)$ – thickness of the electrolyte layer, $h_{1\vec{k}} = \int_{-\infty}^{\infty} \exp(-ik_y y) \int_{-\infty}^{\infty} \exp(-ik_x x) h_1 dxdy$, $h_{2\vec{k}} = \int_{-\infty}^{\infty} \exp(-ik_y y) \int_{-\infty}^{\infty} \exp(-ik_x x) h_2 dxdy$.

We consider the solutions of the system (1) for different values kH

I. For kH<<1

$$\begin{cases} \hat{\varphi}_{1\vec{k}} = 0 \\ \hat{\varphi}_{2\vec{k}} = \dfrac{(\alpha-\beta)\sigma_1}{\sigma_1\sigma_2 kH + \sigma_1\sigma + \sigma_2\sigma} \end{cases}$$

II. For kH>>1

$$\begin{cases} \hat{\varphi}_{1\vec{k}} = 0 \\ \hat{\varphi}_{2\vec{k}} = \dfrac{2}{\sigma_2}\left(\dfrac{\alpha}{e^{kH}} - \beta\right) \end{cases}$$

2.2. Calculation of the evolution of perturbations boundaries of alumina, without the consideration of surface diffusion

The rate of change of small perturbations of the boundaries Al-Al_2O_3 and Al_2O_3-electrolyte without the influence of surface diffusion is proportional to the perturbation of the current density at these interfaces $dh/dt = a\sigma\, \partial\varphi/\partial z$.

For Al-Al_2O_3 interface:

$$h_{1\vec{k}}(t) = h_{1\vec{k}}(0) \cdot \exp\left(\frac{a_1 kv \cdot cth(kH_1)}{\vartheta} t\right) \tag{2}$$

For Al2O3-electrolyte interface we consider two cases

I. When kH<<1

Since the conductivity of the electrolyte and alumina is small compared with aluminum we obtain, we obtain:

$$h_{2\vec{k}} = h_{1\vec{k}}(0) \cdot \exp\left(\frac{a_1 kv}{\vartheta} t\right) + \left(h_{2\vec{k}}(0) - h_{1\vec{k}}(0)\right) \exp\left(-\frac{a_2 kv}{\vartheta H} t\right) \tag{3}$$

II. When kH>>1

$$h_{2\vec{k}} = \frac{2h_{1\vec{k}}(0) a_2 \exp\left(\dfrac{a_1 kv}{\vartheta} t\right)}{(a_1+a_2)e^{kH}} + \left(h_{2\vec{k}}(0) - \frac{2h_{1\vec{k}}(0)a_2}{(a_1+a_2)e^{kH}}\right) \exp\left(-\frac{a_2 kv}{\vartheta} t\right) \tag{4}$$

From (3-4) we see, that in this approximation, the perturbation on the interface of Al_2O_3-electrolyte increases indefinitely with time.

3.3. Calculation of the evolution of perturbations boundaries of alumina, with the consideration of surface diffusion

The rate of change of small perturbations of the interfaces Al-Al_2O_3 and Al_2O_3-electrolyte with the influence of surface diffusion is described by the relation $\partial h/\partial t = V + D\Delta^2 h$, where D – is the surface diffusion coefficient.

ForAl-Al_2O_3interface:

$$V_1 = \left(\frac{a_1 v}{\vartheta} h_{1\vec{k}} \right)$$

$$h_{1\vec{k}} = h_{1\vec{k}}(0) \exp\left(\left(\frac{a_1 v k}{\vartheta} - D_1 k^4 \right) t \right) \tag{5}$$

From (5) we get the wavelength corresponding to the limit of stability for Al-Al_2O_3 interface (figure 3):

$$\lambda_1 = 2\pi \sqrt[3]{\frac{\vartheta D_1}{a_1 v}}. \tag{6}$$

All perturbations of this interface with wavelength $\lambda > \lambda_1$ are unstable.

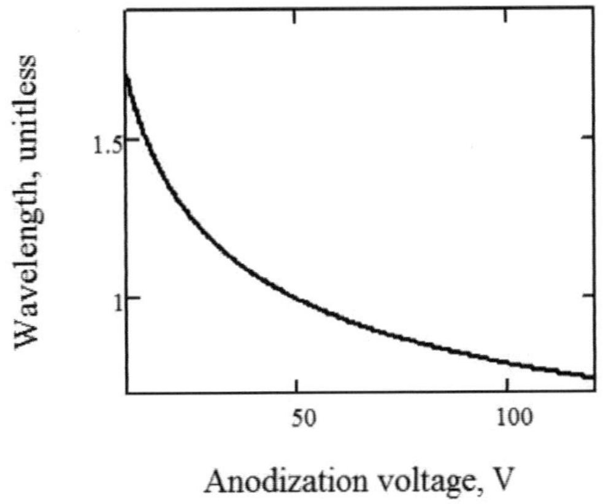

Figure 3. Change of the wavelength corresponding to the limit of stability with anodization voltage

For Al_2O_3-electrolyte interface:

For kH<<1
$$V_2 = \frac{v k a_2}{\vartheta H} \left(h_{1\vec{k}}(0) \cdot \exp\left(\frac{a_1 k v \cdot cth(kH_1)}{\vartheta} t \right) - h_{2\vec{k}} \right) \tag{7}$$

For kH>>1
$$V_2 = \sigma_2 k a_2 \left(\frac{2}{\sigma_2} \left(\frac{\alpha}{e^{kH}} - \beta \right) + \frac{v}{\sigma_2 \vartheta} h_{2\vec{k}} \right) \tag{8}$$

Then, as in (7-8) we obtain the expression, taking into account the surface diffusion:

For kH<<1
$$h_{2\vec{k}} = h_{1\vec{k}}(0) \cdot \exp\left(\left(\frac{a_1 k v}{\vartheta} - D_2 k^4 \right) \cdot t \right) \tag{9}$$

For kH>>1
$$h_{2\vec{k}} = \frac{2 h_{1\vec{k}}(0) a_2}{(a_1 + a_2) e^{kH}} \cdot \exp\left(\left(\frac{a_1 k v}{\vartheta} - D_2 k^4 \right) \cdot t \right) \tag{10}$$

From (9-10) we get the wavelength corresponding to the limit of stability for Al_2O_3-electrolyte interface:

$$\lambda_2 = 2\pi \sqrt[3]{\frac{\vartheta D_2}{a_1 k v}}. \quad (11)$$

All perturbations of this interface with wavelength $\lambda > \lambda_2$ are unstable.

We assume that surface diffusion coefficient D_2 on the Al_2O_3-electrolyte interface is larger than the same coefficient D_1 on the Al-Al_2O_3 interface. According to (6), (11), the correspondent critical wavelengths satisfy of inequality $\lambda_2 < \lambda_1$. Therefore, for perturbations with wavelength λ in interval $\lambda_2 < \lambda < \lambda_1$ the upper boundary of alumina layer is unstable and bottom boundary is stable. The perturbations of upper boundary in this interval of wavelength are the source of porous formation. For wavelength $\lambda > \lambda_1$ the bottom boundary of alumina is unstable. The evolution of perturbations in this wavelength region determines the irregularities of porous structure.

3. Conclusion

In this work we consider the solutions of the system (1) for different values kH. And add the influence of surface diffusion on the process.

As a result of the developed model we obtained the minimum distance $2\pi\left(\vartheta D_1/a_1 v\right)^{1/3}$ between centres of alumina pores in the beginning of anodizing process. The irregularities of porous structure contains in the spectral interval $k_2 < \left(a_1 k v/\vartheta D_2\right)^{1/3}$.

4. Acknowledgments

This work was financially supported by Government of Russian Federation, Grant 074-U01

References

[1] Limonov A G 2010 *Matem. Mod.* **22** 97–108
[2] Petukhov D I, Napolskii K S and Eliseev A A 2012 *Nanotechnology* **23** 335601
[3] Kim K H, Lefeuvre E, Gohier A, Chatelet M, Pribat D, Kim B S and Cojocaru C S 2010 *Proc. SPIE* **7761** 776109
[4] Feofilov S P and Kaplyanskii A A 1996 *Proc. SPIE* **2706** 128
[5] Maat J, Regeling R, Ingham C J, Weijers C, Giesbers M, Vos W M and Zuilhof H 2011 *Langmuir* **27** 13606–17

Self-formation of lead telluride nanostructures during argon plasma etching of single-crystal wafers

G A Dubov[1], S P Zimin[1], E S Gorlachev[1,2], I I Amirov[2], V V Naumov[2] and G Z Bagiyeva[3]

[1] Microelectronics Department, Yaroslavl State University, Sovetskaya Street 14, Yaroslavl 150000, Russia
[2] Yaroslavl Branch of the Institute of Physics and Technology, Russian Academy of Sciences, Universitetskaya Street 21, Yaroslavl 150007, Russia
[3] Abdullaev Institute of Physics, National Academy of Sciences of Azerbaijan, H. Javid Avenue 33, Az-1143 Baku, Azerbaijan

E-mail: gdubov@gmail.com

Abstract. In this work we report a phenomenon of PbTe nano- and microstructure self-formation that takes place during dry etching of lead telluride (111)-oriented single-crystal wafers with 4 at.% excess tellurium in RF high-density low-pressure inductively coupled argon plasma. Using scanning electron microscopy and energy dispersive X-ray spectroscopy investigations it is established that during the plasma treatment high sputtering rates are accompanied by active redeposition processes on crystal surface and on the inner side of the Si etching mask resulting in a nanostructure arrays self-formation via vapour-liquid-solid mechanism.

1. Experimental

PbTe single-crystal ingots were grown in (111) direction using Bridgman method according to a well-established technique [1]. Synthesized composition had excess tellurium relative to stoichiometry of 4 at.%. The grown crystals were cut perpendicular to the growth axis in order to obtain disk wafers with the thickness of 3-5 mm, whose surface was polished with diamond paste with chemical polishing finish. Plasma sputtering (dry etching) of the surface of PbTe(Te) disks was performed using a radio-frequency high-density low-pressure inductively coupled plasma (RF ICP) treatment [2-4]. Power applied to the inductor of the plasma reactor was 800 W, RF bias power at the substrate was varied in the range of 100-300 W, self-bias negative potential was 85-195 V, argon flow rate was 10 sccm. Series of consecutive surface treatments were carried out with a change in the duration (10-60 s) and bias power. During treatment stencil Si masks with native oxide and 0.6 μm thick SiO_2 layer were used, with a typical microgap between the sample and Si surfaces being 1-10 μm. The morphology was studied with scanning electron microscopy (SEM) using Supra 40 Carl Zeiss microscope and local elemental composition was determined from energy dispersive X-ray (EDX) data using INCAx-act Oxford Instruments spectrometer.

2. Results and discussions

The etching process for PbTe(Te) crystals was very intense, with maximum sputtering rate reaching 25 nm·s^{-1}, while for the (111)-oriented epitaxial PbTe films under the same treatment conditions we observed sputtering rate of ~15 nm·s^{-1} [3]. This phenomenon is explained with the defect structure of the PbTe(Te) crystals, since it is known that an excess tellurium atoms in the crystal lattice result in Pb atoms vacancies, which in turn result in a sublimation energy decrease.

Plasma treatment processes were accompanied by the formation of a developed surface microrelief of PbTe(Te) crystals. It was characterized by the presence of an array of homogeneous cone-shaped microhillocks with lateral dimensions of 1000-2000 nm and a height of up to 1500 nm and of an array of smaller nanohillocks with an average size of 50-200 nm. The density of both micro- and nanohillocks depended on the ion energy, with a highest density of large microhillocks not exceeding the density of threading dislocations in the wafer, which typically was (1-2) × 10^8 cm^{-2}. Formation of such large hillocks can take place due to the micromasking of dislocations, as it was previously systematically observed by us for lead chalcogenide films on silicon and barium fluoride substrates [2]. The density of nanohillocks (figure 1) was typically ~10^9 cm^{-2} and their formation took place due to redeposition processes on the sputtered surface.

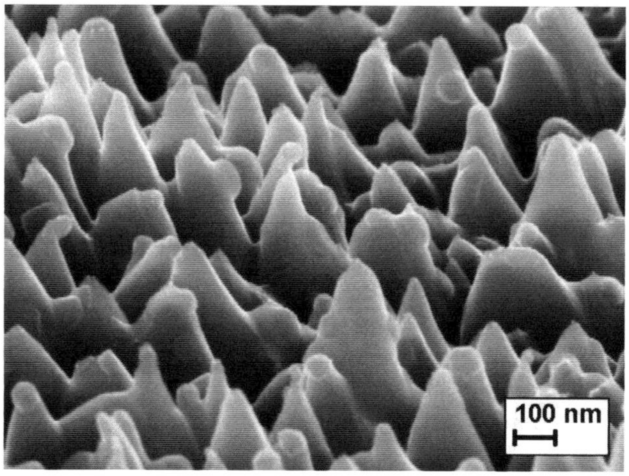

Figure 1. Scanning electron microscopy image of nanohillocks grown on PbTe single crystal surface after consecutive treatments with 200 and 300 W RF bias power.

A newly observed phenomenon connected to the intense plasma sputtering and redeposition of lead telluride was that the formation of nanostructures occurred not only over the sputtered PbTe crystal surface but also on the inner side of the Si stencil mask. This phenomenon has been used by us to develop a new approach to the formation of nanostructures PbTe directly on Si.

The formation of nanostructures takes place during the penetration of molecules and clusters of the sputtered PbTe single crystal to the microgap near the etching step and the redeposition of this material on the substrate inside microgap. Previously [5], we reported a similar phenomenon of PbSe nanowires growth according to a vapour-liquid-solid (VLS) mechanism [6] in the microgap between a stencil mask and a lead selenide epitaxial film directly on the surface of PbSe. A principal feature of the new approach reported in this work is the formation of PbTe nanostructures on the inner side of the silicon mask that is covering a part of the lead telluride sample surface during plasma processing.

As a result of our experiments, it was found that the on the Si wafer surface due to reprecipitation of the sputtered material during the plasma treatment there takes place a formation of various types of PbTe nanostructures. Morphologically the nanostructures could be divided into several basic types: nanopyramids, nanocubes and nanowires (quantum wires). Dimensions and density for all the types of these structures in this work depended on the treatment regime (RF bias power, the duration of etching) and the distance from the edge of the mask (etching step). The most advanced array of pyramidal structures was grown under the regime of processing with 100 W bias RF power. A typical size of the nanopyramid faces at the base was 150 nm, typical height – 700 nm, their density was about 10^9 cm^{-2}. Moreover, for some of these structures there was observed a growth of perpendicular nanoscale "branches". Nanocubes with edge length of 40 nm to 200 nm were grown at higher RF bias power of 200 W and 300 W, which provided a higher energy of the bombarding ions. Under these conditions, inside the microgap a higher concentration of sputtered lead telluride is gathered, which leads to the growth of cubic rather than pyramidal structures. A typical SEM image of cubical structures is shown in figure 2.

Figure 2. Scanning electron microscopy image of PbTe nanocubes grown on Si substrate after plasma etching with 200 W RF bias power.

The third pronounced type of lead telluride nanostructures formed on silicon were cylindrical nanowires. The structures were formed on the mask at relatively large distances from the etching step constituting 20-200 μm, with an RF bias power of 150-200 W. The dependence of the geometric dimensions of the grown nanowires on the distance from the etching step is shown in Table 1. A typical SEM image of the nanowires obtained with RF bias power of 200 W is shown in figure 3.

Table 1. Average geometrical parameters of PbTe nanowires.

Distance from the etch step, μm	Diameter, nm	Length, nm
25	20-30	105
50	20	275
80	20	420
160	20	410

Nanowire growth occurs at large distances from the etching step, where the concentration of sputtered lead telluride is relatively small. The nanowire self-formation takes place due to equilibrium redeposition of physically sputtered PbTe molecules and clusters from etched crystal surface into the microgap, which can take place via various physical mechanisms. The mostly observed mechanism, which is responsible for nanopyramid and nanowire formation in particular, is VLS growth, which is indirectly confirmed by the characteristic form of vertices with droplets (figure 3). In accordance with the results of our previous studies [5], these droplets may be sublimated metallic lead nanocrystals. EDX analysis showed that the composition of self-grown nanostructures on Si on the scale of ~1 μm corresponds to the initial PbTe(Te) crystal. The observed phenomenon has extremely promising applied prospects as a basis for a novel method of lead chalcogenide low-dimensional objects growth with predetermined shape, size, and density on silicon substrates for use in the optoelectronic and thermoelectric device structures.

Figure 3. Scanning electron microscopy image of PbTe high aspect ratio nanowires grown on Si substrate after plasma etching with 200 W RF bias power.

Acknowledgments
This work is financially supported by the Russian Foundation for Basic Research (grant 13-02-00381). SEM and EDX investigations were carried out at the Facilities Sharing Centre "Diagnostics of Micro- and Nanostructures" with the financial support of the Ministry of Education and Science of the Russian Federation.

References
[1] Bagiyeva G Z, Mustafayev N B, Abdinova G Dj and Abdinov D Sh 2011 *Semicond.* **45** 1391
[2] Zimin S P, Gorlachev E S, Amirov I I and Zogg H 2009 *J. Phys. D: Appl. Phys.* **42** 165205
[3] Zimin S P, Gorlachev E S, Amirov I I, Zogg H, Abramof E and Rappl P H O 2011 *Semicond. Sci. Technol.* **26** 105003
[4] Zimin S P, Gorlachev E S, Amirov I I, Naumov V V, Dubov G A, Gremenok V F and Bashkirov S A 2014 *Semicond. Sci. Technol.* **29** 015009
[5] Zimin S P, Gorlachev E S, Amirov I I and Naumov V V 2011 *Tech. Phys. Lett.* **37** 929
[6] Dubrovskii V G, Timofeeva M A, Tchernycheva M and Bolshakov A D 2013 *Semicond.* **47** 50

Resonance properties of metallic nanocantilevers

Ilya V Uvarov, Victor V Naumov and Ildar I Amirov

Yaroslavl Branch of the Institute of Physics and Technology, Institution of Russian Academy of Sciences, 150007, Universitetskaya Street 21, Yaroslavl, Russia

E-mail: ilnik88@mail.ru

Abstract. Resonant properties of the three-layer metallic cantilevers with 40-120 nm thickness were investigated. Two types of cantilevers were fabricated: Cr/Al/Cr and Ti/Al/Ti. Dependencies of the fundamental resonant frequency and Q-factor on the cantilever length, width and thickness were experimentally obtained. The experimental data analysis and the comparison with the theoretical predictions were performed. Dependence of material properties on the cantilever thickness was not observed. Energy losses in the surface layer of the cantilever were the dominant energy loss mechanism.

1. Introduction

NEMS resonators are used in a variety of technological and scientific applications, such as ultra fast sensors, actuators and signal processing components [1]. With decreasing the resonator size the fraction of surface layer in total resonator volume increases. Atoms of the surface and surface layer of the resonator have altered energy state compared with atoms in the bulk of the material. Therefore surface layer has specific elastic properties which affect the resonance characteristics of the device. In a number of studies the decrease in the Young's modulus of the material with decreasing the thickness of the vibrating cantilever is experimentally observed [2]. Also the rise of energy losses due to surface layer defects is observed with decreasing thickness of the cantilever [3]. However, available experimental data are often discrepant and require verification. In this study we investigate resonance properties of metallic cantilevers having a nano-sized thickness and micron length and width, and hence having ultra high surface area to volume ratio.

2. Cantilever fabrication technology and resonance properties measurement setup

Cantilevers having length from 10 to 100 μm, width from 2 to 10 μm and thickness from 40 to 120 nm were fabricated. To avoid bending of the cantilever by residual stresses a three-layered structure was implemented. Two types of cantilevers were fabricated: Cr/Al/Cr and Ti/Al/Ti, the layer thickness ratio was 1/6/1. The outer layers were made of the material with a higher Young's modulus than the material of the inner layer. They served to compensate for the residual stress gradient occurring in the inner layer of the cantilever during the deposition.

Detailed description of the cantilever fabrication technology can be found in our previous work [4], here the process of fabrication is described briefly. Thermally oxidized silicon wafer (thickness of SiO_2 layer was approximately 0.9 μm) was covered by the amorphous silicon (a-Si) sacrificial layer with 2 μm thickness. Next the wafer was covered by the positive photoresist layer, and the cantilevers were patterned by the contact photolithography. Structural layers of Cr/Al/Cr or Ti/Al/Ti with total thickness of 40-120 nm were deposited by the magnetron sputtering method. Then the lift-off process was performed. After that the wafer was covered by the photoresist layer again, the contact pads were patterned by the contact photolithography. Dry etching of a-Si in SF_6 plasma was performed, followed by the wet etching of SiO_2 in HF/NH_4F solution. Then the Al layer of 100 nm thickness was magnetron sputtered and the lift-off process was performed. Release of the cantilevers was performed by SF_6 plasma etching of the sacrificial a-Si layer. The gap between the cantilevers and SiO_2 layer was 2 µm. SEM image of the fabricated nanocantilevers is shown on figure 1. Cantilevers had ultrahigh length to thickness ratio that reached record high value of 1500. Fairly wide range of geometric dimensions allowed identifying the features of presented cantilevers.

Cantilever vibrations were excited by the electrostatic force. Tungsten needles were placed to the cantilever contact pad and to the silicon wafer contact pad of the sample (figure 2). The needles were connected to the precision signal generator. Voltage was applied between the cantilever and the silicon wafer, so the cantilever was attracted to the wafer. SiO_2 layer served as an insulator. Excitation was performed by AC voltage, DC voltage was not applied. The cantilever was attracted to the wafer twice a period of an applied voltage, so the cantilever vibration frequency was two times larger than the driving signal frequency.

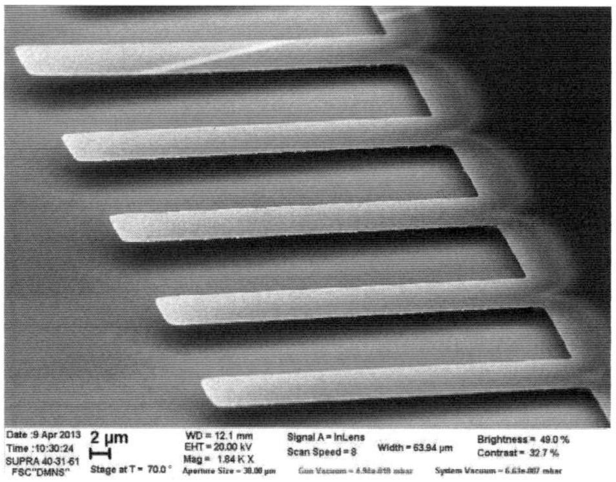

Figure 1. The example of fabricated cantilevers with dimensions 40×8×0.08 µm.

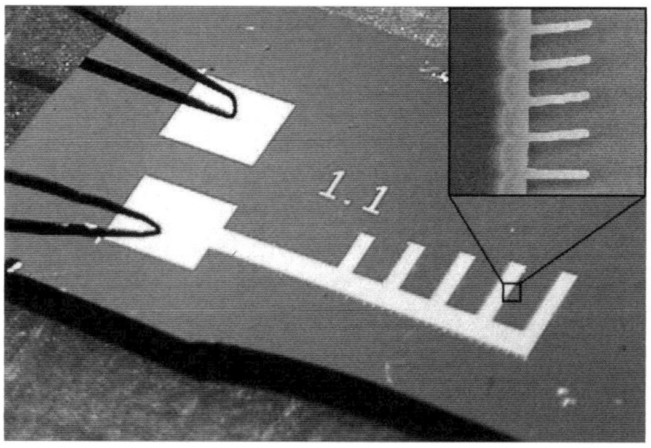

Figure 2. The fabricated sample with tungsten needles placed to the contact pads.

Registration of the cantilever motion was performed by the optical lever method. Laser beam from He-Ne laser was focused by the objective to the cantilever surface. Reflected beam was going to the two-segment photodiode. The dependence of the cantilever vibration amplitude on the frequency of driving voltage was measured. Fundamental resonant frequency and Q-factor of the cantilever were determined from the obtained dependence. Measurements were performed in air at pressure of 10^{-2} mbar. Detailed description of the resonance properties measurements method can be found elsewhere [4].

3. Results and discussion

Calculation of the fundamental natural frequency of the cantilever was carried out in accordance with Bernoulli-Euler beam theory, applied to the multi-layer cantilever with length L, thickness t and consisting in z direction of N layers with thicknesses t_i and mass densities ρ_i [5]:

$$f_0 = \frac{(1.875)^2}{2\pi L^2}\left(\frac{\int_t E(z - z_0)^2\, dz}{\sum_{i=1}^{N} t_i \rho_i}\right)^{\frac{1}{2}} \tag{1}$$

where E – Young's modulus of the material, z_0 – coordinate of the neutral axis of the cantilever. For the three-layer cantilever with a layer thickness ratio of 1/6/1 equation (1) may be converted to the following form:

$$f_0 = \frac{(1.875)^2}{16\pi}\frac{t}{L^2}\left(\frac{12.33E_{ext}+9E_{int}}{\rho_{ext}+3\rho_{int}}\right)^{\frac{1}{2}} \qquad (2)$$

where E_{ext}, E_{int}, ρ_{ext}, ρ_{int} – Young's moduli and densities of the material of external and internal cantilever layers, respectively. Bulk material values of Young's modulus and density were used in the calculations: $E_{Cr} = 279$ GPa, $E_{Ti} = 112$ GPa, $E_{Al} = 70$ GPa, $\rho_{Cr} = 7190$ kg/m^3, $\rho_{Ti} = 4510$ kg/m^3, $\rho_{Al} = 2700$ kg/m^3.

Experimentally obtained values of the resonant frequency of the fabricated cantilevers were in the range of 10-1700 kHz. Analysis of the dependence of the cantilever resonant frequency on the cantilever length showed that there was a qualitative agreement between the experimental data and the calculations. The main reason of the quantitative mismatch was the effect of undercut or, vice versa, the incomplete removal of the sacrificial layer from under the cantilever. This effect changed the effective cantilever length and therefore affected the resonant frequency [6]. Dependence of the resonant frequency on the cantilever width was not observed, which corresponded to the theoretical predictions.

According to the Bernoulli-Euler model, the resonant frequency of the cantilever should vary in proportion to the ratio of thickness to length squared (equation (2)). Proportionality factor is determined by the values of Young's modulus and density of the material. The graph of the dependence of cantilever resonant frequency on t/L^2 is shown on figure 3. Lines denoted the results of calculations for cantilevers Cr/Al/Cr and Ti/Al/Ti. Experimental values of the resonant frequency, except for the values of the cantilevers made with the undercut more than 1 µm, are also plotted on the graph. Discrepancy between the experimental data and theory does not exceed 10%. It is important that a linear dependence of the cantilever resonant frequency on the thickness to the length squared ratio was observed in the experiment (see figure 3). This indicated that the dependence of the material properties on the cantilever thickness until a thickness of 40 nm was not observed, and the classical beam theory with the bulk material values of Young's modulus and density was appropriate for the calculation of the natural frequency.

Cantilevers had a Q-factor from 100 to 2100. There was a significant variation of Q-factor, but it was possible to identify a number of regularities. First, there was no dependence of Q on the length and the width of the cantilever. Second, the quality factor decreased with decreasing thickness of the cantilever in a manner close to linear. The dependence of the quality factor on the cantilever thickness is shown on figure 4, approximating line is drawn through the mean values. Third, Q-factor of Ti/Al/Ti cantilevers was lower by the order than that of cantilevers Cr/Al/Cr. Fourth, there was a decrease of Q over time: in the case of cantilevers Cr/Al/Cr quality factor decreased by 20% over 3 months in air under normal conditions.

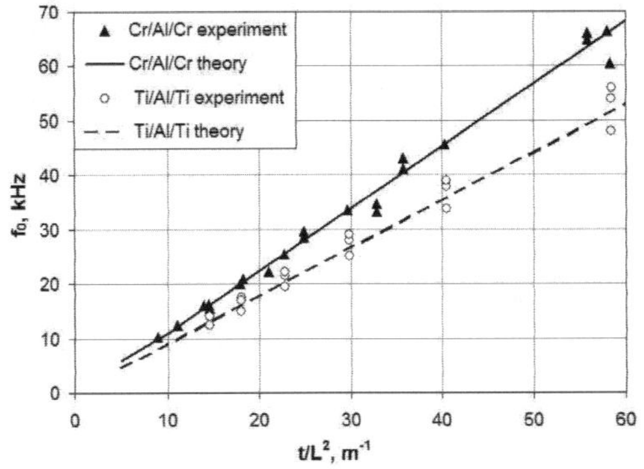

Figure 3. Dependence of the fundamental resonant frequency on t/L^2.

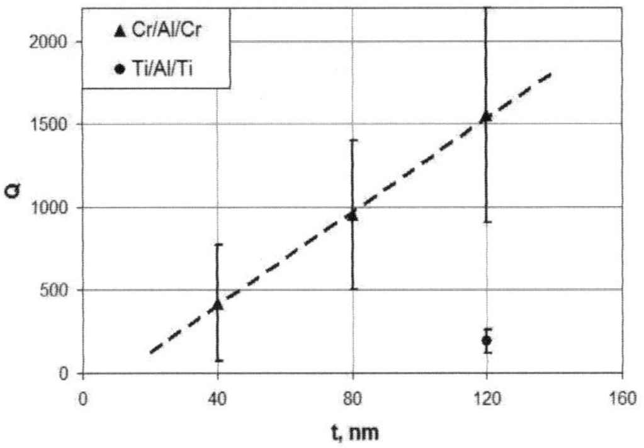

Figure 4. Dependence of the quality factor on the cantilever thickness.

Quality factor of NEMS resonator is determined by several energy loss mechanisms: air damping (Q_{air}), cantilever vibration energy transfer to the substrate through the clamping region (Q_{clamping}), thermoelastic damping (Q_{TED}) and the energy loss due to defects in the bulk (Q_{volume}) and in the surface layer of material (Q_{surface}). These sources operate independently of each other, and the resulting quality factor of the resonator is given by:

$$\frac{1}{Q}=\frac{1}{Q_{\text{air}}}+\frac{1}{Q_{\text{clamping}}}+\frac{1}{Q_{\text{TED}}}+\frac{1}{Q_{\text{volume}}}+\frac{1}{Q_{\text{surface}}} \quad (3)$$

Measurements of resonance characteristics of cantilevers were performed in the intrinsic damping regime, when Q-factor of the cantilever reached a maximum value and was not dependent on the air pressure. Therefore, air damping assumed negligible ($Q_{\text{air}} = \infty$).

Quality factor for the clamping losses in the cantilever with flexural vibrations is determined by the following formula [7]:

$$Q_{\text{clamping}}=\left[\frac{0{,}24(1+\upsilon)}{(1-\upsilon)\psi}\right]\frac{1}{(\gamma_n\chi_n)^2}\left(\frac{L}{t}\right)^3 \quad (4)$$

where υ – Poisson's coefficient of the material, ψ – constant, equal 0.336, γ_n and χ_n – constants for the n-th vibration mode (for the fundamental mode $\gamma_n = 0.597$ and $\chi_n = -0.734$). According to equation (4), Q-factor of the cantilever is proportional to the length to thickness ratio cubed. Therefore, the cantilevers having a length to thickness ratio of about 100 should have $Q \sim 10^6$ for the first vibration mode. If the length to thickness ratio is of about 1000, the calculated value of the quality factor is on the order of 10^9. At the same time, the highest measured Q-factor value of the fabricated cantilevers was of about 2000. Therefore, the energy losses through the clamping point were not the dominant source of energy loss, and the quality factor of cantilevers was limited by other damping mechanisms.

Thermoelastic damping Q-factor is determined by the equation [8]:

$$Q_{\text{TED}}=\frac{\rho C}{E\alpha^2 T}\frac{1+(\omega\tau_{\text{R}})^2}{\omega\tau_{\text{R}}} \quad (5)$$

where C – specific heat, α – linear thermal expansion coefficient, T – absolute temperature, ω – circular frequency of the cantilever oscillations, τ_{R} – relaxation time, given by the equation:

$$\tau_{\text{R}}=\frac{t^2\rho C}{\pi^2\kappa} \quad (6)$$

where κ – heat conductivity. Q-factor due to thermoelastic damping was calculated in accordance with equation (5). Calculations were performed for the cantilever made of aluminium, having a thickness of 100 nm and a resonant frequency of 100 kHz. The following values of constants and variables were used: $\kappa_{\text{Al}} = 237$ W/(m·K), $C_{\text{Al}} = 930$ J/(kg·K), $\alpha_{\text{Al}} = 23.1\cdot10^{-6}$ 1/K, $T = 300$ K. For the cantilever with this parameters relaxation time is $1.1\cdot10^{-11}$ s. At 100 kHz cantilever oscillation period is 10^{-5} s. Thus, the relaxation time is much smaller than the oscillation period. Consequently, the thermoelastic damping is small, and, according to equation (5), the Q-factor of the cantilever is $2.09\cdot10^8$. For the cantilever made of chromium ($\kappa_{\text{Cr}} = 93.9$ W/(m·K), $C_{\text{Cr}} = 461$ J/(kg·K), $\alpha_{\text{Cr}} = 8.2\cdot10^{-6}$ 1/K) relaxation time is $3.6\cdot10^{-11}$ s and Q-factor is $1.63\cdot10^8$. Thus, thermoelastic damping, as well as the clamping losses, was not the dominant source of energy loss in the fabricated cantilevers.

Energy loss occurring at the defects in the bulk and the surface layer of material are due to the interaction of mechanical waves, propagating in a vibrating cantilever, with various defects. These defects are grain boundaries, foreign atoms in the lattice, surface roughness, surface oxide, various adsorbates, etc. In case of nano-sized resonator the surface layer has sufficient part of the total resonator volume, and considerable part of defects is located in the surface layer. According to the applied model [3], energy loss due to defects in the bulk of the cantilever does not depend on its geometry and defined only by properties of the material. Q-factor due to energy losses at the defects of the surface layer is determined by the following formula:

$$Q_{\text{surface}} = \frac{t}{6t_S}\frac{E}{E_d^S} \quad (7)$$

where t_S – thickness of the surface layer of the cantilever, E_d^S – dissipative part of the Young's modulus of the surface layer. Equation (7) does not allow to calculate the quality factor, because it contains unknown quantities. However, it predicts the dependence of the quality factor on the geometry of the cantilever: Q-factor of the cantilever having nanoscale thickness should decrease with decreasing thickness in a linear manner, and should not depend on the length and width.

On the basis of experimentally obtained dependencies of Q on the geometrical dimensions of the cantilever, it was concluded that the quality factor was mainly determined by the energy losses occurring at the surface layer defects of the cantilever. It was noted above that Ti/Al/Ti cantilevers had a lower quality factor (100-250) than the Cr/Al/Cr cantilevers (900-2100). The difference in Q-factor of Cr/Al/Cr and Ti/Al/Ti cantilevers was presumably caused by high getter ability of titanium. The surface of Ti/Al/Ti cantilever actively absorbed various substances from the environment, which led to the large number of defects and hence a large number of energy loss sources. Chromium does not have high getter ability. The reason of decrease in the quality factor over time most likely was the accumulation of various adsorbates on the cantilever surface. Thus, the experimental data supported the conclusion that the Q-factor of metallic nanocantilevers was mainly determined by the surface layer defects.

4. Conclusions

Resonance properties of the three-layer metallic cantilevers having 40-120 nm thickness and ultra high length to thickness ratio were investigated. Two types of cantilevers were fabricated: Cr/Al/Cr and Ti/Al/Ti. Three-layered structure was implemented to avoid bending of the cantilever by residual stresses. Dependencies of the fundamental resonant frequency and Q-factor on the cantilever length, width and thickness were experimentally obtained. The experimental data analysis and the comparison with the theoretical predictions were performed. Size effects of Young's modulus and density of the material was not observed. The resonant frequency of the cantilevers was determined with a high degree of accuracy by Bernoulli-Euler beam theory for values of Young's modulus and density of the bulk material. The quality factor of cantilevers in the absence of air damping was mainly determined by energy losses occurring at the defects of the surface layer of the cantilever.

The reported study was supported by RFBR, research project No. 14-07-31156 mol_a, by Facilities Sharing Centre "Diagnostics of Micro- and Nanostructures" and by the Ministry of education and science of Russian Federation.

References

[1] Greenberg Y, Pashkin Y A and Il'ichev E 2012 *Phys.-Usp.* **55** 382
[2] Nilsson S G, Borrisé X and Montelius L 2004 *Appl. Phys. Lett.* **85** 3555
[3] Yasumura K Y, Stowe T D, Chow E M, Pfafman T, Kenny T W, Stipe B C and Rugar D 2000 *J. Microelectromech. Syst.* **9** 117
[4] Uvarov I V, Naumov V V and Amirov I I 2012 *Proc. SPIE* **8700** 87000S-1
[5] Sandberg R, Svendsen W, Mølhave K and Boisen A 2005 *J. Micromech. Microeng.* **15** 1454
[6] Guillon S, Saya D, Mazenq L, Perisanu S, Vincent P, Lazarus A, Thomas O and Nicu L 2011 *Nanotechnology* **22** 245501
[7] Hao Z, Erbil A and Ayazi F 2003 *Sens. Actuators A* **109** 156
[8] Zener C 1938 *Phys. Rev.* **53** 90

Transverse optical phonon dispersion for multi-layer graphene

I Kurkina[1] and S Smagulova[2]

[1]Junior Researcher Fellow, graduate student, M.K. Ammosov North-Eastern Federal University, Institute of Physics and Technology, Yakutsk, Russia

[2]Chief Researcher, scientific adviser, M.K. Ammosov North-Eastern Federal University, Institute of Physics and Technology, Yakutsk, Russia

E-mail: volkiraly@mail.ru

Abstract. The Raman spectra were measured with different wavelength excitation lasers for multi-layer graphene samples obtained by mechanical and electrostatic exfoliation of highly oriented pyrolytic graphite on a silicon substrate. Phonon frequencies depending on the laser energy value are constructed, and the average distance between defects in the resulting multi-layer graphene calculated using G and D lines intensity ratio.

1. Introduction

It is known, that the number of graphene studies is growing every year. Graphene is interesting due to its electronic [1], mechanical [2] and optical properties [3, 4]. Raman scattering spectrum is an individual optical characteristic of the substance molecules, as a result of Raman scattering spectrum gives reliable identification of the scattering molecules. The most significant for qualitative analysis is a determination of intramolecular vibrations frequencies, which correspond to certain Raman bands. A quantitative analysis requires the band intensities values. Raman spectroscopy had historically played an important role for the structural description of graphite materials, and it is the primary tool for graphene identification and electron-phonon interaction understanding, structure disordering calculation and layer's number determination.

2. Experimental details

In present work multi-layer graphene samples obtained by mechanical exfoliation of highly oriented pyrolytic graphite (HOPG) on a silicon substrate using a conventional and masking tapes, as well as electrostatic force were investigated. For graphene flakes obtaining by electrostatic method ~ 1kV voltage is applied to HOPG, resulting in the interlayer bonds are broken under the influence of electrostatic repulsion forces, and graphene sheets are detached from the source material on a silicon substrate.

Using an optical microscope, it was found that the average lateral size of graphene flakes reached values ~ 26 × 7 um, ~ 8 × 4 um and ~ 30 × 3 um for samples prepared by mechanical exfoliation with scotch tape, krepp (masking tape) and electrostatic method, respectively. Graphene sheet fragments on the substrate surface were more for samples prepared by krepp than ones produced by scotch tape. Electrostatic exfoliation results were differ from the previous ones in that the multi-layer graphene films often obtained in the striped form with ~ 5 um width and ~30 um length located near the thick graphite flakes (Fig.1).

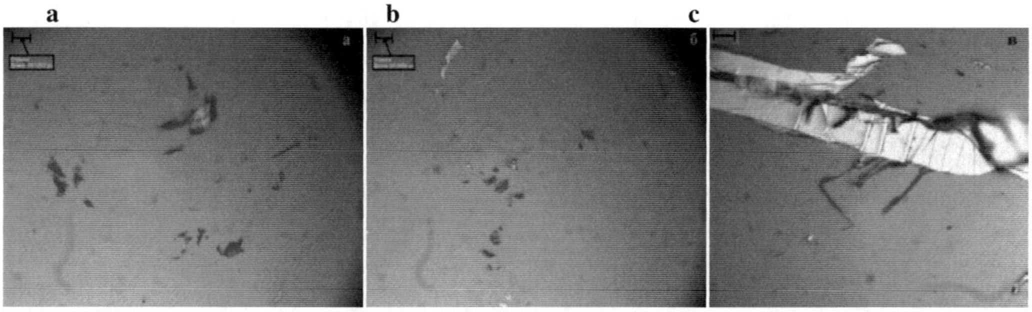

Figure 1. The multi-layer graphene films prepared by scotch (a) and krepp (masking tape) (b) mechanical exfoliation and electrostatic method (c), respectively. Photos were taken at the same magnification. Dimensional scale in the upper left corner is 10 microns

Graphene flakes thickness was in the range of 2 to 10 nm according to the atomic force microscopy. Raman spectra were measured with a installation "Integra spectra" (NT-MDT, Russia) having three lasers with energy (wavelength) of 2.62 eV (473 nm), 2.33 eV (532 nm) and 1.96 eV (632.8 nm). Laser spot diameter on the measured surface was approximately 1 micron.

3. Results and discussion

The experimental Raman spectra of multi-layer graphene flakes contain basic combinational bands G (~ 1580-1590 cm^{-1}), G '(~ 2650-2740 cm^{-1}) and D (1330-1400 cm^{-1}). Typical Raman spectra of the sample obtained by mechanical exfoliation with masking tape, presented in Figure 2.The position (Pos), intensity (I) and width (Γ) of the Raman bands are determined according to the obtained spectrum.

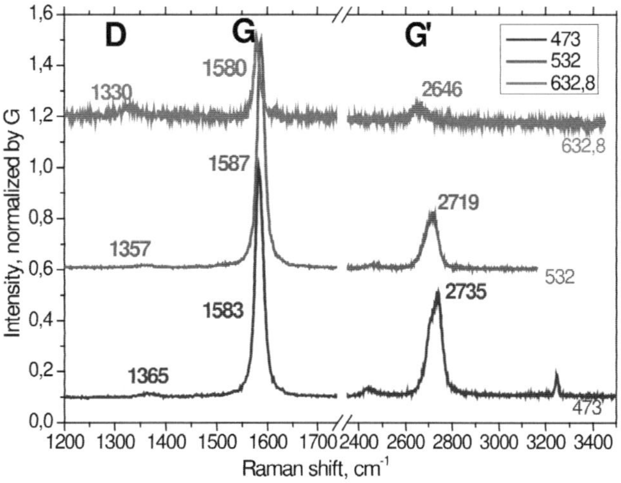

Figure 2. Typical Raman spectra. In this case, the spectra of the sample obtained by mechanical exfoliation using masking tape, when excited by lasers with wavelengths of 473nm, 532 nm, 632.8 nm.

The G band width which is associated with doubly degenerated optical phonon modes in the center of the Brillouin zone (point Γ) is mainly determined by the electron-phonon interaction. A typical value for the G band width is ~ 14 cm^{-1} for graphene obtained by mechanical exfoliation [5]. Phonon scattering by defects should lead to their corresponding attenuation and broadening of the Raman bands. All samples spectra have the D band, which is appearing due to graphene lattice disordering, and it originated from the inter-valley double resonance process involving one transverse optical phonon and defect, i.e. process connecting point belonging to a cone around the Dirac point K and point belonging to a cone around the point K' [6]. G band width had values between 14 and 18 cm^{-1} in the measured spectra (Table 1).

The ratio of D to G bands intensities is determined by Raman spectra. There is a distance between defects L_D dependence on the relationship I_D/I_G [5], which is described by the following empirical formula: $L_D^2(nm^2) = \frac{(4.3\pm1.3)\times10^3}{E_L^4}\left(\frac{I_D}{I_G}\right)^{-1}$. According to this formula the distances between defects were calculated in these films, and the values are presented in Table 1. As one can see, the smallest distance between defects is for samples prepared by mechanical exfoliation with adhesive tape, i.e. they are the most defective.

Table 1. The G band position, G band width and distance between defects L_D calculated by the formula from [5], determined from the Raman spectra.

	Pos(G), $см^{-1}$	Г(G), $см^{-1}$	L_D, нм		
			Scotch (adhesive tape)	krepp (masking tape)	electrostatic
Ср. знач.	1578 - 1590	14 - 18	32-51	42-150	62-123
2,62 эВ	1584	~18	32-44	60-80	62-85
2,33 эВ	1588	~17	32-44	110-150	90-123
1,92 эВ	1581	~14	38-51	42-57	72-98

D and G' band position in graphene always depends on the energy of the exciting light. Figure 3a shows the dependence of the G' band position on the laser energy. Dispersion velocity Disp (G') = $\Delta\omega_{G'}/\Delta E_L$ from this graph is equal to 90-140 sm^{-1}/eV. Figure 3b shows the dependence of the phonon energy on the excitation energy, calculated from the formula $E_{ph}=\hbar\omega(q)$. Phonon frequency $\omega(q)$, as shown in Figure 3a, depends on the laser energy.

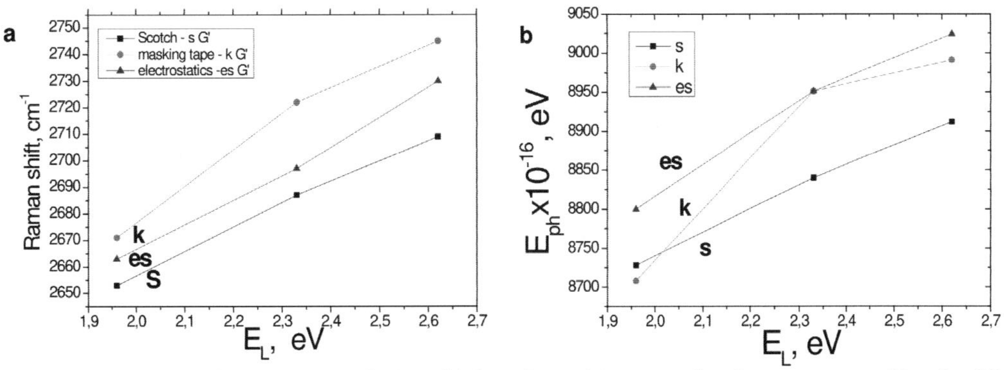

Figure 3. a. The dependence of the G' band position on the laser energy E_L; b. The dependence of the phonon energy E_{ph} on the excitation energy E_L.

4. Conclusion

The Raman spectra of multi-layer graphene obtained by mechanical and electrostatic exfoliation of highly oriented pyrolytic graphite on a silicon substrate, showed G, G' and D characteristic features at excitation with three different lasers. It is shown that the G band width depends on the presence of defects in the graphene structure. The appearing of G' and D bands depending on the electron scattering with transverse optical phonons, have significant dispersion. The dependence of the phonon energy on the exciting light energy is demonstrated. Because of D to G band intensity ratio was defined from the experimental spectra, the distance between the defects was calculated in the measured multi-layer graphene flakes.

References

[1] A.H. Castro Neto et al. The electronic properties of graphene // Rev. Mod. Phys. 81, 109, 2009, P. 109-162

[2] C. Lee et al. Measurement of the Elastic Properties and Intrinsic Strength of Monolayer Graphene // Science. 321, 385, 2008, P. 358-388

[3] J.I. Inoue. Valley-contrastive selection rules of a nonlinear optical transition in graphene with an energy gap // Phys. Rev. B 83, 2011, P. 205404-1 - 205404-1

[4] V.M. Apalkov, T. Chakraborty. Phys. Rev. B 84, 2011, P. 033408-1 - 033408-4

[5] L. G. Cancado et al. Quantifying Defects in Graphene via Raman Spectroscopy at Different Excitation Energies // Nano Lett., vol. 11, 2011, P. 3190–3196

[6] L.M. Malard et al. Raman spectroscopy in graphene // Physics Reports, vol. 473, 2009, P. 51-87

Studying fringe field effect of a field emitter array

M F Sayfullin and K A Nikiforov

Saint Petersburg State University, 7-9, Universitetskaya nab., St. Petersburg, 199034, Russia

E-mail: knikiforov@cc.spbu.ru

Abstract. Field emitter arrays on heavy As-doped Si wafer are studied in vacuum nanoelectronics diode configuration. Different shapes of emitters are considered: cone-shaped point-emitters and cylinder-shaped sharp-edge-emitters are compared. Micro scale field enhancement factor on the edge of cylindrical emitter was calculated via home-developed Matlab application and the results are presented. Two types of anode geometry are proposed: plane anode and spherical anode. Experimental and modelling results of surface electric field distribution are presented. The spherical shape of anode allows higher voltage (and higher field emission current) without destructive arcs risk.

1. Introduction

The standard emission mechanisms of the cathodes used in many applications are photoemission and thermionic emission. One alternative technology is field-emitter arrays (FEA) where electrons are emitted with energies close to the Fermi level. Such cathodes have potentially a lower mean transverse kinetic energy of the produced electron beam, which is mainly determined by the geometry of the electric field lines. There are 2 main architectures of field-emission cathodes: the needle cathode, i.e. one single sharpened tip and FEA which are arrangements of field emitters on a periodically spaced lattice. We have presented first measurements on edge-type FEA in atmosphere pulsed low-voltage operation and DC high-vacuum high-voltage regime as well as numerical simulations of the electric field near the cathode surface [1]. It was shown fringe enhancement of electric field on the boundary of FEA plays negative role on destabilization of emission current.

On the contrary field enhancement is necessary near the curved emission surface on the top of nanostructure emitters. This effect was experimentally studied in case of cylindrical-shaped field emitters (figure 1) on heavy As-doped Si wafer [1] and computer simulation was used to study in this paper.

Journal of Physics: Conference Series **541** (2014) 012020 doi:10.1088/1742-6596/541/1/012020

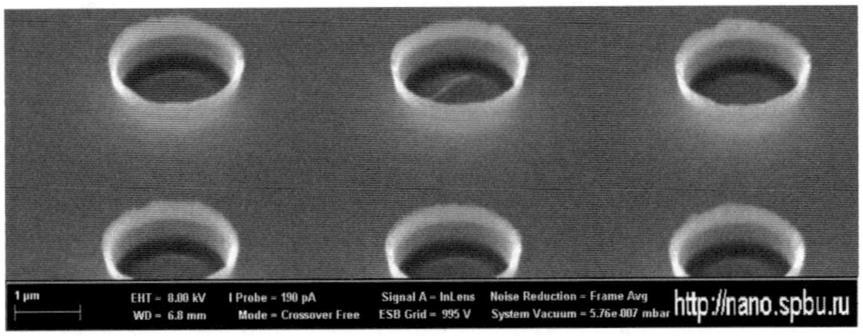

Figure 1. FEA surface with cylinder-shape emitters.

2. Finite element solution

Study of micro-scale non-uniformity of electric field with a software complex developed in Matlab [2] has shown that field enhancement factor depends of the distance at which the field is applied to the sharp margin of cylindrical microemitter (figure 2). As this distance grows, the dependence becomes less significant, and at about 6 times the height of emitter protrusion becomes negligible. Figure 3 shows dependence between field strength and absolute value of the applied uniform field strength.

Numerical solutions of corresponding electrostatic boundary problems were carried out with Matlab and Matlab PDE Toolbox. Software implementation should take into account the specifics of emission systems:

- Calculation are of complex shape includes the emitter boundary with high curvature and multi-scale size, which leads to considerable diversity of characteristic measurements in the same geometric configuration.
- The exponential dependence of current density on field strength requires increased precision when considering boundary conditions on the emitter.

The solution has a rapidly changing gradient in the emission area (on edge of the cathode), thus the finite-element mesh has to become tighter in the neighborhood of the emitter edge to avoid reduction of speed of convergence towards the exact solution and increase of the number of variables (i.e. the dimension of the finite-element system).

With adaptable mesh it is natural to employ an error indicator, that would include the residual norm of the equation and fluctuations of the gradient of finite-element solution, as they are connected with one of the basic values given for this problem – that is, with the electric field strength. This indicator is implemented in pdejumps function which is called by adaptmesh function aimed for the adaptive solution of the problem. For making the mesh more fine a way of splitting finite elements was chosen, such that they are split along their longest side (rmethod=longest), i.e. a triangle (based on the value of error indicator) is replaced with two smaller triangles by splitting its longest side in half.

For the case of conical-shape emitters the macro-scale field was calculated by Comsol Multiphysics coupling with Matlab via LiveLink.

3. Results

It was found the surface field and emission current of emitters in the center of the array is less dependent on the emitter to emitter spacing than on emitters on the boundary of array.

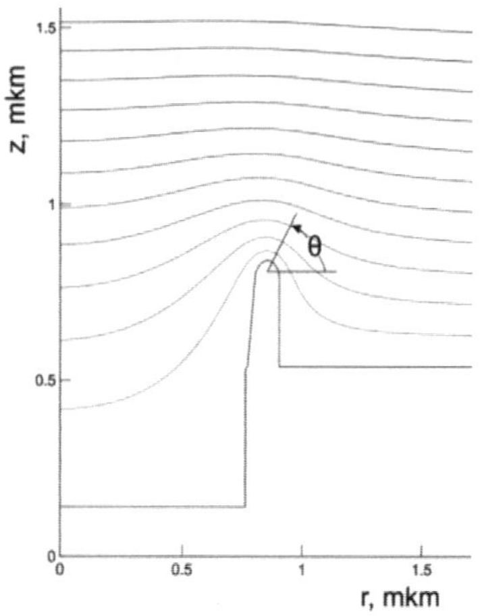

Figure 2. Equipotential lines (with step 1 V) of electric field near the edge of nanosized wall cylinder of microemitter. One half cross section is shown in cylindrical coordinates

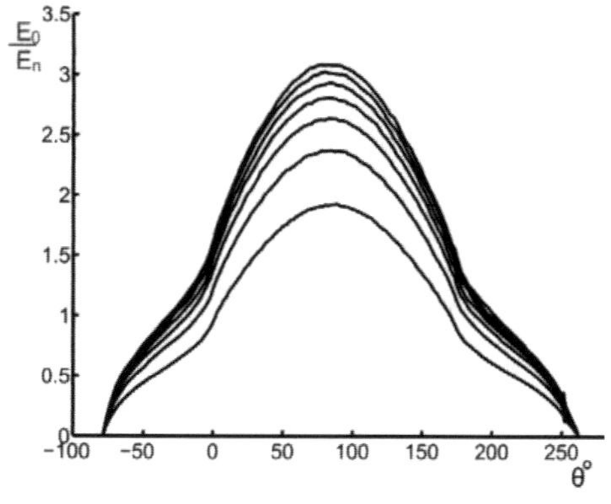

Figure 3. Angle distribution of the field enhancement factor. The uniform field $E_0 = 1.43 \cdot 10^7$ V/m is applied on the distances 0.2-1.4 μm with step 0.2 μm (the lower line corresponds 0.2 μm, the upper line corresponds 1.4 μm).

The shape not only of emitters, but also of anode is important in case of macro-scale uniformity of electric field distribution over the array surface. The fringe enhancement effects of macro-scale field distribution can be reduced due to spherical (see figure 5 instead of plane as on figure 4) anode more than two times. After parameter optimization procedure (in mathematical simulation) it was found for case of square (1 by 1 cm) FEA the best diameter of spherical anode is also 1 cm, and gap between electrodes can be increased to 1-3 mm with applied voltage up to 20 kV.

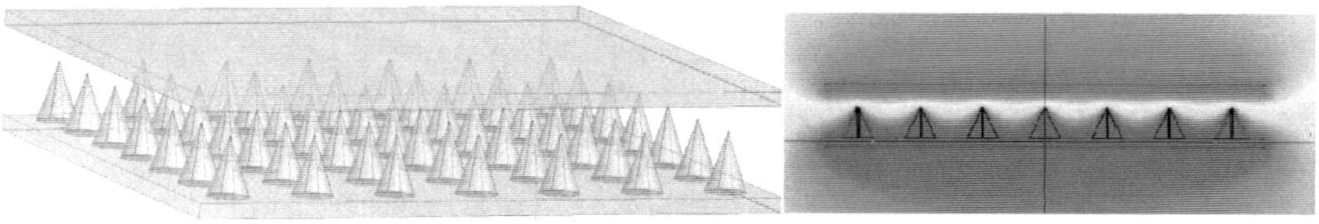

Figure 4. Electrostatic potential distribution with fringe effect non-uniformity between electrodes in case of plane anode and conical emitters.

Figure 5. Spherical anode and square FEA cathode.

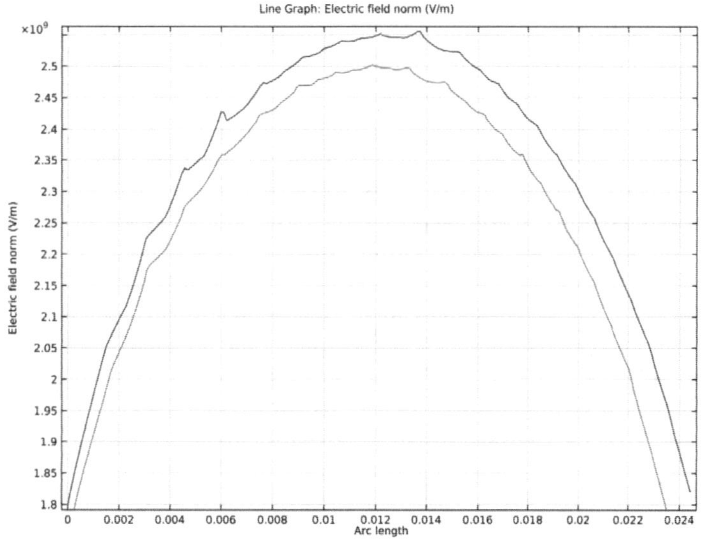

Figure 6. Electric field distribution over the surface of conical emitter top at FEA boundary in two distances to the boundary. The upper line – for distance of half of lattice step in FEA. The lower line – for large distance (half of lattice parameter plus 3 mkm).

4. Conclusion

This study includes multi-scale modeling of electric field distribution on nanostructured cylinder and conical emitter surface in FEA.

Acknowledgments

Financial support has been obtained from RFBR (13-01-00150) and partially from SPbSU (9.38.673.2013). Research was carried out using computational resources provided by Resource Center "Computer Center of SPbU" (http://cc.spbu.ru/en) and using experimental equipment of the Interdisciplinary Resource Center for Nanotechnology of St. Petersburg State University.

[1] Nikiforov K A, Antonova L I, Egorov N V, Trofimov V V, Makeev V V and Ogurtsov O F 2012 Non-gated field emission array as low-energy electron source: Experiment and simulation *RuPAC 2012 Contributions to the Proc. of 23 Russian Particle Accelerator Conf. (St. Petersburg)*, (Cern: JACoW) 218–20

[2] Nikiforov K A and Egorov N V 2012 Program complex for vacuum nanoelectronics finite element simulations *RuPAC 2012 Contributions to the Proc. of 23 Russian Particle Accelerator Conf. (St. Petersburg)*, (Cern: JACoW) 409–11

Optical spectroscopy of organic materials based on C_{60}<A_2B_6>

M. Elistratova[1], N. Romanov[2], I. Zakharova[1] and O. Kvyatkovskii[1,3]

[1]St. Petersburg State Polytechnic University, 195251, St. Petersburg, Russia
[2]Lappeenranta University of Technology, 53850, Lappeenranta, Finland
[3]Ioffe Institute, 194021, St.Petersburg, Russia

E-mail: elistratovamari@yandex.ru

Abstract. The structure, optical and photoelectrical properties of thin films based on C_{60}<A_2B_6> materials are studied. Reproducible vacuum method of thin fullerene films production with Cd<Te;S> impurity on Si, glass and mica surfaces are developed. Surface morphology of the films are investigated by AFM and SEM methods. The *ab initio* quantum - chemical calculations of the geometry, total energy, excited states energy of supramolecules complex fullerene- cadmium telluride are performed. Optical absorption spectra in UV and visible regions, Raman and photoluminescence spectra are studied. Comparing to a pure fullerene, the additional absorbance and luminescence peak at 600 nm are observed. The intensity of additional peaks are defined as the charge composition as a type of substrate. These results are interpreted as an appearance of the dipole-allowed transitions in the fullerene excited singlet states spectrum due to an interference with cadmium telluride.

1. Introduction

Organic/inorganic mixed nanocomposites promise to be valuable to solar energy conversion, photovoltaics, specifically to novel light-driven systems [1]. New composite materials based on C_{60} fullerene as well as nonorganic semiconductor nanoparticles A_2B_6 are a great interest [2] The properties of hybrid materials can be regulated via a variety of functional groups, which can be incorporated in the organic C_{60} matrix. Photovoltaic devices, sensors, opto-electronics, etc. might represent some of the most important applications of these novel materials. In this work thin films of new hybrid organic-inorganic semiconductors, exhibiting photoconductive properties have been produced by simple vacuum method. Molecular semiconductor materials have been studied for more than a half century because they have high potential for unique device applications such as light-emitting diodes, solar cells, and field-effect transistors with low cost, low weight, and mechanically flexible electronics. Therefore development of composite electronic and photoelectronic materials and structures based on C_{60} are relevant and important. In the last years many studies have been focused on optical characteristics of both C_{60} molecules and solid fullerite. Frenckel exiton formation takes place as a result of light quant absorption by C_{60} molecule. It is necessary to separate exitons to get photovoltage. Geterojunction of donor and acceptor materials may observe this function very effective, because it provide with charge transfer mechanism between molecules. It was found that fullerene C_{60} is a strong electron acceptor. Conjugated polymers, phthalocyanines, materials from porphyrin class and also nonorganic semiconductor from A_2B_6 group are effective donors. Therefore nanostructured material contacts with fullerene are considered as a possible material for solar cell creation. Nevertheless, the main property which has to be taken into account when we use organic structures is their quick degradation under the influence of oxygen or H_2O vapor. This aspect reduces to comparatively short useful time of those devices. Therefore intensive research in the field of composite organic and nonorganic semiconductor, which includes C_{60}-CdTe films, structures is underway. It is done to increase the lifetime of organic solar cells. CdTe is a well-known material for photovoltaic because. It is a direct-gap semiconductor with 1.5 eV energy band gap. This band gap is optimal for absorption of sunlight, because increasing band gap lead to decrease covered the solar spectrum. Decreasing bang gap leads to leaving of a portion of absorbed photons to intraband scattering and solar cell heating.

2. Methods of obtaining and experimental research of samples

Thin nanocomposite films from composite mixture of fullerene C60 (99.98%) and inorganic donor CdTe(S) were prepared by vacuum evaporation method on a pre-heated in a height vacuum up to 150°C glass/indium tin oxide, silicon (111) and KBr substrates [3]. Two different techniques - quasi-closed volume (QCV) and Knudsen cell are used, films thickness was 200-500 nm. In the case of QCV method it was provided a full initial charge mixture transfer to the substrate in the process of sublimation. A Knudsen cell evaporation method is a sophisticated method of evaporation from molecular beam. In comparison with the initial charge it can lead to a significant changing of a film composition. Obtaining a large area film possibility is an advantage of the Knudsen cell. Evaporation was produced on a pre-heated in a height vacuum up to 150°C glass/indium tin oxide, silicon (111) and KBr substrates in both methods. Then substrates were cooled to 40°C. The result is random distribution of acceptor and donor materials in amorphous films. Films prepared at 500-550C of the evaporator temperature from C_{60} and CdTe mixture in various percentage (10 – 50%). The films composition and surface morphology was studied SEM-EDAX and AFM methods. A scanning electron microscope Jeol JSM-6390 was used to study the surfaces' morphology. It worked with resolution of 3 nm. The films composition in the selected area was measured by energy dispersive microanalysis console "Oxford INCA Energy" with the utmost sensitivity 0.1%. Films' surfaces and composition of components were studied by electrons with 8 KeV energy. Raman spectra were measured by "Mikroraman" from Renishaw device at 512 nm wavelength. The power of exiting beam was set of 0.1 mW/mcm. It was done to account an oxidation and photopolymerisation effects in films. Measuring time (and the laser irradiation time accordingly) were varied from 30 sec to 30 min. Photoluminescence spectra were measured by using automated installation based on Horiba Jobin Yvon spectrometer. It is composed of FHR 640 monochromator with a grating of 1200 mm^{-1} and the Symphony II (1024*256) Cryogenic Open - Electrode CCD detector. . Excitation of photoluminescence was produced by continuous semiconductor laser with 410 nm wavelength, 50 mW powerty. Filter was installed before the slit of the monochromator to avoid scattering laser radiation falling into the monochromator. CCD chamber was maintained at the temperature of 77 K. Ellipsometry spectra were obtained using an Alpha - SE Wolland ellipsometer with fixed incidence angle of 70 ° and a wavelength from 380 to 890 nm. Inverse problem solution was performed using the CompleteEASE considering the surface roughness and the Kramers-Kronig method B - spline.

3. Experimental results and discussion

The surface topography of all samples was measured by scanning electron microscope. Irregularities in films composition (phase contrast) weren't detected until the limit of the microscope resolution (10 nm). Typical surface topography is shown in fig. 1. The film surface generally smooth with unique (fraction 2 and 6) in the field 100×100 mkm. Also chemical compositions spectra were measured by energy dispersive consoles (table 1). The silicon substrate is present in the spectrum due to the fact that the absorption of electrons and output of x-ray radiation is possible from the substrate when the films thickness is about 200 nm. Spectrum doesn't contain CdTe in the center of the defect on the point 6. The defect size is about 3 mkm. It can be the inclusion of fullerene mixture with a thickness about 2mkm. Point 2 is partly gone from the fullerene crystallite and contains some CdTe. Other points (3, 4, 5) selected on a smooth homogeneous surface show sufficiently good uniformity of composition with an average content value of Cd – 1.60, Te – 1.76%.

Figure 1. SEM surface image of sample containing 15% of CdTe

Table 1. Films homogeneity composition analysis based on EDAX results.

Spectra	C	O	Si	Cd	Te	Total
Spectrum 1	40,49	0,93	55,31	1,00	2,27	100,00
Spectrum 2	77,09	1,22	19,83	0,74	1,12	100,00
Spectrum 3	59,09	0,92	36,38	1,65	1,96	100,00
Spectrum 4	59,12	0,77	37,1	1,44	1,58	100,00
Spectrum 5	59,41	0,79	36,37	1,68	1,74	100,00
Spectrum 6	88,3	1,53	10,18			

Given the molar mass of Cd and Te, it is possible to conclude that CdTe generally is included in the composition the molecular form with some depletion of Cd.

Table 2. Films composition in comparison with the initial charge composition based on EDAX results.

Evaporation method	Knudsen cell				QCV
CdTe weight percents **in initial charge**, %	67	50	33	20	50
CdTe weight percents **in thin films**, %	27	13	5-7	3	50

Comparative characteristics of different samples are shown in table 2. The weight percentages composition of CdTe was determined from measured chemical composition data by subtracting the effect of silicon. It may be concluded that Knudcen cell evaporation leads to significant decreasing content of CdTe in the films in comparison with the original composition. This fact can be explained by difference in condensation and re-evaporation coefficients our materials, because the saturated vapor pressure of CdTe and C_{60} in this temperature range is the almost same. On the other hand evaporation in a quasi-closed volume (QCV-method) leads to a complete preservation of original charge. Moreover this method allows to obtain thicker films. Different authors observed a wide range of C_{60} photoluminescence with some maxima with different relative intensity: about 720 nm (1.7 eV) and about 800 nm (1.55 eV). This range depends on rather experimental conditions which described in numerous works devoted to study of C_{60} PL. Fig. 2 shows obtained PL spectra of nanocomposite fullerene films with different CdTe content. It is seen that the luminescence peak, which can be correspond to a single phase of CdTe (800 - 850 nm), is not observed. There are wide peaks, which overlaps the region 700 – 800 nm are accordingly include fullerene peaks at 720 and 800 nm.

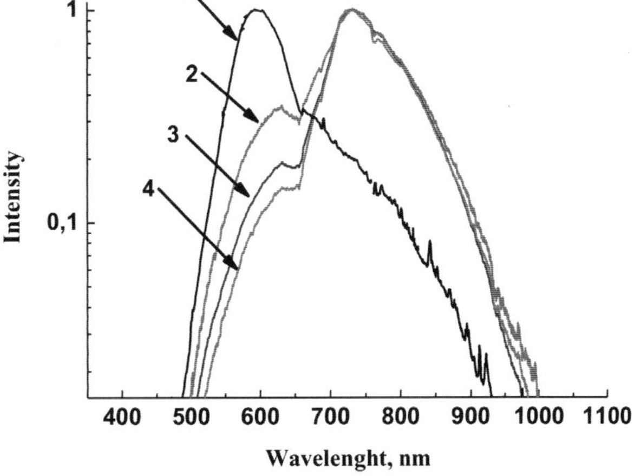

Figure 2. Photoluminescence specta of C_{60}(CdTe) samples containing various weight persentage of CdTe: 1 - 50%, 2 – 27%, 3 – 12%, 4 – 5-7%.

These peaks are usually associated with $T_1 \rightarrow S_0$ transition and it's vibrations sublevels. Also there is increasing relative emission intensity at wavelength of 580 - 620nm with the CdTe content increasing. Thereby the presence of the additional luminescence peak can be explained by change in the fullerene electron structure in the interaction with CdTe. In our opinion this peak is S_1-S_0 energy transition of fullerene. We can assume that with the CdTe addition in fullerene matrix due to decrease in symmetry of fullerene molecules, and therefore partially lifted a ban on the transitions S_1 - S_0. This possible explanation is based on the results of quantum chemical calculations, which allowed singlet-singlet transitions for complex C_{60}CdTe have energy 1.9 - 2.2 eV. Influence of gamma and X-ray irradiation on the properties, composition and structure of the samples were studied.

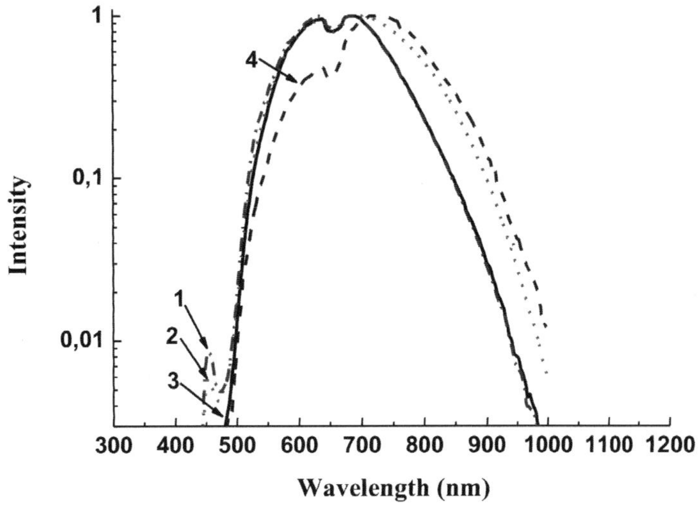

Figure 3. Photoluminescence specta of C_{60}(CdTe) samples containing 20% of CdTe on various substrates: 1,2 – on mica, 3,4 – on silicon, 2,4 – after gamma rays irradiated

In comparison to the pure C_{60} samples additional absorption and PL areas appeared in C_{60}<CdTe> films spectra with 1.9-2.1 eV and 2.75 eV photon energy (fig.3). The intensity of the lines depends on the content CdTe (CdS) and substrate type. Ellipsometry data allow to obtain the extinction coefficient and refractive index spectra. Fig. 4 (below) shows the extinction coefficient curves for pure fullerene sample and the sample with a high concentration of CdTe (50 mol. %). It can be seen that the sample with a high content of CdTe has an additional absorption peak at a wavelength of 630 nm, which well agrees with photoluminescence results. Increase the refractive index (Fig. 4, top) due to the fact that cadmium telluride has a large dipole moment, which polarizes the fullerene molecule from the nearest environment.

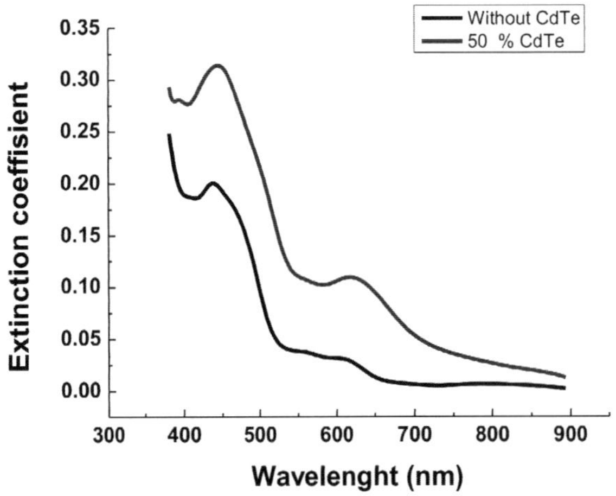

Figure 4. Extinction coefficient curves for clear fullerene sample (up) and sample with a high concentration of cadmium telluride (50 mol.%) (down).

4. Quantum-chemical calculations of C_{60}<A_2B_6> complexes

Quantum-chemical calculations of optimized geometry and electronic structures of C_{60}-CdTe complexes were performed. It was done in frame of density functional theory (DFT) with the hybrid functional B3LYP using MOLCAO-SCF method of GAUSSIAN 03 program package. The calculations were performed using the following sets of atomic Gaussian basis functions: 3-21G for carbon atoms, SBKJC VDZ ECP for cadmium and SBKJC Polarized (p, 2d)-LFK for tellurium. Many configurations, differing in initial CdTe molecules position in fullerite are considered. Calculations results for cluster C_{60}CdTe show that the structure has the energy minimum when tellurium atom is located above the bond 6-6 fullerene molecule at a distance of 0.22 nm. Mayer bond order for tellurium atom with the nearest carbon atoms is 0.54 and 0.60, and on the cadmium atom about 0.16 (for comparison, the order C-C bond in the fullerene molecule is equal to 1.12 for 5-6 bond, 1.38 for 6-6 bond). Energy gain in the total formation energy of such complex is 0.57 eV, it is sufficiently high to ensure the formation of the complex at room temperature. HOMO-LUMO gap is equal to 2.4 eV (for comparison, the calculation gives a molecule C_{60} HOMO-LUMO = 2.9 eV, and for the molecule CdTe Eg = 1.3 eV). Energy of low-lying excited states of the electron spectrum for C_{60} as and complexes of C_{60}-CdTe are calculated using TDDFT method. The results show that for the complexes of C_{60}-CdTe spectrum of excited states is considerably complicated compared to the fullerene molecule and at the same time dipole-allowed singlet transitions with energies of 1.9 - 2.2 eV appear in the spectrum, which may explain the data of optical spectroscopy and luminescence.

5. Conclusions

The novel technology of thin-film composite structures based on C_{60}-CdTe is developed. .. Evaporation from the Knudsen cell leads to significant depletion of the CdTe-component in compare with the initial mixture. Method QCV provide completely transfer of mixture composition. SEM-EDAX investigations confirmed the presence of CdTe in all the deposited layers and consequently the preparation of a hybrid system. Comprehensive study of structure, composite and optical properties this sample led to conclusion that this method leads to molecular dispersivity of CdTe in fullerite matrix. Molecule of CdTe actively interacts with the fullerene molecules and distorts it's symmetry. It manifests in the additional processes of emission and absorption between levels S_1 - S_0 (2.0-2.75 eV). It is assumed that this interaction may increase the probability of photoinduced charge transfer in such composite materials. The CdTe presence in these films also significantly reduces oxidation and polymerization of fullerene, which reduces degradation of the samples in the interaction with the atmosphere. Quantum-chemical calculations by the TD DFT (B3LYP) method was used to explain these results. Energy spectra of the excited triplet and singlet states for complexes C_{60}CdTe (CdS), as well as the supramolecular complexes simulating the position of the impurity molecules in the octahedral and tetrahedral interstices of the crystal lattice of fullerite are calculated. Obtained results are interpreted as an appearance of the dipole-allowed transitions in the fullerene excited singlet states spectrum due to distortions caused by intercalation CdTe in fullerite.

References

[1] Biebersdorf A, Dietmüller R, Susha A S, Rogach A L, Poznyak S K, Talapin D V, Weller H, Klar T A, Feldmann J 2006 Semiconductor nanocrystals photosensitize C_{60} crystals *Nano Lett.* **6** 1559

[2] Guldi D M, Zilbermann I, Anderson G, Kotov N A, Tagmatarchis N, Prato M 2004 Versatile Organic (Fullerene)-Inorganic (CdTe Nanoparticle) Nanoensembles *J. Am. Chem. Soc.* **126** 14340

[3] Zakharova I B, Ziminov V M, Nashchekin A V, Vainshtein Yu S, Aleshin A N 2013 Optical spectroscopy of thin C_{60}:CdS composite films *Semiconductors* **47** 105

Structural and optical properties of the bismuth-containing quartz-like glasses

M. A. Girsova[1], S. V. Firstov[2], T.V. Antropova[1]

[1]Grebenshchikov Institute of Silicate Chemistry, Russian Academy of Sciences, nab. Makarova 2, 199034 Saint-Petersburg, Russia
[2]Fiber Optics Research Center of the Russian Academy of Sciences, 38 Vavilov Street, 119333 Moscow, Russia

E-mail: girsovamarina@rambler.ru

Abstract. Bismuth-containing quartz-like glasses (BCQGs) were synthesized by an impregnation of the silica porous glasses (PGs) with $Bi(NO_3)_3$ aqueous solution. Then the samples were sintered at the temperatures T from 700 °C to ≥ 1500 °C up to closing the pores. Characterization of BCQGs was carried out using the UV-VIS-NIR spectroscopy and IR spectroscopy. The absorption and luminescence properties of the BCQGs are discussed. It is shown that the growth of absorption in the short-wave region is characteristic for BCQGs. IR spectra confirmed the presence $[BiO_3]$ and $[BiO_6]$ units and $\alpha - Bi_2O_3$ phase in BCQGs. The spectral dependences of the luminescence and excitation of the luminescence of the synthesized glass depending on their thermal prehistory were investigated. For the first time, 3-dimensional luminescence spectra have been obtained for BCQG.

1. Introduction

Bismuth-doped glasses are new active optical materials having a broad luminescence spectrum in the spectral range 1000-1700 nm. IR luminescence lifetime of such glasses is amounted to 0.1-1 ms [1, 2]. Bismuth-activated glasses can be used for development multifunctional 3D sources [3, 4], fibers doped with bismuth that are basis for superluminescent light source, tunable high-power lasers, and broadband optical amplifiers for spectral region of 1300–1530 nm [5-7]. Among them the bismuth-containing quartz-like glasses produced on porous glass matrices are of particular interest because of the resource-saving technology [4, 8-10].

2. Experimental

Bismuth-doped glasses (BCQGs) were fabricated by an impregnation of the silica porous glasses (PGs) with $Bi(NO_3)_3$ aqueous solution. Initially, PG samples (in the form of the rectangular plane-parallel plates in the size (1.5±0.03) x (10–15) x (15–25) mm^3) were placed in $Bi(NO_3)_3$ aqueous solution. Then obtained composite material was sintered at 700-900 °C under air atmosphere for 10-30 min, heating rate 1-4 °C/min only or additionally heat-treated at T ≥ 1500 °C. In detail synthesis technuque of BCQGs was described in [9, 10]. The glasses under study have following compositions (as analyzed, wt. %): 0.30 Na_2O, 3.14 B_2O_3, 0.11 Al_2O_3, 96.45 SiO_2 (PGs); ≤ 0.2 Na_2O, (2.7–3.2) B_2O_3, (92.7–96.1) SiO_2, (0.9–1.6) Bi_2O_3, ≤0.10 Al_2O_3 (BCQGs).

Transmission spectra of synthesized glasses were measured by SF-2000 UV / VIS spectrophotometer in the wavelength range of 190–1100 nm with a step of 0.115 nm. IR transmission spectra in the range of 4000 to 400 cm^{-1} at the spectral resolution of 4 cm^{-1} were recorded wih SPECORD M-80 spectrophotometer. For measurements of IR spectra the glasses were powdered and mixed with KBr in order to obtain thin pellets by vacuum pressing. Luminescence spectra of the BCQGs were collected by FLSP920 optical spectrofluorometer. All measurements were conducted at room temperature.

3. Results and discussion

According to the transmission spectra (Fig. 1) and optical loss spectra [10] of the glasses under study the strong absorption in UV spectral region is characteristic of them. There are seven peaks at below 240, 471, 522, 713 nm (Fig. 1a, curve 1); 204, 497, 568 nm (Fig. 1a, curve 2) which assigned to the glass matrix. The UV-VIS transmission spectra of bismuth-containing glasses are shown in Fig. 1b. It was found that bismuth-containing glasses have two bands at <280 nm and around 400 nm. Presence of UV absorption of Bismuth-doped glasses (here, <280 nm) could be associated with different states of bismuth oxidation: Bi^{3+} [11], Bi^{2+} [2], colloidal bismuth [12]. Emission properties of BCQGs confirmed that UV absorption band is associated to Bi^{3+} ions. The band at 405-408 nm corresponds to absorbance band of Bi^0 nanoparticles (NPs) [9]. In our case an origin of this band was still unknown.

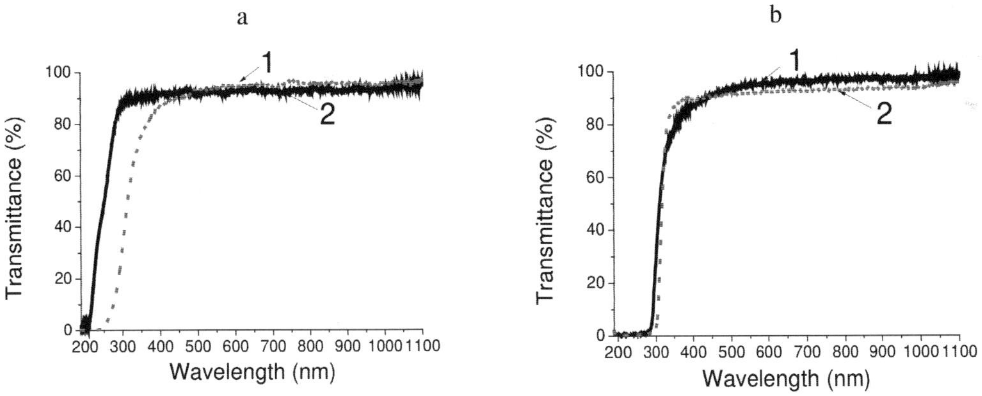

Figure 1. Transmission spectra of the glasses: (a) 1 – silica porous glass without bismuth, 2 – quartz-like glasses without bismuth (T ≥ 870 °C); (b) 1 – bismuth-containing porous glass (T ≤ 700 °C), 2 – bismuth-containing quartz-like glass (T ≥ 870 °C).

IR transmission spectra of glasses under study are shown in Fig. 2. It was found that all types of glasses have eight fundamental absorption bands at 3684–3676, 3496–3472, 3432–3416, 2932–2908, 2824–2808, 1664–1644, 1400–1380, 1108–1084 cm^{-1}. Using published data [13-21] the obtained IR bands have been identified. Broad bands placed at 3684–3676, 3496–3472, 3432–3416, 1664–1644 cm^{-1} are assigned to stretching vibrations of hydroxyl and water groups. Two bands at 2932–2908 and 2824–2808 cm^{-1} are associated with hydrogen bonds. The band at 1400–1380 cm^{-1} is connected with asymmetrical stretching vibrations of B-O and B-O$^-$ bond in trigonal BO_3 units and is characteristic of =B-O-B≡ linkage with one of the boron in tetrahedral coordination. Asymmetrical vibrations of [SiO_4] tetrahedral in a polymerized network with dominated Si–O–Si linkages and presence of tri-, tetra-, pentaborate and diborate groups belonging to BO_4 groups are responsible for band at 1108–1084 cm^{-1}. Apart from above-mentioned bands, it was observed the additional bands in IR transmission spectra of bismuth-activated glasses. The spectra analysis showed that BiO_3 and BiO_6 units formed in glasses. The band at 944-924 cm^{-1} (Fig. 2b, curves 2, 3) is caused by the increasing of the degree of cross-linking by Bi-O-Si linkages. It was revealed that band at 940 nm appeared after annealing T>870 °C. The band at 676–652 cм$^{-1}$ (Fig. 2b, curves 1, 3) may be assigned to Bi-O bending vibrations in BiO_6 units and indicates the presence of $\alpha - Bi_2O_3$ phase also. The peaks around 576–

564 см$^{-1}$ (Figs 2b, curve 2) are ascribed to doubly degenerate stretching vibrations of Bi–O bonds in BiO_3 and BiO_6 units, Bi-O-Bi and Bi-O bending vibrations in BiO_6 units.

The luminescence spectra of the glasses are given in Fig. 3. The obtained luminescence spectrum of initial matrix without bismuth consists of one band in the UV region (close to 315 nm) (Fig. 3, curve 1). It was revealed that heat treatment process of BCQGs at T < 870 °C initiated blue-green luminescence with the maximum at 450 nm (excitation at 250 nm), which is originated from Bi^{3+} ions

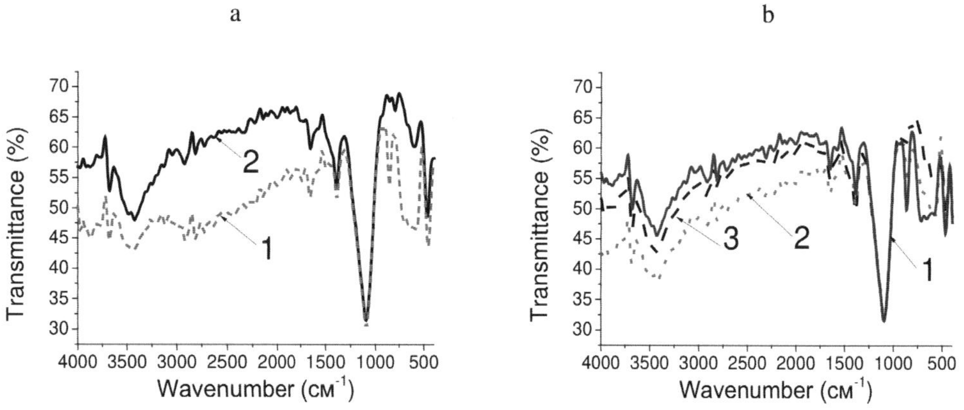

Figure 2. IR transmission spectra of the glasses: (a) 1 – silica porous glass without bismuth, 2 – quartz-like glasses without bismuth (T ≥ 870 °C); (b) 1 – bismuth-containing porous glass (T ≤ 700 °C), 2 – bismuth-containing quartz-like glass (T ≥ 870 °C), 3 – bismuth-containing quartz-like glass (T ≥ 1500 °C).

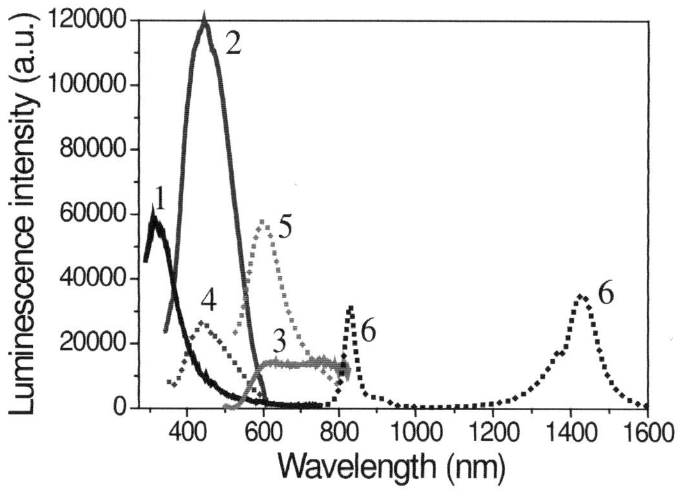

Figure 3. Luminescence spectra of the studied glasses with the different thermal prehistory: (1) quartz-like glass without bismuth (T ≥ 870 C; excited at 280 nm); (2) bismuth-containing porous glass (T ≤ 700 °C; excited at 280 nm); (3, 4) the bismuth-containing quartz-like glass, heat-treated at T ≥ 870 °C (3 – excited at 500 nm, 4 – excited at 300 nm); (5, 6) the bismuth-containing quartz-like glass, additionally heat-treated at T ≥ 1500 °C (5 – excited at 500 nm and 6 – excited at 420 nm).

in the glass (Fig. 3, curve 2). It is determined that heat treatment of BCQGs at T ≥ 870 °C leads to decrease in the intensity of blue-green luminescence (Fig. 3, curve 4). The observed changes are caused by reduction of the Bi^{3+} to Bi^{2+}. Bi^{2+} ions are responsible for red luminescence with the maxima at 610 and 750 nm at excitation at 480 nm. These emission bands can be observed after annealing process at temperature higher than 870 °C (Fig. 3, curve 3). Re-annealing of BCQGs at high temperatures (T ≥ 1500 °C) leads to an increase in red luminescence intensity owing to the growth of divalent bismuth (Fig. 3, curve 5). Moreover two bands of infrared luminescence with maxima about 830 nm and 1410 nm can be seen also (Fig. 3, curve 6).

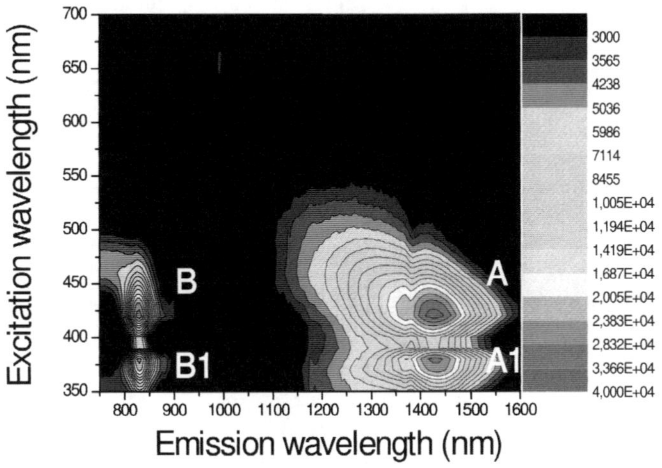

Figure 4. IR luminescence intensity distribution vs the luminescence and excitation wavelengths in the bismuth-containing quartz-like glass heat-treated at T ≥ 1500 °C.

Fig. 4 shows the distribution of the intensity of IR luminescence depending on the wavelengths of excitation and luminescence. In this diagram four bright peaks A, A1, B, and B1 are distinctly seen. Earlier the similar bands in silicate glass were obtained in [22]. These bands are inherent to IR active bismuth centres formed in silica-based glass.

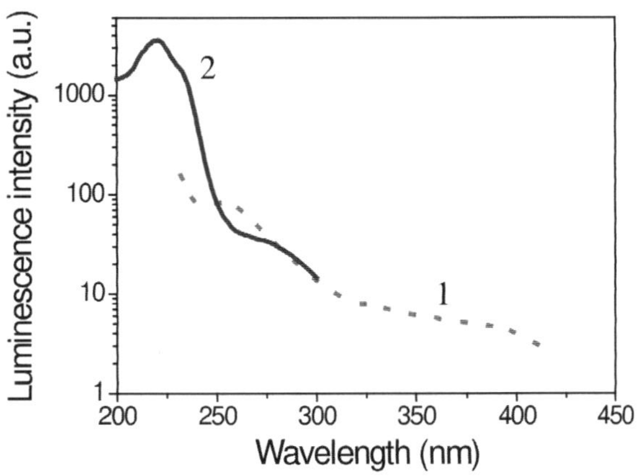

Figure 5. Excitation spectra of blue luminescence in the bismuth-containing quartz-like glass heat-treated at: 1 – T ≥ 870 °C (λ_{em} = 450 nm); 2 – T ≥ 1500 °C (λ_{em} = 450 nm).

Fig. 5 demonstrates excitation spectra of BCQG annealed at ≥ 870 °C (1) and >1500 °C (2). We have observed the transformation of the excitation spectra of blue luminescence at the increasing of treatment temperature. Main change is an appearance of additional excitation band at 220 nm at higher temperature. Recently this band was also observed in the absorption spectrum of the fibre preforms from pure silica glass doped with bismuth ions and was attributed to Bi^{3+} [23]

Thus, the appearance of the red and infrared luminescence is accompanied by structural changes (increasing of the degree of cross-linking by Bi-O-Si linkages) in the glass network.

Conclusion

In conclusion, fabrication technique for new bismuth-containing quartz-like glasses has been developed. Series of BCSG on the base of silica porous glass matrices have synthesized. Their structural, optical and luminescent properties have been investigated.

It has been shown that synthesized glasses can emit blue-green, red and also infrared luminescence owing to presence of different bismuth active centres. Formation of red and IR active centers is accompanied by glass network modification at annealing process.

Acknowledgements

The work was supported by the Presidium of the Russian Academy of Sciences (Program of Basic Researches no. 24).

The authors thank T. G. Kostyreva, L. N. Kurilenko, and L. F. Dikaya for chemical analysis of the glasses, and G. F. Golovina for help in conducting of IR spectroscopic studies.

4. References

[1] Fujimoto Y, Nakatsuka M 2001 *Jpn. J. Appl. Phys.* **40** L 279.

[2] Firstov S V, Khopin V F, Bufetov I A, Firstova E G, Medvedkov O I, Guryanov A N, Dianov E M 2011 *Optics Express* **19** 19551.

[3] Ruan J, Su L, Qiu J, Chen D, Xu J 2009 *Optics Express* **17** 5163.

[4] Zhou S, Lei W, Jiang N, Hao J, Wu E, Zeng H, Qiu J 2009 *J. Mater. Chem.* **19** 4603.

[5] Riumkin K E, Melkumov M A, Bufetov I A, Shubin A V, Firstov S V, Khopin V F, Guryanov A N, Dianov E M 2012 *Optics Letters* **37** 4817.

[6] Shubin A V, Bufetov I A, Melkumov M A, Firstov S V, Medvedkov O I, Khopin V F, Guryanov A N, Dianov E M 2012 *Optics Letters* **37** 2589.

[7] Bufetov I A, Firstov S V, Khopin V F, Medvedkov O I, Guryanov A N, Dianov E M 2008 *Optics Letters* **33** 2227.

[8] Zhou S, Jiang N, Zhu B, Yang H, Ye S, Lakshminarayana G, Hao J, Qiu J 2008 *Adv. Funct. Mater.* **18** 1407.

[9] Girsova M A, Firstov S V, Anfimova I N, Golovina G F, Kurilenko L N, Kostyreva T G, Polyakova I G, Antropova T V 2012 *Fizika i Khimia Stekla* **38** 861 (in Russian).

[10] Firstov S V, Girsova M A, Dianov E M, Antropova T V 2014 *Glass Physics and Chemistry* **40** (in press).

[11] Duffy J A, Ingram M D 1970 *J. Chem.Phys.* **52** 3752.

[12] Singh S P, Karmakar B 2011 *Optical Materials* **33** 1760.

[13] Ramadevudu G, Rao S L S, Shareefuddin MD A H, Chary M N 2011 *International Journal of Engineering Science and Technology* **3** 6998.

[14] Kundu V, Dhiman R L, Maan A S, Goyal D R 2009 *Journal of optoelectronics and advanced materials* **11** 1595.

[15] Qiao W, Chen P 2010 *Glass Physics and Chemistry* **36** 304.

[16] Dimitriev Y, Krupchanska M, Ivanova Y, Staneva A 2010 *Journal of the University of Chemical Technology and Metallurgy* **45** 39.

[17] Ardelean I, Cora S 2008 *Journal of Materials Science: Materials in Electronics* **19** 584.

[18] Krupchanska M, Iordanova R, Dimitriev Y, Bachvarova-Nedelcheva A 2010 *Journal of*

Optoelectronics and Advanced Materials **12** 1692.
[19] Doweidar H, Saddeek Y B 2009 *Journal of Non-Crystalline Solids* **355** 348.
[20] Baia L, Stefan R, Kiefer W, Simon S 2005 *Journal of Raman Spectroscopy* **36** 262.
[21] Ardelean I, Rusu D 2008 *Journal of optoelectronics and advanced materials* **10** 66.
[22] Bufetov I A, Melkumov M A, Firstov S V, Shubin A V, Semenov S L, Vel'miskin V V, Levchenko A E, Firstova E G, Dianov E M 2011 *Optics Letters* **36** 166.
[23] Bufetov I A, Semjonov S L, Vel'miskin V V, Firstov S V, Bufetova G A, Dianov E M 2010 Quantum Electronics **40** 639.

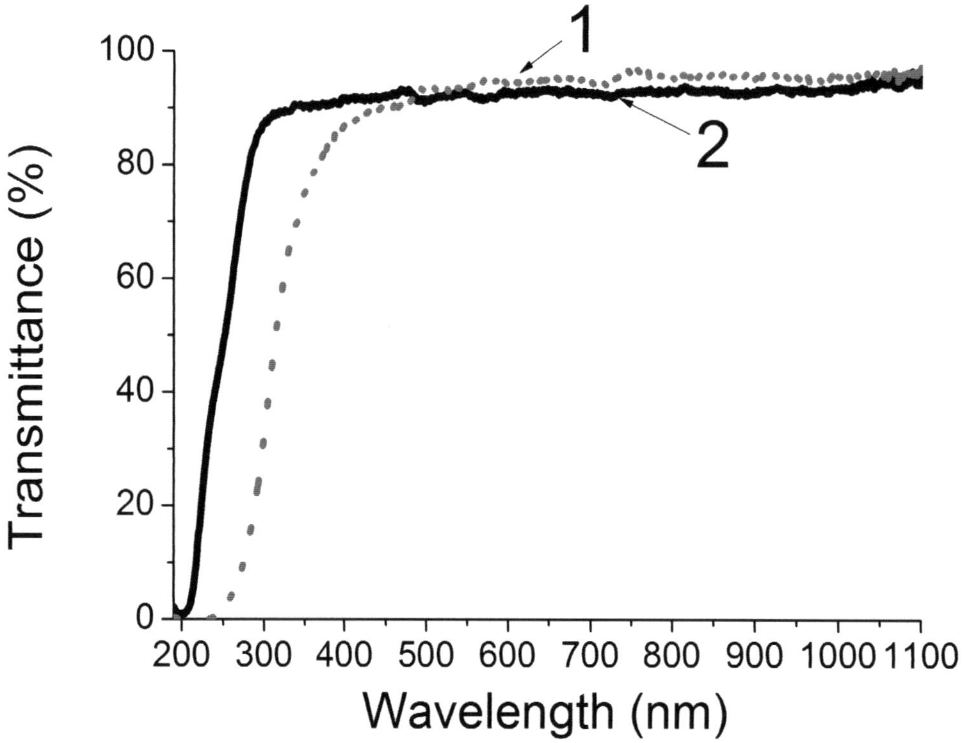

1
2
Transmittance (%)
100
80
60
40
20
0
200
300
400
500
600
700
800
900
1000
1100
Wavelength (nm)

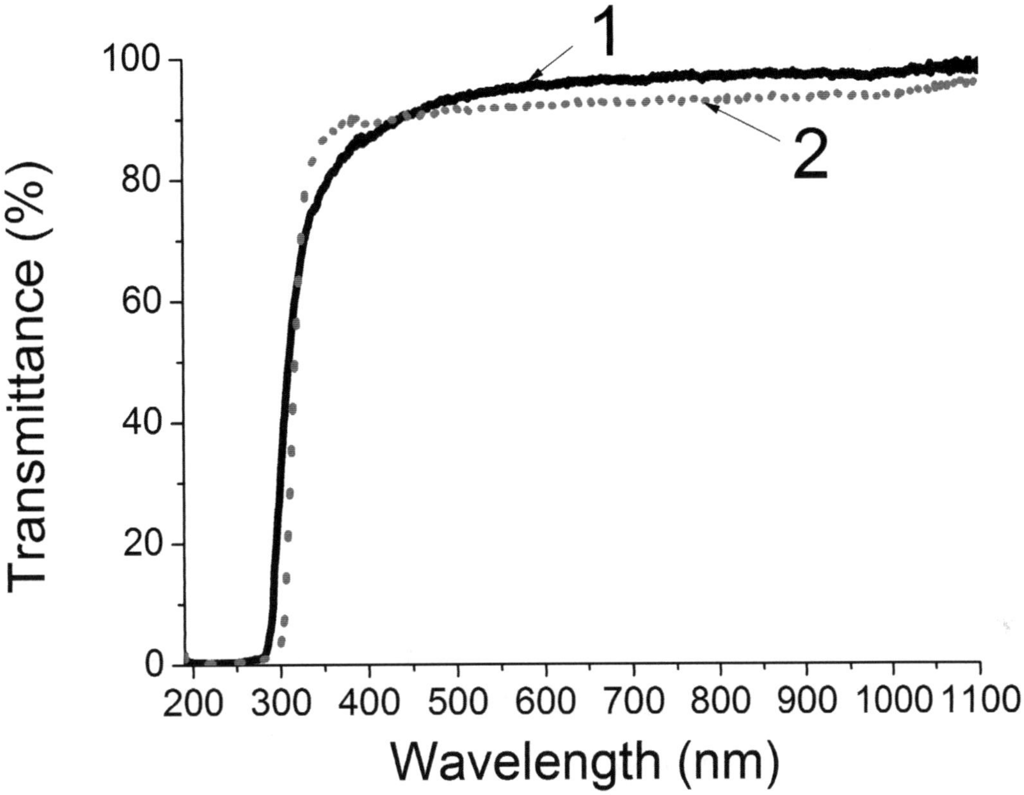
1
2
100
80
60
40
20
0
Transmittance (%)
200 300 400 500 600 700 800 900 1000 1100
Wavelength (nm)

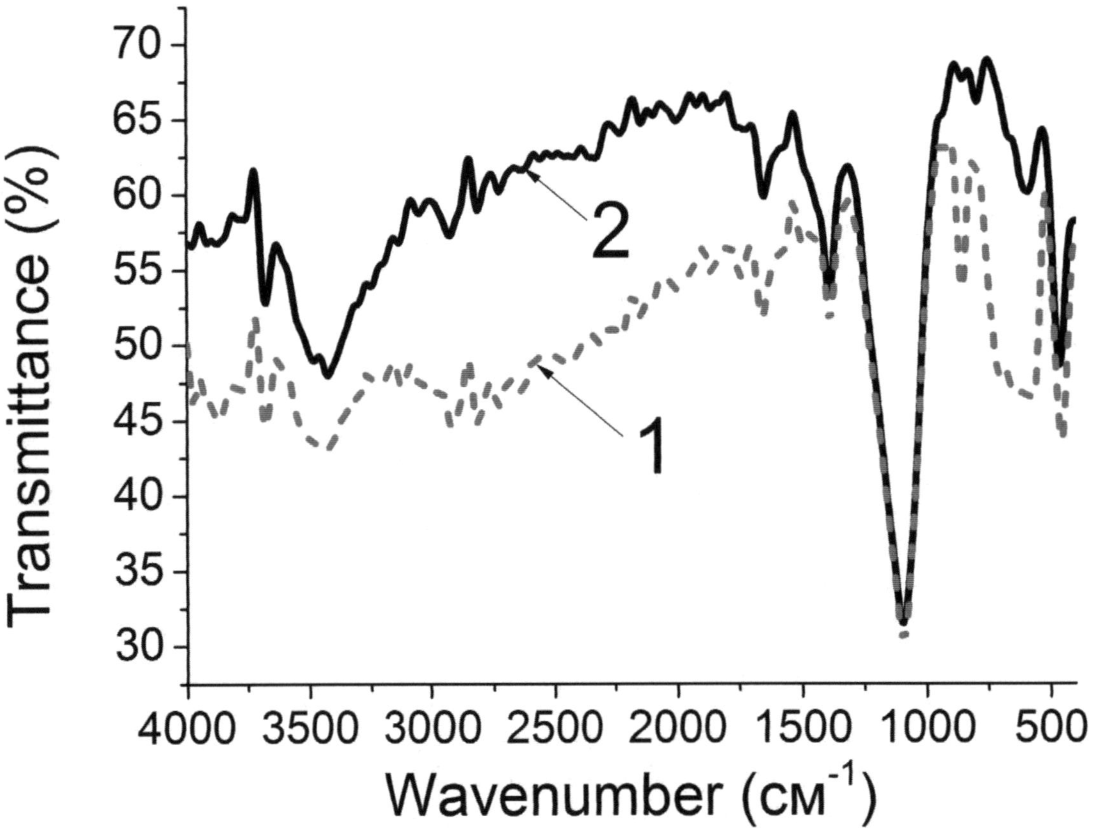

Transmittance (%)
70
65
60
55
50
45
40
35
30
4000
3500
3000
2500
2000
1500
1000
500
Wavenumber (см-1)
2
1

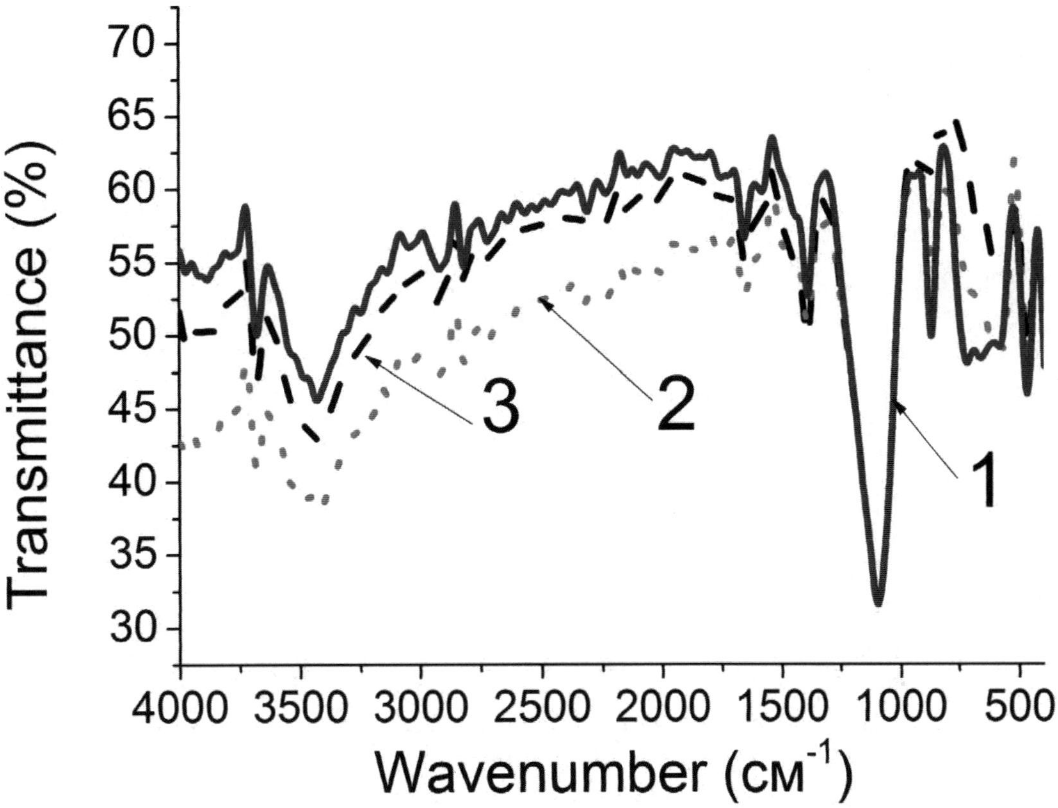

Transmittance (%)
70
65
60
55
50
45
40
35
30
4000
3500
3000
2500
2000
1500
1000
500
Wavenumber (см$^{-1}$)
1
2
3

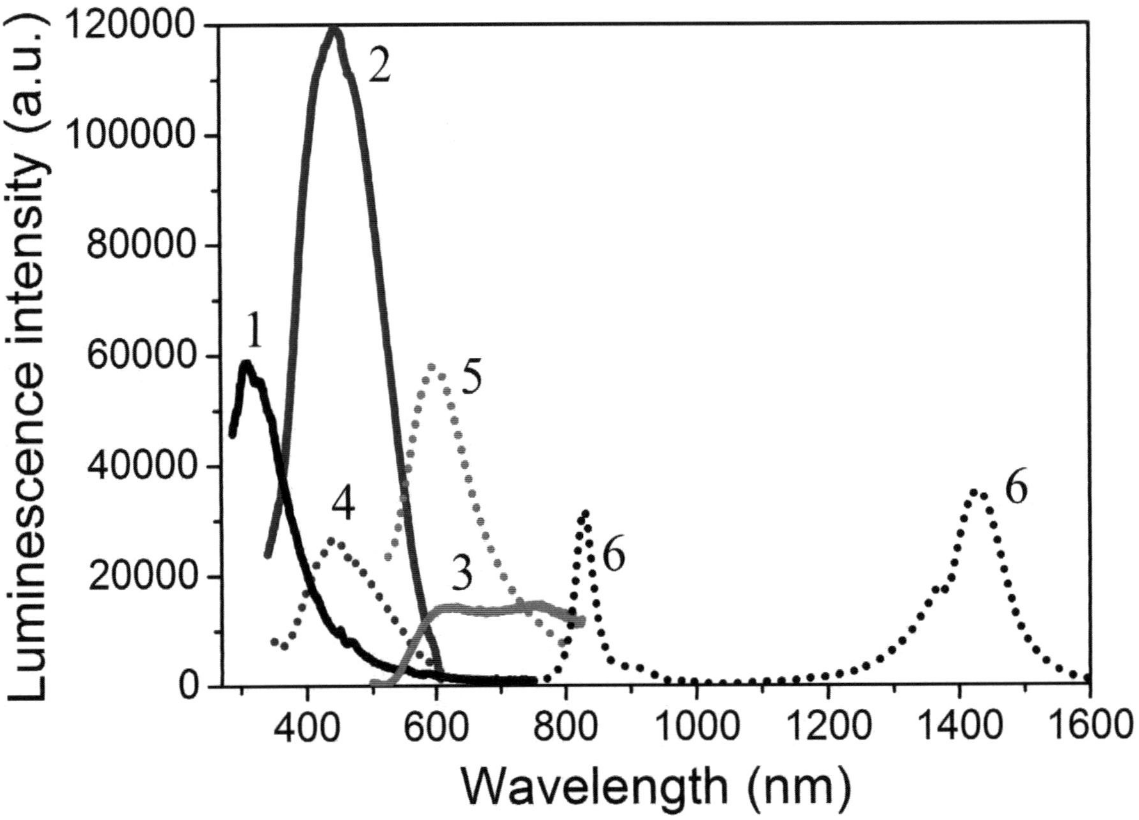
Luminescence intensity (a.u.)
120000
100000
80000
60000
40000
20000
0
400
600
800
1000
1200
1400
1600
Wavelength (nm)
1
2
3
4
5
6
6

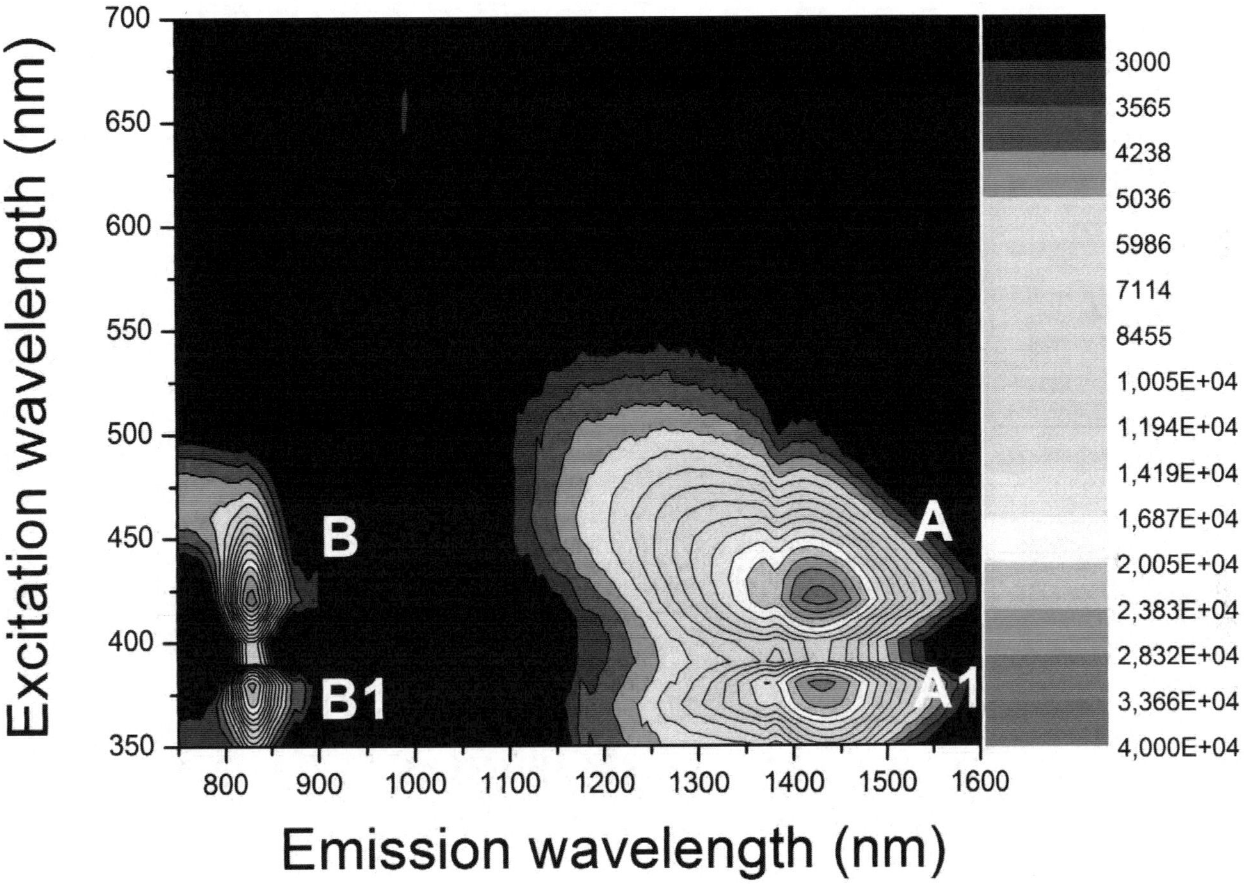

Excitation wavelength (nm)
700
650
600
550
500
450
400
350
800
900
1000
1100
1200
1300
1400
1500
1600
Emission wavelength (nm)
A
A1
B
B1
3000
3565
4238
5036
5986
7114
8455
1,005E+04
1,194E+04
1,419E+04
1,687E+04
2,005E+04
2,383E+04
2,832E+04
3,366E+04
4,000E+04

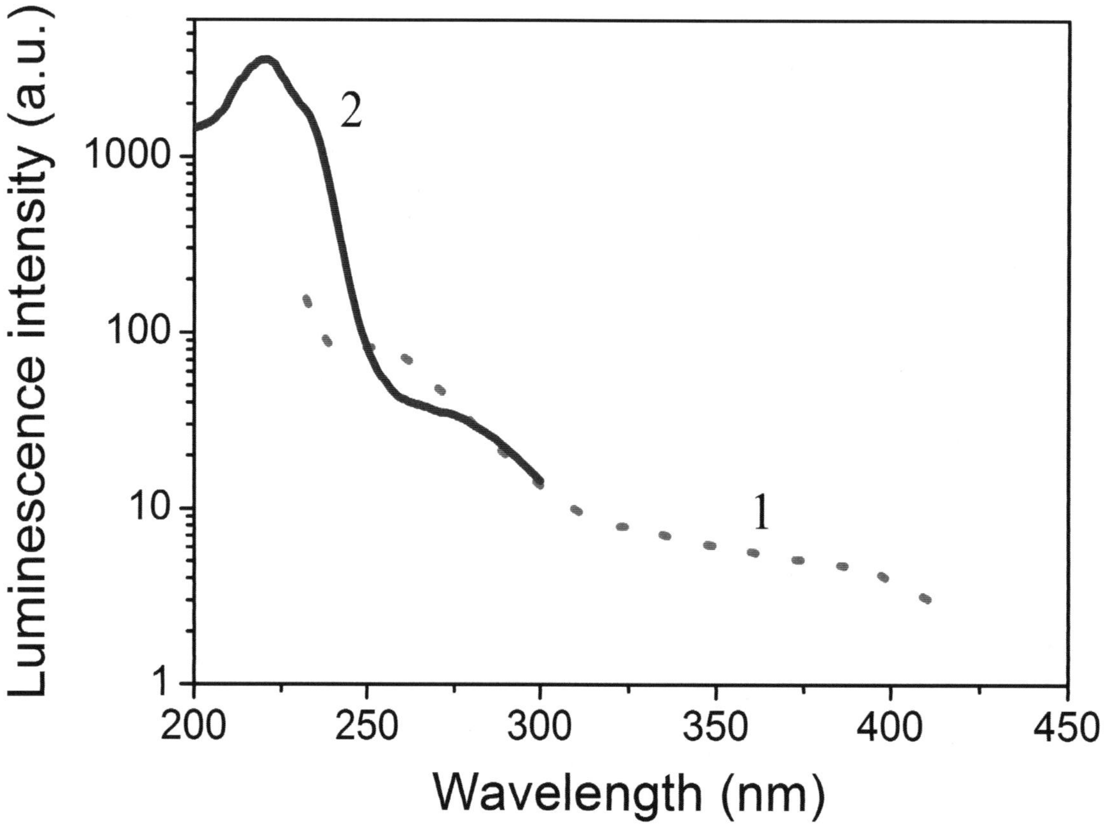
Luminescence intensity (a.u.)
1000
100
10
1
200
250
300
350
400
450
Wavelength (nm)
2
1

Permittivity of anisotropic dielectric near surface with local field effects

M N Anokhin, A A Tishchenko, M I Ryazanov and M N Strikhanov
National Research Nuclear University MEPhI, Moscow, Russia

E-mail: MNAnokhin@mephi.ru, tishchenko@mephi.ru

Abstract. We generalize Clausius-Mossotti relation for the case of anisotropic dielectric interface and obtain dielectric permittivity with local-field effects.

1. Introduction

It is known that the dielectric properties change near the interface [1]. One of the reasons for this phenomenon is the natural anisotropy of the medium near the surface. Because of this anisotropy the dielectric permittivity turns to tensorial form. The natural changing of dielectric properties near in the thin near-surface layer leads to new conditions for existing of surface electromagnetic waves [1]. Moreover, it affects the calculation of injection of charge carriers from the electrode into the layer of organic material [2], [3]. So, the effect of the changing of the dielectric permittivity at interfaces is of importance for physics of light-emitting diodes and other multilayered devices.

2. Macroscopic and local fields

At first let us consider the amorphous substance composed of uniformly located isotropic molecules.
We use the time and coordinates Fourier transform for Maxwell's equations:

$$\begin{cases} i[\mathbf{l},\mathbf{H}(\mathbf{l},\omega)] = \dfrac{4\pi}{c}\mathbf{j}(\mathbf{l},\omega) - i\dfrac{\omega}{c}\mathbf{E}(\mathbf{l},\omega) \\ i[\mathbf{l},\mathbf{E}(\mathbf{l},\omega)] = i\dfrac{\omega}{c}\mathbf{H}(\mathbf{l},\omega) \\ i(\mathbf{l},\mathbf{E}(\mathbf{l},\omega)) = 4\pi\rho(\mathbf{l},\omega) \\ i(\mathbf{l},\mathbf{H}(\mathbf{l},\omega)) = 0 \end{cases} . \tag{1}$$

From this equations we can find connection between the electric field and the density of microcurrents, i.e. $\mathbf{E} = f(\mathbf{j})$:

$$E_i(\mathbf{l},\omega) = \frac{4\pi i}{\omega} S_{ij}(\mathbf{l},\omega) j_j(\mathbf{l},\omega), \tag{2}$$

where

$$S_{ij}(\mathbf{l},\omega) = \frac{(\omega/c)^2 \delta_{ij} - l_i l_j}{l^2 - (\omega/c)^2 - i0}. \tag{3}$$

Then let us consider the amorphous substance occupying half-space $z > 0$. It is composed of N uniformly located anisotropic molecules. Let all the molecules of the material have the polarizability $\alpha_{ij}(\omega)$. Let the external field $\mathbf{E}^0$ act upon this material. Because of linearity of Maxwell's equations we can write a microscopic field as sum of a primary field $\mathbf{E}^0$ and the secondary fields created by all the molecules.

Let all the atomic electrons be non-relativistic. Thus, we can use the dipole approximation for the density of microcurrents:

$$\mathbf{j}^{mic}(\mathbf{r},\omega) = (-i\omega)\sum_b \mathbf{d}(\mathbf{R}_b,\omega)\delta(\mathbf{r}-\mathbf{R}_b), \quad (4)$$

where $\mathbf{d}(\mathbf{R}_b,\omega)$ is the moment dipolaire of the b-th molecule, $b=\overline{1,N}$, N is the total number of molecules of the substance.

The Fourier transform of the solution of Maxwell's equations in a medium is given by:

$$E_i^{mic}(\mathbf{r},\omega) = E_i^0(\mathbf{r},\omega) + \frac{1}{2\pi^2}\int d^3 l S_{ij}(\mathbf{l},\omega)\sum_b d_j(\mathbf{R}_b,\omega)\exp\{-i\mathbf{l}(\mathbf{R}_b-\mathbf{r})\}. \quad (5)$$

Whereas

$$d_j(\mathbf{R}_b,\omega) = \alpha_{jk}(\omega)E_k^{mic}(\mathbf{R}_b,\omega). \quad (6)$$

If we put $r = R_a$ in Eq. (5), we will obtain a system of N equations for acting on the a-th molecule field. The number N is macroscopically large, and such a system can be solved only approximately.

$$E_i^{mic}(\mathbf{R}_a,\omega) = E_i^0(\mathbf{R}_a,\omega) + \frac{1}{2\pi^2}\int d^3 l S_{ij}(\mathbf{l},\omega)\alpha_{jk}(\omega)\sum_b E_k^{loc}(\mathbf{R}_b,\omega)\exp\{-i\mathbf{l}(\mathbf{R}_b-\mathbf{R}_a)\}, \quad (7)$$

where index b numbers all molecules of the dielectric.

If we average the microscopic field E_i^{mic} over the location of other molecules, we will find the local field E_i^{loc}. As a first approximation we replace the exact field which affects a molecule by the local field:

$$E_i^{loc}(\mathbf{R}_a,\omega) = E_i^0(\mathbf{R}_a,\omega) + \frac{1}{2\pi^2}\int d^3 l S_{ij}(\mathbf{l},\omega)\alpha_{jk}(\omega)\int d^3 R_{ba} N w(\mathbf{R}_{ba}) E_k^{loc}(\mathbf{R}_a+\mathbf{R}_{ba},\omega)\exp\{-i\mathbf{l}\mathbf{R}_{ba}\}, \quad (8)$$

where $w(\mathbf{R}_{ba})$ is the probability density of finding the b-th molecule at the distance $\mathbf{R}_{ba} = \mathbf{R}_b - \mathbf{R}_a$ $\left(k=\overline{1,N-1}\right)$.

Unlike microscopic field, the local field is independent of the coordinates of neighbouring molecules. Nevertheless, dependence on the structure of material remains and it is determined by the function $w(\mathbf{R}_{ba})$.

For integration by the coordinates we take into account that the molecules of layer are located only in the region Z_b, i.e. $Z_{ba} > -Z_a$.

We average over the coordinates of all the molecules of the substance and obtain the macroscopic field:

$$E_i(\mathbf{R},\omega) = E_i^0(\mathbf{R},\omega) + \frac{1}{2\pi^2}\int d^3 R' S_{ij}(-\mathbf{R}',\omega)P_j(\mathbf{R}+\mathbf{R}',\omega), \quad (9)$$

where

$$S_{ij}(-\mathbf{R}',\omega) = \int d^3 q S_{ij}(\mathbf{l},\omega)\exp\{-i\mathbf{l}\mathbf{R}'\}, \quad (10)$$

$$P_j(\mathbf{R},\omega) = n\alpha_{jk}(\omega)E_k^{loc}(\mathbf{R},\omega). \quad (11)$$

We represent the probability density distribution of molecules in the form:

$$w(\mathbf{R}) = \frac{1}{V}\left[1 - f(\mathbf{R})\right]. \quad (12)$$

Now we can obtain a relation between the macroscopic field $\mathbf{E}$ and the local field $\mathbf{E}^{loc}$:

$$E_i(\mathbf{R},\omega) = E_i^{loc}(\mathbf{R},\omega) + \frac{1}{2\pi^2}\int d^3 R' S_{ij}(-\mathbf{R}',\omega) f(\mathbf{R}') P_j(\mathbf{R}+\mathbf{R}',\omega). \quad (13)$$

We write this equation in the coordinates $(\mathbf{q},z,\omega)$:

$$E_i(\mathbf{q},z,\omega) = E_i^{loc}(\mathbf{q},z,\omega) + 4\pi n \int_{z'>-z} \frac{d^3R'}{(2\pi)^3}\exp\{i\mathbf{q}\mathbf{R}'\} f(\mathbf{R}') S_{ij}(-\mathbf{R}',\omega)\alpha_{jk}(\omega)E_k^{loc}(\mathbf{q},z,\omega). \quad (14)$$

Equation which connects the Fourier transforms of the local and macroscopic fields:

$$t_{ik}(\mathbf{q},z,\omega)E_k^{loc}(\mathbf{q},z,\omega) = E_i(\mathbf{q},z,\omega), \quad (15)$$

where

$$t_{ik}(\mathbf{q},z,\omega) = \delta_{ik} + 4\pi n\int \frac{d^3R'}{(2\pi)^3}\exp\{i\mathbf{q}\mathbf{R}'\} f(\mathbf{R}') S_{ij}(-\mathbf{R}',\omega)\alpha_{jk}(\omega). \quad (16)$$

3. Dielectric function

Taking into account that

$$P_i(\mathbf{q},z,\omega) = n\alpha_{ij}(\omega)E_j^{loc}(\mathbf{q},z,\omega) \tag{17}$$

and on the other hand

$$4\pi P_i(\mathbf{q},z,\omega) = \left[\varepsilon_{ij}(\mathbf{q},z,\omega) - \delta_{ij}\right]E_j(\mathbf{q},z,\omega), \tag{18}$$

using Eq. (15) we can obtain the dielectric permittivity:

$$\varepsilon_{ik}(\mathbf{q},z,\omega) = \delta_{ik} + 4\pi n\alpha_{ij}(\omega)t_{jk}^{-1}(\mathbf{q},z,\omega), \tag{19}$$

In the long-wave approximation

$$n^{-1/3} \ll q^{-1} \ll c/\omega: \tag{20}$$

$$t_{ik}(z,\omega) = (\delta_{ik} - e_i e_k)\left\{1 - 4\pi n\alpha_{\perp}(\omega)\left(\frac{1}{3} + c(z)\right)\right\} + \left\{1 - 4\pi n\alpha_{\parallel}(\omega)\left(\frac{1}{3} + c(z)\right)\right\}e_i e_k \tag{21}$$

From Eqs. (21) we find t_{ik}^{-1}:

$$t_{ik}^{-1}(z,\omega) = \frac{1}{1 - 4\pi n\alpha_{\perp}(\omega)\left(\frac{1}{3} + c(z)\right)}(\delta_{ik} - e_i e_k) + \frac{1}{1 - 4\pi n\alpha_{\parallel}(\omega)\left(\frac{1}{3} + c(z)\right)}e_i e_k, \tag{22}$$

where

$$c(z) = \int d^3 l \int\limits_{Z'>z} \frac{dZ'}{2\pi} \exp\{il_z Z'\} f(\mathbf{l}_{\perp}, -Z'). \tag{23}$$

One can obtain the dielectric permittivity

$$\varepsilon_{ik}(z,\omega) = \varepsilon_{\perp}(z,\omega)(\delta_{ik} - e_i e_k) + \varepsilon_{\parallel}(z,\omega)e_i e_k, \tag{24}$$

where

$$\varepsilon_{\perp}(z,\omega) = \frac{1 + 4\pi n\alpha_{\perp}(\omega)(2/3 - c(z))}{1 - 4\pi n\alpha_{\perp}(\omega)(1/3 + c(z))}, \quad \varepsilon_{\parallel}(z,\omega) = \frac{1 + 4\pi n\alpha_{\parallel}(\omega)(2/3 - c(z))}{1 - 4\pi n\alpha_{\parallel}(\omega)(1/3 + c(z))}, \tag{25}$$

There is a dependence on z in the denominator of the Eqs. (25). When the function $c(z)$ is such that the denominator is minimal, the local field effect is strong.
The conditions of the maxima of the dielectric permittivity taking into account Eqs. (25) can be written in form:

$$\begin{aligned} &\mathrm{Re}\left\{\frac{4\pi}{3}n\alpha_{\perp}(\omega) + 4\pi n\alpha_{\perp}(\omega)\int d^3 l \int\limits_{Z'>z} \frac{dZ'}{2\pi} \exp\{il_z Z'\} f(\mathbf{l}_{\perp}, -Z')\right\} = 1 \\ &\mathrm{Re}\left\{\frac{4\pi}{3}n\alpha_{\parallel}(\omega) + 4\pi n\alpha_{\parallel}(\omega)\int d^3 l \int\limits_{Z'>z} \frac{dZ'}{2\pi} \exp\{il_z Z'\} f(\mathbf{l}_{\perp}, -Z')\right\} = 1 \end{aligned} \tag{26}$$

The expressions obtained in the limiting cases explored before go into known results. Indeed, in case of infinite substrate the results obtained turn into results of the article [1]. In the limiting case of infinite and isotropic media the expressions turn into the widely known Clausius–Mossotti relation:

$$\varepsilon(\omega) = \frac{1 + \frac{8\pi}{3}n\alpha(\omega)}{1 - \frac{4\pi}{3}n\alpha(\omega)}. \tag{27}$$

Thus, we report here for the first time of the new result – generalized Clausius-Mossotti relation. We have obtained dielectric permittivity $\varepsilon_{ik}(z,\omega)$ - the macroscopic physical quantity expressed through microscopic polarizability of single molecules and the function describing their mutual distribution. In conclusion we would like to note, that the theory constructed here is useful not only in solid state physics. All the problems connected with scattering of different electromagnetic waves in matter – we mean light, the electromagnetic field of charged particle beams moving through the target or in its vicinity – all of them are strongly dependent on the local field effects [4], [5]. For two dimensional structures we managed to find the expressions linked $\mathbf{E}^{loc}$ and $\mathbf{E}^0$ [6]; for the near-surface layer it has not been done yet, and this our work is a very good step in the way of solving this great problem, which is of real use and interest for numerous applications radiation physics of charged particles and in optics, along with solid state physics.

The work was supported by the Ministry of Education and Science of the Russian Federation in the frames of Competitiveness Growth Program of National Research Nuclear University MEPhI, Agreement 02.A03.21.0005.

References

[1] Ryazanov M I The effect of the natural variation in the polarization of a near-surface layer onelectromagnetic surface waves 1996 *JETP* **83** 529
[2] Garstein Yu N and Conwell E M Field-dependent thermal injection into a disordered molecular insulator 1996 *Chem. Phys. Lett.* **255** 93
[3] Arkhipov V I, Emelianova E V, Tak Y H and Bossler H Charge injection into light-emitting diodes: Theory and experiment 1998 *J. Appl. Phys.* **84** 848
[4] Ryazanov M I and Tishchenko A A Emission of a Fast Charged Particle Passing through a Monomolecular Film 2002 *Laser Physics* **12**, 1442
[5] Ryazanov M I, Strikhanov M N and Tishchenko A A Local field effect in diffraction radiation from a periodical system of dielectric spheres 2008 *Nucl. Instr. and Meth. B* **266**, 3811
[6] Ryazanov M I and Tishchenko A A Clausius-Mossotti-Type relation for planar monolayers 2006 *JETP* **103** 539 [*ZhETF* **130** 621]

Local field effects in anisotropic metamaterials

O V Porvatkina, A A Tishchenko and M N Strikhanov

National Research Nuclear University MEPhI, Moscow, Russia

E-mail: OVPorvatkina@mephi.ru, tishchenko@mephi.ru

Abstract. We consider dielectric and magnetic properties of metamaterials consisting of single particles. Local field effects are taken into account. The so called method of multiscale consideration is realized: the macroscopic quantities of permittivity and permeability are expressed through the microscopic polarizability and magnetization of single particles. We obtain the generalized Clausius-Mossotti relations for the case of magnetic medium and with anisotropic particles in the long wave approximation. In cases of isotropic magnetic and nonmagnetic particles our results coincide with published results of other authors.

1. Introduction

In the frames of macroscopic electrodynamics there are two ways to determine the constitutive equations. The first way is that we consider only three macroscopic fields: $\mathbf{E}$, $\mathbf{D}$, $\mathbf{B}$. In this case, permittivity $\varepsilon(\omega)$ describes both electric and magnetic response of medium. This approach suggests the attractive opportunity to take into account the effects of spatial dispersion, see the article of V.M. Agranovich [1]. The second or "symmetric" way is based on the use of four macroscopic quantities: $\mathbf{E}$, $\mathbf{D}$, $\mathbf{B}$, $\mathbf{H}$. We investigate permittivity $\varepsilon(\omega)$ and permeability $\mu(\omega)$ by second way, meaning to develop the theory valid for description of metamaterials with the so called local field effects that are due to interaction of single particles the medium consist of. These particles can be atoms, molecules, nanoparticles or macroscopic particles.

2. Permittivity and permeability with local field effects

Let us consider homogenous magnetic medium which consists of N anisotropic particles. The microscopic field $\mathbf{E}^{mic}$ in a medium is different from the incident field $\mathbf{E}^{0}$ generated by an external source in vacuum. Relations between the external field $\mathbf{E}^{0}$, the macroscopic field $\mathbf{E}$, and the local field $\mathbf{E}^{loc}$ that lead to the Clausius–Mossotti formula for three-dimensional systems are obtained by using a pair distribution function for molecules [2].

Equation for the density of micro-currents in homogenous magnetic medium can be written as:

$$j_i^{mic}(\mathbf{r},\omega) = -i\omega P_i(\mathbf{r},\omega) + c\left(rot\mathbf{M}(\mathbf{r},\omega)\right)_i, \tag{1}$$

where $\mathbf{P}(\mathbf{r},\omega)$ and $\mathbf{M}(\mathbf{r},\omega)$ are the average density of the electric and magnetic moments, respectively.

The relationship between the average density of the electric moment and microscopic electric field, the average density of the magnetic moment and the microscopic magnetic field given by the expressions:

$$P_i(\mathbf{r},\omega) = \sum_n \alpha_{ij}^{e}(\omega) E_j^{mic}(\mathbf{R}_n,\omega)\delta(\mathbf{r}-\mathbf{R}_n), \tag{2}$$

$$M_i(\mathbf{r},\omega) = \sum_n \alpha_{ij}^{m}(\omega) H_j^{mic}(\mathbf{R}_n,\omega)\delta(\mathbf{r}-\mathbf{R}_n), \tag{3}$$

where $\alpha^e_{ij}(\omega)$, $\alpha^m_{ij}(\omega)$ are polarizability tensor and magnetization tensor, respectively. These tensors can be written in the form

$$\alpha^e_{ij}(\omega)=\alpha^e_{\perp}(\omega)\left(\delta_{ij}-e_ie_j\right)+\alpha^e_{\parallel}(\omega)e_ie_j\ , \tag{4}$$

$$\alpha^m_{ij}(\omega)=\alpha^m_{\perp}(\omega)\left(\delta_{it}-e_ie_j\right)+\alpha^m_{\parallel}(\omega)e_ie_j\ . \tag{5}$$

The microscopic field in medium is the sum of the primary field $\mathbf{E}^0(\mathbf{r},\omega)$, generated by external charges, and secondary fields, produced by all particles of the medium. Maxwell's equations for Fourier transform of the microscopic fields acting on a particle of medium placed at a point $\mathbf{r}$ give [2]:

$$\mathbf{E}^{mic}(\mathbf{r},\omega)=\mathbf{E}^0(\mathbf{r},\omega)+\frac{4\pi i}{\omega}\int d^3qe^{i\mathbf{qr}}\frac{k^2\mathbf{j}^{mic}(\mathbf{q},\omega)-\mathbf{q}\left(\mathbf{q},\mathbf{j}^{mic}(\mathbf{q},\omega)\right)}{q^2-k^2}, \tag{6}$$

$$\mathbf{H}^{mic}(\mathbf{r},\omega)=\mathbf{H}^0(\mathbf{r},\omega)+\frac{4\pi i}{c}\int d^3qe^{i\mathbf{qr}}\frac{[\mathbf{q},\mathbf{j}^{mic}(\mathbf{q},\omega)]}{q^2-k^2}, \tag{7}$$

where $k^2=\dfrac{\omega^2}{c^2}$.

The effective field acting on a molecule is formed by adding the fields of many molecules lying in some volume of medium with linear dimensions L that are large by the comparison with intermolecular distance $n^{-1/3}$, but small by the comparison with wavelength (long-wave limit) . When the inequality

$$n^{-1/3}=L=\frac{c}{w} \tag{8}$$

is valid, the values of the microscopic electric and magnetic fields acting on a molecule are close to its values averaged over the positions of the over molecules, so-called the local fields. The local field is independent of the positions of other molecules, whereas its dependence on the structure of the medium is preserved by the function of distribution $dw(\mathbf{R}_{an})$.

Averaging Eqs. (6) and (7) with two-particle function of distribution

$$dw(\mathbf{R}_{an})\equiv\frac{1}{V}\{1-f(\mathbf{R}_{an})\}d^3R_{an} \tag{9}$$

(here, V is the volume of the medium) i.e. over the positions of all particles in respect to the single particle (see [2] and [3]), one can get equations for the local electric and magnetic fields acting on a molecule located at a point with radius-vector $\mathbf{R}_a$:

$$E^{loc}_i(\mathbf{R}_a,\omega)=E^0_i(\mathbf{R}_a,\omega)+\frac{n}{2\omega\pi^2}\int d^3q\frac{k^2\delta_{ij}-q_iq_j}{q^2-k^2}\int d^3R_{an}e^{i\mathbf{qR}_{an}}$$
$$\times\left(1-f(R_{an})\right)\left(\omega\alpha^e_{jk}(\omega)E^{loc}_k(\mathbf{R}_a-\mathbf{R}_{an},\omega)-c\alpha^m_{jk}(\omega)\left[\mathbf{p},\mathbf{B}^{loc}(\mathbf{R}_a-\mathbf{R}_{an},\omega)\right]_k\right) \tag{10}$$

$$B^{loc}_i(\mathbf{R}_a,\omega)=B^0_i(\mathbf{R}_a,\omega)+\frac{n}{2c\pi^2}\int d^3q\frac{1}{q^2-k^2}\int d^3R_{an}e^{i\mathbf{qR}_{an}}\left(1-f(R_{an})\right)$$
$$\times\left(\omega\alpha^e_{ij}(\omega)\left[\mathbf{q},\mathbf{E}^{loc}(\mathbf{R}_a-\mathbf{R}_{an},\omega)\right]_j-c\alpha^m_{ij}(\omega)\left[\mathbf{q},\left[\mathbf{q},\mathbf{B}^{loc}(\mathbf{R}_a-\mathbf{R}_{an},\omega)\right]\right]_j\right) \tag{11}$$

where $R_{an}=R_a-R_n$ $\left(k=\overline{1,N-1}\right)$ is the distance between molecules a and n.

In these equations $f(R_{an})$ can be found from equation for the radial distributional function, which in turn can be found from independent experiment, e.g., using x-ray diffraction.

On the other hand, averaging equations Eqs. (6) and (7) over the positions of all particles one get equations for usual average macroscopic fields:

$$E_i(\mathbf{r},\omega)=E_i^0(\mathbf{r},\omega)+\frac{n}{2\omega\pi^2}\int d^3qe^{i\mathbf{qr}}\frac{k^2\delta_{ij}-q_iq_j}{q^2-k^2}$$
$$\times\left(\omega\alpha_{jk}^e(\omega)E_k^{loc}(\mathbf{q},\omega)-c\alpha_{jk}^m(\omega)[\mathbf{q},\mathbf{B}^{loc}(\mathbf{q},\omega)]_k\right), \quad (12)$$

$$B_i(\mathbf{r},\omega)=B_i^0(\mathbf{r},\omega)+\frac{n}{2c\pi^2}\int d^3qe^{i\mathbf{qr}}\frac{1}{q^2-k^2}$$
$$\times\left(\omega\alpha_{ij}^e(\omega)\left[\mathbf{q},\mathbf{E}^{loc}(\mathbf{q},\omega)\right]_j-c\alpha_{ij}^m(\omega)\left[\mathbf{q},\left[\mathbf{q},\mathbf{B}^{loc}(\mathbf{q},\omega)\right]\right]_j\right). \quad (13)$$

Substituting vector $\mathbf{R}_a$ of molecule instead vector $\mathbf{r}$ in Eqs. (12),(13) gives expressions for macroscopic fields at a point $\mathbf{R}_a$. Combining Eqs. (10) - (13) ,in the long-wave limit one can obtain relations between the local field, i.e. locally acting fields, and the average macroscopic field in medium:

$$E_i(\mathbf{r},\omega)-E_i^{loc}(\mathbf{r},\omega)=-(4\pi n/3)\alpha_{ik}^e(\omega)E_k^{loc}(\mathbf{r},\omega), \quad (14)$$
$$B_i(\mathbf{r},\omega)-B_i^{loc}(\mathbf{r},\omega)=(8\pi n/3)\alpha_{ik}^m(\omega)B_k^{loc}(\mathbf{r},\omega). \quad (15)$$

To obtain Eqs. (14), (15) we used properties of the function $f(R_{an})$. On the one hand, the probability of unlimited approachment between the two molecules is negligibly small, and therefore $f(R_{an})=1$ at $R_{an}=n^{-1/3}$. On the other hand, there is no correlation between distant molecules, therefore $f(R_{an})=0$ at R_{an} ? $n^{-1/3}$ [2].

Polarization of material can be expressed through the macroscopic field and on the other hand is expressed through the local field:

$$E_k(\mathbf{r},\omega)(\varepsilon_{ik}(\omega)-\delta_{ik})=4\pi n\alpha_{ij}^e(\omega)E_j^{loc}(\mathbf{r},\omega), \quad (16)$$
$$(\mu^{-1}_{ij}(\omega)-\delta_{ij})B_j(\mathbf{r},\omega)=4\pi n\alpha_{ij}^m(\omega)B_j^{loc}(\mathbf{r},\omega). \quad (17)$$

Eqs. (14), (15) and (16), (17) let us obtain the dependence of the permittivity tensor $\varepsilon_{ik}(\omega)$ from the polarizability tensor $\alpha_{ij}^e(\omega)$ and dependence of the permeability tensor $\mu_{ij}(\omega)$ from the magnetization tensor $\alpha_{ij}^m(\omega)$:

$$\varepsilon_{ik}(\omega)=\varepsilon^{\perp}(\omega)(\delta_{ik}-e_ie_k)+\varepsilon^{\parallel}(\omega)e_ie_k, \quad (18)$$
$$\mu_{ij}(\omega)=\mu^{\perp}(\omega)(\delta_{ij}-e_ie_j)+\mu^{\parallel}(\omega)e_ie_j, \quad (19)$$

where

$$\varepsilon^{\perp}(\omega)=\frac{1+(8\pi/3)n\alpha_{\perp}^e(\omega)}{1-(4\pi/3)n\alpha_{\perp}^e(\omega)},\quad \varepsilon^{\parallel}(\omega)=\frac{1+(8\pi/3)n\alpha_{\parallel}^e(\omega)}{1-(4\pi/3)n\alpha_{\parallel}^e(\omega)} \quad (20)$$

$$\mu^{\perp}(\omega)=\frac{1+(8\pi/3)n\alpha_{\perp}^m(\omega)}{1-(4\pi/3)n\alpha_{\perp}^m(\omega)},\quad \mu^{\parallel}(\omega)=\frac{1+(8\pi/3)n\alpha_{\parallel}^m(\omega)}{1-(4\pi/3)n\alpha_{\parallel}^m(\omega)}. \quad (21)$$

In Eqs. (18) - (20) denominators are not vanish because polarizability and magnetization are complex quantities. In particular case of isotropic medium Eqs. (18), (19) give

$$\varepsilon(\omega)=\frac{1+(8\pi/3)n\alpha_e(\omega)}{1-(4\pi/3)n\alpha_e(\omega)}, \tag{22}$$

$$\mu(\omega)=\frac{1+(8\pi/3)n\alpha_m(\omega)}{1-(4\pi/3)n\alpha_m(\omega)}. \tag{23}$$

Eqs. (22),(23) coincides with ones obtained in article [4] with the help of another method. For nonmagnetic material our results go to the well-known Clausius-Mossotti relations. The main approximation we use is that the wavelength of the field must exceed considerably the size of the particles. So, we have obtained the generalized Clausius-Mossotti relations for magnetic media in general and metamaterials in particular.

The conditions when denominators in Eqs. (20) - (23) are small implies the effects of difference between local fields and the macroscopic fields. The strong enhancement of local fields means sharp enhancement of all electromagnetic processes and can be of great interest for numerous applications in physics of solid state, spectroscopy, optics, including nonlinear optics [5], and for artificial periodic structures, in particularly, in metamaterials [6].

The work was supported by the Ministry of Education and Science of the Russian Federation in the frames of Competitiveness Growth Program of National Research Nuclear University MEPhI, Agreement 02.A03.21.0005.

References

[1] Agranovich V M and Gartstein Yu N Spatial dispersion and negative refraction of light 2006 *Physics-Uspekhi* **49** 1029 [UFN **176** 1051]

[2] Ryazanov M I and Tishchenko A A Clausius-Mossotti-type relation for planar monolayers 2006 *JETP* **103** 539 [*ZhETF* **130** 621]

[3] Ryazanov M I *Electrodynamics of Condensed Matter* [in Russian] (Nauka, Moscow, 1984)

[4] Kästel J and Fleischhauer M Local-field effects in magnetodielectric medium: Negative refraction and absorption reduction 2007 Physical Review A **76** 062509

[5] Gorkunov M V and Ryazanov M I The effect of a local field on Raman scattering in a uniaxial crystal 1997 *JETP* **85** 97 [*ZhETF* **112** 180]

[6] Pendry J B Negative refraction makes a perfect lens 2000 *Phys. Rev. Lett.* **85** 39

Modelling of ion convection during electrochemical oxidation of aluminium

Pavel Zun[1], Andrey Svitenkov
University ITMO, Saint-Petersburg

Abstract. Mesoscopic electrically induced ionic convection in the electrolyte surrounding the anode is modelled. Dissipative particle dynamics approach is used for the modelling. Simulation shows that the relation between forming voltage and convection cell size is likely nonlinear. Interactions between this nonlinearity and modes in the oxide layer can be the cause of irregularities in the resulting structures.

1. Introduction

Anodizing of aluminium is a process that's been used in the industry for at least the last 50 years. More recently, nano-porous anodic aluminium oxide membranes have been widely used as templates in the synthesis of one-dimensional nanomaterials and quantum dot arrays [1], but some microscopic characteristics of it are not yet fully understood.

For example, there are different theories explaining the ordering of pores in the oxide layer [2, 3]. Empirically a linear dependence has been found between the inter-pore distance and the voltage applied [4]. The most common explanation involves the stress in the oxide layer [2] as the reason for pore ordering and diameter. This explanation results in a close to linear dependence for most common voltages, but it does not completely explain why the quality of structures varies dramatically for different regimes.

However, convective flow inside the electrolyte is another factor that influences pore structure. Interactions between it and the processes in the oxide layer can explain irregularities in the resulting structures. There have been few attempts to model the convective flow for this process. The only known to us attempt is [3], and it has serious drawbacks, which are discussed in more detail in the next section. The aim of this work is to clarify the influence of convective flows on inter-pore distance.

In this work the initial phase of anodic oxidation is modelled. For modelling the flows in the electrolyte a mesoscopic dissipative particle dynamics approach [5] is used.

2. Theory and related work

The process of anodizing consists of submerging an aluminium component (anode) and a cathode in water-acid solution (oxalic, sulphuric and phosphoric acids are commonly used) and applying voltage to them. This leads to electrochemical oxidation of the aluminium.

Negatively charged anions, i.e. hydroxide, acid residual and maybe oxide ions move to the anode. The electrical charge in the circuit causes positively charged aluminium ions (Al^{3+}) to be generated in

[1] E-mail: pavel.zun@gmail.com

the anode and move toward the cathode. At the anode surface they react with the oxide/hydroxide ions to form aluminium oxide (in the case of the hydroxide ion, hydrogen ions are released into the solution). The following reactions are believed to occur on the anode:

$$Al \rightarrow Al^{3+} + 3e^-$$

$$2Al^{3+} + 3O^{2-} \rightarrow Al_2O_3$$

$$2Al^{3+} + 3OH^- \rightarrow Al_2O_3 + 3H^+$$

For which the overall equation is

$$2Al + 3H_2O \rightarrow Al_2O_3 + 6H^+ + 6e^-$$

The rate of the oxidation depends on the concentration of negatively charged ions, which are brought to the surface by the electrolyte flow. Since the characteristic inter-pore distances are on the order of 100 nm, the distribution of these ions is affected by both hydrodynamic and molecular kinetic laws. Because of that, it is reasonable to use a mesoscale approach, which accounts for both of these factors. Dissipative particle dynamics (DPD) is used for the simulation.

The main driving force of the convection is the external electric field. The field and the flow is also significantly affected by the uneven distribution of positive and negative charges in the electrolyte, for example, by the anodic shielding.

There have been attempts [3] to model this process before, but the method used in [3] has severe shortcomings.

Firstly, a continual model was used, which did not account for the kinetic processes governing the charge distribution. For example, equilibrium solution of their system of equations does not lead to anodic shielding. Using their model, charge concentration drops linearly instead of exponentially.

Secondly, the convective cells took up all of the simulation area, and because of that the result was dependent upon the distance from the anode to the arbitrarily chosen upper bound of the simulation box. This upper bound does not correspond to any physical length. The model from [3] results in a linear relation between voltage and convection cell size, which is in accordance with the experiments.

Our hypothesis is that the ionic convection near the anode is an important factor for the formation of irregularities in the oxide layer. Nonlinear dependence of inter-pore distance on the applied voltage would be an indirect proof of this hypothesis, since the superposition of the effects of convection and stress in the oxide can cause the pores to distort.

3. Model, implementation and results

We implemented the model in the LAMMPS (Large-scale Atomic/Molecular Massively Parallel Simulator) package [6]. The area directly neighbouring the anode is modelled, with periodic boundary conditions on x and y axes. In the uppermost part of the simulation box electroneutrality is enforced in the electrolyte. For short-range interactions the Coulomb forces are calculated directly and for long-range interactions a k-space PPPM solver is utilized. Long-range interactions ensure that the model is self-consistent.

For the modelling, the following parameters were used: simulation area 360×360×2 nm, temperature equal to 40° C, density of 1 g/cm^3, dynamic viscosity of 0,7 mPa · s (viscosity of water at that temperature). The system was calibrated so that for the forming voltage 48 V (voltage for one of the stable modes) the cell size is equal to half experimental inter-pore distance.

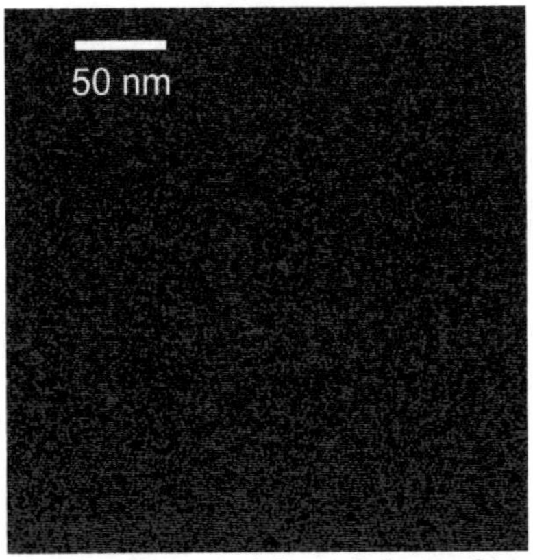

Fig. 1: *Visualization of the simulation box. Positive charges in blue, negative charges in red. Debye shielding of the anode is visible in the lower part*

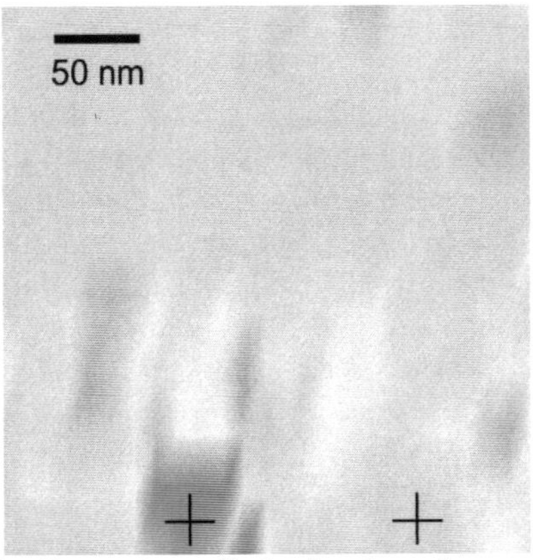

Fig. 2: *Convective cells in the electrolyte. Crosses mark conjugate streams. Colour indicates charge density — orange areas are positively charged*

Figure 1 shows a steady-state solution of the model. Red points indicate negatively charged particles, and blue points indicate positively charged ones. The anode is situated directly below the simulated area, and Debye shielding can be observed in the lower part of the area.

For our simulation we use a constantly-sized square box. Only the lower part of the electrolyte is involved in the convective flow, which guarantees independence of our results from the geometry of the simulation area. For higher voltages the whole mass of the liquid gets involved in the convection. This marks the limits of applicability of our model.

Our main point of interest is to determine whether the dependence of inter-pore distance on the applied voltage is linear or not. The distance between the opposite charged electrolyte streams is equal to half inter-pore distance. The convective cells are shown in Figure 2, where color indicates electrolyte charge density. The crosses mark the centers of two conjugate streams.

The plot of the cell size against the voltage is given in Figure 3. Lowest, highest and average cell sizes are marked. Note that the cell size (distance between two conjugate streams) is equal to half inter-pore distance.

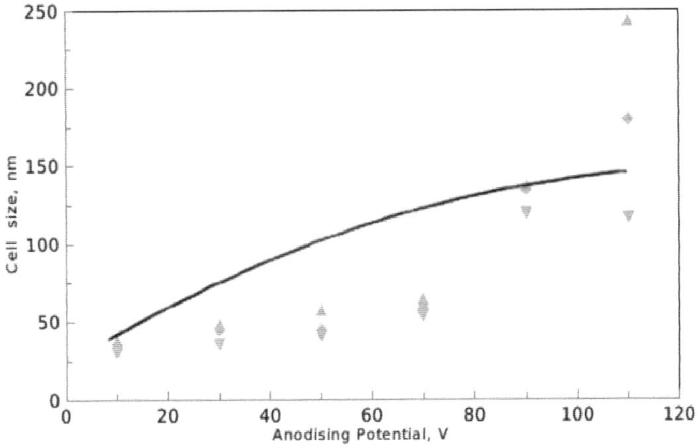

Fig. 3: *The plot of cell size (half inter-pore distance) against forming voltage. Lowest, highest and average cell sizes are marked.*

The black line shows wavelength of the most rapidly growing mode in oxide layer as a function of the applied voltage [2]

4. Conclusions

In this work electrically induced mesoscale convection was modelled. The steady-state solution of the model shows Debye shielding. In the dynamic solution convective cells can be observed.

The dependence between the applied voltage and the cell size is likely non-linear (see Fig. 3). This is more noticeable for higher voltages, where the cell size growth rate in relation to voltage is faster than linear. This, combined with the slower than linear rate for fastest-growing mode in oxide, derived in [2], (shown in Figure 3 with the black line), can lead to instabilities and irregularities in cell structure. In the regimes where cell size and the mode wavelength match, good quality structures can be produced, and in the other regimes pores can be distorted because of interference.

Hence, this supports our hypothesis about the possibility of the interactions between ionic convection and elastic stress in the oxide layer playing a key role in the formation of irregularities in the resulting structures.

5. References

[1] Shingubara S 2003 *J. Nanopart. Res.* **5** 17
[2] Singh G K, Golovin A A and Aranson I S 2000 *Phys. Rev.* B **73** 205422
[3] Lu S, Su Z, Sha J and Zhou W 2009 *Chem. Commun.* 5639–5641
[4] Zhang F, Liu X, Pan C and Zhu J 2007 *Nanotechnology* **18** 345302
[5] Español P and Warren P B 1995 *Europhysics Letters* **30(4)** 191–196
[6] Plimpton S 1995 *J Comp Phys* **117** 1-19

Low temperature growth of the epitaxial Ge layers on Si(100) by Hot Wire Chemical Vapor Deposition

S A Matveev[1,2,3], S A Denisov[1], D V Guseinov[1], V N Trushin[1], A V Nezhdanov[2], D O Filatov[1], and V G Shengurov[1]

[1] Physico-Technical Research Institute, Nizhniy Novgorod State University, Nizhniy Novgorod, 23/3 Gagarin Ave., Nizhny Novgorod 603950 Russia
[2] Physics Department, Nizhniy Novgorod State University, Nizhniy Novgorod, 23/3 Gagarin Ave., Nizhny Novgorod 603950 Russia

E-mail: shengurov@phys.unn.ru

Abstract. In the present paper, we report on the growing of thick (~0.2-3.0 μm) epitaxial Ge/ Si(100) layers by Hot Wire Chemical Vapor Deposition (HWCVD) at low growth temperatures (350°C). The single crystal epitaxial Ge layers with low threading dislocation density (~10^5 cm^{-2}) and surface roughness (< 0.5 nm) have been obtained.

1. Introduction

The possibility of heteroepitaxial growth of pure Ge on Si(100) substrates has generated a high interest because of the prospects for their potential applications in various novel electronic and photonic devices [1]. The high electron mobility, small direct bandgap of Ge (≈ 0.8 eV at room temperature), and the compatibility with well established silicon-based complimentary-symmetry metal-oxide-semiconductor (CMOS) technology makes it possible to design efficient photodetectors operating in the low optical fiber loss window (the light wavelength λ ranging from 1.3 to 1.55 μm) [2]. However, the key issue in growing the Ge films on Si for device applications is overcoming the negative effects of the large lattice mismatch (4.2% at 300K) between these two materials, which results in a high misfit dislocation density at the Ge/Si interface and a high threading dislocations density (TDD) in the Ge layers along with a high surface roughness due to island growth. To reduce the misfit dislocation density, various approaches have been proposed, e. g. a cyclic thermal annealing at the temperatures varied from 700 to 800°C and back [3]. However, for the integration of Ge/Si photodetectors into standard CMOS integrated circuits (ICs), low temperature growth techniques of the Ge layers are desired highly [4].

Hot Wire Chemical Vapour Deposition (HWCVD) is one of the promising candidates for the low-temperature Ge/Si heterostructure growth providing the abrupt heterointerfaces [5]. In this paper, we report on the growing of thick (up to 3 μm) single crystal epitaxial Ge layers on Si(100) using HWCVD method.

[3] To whom any correspondence should be addressed.

2. Experiment

The epitaxial growth of Ge films on Si(100) was carried out in a home made Ultra High Vacuum (UHV) system for Molecular Beam Epitaxy (MBE) of Si and Ge shown schematically in Figure 1. The base residual pressure in the growth chamber was less than 10^{-7} Torr.

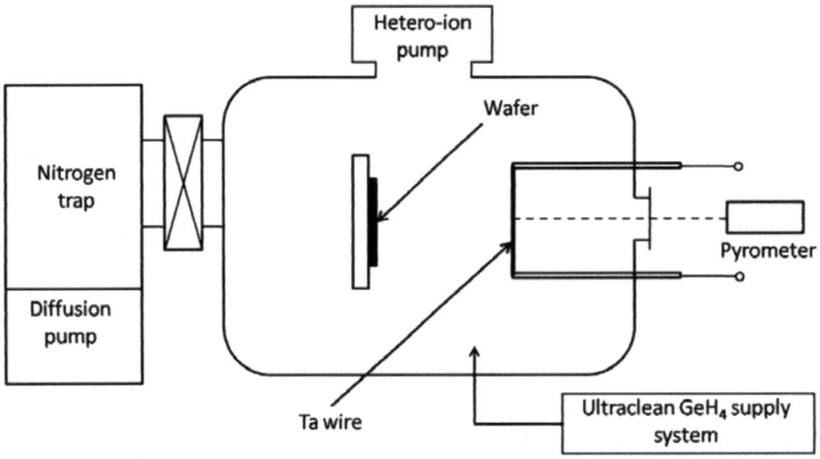

Figure 1. Schematic representation of HWCVD setup for the epitaxial growth of Ge layers on Si(100) substrates.

The Ge layers were grown on n-Si(100) substrates (the specific resistivity of the substrate material was 4.5 and 0.005 Ω·cm) with a epiready grade polished (100) surface. The Si buffer layers were grown from a sublimation Si source at the substrate temperature ≈ 800°C. The deposition of Ge was carried out by letting purified germane (GeH_4) inside the growth chamber through a leak valve. Germane was cracked pyrolitically on a tantalum (Ta) filament located 3 cm away from the substrate. The filament was heated up to 1200 ÷ 1500°C by a direct electrical current. The filament temperature was measured by an optical pyrometer. The partial pressure of germane inside the growth chamber was $(1 \div 4)\cdot 10^{-4}$ Torr, the substrate temperature during the deposition of Ge was ≈ 350°C.

The surface morphology and roughness of deposited Ge films were examined by Atomic Force Microscopy (AFM). The crystal structure of the Ge films was examined by Double crystal X-Ray Diffractometry (DXRD) and by Reflection Electron Diffraction. In addition, the elastic strain in the Ge/Si(100) epilayers was examined by Raman Spectroscopy. The threading dislocation density in the Ge/Si(100) films was determined by standard etch pit counting method. To study the depth profiles of the impurity concentrations in the Ge layers Secondary Ion Mass Spectrometry (SIMS) has been applied. The carrier concentration and mobility in the Ge/Si(001) films were measured by standard van-der-Pauw method at room temperature using the samples grown on low conductive substrates.

3. Results and discussion

The growth rate of the Ge layers was found to depend on the temperature of the Ta filament linearly (see Figure 2). Evidently, increasing the filament temperature increases the decomposition rate of germane. As a result, the Ge flux density onto the substrate increases with increasing Ta filament temperature.

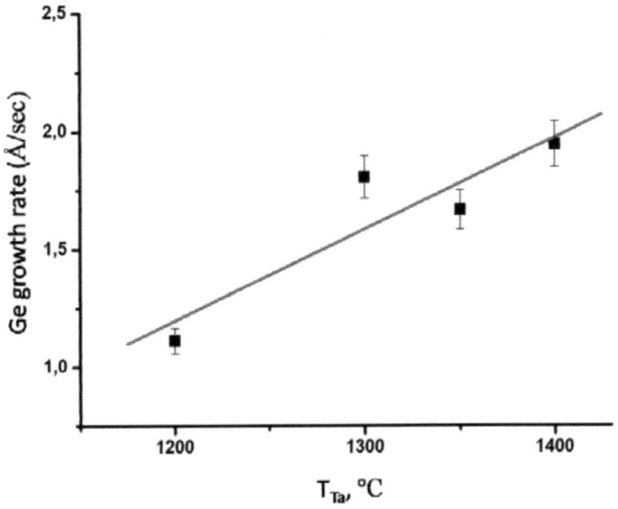

Figure 2. The Ge growth rate (Å/sec) as a function of the Ta filament temperature.

The Reflection Electron Diffraction pattern of a Ge film on Si(100) demonstrated a typical single crystal diffraction pattern with Kikuchi lines expressed clearly (Figure 3) that points to the single crystal structure of the Ge film.

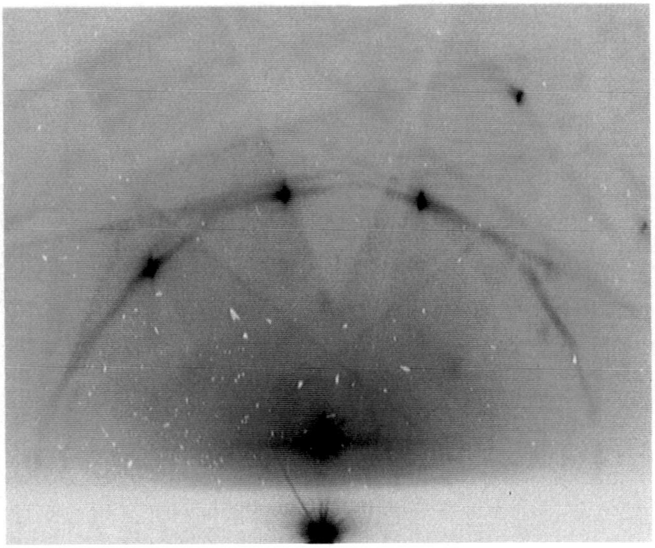

Figure 3. Reflection Electron Diffraction pattern of a Ge film on Si(100).

The dependence of the full width at half maximum (FWHM) of the Ge(004) peak in the X-ray diffraction curves of the Ge/Si(100) films on the Ta filament temperature is shown in Figure 4. It should be noted that the growth times for the Ge layers were the same for all samples, therefore, the higher the Ta filament temperature, the thicker the Ge layer.

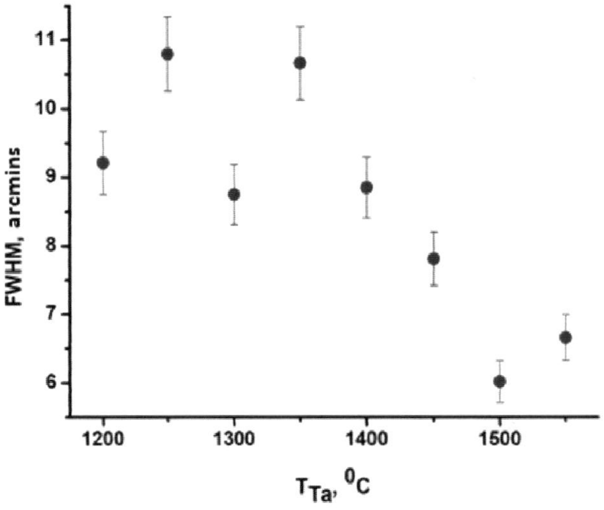

Figure 4. The dependence of the FWHM of Ge(004) peak in the X-ray diffraction curves of the Ge layers on the Ta filament temperature.

The overall surface morphology pattern of the Ge/Si(100) layers surface was found to be a step(terrace)-wise pointing to two-dimensional layer-to-layer growth mode. The averaged root-mean-square (RMS) surface roughness of the Ge layers over a scan area of 2.5 mkm^2 was ≈ 0.37 nm that satisfies well the requirements of modern CMOS technology.

Raman spectroscopy (Figure 5) revealed a red shift of the Ge peak form the Ge/Si(100) epitaxial films (302.02 cm^{-2}) relative to the bulk Ge reference (300.7 cm^{-7}) indicating the tensile strain of the Ge layers.

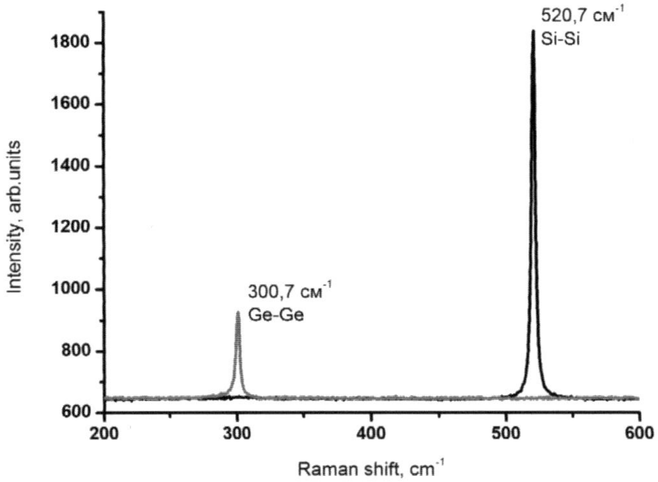

Figure 5. A Raman scattering spectrum of a Ge epitaxial film on Si(100) substrate.

The SIMS impurity depth profiles in the Ge/Si(100) films shown in Figure 6 demonstrate clearly that the carbon (12 C) and oxygen (16 O) concentrations in the epitaxial Ge film is lower than those in the Si substrate. A considerably long decay tail of the Ge concentrations profile into the Si substrate could be attributed to the knock-in effect.

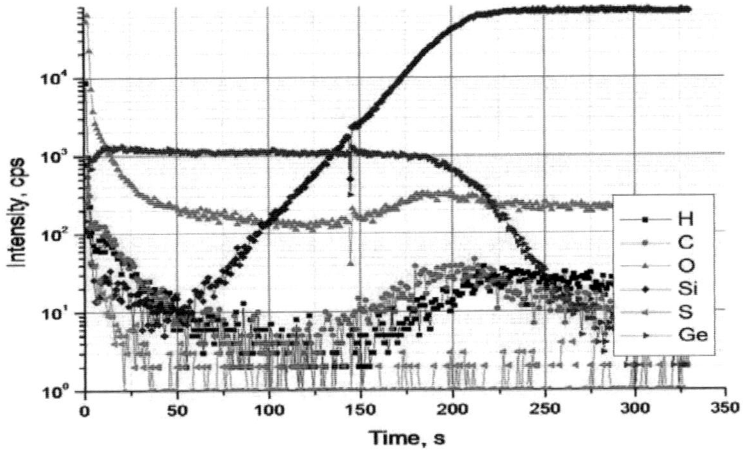

Figure 6. SIMS impurity profiles for carbon (C), oxygen (O), hydrogen (H), silicon (Si) and germanium (Ge) in a Ge/Si(100) epitaxial film.

The treading dislocation density (TDD) in the Ge/Si(100) layers was ~ 10^5 cm^{-2}. This value is among the lowest ones reported in the literature [6].

The Hall measurements performed on the samples grown on low conductive substrates revealed the Ge film material conductivity to be the p-type one. The hole concentration and mobility at room temperature were $(1 \div 3)\cdot10^{17}$ cm^{-3} and $\approx$ 900 cm^2/V·sec, respectively that is close to the respective values for bulk Ge [7].

4. Conclusion

The results of the present studies demonstrate the capabilities of Hot Wire Chemical Vapor Deposition in low-temperature growing thick (up to 3 μm) single crystal epitaxial Ge films on Si(100) substrates featured by low threading dislocation density, low surface roughness, and good electrical parameters. Such Ge/Si(100) films are promising for various device application in microelectronics and photonics, particularly, for the infrared photodetectors integrated into the Si-based CMOS ICs.

Acknowledgements

The authours greatfully acknowledge the financial support from Russian Foundation for Basic Research (RFBR), project No. 14-02-31370_mol_a.

References

[1] Pavesi L and Lockwood D J 2004 *Silicon Photonics* Springer
[2] Colace L and Assanto G 2009 *IEEE Photonics J.* **1** 69
[3] Michel J, Liu J and Kimerling L C 2010 *Nat. Photonics* **4** 527
[4] Masini C, Calace L, Assanto G, Luan Hsin-Chiao and Kimerling L C 2001 *IEEE Trans. Electron Dev.* **48** 1092
[5] Mukherjee C, Seitz H and Schroder B 2001 *Appl. Phys. Lett.* **78** 3457
[6] Wang J and Lee S 2011 *Sensors* **11** 696
[7] Golikova O A, Moizhez B Ya and Stilbans L S 1962 *Sov. Phys. Solid State* **3** 2259

Journal of Physics: Conference Series **541** (2014) 012027 doi:10.1088/1742-6596/541/1/012027

Transparent conductor based on aluminum nanomesh

B Kazarkin, A S Mohammed, A Stsiapanau, S Zhuk, Y Satskevich and A Smirnov
Belorussian State University of Informatics and Radioelectronics (BSUIR), Minsk, 220013, Belarus

E-mail: kazarkin_boris@mail.ru

Abstract. We report a transparent conductor based on Al nanomesh, which was fabricated through Al anodization and etching processes. The Al anodization was performed at low temperature condition to slow down the anodization rate to achieve the well-controlled thickness of an Al nanomesh. By careful controlling of the anodization process, we can fabricate Al nanomesh transparent conductors with different sheet resistance and optical transparency in the visible spectrum range. We shall show that Al nanomesh transparent conductor is a strong contender for a transparent conductor dominated by ITO.

1.Introduction

The main interest in considering of porous aluminum oxide is the technology of its manufacturing. Electrochemical anodizing of Al results in a relatively ordered structure of aluminum oxide. This technology doesn't require photolithography processing, therefore it's could be quite cheap [1-2].

To obtain nanomesh aluminum structure it's necessary to use features of growth of aluminum oxide. This structure is optically transparent and electrically conductive. These parameters depend on geometrical structure of porous alumina films [3].

In addition to good optical transparency and almost metallic conductivity, ultrathin aluminum nanomesh film is flexible and durable. In order to obtain an aluminum nanomesh with various pores size, it's necessary to change anodizing voltages, content and concentration of an electrolyte. It allows one to change the basic electro optical properties, such as optical transparency and electrical conductivity.

Aluminum nanomesh films can be used also as a substitute for ITO in light emitting devices, solar cells and other optoelectronic applications. It is possible also to fabricate a high resistance transparent ultrathin film if needed.

Journal of Physics: Conference Series **541** (2014) 012027 doi:10.1088/1742-6596/541/1/012027

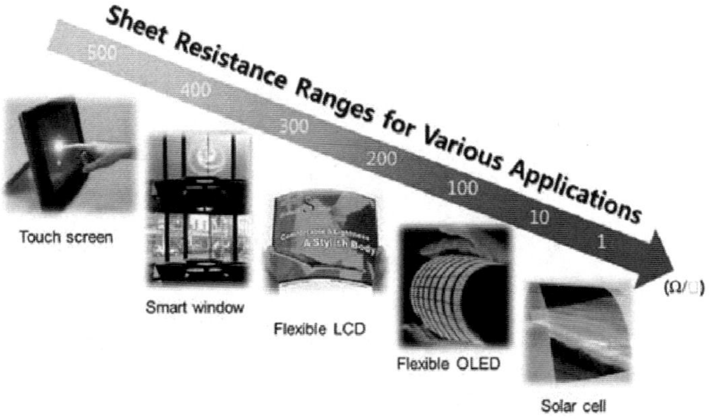

	Sheet resistance	Optical transmission	Ease of customisation	Haze	Ease of pattering	Thickness	Stability	Flexibility	Reflection	Low cost
ITO-on-glass	3	3	3	3	3	1	4	1	1	1
ITO-on-pet	3	3	3	3	3	1	4	2	1	1
Silver Nanowires	4	4	3	1	3	4	1	5	4	3
Graphene	1	5	2	5	3	5	2	5	5	1
Carbon Nanotubes	3	4	3	5	3	3	5	5	5	2
PDOT:PSS	1	3	3	1	4	2	1	5	3	5
Micro Fine Wire	4	1	5	2	5	1	2	5	3	5
Metal Mesh (emboss)	5	3	1	2	5	5	4	4	3	5
Metal Mesh (direct print)	5	1	1	2	5	2	2	4	3	5
Metal Mesh(etching)	5	3	2	2	5	2	2	4	3	5
Other nanotechnology	5	4	5	2	5	4	3	5	2	3

Excellent	5
Good	4
Moderate	3
Moderate-to-Poor	2
Poor	1

2.Methodology

It is possible to achieve many useful structures by anodizing the bi-layer structures, where top porous anodic oxide is used as a template for anodizing or etching the bottom layer. Masking makes easier the pore nucleation process in the bottom layer, therefore more regular porous structures in less aggressive electrolytes can be formed in the bottom layer.

In this paper we will describe how to get a transparent conductive aluminum nanomesh film on a glass substrate using low temperature non lithography low cost anodizing technology. We propose to change anodizing parameters in the final stage of the process, which allows to open transparent windows and leave a nanomesh from aluminum residues.

The technology for the fabrication of aluminum nanomesh films with needed parametes is presented in Fig. 1

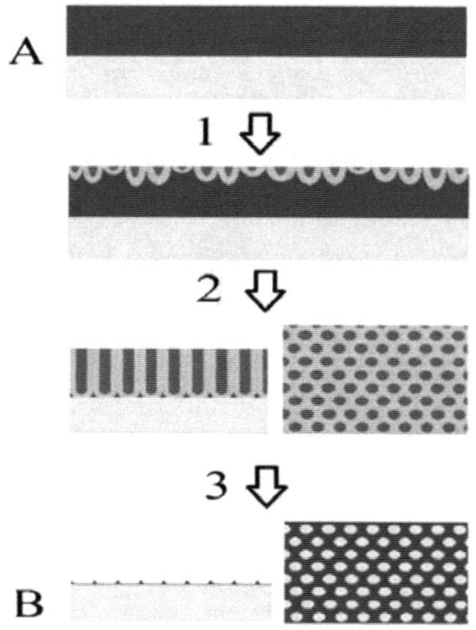

Figure 1. Al nanomesh manufacturing process: A - a glass substrate with initial aluminum layer; 1- start of the anodizing process; 2- stop anodizing; 3-etching of porous Al_2O_3; B- aluminum nanomesh on a glass substrate.

3.Results

In our experiments we used 1x1 cm^2 glass substrates coved with magnetron sputtered aluminum film of 200 nm thick. Aluminum nanomesh was obtained by anodic oxidation of aluminum in anodizing cell (Fig. 2) at a temperature of 10^0C in 2% solution of phosphoric acid at anodizing voltage of 100 V. Chromic acid solution (potassium dichromate) was used to etch of porous aluminum oxide. Images of obtained samples are shown in Fig. 3.

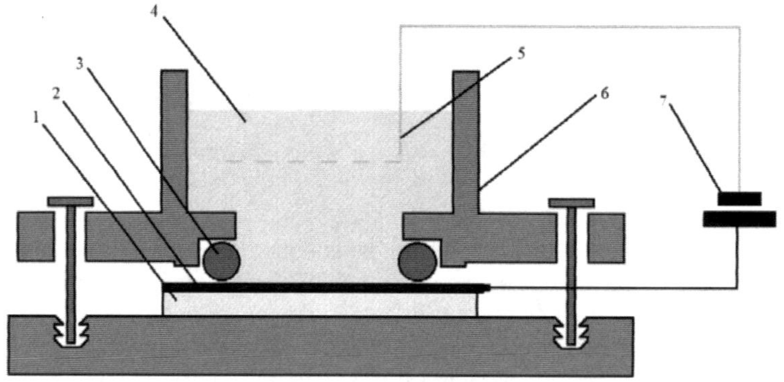

Figure 2. Electrochemical cell: 1- glass wafer; 2- deposited aluminum; 3- O-ring; 4- electrolyte;5- platinum electrode with platinum net;6- cell body;7 – power source.

Journal of Physics: Conference Series **541** (2014) 012027 doi:10.1088/1742-6596/541/1/012027

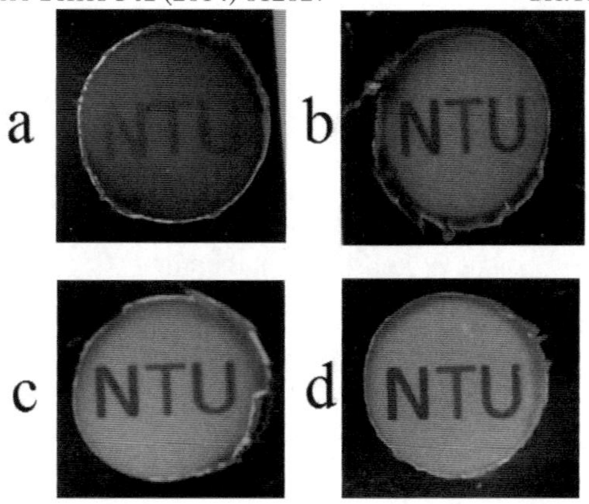

Figure 3. Obtained samples aluminum nanomesh on glass wafers: a- Rs=3 Ohm/sq; b- Rs=16 Ohm/sq; c- Rs=1 KOhm/sq; d- Rs=17 KOhm/sq.

Morphology of obtained nanomesh structures was obtained with SEM SUPRA-55WDS(Fig.4-5).

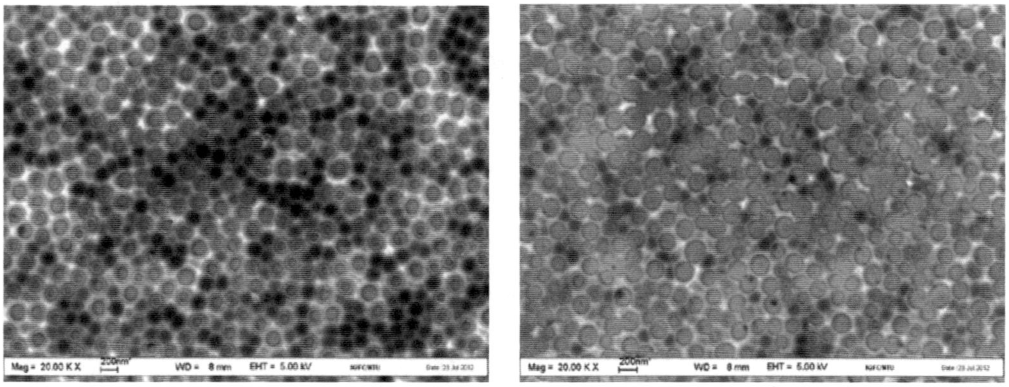

a b

Figure 4. SEM images of nanomesh: a - Rs=3 $\Omega/\square$; transparency=17%, b - Rs=16 $\Omega/\square$; transparency=39%.

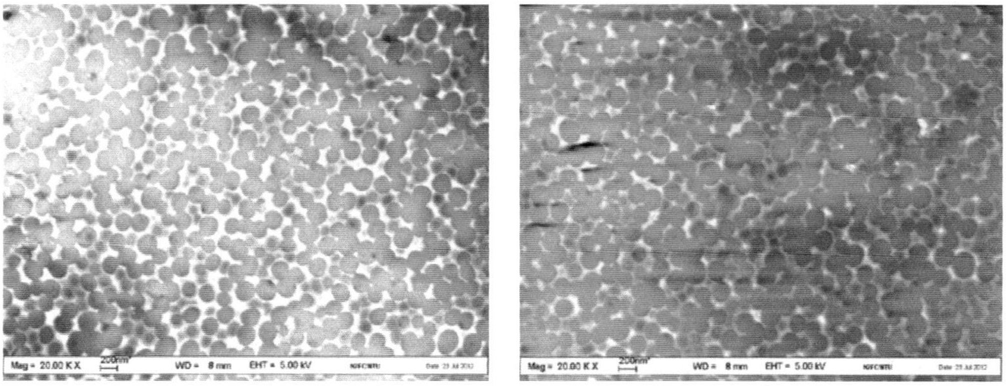

c d

Figure 5. SEM images of nanomesh: c - Rs=1 K $\Omega/\square$; transparency=57%, d - Rs=17 K $\Omega/\square$; transparency=63%.

Through the method of anodic oxidation of aluminum, we were able to obtain an aluminum nanomesh with different values of transparency (from 17% to 63%) and sheet resistance (from 3 *Ω*/□ to 17K*Ω*/□). SEM-images (Fig. 1 a-d) clearly show the structure of the obtained nanomeshes. The sizes of the porous (150-200 nm) and inter porous distances (200-250 nm) are almost equel.

The spectral characteristics of the samples are shown in the Figure 3.

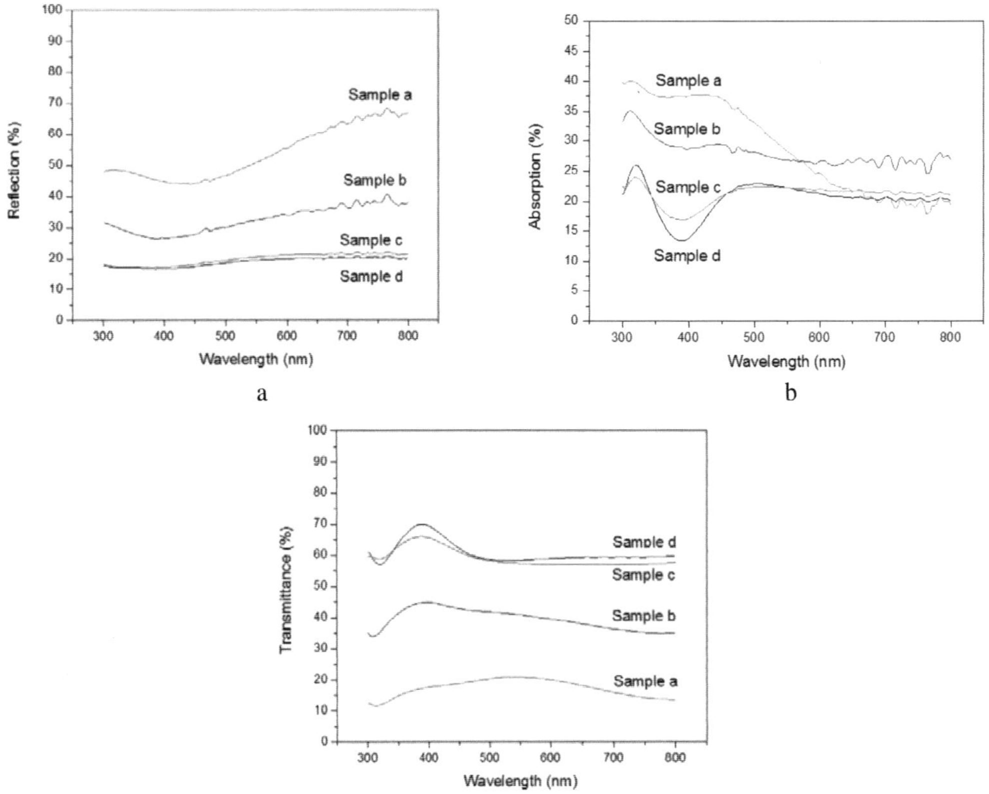

c

Figure 3. Spectra for obtained samples: a – reflection spectra, b - absorption spectra, c - transmittance spectra. Sample a- Rs=3 *Ω*/□; sample b- Rs=16 *Ω*/□; sample c- Rs=1 K *Ω*/□; sample d - Rs=17 K*Ω*/□.

4. Conclusions

Using the features of the porous alumina growth allowed us to obtain a cheap ultra-thin material with good optical properties and conductivity of a metal. Aluminum nanomeshes are cheaper than ITO, and it gives the new advantages in manufacturing areas. We believe that the unique structure, optical and electrical properties of nanomesh films can be used in various research objectives and production purposes.

References

[1] Smirnov A, Jaguiro P, Stsiapanau A, Martinovich A, Maksimov A, Tarasevich S, Lapanik V, Kwok H S 2009 Electrochemical fabrication of alignment and multifunctional nanostructured layers for LCD *Proc. 29th Int. Display Research Conf.* 548-550

[2] Smirnov A, Stsiapanau A, Mohammed A, Mukha E, Kwok H S, Murauski A 2011 Combined nanostructured layers for display applications *Proc. SID Symposium Display Week-2011* 1385-1387

[3] Jaguiro P, Stsiapanau A, Hubarevich A, Mukha Y, Smirnov A 2010 Self-organized nanostructured anodic oxides for display applications *Semiconductor Physics, Quantum Electronics & Optoelectronics* **13** 305-308

[4] Du Q G, Sathiyamoorthy K, Zhang L P, Demir H V, Kam C H, Sun X W 2012 A two-dimensional nanopatterned thin metallic transparent conductor with high transparency from the ultraviolet to the infrared *Appl. Phys. Lett.* **101** 181112

http://www.idtechex.com/research/reports/transparent-conductive-films-tcf-2014-2024-forecasts-markets-technologies-000366.asp

Journal of Physics: Conference Series **541** (2014) 012028 doi:10.1088/1742-6596/541/1/012028

Drift mobility in thin organic layers: joint Monte-Carlo and analytic modeling

V R Nikitenko, V M Sukharev, N A Sannikova
National Research Nuclear University MEPhI (Moscow Engineering Physics Institute), Moscow, Russia

E-mail: VMSukharev@mephi.ru

Abstract. Hopping transport of charge carriers in thin (up to 100 nm) organic layers, which are suitable for organic light-emitting diodes and solar cells, is modeled in the framework of Gaussian disorder model. Monte-Carlo simulations yield considerable decrease of drift mobility with thickness of the layer, which is in good agreement with the developed analytic model.

1. Introduction

Organic electronics devices (light-emitting diodes, photovoltaic elements, etc.) consist of thin (in the range of 10 - 100 nm) active layers. Predictive modelling of operation of such devices unavoidably needs correct characterization of charge transport in thin organic layers. Charge transport, being of hopping nature, can be described in the framework of Gaussian disorder model (GDM) [1]. Values of mobility, which have been obtained from experiments for thicker layers (> 1 micrometre), are often used for modelling the light-emitting diodes. However, one has to expect, that mobility is size dependent on a spatial scale of tens and even hundreds nanometers (irrespective to non-equilibrium or quasi-equilibrium regime of charge generation), in accordance with recent Monte-Carlo (MC) simulations [2,3]. The reason is that percolation paths, which are interrupted in thick films, determine conductivity in thin disordered films, providing that a thickness L is less or comparable with a scale of percolation network [4]. In other words, a carrier, passing through a thin film, has small probability to meet rather deep states, which control mobility. In this work this statement is supported by joint MC and analytic modeling.

2. Monte-Carlo procedure and analytic model.

In MC simulations, we use a conventional model, where hopping centres (localized states) are located at sites of simple cubic lattice with a lattice constant a_0. Number of sites along any transversal direction exceeds 10 times a number of sites along the electric field direction. Energies of sites are distributed randomly according to Gaussian distribution $g(E)$ with the variance σ:

$$g(E)=\left(1/\sqrt{2\pi\sigma^2}\right)\exp\left[-E^2/2\sigma^2\right]. \quad (1)$$

Probability of hopping from site i to site j is $P_{ij} = \nu_{ij}/\nu_i$, $\nu_i = \sum \nu_{il}$ $(i\neq l)$. Due to exponential decreasing of hopping rate with increasing of distance r, we take into account only those sites whose distance from point i is less than $3a_0$ along any direction.

The final state j is determined on each step i by the condition, that a random number $0 \le x_i \le 1$ is between P_{ij-1} and P_{ij} ($P_{i0} = 0$). Waiting time before jump, $t_i = y_i/v_i$, is defined by another random number y_i, according to exponential distribution $P_{ii} = \exp(-v_i t)$. Hopping rates are determined by usual Miller-Abraham's expression:

$$v_{ij} = \omega_0 \exp\left[-2\gamma r_{ij} - \left(\Delta E_{ij} + \left|\Delta E_{ij}\right|\right)/2kT\right], \tag{2}$$

where ω_0 is a frequency factor, γ is inverse localization radius, r_{ij} is a distance between sites i and j, $\Delta E_{ij} = E_i - E_j - eFr_{ij}\cos\theta$, T is absolute temperature, k is Boltzmann constant, and θ is angle of directions of electric field, **F**, and of a jump, $\mathbf{r_{ij}}$. The value $2\gamma a_0 = 10$ is assumed. Quasi-equilibrium initial distribution is considered in order to omit effects of dispersive transport, $g_{occ}(E) \propto g(E)\exp(-E/kT)$. An event (i.e. random walk of a carrier across the volume) starts with an electron being born in the l-th layer, where the number l satisfies to the condition: $leFa_0 \gg 1$, hence visit of a carrier to the left boundary is improbable. It allows analysing the mobility, depending on the medium properties, without the influence of the electrode (dummy injecting electrode). If the electron appears in the nearest to right electrode layer, event is over. Drift mobility is calculated from the average drift velocity, i.e. from an average (over 10000 events) inverse transit time t_{tr},

$$\mu = (L/F)\langle 1/t_{tr}\rangle \tag{3}$$

The basic idea of the analytic model is the following: probability to meet a rather deep state (releasing from this state determines drift mobility) decreases along with the number of states N_0, visited by a carrier. Average (over the ensemble of drifting carriers) value $\langle N_0\rangle$, obviously, decreases along with decreasing of film thickness or with increasing of field strength F. One can expect increase of probability to meet a deep state of a given energy, hence decrease of drift mobility, along with increasing of layer thickness. The simplest approach to quantify this dependence of mobility is to "cut off" deep energies from the Gaussian density of states (DOS) (1), hence to introduce an effective DOS $g_{eff}(E)$:

$$g_{eff}(E) = g(E)\eta(E - E_*), \tag{4}$$

where $\eta(E - E_*)$ is the step function. The average number $\langle N\rangle$ of carrier's captures by states with energies $E < E_*$, is less than unity, consequently such capture is improbable:

$$\langle N\rangle = \langle N_0\rangle \varphi(E_*) = 1, \tag{5}$$

where $\varphi(E_*)$ is a probability of a carrier to be captured after the hop to a state of energy $E < E_*$. Obviously, the value of $\langle N_0\rangle$ increases along with the layer thickness L at a given field strength, while $\varphi(E_*)$ decreases along with decreasing of the energy E_*, hence this energy decreases along with increasing L. The magnitude of mobility is determined by the states of energies around $-\sigma^2/kT$ [1]. Consequently, mobility approaches to its infinite-medium limit under the condition $E_*(L) < -\sigma^2/kT$.

The model for calculations of mobility rests upon transport level concept [5], which has been modified recently [6], introducing a disorder-dependent percolation factor $B>1$ on the base of MC simulations of a random walk of a carrier around deep initial state. The concept of transport level (or transport energy) is known to be a useful approximation for analytic modeling, because it can greatly simplify the description of hopping transport by its reduction to the formalism of multiple trapping model. Real and formal (effective) transport levels, E_{trans} and E_{tr}, and typical hopping distance, a, are defined by the following equations [6]:

$$\frac{4\pi}{3} N_t \int_{E_{trans}}^{E_{tr}} dE' g(E')\left[(E_{tr} - E')/2\gamma kT\right]^3 = B/2, \tag{6}$$

$$\frac{4\pi}{3} N_t \int_{-\infty}^{E_{tr}} dE' g(E')\left[(E_{tr} - E')/2\gamma kT\right]^3 = B, \tag{7}$$

$$2\gamma a + E_{trans}/kT = E_{tr}/kT\ , \tag{8}$$

where N_t is a spatial density of localized states. Integration in Eq. (7) performs over E-r region around deep initial state of energy $E \to -\infty$, which includes $B>1$ neighbour states with transition rate $\nu \geq \omega_0 \exp\left[-\left(E_{tr} - E\right)/kT\right]$. Eq. (6) means that upward and downward jumps from a state of energy E_{trans} occur with equal probabilities. Percolation factor B was determined from MC simulations [6]:

$$B = 2.2 + 0.2(\sigma / kT - 1.5),\ \sqrt{2} \leq \sigma / kT \leq 4\ . \tag{9}$$

Mobility is calculated in analogy with multiple trapping model [6],

$$\mu = \mu_C \nu_0 \tau_0 \Big/ \int_{E_*(L,F)}^{E_{trans}} dEg\left(E\right)\exp\left[\left(E_{trans} - E\right)/kT\right], \tag{10}$$

by the use of effective density of states, $g_{eff}\left(E\right)$, see Eq. (4). In Eq. (9) $\mu_C = \omega_0 \exp\left(-2\gamma a\right)\left(e/kT\right)a^2/6$ is the characteristic mobility in "conducting" states (near transport level), $\nu_0 \tau_0 \approx kT/\sigma$ [6]. The energy E_* is defined by Eq. (5), where the probability $\varphi(E_*)$ is also defined on the base of transport level. Transport level does not depend on E_*, as long as E_*)<<E_{trans}, due to sharp decreasing of $g\left(E\right)$, $E \to -\infty$.

Results and discussion

Figure 1 shows, that the analytical model yields good agreement with MC simulations, up to field strength $F = 5{\cdot}10^5$ *V/cm*. Mobility is normalized by the value $\mu_0 = \omega_0 \exp\left(-2\gamma a_0\right)\left(e/kT\right)a_0{}^2$. Mobility decreases with increasing thickness as a power law [2], approaching to the infinite-media limit μ_∞,

$$\mu = \mu_\infty \left[1 + \left(L/L_0\right)^{-\beta}\right],\ L \geq 20\,\text{nm}. \tag{11}$$

Values of constants μ_∞, L_0 and β are defined by fitting of MC-data to Eq. (11). The power-law exponent β increases slightly along with increasing of disorder and decreases along with increasing of field strength. It increases from 1.0 to 1.3 and from 0.6 to 1.0 along with the increase of σ/*kT* from 3 to 5 at $F = 1 \cdot 10^5$ and $F = 5 \cdot 10^5$ *V/cm*, respectively.

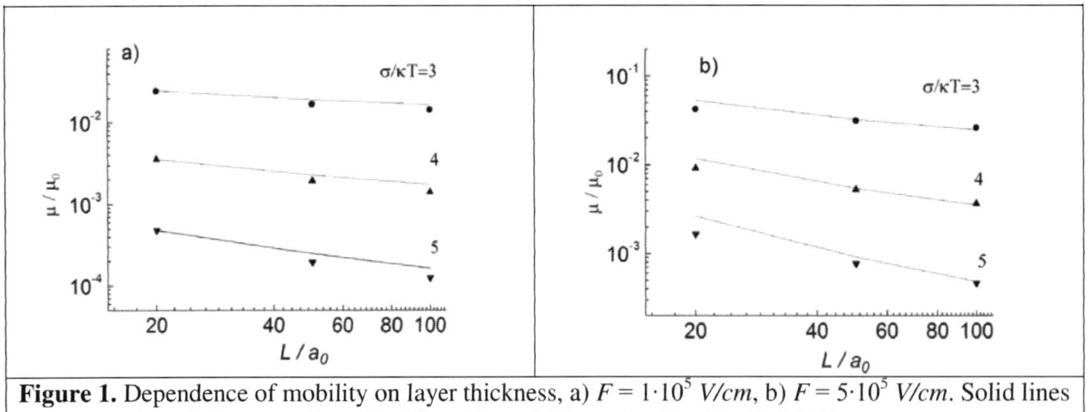

Figure 1. Dependence of mobility on layer thickness, a) $F = 1{\cdot}10^5$ *V/cm*, b) $F = 5{\cdot}10^5$ *V/cm*. Solid lines – analytical modelling, see Eqs. (5)-(10), points – MC results, Eq. (3). *T=295K*.

Remarkably, the analytic model is able to fit MC data for mobility in thin films for a broad range of disorder parameter σ/*kT* and electric field strength, by the use the same disorder-dependent (but field-independent) percolation factor *B*, see Eq. (9), which has been used successfully [6] in modeling of low-field mobility in infinite media. One has to consider, however, field dependence of transport level to achieve quantitative accuracy at high-field and high-disorder limit.

Figure 2 shows considerable impact of film thickness on disorder dependence of drift mobility. Temperature dependence in thick (> 1 micrometer) films, referred here as infinite media, describes ubiquitously as follows [1]:

$$\mu_\infty(T)\propto\exp\left[-A(F)(\sigma/kT)^2\right],\ A(F)=C_\infty-\sqrt{F/F_\infty}, \qquad (12)$$

where $C_\infty\approx 0.44$ and $F_\infty\approx 1.2\cdot 10^7\ V/cm$ at $2\gamma a_0$ = 10. Figure 2 shows progressive deviations from the law (11) with increasing disorder and decreasing *L*. Fit of the low-disorder pattern by eq. (3), see dashed lines in the Figure 2, yields considerable decrease of the value of *A* along with the decrease of thickness, namely *A* = 0.37, 0.33, 0.31 and 0.27 at *L* = ∞, 100, 50 and 20 nm, respectively. The first value of *A* is in excellent agreement with Eq. (11) at $F=1\cdot 10^5\ V/cm$.

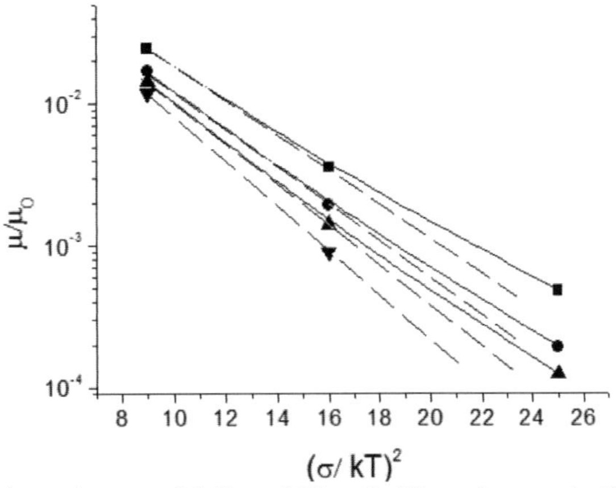

Figure 2. Temperature dependences of drift mobility: *L*=20 nm (squares), 50 nm (circles), 100 nm (up triangles), infinity (down triangles), see μ_∞ in Eq. (11). Solid lines – interpolation of MC data, dashed line – low-disorder fit in accord with Eq. (12). $F=1\cdot 10^5$ *V/cm, T=295K.*

Size dependence of mobility in thin organic layers has to be accounted for predictive modeling of charge transport in organic light-emitting diodes. As a first approximation, one can use thickness-dependent spatially-uniform mobility, see Eq. (11).

The work was supported by Russian Ministry of Education and Science in the frames of Competitiveness Growth Program of National Research Nuclear University MEPhI, Agreement 02.A03.21.0005.

References

[1] Bässler H, 1993 *Phys. Stat. Sol. (b).* **175** 15
[2] Korolev N A, Nikitenko V R, and Ivanov D V 2011 *Semiconductors* **45** 230
[3] S Raj Mohan, Manoranjan P Singh, M P Joshi 2010 *Organic Electronics* **11** 1642
[4] Shklovskii B I and Efros A L 1984 *Electronic Properties of Doped Semiconductors* (Heidelberg: Springer)
[5] Arkhipov V I, Emelianova E V and Adriaenssens G I 2001 *Phys. Rev. B* **64** 125125
[6] Nikitenko V R, Strikhanov M N 2014 *J. Appl. Phys.* **115** 073704

Photoelectrical characteristics of Ga_2O_3-GaAs structures

V.V. Vishnikina, V. M. Kalygina, Yu. S. Petrova, I. A. Prudaev

National Research Tomsk State University, 36 Lenin Prospekt, Tomsk, Russia

E-mail: vishnikina.vera@rambler.ru

Abstract. The influence of thermal annealing and exposure in oxygen plasma on photoelectrical characteristics of Ga_2O_3-GaAs structures in visible and UV-region is investigated. Thermal annealing enhances the transparence of Ga_2O_3 films in visible range and leads to the appearance of photocurrent in Ga_2O_3-GaAs-structures. Ga_2O_3 films absorb UV-radiation beginning with 240 nm and lower.

1. Introduction

Thin Ga_2O_3 films have a wide variety of applications including the fabrication of gas sensors, solar-blind photodetectors, phosphors and other devices. Also such films are applied as transparent conductive electrodes and antireflection coatings. It is known that films' properties are determined by conditions of obtaining and processing. So the purpose of this work is investigation of how high-temperature annealing and treatment in oxygen plasma influence on photoelectrical characteristics of MOS-structures. Ga_2O_3 can be found in five different phases which differ from each other by crystalline structure. β-phase is the most interesting. It is the most chemically and thermally stable crystalline modification. It is known from literature that Ga_2O_3 films are transparent enough to visible light. And such films can be used as transparent conductive electrodes in optoelectronic devices. It is interesting to investigate how high-temperature annealing and O_2 exposure influence on photoelectric phenomena in such structures.

2. Experiment details

All samples were fabricated on GaAs epitaxial layers (N_d = 8.9·1015 cm^{-3}) grown on n-type GaAs substrates ($n_0 \approx 10^{18}$ cm^{-3}). In the first stage standard chemical purification is conducted. Ga_2O_3 films were grown by photoelectrochemical oxidation method (or anodic oxidation). It is the most sparing method for obtaining thin films. As the result oxide of gallium and oxide of arsenic are formed. In order to remove an easily volatile oxide of arsenic the annealing in atmosphere of hydrogen is conducted (300 °C, 10 min). In order to obtain β-phase the annealing in argon at 900 °C (30 min) is carried out. After that the samples were exposed to the oxygen plasma (20, 30, 50 min, 90 °C). The phase composition of our films was determined by X-ray diffraction analysis (XRD). The surface morphology was investigated by an atomic force microscope (AFM).

3. Results and discussion

It is important to notice that as-grown films were amorphous. According to the results of XRD (**Figure 1**) it can be assumed that introduction of oxide atoms in Ga_2O_3 lattice leads to the formation of β-phase crystallites with different crystallographic orientation [1, 2].

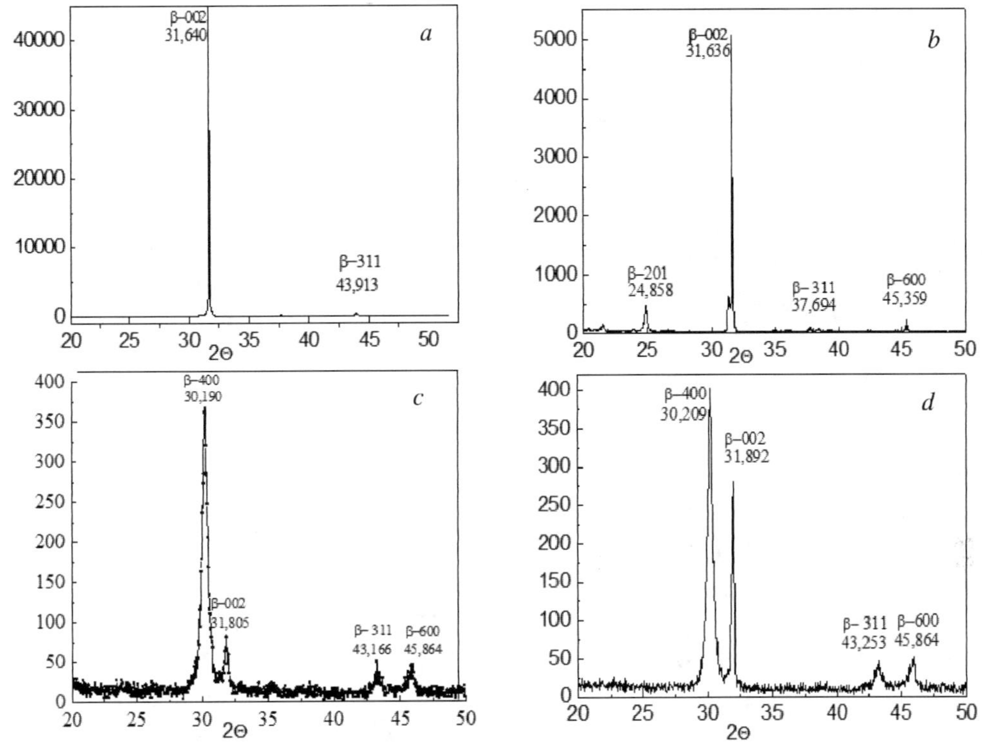

Figure 1. Spectra of X-ray diffraction: *a*) – treatment in H_2 and 20' in O_2; *b*) – treatment in O_2 20', annealing in Ar (900 °C, 30'); *c*) – treatment in O_2 50' , annealing in Ar (900 °C, 30'); *d*) – treatment in O_2 20' after annealing in Ar (900 °C, 30')

From the AFM-analysis it is evident that annealing led to a change in the microstructure from an amorphous to a polycrystalline phase (**Figure 2-4**).

The reduction of charge carriers concentration takes place in GaAs after the growth of Ga_2O_3 film. Such effect depends on the duration of O_2 exposure (**Figure 5**).

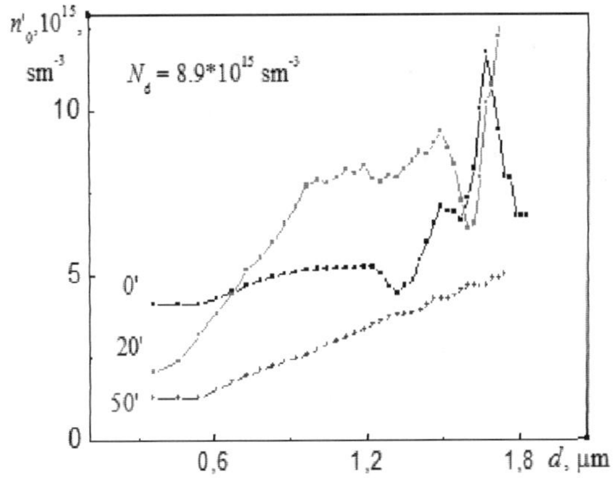

Figure 5. The concentration of charge carriers in n-GaAs after high temperature annealing at different duration of O_2 exposure

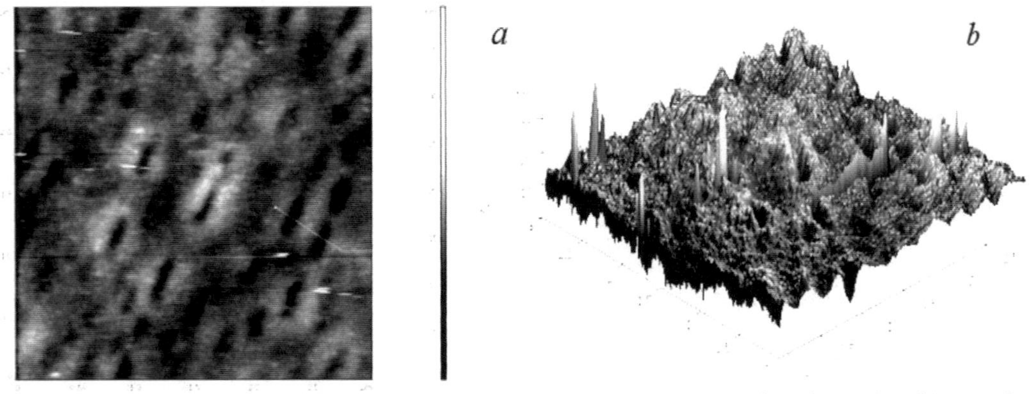

Figure 2. Two-dimensional image (*a*) and three-dimensional image (*b*) of Ga_2O_3 film surface (without treatment in oxygen plasma and annealing in Ar)

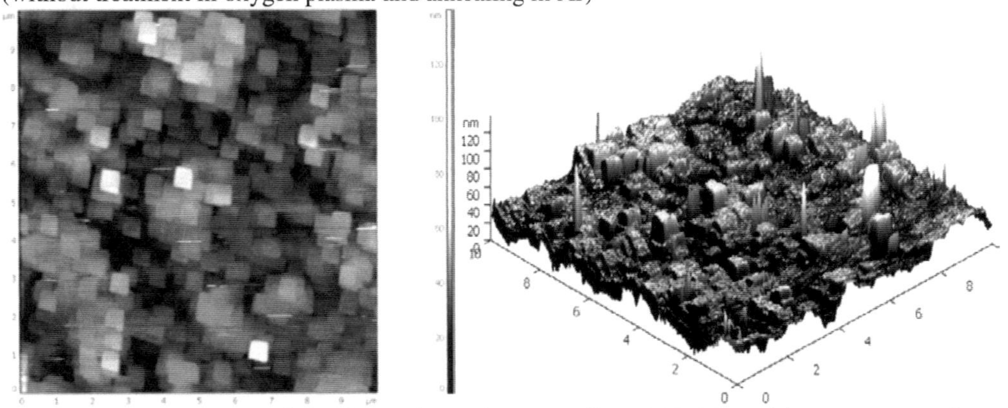

Figure 3. AFM picture of Ga_2O_3 surface after annealing (900 °C), without treatment in oxygen plasma - two-dimensional (*a*) and three-dimensional (*b*)

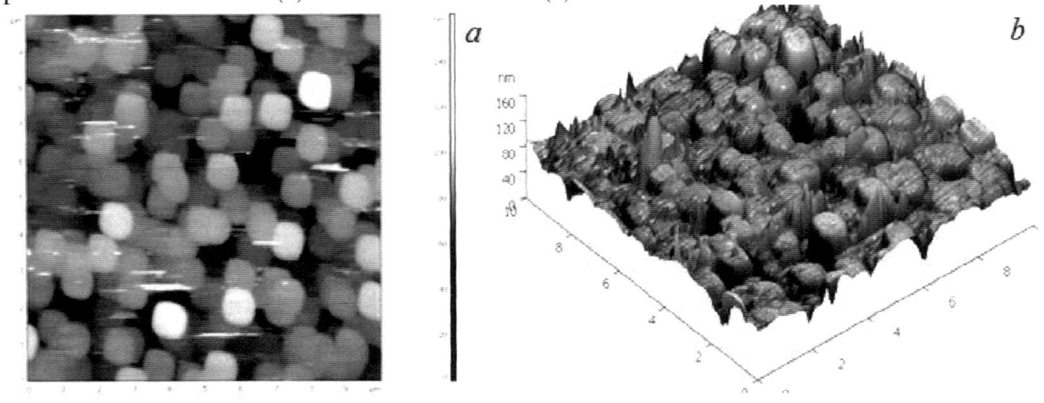

Figure 4. AFM picture of Ga_2O_3 surface after annealing (900 °C), treatment in O_2 before annealing (50 min.) – two-dimensional (*a*) and three-dimensional (*b*)

Photoresponse is observed for samples after only high-temperature annealing. Photocurrent is explained by the generation of charge carriers in the space charge region of the semiconductor. After 20 minutes of O_2 exposure we obtain the most transparent films for visible light (**Figure 6**).

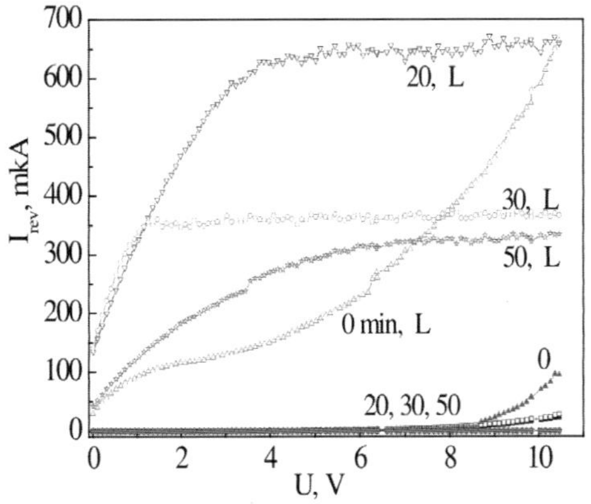

Figure 6. Reverse dark and light current-voltage curves for samples after thermal annealing and O_2 exposure

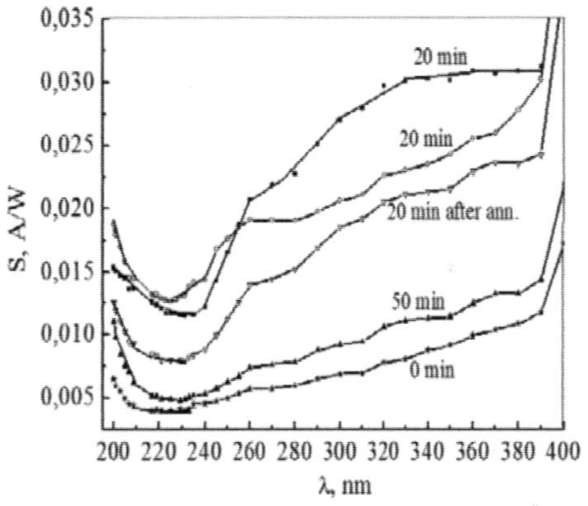

Figure 7. The spectral dependence of the sensitivity of Ga_2O_3-GaAs structures according to the duration of O_2 exposure

Similar with the visible range the influence of UV-radiation on structures without high-temperature annealing isn't observed. Maximum sensitivity is observed for samples after annealing at 900 °C which were exposed to the oxygen plasma during 20 minutes. In our experiment sensitivity increases from 240 nm to 200 nm – in this region of wavelength gallium oxide films are not transparent and absorb the radiation (**Figure 7**).

4. Conclusion

The connection between the duration of oxygen plasma treatment and the emergence of β-phase crystallites with different crystallographic orientation is established. It is shown that after high-temperature annealing the films become transparent in visible range and photocurrent is explained by the generation of charge carriers in the space charge region of the semiconductor. Ga_2O_3 films absorb UV-radiation from 240 nm.

The work was supported by the Ministry of Education and Science of Russian Federation (Project 1975).

References

[1] Kalygina V M, Zarubin A N, Naiden E P, Novikov V A, Petrova Yu S, Tolbanov O P, Tyazhev A V, Yaskevich T M. Semiconductors.2012. 46 (2). P. 267-273.

[2] Kalygina V M, Zarubin A N, Naiden E P, Novikov V A, Petrova Yu S, Skakunov M S, Tolbanov O P, Tyazhev A V, Yaskevich T M. Semiconductors.2011. 45 (8). P. 1097-1102.

Features and benefits of III-N growth by ammonia-MBE and plasma assisted MBE.

A N Alexeev[1], V P Chaly[2], D M Krasovitsky[2], V V Mamaev[1,3,*], S I Petrov[1], V G Sidorov[3]

[1]SemiTEq JSC, Saint-Petersburg, Russia

[2]Svetlana-Rost JSC, Saint-Petersburg, Russia

[3] Saint Petersburg State Polytechnical University, Saint-Petersburg, Russia

*toodayl88@gmail.com

Abstract. Features of III-N heterostructures growth by ammonia and plasma MBE are discussed. Use of ammonia MBE for growth high temperature AlN/AlGaN buffer layer at the initial stage allows to improve the structural quality and reduce the dislocation density in the GaN layers down to $9x10^8$-$1x10^9$ cm^{-2}. However, to produce an effective p-doping, as well as growth of In-containing layers (InGaN with a high content of In and InAlN, lattice matched with GaN) is necessary to use a method that has no limits at low temperatures–plasma-assisted MBE.

[1] Introduction

At present time, an increasing number of researchers choose plasma-assisted MBE as it is simpler in service and also has a number of features, for example, such as possibility of low-temperature growth and absence of hydrogen on the growing surface. These features are very useful for InN and InGaN with high In content as well as InAlN lattice matched to GaN growth and effective p-doping.
Besides, this choice is caused by improvement of plasma sources and development of technology of plasma-assisted MBE in last years that, in particular, allowed to increase considerably growth rate (up to 1 micron/hour). In recent years it was shown that by optimization of growth conditions at the initial stage and during the growth process by plasma-assisted MBE it is possible to improve significantly quality of grown material[1,2], for example, using the technique of migration enhanced epitaxy (MEE).
However unlike to ammonia MBE this method doesn't allow to increase noticeably the growth temperature, and thus improve the quality of the material by increasing the surface mobility of atoms.
In this paper we present the results of using of both types of MBE (plasma-assisted and ammonia), obtained in STE3N MBE System (SemiTEq).

[2] Features and benefits

One of the main problems in manufacturing of GaN-based devices up to date is the lack of low cost, lattice matched substrates to III-nitrides. In spite of using special procedures at the initial growth stage, heteroepitaxy of III-nitrides on mismatched substrates usually yields high dislocation density, which affects device quality and reliability. Typical values are in the range of 10^9-10^{10} cm^{-2} for MBE and 10^8-10^9 cm^{-2} for MOCVD (10^7 cm^{-2} using ELOG).

Moreover, typical growth temperatures in MBE are much lower as compared with MOCVD. It leads to insufficient surface mobility of adatoms, worse coalescence of nucleation blocks on initial growth stage, and, as a result, to high dislocation density.

But on the other hand the MBE method has many advantages compared with MOVPE. Namely it enables in-situ diagnostics of grown layers at the one monolayer level using RHEED, gives sharper heterojunctions, provides a higher purity of the growth chamber and respectively grown material, enables the construction of high vacuum cluster systems using the combination with other high vacuum growth and research chambers, more safe and cost effective due to less consumption of precursors and other. So the task of reducing the dislocation density and improvement of structural quality of the grown material by increasing the growth temperature in MBE technology is very urgent.

MBE using ammonia is characterized by the tendency to increase V/III ratio and growth temperature (typical growth temperatures in ammonia MBE are 100-200C higher than in PAMBE) that allows to grow at higher growth rate (more than 1 μm/hour) and improve structural quality as it was showed by Dr. Webb et al from Institute for Microstructural Sciences (Canada) [3] and Dr. Semond et al from CNRS (France) [4]. However the condition of high vacuum maintenance in MBE doesn't allow to increase considerably ammonia flow (V/III ratio) and so to increase temperature of GaN growth to values comparable with MOCVD (more than 1000^0C). It leads to the fact that the density of dislocations in GaN layers grown by both MBE methods on mismatched substrates differs not strongly and 1.5-2 orders more than in MOCVD.

Most effective way to significant reduction of the dislocation density is the growth of thermally stable AlN buffer layers at extremely high for MBE, comparable to MOCVD, substrate temperature (more than 1000-1100^0C). The use of such buffer layers may allow to drastically reduce dislocation density by the increasing of nucleation blocks coalescence and quickly realize transition to two-dimensional growth mode.

The main advantage of ammonia MBE compared with plasma-assisted MBE is just the possibility of growth of AlN and AlGaN layer with high aluminum content at increased substrate temperature and V/III ratio more than 1.

Growth at such temperatures in the plasma-assisted MBE is very difficult to realize since Al-rich mode is necessary for two-dimensional growth of AlN and desorption of aluminum becomes significant at substrate temperature more than 900^0C.

Realization of this approach required development and creation of the MBE system allow to provide substrate temperature for growth of AlN layers comparable with MOCVD. In this case, the fact that less ammonia flow is required for optimal growth mode of AlN than for GaN and also use of cryopanels with increased area allows to provide vacuum typical for MBE even at extremely high substrate temperature in the gas-load mode. In turn, high vacuum and use in MBE as aluminum precursor of effusion cell with metal aluminum allow to avoid parasitic reactions which occur in the MOCVD reactor between trimethylaluminum and ammonia. Meanwhile, the solution to this problem requires a special effort in growth of AlN and AlGaN layers with high aluminum content by MOCVD. Thus, high temperature ammonia MBE opens the way for the growth of high quality structures for micro and optoelectronics with an active region grown further by ammonia or plasma-assisted MBE in one growth run. In particular this approach is perspective for growth of heterostructures for UV devices and microwave transistors in which high quality layers of AlN and AlGaN with high aluminum content are required.

Using of this approach and such unique, high temperature MBE System allowed our team to develop very special AlN/AlGaN complex buffer layer providing extremely low for MBE dislocation density growing on several types of mismatched substrates (sapphire, SiC, Si etc.).

[3] Results

We demonstrate low-dislocation-density and high-electron-mobility GaN layers grown by ammonia-MBE on c-Al_2O_3 under the ultimate for MBE growth temperature as high as T_S=1150ºC. The buffer layer sequence from c-sapphire substrate involved a 200-nm-thick AlN layer (T_S=1100–1150ºC), AlGaN/AlN superlattice (SL), and $Al_xGa_{1-x}N$ (x=0.1–0.3) transition layers (900–920ºC). The scanning transmission electron microscopy (STEM) study showed gradual decrease of threading dislocation density from (2-4)·10^{10} cm^{-2} in AlN to (4-6)·10^9 cm^{-2} in $Al_{0.1}Ga_{0.9}N$ grown after SL (Fig. 1). Finally, the dislocation density in the top GaN active layer was reduced down to (9-10)·10^8 cm^{-2} that comparable with MOCVD GaN grown on sapphire and several times lower than in conventional NH3-MBE. Detailed study by TEM allowed separate evaluation of the density of screw (~2·10^8 cm^{-2}), edge (~4·10^8 cm^{-2}), and mixed (~4·10^8 cm^{-2}) dislocations in the top GaN layer. The improvement of structural quality resulted in substantial increase in electron mobility which reached the maximum value 600–650 cm^2/V·s in a 1.5-μm-thick GaN top layer lightly doped with silicon up to n=(3–5)·10^{16} cm^{-3}, that corresponds to a good quality MOCVD GaN. The experimental values of mobility and dislocation density are in good agreement with calculations.[5]
Importantly, growth of AlN at 1150^{0}C is difficult to realize in PA-MBE since Al-rich mode is necessary for 2D-growth, while desorption of Al becomes significant at T>900^{0}C.

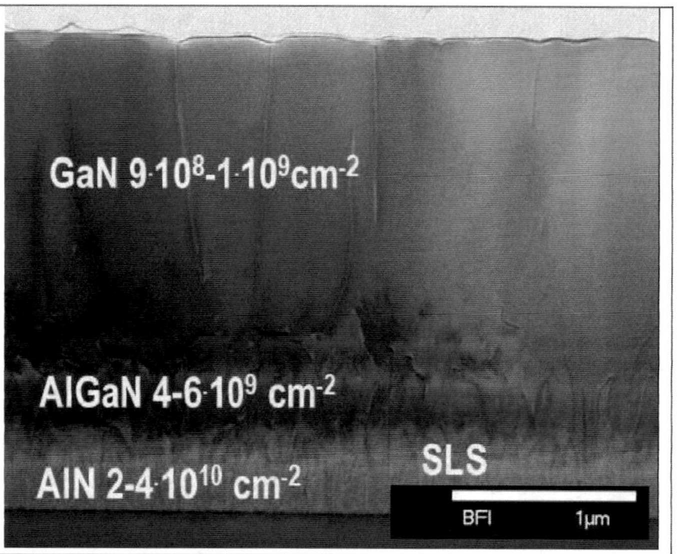

Figure 1. STEM image of AlN/AlGaN/GaN heterostructure..

Employing such a GaN layer in a double heterostructure (DH) with the cap $Al_xGa_{1-x}N$ barrier layer of variable composition (x=0.25-0.4) allows to change the electron sheet density and mobility in a two dimensional electron gas (2DEG) in the range of 1300-1700 cm^2/V·s and (1.0-1.8)×10^{13} cm^{-2}, respectively, providing control of a 2DEG channel sheet resistance in the range of 230-400 Ω/□.
Performance and reliability of nitride HEMT devices can be improved by replacing the barrier AlGaN layer by InAlN lattice matched to GaN. Indium content in this composition is 17 percent. Thus, at this composition stress and piezopolarisation are not present, potentially improving the stability of the heterostructure. Historically, only PAMBE could be used to grow the high-quality, lattice-matched

InAlN layer, which requires an indium content of 17%. The lower temperature limit for ammonia MBE is determined by ammonia cracking efficiency, and this is negligible at temperatures lower than 500 °C. Increase the temperature above this value and typical ammonia flow leads to very low indium content – it is insufficient for growing InAlN lattice-matched to GaN.
However, Speck's group has shown recently that if the ammonia flow is increased to 1000 sccm, growth of InAlN that is lattice-matched to GaN is possible [6]. Motivated by this result, we have tried to grow such structures by PAMBE and ammonia MBE, realizing good results in both cases. It is worth noting that thanks to the special cryopanels and pumping system design, the vacuum levels are almost unchanged when ammonia flow hits 1000 sccm.
Also employing of lattice-matched to GaN barrier layers InAlN in a double heterostructure (DH) allows we have achieved electron sheet density and mobility 1100-1200 $cm^2/V{\cdot}s$ and $(2.3\text{-}2.5)\times10^{13}$ cm^{-2}. InAlN layers were grown both by PA MBE and ammonia MBE. It is interesting to note that due to the special cryopanels vacuum in growth chamber was practically unchanged with increasing ammonia flow up to 1000 sccm.

Thus, both methods have their own advantages, and to obtain effective opto-and microelectronic devices it is necessary to combine both techniques in a single MBE system.

[1] Kaun, S. W., Wong, M. H., Mishra, U. K., & Speck, J. S. (2012). Correlation between threading dislocation density and sheet resistance of AlGaN/AlN/GaN heterostructures grown by plasma-assisted molecular beam epitaxy. *Applied Physics Letters*, *100*(26), 262102. .

[2] Nechaev, D. V., Aseev, P. A., Jmerik, V. N., Brunkov, P. N., Kuznetsova, Y. V., Sitnikova, A. A., ... & Ivanov, S. V. (2013). Control of threading dislocation density at the initial growth stage of AlN on c-sapphire in plasma-assisted MBE. *Journal of Crystal Growth*, *378*, 319-322.

[3] Webb, J. B., Tang, H., Bardwell, J. A., Moisa, S., Peters, C., & MacElwee, T. (2001). Defect reduction in GaN epilayers and HFET structures grown on (0001) sapphire by ammonia MBE. *Journal of crystal growth*, *230*(3), 584-589.

[4] Semond, F., Lorenzini, P., Grandjean, N., & Massies, J. (2001). High-electron-mobility AlGaN/GaN heterostructures grown on Si (111) by molecular-beam epitaxy. *Applied Physics Letters*, *78*(3), 335-337.

[5] Ng, H. M., Doppalapudi, D., Moustakas, T. D., Weimann, N. G., & Eastman, L. F. (1998). The role of dislocation scattering in n-type GaN films. *Applied physics letters*, *73*(6), 821-823.

[6] Wong, M. H., Wu, F., Hurni, C. A., Choi, S., Speck, J. S., & Mishra, U. K. (2012). Molecular beam epitaxy of InAlN lattice-matched to GaN with homogeneous composition using ammonia as nitrogen source. *Applied Physics Letters*, *100*(7), 072107.

Synthesis and characterization of ZnO nanorods obtained by catalyst-free thermal technique

V. I. Pushkariov [(1,2)], **A. L. Nikolaev**[(1,2)], **E.M. Kaidashev**[(1,2)]

[1]Vorovich Mechanics and Applied Mathematics Research Institute, Southern Federal University, Russia
[2]Physics Department, Southern Federal University, Russia

E-mail: velascas@rambler.ru ; andreynicolaev@eurosites.ru ; kaidashev@mail.ru

Abstract. ZnO nanorod arrays have been synthesized on silicon wafers by catalyst free thermal vapor deposition with different temperature values. Vertical-aligned ZnO nanostructures were grown at the quite low temperatures by this method. Modification of the method allowed to place silicon substrates in areas with different ratios of concentrations Ar and O_2 molecules in the vapor phase. Photoluminescence spectroscopy and electron microscopy studies have shown that morphologies and optical properties of nanorods depends on different ratios of concentrations of the molecules in the vapor phase.

1. Introduction

Zinc oxide, is a promising semiconducting and piezoelectric material with direct wide-band gap (3.37 eV at room temperature) and large exciton binding energy (60 meV). During the last years one-dimension nanocrystalline ZnO structures such as nanowire, nanorods, nanobelts and nanocables attract much attention of various research groups, due to their unique physical properties and potential applications in nanoscale devices. Of particular interest are well-oriented arrays of ZnO nanorods in connection with the combination of excellent electronic and optoelectronic properties of each nanorod. These ZnO nanorods arrays can be used as active elements of nanodevices, such as LEDs [1], nanosensors [2], UV lasers [3] photodetectors [4], etc. For practical applications the main objective is to obtain arrays of nanostructures ZnO, with the necessary morphology and physical properties.

There are various methods of producing arrays of nanorods ZnO: hydrothermal method, pulsed laser deposition, chemical vapour deposition and metalorganic chemical vapour deposition, thermal evaporation and etc. One of the promising methods for the growth of ZnO nanorods is the thermal technique of synthesis from Zn vapor, which implemented in this paper. In our work we obtained ZnO nanorod arrays by means of catalyst free thermal evaporation technique under different temperature values and investigated properties of the structures.

2. Experiment

The technique used during the synthesis of ZnO nanorod arrays have been based on thermal evaporation of metallic granule Zn in Ar and O_2 gas mixture, transferring vapor precursors to the substrate and subsequent deposition of ZnO nanorods on the substrate from the vapor phase without the use of any metal particles as catalysts.

ZnO nanorods grew at a quite low temperature without any catalysts on the ZnO (100nm) / (100) silicon substrates 10x10 mm^2 in size. The thin ZnO buffer layer was prepared by pulsed laser deposition technique on the Si substrate. Laser radiation of KrF laser (λ = 248 nm, E = 300 mJ) was focused on the surface of rotating ZnO ceramic. Distance between target and substrate was 50 mm. The repetition frequency of the laser pulses was 10 Hz. The substrate temperature was 500°C. ZnO film was deposited at 2,000 laser pulses, the oxygen pressure was 2×10^{-2} mbar.

It was found the morphology and optical properties of ZnO rods on the substrates are very sensitive to the Zn vapor supersaturation. Our reactor used for the synthesis comprised a horizontal tube furnace, a pumped quartz tube and an argon and oxygen gas supply. In our work, the precursor (Zn granule) and substrates placed in a small quartz tube. In contrast to works [5-7], we use a small quartz tube opened on both sides. This tube is placed in a reaction chamber - a quartz tube 32 mm in diameter, located in a resistive heater. Using this modification could be achieved different ratio values of concentrations of the zinc and oxygen molecules in the vapor phase in different areas of reactor. Pressure value in the experimental setup was controlled by the means of changing pumping rate and flow of Ar buffer gas.

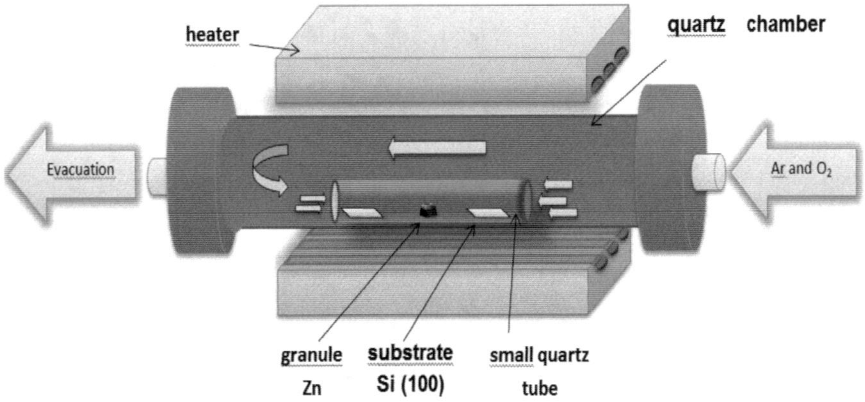

Figure 1. Experimental setup used for the growth of nanostructures.

In the first series of experiments zinc granule (purity 99.9999%) weighing 500 mg was placed in the center of the small quartz tube. Substrate ZnO / (100) Si was placed on a distance of 0.5 cm from the Zn granule between zinc precursor and input aperture of gas mixture. After the quartz chamber was pumped to 2 mbar, Ar gas was flowed at a flow rate of 240 sccm, and the deposition pressure was set 11 mbar. Next the substrate and Zn granule are heated to a synthesis temperature in 35 minutes. When the necessary temperature was achieved, gas also started flow in the reactor at 8 sccm. Well-aligned ZnO nanorod arrays have been grown at the furnace temperature 600°C in 15 minutes.

Scanning electron microscopy (SEM) and photoluminescence (PL) was used to characterize optical and morphological properties of the ZnO nanorod arrays. SEM images and PL spectra of synthesized nanorods shown in Figure 2. SEM and PL researches was carried out on different areas of the substrate to study the effect of the atmosphere supersaturation by Zn vapor on the properties of produced zinc nanorods. The size of crystals depends on distance to the edge of a small quartz tube that shown in scanning electron microscope images. It is found that the ratio values of zinc and oxygen molecules concentrations in the vapor phase influences on size and optical properties of ZnO nanostructures. Average diameter of the rods is about 50 nm. Average height of rods increases with increasing substrate-to-source distance from 200 nm to 800 nm. Nanorods from various areas had different concentrations of point defects related to oxygen vacancies according to the red-shift of the PL peak corresponding to exciton transitions.

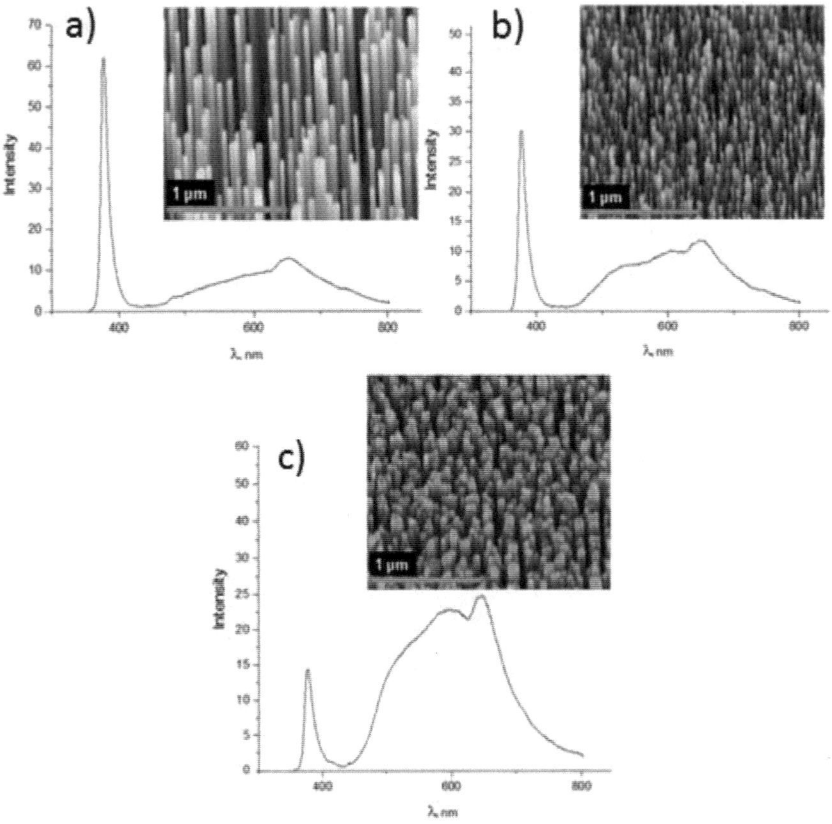

Figure 2. SEM images and PL spectra of nanorods obtained from areas that are at different distances from the open end of the tube: a=35 mm, b=40 mm and c=45mm.

In the second series of experiments substrate was placed on a distance of 4.5 cm from the zinc granule downstream of gas mixture. The deposition pressure was set 11 mbar. Next the substrate and Zn granule are heated to a synthesis temperature in 25 minutes. When the necessary temperature was achieved, gas also started flow in the reactor at 5 sccm. At the final stage of the synthesis process the pressure in the chamber was rapidly increased from 10 mbar to 50 mbar by changing the pumping rate. ZnO nanorod arrays with sharp tips have been grown at the furnace temperature 600°C in 30 minutes.

SEM and PL researches was carried out on different areas of the substrate to study the effect of the Zn vapor supersaturation on the properties of obtained zinc nanorods. The sharpened form of nanorods (Fig. 3) is similar to the shape of the nanostructures obtained in work [8]. The rapid increase of pressure value leads to a change in the rate of generation of zinc vapor and molecular mobility. Consequently were grown the rods with various form of the tips in different areas of the reactor. Photoluminescence spectra show that obtained structures have a high intensity peak in the ultraviolet spectrum range at a wavelength of ~ 380nm, and a low intensity peak in the visible spectrum, which corresponds to low concentrations of point defects related to oxygen vacancies.

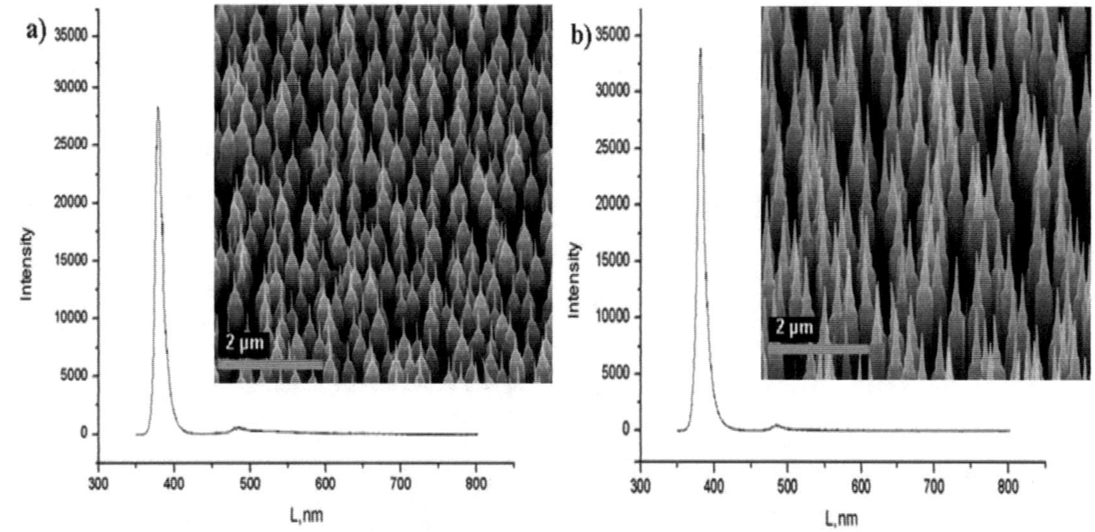

Figure 3. SEM images and PL spectra of nanorods obtained by changing pressure values during the synthesis from areas with different Zn vapor supersaturation values (distances from distances from the granule Zn : a=40 mm and b=50 mm).

3. Conclusion

A method of synthesis one-dimensional ZnO nanostructures by thermal evaporation technique at the quite low temperature (550°C -600°C) is proposed. The configuration used in this work allows us to localize the substrate in areas of reactor with different ratio values of concentrations of the zinc and oxygen molecules in the vapor phase. SEM images and PL spectra have shown that physical properties of synthesized structures depend on Zn vapor supersaturation values. Using our modification presented in this work, makes it possible to obtain the nanocrystals with the necessary morphological and excellent optical properties. Such rods are very interesting in quantum optoelectronics. Particularly they can be used in ultraviolet photodetectors and light-emitting diodes.

4. Literature

[1] Sun X W, Huang J Z, Wang J X, Xu Z 2008 *Nano Lett.* **8** №4 1219–1223
[2] Kim D Y, Son J Y 2009 *Electrochemical and Solid-State Letters* **12** №12 109-111
[3] Huang M H, Mao S, Feick H, Yan H, Wu Y, Kind H, Weber E, Russo R, Yang P 2001 *Science* 1897-1899
[4] Chien Y L, Shoou-J C, Sheng P C, Ching T L, Che F K, Hong M C 2006 *Applied Physics Letters* **89** 153101
[5] Park J, Choi H, Siebein K, Singh R 2003 *Journal of Crystal Growth* **258** 342–348
[6] Chang P, Fan Z, Wang D, Tseng W, Chiou W, Hong J, Lu J 2004 *Chem. Mater* **16** 5133-5137
[7] Red'kin A N, Makovei Z I, Gruzintsev A N, Dubonos S V, Yakimov E E 2007 *Inorganic Materials* **43** №3 253-257
[8] Pan N, Wang X, Zhang K, Hu H, Xu B, Li F, Hou J 2005 *Nanotechnology* **16** 1069–1072

Investigation of defects formation on different stages of sapphire crystals growth

S P Malyukov and Y V Klunnikova

Department of Electronic Apparatuses Design, Southern Federal University, Taganrog, Nekrasovsky, 44, Russia

E-mail: jklunnikova@rambler.ru

Abstract. The mathematical simulations and calculations allow to estimate the large number of factors influence on crystals growth and quality and to determine the main cause of the defect structures formation in the solid phase. The research results allow making the quantity estimation of defects formation in sapphire crystals. The simulation results can help to define defect formation zones and can be used for recommendations about crystals quality improvement.

1. Introduction

The SOI technology is one of the most progressive manufacturing techniques of inertial MEMS-sensors [1]. The various poly- and monocrystal materials can be applied as a dielectric. The sapphire crystal is one of perspective materials especially for MEMS in which sensitive elements work in the gigahertz ranges. The quality of all MEMS-sensor operation depends on quality and a minimum of such crystal superficial defects. So the problem of defects formation on different stages of sapphire crystals growth is actual one now.

The big size sapphire crystals growth with improved structural characteristics became a priority of modern science and equipment. In the last decades the Czochralsky's method, Kiropoulos's method, the horizontal directed crystallization (HDC) method and Stepanov's method were applied for big size monocrystals growth. These crystals were widely used in electronic technique [2].

The sapphire crystal quality can be characterized by the orientation, density and the nature of dot defects distribution, density of single dislocations, extension of blocks borders and their disorientation, the value of residual tension, chemical purity and impurity heterogeneity [2-4].

2. Mathematical simulation of defects formation during sapphire crystals growth

The HDC method has some advantages in comparison with other methods. First of all it is because the melt height and its square are constant during all process. The large melt volume allows to provide the effective impurities evaporation. The fuse process and the border of phase division can be controlled visually or by means of optical devices.

The bubbles formation in the melt and their capture by the crystallization front are one of the problems during sapphire crystals growth by the HDC method. These bubbles can be the cause of pores emergence in crystal which being tension concentrators are an additional source of dislocations and, as a result, a source of crystal defects in the form of blocks.

It is noticed [5,6] that at the different stages of crystal growth the position and form of bubbles congestions are various. The largest problem for bubbles congestions removal is made by the bubbles located in the benthonic zone of sapphire crystal growth furnace SZVN 155.320 crucible beginning, approximately, at the distance 250 mm from the fuse with the container length 320 mm. It is shown schematically in Figure 1. It is also shown in the photographs of crystals taken at lighting, presented in Figure 2.

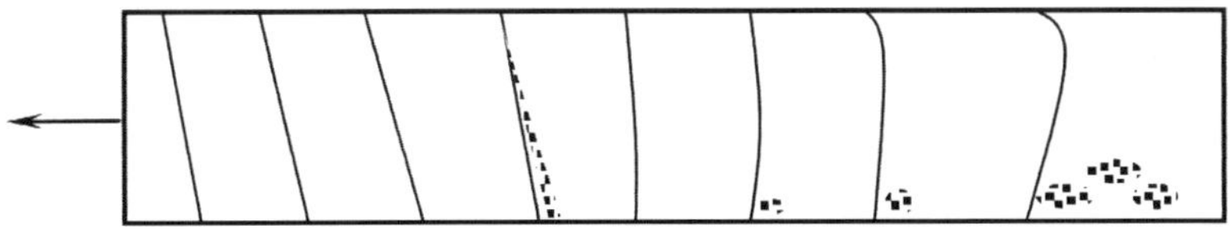

Figure 1. The scheme of position of the crystallization front and bubbles congestion at various stages of growth (the arrow showed the direction of crucible movement)

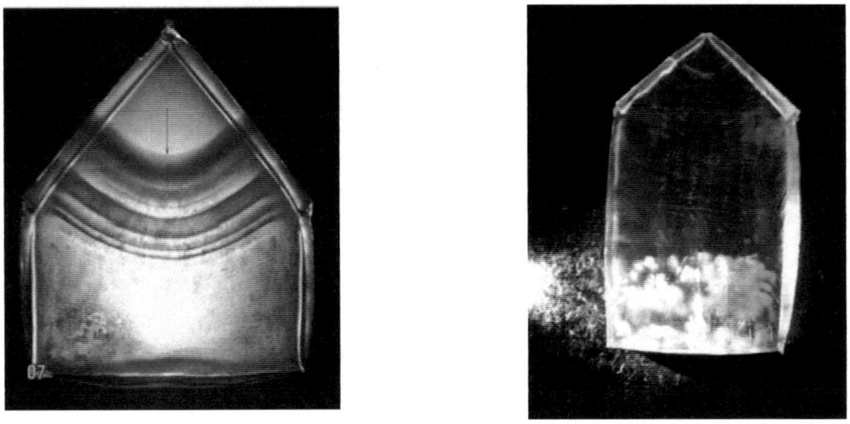

a) b)

Figure 2. Inclusions in sapphire crystals: a) at the distance 60 mm from the fuse, b) at the distance 250 mm from the fuse

In the case of horizontal directed crystallization, the projection of Archimedean velocity on the direction of front movement is zero, and the direction of the temperature gradient in the melt ∇T coincides with the direction of growth front velocity.

Considering that the thermocapillary force in the stationary mode [7,8]

$$F_k = -2\pi \cdot r^2 \cdot \frac{\partial \alpha}{\partial T} \cdot \nabla T, \tag{1}$$

where ∇T is the direction of the temperature gradient in the melt; r is the radius of gas bubble; α is the surface tension between gas and liquid boundary must be equal to the force of the medium viscous resistance [9]

$$F_c = 6\pi \cdot r \cdot \mu \cdot V_k. \tag{2}$$

The rate of bubble thermocapillary movement relatively to the liquid can be determined by the formula

$$V_k = \frac{r}{3 \cdot \mu} \cdot \frac{\partial \alpha}{\partial T} \cdot \nabla T, \tag{3}$$

where μ is the dynamic viscosity of liquid. It is also considered that for the majority of liquids the derivative $\partial\alpha/\partial T \prec 0$. At the rate $V_a = 0$ the velocity of the bubble relatively the growth front is defined by equality

$$V_p = \frac{r}{3\cdot\mu}\cdot\frac{\partial\alpha}{\partial T}\cdot\nabla T - V_f . \quad (4)$$

As we can see from (4) in the bubbles ensemble in the melt there are bubbles of critical size r_k at which the bubble motion direction relatively the crystal growth surface is changed. Gas bubbles with size

$$r > r^* = \frac{3\cdot\mu}{\nabla T}\cdot\frac{V_f}{\partial\alpha/\partial T} \quad (5)$$

will move in the direction from the front, and smaller bubbles $r < r^*$ will move in the opposite direction.

Bubbles with radius $r < r^*$ appeared in surface of phase transformation during crystallization will be grasped by the growth front. So they create system of steady defects in material volume.

It is necessary to calculate the temperature distribution in sapphire crystals growth equipment by the HDC method for defects diagnostic. The temperature distribution determination in crystal-melt-crystal powder system consists of heat conductivity equations solution [10]:

$$\frac{\partial T_i(x,y,z,\tau)}{\partial \tau} = \left(\frac{\partial}{\partial x}a_i\frac{\partial T_i(x,y,z,\tau)}{\partial x} + \frac{\partial}{\partial y}a_i\frac{\partial T_i(x,y,z,\tau)}{\partial y} + \frac{\partial}{\partial z}a_i\frac{\partial T_i(x,y,z,\tau)}{\partial z}\right) - W\frac{\partial T_i(x,y,z,\tau)}{\partial x}, \quad (6)$$

where $i = 1,2,3$ is the crystal, melt and crystal powder accordingly; a_i is the heat diffusivity coefficients ($a_i = \frac{\lambda_i}{\rho_i \cdot C_i}$, where λ_i is the thermal conductivity coefficient; ρ_i is the material density; C_i is the specific heat); W is the container movement speed.

The container movement speed is small enough, so we can conclude that the process is in quasisteady state and the temperature distribution can be found by the following formula [11]

$$div(\lambda_i \, gradT_i(x,y,z)) = 0. \quad (7)$$

The boundary conditions for system of the equations (7), reflecting continuity of thermal fields and streams on limits of environments section, can be found by the following ratios:

$$\lambda_1\frac{\partial T_1(x_T y,z)}{\partial x} = \lambda_2\frac{\partial T_2(x_T,y,z)}{\partial x}, \quad (8)$$

$$\lambda_2\frac{\partial T_2(x_T+\Delta x,y,z)}{\partial x} = \lambda_3\frac{\partial T_3(x_T+\Delta x,y,z)}{\partial x}, \quad (9)$$

$$q_{s_1} = q_{s_2} = q_{s_3} = \sigma\beta(T^4 - T_{hot}^4), \quad (10)$$

where σ is the Stephane-Boltsmana constant; β is the radiation coefficient; T_{hot} is the function with the heat temperature distribution.

The final volume method on the unstructured grid was used for simulation. The temperature fields in crystal-liquid-powder system are showed in Figure 2, a. The temperature fields in vertical crystal cut are presented in Figure 2, b.

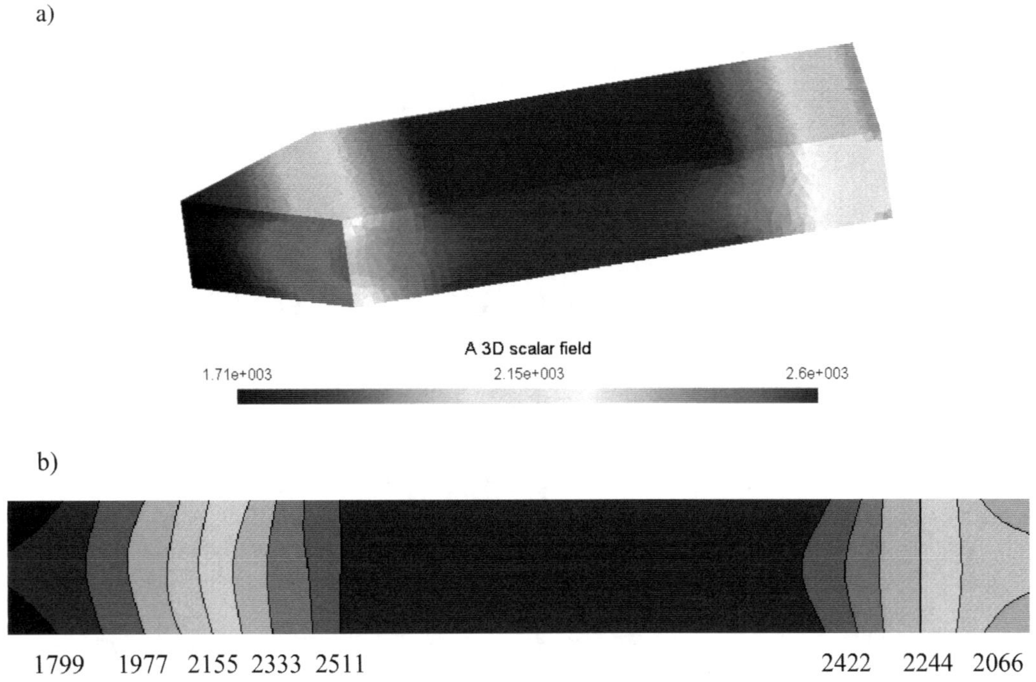

Figure 2. Temperature distribution in crystal-liquid-powder system

The temperatures gradients allow to estimate the sizes of forming vapour bubbles. So it is known that the crystal growth rate influence on the capture of impurity and foreign inclusions by the front of crystallization. At bubble capture the ratio of crystal growth and bubble rates defines a final form of inclusion. The received dependences of the bubbles average size from the growth speed can be explained [10]. At the set crystallization speed the bubbles with certain radius can be characterized by the greatest probability of capture. The increase in growth rate will lead to increase the probability of larger bubbles capture.

The dependence of segregation effective coefficient on the interface of phases from growth rate can be determined [2,5,10] as

$$K_{eff} = \frac{K_0}{K_0 + (1 - K_0)e^{-\frac{V\delta}{D}}}, \tag{11}$$

where K_0 is the equilibrium segregation coefficient; δ is the depth of the area with raised or lowered concentration of impurity at the crystallization front; D is the diffusion coefficient of impurity in melt; V is the crystal growth rate. At the intensive interfusion $\delta \to 0$ or at the small crystallization speed $V \to 0$, $K_{eff} \to K_0$ the impurities and bubbles are taking away in melt. When the growth speed $V >> D/\delta$ ($e^{-\frac{V\delta}{D}} \to 0$), $K_{eff}=1$ it means that all bubbles and impurity are taken by crystal.

The growth rate depends on position of the container relatively the heater and increases throughout the process of sapphire crystals growth in the conditions of large temperature gradients [5]. The effective coefficient dependence at the various stages of growth is shown in Figure 3.

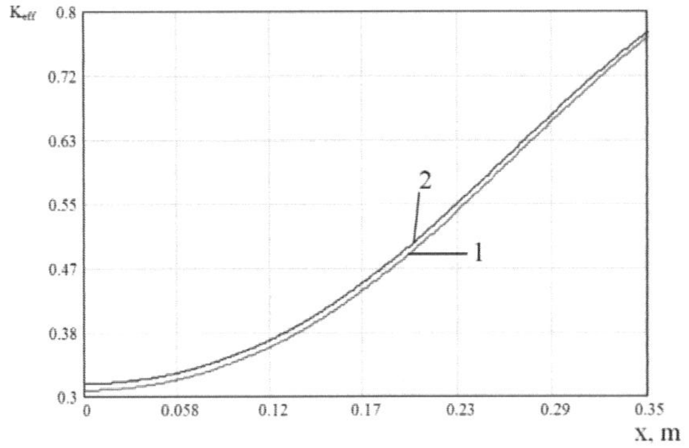

Figure 3. The effective coefficient of segregation depending on the crucible position: 1 – the gradient in melt is 1000 K/m, 2 – the gradient in melt is 500K/m

From Figure 3 it is visible that only at the initial stage (x=0.1m) the effective coefficient of segregation is close to equilibrium one. It is defined by small growth rate. Thus the crystal grows without foreign inclusions. As you can see from the figure the effective coefficient of segregation doesn't depend on temperature gradient in melt practically.

3. The investigation results

The results showed that the melt width increased with heat temperature increase on 300 K and as a result the crystallization front location changed. So we can conclude that we have possibility to control the crystallization front location by heat temperature increase and it can lead to defects redistribution. The suggested model describes adequately the processes of bubbles formation in sapphire and can be used at the decision of various scientific and practical problems for crystals defects investigation.

The results of research presented in this paper were obtained with the financial support of the Ministry of Education of the Russian Federation within the framework of the project "Creation of a plant for manufacture of complex reconfigurable high-precision positioning of objects on the basis of satellite navigation systems, local area networks laser and microwave beacons and MEMS Technology" by government decree № 218 of 09.04.2010, the research was carried out in the FSAEI HPE SFU (SFedU).

References

[1] Skorupa W 1995 *J. NATO ASI Series 3. High Technology* vol 4 pp 39-54
[2] Dobrovinskaya E R, Lytvynov L A and Pishchik V V 2009 *Sapphire. Material, Manufacturing, Applications* (Berlin: Springer)
[3] Shtukenberg A G and Punin Y O 2004 *Optical Anomalies in Crystals* (St. Petersburg: Science)
[4] Hannanov S C, Nikanorov S P and Baholdin S I 2003 *J. Solids Phys.* vol 45(6) pp 1020-1023
[5] Malyukov S P, Nelina S N and Stefanovich V A 2012 *Physicotechnological Aspects of Sapphire Production* (Saarbrucken:Lap Lambert Academic Publishing)
[6] Cherednichenko D I, Malyukov S P, and Klunnikova Y V 2013 *Sapphire: Structure, Technology and Applications* (USA: Nova Science Publishers) pp 101-118
[7] Dzyuba A S, Iong Zu 1985 *Journal of Crystallography* vol 30(6) pp 1177-1180
[8] Geguzin Y E, Dzyuba A S, Kononenko N V 1981 *Journal of Crystallography* vol 26 pp 577-581
[9] Frenkel Y I 1975 *Kinetic Theory of Liquids* (Leningrad: Science)
[10] Malyukov S P and Klunnikova Y V 2013 *Nano- and Piezoelectric Technologies, Materials and Devices* (USA: Nova Science Publishers) pp 133-150
[11] Lykov A V 1967 *Heat Conductivity Theory* (Moscow: Higher School)

Photoelectrical properties of TiO_2-Si structures

Yu S Petrova, A N Zarubin, V M Kalygina, S Yu Zupiy

Tomsk State University, 36 Lenin Avenue, Tomsk, Russia

E-mail: petrovays@mail.ru

Abstract. The effect of thermal annealing at 500 and 750°C as well as treatment in oxygen plasma on electrical and photoelectrical characteristics of TiO_2-Si structures were investigated. TiO_2 films were deposited on n-Si substrates by magnetron sputtering. It was found that in the structures annealed at 500°C and treated in oxygen plasma a substantial part of the enhanced photocurrent was observed over a long time after removing of illumination with $\lambda = 400$ nm.

1. Introduction

Anatase and rutile are two stable crystalline phases of TiO_2 that have a vast variety of applications. Titanium oxides are chemically stable, relatively hard and transparent. Its particularities are high dielectric constant and photocatalytic activity. Due to high refractive index TiO_2 is interesting for optical applications, e.g. antireflection coatings, narrow band filters, and optical waveguides. TiO_2 phases are wide gap semiconductors with a large range of electrical properties depending on chemical composition. In this paper the results of effects of thermal annealing and oxygen plasma treatment on electrical and photoelectrical characteristics of TiO_2-n-Si structures are presented.

2. Experiment details

TiO_2 films with thickness 90 – 100 nm were fabricated by magnetron depositing on n-Si substrates. TiO_2 was used as a target. TiO_2-n-Si structures were annealed in argon at 500 or 750°C. Some part of the samples were treated with oxygen plasma during 20 min at 50°C before annealing. Structural characteristics of the samples were measured by X-ray diffraction and atomic force microscopy (AFM). Metal Ni/V contacts were deposited on TiO_2 and Si substrate. For investigation of photoelectrical characteristics LED with $\lambda = 400$nm was used. Light power P was changed in 20 – 520mkW interval.

3. Results and discussion

As-deposited TiO_2 films had amorphous structure. Figure 1 indicates the results for films after annealing at 500°C. XRD spectra contain amorphous matrix peaks at $2\,\Theta = 25.2^{\circ}$, 27.4° and 56.7° which belong to anatase (101) crystalline as well as rutile (110) and (211) crystalline respectively. If before annealing TiO_2 films had been treated in oxygen plasma at 50°C the XRD patterns show two new peaks: at $2\theta = 38^{\circ}$ which corresponds to anatase crystalline (211) and $2\theta = 44.1^{\circ}$ corresponds to rutile crystalline (210) (**Figure 1b**). Besides, effect of oxygen plasma leads to increasing of antecedent peaks about two times in XRD spectra of TiO_2 films.

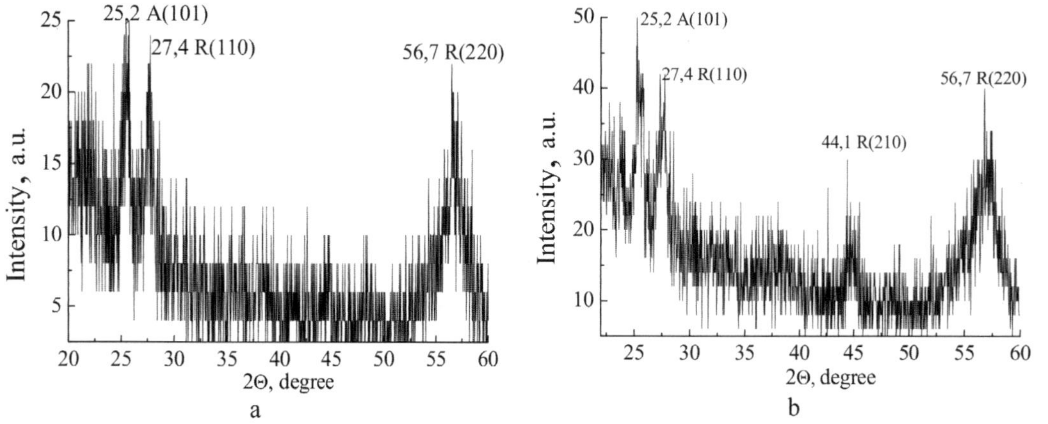

Figure 1. XRD patterns of TiO_2 film after annealing in argon at 500°C: a – without treatment in oxygen plasma; b – with treatment in oxygen plasma before annealing

After annealing at 750°C amorphous phase vanishes and the film becomes polycrystalline (**Figure 2a**). Anatase phase disappears and in TiO_2 film only rutile crystalline remain mainly with crystal orientation (220) – peak at 2Θ = 56.7 degree. Insert in figure 2a shows detailed XRD pattern of TiO_2 film and it can be seen the peaks at 2Θ = 27.4 and 44.5 degree which belongs rutile crystalline. Preliminary expose of TiO_2 films to oxygen plasma before annealing at 750°C does not change their structure (**Figure 2b**).

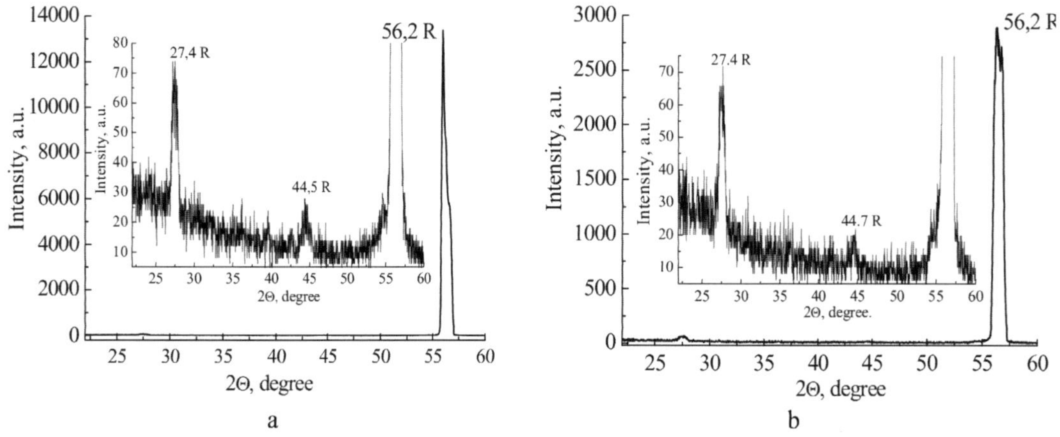

Figure 2. XRD patterns of TiO_2 film after annealing in argon at 750°C: a – without treatment in oxygen plasma; b – with treatment in oxygen plasma before annealing

According to AFM images the roughness of TiO_2 films is 8 – 10 nm after annealing at 500°C (**Figure 3a, 3b**). The morphology of films significantly varied as result of their treatment in oxygen plasma. The influence of oxygen on TiO_2 films results in merging of irregularities and formation of some grain-like structure (**Figure 3c, 3d**). The size of grains increases at substrate plane after influence of oxygen plasma and it is about 100 nm.

It is evident from figure 4a and 4b that grain size increases with rise of annealing temperature. As a rule annealing of TiO_2 films at 750°C results in increasing of grains perpendicularly to substrate. The effect of growth temperature is also reflected in the case of root-mean-square surface roughness and their values increases to 20 nm (**Figure 4b**). In addition new crystalline particles nucleate with size of grains equal 20 – 50 nm (**Figure 4a**).

After oxygen exposure on TiO_2 films and subsequent annealing in argon at 750°C it can be seen intergrowth of crystalline and formation of large particulars with size 150-200 nm (**Figure 4c**). The surface roughness decreases to 10 nm.

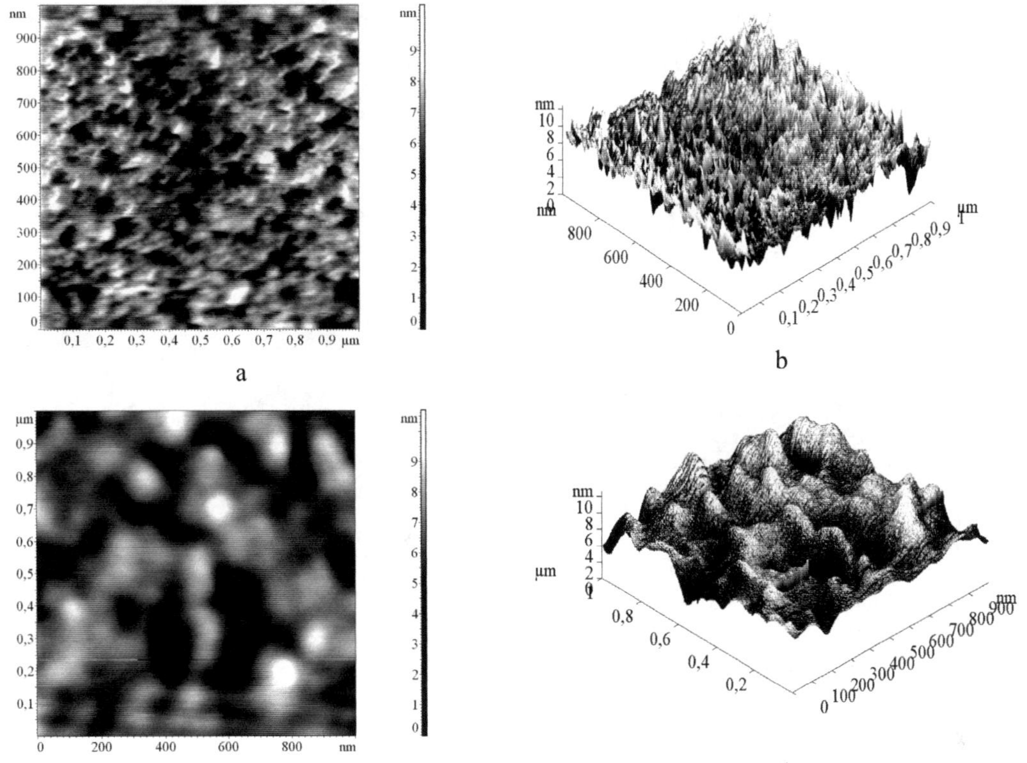

a b

c d

Figure 3. AFM images of TiO_2 films formed on Si substrates: a, b– after annealing at 500°C; c, d – treatment in oxygen plasma before annealing at 500°C

The TiO_2-n-Si structures showed rectifying properties. At positive potentials on the gate direct current-voltage (I-U) characteristics changed very slightly even under illumination with light power P= 520 mkW. Under illumination at negative potential on a gate the photocurrent appeared. Its value depended on optical power, voltage (U) at structure and treatment regime (**Figure 5**). Enhancement of photocurrent with increasing voltage was observed at $U \geq 2V$ for samples annealed at 500°C.

An important feature of structures behavior is continuous photocurrent decay after switch of light (**Figure 6**). Decrease of photocurrent as a function of time is described by exponential law. Time constant does not depend on voltage at $U \geq 2V$ and equals 22 ± 2 min at room temperature. Persistent photocurrent is absent in structures annealed at 750°C. After illumination with light power 520 mkW dark reverse current of the structures annealed at 750°C in argon without a treatment in oxygen plasma decreased and its value was about two times less than initial dark current. This effect did not monitored in structures treated in oxygen plasma before annealing at 750°C. Experimental data are explained under consideration of phase and structural changing in TiO_2 films and the interface Si-TiO_2.

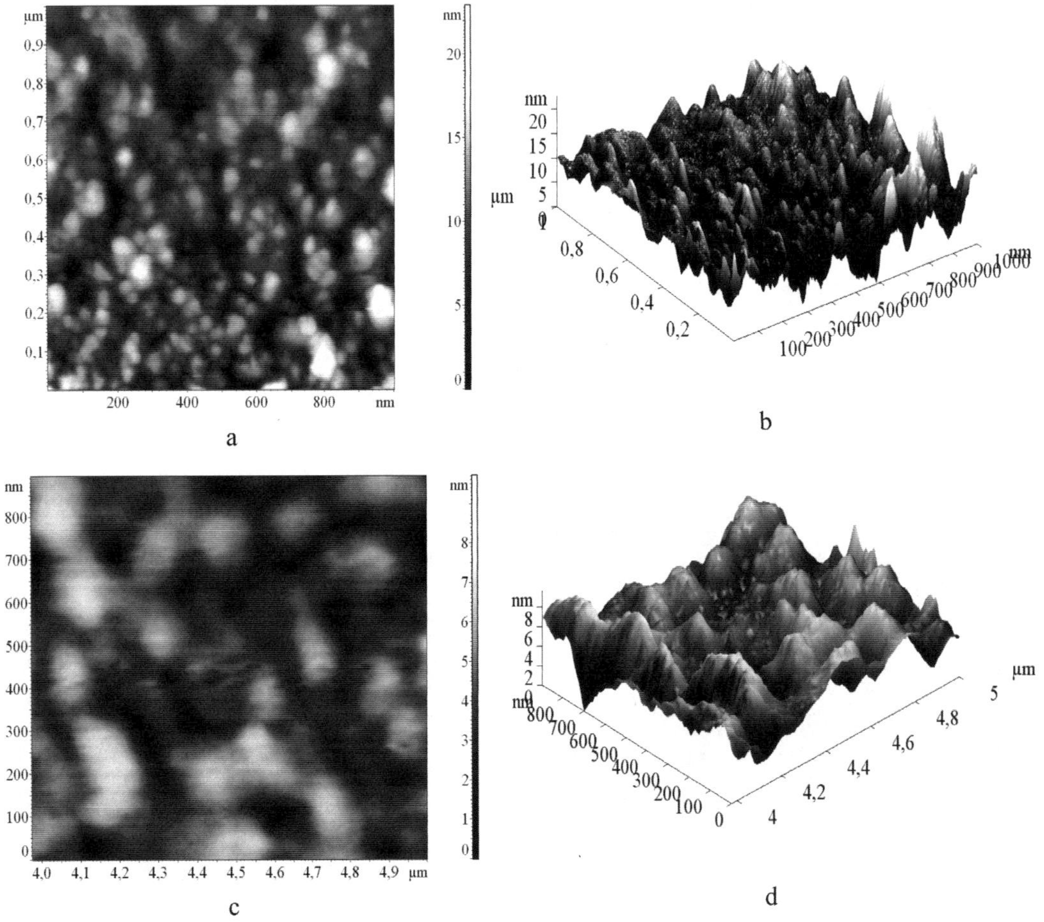

Figure 4. AFM images of TiO_2 films formed on Si substrates: a, b– after annealing at 750ºC; c, d – treatment in oxygen plasma before annealing at 750ºC

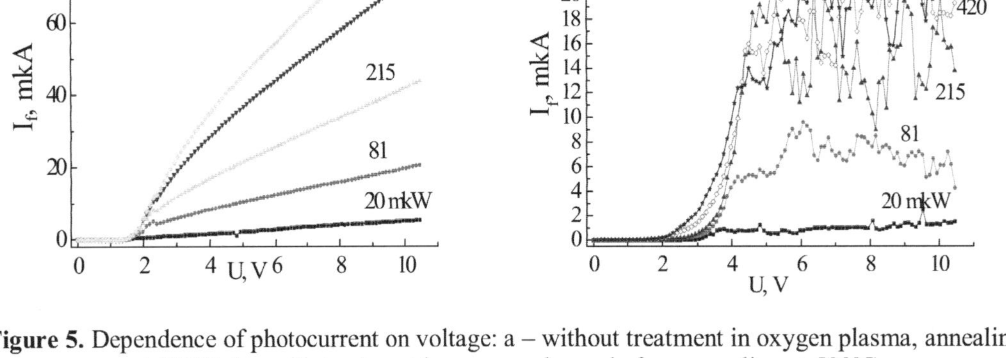

Figure 5. Dependence of photocurrent on voltage: a – without treatment in oxygen plasma, annealing at 500ºC; b – with treatment in oxygen plasma before annealing at 500ºC

Photocurrent is absent in $0 \leq U \leq 1$ V interval. It is explained by forming of depletion regime and appearance of inversion layer. At very low frequency (continuous current) all bias voltage is applied on TiO_2 and photocurrent is defined by band to band generation of electron-hole pairs in dielectric film.

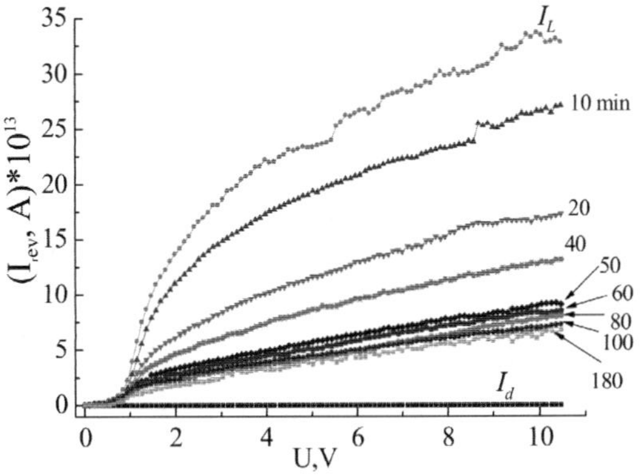

Figure 6. Change of dark reverse I-U characteristics of the structure (20min, 500 °C) after switch off light with P = 520 mkW at different time intervals; I_L is current at P = 520 mkW ; I_d – initial dark current.

4. Conclusion

The influence of thermal annealing and treatment in oxygen plasma on electrical and photoelectrical characteristics of TiO_2-Si structures have been investigated. It has been shown connection of structure and phase composition of TiO_2 films with photoelectrical characteristics of TiO_2-Si structures. Slow decreasing of reverse current after switch off light is explained by persistent photoconductivity which takes place in TiO_2 – n-Si structures only after define technological processing.

Journal of Physics: Conference Series **541** (2014) 012034 doi:10.1088/1742-6596/541/1/012034

The application of the modified Blumlein scheme for electrodischarge obtainment of nanopowders in liquids

Y Izotov[1] **and A Tikhomirov**[2]
[1]Petrozavodsk State University, Department of Physics- Engineering, Undergraduate student, Lenin Avenue 33, Russia.
[2]Petrozavodsk State University, Department of Physics- Engineering, Researcher, Lenin Avenue 33, Russia.

E-mail: izotov93@yandex.ru

Abstract. The formation of metal-carbon structures in electro-impulse discharge is investigated. The method of decrease of discharge voltage is proposed.

1. Introduction

Electro-impulse technology allows to obtain nanoparticles of the electrode material at low-powered electric impulse discharges in water, gasoline, ethanol, liquid nitrogen. Discharge impulse method has some essential advantages: cleanliness and purity, isomerism and preservation of crystal planes of the resulting product that enable to select the material effectively and to obtain new surface, physical and chemical and technological properties.

2. Experimental set-up

This work considers the electro-impulse method based on the so-called Blumlein scheme but modified to some extent and electrodes with the geometric pattern "needle-plane". Such geometric configuration apparently seems to be the most appropriate for voltage decrease at the discharge gap in connection with the local electric field enhancement near the needle. The effective mechanism to obtain powder materials in such discharge is the impulse plasma in the discharge gap, the formation of which is accompanied by strong heating of the discharge channel and by hydraulic impact.

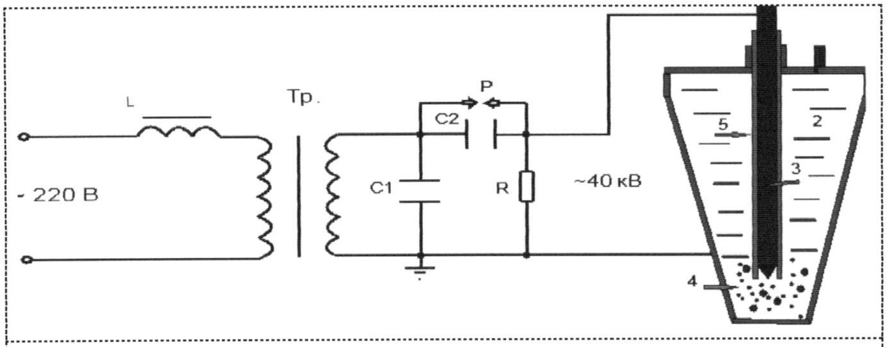

Figure 1. Scheme of the experimental set-up to obtain metal-carbon structures.

1. Cone electrode 2. Liquid (distillated water) 3. Pointed electrode (carbon) 4. Plate electrode (molybdenum) 5.Insulator Tp – high-voltage transformer L – choke P – commutator (air gap) R – resistor C1, C2 – capacitors (0.025 μF)

In this scheme electric capacity C1 is charged from high-voltage transformer Tp through ohmic resistance of the discharge gap between electrodes 3,4 which results in voltage increase on commutator P. When the critical voltage on commutator P is reached, capacity C1 turns out to be connected to the discharge gap and leads to plasma ignition.

3. Results and Discussion

The obtained powders were examined by X-ray analysis and scanning microscopy. The method of ACM-analysis is described below. First of all, the series of surface images of the samples in the area ≈ 3 μm / 3 μm were obtained. Then the most perspective sample of the smaller size was chosen out of them. It was scanned at a higher resolution once again. After that the spatial resolution of the microscope was increased and the images were obtained in the area ≈ 1 nm / 1 nm. Next, the analysis of the obtained images including the estimation of the sizes of the detected fractions, the statistical data processing and comparison of the given results with the results of other types of microscopic researches.

In all images not only separate particles but also multiple groups are easily determined. The smallest particles have the sizes about 150 nm, height 20 nm. The average size of particles is 200-300 nm, height 40-50 nm. In figure 2 the profile of a typical particles is given, its dimensions are in accordance with the average ones.

Figure 2. Profile of typical particles

It should be mentioned that one can notice bigger particles as well. As a rule, they have the stretched forms. But the thorough analysis of this profile showed that it is multiple groups of smaller particles. Such groups of particles make determination of size distribution of particles more difficult. But the analysis of the images revealed that this distribution will not be significantly wide.

Almost all particles have the sizes of the same order – hundreds of nanometers. In the process of electrode erosion not only the molybdenum plate but also the copper electrode was destructed to the same extent.

X-ray diagrams were registered from two areas of the sample: the area having a porous black layer on its surface and the light area. The corresponding x-ray diagrams are presented on figure 3.

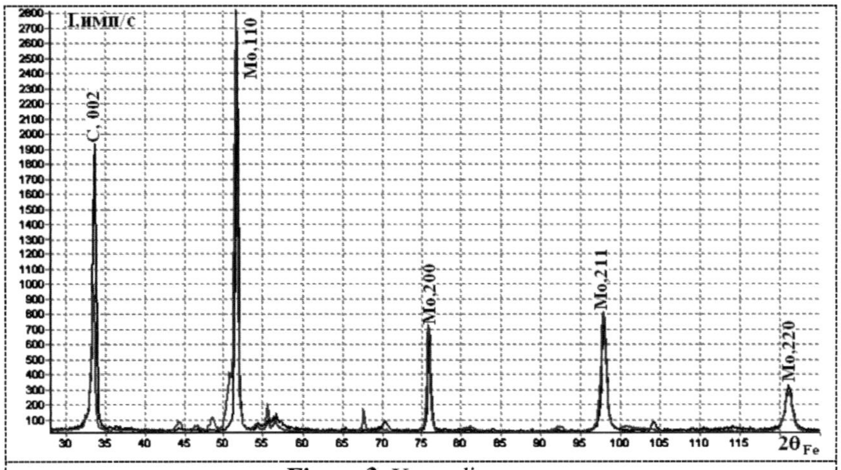

Figure 3. X-ray diagrams

According to figure 3 one can see that both intensive and weak reflections are observed in x-ray diagrams. The major difference of x-ray diagrams is noticed in the area of weak reflections.

Quality phase analysis was done by the way of comparison of experimental x-ray diagrams with the x-ray diagrams theoretically modeled on the basis of data on the atomic structure of molybdenum, molybdenum carbides of different compositions and carbon allotropic phases presented in the base of crystallographic characteristics of inorganic compounds ICSD.

It was found out that strong lines are the reflection of (002) graphite and lines (200), (211), (220) of molybdenum. The further analysis of diffraction lines was done by Ritweld method. According to figure 4 the result of two stages of calculation of x-ray diagram is presented.

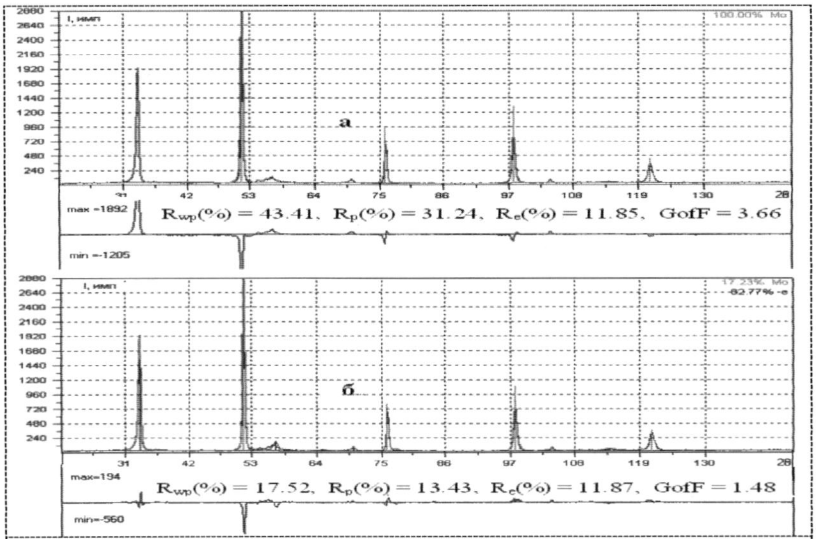

Figure 4. Calculated X-ray diagrams

At the first stage only the molybdenum phase with volume-centered cubic lattice (space group of symmetry Im 3 m) was taken into account. Profile factor weight of unreliability R_{wp} is 43% (according to 4 (a)). Having taken into consideration the contribution made by dissipation of hexagonal graphite

(space group of symmetry $P6_3mc$), profile factor weight of unreliability R_{wp} decreases to 17.5% (according to figure 4(b)).
According to figure 5, all lines of the experimental x-ray diagram are described by the x-ray diagram of molybdenum substrate and the layer of hexagonal graphite on it 83: 17. In the process of calculation all profile characteristics of x-ray diagrams of both phases were defined more precisely. Values of periods of elementary cell correspond to literature data: for Mo a=3.147(1); for graphite a=2.462(1), b=6.727(1) Å.

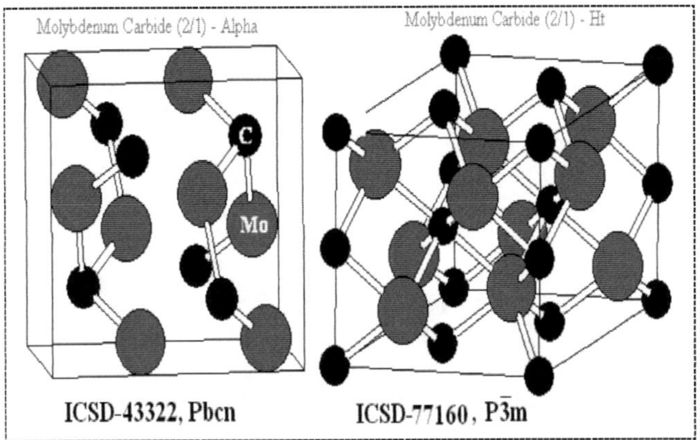

Figure 5. Structure of powder

The studies of formation of metal-carbon structures in electric impulse discharge by X-ray analysis have been conducted. It was stated that strong lines are the reflection of graphite and lines are the reflection of molybdenum.
The analysis of diffraction patterns showed that carbide film is formed through the reaction with carbon on the molybdenum surface.
This work was supported by program of strategic development of Petrsu 2009-2013

References:

[1] Oswald R.B., Schallhorn D.R., Eiasen H.A. IEEE Trans. Nucl. Sci. Vol. NS-15. N 6. 1968.
[2] Morozov N., Petrov Y. Dynamics of Fracture. B.; Heidelberg; N.Y.:Springer, 2000.
[3] Patent RU № 107010 27.07.2011

Model of electron transport in cell of a thin-film vacuum nanotriode

Z R Gallyamov and K A Nikiforov

Saint Petersburg State University, 7-9, Universitetskaya nab., St. Petersburg, 199034, Russia

E-mail: knikiforov@cc.spbu.ru

Abstract. Thin-film vacuum nanotriode with cone-shaped emitters is considered. Each emitter has height 200 nm with 1 nm radius of surface curvature on the top. Current-voltage characteristics are calculated for single cylindrical triode cell (655 nm in diameter). It is shown the value of the emission current comes to microamperes, when the voltage is less than 20 V applied on the anode-emitter gap 500 nm. Anode-cathode voltage 16-20 V can generate more than 0.5 mkA emission current flown to the anode when gate-cathode voltage is less than 14 V and gate diameter is 150 nm.

The modulator or extractor electrode is often used to control field emission process from cold cathodes in height-power THz applications. Although simple vacuum diode system is sufficiently to start electron emission, triode configuration with field emitter as cathode, extractor electrode as gate and distant anode as collector is more stable. This configuration (figure 1 (a)) is selected for modeling of thin-film field-emission Spindt-type cathode [1]. One can say that the whole field of vacuum electronics developed from the production technology for a thin-film field emission cathode developed by Charles Spindt et al [1]. For an instance, figure 1 (a) schematically shows a field emission triode cell based on Spindt cathode. On substrate plane there is a conic metal emitter whose tip protrudes inside a hole in the controlling electrode film acting as a gate. Height of the cone, radius of the tip and of the gate hole can vary depending on the technological process requirements. Besides, there is a remote anode that acts as an electron collector.

The function of a vacuum thin-film nanotriode and other devices of vacuum nanoelectronics is based on several mutually interacting processes:

- Base factor is appearance of an electric field within the structures.
- Some particular values of field strength at the emitter surface cause field electron emission to appear. The process of field electron emission of electrons from metal is sufficiently precisely described by the well known Fawler-Nordheim equation:

$$j(E) = \frac{AE^2}{\Phi t^2(y)} \exp\left[-B\frac{\Phi^{3/2}}{E} v(y)\right] \quad (1)$$

Here j stands for the emission current density, while A and B are constants connected to fundamental physical values via equations $A = \frac{e^3}{8\pi h}$, $B = \frac{8\pi\sqrt{2m}}{3eh}$, where h is the Planck constant, e is the electron charge and, m is the rest mass of a free electron, $t(y)$ and $v(y)$ are elliptic Nordheim functions of

Journal of Physics: Conference Series **541** (2014) 012035 doi:10.1088/1742-6596/541/1/012035

$y = \sqrt{\frac{e^3}{4\pi\varepsilon_0}\frac{E^{1/2}}{\Phi}}$, approximated by $t^2(y) \approx 11619$ and $v(y) \approx 0{,}95 - y^2$, ε_0 is an electric constant, E is a value of external electron field strength, Φ is the work function of the material used as a measure of energy of bond between the electrons and the solid.

- Because of the emission, electronic gas (plasma) is formed inside the structures. It can move under the influence of applied field.
- A space charge inside the structures that appears because of that acts as a source of secondary electric field that superimposes itself onto the base field.
- Processes that occur in parallel to this are secondary electron emission, polarisation and emergence of surface charge on the insulant, ionisation of remaining gas molecules.

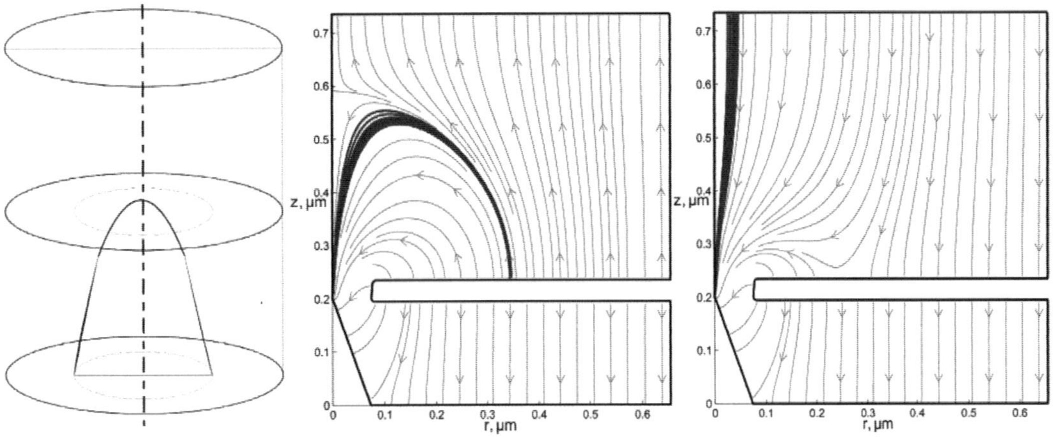

(a) (b) (c)

Figure 1. Scheme of the thin film vacuum field emission nanotriode cell and electron trajectories with electric force lines in the computer model.

It is well-known that the space charge created by emitted electrons over the emitter surface can have influence on field electron emission. In [2] it is shown that for objects with emitter surface curvature radius of $1 \div 10$ nm non-negligible influence of space charge (change in the field by $1 \div 10$ %) begin with current density of 10^{13} A/m, while large influence (about 50%, i.e. for the same current density the voltage has to be increased two-fold or more relative to calculated values) appears with current density 10^{14} A/m.

On the other hand, for emission current of approx. 10^{-7} A and distance between cathode and anode of about 500 nm every electron in the interelectrode space can be regarded as not interacting with other electrons, as the time of passing of particular electrons is smaller than their average time interval between two consecutive exits into the vacuum.

Besides, the emitted electron trajectories can be roughly regarded the same as the electric field force lines. This approach is called the assumption of "zero mass", and it is the more precise the closer the force lines of the field are to straight lines. This approach is known also as electron massless approximation and used for electrostatic simulation of vacuum emission nanoelectronics devices. In vacuum nanoelectronics due to potential character of the electric field, integral of the vector $\mathbf{E}$ along an arbitary force line between two conductors is equal to the difference of their potentials, i.e. it is invariant. Thus, force lines with less curvature correspond to higher electric field strength. On the other hand, because of considerable non-linearity of Fawler-Nordheim formula, greater part of emission current is formed in areas with maximum field strength. Thus electron trajectories inside the whole area of emission can be regarded the same as force lines of vector $\mathbf{E}$.

Trajectory analysis in massless approximation is performed by introducing the current flow function, well known in hydrodynamics [3]. The model supplies current-voltage characteristics including the effects of grid leakage current and the Fowler-Nordheim equation is used to calculate cathode current density.

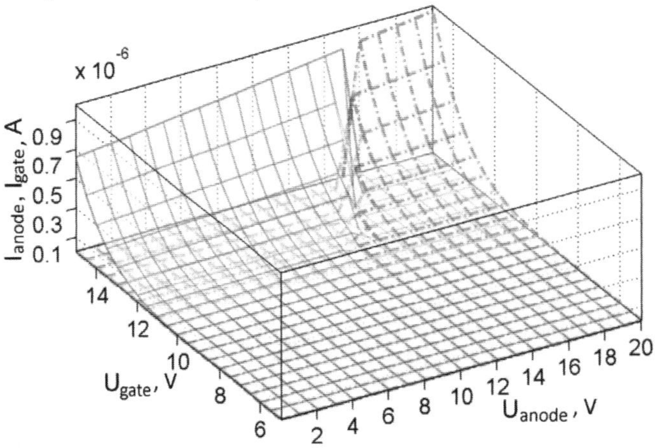

Figure 2. Current-voltage characteristics of the triode cell. The anode current is dotted line, the gate current – solid line.

The thin-film field emitters produced by the technology of electron-beam evaporation are naturally polycrystallic. Cathodes produced this way have the same multiple shape as all thin-film electrodes and the crystal structure of its surface may vary between samples depending on crystal orientation, shape and contents of the substance.

Consequently, physical conditions on the emission surface of cathode are often non-uniform and prone to variance because of asperities, changes in the work function (incl. due to adsorbents) etc. This leads to emission not being observed from all the surface of cathode. For Spindt cathode the real emission surface is the surface of hemisphere with a radius equal to curvature radius of the emitter tip. For example, for radius of 30 nm the emission surface is about 565 sq. nm, and for a multitude of small areas of atomic size the combined emission surface would range from 0.13 sq. nm to 0.2 sq. nm [4] (other sources mention values up to 3 sq. nm [5]).

Accordingly, a considerably smaller voltage is required for a current to appear than one would expect in case of a physically and geometrically uniform emission surface.

Accounting for such minute geometrical details would make modelling of emitters considerably more complex (there are works in that direction: [6-10]). On the other hand, even then the Fawler-Nordheim equation would still describe the Spindt cathode emission current with sufficient confidence, despite that it assumes planar geometry.

Both those facts call for using the Fawler-Nordheim model. To adapt it for our needs, we introduce the local field enchantment factor

$$\gamma(\mathbf{s}) = \frac{E_n(\mathbf{s})}{E_M(\mathbf{s})} \tag{2}$$

where $E_n(\mathbf{s})$ is the normal component of local electric field at point $\mathbf{s}$ of the surface, $E_M(\mathbf{s})$ is the value of surface electric field strength in approximation of flat geometry (i.e. without nano-protrusions and other non-uniformities).

Regarding physical conditions of the surface, we make the following assumptions:

- Work function is constant throughout the entirety of cathode surface (influence of the local changes in work function on the emission current is difficult to separate from field enhancement effect).
- The variable $\gamma(s)$ is replaced with an effective constant γ_{eff}:

$$\int_S \frac{A(\gamma(s)E_M(s))^2}{\Phi t^2(y)} exp\left(-B\frac{\Phi^{3/2}}{\gamma(s)E_M(s)}v(y)\right)ds = \int_S \frac{A(\gamma_{eff}E_M(s))^2}{\Phi t^2(y)} exp\left(-B\frac{\Phi^{3/2}}{\gamma_{eff}E_M(s)}v(y)\right)ds$$

where S is the total surface of cathode, all the other values are the same as in equation (1).

Effective values of local field enchantment factor, that can be experimentally measured by the slope of Fawler-Nordheim line for flat electrodes in vacuum usually vary from several dozens to several thousands of units.

This work is based on modeling of nanotriode with a matrix of classical Spindt cathodes produced by the MIT (USA) [11]. This nanotriode is special in that radius of curvature at the tip of a cathode is just about 1 nm, thus influence of microstructure and nanoprotrusions of emission surface in negligible, and values of γ_{eff} must be close to one. The microtriode matrix consists of 975 emitter cells. Modeling of one cylindrical cell used the following initial parameters: distance between the gate and the anode 500 nm, height of the tip 200 nm, tip curvature radius 1 nm, diameter of gate diaphragm 150 nm, thickness of gate 40 nm, distance between the gate and the base of the cell 195 nm, radius of a cylindrical cell 655 nm, work function value $\Phi=4.2$ eV.

Calculated integral current characteristics are presented on figure 2. In the same time, current characteristic for γ_{eff}=1,15 is the same as on the experimental line presented in [11]. The value of the emission current comes to microamperes, when the voltage is less than 20 V. In dependence of the gate and the anode voltages current of emitted electrons can flow to the gate film (figure 1 (b)) or to the anode film (figure 1 (c)). As a result we have found voltage range corresponding to switching regime of electron current between the anode and gate electrodes. A comparison with edged cylinder-shape emitter cell [12] is next step of the study.

Acknowledgments

Financial support has been obtained from RFBR (13-01-00150) and partially from SPbSU (9.38.673.2013). Research was carried out using computational resources provided by Resource Center "Computer Center of SPbU" (http://cc.spbu.ru/en) and using experimental equipment of the Interdisciplinary Resource Center for Nanotechnology of St. Petersburg State University.

[1] Spindt C A 1968 *J. Appl. Phys.* **39** 3504
[2] Pavlov V G 2004 *Technical Physics* **49** 1610
[3] Nikiforov K A and Egorov N V 2012 Program complex for vacuum nanoelectronics finite element simulations *RuPAC 2012 Contributions to the Proc. of 23 Russian Particle Accelerator Conf. (St. Petersburg)*, (Cern: JACoW) 409–11
[4] Spindt C A, Brodie I, Humphrey L and Westerberg E R 1976 Physical properties of thin-film field emission cathodes with molybdenum cones *J. Appl. Phys.* **47** 5248
[5] Desieres Y, Nicolas P, Sermet F, Geffraye F and Gidon S 2005 Electron optics and electrical studies of a single Spindt-type field emitter *J. Vac. Sci. Tech. B* **23** 640
[6] Mesa G and Saenz J J 1996 Three-dimensional image interaction for nonsmooth emitters *Appl.Phys.Lett.* **69** 1169

[7] Mouton R, Adessi C, Semet V nad Binh V T 2007 Field emission characteristics from patchwork cathodes: a numerical approach *Technical Digest of 20-th International Vacuum Nanoelectronics Conference* 54

[8] Jensen K L, Petillo J J, Feldman D W et al. 2007 Application of a General Electron Emission Equation to Surface Non-Uniformity and Current Density Variation *Technical Digest of 20-th International Vacuum Nanoelectronics Conference* 167

[9] Nikiforov K A, Egorov N V and Shen C-C 2009 Surface reconstruction of a field electron emitter *Surface Investigation X-Ray, Synchrotron and Neutron Techniques* **5** 833–39

[10] Plšek, J, Hrubý P, Nikiforov K and Knor Z 2005 Properties of physisorbed water layers on gold revealed in a FEM study *Applied Surface Science* **252** 1553–60

[11] Jensen K L, Zaidman E G, Kodis M A et al. 1995 Analytical and semi- numerical models for gated FEA's *Technical Digest of 8-th International Vacuum Microelectronics Conference* 216–220

[12] Nikiforov K A, Antonova L I, Egorov N V, Trofimov V V, Makeev V V and Ogurtsov O F 2012 Non-gated field emission array as low-energy electron source: Experiment and simulation *RuPAC 2012 Contributions to the Proc. of 23 Russian Particle Accelerator Conf. (St. Petersburg)*, (Cern: JACoW) 218–20

Localization microscopy study of FtsZ structures in *E. coli* cells during SOS-response

A D Vedyaykin[1, a], A V Sabantsev[1], I E Vishnyakov[2], S N Borchsenius[2], Y V Fedorova[1], A S Melnikov[1], P Yu Serdobintsev[1] and M A Khodorkovskii[1]
[1] St. Petersburg State Polytechnical University, Institute of Nanobiotechnologies, St. Petersburg, Russia
[2] Institute of Cytology, Russian Academy of Sciences, St. Petersburg

E-mail: god-fish@mail.ru

Abstract. Localization microscopy allows visualization of biological structures with resolution well below the diffraction limit. This is achieved by temporal separation of single fluorophore molecules emission and subsequent localization of them with the precision of few tens of nanometers. This method was previously successfully used to obtain images of FtsZ structures in Escherichia coli cells using FtsZ fusion with fluorescent protein mEos2. In this work we obtained superresolution images of FtsZ structures in fixed *E. coli* cells using immunocytochemical labeling. Comparison of superresolution FtsZ structures in cells undergoing SOS-response and "healthy" cells shows that FtsZ structures are partially disassembled during SOS-response, but still retain some periodicity.

1. Introduction.

Localization microscopy is a novel fluorescence microscopy technique, surpassing the diffraction limit of resolution by an order of magnitude [1]. The increased resolution is extremely valuable in the study of biological structures. This method is based on creating conditions that render only small fraction of fluorescent molecules in the sample visible at the same time so that images of individual molecules are separated. This allows these molecules to be localized up to few tens of nanometers. A series of sample images (typically from few thousands to few tens of thousands) is collected and molecule positions are determined by computer algorithm. The principle of localization microscopy is illustrated by figure 1. One of the most convenient ways to control the number of fluorescent molecules in the sample is to use photoactivatable or photoconvertible dye to lable the structure of interest and adjust the power of activating/converting laser to achieve desired fluorescent molecule density. One of the most promising areas of localization microscopy application is microbiology as most of the prokaryotic cells have the sizes comparable to the resolution limit of conventional fluorescent microscopy, which renders this method largely inapplicable to the study of inner structures of these tiny cells [2].

[a] To whom any correspondence should be addressed.

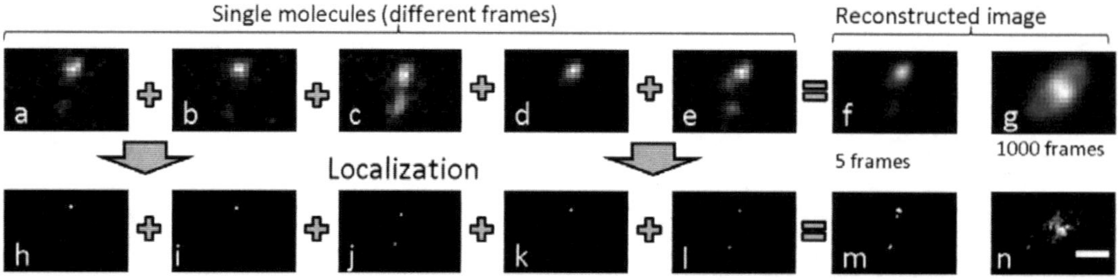

Figure 1. Localization microscopy scheme. Time separated conventional fluorescent images of single molecules (a-e) are processed using software, this results in localized images (h-l). Next these localized images are combined into image with improved resolution: m – image reconstructed from 5 frames, n – from 1000 frames. There are conventional fluorescent images of these 5 and 1000 frames on figures f and g respectively. Scale bar – 1 um.

FtsZ is a prokaryotic tubulin homolog that plays a crucial role in cell division, participating in the formation of the septum between two dividing cells [3, 4]. Localization microscopy was previously used to visualize the structure of the Z-ring formed by FtsZ at the septum using FtsZ fusion with fluorescent protein [5]. Furthermore, well before FtsZ was shown to form spiral-like structures in mutant *Escherichia coli* cells [6]. Fluorescent labeling of proteins can introduce artifacts both when fluorescent proteins or antibodies are used. It is therefore important to compare structured obtained using different labeling techniques. It is also known that FtsZ polymerization is one of the main regulation points in cell-division arrest during SOS-response [7]. The aim of this work was to compare FtsZ superresolution structures in "healthy" *E. coli* cells and cells undergoing SOS-response and also to compare superresolution FtsZ structures obtained using fusion proteins and immunocytochemical labeling.

2. Materials and methods.

2.1. Bacterial strain and immunofluorescent staining.

For experiments with normally dividing cells *E. coli* Top10 strain was used. For SOS-response experiments *E. coli* strain KD 403 with CRISPR-system (genes casA,B,C,D,E,1 and 2 under the control of arabinose-inducable promoter), carrying a spacer, complimentary to the site in bacteria genome was used. Induction of casA-E,1,2 expression resulted in bacterial DNA double-stranded brakes and caused SOS-response.

Four hours after the induction cells were immobilized on poly-L-lysyne coated coverslips with nonspecifically adsorbed polystyrene microspheres (2.35 μm, Bangs Laboratories). Then cells were fixed with methanol-water-acetic acid solution (3:4:1) for 10 minutes. Next bacteria were washed 3 times with PBS and permeabilized with lysozyme in GTE buffer (glucose 50 mM, Tris 32.5 mM pH=7.5, EDTA 10 mM). After blocking in 2% w/v BSA for 1 hour, cells were incubated with anti-FtsZ rabbit primary antibodies (Agrisera, 1 to 200) overnight. Then bacteria were washed 5x in 0.05 % Tween and stained with Cage 635 Goat anti-Rabbit secondary antibodies (Abberior, 1 to 100).

2.2. Experimental setup and data processing.

The observation of samples were performed using experimental setup based on Axioimager.Z1 (Carl Zeiss) upright microscope (see figure 3) with extended facilities (including quadrant back focal plane detection and multicolor fluorescent microscopy with laser excitation) [8].

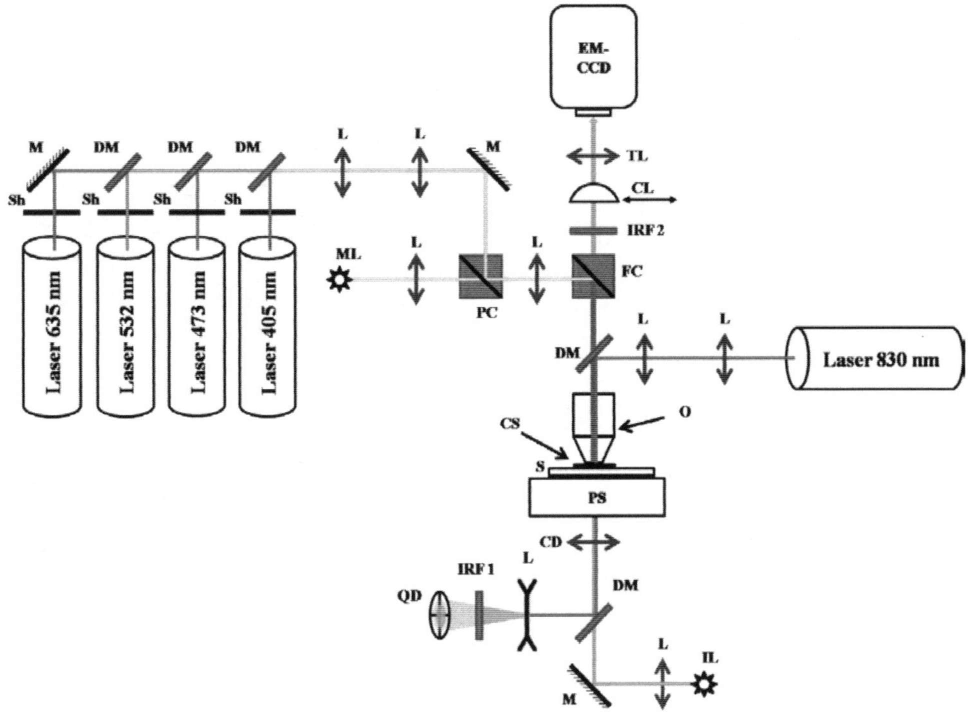

Figure 2. Experimental setup. *CD – microscope condenser, CL – cylindrical lens, CS – coverslip, DM – dichroic mirror, EM-CCD – microscope camera, FC – fluorescence filter cube, IL – transmitted light lamp, IRF 1 – infrared filter, passing 830 nm, IRF 2 – infrared filter, blocking 830 nm, L- lens, M – mirror, ML – mercury lamp for fluorescence excitation, O – microscope objective, PC - polarizing cube, PS – piezo stage, QD – quadrant detector, S – sample, Sh – shutter, TL – microscope tube lens.*

Cage 635 was excited by 635 nm diode laser (power density about 10 kW/cm^2 in focal plane) and constantly activated by 405 nm diode laser (100 W/cm^2). To reconstruct a superresolution image 1000 frames were collected and analyzed using QuickPALM [9] plug-in for ImageJ [10]. Due to increased resolution and long acquisition time localization microscopy is extremely sensitive to the sample drift. To overcome this problem we use an active sample stabilization system based on quadrant detection of 2.35 μm polystyrene microsphere position. It can achieve precision sufficient to render drift effect on localization microscopy resolution neglectable in combination with high refresh rate (up to few hundred Hz) which allows it to compensate higher-frequency fluctuations of sample position compared to CCD-based stabilization. Low requirements for sample preparation allows this method to be easily incorporated into existing microscopy sample preparation protocols.

3. Results and discussion.

Localization microscopy of *E. coli* Top10 cells undergoing normal cell division (figure 3) showed that FtsZ forms various structures including Z-ring between two dividing cells (figure 3, column 4,5) and spiral-like structures along the cell membrane (figure 3, column 3). These observations are consistent with previous data obtained using fluorescent protein mEos2 [5].

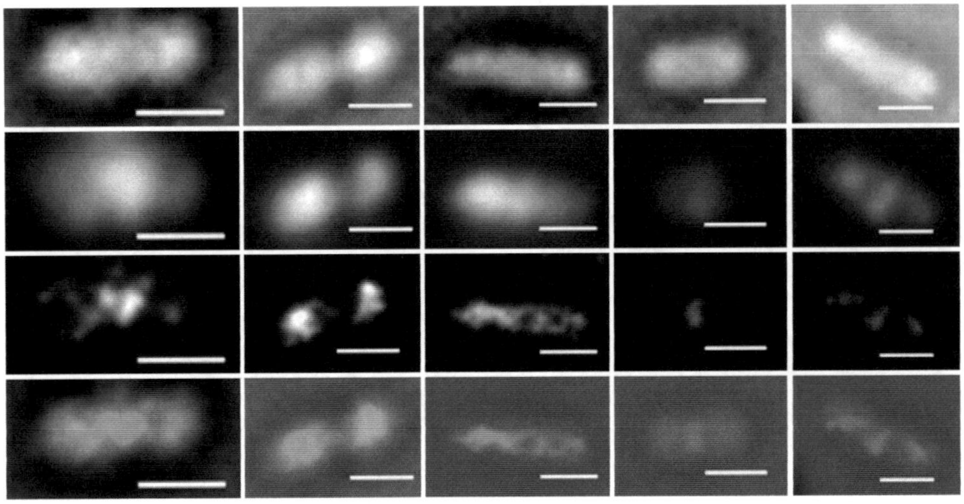

Figure 3. Structures formed by FtsZ in "ordinary" cells. Transmitted light (a), "conventional" fluorescence (b), localization microscopy (c) and composite of a and c images (d). For clarity, the glow of the fluorophore on the composite image is shown in red. Scale bar – 1 μm.

Figure 4 shows a localization microscopy images of FtsZ in *E. coli* cell during SOS-response (b), corresponding transmitted light image (a) and "conventional" fluorescent image (c) obtained by averaging 2000 frames of the sample that were used for the reconstruction of a superresolution image. This images show evidence of the regular structure with the period of approximately 1 μm, but no ring-like structures are formed which could explain the division arrest in these cells. It was shown that SulA protein induced during SOS-response can partially inhibit FtsZ polymerization [7]. To test whether SulA is responsible for the observed change in FtsZ pattern either SulA inhibition or *sulA* gene deletion/knock-out mutation can be used. Three-dimensional localization microscopy should allow us to obtain a better view of the structures formed by FtsZ.

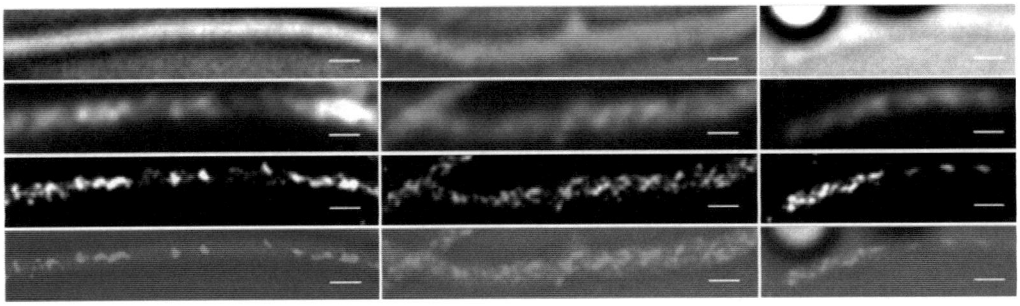

Fig. 4. FtsZ structure in *E. coli* during SOS-response. Transmitted light (a), "conventional" fluorescence (b), localization microscopy (c) and composite of a and c images (d) of a bacteria in SOS-response with fluorescently-labeled FtsZ. Bar – 1 um.

The authors would like to thank Dr. Konstantin Severinov and Dr. Kirill Datsenko for the provided *E. coli* KD 403 strain. Part of the work was carried out using scientific equipment of the Center of Shared Usage "The analytical center of nano- and biotechnologies of SPbSPU" with financial support from the Russian Foundation for Basic Research (mol_a 12-04-31536).

4. References

1. M.J. Rust, M. Bates, X. Zhuang. Sub-diffraction-limit imaging by stochastic optical reconstruction microscopy (STORM). // Nat Methods, 2006. 3(10): p. 793-5.
2. C. Coltharp, J. Xiao. Superresolution microscopy for microbiology. // Cell Microbiol, 2012. 14(12): p. 1808-18.
3. S.G. Addinall, E. Bi, J. Lutkenhaus. FtsZ ring formation in fts mutants. // J Bacteriol, 1996. 178(13): p. 3877-84.
4. P. de Boer, R. Crossley, L. Rothfield. The essential bacterial cell-division protein FtsZ is a GTPase. // Nature, 1992. 359(6392): p. 254-6.
5. G. Fu, T. Huang, J. Buss, C. Coltharp, Z. Hensel, J. Xiao. In vivo structure of the E. coli FtsZ-ring revealed by photoactivated localization microscopy (PALM). // PLoS One, 2010. 5(9): p. e12682.
6. E. Mileykovskaya, Q. Sun, W. Margolin, W. Dowhan. Localization and function of early cell division proteins in filamentous Escherichia coli cells lacking phosphatidylethanolamine. // J Bacteriol, 1998. 180(16): p. 4252-7.
7. D. Trusca, S. Scott, C. Thompson, D. Bramhill. Bacterial SOS checkpoint protein SulA inhibits polymerization of purified FtsZ cell division protein. // J Bacteriol, 1998. 180(15): p. 3946-53.
8. A. Sabantsev, I. Vishnyakov, A. Vedyaykin, G. Pobegalov, S. Borchsenius, M. Khodorkovskiy. «Study of structures formed by FtsZ in Escherichia coli and Mycoplasma hominis cells». // FEBS Journal, Special Issue: 38th FEBS Congress, Saint Petersburg, Russia, 2013. 280(Sup. s1): p. 469.
9. R. Henriques, M. Lelek, E.F. Fornasiero, F. Valtorta, C. Zimmer, M.M. Mhlanga. QuickPALM: 3D real-time photoactivation nanoscopy image processing in ImageJ. // Nat Methods, 2010. 7(5): p. 339-40.
10. T.J. Collins. ImageJ for microscopy. // Biotechniques, 2007. 43(1 Suppl): p. 25-30.

Complexes of carbon nanotubes with oligonucleotides in thin Langmuir–Blodgett films to detect electrochemically hybridization

A S Egorov[1], V P Egorova[2], H V Krylova[3,5], I V Lipnevich[3], T I Orekhovskaya[4], A A Veligura[2], M I Govorov[1] and B G Shulitsky[4]

[1] Institute of Physiology, Nat. Acad. Sci. of Belarus, 28 Academicheskaya Str, 220072 Minsk, Belarus
[2] Belarusian State Pedagogical University, 18 Sovetskaya Str, 200005 Minsk, Belarus
[3] Physics Department, Belarusan State University, 4 Nezavisimosti Ave., 200030 Minsk, Belarus
[4] Belarusian State University of Informatics and Radioelectronics, 6 P. Brovki Str, 200013 Minsk, Belarus

Abstract. Self-assembled complexes consisting of thin multi-walled carbon nanotubes (MWCNTs) and DNA-oligonucleotides which are able to a cooperative binding to complementary oligonucleotides have been investigated. It was establised a high-performance charge transport in nanostructured Langmuir–Blodgett complexes thin MWCNTs/DNA. A method to electrochemically detect DNA hybridization on the self-organized structures has been proposed.

1. Introduction

An unique property of nucleotide bases is to be arranged by stacking interactions on a surface of carbon nanotubes (CNTs). CNT can be complexified with DNA due to the property. Carbon nanostructures such as graphene and CNTs are characterized by very weak scattering of charge carriers. Langmuir–Blodgett (LB) technique is a nanotechnology which allows to fabricate highly ordered two-dimensional (2D) clusters of functionalized CNTs. These clusters can be spontaneously polarized by virtue localization of the carriers ballistically moving in "zig-zag"-CNT [1]. In [2] we have proposed a method to synthesize new DNA-oligonucleotide complexes of few-walled (thin MWCNTs with diameter d ~ 2–5 nm) and multi-walled carbon nanotubes (thick MWCNTs with diameter of 10 nm) under conditions of self-organization.

One needs bioactive ligand-binding DNA-containing model structures with carboxylated carbon nanotubes to carry out an effective nanobiodetection of DNA hybridization. It is possible to develop a new generation of biosensors – bionanodetectors when fabricating of synthetic carbon nanotubes with low dispersion. Bionanodetecting of DNA – genotyping of SNP (Single Nucleotide Polymorphism) should combine mutually exclusive properties. On the one hand, the oligonucleotides must interact with detector high sensitivity ("reporter") sites more strongly than with nonspecific molecular agents, but, on the other hand – form a specific complex with DNA targets through weak hydrogen bonds. As

[5] Corresponding author, E-mail: grushevskaja@bsu.by

known [3, 4], methods of covalent binding "reporter" groups distort significantly the conformation of the DNA, and, respectively, the hybridization process is not sufficiently specific (ambiguous). It was shown [5] that pyrenyl conjugates of oligonucleotides can "coagulate" on the single-walled nanotubes surface into self-organized electron-dense nanostructures. In this case, the self-assembly was a side effect, which prevents the molecular recognition. However, we will represent self-assembled complexes MWCNTs/DNA, which are able to a cooperative binding to complementary oligonucleotides.

A goal of the paper is to establish a high-performance charge transport in nanostructured LB-complexes thin MWCNTs/DNA and show the ability of such heterostructure to efficient DNA hybridization.

2. Materials and methods

Oligonucleotide ON1 (structure sequence: 5'-GCCATATACTCTCCTTGGTGACA-3'), oligonucleotides ON2 and ON4, complementary and non-complementary to the ON1, as well as oligonucleotides ON5 and ON3 with one and three non-complementary bases, respectively, were synthesized at «Praymteh" (Minsk, Belarus). We will utilize MWCNTs obtained by chemical vapor deposition method with subsequent covalent modification of their surface by carboxyl groups.

A complexification is performed by two methods. First method is ultrasound treatment of a mixture from ON1 with thin MWCNTs, suspended in TE buffer at various concentrations. Second method is to form MWCNT/oligonucleotide complexes by LB-technique with additional non-covalent functionalization of MWCNTs by stearic acid molecules [6].

Electron microscopy images are obtained with help of a transmission electron microscope (TEM) JEM-100CX TEM (JEOL, Japan) at an accelerating voltage of 100 kV.

Electrochemical detection was carried out by the method [7]. A sensor has an interdigital structure from the N pairs of twisted aluminum electrodes, which are arranged on the pyroceramics substrate and have octagonal shape. One utilizes the electrodes system with N = 10, 15. Each pair of electrodes is an capacity of "open"-type. A nanocomposite dielectric coating of electrodes is a metal-containing LB film of thiophene-pyrrole series oligomer on sublayer of nanoporous anodic oxide [8].

Impedance analysis has been performed in deionized (resistivity>18.0 M Ω cm at 25 °C).

3. Results and discussion

3.1. Structural analysis

MWCNTs have open ends (see figure 1a). Figure 1b demonstrates that flexible ultrathin CNTs are helically twisted with a pitch of helix ~ $1/d$.

(a) (b)

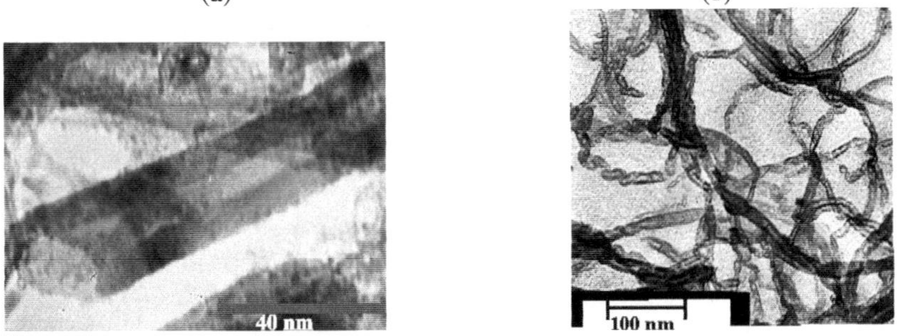

Figure 1. TEM-images of MWCNTs.

A distribution of ss-DNA-oligonucleotide / MWCNT (ss-DNA/MWCNT) in TE buffer after the ultrasonic treatment is shown in figure 2. ss-DNA/MWCNT complexes hold three-dimensional (3D) structure of cable type (figure 2), a core of which is the conductive MWCNT, and a shell ("casing") – 3D self-organized oligonucleotide layer dark-colored in comparison with bright white CNT-image with sharply delineated dark edge. The dark color indicates a high density of the nucleotide self-organized on the surface of CNT, and, hence, the complexes under investigation are электронно-плотные наноструктуры. The stacking interactions decrease a polarization of the CNT in direction transverse to CNT surface (spontaneous polarization) and stiffen CNT walls, whereby the pitch and the radius of helix becomes very large (figure 2b).

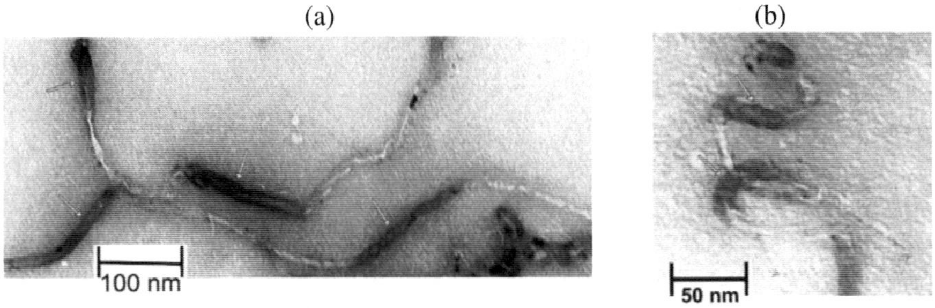

Figure 2. TEM-images of complexes ss-DNA/MWCNT in TE buffer.

3.2. Electro-chemical properties

A electrochemical detecting is performed for two frequency ranges : low-frequency (~ 100 kHz) and high frequency (~ 400 – 800 kHz) ones. A Maxwell polarization of conducting inclusions gives main contribution to polarization in the low-frequency region. A dipole polarization vector of molecules or molecular groups relaxes in the high-frequency region. 2D LB-CNT-clusters effectively shield the electrodes due to charge ballistic transport in MWCNTs. Therefore, ionization of water molecules in an electric field of the electrodes is suppressed, and, respectively, the capacitance C_{CNT} of the sensor with 2D LB-CNT-clusters is smaller than the capacitance of the pure sensor. A decrease of C_{CNT} in the high frequency region is $\Delta C_{CNT} \sim -1 \div -1.5$ pF and practically does not depend on the geometry of the sensor. The dielectric permittivity of the oligonucleotide shell in ss-DNA/MWCNTs LB-complexes is high, and a shielding degree is considerably lowered because the charge transport happens mainly transversely to helix axes. Therefore, an increase of capacitance $C_{ss\text{-}DNA/CNT}$ by $\Delta C_{ss\text{-}DNA/CNT} > 0$ takes place. $\Delta C_{ss\text{-}DNA/CNT}$ in the high frequency region is practically independent on the sensor geometry also, but has a larger range of values: from +2 to +6 pF.

The sensors with ss-DNA/MWCNT LB-complexes were incubated about 10 minutes in solutions of different nucleotides. Then, free oligonucleotide is washed off 10 min. from the sensor surface in water at a temperature of 39 °C.

Recordings in the low frequency region allow to estimate an increase of the conductive 2D LB-CNT-clusters portion when releasing ON1 from ss-DNA/MWCNTs to form duplexes ds-DNA. One can observe in figure 3 that the recognized oligonucleotides forming duplexes ds-DNA, partially liberate CNTs. This leads to a capacitance decrease towards to capacitance values for the sensor with conductive 2D LB-CNT-clusters only. Figure 3 demonstrates *S*-shaped dependence of low-frequency module $|\Delta C|$, $\Delta C < 0$ of electrical capacitance displacement due to the cooperative binding of ON2 with ON1. The value of the capacitance of the sensor with the homoduplexes ON1/ON2 is closest to $C_{ss\text{-}DNA/CNT}$ owing to the effect of phase separation in the phase transition from ds-DNA/MWCNTs complexes to double-stranded DNA. This effect is absent when forming of heteroduplexes ON1/ ONi, $i = 3, 4, 5$.

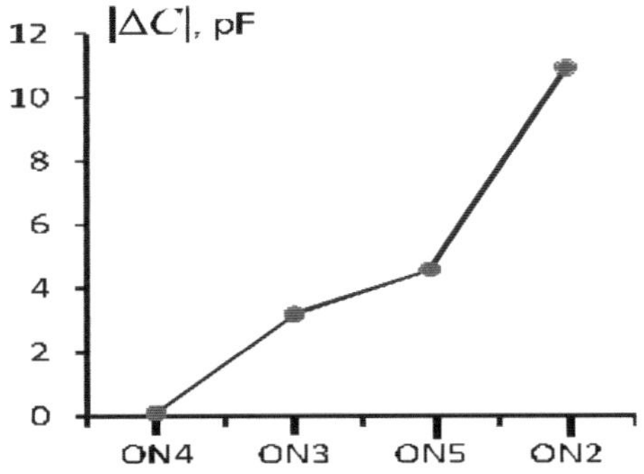

Figure 3. A low-frequency electrical capacitance displacement module $|\Delta C|$, $\Delta C < 0$ due to the interaction of ON1 with oligonucleotides ON2, ON3, ON4, ON5.

4. Conclusion

Thus, a highly effective transport of electric charge holds in such nanoheterostructures as LB-complexes thin MWCNTs/DNA-oligonucleotides. A method to electrochemically detect DNA hybridization on the self-organized structures has been proposed.

5. References

[1] Grushevskaya H V, Hrushevsky V V, Krylova N G, Kulahava T A, Lipnevich I V, Orekhovskaya T I, Semenkova G N and Shulitsky B G 2010 *J. Nonlin. Phen. in Complex Sys.* **13** 396

[2] Veligura A A, Egorov A S, Egorova V P, Krylova H V, Lipnevich I V and Shulitsky B G 2013 *Scientific Herald of the* Uzhgorod University. Series Physics. **32** 94

[3] Ha T 2001 *Curr. Opin. Struct. Biol.* **11** 287

[4] Van Hooijdonk E, Bittencourt C, Snyders R 2013 *Beilstein J. Nanotechnol.* **4** 129

[5] Aparzin E K, Novopashina D S, Nastaushev Yu V and Venyaminova A G 2012 *Russian Nanotechnologies* **7** 38

[6] Egorov A S, Krylova H V, Lipnevich I V, Shulitsky B G, Baran L V, Gusakova S V and Govorov M I 2012 *J. Nonlin. Phen. in Complex Sys.* **15** 121

[7] Abramov I I , Hrushevsky V V, Krylov G G, Krylova H V, Lipnevich I V and Orekhovskaya T I 2012 *Petersburg J. Electronics.* **4 (73)** 59

[8] Abramov I I, Krylova H V, Lipnevich I V and Orekhovskaya T I 2013 Aluminum oxide based nanostructures for impedance nanoelectronics *Microwave & Telecommunication Thechnology, Electrochemical detection: Proc. 23rd Int. Crimean Conf. (Sevastopol, Crimea, Ukraine, 8–13 September 2013)* (Moscow, Kiev: Wb) pp 851–852.

Photoelectric properties of MSM structure based on ZnO nanorods, received by thermal evaporation and carbothermal syntesis.

D A Zhilin[1,2], N V Lyanguzov[2], L A Nikolaev[1,2], V I Pushkariov[1,2] and E M Kaidashev[1,2].

[1] Vorovich Mechanics and Applied Mathematics Research Institute, Southern Federal University, Russia, Rostov-on-Don, Stachki street 200/1

[2] Physics Department, Southern Federal University, Russia, Rostov-on-Don, Zorge street 5

E-mail d.a.zhilin@mail.ru

Abstract. Photoelectric characteristics of metal-semiconductor-metal (MSM) structures: Au/ZnO(nanorods)/ZnO(film)/ZnO(nanorods)/Au were investigated. Synthesis of ZnO nanorods (NR's) was carried out by two different methods, such as catalyst-free carbothermal synthesis (t = 950°C, precursor - ZnO:C) and catalyst-free thermal evaporation of metallic Zn (t = 600°C, precursor Zn). Photoluminescence spectra have shown that the ZnO NR's obtained by means of thermal evaporation technique have a lower concentration of point defects related with oxygen vacancies, than ZnO NR's prepared by carbothermal synthesis. MSM structure with ZnO NR's obtained by thermal evaporation technique have photosensitivity 97 mA/W at ~ 325 nm illumination and 44 mA/W at ~ 518 nm illumination at a bias 5V. For MSM structure with ZnO NR's obtained by method of carbothermal synthesis, the photosensitivity values were 22 mA/W, and 103 mA/W, respectively.

1. Introduction

Recently, interest to a one-dimensional ZnO nanostructures for fundamental research has increased due to their specific characteristics, such as the dimensional quantum effects and large surface area of zinc oxide nanostructures. Also, there is a huge potential for the use of such structures in nanoelectronics [1, 2]. Furthermore ZnO has several special properties: high adsorption capacity, radiation resistance, transparency in the visible range of the electromagnetic radiation and direct wide band gap (3.37 eV) [3, 4]. ZnO can form a Schottky barrier with metals such as Pt and Au [5], and as is known, such photodiode structures have higher performance than the conventional p-n junction. All of these properties are very attractive for creation high performance photosensitive structure based on Schottky barrier Au/ZnO, and operating in the ultraviolet (UV) spectrum.

The main aim of this work was studying of the photovoltaic curves of metal-semiconductor-metal (MSM) structure: Au/ZnO(NR's)/ZnO(film)/ZnO(NR's)/Au. These photodetectors can be used in areas where the control of UV radiation is needs, for example, to monitor solar radiation, UV radiation in air and water disinfection systems.

2. Experiment

MSM structures were produced on the Si (100) substrate with thin oxide layer SiO_2. Deposition of ZnO thin film was carried out by pulsed laser deposition (PLD) in an oxygen atmosphere, $P(O_2) = 2\times10^{-2}$ mbar at a substrate temperature of 500°C, the distance between the target and the substrate was 50 mm. For the ablation of ZnO target was used KrF-excimer laser with wavelength of 248 nm and frequency f = 10Hz. The flux density of the laser pulse energy was ~ 2.3 J/cm^2. ZnO film used as a sublayer reducing the barrier nucleation during the growth of nanorods ZnO. ZnO nanorods growth was carried out by two different methods: catalyst-free vapor phase transport growth (thermal evaporation of metallic Zn) and catalyst-free carbothermal synthesis. In the vapor phase transport synthesis [6], evaporation of zinc produced in the argon and oxygen atmosphere at 600°C. The argon flow was 240 sccm, oxygen flow was 8 sccm and synthesis pressure was 15 mbar (scheme of the reaction chamber shown in Fig. 1a). In the carbothermal synthesis (scheme of the reaction chamber shown in Fig. 1b) [7, 8] was used pressed tablet of ZnO and C as a precursor (1:1 on the molar mass). The synthesis temperature was 950°C. The argon flow is 200 sccm and synthesis pressure is 20 mbar.

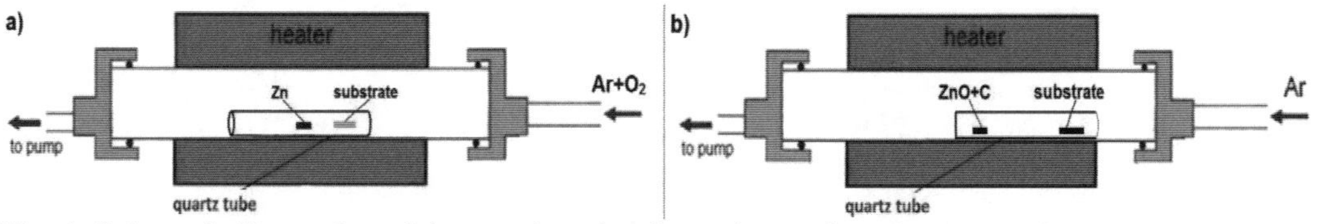

Fig. 1. Schematic illustration of the reaction chambers: a) - catalyst-free vapor phase transport growth (in this case Zn precursor is located in the center of a quartz tube, which is open on both sides) and b) - catalyst-free carbothermal synthesis (it uses a quartz tube with one closed (soldered) end).

After the synthesis of ZnO nanorods, Au contacts were deposited on the nanorods surface through a mask by PLD at room temperature under vacuum value 2×10^{-5} mbar, a distance between the target and the substrate was 50 mm, the laser wavelength λ = 248 nm, frequency f = 10 Hz, the energy density was 3.7 J/cm^2. During the deposition structure was located at 45° to the axis of the plasma torch, for uniform deposition of gold on the surface of the nanorods ZnO. Thus was obtained a structure based on a double Schottky barrier: Au/ZnO(NR's)/ZnO(film)/ZnO(NR's)/Au (Fig. 2).

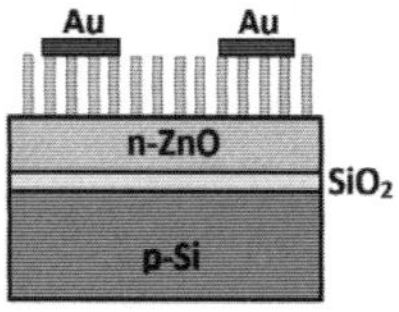

Fig. 2. Schematic view of the MSM structure.

3. Experimental results and discussion

The main difference between two techniques is a lower temperature of synthesis at vapor phase transport method (600°C) than at the carbothermal synthesis (950°C), resulting in a lower defect concentration in ZnO NR's [9]. This fact is proved by photoluminescence spectra (Fig. 3), which show that the ZnO NR's produced by vapor phase transport synthesis have a high peak in the UV region of the spectrum with a peak wavelength of ~ 380 nm, and low peak in the green region of the spectrum, at wavelength ~ 525 nm (Fig. 3a). This fact can indicate low concentration of point defects related with oxygen vacancies in samples. Opposite situation is observed for ZnO nanorods, obtained by carbothermal method. (Fig. 3b).

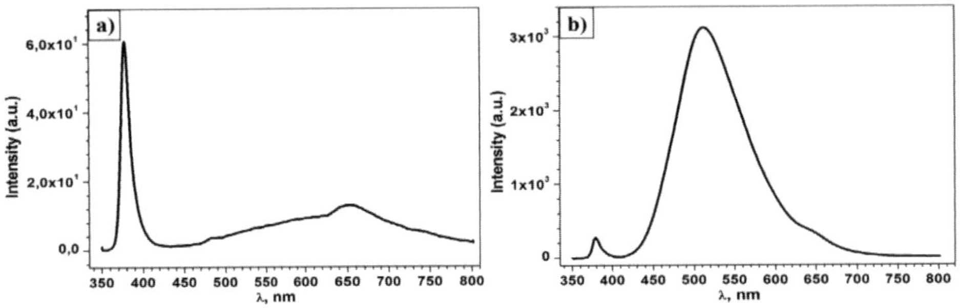

Fig. 3. Photoluminescence spectra of ZnO NR's, obtained a) - catalyst-free vapor phase transport growth and b) - catalyst-free carbothermal synthesis.

I-U curves (Fig. 4a, 4b) of MSM photodetectors (Au/ZnO(NR's)/ZnO(film)/ZnO(NR's)/Au) shows that studied structures have sensitivity in the visible and ultraviolet region of the spectrum. MSM structure with ZnO NR's prepared by the method of thermal evaporation, has a high photosensitivity in the ultraviolet region of the spectrum (~ 325 nm) - 97 mA/W, and low photosensitivity in the green region (~ 518 nm) - 44 mA/W (Fig. 4b). For structures prepared by carbothermal synthesis photosensitivity values were 22 mA/W and 103 mA/W respectively at a bias 5V (Fig. 4a). It is caused by low concentration of defects in ZnO NR's produced by vapor phase transport synthesis.

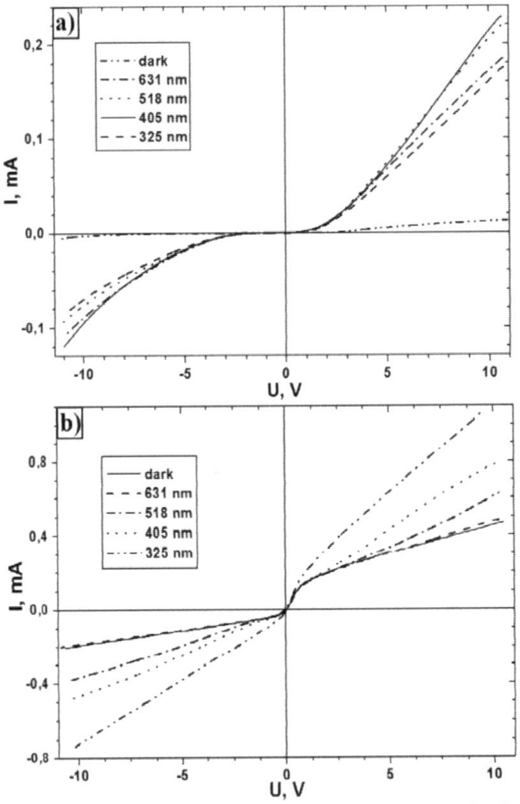

Fig. 4. I-U curves of MSM structure: a) – ZnO NR's produced by carbothermal method; b) – ZnO NR's produced by thermal evaporation.

Bandgap of ZnO is 3.37 eV (corresponding to the energy of photons with a wavelength ~ 375 nm), and hence the sensitivity at a wavelength of ~ 325 nm is easily explained by the intrinsic absorption of ZnO. Thus photons have a higher energy than the band gap of a semiconductor, so the optical excitation occurs from the valence band to the conduction band. Then charge separation occurs at the Schottky barrier. In addition, these structures has a photosensitivity in the visible region of the spectrum, as seen from Fig. 4a and 4b. This is due to the presence of point defects related with oxygen vacancies in ZnO. So, if the donor defect levels are in the band gap of ZnO, then electron transfer between these levels and bands requires minimal energy. In this case the generation of charge carriers will occur even when excited by light with photon energies below the band gap of zinc oxide.

The temporal response (Fig. 5) of MSM structure for ZnO NR's, produced by the method of thermal evaporation, was received by laser pulses excitation with 10 ns pulse length (Nd-YAG laser, λ = 355 nm) at a bias 5V. Rise time was ~ 1 μs. The fall time at full-width at half-maximum was ~ 10 μs.

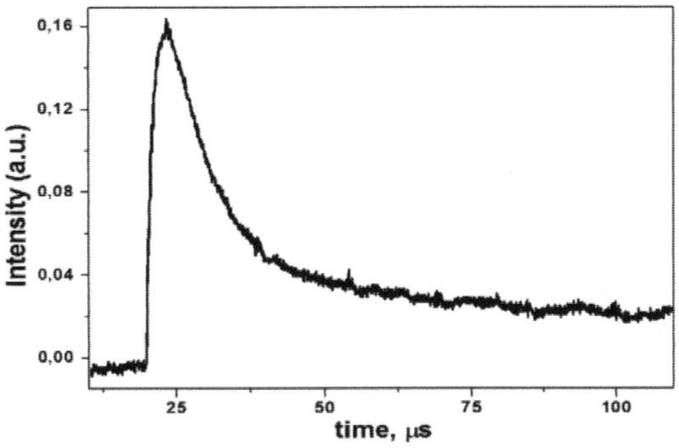

Fig. 5. a) - temporal response of MSM detectors based ZnO NR's excited by Nd-YAG laser (λ = 355 nm, 10 ns).

Due to good radiation durability, high photosensitivity and performance our photodetectors could find wide application in many areas where the control of UV radiation needs, for example, to monitor the UV solar radiation, UV radiation monitoring in air and water disinfection devices, etc [10].

References

[1] Logeeswaran V J, Jinyong Oh, Avinash P N, Aaron M K, Kristin H G, Sonia G, Nobuhiko P K, Shih-Yuan Wang and et al. 2011 IEEE *Journal of Selected Topics in Quantum Electronics* **17** №4 1002-1032
[2] Heo Y W, Norton D P, Tien L C, Kwon Y, Kang B S, Ren F, Pearton S J, LaRoche J R 2004 *Materials Science and Engineering R* **47** 1-47
[3] Mridha S and Basak D J 2007 *Applied Physics* **101** 08102
[4] Cheng-Pin Chen, Pei-Hsuan Lin, Liang-Yi Chen, Min-Yung Ke, Yun-Wei Cheng and JianJang Huang 2009 *Nanotechnology* **20** 245204
[5] Leonard J Brillson and Yicheng Lu 2011 *Journal Of Applied Physics* **109** 121301
[6] Pushkariov V I, Lyanguzov N V, Kaidashev E M 2013 *Inženernyj vestnik Dona* № 4
[7] Lyanguzov N V, Kaidashev E M, Zakharchenko I N, Bunina O A 2013 *Letters to the Journal of Technical Physics* **39** 17 27-34
[8] Lyanguzov N V, Zhilin D A, Kaidashev E M 2014 *Advanced Materials Physics, Mechanics and Applications, Series: Springer Proceedings in Physics* **152**
[9] Alexey A S, Samuel A F, Stefan T B, Richard A C, Huub J J van Dam and Paul S 2007 *Faraday Discussions* **134** 267-282
[10] Abdulgafour H I, Hassan Z, Ahmed N M, and Yam F K 2012 *Journal of Applied Physics* **112** 074510 1-9

Optoelectronic method for analysis of biomolecular interaction dynamics

E. Nepomnyashchaya, E. Velichko, E. Aksenov, T. Bogomaz
Saint-Petersburg State Polytechnical University

E-mail: elina.nep@gmail.com

Abstract. Optoelectronic method of laser correlation spectroscopy for study of intermolecular interaction in biomolecular suspension is presented. The method of laser correlation spectroscopy is integrated with orthogonal laser light scattering and ultramicroscopy technique for visual control of biomolecular interactions. The capabilities of the method for analysis of biomolecular conglomerates dynamics are considered.

1. Introduction

A number of pathological processes in human body are accompanied by molecular parameters shift and changes in biological fluids [1-2]. For instance, autoimmune deceases cause changes in work of regulatory proteins of immune system and complement system, and in some cases the complement inhibitor introduction is necessary. So, express analysis of inhibitors and their interaction with complement system is important task in medicine [3-4].

Complement system should be tested in liquid medium with blood plasma that limits use of electron, tunnel and atomic-force microscopy. One of the effective optoelectronic methods of biological fluid media investigation is laser correlation spectroscopy. Recently the laser correlation spectroscopy method became a standard technique for measurement of such static characteristics as size, charge and diffusion constant of particles in range of 0.5-100 nm [5-6]. But resources of the method in investigation of biomolecular interaction dynamics are not studied enough [7].

The aim of our work is the development of laboratory model of small-sized portable laser correlation spectrometer with one of the measuring channels based on orthogonal light scattering with visual control for analysis of biomolecular dynamics in water suspensions. The light scattering was chosen as a complementary method of investigation due to its simplicity and measurement capabilities of evaluation of biomolecular interaction characteristics, affording solution-based measurements of dynamic equilibrium properties and kinetic phenomena [5].

2. Experimental setup and samples

2.1. Experimental setup

The procedure of measurements combined two methods – light scattering analysis (which includes laser correlation spectroscopy and orthogonal light scattering techniques) and ultramicroscopy technique. The scheme of the experimental setup is presented in figure 1.

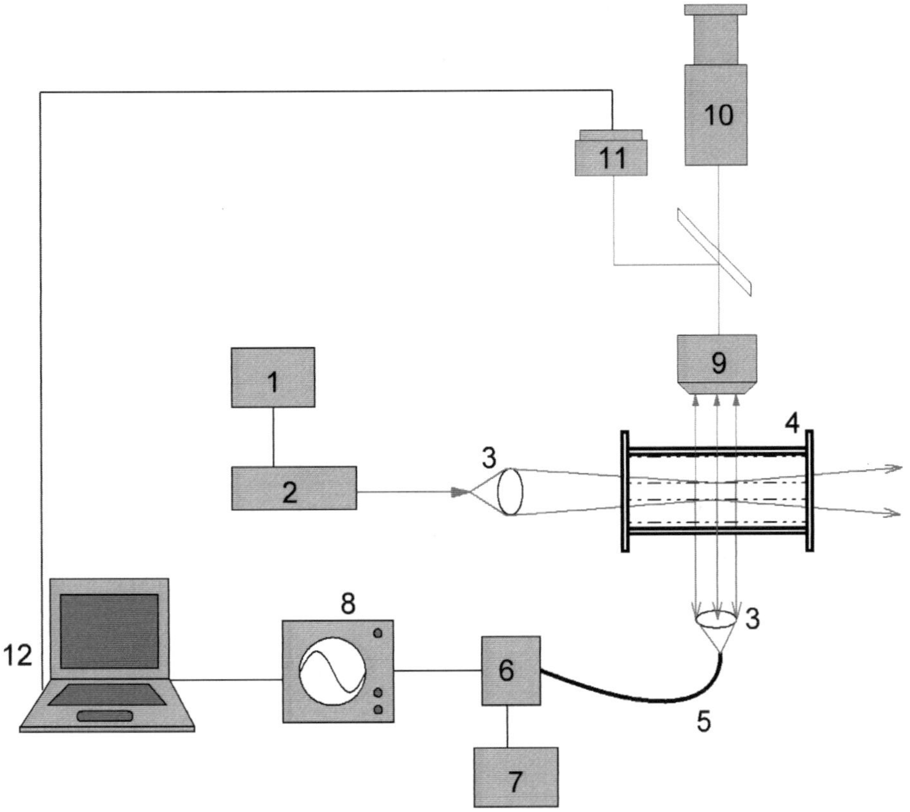

Figure 1. Experimental setup for investigation of biomolecular dynamics in solution: 1 – laser power supply; 2 – junction laser; 3 – focusing lenses; 4 – optical cell with a solution; 5 – optical fiber; 6 – photomultiplier tube; 7 – power supply; 8 – oscilloscope; 9 –microscope lens; 10 – microscope eyepiece; 11 – CCD camera; 12 – computer.

Light from the junction laser (KLM-G650-13-5) (2) with wavelength 650 nanometers, emissive power 13 mW and diameter of a ray 5 mm passed through the focusing lens (3) with focal distance 1.4 mm. The light was focused on the optical cell with water suspension of biomolecules (4) and the beam with the maximum power density passed through the transparent cuvette completely filled with anolyte solution. The scattered light was registered by photomultiplier tube (H11706-01 Hamamatsu) (6) which may be positioned in different ways for registration of light scattered on angle degree in a range from 30^0 to 180^0. .The signal from photodetector was transmitted to the oscilloscope (8) and to the computer (12) with software for correlation and statistical analysis. The optical cell with solution was located directly under a lens of a microscope (9), the angle between the laser ray and optical axis of microscope was 90^0. The scattered light image was visualized via microscope (10), registered by a CCD camera (11) and also transmitted to the computer (12) for further treatment.

Additional visualization of orthogonal scattered light allowed us to analyze character of conformation changes of biomolecular conglomerates structure and their relocations within the volume under the influence of external factors in real time. Such visual control gives additional information which can be very helpful in medical applications.

2.2. Samples
In our preliminary experiments the samples of distilled water suspensions of albumin, glycine and insulin at concentrations 0.001, 0.01, 0.05, 0.1 and 1 % were measured. The solution volume in the optical cell was equal to 0.3 ml. Measurements were carried out in standard conditions (room temperature was 23^0 C, relative humidity was 48 %, normal atmospheric pressure).

3. Results and discussion
Experimental results on measurement of scattered light characteristics of biomolecular suspensions in dependence on concentration, acidity of solution, ionic force of the solution, conformational structure and other factors were conducted. Albumin is one of the main blood and liquor proteins, so in this work we present results received on albumin solutions.

In figure 2 the dependence of average intensity of orthogonal scattered light on concentration of albumin in distilled water solution is presented. The received data revealed that average intensity of scattered light rises with increasing concentration which was expected [8].

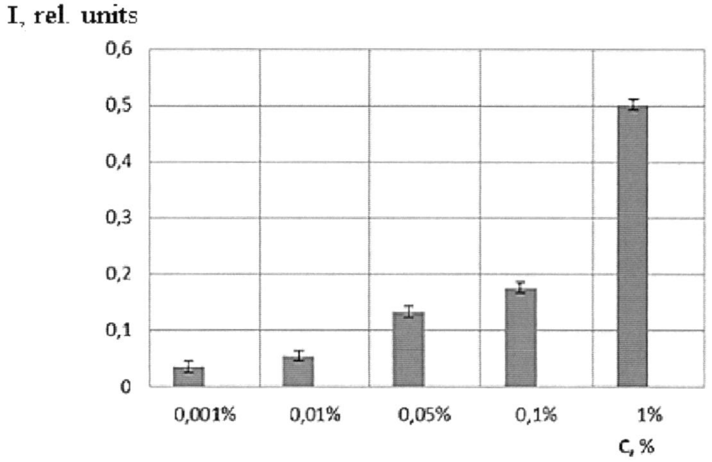

Figure 2. Dependence of orthogonal scattered light intensity on concentration of albumin in solution.

For investigation of biomolecular dynamics/kinetics in peptide solutions the influence of media acidity and ionic force on conglomerate formation was studied. The medium acidity dependence was investigated for distilled water albumin solution (at concentration 1 %). In case the hydrochloric acid concentration reached the value 2 % the size of diffusers in solution (which was clearly seen due to visualization technique) and registered intensity of scattered light significantly increased. The rise of intensity of scattered light continued with acidity increasing till the acidity concentration was 4 %. Further increasing of solution acidity had no influence on scattered light intensity and we could observe the saturation region on the graph. Addition of electrolyte (NaCl) in initial solution caused the increase in the amount of scattered particles at lower acid concentrations (0.3 %), which was in agreement with work [9]. We measured solutions with ionic force μ from 0.0025 to 0.85 mol/L. The received data revealed that increasing of solution ionic force affects the conglomeration process and protein aggregation ability. The dynamic of conglomerates formation was observed by CCD camera which allowed us to evaluate correspondence between intensity values and aggregation processes in solution.

More detailed numerical data were received by mathematical treatment of the results of laser correlation spectroscopy method – by registration of the signal via photomultiplier tube and oscilloscope and further correlation analysis. Correlation function of the signal scattered on solution with different refractive indexes includes information about spectral structure of the signal, which is defined by the Doppler shifts caused by Brownian motion of particles. Spectral shift depends on the articles velocity, and according to Stokes's – Einstein equation (3), is directly connected with particle sizes. Thus, the evaluation of scattered light correlation function allows us to receive data about particle sizes and interaction level.

In the experiment polydisperse systems were investigated, and the correlation functions were on the form of a descending exponential curves superposition:

$$G(\tau) = \int_0^{\infty} A(\Gamma)e^{-\Gamma\tau}d\Gamma\,, \tag{1}$$

where τ is the time, $A(\Gamma)$ is distribution of the rate of decay.

According to cumulants method we consider equation (1) as the Taylor series expansion of $G(\tau)$ in small τ:

$$G(\tau) = G(0)\exp\left[-\sum_{n=1}^{\infty} K_n \tau^n \frac{1}{n!}\right], \tag{2}$$

where K_n are cumulants which can be expressed through the central moments.

In the solution of these equations only the first cumulants are considered, and the minimal residual method is used. Taking into consideration that $\Gamma = Dq^2$ (where D is the diffusion coefficient) and Stokes-Einstein's formula:

$$D = \frac{k_b T}{6\pi\eta R}, \tag{3}$$

where T is the temperature, η is the viscosity, we can calculate the hydrodynamic radius of studied particles – R.

In figure 3 the example of the received signal of dispersion on solution of the bull serumal albumine (BSA) in distilled water is presented.

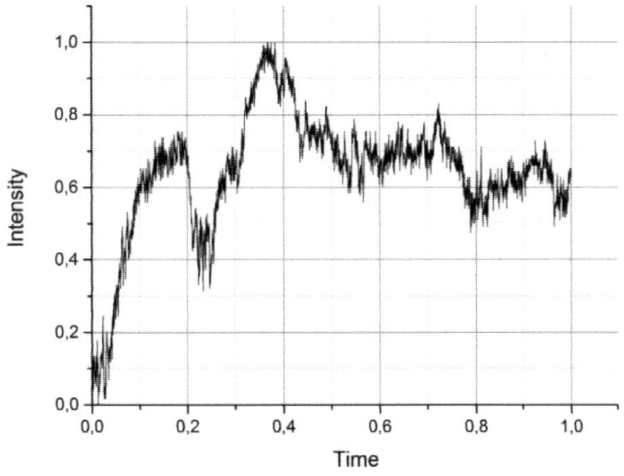

Figure 3. Example of the registered signal of scattered light from distilled water albumin solution (angle was 90^0).

Taking into consideration that the solution was a monodispersed, we calculated the value of diffusion coefficient of albumine molecules $D = 7{\cdot}10^{-7}$ cm^2/s. Using Stokes-Einstein's formula we also calculated the average hydrodynamic radius of a particle R = 5 nm. According to tabular data the BSA molecules has an ellipsoid form with half radiuses of 1,7 nm and 4,2 nm, which are in good agreement with the received results.

In the case the concentration of the hydrochloric acid in solution was 3% the diffusion coefficient was $D = 4{\cdot}10^{-7}$ cm^2/s, and the average hydrodynamic radius of particles increased ($R = 7$ nm).

Received data demonstrated that introduced method of measurements of scattered light characteristics allows investigating structure, interactions and kinetics of biomolecular systems in liquid mediums.

4. Conclusion

The utility of introduced optoelectronic method of investigation, which includes both laser correlation spectroscopy and visual control, was confirmed. Subsequently our approach will be realized as a model of small-sized portable laser correlation spectrometer with one of the measuring channels providing visual control. This spectrometer can be used for diagnostics of biomolecular solutions in diagnostics of such socially significant diseases as oncology, diabetes, cardio-vascular and autoimmune diseases and for assistance in therapy decision.

References

[1] Booth C S 2012 *Chemical & Biomolecular Engineering Theses, Dissertations, & Student Research.* **12** 147

[2] Liu R, Li M, Zhi-Ping L, Wu J, Luonan C and Kazuyuki A 2012 *Scientific Reports* **2** 813

[3] Hanlon A D, Larkin M I and Reddick R M 2010 *Biophysical J.* **98** 297–304

[4] Le T M, Paul J S, Al-Nashash H, Tan A, Luft A R, Sheu F S and Ong S H 2007 *IEEE transactions on medical imaging* **26** 833

[5] Some D 2013 *J. Biophys. Rev.* **5** 147

[6] Minton A P 2007 *Biophysical J.* **93** 1321

[7] Some D and Kenrick S 2012 Characterization of protein-protein interactions via static and dynamic light scattering *Protein interactions* ed J Cai (USA: Intech) chapter 20 pp 402–426

[8] Petrova G P, Petrysevich U M and Ten D I 2002 *J. Quantum electronics* **32** 10

[9] Petrova G P, Petrysevich U M and Evseevicheva A N 1998 *J. Vestnik of the Moscow university. Series 3. Physics. Astronomy* **4**

Different photodynamic effect between continuous wave and pulsed laser irradiation modes in k562 cells in vitro.

V.V. Klimenko[1], A.A. Bogdanov[1], N.A. Knyazev[1], A.A. Rusanov[2], M.V. Dubina[1]

[1] St. Petersburg Academic University, St. Petersburg, Russia
[2] First Pavlov State Medical University of St. Petersburg, St. Petersburg, Russia

E-mail: klimenko.vova@gmail.com

Abstract. Photodynamic therapy is a cancer treatment method is used primarily continuous mode laser radiation. At high power density irradiation occurs intense consumption of molecular oxygen and this caused hypoxic tumor tissue, which leads to inefficiency PDT. In this paper, pulsed and continuous irradiation modes during PDT photosensitizer Radachlorin were compared. A mathematical model for the generation of singlet oxygen $^1O_2$ in tumor cells during photodynamic therapy with tissue oxygenation was developed. Our study theoretically and experimentally demonstrates the increased singlet oxygen generation efficiency in a pulsed irradiation mode compared to continuous wave mode with the same power density 20mW/cm^2. Experimental in vitro showed that pulsed irradiation mode mostly induces apoptosis k562 tumor cells at irradiation doses of k562 1.25 - 2.5J/cm^2 while the continuous mode induced necrosis.

1. Introduction

Photodynamic therapy (PDT) has now reached the level of being an accepted treatment for several types of cancer. PDT is based on production of singlet oxygen ($^1O_2$) by energy transfer from light-excited photosensitizer molecules in target tissue. Singlet oxygen effectively oxidizes many kinds of biomolecules, leading to damage and cell death. The introduction of semiconductor heterostructures and development of semiconductor-based diode lasers [1] have significantly accelerated the evolution of PDT. Today's preferential use of semiconductor lasers is primarily owing to their low cost, ease of treatment and the ability to match their emission peak with absorption of any particular photosensitizer. High power light irradiation in continuous mode and porphyrin photosensitizers (absorption wavelength 600-700 nm) are mainly applied in clinic now [2, 3] to achieve an impactful penetration depth in tumor tissue. Despite the fact that such treatment is successful enough it has some critical disadvantages connected with tissue decreased oxygenation, overheating and photosensitizer bleaching. Moreover, continuous mode PDT typically results in necrotic cell death with further inflammatory response. Semiconductor lasers give us opportunity to apply different regulations including easy light power switching and pulse mode irradiation. There is some literature data that pulse mode PDT in vitro leads to apoptotic cell death mechanism [4, 5]. In present work we have investigated the difference of continuous and pulse irradiation modes and their biological effect in vitro.

2. Theoretical PDT model

In our study we used macroscopic PDT theoretical model adapted from [6]. Briefly, we exploited differential system eq. (1)-(4) to calculate cumulative concentration of singlet oxygen. This model is good for quality description of triplet oxygen flux impact on singlet oxygen generation in PDT.

$$\frac{d[S_0]}{dt}+\left(\xi\sigma\frac{I([S_0]+\delta)[{}^3O_2]}{[{}^3O_2]+\beta}\right)[S_0]=0 \tag{1}$$

$$\frac{d[{}^3O_2]}{dt}+\left(\xi\frac{I[S_0]}{[{}^3O_2]+\beta}\right)[{}^3O_2]-g\left(1-\frac{[{}^3O_2]}{[{}^3O_2](t=0)}\right)=0 \tag{2}$$

$$\frac{d[{}^1O_2]}{dt}-\left(\xi\frac{I[S_0]}{[{}^3O_2]+\beta}\right)[{}^3O_2]=0 \tag{3}$$

$$[{}^1O_2]_{cumulative}=\int_0^t\left(\xi\frac{I[S_0]}{[{}^3O_2]+\beta}\right)[{}^3O_2]dt \tag{4}$$

, where $[S_0]$ is the ground state sensitizer concentration, $[{}^3O_2]$ and $[{}^1O_2]$ are the ground triplet and excited singlet state oxygen concentration, respectively, I is intensity of laser irradiation, g is maximum oxygen supply rate, δ is a low photosensitizer concentration correction term, I is intensity of laser irradiation. Simulation PDT process was performed using the photochemical parameters shown in Table 1. For continuous wave and pulse modes were applied parameters: intensity of laser irradiation - 20mW/cm^2, pulse duration – 200 ms, interval between pulse – 500 ms. These pulse mode parameters satisfy maintaining a high level of triplet oxygen concentration.

Table 1. Photochemical parameters for Photofrin at 630 nm and initial conditions used for the macroscopic model.

Symbol	Value	References
$\xi(cm^2mW^{-1}s^{-1})$	3.7 x 10^{-3}	[7]
$\beta(\mu M)$	11.9	[7]
$\sigma(\mu M^{-1})$	7.6 x 10^{-5}	[7]
$g(\mu M/s)$	0.7	[6]
$[{}^3O_2](\mu M)$	100	[6]
$[S_0](\mu M)$	7	[6]

We can see from figure 1 and figure 2 that cumulative singlet oxygen concentration is higher in a pulse mode than in continuous one. Moreover, we observe that efficiency of singlet oxygen generation is higher in pulse mode. On the other hand the average singlet oxygen generation rate at pulse mode is several times smaller.

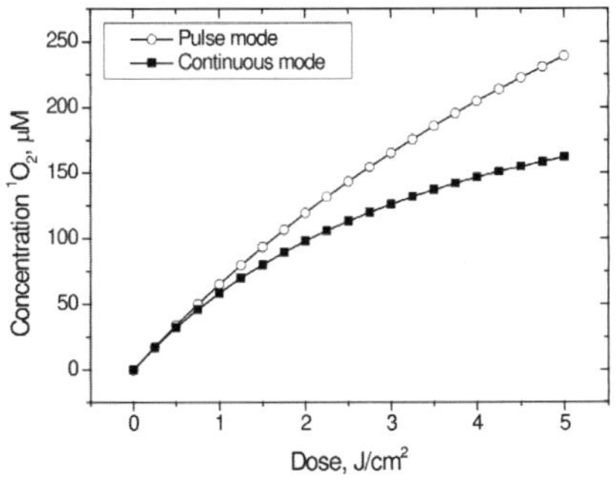

Figure 1. The dependence of singlet oxygen cumulative concentration on irradiation dose applied for pulse and continuous wave irradiation modes.

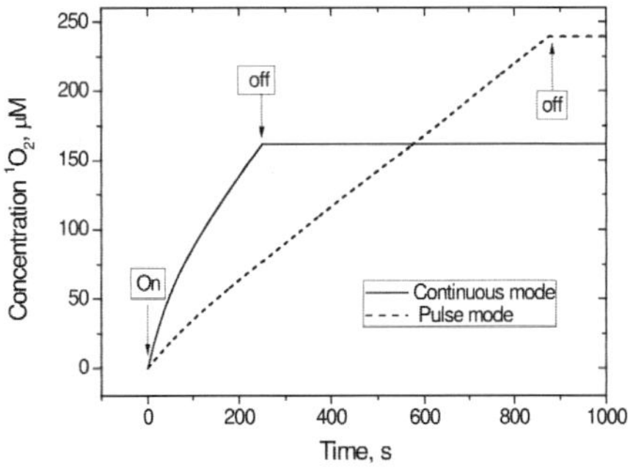

Figure 2. The dependence of singlet oxygen cumulative concentration on time for pulse and continuous wave irradiation modes.

3. PDT experiment in vitro.

In experimental studies in vitro cancer cell line k562 (Bank of Cell Cultures, Institute of Cytology, RAS, St. Petersburg, Russia), Radachlorin® (RADA-PHARMA Co, Ltd., Moscow, Russia) photosensitizer (PS) and semiconductor laser for PDT «LAHTA – MILON» (MILON Laser, LLC, St. Petersburg, Russia) were used. $1\cdot 10^6$ cells per well k562 cells were seeded in 6-well plates and incubated during 12 hours with PS in 5 μg/ml concentration. After incubation cells were washed with PBS buffer and irradiated with laser source in different modes. For irradiation in pulse mode we applied parameters described above. Estimation of cell viability was performed before irradiation, 2 h and 24 h after irradiation using an EPICS XL flow cytometer (Backman Coulter, United States) and standard protocols. Nonirradiated cells served as a control.

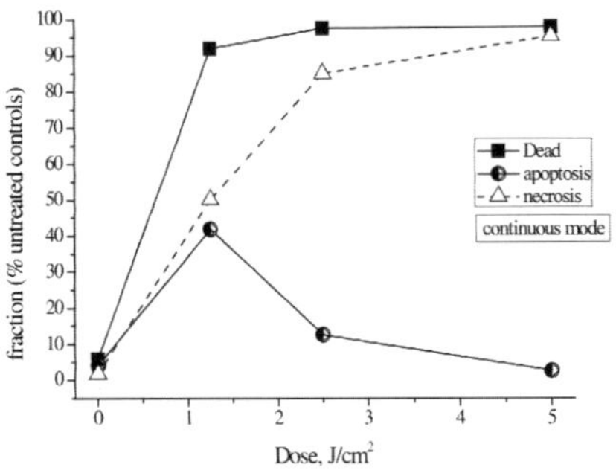

Figure 3. Dose-dependent distribution apoptosis and necrosis fraction k562 cells in total count of dead cells at continuous mode irradiation 20mW/cm^2 with 5μg/ml Radachlorin concentration.

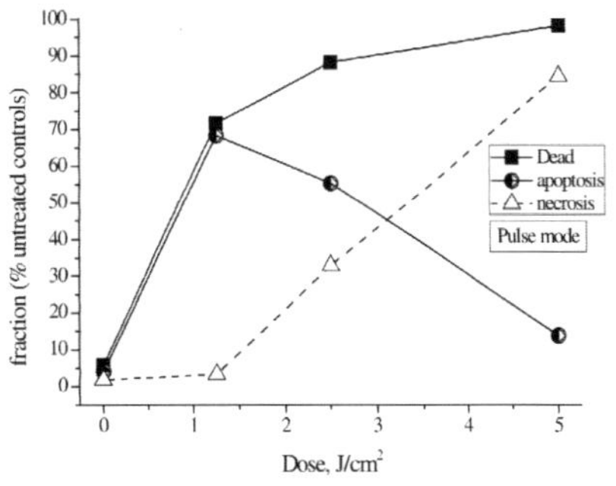

Figure 4. Dose-dependent distribution apoptosis and necrosis fraction k562 cells in total count of dead cells at pulse mode irradiation 20mW/cm^2 with 5μg/ml Radachlorin concentration.

The figure 3 and figure 4 shows that fraction of necrosis cells at all doses in case of continuous mode is greater than the same one in pulse mode. On contrary in pulse mode we have the convenience of apoptosis except 5J/cm^2. The total count of dead cells is almost equal for both modes. We can consider irradiation doses used as penetrating doses in different depth of tumor tissue. In that case we make conclusion that pulse mode can provide the same cell destructive potency with preferable biological effect – apoptosis.

4. Discussion

Taking into account all data obtained, we can conclude that the main factor in biological effect during PDT is the speed of singlet oxygen generation. Low speed of singlet oxygen generation leads to apoptosis and high speed – to necrosis. The next factor facilitating cell death in PDT is cumulative singlet oxygen concentration, which directly connected with total molecular damages. Therefore, results given above prove the hypothesis about major efficiency of pulse mode in comparison with continuous one in terms of singlet oxygen generation and apoptosis initiation.

References

[1] Zhores A 2000 Double Heterostructure Lasers: Early Days and Future Perspectives *IEEE J. Sel. Top. Quant. Electron.* vol 6 n 6

[2] Agostinis P *et al* 2011 Photodynamic Therapy of Cancer: An Update. *CA Cancer J Clin.* **61**(4) pp 250-81

[3] Oleinick N, Morris R and Belichenko I 2002 The role of apoptosis in response to photodynamic therapy: what, where, why, and how *Photochem. Photobiol. Sci* **1** pp 1–21.

[4] Yuuichi M, Yukihiro U and Tsuyoshi N 1999 Comparison of phototoxicity mechanism between pulsed and continuous wave irradiation in photodynamic therapy *J. Photochem. Photobiol. B: Biol.* **53** pp 53–59

[5] Kawauchi S, Morimoto Y, Sato S, Arai T, Seguchi K, Asanuma H and Kikuchi M 2004 Differences between cytotoxicity in photodynamic therapy using a pulsed laser and a continuous wave laser: study of oxygen consumption and photobleaching *Lasers in Medical Science* **18** pp 179–183

[6] Wang K, Finlay J, Busch T, Hahn S and Zhu T 2010 Explicit dosimetry for photodynamic therapy: macroscopic singlet oxygen modeling. *J Biophotonics* **5-6** pp 304-18

[7] Georgakoudil I, M G Nichols and Foster T H 1997 The Mechanism of Photofrins Photobleaching and Its Consequences for Photodynamic Dosimetry *Photochemistry and Photobiology* **65**(1) pp 135-144

Journal of Physics: Conference Series 541 (2014) 012041 doi:10.1088/1742-6596/541/1/012041

Angular measurements of light scattered by the glucose containing biological tissues and their phantoms

K G Domnin, G A Cherevatenko, E T Aksenov

Department of Quantum Electronics, Institute of Physics, Nanotechnology and Telecommunications, Saint-Petersburg State Polytechnical University, Polytechnicheskaya, 29, St. Petersburg 195251 Russia.

E-mail: gcherevatenko@gmail.com

Abstract. Method of optical polarimetry is well known, but earlier it was mainly used to low scattering media. In this paper we consider the possibility of measuring the glucose concentration by detecting polarization of the backscattered laser light. Dependencies of the degree of polarization and of the optical rotation on the registration angle were received. Furthermore, the significant impact of the wavelength and the power of probing beam on the degree of polarization was shown.

1. Introduction

The development of noninvasive and continuous or repeated methods of monitoring blood glucose concentration is one of the priorities of modern biomedical diagnostics. Due to the urgency of the problem, a number of methods of non-invasive measurement of glucose concentration in the blood are developing. However, none of the existing methods has not yet reached the stage of commercial production, so research in this area continues.

Recently, several key areas in the optical measurement of glucose concentration in tissues have seen progress, such as: infrared spectroscopy; Raman spectroscopy, polarimetry and fluorescence spectroscopy. Furthermore, the preliminary positive results were obtained by the methods of optoacoustics and optical coherence tomography [1]. One of the most perspective methods is polarimetry of backscattered light. Polarimetry has been used for investigation the optically homogeneous mediums, earlier polarimetric measurements in turbid media considered to be impossible. However, a number of researchers [2, 3] have shown that laser light backscattered by turbid media (which includes most of biological tissues) can have sufficient degree of polarization for the corresponding measurements. Furthermore it is known that change in the glucose concentration in biological tissues affects their optical properties. One of the dominant effects is reducing of the scattering coefficient with increasing glucose concentration. The scattering coefficient of biological tissue is determined by the difference in refractive index of the extracellular fluid and the cellular membranes. Increasing the concentration of glucose in the extracellular fluid leads to an increase its refractive index, which in turn reduces the total scattering coefficient of the tissue. As a result, the photon path length varies, and hence the polarization of the scattered light is also changed. In connection with this, we determined the objectives of this work: to develop a method of measurement of glucose concentration in biological tissue and their phantoms based on turbid polarimetry.

2. Methodology

The light entering a turbid medium, including the majority of biological tissues, is repeatedly scattered and absorbed, which results in the broadening of the laser beam and its attenuation during propagation through the medium. Volume scattering is the main cause of the significant diffuse light propagation in the backward direction [4]. For each observation point of the scattered light at the tissue surface, the ensemble of the effective optical paths that light travels from the source to this point was characterized by different optical paths. The corresponding trajectories occupy a spatial region, having a «banana» shape. The analytical expression for the photon path distribution of such shape was derived [5]

$$P(x,y,z)=\frac{z^2\exp[-k((x^2+y^2+z^2)^{\frac{1}{2}}+((r-x)^2+y^2+z^2))^{\frac{1}{2}}]}{[x^2+y^2+z^2]^{3/2}[(r-x)^2+y^2+z^2]^{3/2}}\times \\ \times(k(x^2+y^2+z^2)^{\frac{1}{2}}+1)(k((r-x)^2+y^2+z^2))^{\frac{1}{2}}+1) \quad (1)$$

where the source is placed at the origin (0, 0, 0) and the detector is placed at (r, 0, 0), x and y are directed along the surface, z – normal to the surface, $k=\sqrt{3\mu_a\mu_s'}$, μ_a – absorption coefficient, μ_s' – reduced scattering coefficient, $\mu_s'=\mu_s(1-g)$, g – anisotropy factor. Particularly, the most likely path (center line of the "banana" region) in the case of the weak absorption limit, $kr<<1$, can be defined as

$$z_0(x)\approx\left(\frac{1}{8}\left[(x^2+(r-x)^2)^2+32x^2(r-x)^2-x^2-(r-x)^2\right]^{\frac{1}{2}}\right)^{\frac{1}{2}}. \quad (2)$$

At $x = r/2$, the central line of the "banana" region reaches a maximum depth of $z_0^{\max}\approx\frac{\sqrt{2}r}{4}$. For the strong absorption limit $kr>>1$

$$z_{\mathrm{mod}}(x)\approx\left(\frac{2x(r-x)}{(3\mu_a\mu_s(1-g)r)^{\frac{1}{2}}}\right)^{\frac{1}{2}}, \quad (3)$$

reaching the maximum value $z_{\mathrm{mod}}^{\max}\approx\sqrt{r/2k}$ at $x=r/2$.

On the other hand, if we consider the vector nature of the light and characterize the photon by its Stokes vector $S=(I, Q, U, V)$, then the motion of such a photon in the scattering media can be described by a sequence of variation in the Stokes vector associated with the photon. Each scattering event is accompanied by a variation of the Stokes vector for this photon. The new Stokes vector $\boldsymbol{S}_{n+1}$ can be defined as

$$S_{n+1}=M_k(\theta)M_n(\phi)S_n, \quad (4)$$

where $M_k(\theta)$ – Muller matrix of the scattering particle, $M_n(\phi)$ – matrix, that describes the rotation of the Stokes vector around the axis, specifying the direction of the photon's propagation before interaction [6]. Thus, the plane of polarization of the photon is rotated. Since this rotation is relatively random, the detected photon will be randomly polarized and, hence, the detected light will be partially depolarized, where the degree of polarization will decrease with the increasing order scattering.

The change in the concentration of glucose in turbid media leads to a number of processes, that affect its light-scattering properties, herewith, the scattering coefficient and the polarization of the scattered light vary. Below, there are the dependencies of the refractive index (5), scattering coefficient (6) and the anisotropy factor (7) on the glucose concentration in the intralipid (sterile fat emulsion for intravenous infusion) solution, which along with suspension of the microspheres and the milk solution used to simulate biological tissues [7]

$$n = 1.325 + 1.515 \cdot 10^{-6} \cdot C_{gl}, \tag{5}$$

$$\mu_s = (1 - 0.0022 \cdot C_{gl} / 18) \cdot \mu_{sm}, \tag{6}$$

$$g = (1 + 0.000007 \cdot C_{gl} / 18) \cdot g_m, \tag{7}$$

where C_{gl} [mg/dl] – glucose concentration, g_m и μ_{sm} – scattering anisotropy factor and scattering coefficient of intralipid solution, respectively, before the glucose is added. Apparently, the increase in the glucose concentration of such media leads to a decrease in the difference between the refractive index of scatterers and the media $\Delta n = n_s - n_{medium}$, which in turn will decrease the reduced scattering coefficient: $\mu_s' \downarrow$.

On the other hand, the addition of glucose to human blood leads to a slightly different effect. The optics of whole blood is mainly determined by the optical properties of the red blood cells and plasma. Increasing the concentration of glucose in the blood leads to an increase in the reduced scattering coefficient $\mu_s' \uparrow$. This effect is related to the penetration of glucose into the red blood cells and its reaction with hemoglobin inside cells [8].

Furthermore, glucose, due to its chirality, has the ability to rotate the polarization plane of the linearly polarized light passing through the solution. The angle of rotation is directly proportional to the concentration of the optically active molecules, the length of the interaction layer and a constant of the substance, called the specific rotation. The resulting angle is defined as

$$\varphi = \alpha_\lambda L C_{gl}, \tag{8}$$

where α_λ has the dimension [deg•dm^{-1}•g^{-1}•l] and represents the specific rotation at wavelength λ, L – length of the sample in dm, C_{gl} – concentration in g/l. This angle becomes significant, due to the fact that the photon pathlength in turbid media is considerable (1).

Thus, the growth of the glucose concentration in turbid media will alter the polarization of the scattered light, as a result of changing the scattering coefficient and the optical chirality of the glucose molecules. Moreover, it was shown in the Ref. 9 that for different detection angles the optical rotation and the degree of polarization would be different for the same glucose concentration. The experimental study [9] has shown that in the forward direction glucose chirality makes the greatest contribution to the optical activity, and in other directions optical activity is dominated by scattering effects. Therefore we can discuss the optimal choice for the detection angle of the scattered light from the standpoint of the maximum sensitivity to changes in glucose concentration.

3. Experimental setup

The spatial distribution of the light backscattered by turbid media were investigated. To conduct this study we developed a laboratory set (figure 1), consisting of: laser with a polarizer fixed on it, and the cuvette or a holder for samples, photodetector, which position could change relative to the direction of probe beam, rotating analyzer was installed on the photodetector. Furthermore, the significant impact of the wavelength and power of probing beam on the degree of polarization was noticed. Therefore, a semiconductor laser with a wavelength 650 nm was used as the light source, which power was regulated to achieve a maximum degree of polarization.

Experiments were conducted on solutions of milk, human blood and the human finger pad. The cow's milk was used for creation of the human skin phantom. In milk the fat globules and colloidal protein particles (casein micelles) act as scatterers. Such media can be used as a biological tissue phantom, as is strongly scattering with scattering anisotropy factor g> 0,7, i.e. close to tissues.

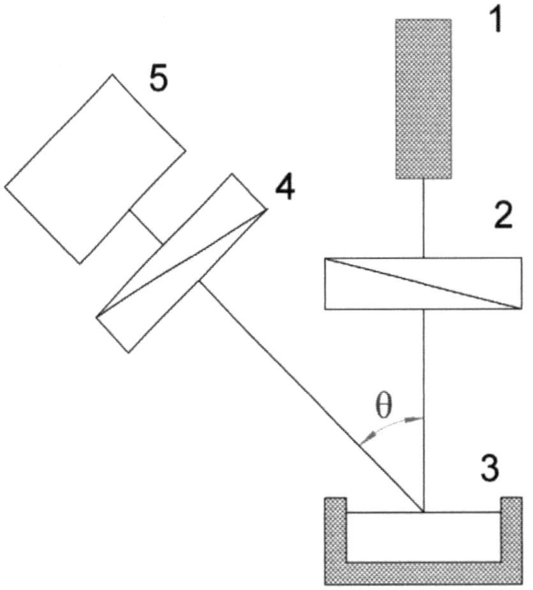

Figure 1. Scheme of experiment setup: 1 – laser, 2 – polarizer, 3 – cuvette with sample, 4 – rotating polarizer, 5 – photodiode.

4. Results

Experiments with solutions of milk and human blood shown that the dependences of the parameters of the scattered light on the registration angle θ are symmetric with respect the probe beam axis. The figure 2 shows the dependences of the light intensity (voltage on the photodiode), degree of polarization (DOP) and optical rotation of light scattered by milk solution (1:4 with distilled water) and splution of human blood (1:1 with NaCl solution) on the registration angle for different glucose concentrations in the solutions.

The experimental results show that the intensity (figure 2 a, d) and degree of polarization (figure 2 b, e) of the light decrease with increasing angle θ, and the optical rotation increases (figure 2 c, f) for both samples. Increasing the concentration of glucose causes a slight change in the intensity for the solution of milk (figure 2 a) and a substantial increase for the solution of human blood for all registration angles (figure 2 d), which is the result of reduced scattering coefficient increased. The degree of polarization evenly decreases with increasing glucose concentration for milk (figure 2 b), while for the blood the dependence is more complex (figure 2 e). The optical rotation increases in the case of milk for all registration angles (figure 2 c) and vice versa decreases for the blood at all angles (figure 2 f), but the most significant absolute change for both samples was observed at the minimum values of θ, i.e. closest to the opposite direction.

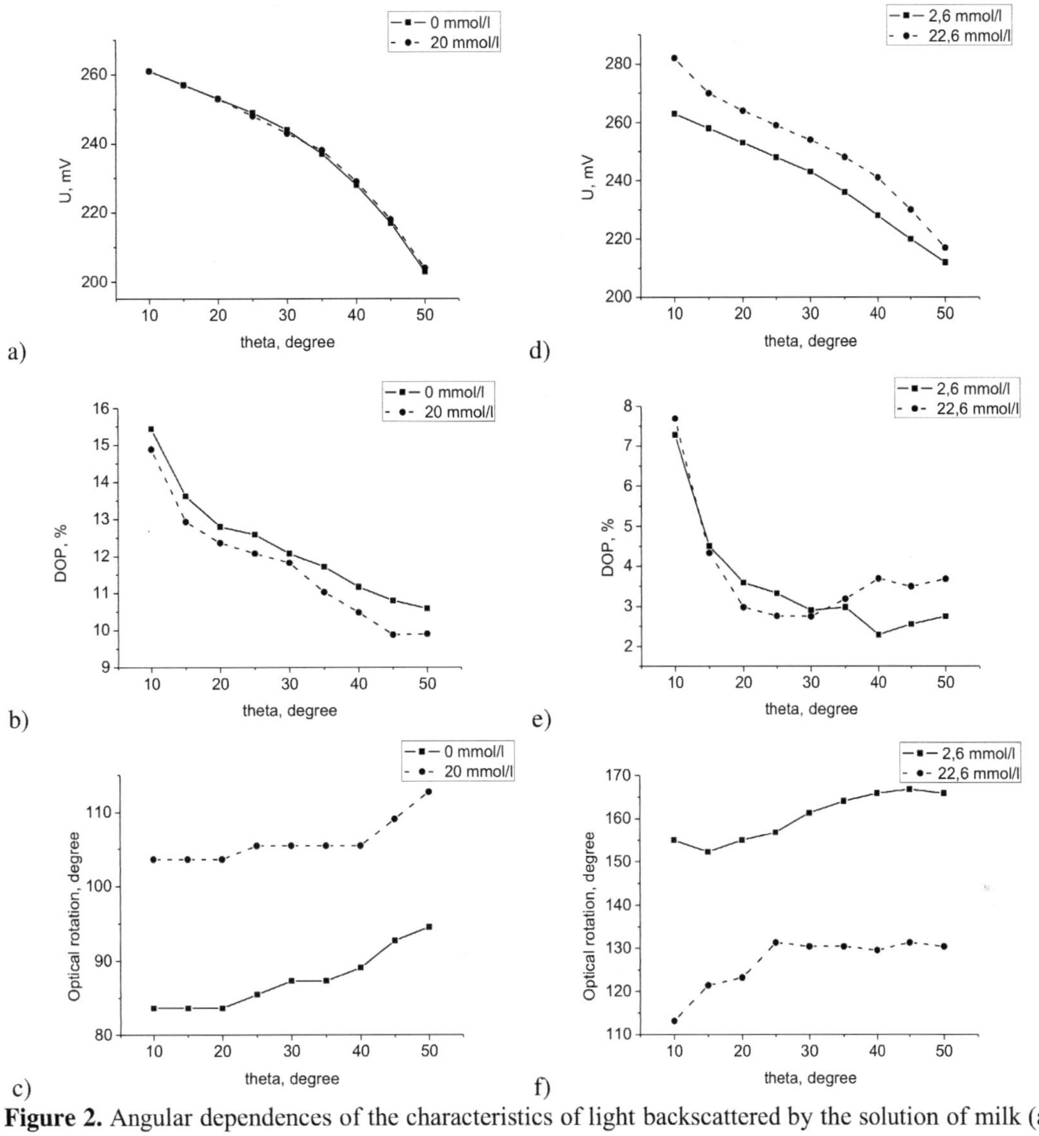

Figure 2. Angular dependences of the characteristics of light backscattered by the solution of milk (a-c) and solution of human blood (d-f) for two glucose concentrations: voltages on the photodiodes versus registration angle θ (a, d), degree of polarization versus registration angle θ (b, e), optical rotation versus registration angle θ (c, f). The symbols are data points; the lines are guide for the eye.

Preliminary measurements on the human finger pad shown that the angular dependences of the intensity, the degree of polarization and optical rotation were asymmetrical relative to the axis of the probe beam and were relatively complex function of the registration angle. However, the greatest degree of polarization was observed closer to the opposite direction as for the samples described above, indicating the prospect of such method of measurement for *in vivo* experiments.

5. Conclusions

In this paper, the angular dependences of the parameters of backscattered polarized light in the the glucose containing biological tissues and their phantoms was considered. In the experiments we studied a 20% solution of milk, a 50% suspension of human blood and the skin on the human finger pad. The dependencies of the intensity, degree of polarization and of the optical rotation on the registration angle were received for different glucose concentrations using created experimental setup. The experimental results shown that the angles closer to the opposite direction are most preferred for measuring the parameters of polarized light scattered by biological tissues. The obtained results should be considered when creating a non-invasive blood glucometer based on optical polarimetry. An example of such a sensor is the differential optical polarimeter, which working principle and test results were described in [10, 11].

References

[1] Tura A, Maran A, Pacini G 2007 *Diabetes Research and Clinical Practice* **77** 16–40
[2] Guo X, Wood M F G, Vitkin A 2007 *J.Applied optics* **46** 4491–4500
[3] Wood M F G, Guo X, Vitkin A J 2007 *J. Biomedical Optics* **12** 1–10
[4] Tuchin V V 2013 *Tissue Optics: Light Scattering Methods and Instruments for Medical Diagnosis* (Moscow:FIZMATLIT)
[5] Feng S, Zeng F, Chance B 1993 *Proc. SPIE* **1888** 78-89
[6] Maksimova I L , Romanov S V, Izotova V F 2002 *Optics and Spectroscopy* **92** 915-23
[7] Kohl M, Essenpreis M, Cope M 1995 *Phys. Med. Biol.* **40** 1267-87
[8] Amerov A K, Chen J, Small G W, Arnold M A 2004 *Proc. SPIE* **5330** 101–10
[9] Guo X, Wood M F G, Vitkin A 2006 *J. Biomedical* Optics **11** 041105
[10] Cherevatenko G A, Aksenov E T 2013 *Proc. SPIE* **9031** 903105
[11] Kafidova G A, Aksenov E T, Petrov V M 2013 **8803** 880306

Specialized probes with nanowhisker structures for scanning probe microscopy

M.V. Zhukov[1], I.V. Kukhtevich[1,2], V.V. Levichev[1], I.S. Mukhin[3], A.O. Golubok[1,4]

[1] Department of material science and nanotechnology, ITMO University, 197101 Kronverkskiy pr., 49, Saint-Petersburg, Russian Federation
[2] Laboratory of information and measurement biosensor and chemosensor microsystems, Institute of Analytical Instrumentation RAS, 198095 Ivan Chernykh, 31-33, Saint-Petersburg, Russian Federation
[3] Laboratory of nanobiotechnology, St Petersburg Academic University RAS, 194021 Khlopina, 8/3, Saint-Petersburg, Russian Federation
[4] Nanotechnology department, Institute of Analytical Instrumentation RAS, 198095 Ivan Chernykh, 31-33, Saint-Petersburg, Russian Federation

E-mail: cloudjyk@yandex.ru

Abstract. Probes with single nanowhiskers (SNWs) and nanoscalpels (NSs) for scanning force microscopy were created and studied. SNWs consisting of an amorphous Pt/C material and amorphous carbon SNWs were grown up at top of standard Si cantilevers with help of focused electron beam technique. By means of test samples it was shown that a SNW probe give more contrast and accurate images of deep channels and steps on a sample surface in comparison with a standard Si cantilever having the same radius of top as SNW. It was revealed that unlike the standard Si probe, the SNW probe allows to visualize fine nanostructure an erythrocyte membrane. The specialized NS probes achieve better results than standard probes when they work in force lithography mode. Moreover, the possibility of applying NS probes for precision movement of nanoparticles and cell nanosurgery are demonstrated.

1. Introduction

The main element that determines quality of the scanning probe microscopy (SPM) images is a probe. Nowadays works at a modification and improvement of a technology of probes fabrication are performed including nanowhisker structures growth on the probe tips. One of the widely used methods of nanowhiskers growth remains the method of molecular beam epitaxy [1] that allows fabricating arrays of nanowhiskers on the tops of probes. However, this method has difficulties with single nanowhiskers growth with strict control of their parameters such as length, thickness, orientation to initial growth surface. These parameters are required to achieve a stable scanning of objects with complex topology of their surface.

Solution of this problem is applying a new method for nanostructures growth under an influence of electron beam in vacuum chamber. This method allows producing structures in form of single nanowhiskers (SNWs) [2] and nanoscalpels (NSs) [3] on the tips of the probes. The SNWs probes are designed for high-precision imaging of nanotopography, as well as the NSs probes are fabricated for stable force lithography and manipulation of nanoparticles.

It should be mentioned that a significant challenge of SPM application is the study of native biological objects immobilized in a liquid [4, 5]. In this case SPM experiments carried out with SNW structures in various environments may become the first step to precision SPM study of native biological objects with complex topology structures.

The aim of this work was to fabricate and study the influence of SNW probes on SPM visualization of objects with different nature in different SPM modes and environments, as well as to determine the influence of the NS probes on structures created during the SPM lithography.

2. Experimental setup

Creation of the nanowhisker structures on the tops of probes and control of their parameters were performed by means of FEI Inspect (USA) and Carl Zeiss CrossBeam Neon 40 (Germany) scanning electron microscopes. Studies of the features of nanowhisker probes were carried out on NT-MDT Ntegra Aura (Russia) scanning probe microscope.

Samples with different nature such as biological (dried erythrocyte cells) and inorganic (calibration grid, silicon structures, polycarbonate and gold substrates, etc.) objects were investigated. Distilled water, phosphate buffered saline (PBS), borate buffer, 0.1 M NaOH and 0.5 M NaOH were selected as test liquids. The SPM measurements were performed both specialized nanowhisker and standard probes.

3. Results and Discussions

SNWs consisting of an amorphous Pt/C material were characterized with following geometrical parameters: 700-800 nm length, 40-60 nm diameter, and 7-12 nm curvature of tip radius. NSs were fabricated as asymmetrical C nanostructures with 60-80 nm width, 300-400 nm depth, and a length not exceeding 400 nm.

SNW probes demonstrated better quality of SPM images in comparison with standard Si probes in the study of biological objects, as well as samples with inorganic nature. It was shown that SNW probes clearly visualize samples surfaces with sharp features compare to standard probes with the same tip radius due to the increased aspect ratio of probe length to diameter. Moreover, SNW probes have better penetrating ability in microchannels and nanopores, as well as provide more contrast and accurate images in case of surfaces with topology in the range of 2-50 nm. Thus, the results of erythrocytes study showed that SNW probes significantly improve image quality and resolution of the scanning probe microscope (Fig. 1, 2). The visualization of PMMA (Polymethyl methacrylate) microchannel obtained with standard and SNW probes is shown in Fig. 3. It is seen that the SNW probe has substantially better penetration ability (up to 100 nm) than the standard probe (up to 40 nm).

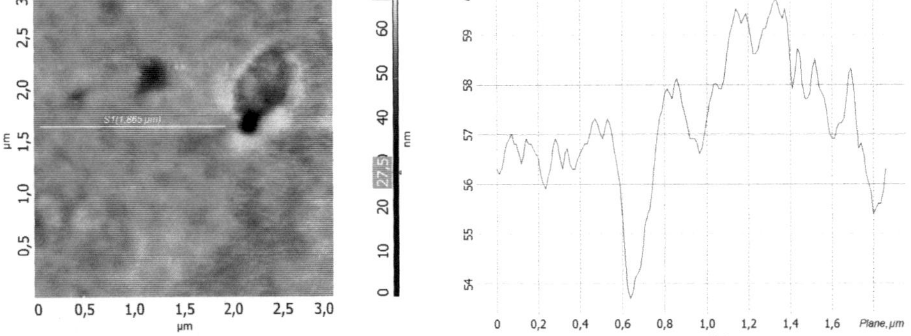

(a) (b)

Fig. 1. SPM image (a) and the cross section (b) of the erythrocyte membrane surface obtained with standard probe.

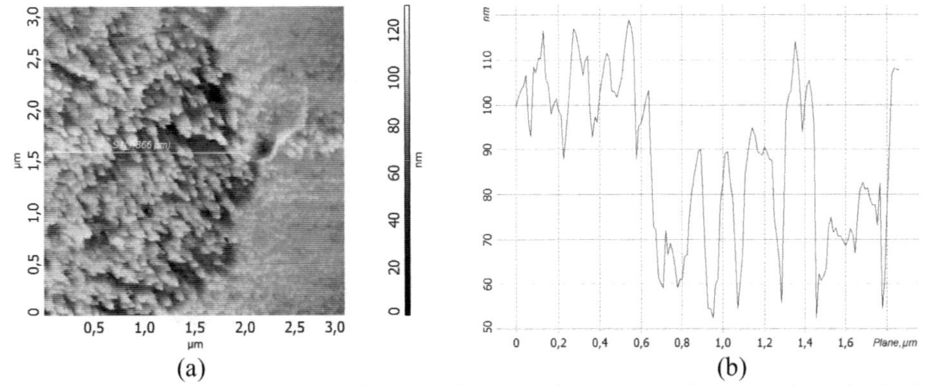

Fig. 2. SPM image (a) and the cross section (b) of the erythrocyte membrane surface obtained with SNW probe.

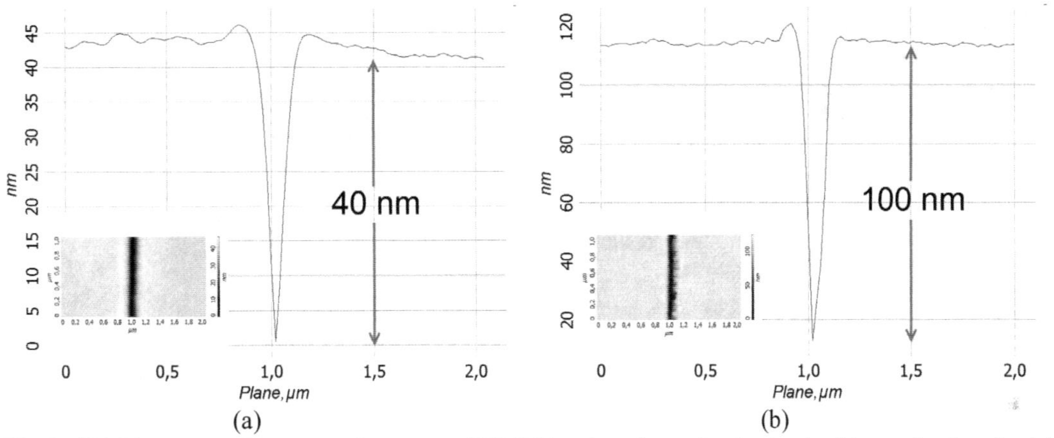

Fig. 3. SPM images and cross section areas of PMMA microchannels obtained with standard probe (a) and SNW probe (b).

To identify the best modes of SPM operating in the liquid the calibration grid NT-MDT TGQ01 (Russia) was investigated. The calibration grid is a set of periodically arranged square-shaped cells with constant height. Study of the calibration grid with help of SNW probes in the liquids showed that PBS is the best liquid according to image quality. The highest resolution was achieved in contact mode while the best accordance with parameters of test structures was observed in tapping mode.

SNW probes are shown significantly improved image quality and resolution of the SPM in liquids in comparison with standard probes (Fig. 4). Thus, in PBS tilt width of grid in X direction has values of about 4 nm for SNW probes and about 20 nm for standard probes, in Y direction has values about 12 nm and 38 nm, correspondingly.

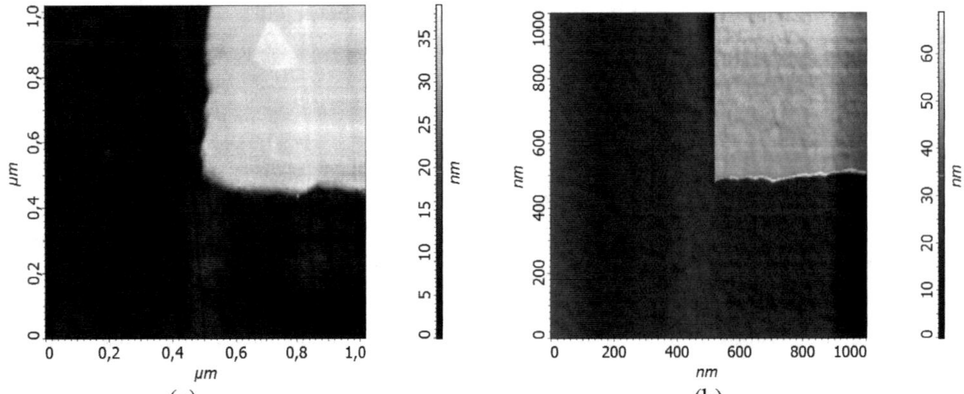

(a) (b)

Fig. 4. SPM images of one cell of TGQ01 grid obtained in PBS with standard probe (a) and SNW probe (b).

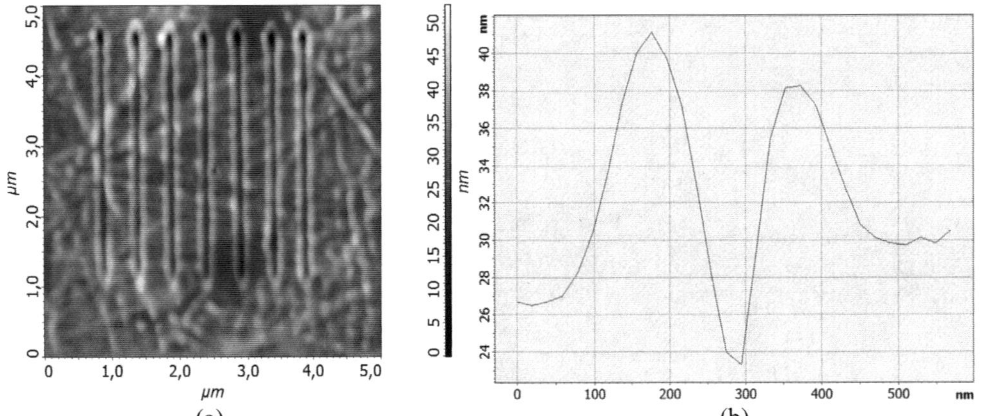

(a) (b)

Fig. 5. SPM image (a) and the cross section (b) of the diffraction grating on the gold layer (10 nm) obtained with standard probe force lithography.

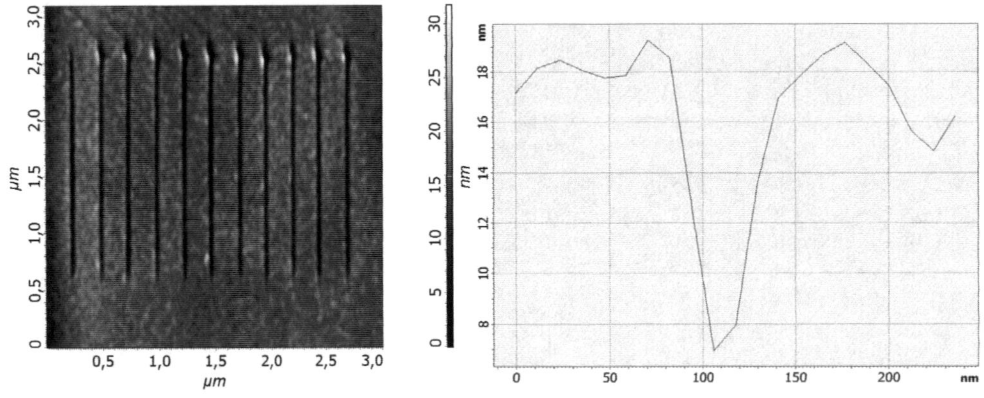

(a) (b)

Fig. 6. SPM image (a) and the cross section (b) of the diffraction grating on the gold layer (10 nm) obtained with NS probe force lithography.

NS probes were characterized with enhanced mechanical stability in force lithography mode and showed better results of produced structures on polycarbonate substrates with a gold coating (10 nm) compare to standard probes (Fig. 5, 6). During SPM lithography with NS probes was found out that the most suitable parameters of diffraction gratings producing (with period from 50 nm to 500 nm) on the gold layers are values of forces in the range of 1-2 μN. The most suitable velocity of the probe movement during lithography process is 2 μm/sec.

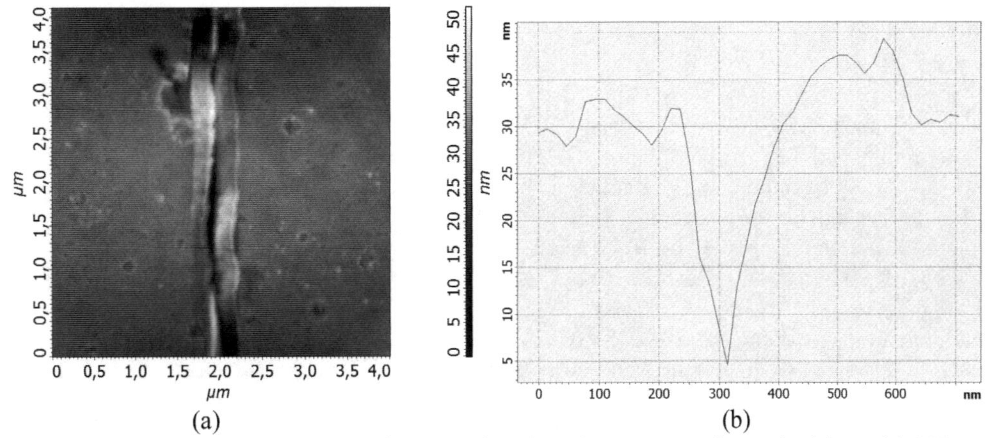

Fig. 7. SPM image (a) and the cross section area (b) of erythrocyte membrane incision with NS probe (6.5 μN force).

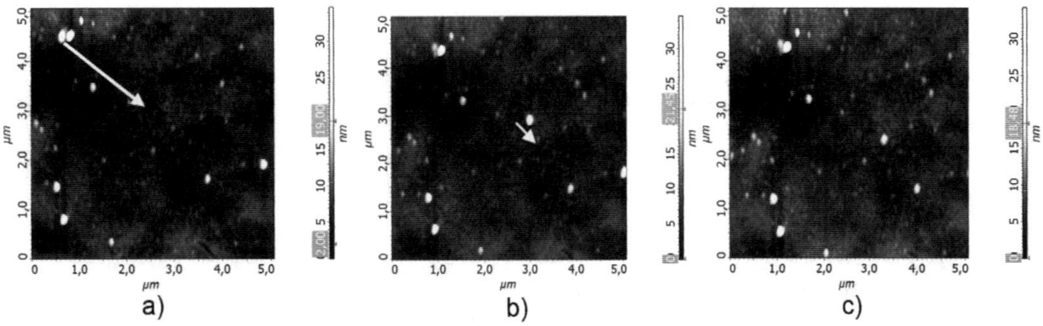

Fig. 8. Manipulation of nanospheres with NS probes: a) the initial position of the particle; b) the initial displacement of the particle, strong impact (about 500 nN); c) repeated displacement of the particle, low impact (about 250 nN).

NS probes also were used for precision cell nanosurgery of dried erythrocyte membranes (Fig. 7). Dissection was carried out by moving the probe relative to the surface of the erythrocyte membrane with a predetermined impact force (0.5 - 7.5 μN) and velocity of about 0.5 μm/sec. In addition the possibility of precise manipulation of nanoparticles (250-300 nm in diameter) was demonstrated by using broad part of NS (Fig. 8).

4. Conclusions

The results showed that the specialized SNW probes gives more reliable and detailed images of surface topology in comparison with standard probes, particularly in the surface features size in range of 2-50 nm. It was revealed that specialized NS probes give better results than standard probes during atomic force lithography. Moreover, the possibility of NS probes applying for precision nanoparticles movements and cell nanosurgery are shown.

These studies have demonstrated the benefit of specialized probes with nanowhisker structures compare to standard probes. Thus, the foundation for the further applications of such structures in the field of nanotechnology and biology was created.

The work was done with support of the Government of St. Petersburg, the U.M.N.I.K. program, Russian Foundation for Basic Research.

References

[1] Dmitry Klinov, Sergei Magonov 2004 True molecular resolution in tapping-mode atomic force microscopy with high-resolution probes *Applied physics letters* vol. 84 issue 14 pp 2697–2699.

[2] V. V. Levichev, M. V. Zhukov, I. S. Mukhin, A. I. Denisyuk and A. O. Golubok 2013 On the operating stability of a scanning force microscope with a nanowhisker at the top of the probe *Technical Physics* vol. 58 issue 7 pp 1043–1047.

[3] J. D. Beard, D. J. Burbridge, A. V. Moskalenko, O. Dudko, P. L. Yarova, S. V. Smirnov and S. N. Gordeev 2009 An atomic force microscope nanoscalpel for nanolithography and biological applications *Nanotechnology* vol. 20 pp 1–10.

[4] Yuanqing Chao, Tong Zhang 2011 Optimization of fixation methods for observation of bacterial cell morphology and surface ultrastructures by atomic force microscopy *Appl. Microbiol. Biotechnol.* vol. 92 issue 2 pp 381-392.

[5] Rikke Louise Meyer, Xingfei Zhou, Lone Tang, Ayyoob Arpanaei, Peter Kingshott, Flemming Besenbacher 2010 Immobilisation of living bacteria for AFM imaging under physiological conditions *Ultramicroscopy* vol. 110 issue 11 pp 1349-1357.

SPbOPEN2014 IOP Publishing
Journal of Physics: Conference Series **541** (2014) 012043 doi:10.1088/1742-6596/541/1/012043

Investigation of blood cell properties using laser diffractometry and kinetic nephelometry

S A Kavalenka, A I Kavalenka, V V Popov, T D L Nguyen and VA Loban

Department of Biophysics, Faculty of Physics, Belarusian State University, Nezavisimosti ave. 4, 220030 Minsk, Belarus

E-mail: kavalenka@bsu.by

Abstract. The portable multi-channel laser nephelometer was developed and applied for quantitative and qualitative researches of blood cells. Obtained findings were analyzed in comparison with results of laser diffractometry and microscopy.

1. Introduction

Coherent optical methods such as laser diffractometry, nephelometry, speckle interferometry, holographic digital microscopy allow to analyze dispersed cellular systems, individual cells and tissues [1-3]. The main advantages of these methods are the lack of contact, the ease to use, high accuracy and ability to automate the measurement.

Laser diffractometry and nephelometry are based on the analysis of light scattering patterns that are results of diffraction, refraction, reflection and absorption of laser light during the laser beam interaction with investigated objects [1-5]. The angular dependence of light scattering intensity called the indicatrix is characteristic for the particle size. Therefore, these methods are the most widely applied for determination of cell sizes and their changes. The most often the Fraunhofer approximation or the Mie theory is used for transforming the measured data to a particle size distribution [4, 5]. The exact theory of scattering (Mie theory, T-matrix method, discrete dipole approximation method and the approximate calculation methods is used to calculate the indicatrix of a single particle. Suspensions of different cells can be analyzed using the multi-mode Laplace transformation for detection and separation of multi-modal particle size distributions.

The researches of human and animal blood cells are required to the development of cellular biological and biomedical technologies. Valuable information can be obtained performing the kinetic studies of geometric and concentration parameters of cell systems. The ability to aggregate and the deformability during stimulated swelling of cells determine the key properties of normal and abnormal blood cells, in particularly, erythrocytes, platelets, neutrophils, hematopoietic progenitors and blasts cells.

The study is aimed to development of the automated multi-channel laser analyzer and its application for quantitative and qualitative researches of blood cells.

2. Results

The automated multi-channel laser nephelometer for researches of cells in static and kinetic regimes was developed. Semiconductor laser diode LFD635-5-3 and high-sensitive opto-electronic integrated circuits OPT301 were used in this device as the laser light source and the scattered light intensity

Journal of Physics: Conference Series **541** (2014) 012043 doi:10.1088/1742-6596/541/1/012043

detectors, respectively. Microcontroller AtMega8535 provides the control of kinetic data registration and processing. Firmware for the microcontroller was written using Atmel AVR-Studio 6.0. As software UniChrom-97 ("New Analytical Systems", Belarus) was used. To support nephelometer driver DLL was developed using driver development kit UniChrom-DDK. Intensities of scattered light I(β) at the angles (β) of 7°, 90°, 187° and intensity of transmitted laser radiation (β = 0°) are measured by this nephelometer.

2.1. Concentration of particles in suspensions

Concentration dependencies I(β, C) for different types of human blood cells (isolated from blood and resuspended in Earle's medium or in NaCl solution) and latex particles were obtained by developed device. Such dependencies are shown in figure 1 for erythrocytes (a), "ghosts" of erythrocytes (b) and latex (c). It was revealed that the concentration dependences are described by distinct non-monotone functions for angles 7°, 90°, 187° and 0°. These dependencies distinguish for particles and cells that have different sizes. Diameter of latex particles was about 600 nm that established by atomic force microscopy. Sizes of investigated cells were determined with the use of small-angle laser diffraction analyzer detecting light scattering indicatrix from 1° to 12° angles and applying Fraunhofer approximation. The different types of investigated cells (erythrocytes and neutrophils) were characterized by sizes from 3 to 12 μm under different conditions of medium. Figure 1 (a) are represented the data for erythrocytes with the mean of diameter about 4.9 μm that were isolated from aged blood and resuspended in Earle's medium (300 mOsm, pH=7.3).

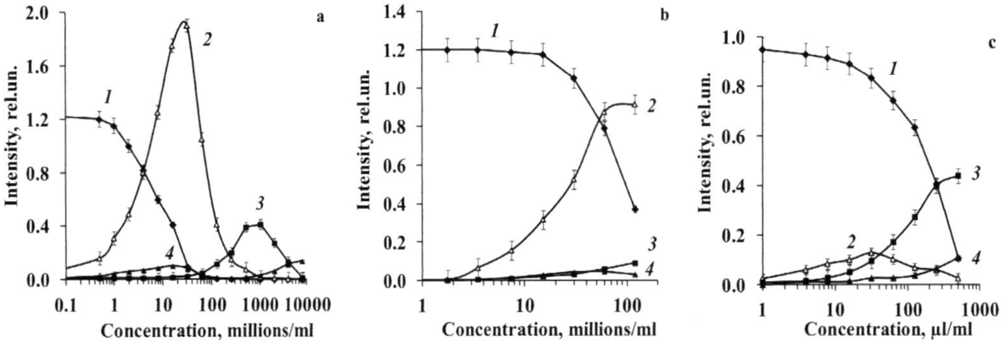

Figure 1. Concentration dependencies of scattered light intensity obtained for erythrocytes (a), "ghosts" of erythrocytes (b) and latex particle (c). ***1*** – light transition (at the angle of 0°), ***2*** – light scattering at the angle of 7°, ***3*** – scattering at the angle of 90°, ***4*** – scattering at the angle of 187°.

To consider the part of volume occupied by scattering particles, parameter V_e/V_0 was calculated as ratio of volume occupied by all cells (V_e) to total volume of suspension (V_0). The parameter V_e/V_0 for erythrocytes is similar to hematocrit that is partial volume of erythrocytes in whole human blood. Normal value of hematocrit is within the range of 0.35-0.48 l/l. The shape of erythrocytes from aged blood can be approximated by equivalent sphere. The mean of diameter of the cells (d) was estimated analyzing small-angle laser diffraction as described higher. Further, value d was used for calculation of volume (V), surface area (S) of the sphere and V_e/V_0 ratio. The data characterizing intensity of light scattering by erythrocytes as dependencies on V_e/V_0 are represented in figure 2. Optical density was calculated as logarithm of light transition intensities ratio of solutions without and with particles. The dependences of intensity of light scattering at small angle (7°) and transmitted light can be approximated by linear functions for narrow ranges of parameter V_e/V_0 and cell concentration (figure 2, a). This effect conforms to single scattering. It was revealed that the range of linearity is wider for smaller particles such as latex particles, and this range is most limited for cells with the biggest diameters. Linear function approaches the dependence describing light scattering at the angle of 90° in the range of middle cell concentrations and V_e/V_0 ratio from 0.005 to 0.035 l/l (figure 2, b). The increase of V_e/V_0 ratio over 0.01 l/l (from 0.01 to 0.6 l/l in the figure 2, b, c) is corresponds to the rise of cell concentration over 100 millions/ml (figure 1, a), and it is accompanied by the loss of light transition and the strong decline of scattering at the angles of 7°. It is obviously that the multiple scattering and amplification of light reflection by particles take place at high values of V_e/V_0. Reflection of light is the main phenomenon in the range of value V_e/V_0 over 0.1-0.2 l/l, and in this case more information can be obtained analyzing the scattering at the large angles such as 90° and 187° (figure 2, c).

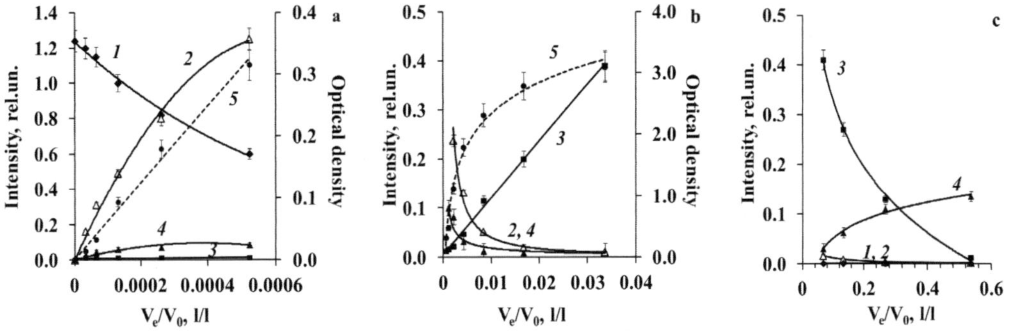

Figure 2. Dependencies of scattered light intensity on parameter V_e/V_0 characterizing the ranges of low (a), middle (b) and high (c) concentrations of erythrocytes. ***1*** – light transition (0°), ***2*** – light scattering in angle 7°, ***3*** – light scattering in angle 90°, ***4*** – light scattering in angle 187°, ***5*** – optical density (dashed line)

2.2. Deformation of cells

To avoid the influence of multiple scattering, the study of cell deformation is performed with the use of low concentration of cells. The concentrations have not to exceed of approximately 5-8 and 2-3 millions/ml for erythrocytes and neutrophils, respectively. If ratio V_e/V_0 is not greater than of 0.0005, that analysis of small-angle scattering and light transition intensities provides the highest both sensitivity and linearity.

Deformation of cells and the increase of their membrane surface area without the loss of cell integrity or aggregation are accompanied by variations that not exceeding of approximately 40 % for erythrocytes and 250 % for neutrophils with respect to normal value. The normal cell surface area in isotonic media with respect to blood plasma (300 mOsm, 0.150±0.004 M NaCl,) is 151±3 μm^2 for discoid blood erythrocytes and 214±30 μm^2 for spherical neutrophils. Surface of the cells can be smooth or rough with folds. Ability of cell membrane to extension depends on the number of factors including initial elastic stress of the membrane, the presence or the absence of folding, membrane lipids and proteins qualitative content, cytoskeleton and other. Membranes of discoid erythrocytes from fresh blood is smooth in isotonic medium (figure 3, a), and their volume is about 98±2 μm^3. In hypertonic media volume of the cells is decreased, and their membrane becomes folding (figure 3, b). In hypotonic media deformation and swelling of the cells are observed (figure 3, c).

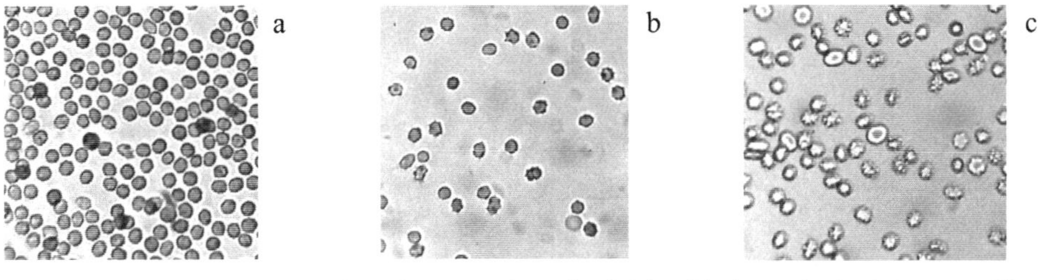

Figure 3. Light microscopy of erythrocytes from fresh blood in isotonic (a, 0.15 M NaCl), hypertonic (b, 0.20 M NaCl) and hypotonic (c, 0.12 M NaCl) media.

Deformation of aged blood erythrocytes (8 millions cells per ml, V_e/V_0 = 0.0005) in hypotonic media was studied during the decrease of NaCl concentration in extracellular medium from 0.150 to 0.015 mM. Dependencies of light scattering intensities on NaCl concentration are shown in figure 4. It was established that light scattering intensity at 7° rises during NaCl concentration decrease from 0.150 to 0.105 mM. Using laser diffractometry analyzer the means of the cell diameters in these media were determined as 4.9 μm in 0.150 mM NaCl, 6.2 μm in 0.135 mM NaCl, 6.7 μm in 0.120 mM NaCl and 7.4 μm in 0.105 mM NaCl. The calculated V and S of equivalent spheres were 62 μm^3 and 76 μm^2 in 0.150 mM NaCl, 122 μm^3 and 119 μm^2 in 0.135 mM NaCl, 159 μm^3 and 142 μm^2 in 0.120 mM NaCl and 211 μm^3 and 171 μm^2 in 0.105 mM NaCl, respectively. In these cases, the means of V correlate linearly with NaCl concentration outside the cells (confidence probability p=0.99). Correlation between V and intensities of light scattering at different angles are illustrated in figure 5 for swelling of the cells. The best linear correlation (p=0.99) is revealed between V and light scattering intensity at 7° (figure 5, ***2***). In concentration range of NaCl from 0.090 to 0.015 mM and osmolarity from 180 to 30 mOsm the transition and scattering of light are reduced in result of lysis of the cells (figure 4, ***1-3***). Extension of cellular membrane during swelling in such conditions exceeds the limit of S about 210 μm^2, and erythrocytes destruct that are confirmed by microscopy.

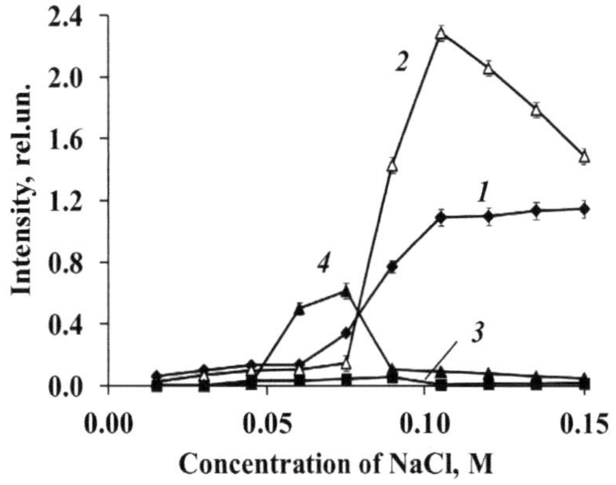

Figure 4. Dependencies of intensities of light transition (*1*) and light scattering at 7° (*2*), 90° (*3*), 187° (*4*) on NaCl concentration for aged blood erythrocytes.

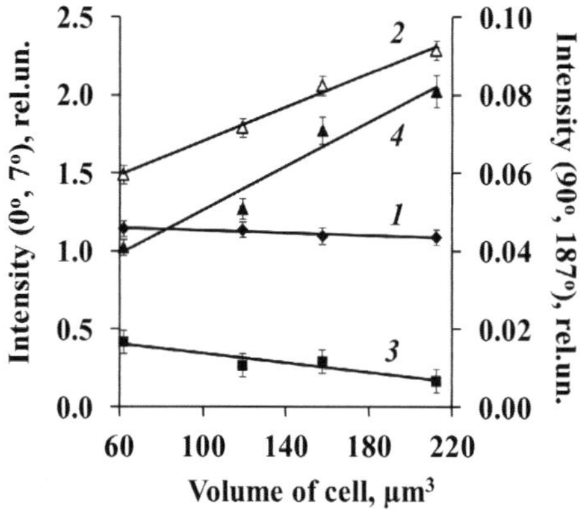

Figure 5. Correlation between volume of the erythrocytes and intensities of light transition (*1*) and light scattering at 7° (*2*), 90° (*3*), 187° (*4*).

One of the methods for evaluation of erythrocyte ability to deform is kinetic analysis of erythrocyte swelling and lysis induced by HCl. We determined the basic characteristic parameters of the processes as shown in figure 6.

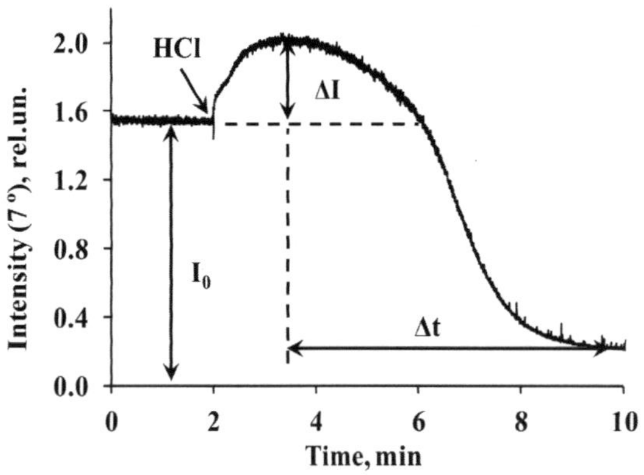

Figure 6. Determination of parameters that describe erythrocyte swelling and destruction using kinetic dependencies of light scattering intensity:
I_0 – intensity concerned by initial volume of cells,
ΔI – the changes of light scattering intensity that characterized the additional swelling of erythrocytes under HCl action before cell disruption,
Δt – duration of cell disruption process.

Kinetics of HCl-induced lysis of erythrocytes swelled in hypotonic conditions are represented in figure 7. It was established that I_0 as well as the initial volume of erythrocytes before addition of HCl correlate linearly with the changes of light scattering intensity ΔI characterizing the additional cell swelling under HCl action before disruption and the duration of cell destruction Δt (figure 8).

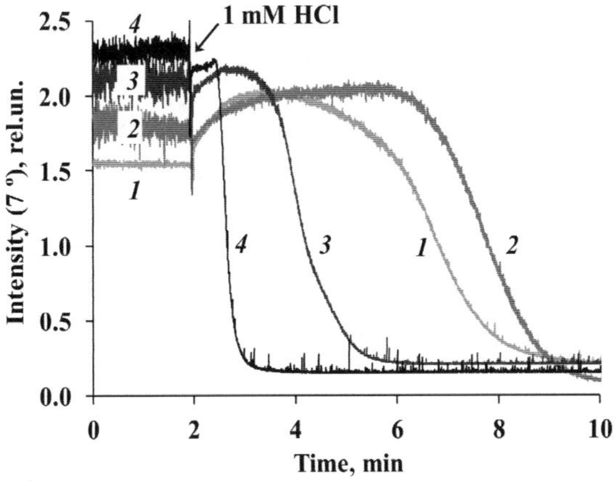

Figure 7. Kinetic dependencies of scattered light intensity (in angle 7°) during HCl-induced lysis of erythrocytes in isotonic (*1*) and hypotonic conditions (*2 – 4*). *1* – 0.150 M NaCl, *2* – 0.135 M NaCl, *3* – 0.120 M NaCl, *4* – 0.105 M NaCl.

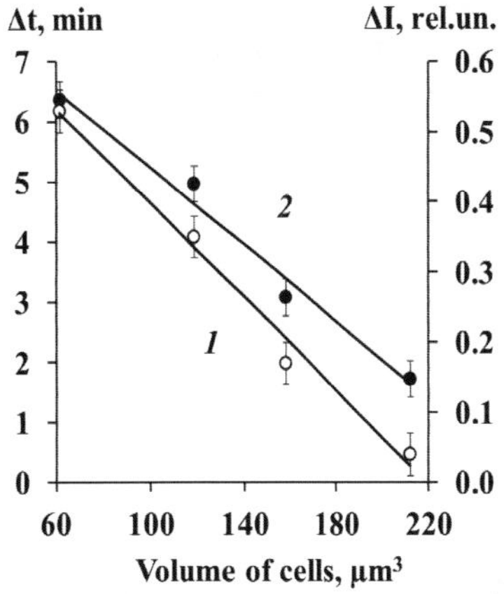

Figure 8. Linear correlation between initial volume of cells and parameters of the following HCl-induced lysis of erythrocytes.
1 – Δt, *2* – ΔI.

2.3. Aggregation of cells

The aggregation processes of erythrocytes and neutrophils from fresh blood under different influences were investigated in kinetic regime. Figure 9 are shown the kinetic dependences of scattered light intensity during erythrocyte aggregation ($5 \cdot 10^6$ cells per ml, $V_e/V_0 = 0.0005$) induced by addition of La^{3+}. The corresponding data from light microscopy are represented in figure 10. During aggregation a quantity of light-scattering centers and total scattering surface are reduced, therefore the intensity of light transition rises (figure 9, *1*) and intensity of light scattering at 7° decreases (figure 9, 2). It is necessary to note that the erythrocytes in aggregates destruct in the presence of La^{3+}, and hemoglobin is released in extracellular medium. Thus, the registering kinetic dependencies are the results of aggregation and destruction of the cells.

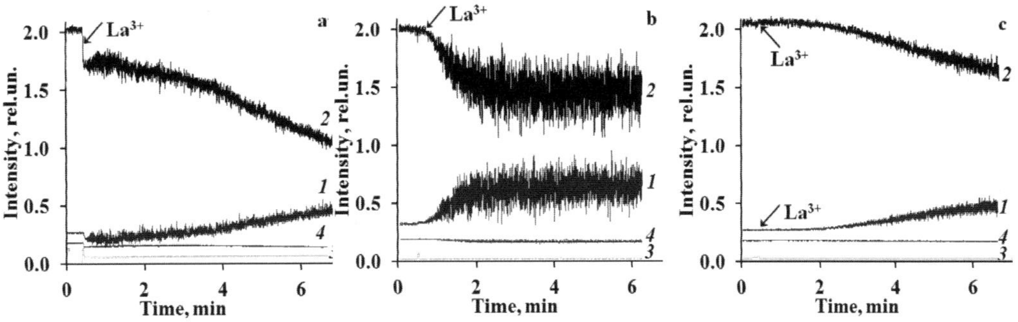

Figure 9. Kinetic dependences of scattered light intensity during erythrocyte aggregation induced by addition of La^{3+} in concentration of 2.5 mM (a), 0.5 mM (b) and 0.125 mM (c). *1* – light transition (0°), *2* – light scattering in angle 7°, *3* – light scattering in angle 90°, *4* – light scattering in angle 187°

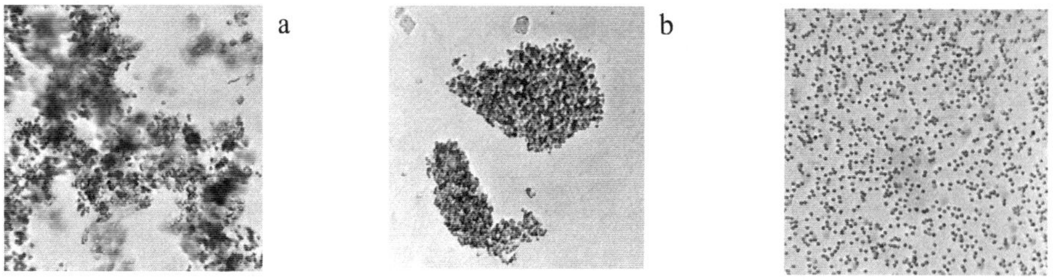

Figure 10. Light microscopy of erythrocyte aggregates. a – 2.5 mM La^{3+}, b – 0.5 mM La^{3+}, c – 0.125 mM La^{3+}.

3. Conclusion
The portable laser nephelometer was developed. This device supports automatic multi-channel registration of kinetic nephelometric data and permits to carry out quantitative and qualitative researches of blood cells. Data about structure of blood cells obtained by nephelometry correlate with data from laser diffractometry and microscopy.

References
[1] Bedner E, Gorczyca W, Melamed M and Darzynkiewicz Z 1997 *Cytometry* **29** 191–6
[2] Shvalov A N, Surovtsev I V, Chernyshev A V, Soini J T and Maltsev V P 1999 *Cytometry* **37** 215–20
[3] Kalashnikov M, Choi W, Yu C, Sung Y, Dasary R R, Badizadegan K and Feld M S 2009 *Optics Express* **17** 19674–81
[4] Doicu A and Wriedt T 1997 *Appl. Opt.* **36** 2971–8
[5] Etzler F M 2004 *American Pharm Review* **7(1)**104–8

Journal of Physics: Conference Series **541** (2014) 012044 doi:10.1088/1742-6596/541/1/012044

Influence of a diffusion mass flux on instability development of surface shape of high-viscosity liquid

T Pogosian and S Chivilikhin

Saint Petersburg National Research University Information Technologies, Mechanics and Optics, St. Petersburg, Russia

E-mail: edo-chan@mail.ru

Abstract. This work is devoted to theoretical investigation of instability development of surface shape of high-viscosity liquid under the influence of diffusion mass flux. If there is a local eminence on the surface it can lead to growth of perturbation of surface shape. Stability and instability of perturbation depends on factors which will prevail. Diffusion causes distortion of surface shape and dominance of capillary forces achieves the opposite effect.

1. Instability of flat shape surface

This work is divided into two parts. First, we consider a surface having the shape of a flat in unperturbed state. The mass stream can be calculated in quasi–steady–state approximation using quasi–steady–state equation of diffusion. Action of capillary forces and gravitation is taken from [1]. Fourier harmonica with wave number k of perturbation of a free surface of liquid change over time under the law:

$$h_k = h_k(0)e^{a(k)t}, \quad a(k) = Vk\,\mathrm{cth}(kH) - \frac{1}{2\mu}\left(\sigma k + \frac{\rho g}{k}\right),$$

where V - upward velocity of the average level of liquid; H - thickness of diffusion boundary layer; μ, ρ and σ - viscosity, density and coefficient of surface tension of liquid, g - acceleration of gravity.

With the wave numbers, where a(k) is less than zero, the perturbation decays, and this corresponds to conditions of stability. Wave numbers, for which a(k) is greater than zero, correspond to the regime of instability.

For the analysis of parameters of system we entered dimensionless quantities:

$$\chi = \frac{k}{k^*}, \qquad k^* = \sqrt{\frac{\rho g}{\sigma}}, \qquad \gamma = k^* H, \qquad \omega = \frac{\sigma}{2\mu V}.$$

In figure 1 areas in space of parameters (γ, ω) corresponding to various modes of perturbation evolution are represented. If wave number is very small we will observe relaxation of the surface at any mode, but if wave number grows there are different options of surface perturbation development: at $\omega>1$ growth of wave number doesn't matter, this is an area of modes of stability, at $\omega<1$ it will lead to distortion of surface shape. Figure A2 and figure B2 visually demonstrate behavior of function $a(\chi)$ at certain values of γ and ω, taken from B and A areas respectively.

There are two specific modes. In the area C instability becomes apparent in the final range of wave numbers (figure C2). It corresponds to formation of wave structure on the surface with constant wavelength.

D area combines stability and instability manifestations in the final range of wave numbers (figure D2). It can be interpreted, as formation of wave structure with wavelength changing over time.

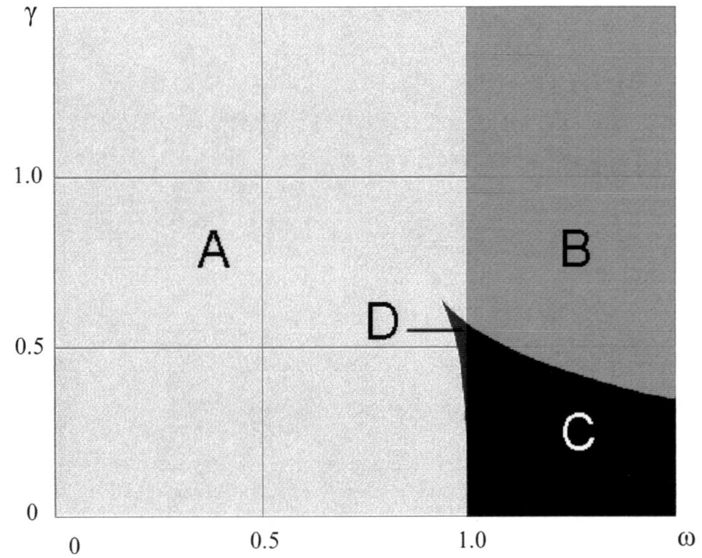

Figure 1. Areas in space of parameters (γ,ω) corresponding to various modes of perturbation evolution: A - mode of instability, B – mode of stability, C – mode of local instability, D - mixed mode.

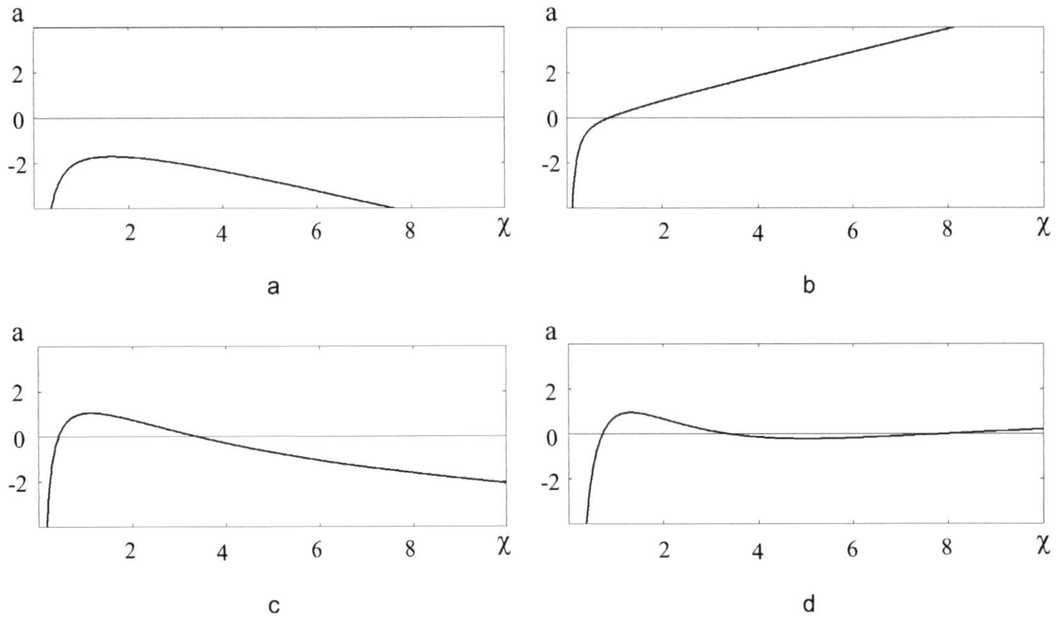

Figure 2. Behavior of function a(χ) at certain values of γ and ω, taken from different areas: a) mode of stability, $\gamma = 1.5$, $\omega = 1.5$; b) mode of instability, $\gamma = 1.5$, $\omega = 0.5$; c) mode of local instability, $\gamma = 0.3$, $\omega = 1.2$; d) mixed mode, $\gamma = 0.5$, $\omega = 0.985$.

2. Instability of spherical shape surface

Second part of work devoted to spherical shape of surface. In particular, we observe the nanoparticle, which has the shape of a sphere in the zero approximation. For the description of a nanoparticle the liquid-drop model is used. The unperturbed surface is considered as an ideal sphere. For the description of a mass stream to a free surface we solved the quasi steady-state diffusion equation where concentration is a solution of the Laplace's equation in spherical coordinates.

Action of capillary forces is taken from [2]. Perturbation of surface is expanded in series of Legendre polynomials:

$$h(\theta,t) = \sum_{n=1}^{\infty} a_n(t) P_n(\cos\theta).$$

The coefficients of expansion change over time and have dependence on indexes n, which is polynomial degree:

$$a_n(t) = a_n(0)\exp\left(\frac{t}{R_0}\left(V(n+1) - \frac{\sigma}{2\mu}\gamma_n\right)\right),$$

$$a_n(0) = \frac{2n+1}{2}\int_0^{\pi} h_0(\theta)\sin\theta\, P_n(\cos\theta)\, d\theta,$$

where V is unperturbed velocity of sphere radius growth; μ and σ - viscosity and coefficient of surface tension of nanoparticle; R_0 is initial radius; $\gamma_n = n(n+2)(2n+1)/(2n^2+4n+3)$; $h_0(\theta)$ initial perturbation.

Big role in research of the equation of evolution of perturbation of a surface form is played by parameters. For the analysis of parameters entering under an exhibitor in coefficients at Legendre polynomials, it is better observe them as ratio σ/2μV.

In the Figure 3 the line of ratio of parameters σ/2μV divided into areas is presented. Left area marked as B is an area of a ratio of parameters at which all harmonicas, since the first, are responsible for growth of perturbation development. Right area marked as A is an area of a ratio of parameters at which all harmonicas are responsible for a relaxation of the perturbation surface. Between these main areas there is located the infinite set of decreasing areas where the part of harmonicas starts being responsible for instability.

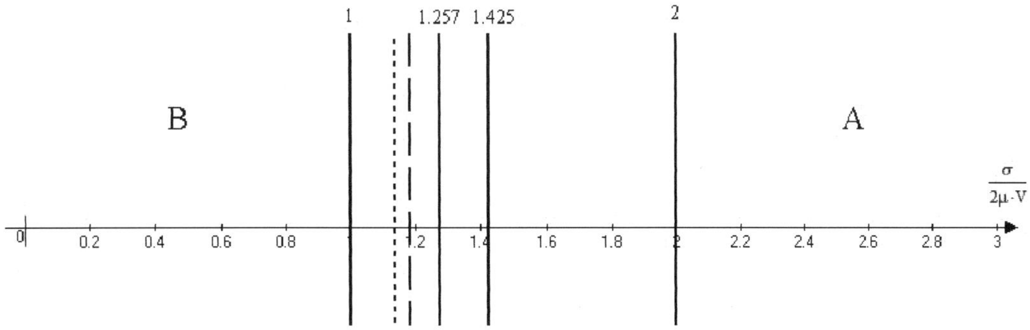

Figure 3. Line of ratio of parameters σ/2μV divided into areas: A is an area of a ratio of parameters at which all harmonicas are responsible for a relaxation of the perturbation surface; B is an area of a ratio of parameters at which all harmonicas are responsible for growth of perturbation development.

The value of a straight line below which the n-harmonica starts making a contribution to instability of perturbation development, can be calculated from:

$$\frac{\sigma}{2\mu V} = \frac{(n+1)(2n^2+4n+3)}{n(n+2)(2n+1)},$$

where n = 1, 2 … ∞.

Thus knowing one of parameters of system it is possible to pick up optimum their ratios for achievement of this or that effect.

It is important to mention that at $n\to\infty$, if $V > \sigma/2\mu$, coefficients of expansion correspond to unstable harmonics and speed of instability development increases with index growth.

3. Summary

In this work the equations of evolution of perturbation of flat and spherical shape surface are received. Various modes of development of the perturbation surface concerning parameters are investigated. There are revealed modes at which the surface loses stability, and modes at which there is a relaxation of perturbation of a free surface.

The most actual is finding conditions for instability mode. Critical velocity must be bigger than surface tension coefficient divided into the doubled effective coefficient of viscosity.

Acknowledgments
This work was financially supported by Government of Russian Federation, Grant 074-U01.

References

[1] Chivilikhin S A 2012 Relaxation of small perturbations of highviscous liquids planar surface *Nanosystems: Physics, chemistry, mathematics* **3**(4) 54–66
[2] Hrma P and Carter C E 1984 Capillarity-induced spheroidization of nearly spherical amorphous bodies by viscous flow *J. of Non-Crystall. Solids* **63** 391-412

Photoinduced modification of quantum dot optical properties affects bacteriorhodopsin photocycle in a (quantum dot)–bacteriorhodopsin hybrid material

V A Krivenkov[1, 2], D O Solovyeva[1], P S Samokhvalov[1], K I Brazhnik[1], G E Kotkovskiy[1,2], A A Chistyakov[1, 2], E P Lukashev[1,3] and I R Nabiev[1,4]
[1] Laboratory of Nano-Bioengineering, National Research Nuclear University MEPhI, Kashirskoye shosse 31, 115409 Moscow, Russian Federation
[2] Department of Micro- and Nanophysics, National Research Nuclear University MEPhI, Kashirskoye shosse 31, 115409 Moscow, Russian Federation
[3] Department of Biophysics, Moscow State University, Vorobyovy Gory, 119992 Moscow, Russian Federation
[4] Laboratoire de Recherche en Nanosciences, LRN-EA4682, Université de Reims Champagne-Ardenne, 51100 Reims, France

E-mail: vkrivenkov@list.ru

Abstract. A method for controlled changes in the radiative properties of quantum dots (QDs) in order to modulate the Förster resonance energy transfer (FRET) rate in nano-hybrid materials is proposed. The mechanism underlying the effect of QDs with optical properties modulated by UV laser irradiation on the photocycle of the photosensitive protein bacteriorhodopsin (bR) in its native purple membranes (PM) isolated from *Halobacterium salinarum* has been studied. The irradiation leads to a twofold decrease in the QD fluorescence quantum yield without changes in the extinction spectrum or the position or shape of the fluorescence spectrum. The bR photocycle is accelerated, which has been shown to be related to the changes of the surface potential of PM upon formation of their complexes with QDs.

1. Introduction

The photosensitive protein bacteriorhodopsin (bR) [1], a photosensitive membrane protein from purple membranes (PMs) of *Halobacterium salinarium*, is a promising potential component of photovoltaic nano–bio hybrid structures [2] which utilizes as little as 0.1–0.5% of the total energy of the solar light optical spectrum [3]. The harvesting and utilization of solar energy by photosensitive proteins can be made substantially more efficient by complexing them with semiconductor nanocrystals or quantum dots (QDs), which are capable of absorbing light in the UV and visual spectral regions and passing the absorbed energy to the photosensitive proteins through the Förster resonance energy transfer (FRET) mechanism [4, 5]. It has been demonstrated that the transfer of additional energy to bR or photoreactive centers can increase the efficiency of proton transport by bR [4] and charge separation during photosynthesis [5]. It was also been shown that the integration of QDs into purple membranes (PMs) containing bR could accelerate the formation of the intermediate M-form of the bR photocycle, although the mechanism of this effect remained unknown [6].

This study was aimed at determining the mechanisms of QD effects on M-form of the bR photocycle. For this purpose, we directionally changed the QD optical properties by means of laser UV irradiation before their integration into PMs in order to estimate the possible contribution of the resonance energy transfer from QDs to bR into the acceleration of the formation of the bR M-form.

UV irradiation allows the FRET efficiency to be changed without affecting the structural organization of the nano-hybrid system. Radiation with a quantum energy greater than the QD band gap is known to cause a decrease in the quantum yield (QY), as well as a blue shift of the absorption and fluorescence spectra of the nanocrystals [7-9]. For preserving the QD structural integrity, the irradiation mode should be so selected that the nanocrystal QY is changed without altering the shape of their absorption spectra. The use of these UV-irradiated QDs could make it possible to modulate the FRET efficiency without affecting the structure of the QD-based nano-hybrid system and determine the mechanism of the QD effect on bR leading to accelerated formation of intermediate forms in the bR photocycle.

2. Theory

FRET is non-radiative energy transfer between donor chromophores in an excited state (D) and acceptor chromophores in their ground state (A) [10]. One of the key parameters determining the efficiency of FRET is the Förster radius (R_0), defined as the distance between D and A for which the FRET efficiency approaches 50%. It can be calculated from the intrinsic optical and spectral parameters of D and A:

$$R_0^6 = \frac{9000(\ln 10) Q_D \kappa^2}{128\pi^5 N n^4} J(\lambda) \tag{1}$$

where Q_D is the donor quantum yield, κ^2 is the transition dipole orientation factor, N is Avogadro's number, n is the refractive index of the medium and $J(\lambda)$ is the normalized overlap integral between the D emission and A absorption. The normalized overlap integral is calculated as follows:

$$J(\lambda) = \int_0^\infty F_D(\lambda)\varepsilon_A(\lambda)\lambda^4 d\lambda \tag{2}$$

where F_D is the normalized emission of D, and ε_A is the molar absorption coefficient of A at the wavelength λ.

The FRET efficiency (E), which reflects the theoretical proportion of absorbed photons transferred to the acceptor at a given distance, can be determined using the following equation:

$$E = \frac{n_A R_0^6}{n_A R_0^6 + r^6} \tag{3}$$

where R_0 is the Förster radius, r is the distance between the donor and the acceptor, and n_A is the number of acceptors accessible for energy transfer from the donor.

3. Experimental

We use bR in its native purple membranes (PM) because the isolation of bR from PMs considerably decreases both the biochemical and thermodynamic stabilities of the protein. PMs were isolated from *H. salinarum* by the standard procedure [11]. The concentration of bR in solutions was measured by light absorbance at 570 nm (the bR molar extinction coefficient was 63000 $M^{-1}cm^{-1}$). The changes in the bR light absorbance were measured using a laboratory flash-photolysis setup with the probing beam passing a double monochromator after excitation of the sample with a Nd-YAG laser (532 nm, 8 ns, 15 mJ).

A chloroform colloid solutions of CdSe/ZnS QDs synthesized as described in [12] and coated with trioctylphosphine oxide (TOPO) were exposed to UV laser irradiation (266 nm, 8 ns, 1 mJ) at a dose of about 1.5×10^{-16} J per particle. The TOPO concentration was about 1 mg/ml. The QDs extinction spectrum, as well as the shape and position of their fluorescence spectrum, remained unchanged after the irradiation, while the QY was two times decreased.

The surface of both irradiated and nonirradiated particles was functionalized with PEG-OH [13] to make them water-soluble and stable in aqueous solutions. The fluorescence peak position of QDs after functionalization was 570 nm. The functionalized QDs were used to obtain the aqueous solutions of QDs, PMs, and their mixtures, the description of which is given in Table 1. The pH of all these solutions was 6.5. The fluorescence intensity and lifetime of the QDs water solutions were measured by means of a FLUOROLOG-3 spectrofluorimeter (Horiba Jobin Yvon) using the TCSPC option (Time-Correlated Single Photon Counting). The lifetimes ($\boldsymbol{\tau}$ obtained from fluorescence and photopotential measurements) were evaluated by multiexponential fits and presented as average lifetimes ($\boldsymbol{\tau}_{\text{avr}}$) in all cases. The $\boldsymbol{\tau}_{\text{avr}}$ value was calculated as the sum of the products of each $\boldsymbol{\tau}$ value by the respective amplitude divided by the sum of all amplitudes.

Table 1. Samples used in experiments.

	Molar concentration of QDs in solution, μM	Molar concentration of bR in solution, μM	QDs condition
Sample 1	1,2	-	nonirradiated
Sample 2	-	7,2	-
Sample 3	1,2	7,2	nonirradiated
Sample 4	1,2	7,2	irradiated

4. Results and discussion

In [6] was found that the integration of QDs into purple membranes (PMs) containing bR could accelerate the formation of the intermediate M-form of the bR photocycle. And this is particularly strong effect was observed for QDs with a fluorescence maximum at a wavelength of 570 nm, almost coinciding with the maximum of the bR absorption band. For such quantum dots FRET to bR is maximal because the overlap integral (equation 2) is maximal [11]. Therefore it is interesting to clarify the role of FRET in bR photocycle acceleration process. It can be done by UV laser irradiating which reduce the quantum yield of QDs without affecting the structure of the QD-PM nano-hybrid system. As follows from equation 1, the reduction in QDs quantum yield should reduce the Förster radius. As follows from equation 3, the reduction in Förster radius should lead to a decrease in FRET efficiency. If the FRET affects the kinetics of the bR M-form formation, then with a decrease in energy transfer efficiency the speed of bR photocycle also changes. We was studied the kinetics of the bR M-form in aqueous mixtures of PMs with both irradiated and non-irradiated QDs.

In order to demonstrate that FRET occurred in the aqueous mixtures of QDs and bR, we plotted the dependence of the nonradiative fluorescence quenching rate for QDs not exposed to UV irradiation on the amount of bR added to the solution (Fig. 1a). The nonradiative fluorescence quenching in QD solutions was estimated as:

$$F = 1 - \eta \frac{I}{I_0} \qquad (4)$$

where F is the nonradiative quenching rate, I is the sample fluorescence intensity in the presence of bR, I_0 is the sample fluorescence intensity in the absence of bR, and η is a correction factor taking into account light energy reabsorption [14]. It has been shown [4] that FRET between the donor (QDs) and acceptor (bR) is the main nonradiative relaxation pathway in aqueous solutions of QDs and bR. Thus, data shown in Fig. 1a demonstrate that the FRET efficiency increases with increasing proportion of bR in the QD–bR system. FRET in this system is also confirmed by a shorter average luminescence decay time in sample 3 compared to sample 1 (Fig. 1b).

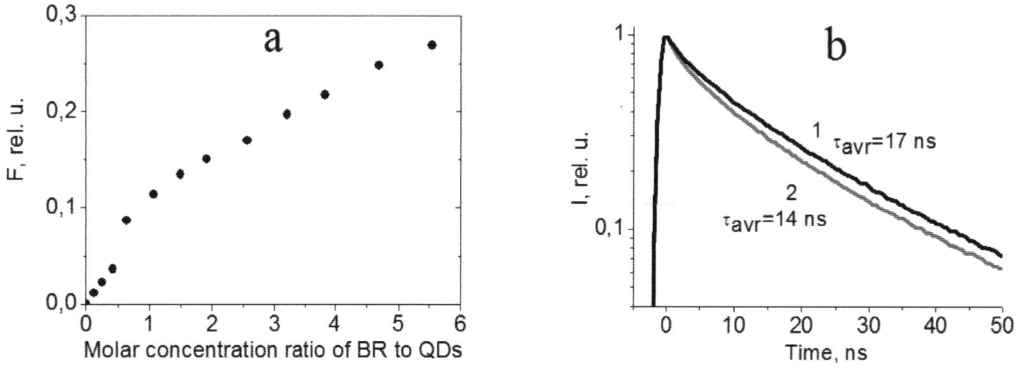

Figure 1. (a) Dependence of the nonradiative fluorescence quenching rate for QDs not exposed to UV irradiation on the bR-to-QD molar ratio in their mixed aqueous solution. (b) Changes in the fluorescence kinetics of QDs in the presence of bR (**1** – nonirradiated QDs water solution; **2** – water mixture of nonirradiated QDs and PMs).

Figure 2a shows the kinetics of the formation of the bR M form (the dependence of the absorbance at a wavelength of 410 nm on time) in the experimental samples 2–4. As can be seen in the figure, this kinetics was the same for the bR–QD mixture with a twofold decreased QY and the mixture of bR and nonirradiated QDs. The average times of the absorption increase are 60±5 μs for PM suspensions and 40±3 μs for all bR–QD mixtures; the average times of the absorption decrease are 4.9±0.4 ms and 3.3±0.2 ms, respectively. Therefore, we may conclude that FRET between bR and QDs does not affect the kinetics of the formation of the bR M-form.

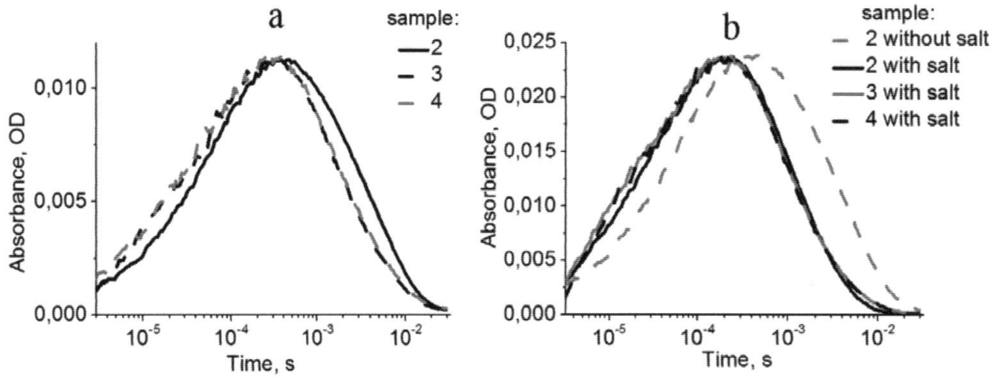

Figure 2. (a) Kinetics of the formation of the bR M-form in samples 2 (PMs water solution), 3 (water mixture of nonirradiated QDs and PMs), and 4 (water mixture of irradiated QDs and PMs). (b) Kinetics of the formation of the bR M-form in samples 2, 3, and 4 with an increased NaCl concentration (100 mM) in comparison with sample 2 without increasing of NaCl.

In addition, we measured the kinetics of the formation of the M form at an increased NaCl concentration in the solutions studied (100 mM; pH remained at 6.5), when charges on the PM surface were screened (Fig. 2b). The results showed that, at a high concentration of the salt, the kinetics of the formation of the bR M-form was the same in the presence and absence of QDs. The average times of the absorption increase and decrease are 37±3 μs and 1.77±0.14 ms respectively for all samples with an increased NaCl concentration. On the other hand, an increased ionic strength accelerated the formation and breakdown of the M-form; i.e., it had the same effect as the addition of QDs to PMs. Thus, the effect of QDs on bR may be accounted for by the modification of the PM surface potential in the QD–PM system.

5. Conclusion

We have developed a method for directional modification of the QD radiative properties in order to control the FRET rate in nano-hybrid materials. The mechanism underlying the effect of QDs with optical properties changed by UV laser irradiation on the photocycle of bR in PMs isolated from *Halobacterium salinarum* has been studied. The mode of laser irradiation has been so adjusted that it caused a twofold decrease in the QD luminescence QY without affecting either their extinction spectrum or the position or shape of their fluorescence spectrum. We have demonstrated that the accelerated formation of the bR photocycle M-form intermediate is related to the changed PM surface potential in the QD–PM complexes rather than FRET between QDs and bR. The method developed offers new opportunities for studying QD-based nano-hybrid systems and makes it possible to improve the characteristics of light-harvesting and light-emitting devices in which these systems are used.

Aknowledgements

This study was supported by the Ministry of Education and Science of the Russian Federation, grant nos. 11.G34.31.0050 and 14.512.11.0034. We thank Vladimir Ushakov for help in the preparation of the manuscript.

References

[1] Oesterhelt D 1998 *Curr. Opin. Struct. Biol.* **8** pp 489–500
[2] Janfaza S Molaeirad A Mohamadpour R Khayati M Mehrvand J 2014 *BioNanoScience* **4** 71–77
[3] Archer M D Barber J 2004 *Imperial College Press* London pp 1–41
[4] Rakovich A et al 2010 *Nano Lett.* **10** pp 2640–2648
[5] Nabiev I et al 2010 *Angew. Chem. Int. Ed.* **49** pp 7217–7221
[6] Zaitsev S Y Lukashev E P Solovyeva D O Chistyakov A A Oleinikov V A 2014 *Colloids Surf. B: Biointerfaces* 6301
[7] Lee S F and Osborne M A 2009 *ChemPhysChem* **10** pp 2174–2191
[8] van Sark W G J H M Frederix P L T M Bol A A Gerritsen H C and Meijerink A 2002 *ChemPhysChem* **3** pp 871-879
[9] Shcherbatyuk V Inman R H and Ghosh S 2011 *J. Appl. Phys* **110** 053518
[10] Lakowicz J R 2006 Principles of Fluorescence Spectroscopy *Springer* New York pp 443-527
[11] Bouchonville N Le Cigne A Sukhanova A Molinari M and Nabiev I 2013 *Laser Phys. Lett.* **10** 085901
[12] Peng Z A and Peng X J 2002 *Am. Chem. Soc.* **124** pp 3343-3353
[13] Sukhanova A et al 2012 *NanoMedicine: NBM* **8** pp 516–525
[14] Borissevitch I E 1999 *J. Lumin.* **81** pp 219–224

Journal of Physics: Conference Series **541** (2014) 012046 doi:10.1088/1742-6596/541/1/012046

Investigation the bit rate of quantum key using Si single photon detectors

A V Kolyako[1,2] I G Neizvestny[1,2], V L Kurochkin[1]

1 Institute of Semiconductor Physics, Siberian Division, Russian Academy of Sciences, Novosibirsk, 630090 Russia
2 Novosibirsk State Technical University, Novosibirsk, 630090 Russia

E-mail: kurochkin@isp.nsc.ru

Abstract. This work reports on the results of experimental investigations bit rate of quantum key distribution on a setup designed for quantum cryptography with single photons. Silicon avalanche photodiodes (C30902S) serve as single-photon detectors. The quantum key is transmitted with pulsed semiconductor lasers by coding polarization states of photons in two alternative nonorthogonal bases over free spase.

1. Introduction

From the practical standpoint, quantum cryptography is today the most mature area of quantum informatics. Quantum cryptography allows for absolutely secure data transfer between legitimate users of communication lines. The security of data transfer, combined with the complete impossibility of unauthorized access, relies on fundamental laws of nature, unlike those cryptography approaches using mathematical methods, where the information basically may be decoded. In accordance with the mathematically proven Shannon statement [1], a message being transferred cannot be decoded if it is encoded by a random one-shot key whose length equals that of the message, provided that the key is known by legitimate users alone. However, the problem in this situation is how to transport the key to remote users. Generally speaking, classical communication methods cannot provide secure transfer of the key over accessible data channels, since there are techniques of inconspicuous listening with subsequent decoding.

The ideas of quantum physics and quantum informatics as applied to long-range data communication may be used to attack the problem of transfer of an absolutely random key through accessible data channels with security assurance. The absolute security provided by quantum cryptography follows from the forbidding of quantum physics that are imposed on a metering device: (i) it is impossible to extract information about nonorthogonal states without disturbance [2] and (ii) it is impossible to clone an unknown quantum state (the no-cloning theorem) [3]. From these forbidding, it follows that, if single quantum objects are employed as data carriers, any intervention into the data transfer process undertaken by an unauthorized person will inevitably cause irreversible changes in the quantum states of the objects, from which the fact of intervention can be established.

Bennet and Brassard in 1984 [4] were the first to justify the principles of quantum cryptography and to devise a communications protocol. The first experimental demonstration of their concept [5] attracted much interest worldwide and gave an impetus to extensive research in this field. In [5], a key dissemination protocol was suggested, which was represented as a secured sequence of zeros and unities enciphered by single photons polarized in two mutually nonorthogonal bases. Later, this protocol was named BB84. In the first experimental setup [5], the distance between the transmitter and receiver (the length of the quantum channel) was 0.3 m. However, later on, rapid progress in the increase in the communication range was observed. At present, the record belongs to the group of the authors of [6], where the results on key distribution over a distance of 144 km are presented.

The important part of the quantum cryptography set up is single photon detector [7]. Single photon detection has large range of applications. It had become essential in broad range of research applications such as atmospheric monitoring, time resolved spectroscopy, quantum communications and investigation of quantum systems like single molecule. It has become an important instrument not only in physics but also in biology and chemistry. Quantum cryptography is the first in the history method which allows to guarantee the data transmission secrecy with the laws of physics. But for the good commercial application it should suit a number of conditions such as distance of communication, bit rate and stability. All these main parameters depend on the single photon detector operation. Specifically, it must offer a high quantum efficiency of measurement, a low noise, and a sufficiently high count rate. For the atmosphere transmission the silicon based avalanche photodiodes are used. This work reports on the results of experimental investigations bit rate of quantum key distribution on a setup designed for quantum cryptography. Silicon avalanche photodiodes (C30902S) serve as single-photon detectors.

2. Experimental setup

To research on the quantum key distribution over free space was create experimental set up [8,9]. The transmitting unit consists of four semiconductor lasers. Each laser generates light pulses with one of the four polarizations 0^0, 45^0, 90^0 and -45°. The laser beams are combined by a system of mirrors into one beam, which is attenuated by an absorbing filter at the exit and is directed through a 70-cm-long air gap toward the receiving unit. The semiconductor lasers with source supplies (modulated in current) operate in a pulsed mode with a pulse length of 8-10 ns. Each laser generates a coherent light pulse when a control pulse from a computer is fed into its power supply. The attenuated laser pulses arrive at the entrance of the receiving unit and are divided by a beam-splitting 50% mirror into two beams. The analysis of photons polarization was with help two Glan prism and four single-photon detectors.

The scheme of the receiving unit permits us to adjust the transmitting unit so that, after the exit attenuator, each laser pulse predominantly involves no more than one photon and the fraction of pulses containing two and more photons is insignificant. Under these conditions, the distribution of photons over the pulses obeys the Poisson statistics. In quantum cryptography, a signal is considered to be a single-photon signal if the mean number n of photons per pulse falls in the range 0.1 - 0.4. In particular, at $n = 0.1$, the fraction of pulses containing two (three) photons is equal to 5% (0.16%) of the number of single-photon pulses. Actually, in this case, nine out of ten pulses contain no photons. A photon (with any polarization) sent by Alice can arrive at three photoreceivers, namely, at one photoreceiver in its own basis (the polarizing splitting prism does not transmit this photon to the second photoreceiver) and at two photoreceivers in the foreign basis with equal probability. In the case when signals from all four photoreceivers are detected simultaneously and, in addition, the number of simultaneous operations of two and more photoreceivers is counted, the fraction of multiphoton pulses in the transmission can be calculated in terms of the Poisson statistics. The required mean number of photons in light pulses of the transmitting unit can be achieved by the sequential tuning of the laser power for each laser.

In our setup, the specially chosen silicon avalanche photodiodes C30902S (EG&G), which are the most sensitive photodiodes in the range of 0.8 μm, served as single-photon detectors. With the aim of counting single photons, the avalanche photodiodes were connected so that they operated in the Geiger mode, in which one photon can induce an avalanche of charge carriers. The photodiodes were connected in a passive avalanche quenching circuit. The ballast resistor (200 kΩ) limited the current in the avalanche photodiode. The signal was picked off the load resistor (50Ω), amplified by an amplifier 6, and fed into a shaper of standard TTL pulses for the interface with a computer. In order to decrease intrinsic noises, the photodiodes were cooled to –20°C with the use of Peltier semiconductor microcoolers. The quantum key was transmitted with pulsed semiconductor lasers by coding polarization states of photons in two alternative nonorthogonal bases (BB84 protocol).

3. Experimental results

In our experiments, the quantum key was generated as follows (protocol BB84). Alice's computer sets the laser pulse repetition rate (clock frequency). For each clock period, a sync (strobe) is generated, which is sent to Bob for transmission–reception synchronization. Simultaneously with the strobe, another pulse is applied at random to one of the four lasers, which generates a 10-nswide light pulse. A random number is created by an intelligent random-number generator, although generally it is preferable to use a noise generator to create a random number [7]. Having received the sync, Bob generates his own 20-ns-wide strobe. The pulses from the photodetectors are recorded only during the strobe application. In this case, a major part of the intrinsic noise of the photodetector is eliminated. For example, at –20°C and a voltage excess of 20 V, the rate of noise pulses is about 3×10^3 per second. At the same time, this value reduces to 100 per 10^6 clock periods when time gating of the signals is employed. After amplification, the duration of the noise pulses and the pulses from the APD was found to be 8–10 ns. Pretuning of the delay between the strobe and the APD pulse (when the diode is triggered by a laser pulse from the transmitter) makes it possible to considerably improve the signal-to-noise ratio and reduce the error rate in the final code.

The pulse from the APD is regarded as bearing information only if it coincides in time with the laser pulse. Intrinsic noise pulses not appearing during the strobe did not fall on the counter. The data from the four photodetectors were read out by Bob's computer under the control of the sync. In our setup, we used the same computer. Without loss in generality, such a design makes the experiment less hardware-intensive. When a pulse from any of the APDs came during the strobe, Bob remembered these data and the number of the sync and generated a signal for Alice by which she remembered the number of the sync and which of the lasers operated in that clock period. Since the mean number of photons in a light pulse was much less than unity, there was no need to remember the transfer process as a whole. Fifty percent mirror at the entrance randomly turns the photon to the vertical–horizontal or diagonal basis for recording. If Alice's and Bob's bases coincided, the measurements were assigned a successive serial number and filed to generate the key. Other wise, they were disregarded. According to the BB84 protocol, such a procedure generates a matched random secured key.

The bit rate of quantum key distribution depends on the clock frequency of laser pulse repetition, the number of photons per pulse, and the frequency characteristics of the avalanche photodiodes used. Sifted key generation rate is calculated by the following formula:

$$R_{sift} = \frac{1}{2} f_L \, n \, \eta \, T$$

Here the 1/2 factor is from sifting the key in the BB84 protocol, f_L is laser repetition rate, n is mean photon number per laser pulse, η is quantum efficiency of SPCM, and T is total transmission of the optical channel from Alice's output to Bob's detectors.

In our experiment, the rate of key generation was limited by the rate of data exchange between the computer and the receiving and transmitting units, which corresponded to a transmission clock frequency of 100 kHz. In the experiment, during the transmission with ~ 0.1 photon per pulse10^6 laser pulses, a key with a length of 12462 bits was generated; among the bits, only 812 (2,0%) were erroneous (the values of bits for Alice and Bob did not coincide –quantum bit error rate (QBER)). Table 1 shows the experimental data according to quantum key value of the average number of photons per pulse at a fixed quantum efficiency of 30%. Table 2 shows the error rate of the average number of photons per pulse at the quantum efficiency of 30%. During the transmission with ~ 0.2, the length of the key was 25336 bits, and QBER 174 bits (0.95%). For the used laser pulse repetition of 100 kHz, this corresponded to a key generation rate of ~1,2 kbits/s and 2,5 kbits/s.

Table 1. Dependence of the quantum key distribution vs the average number of photons per pulse at a fixed quantum efficiency of 30%.

Average number of photons	Amount of quantum key
0.1	12462
0.2	25336
0.4	44884

Table 2. Dependence of quantum bit error rate vs the average number of photons per pulse at 30% quantum efficiency.

Average number of photons	QBER (%)
0.1	2.0
0.2	1.3
0.4	0.9

4. Conclusion

To research on the quantum key distribution over free space was create experimental set up. The specially chosen silicon avalanche photodiodes C30902S (EG&G), which are the most sensitive photodiodes in the range of 0.8 μm, served as single-photon detectors. The key is transferred by pulsed semiconductor lasers, which encode the polarization state of the photons in two mutually nonorthogonal bases. The dependence of the bit rate quantum key and bit error rate vs of the average number of photons per pulse was investigated. When radiation propagates through the atmosphere, its polarization undergoes insignificant changes. Therefore, the polarization method of coding is used for organizing quantum channels over a free space, and, in the long view, the possibility of connection with orbital satellites is considered. The transmittance spectrum of the atmosphere has good transparency windows with wavelengths in the region of 0.8 μm. It is considered that the vertical optical density of the atmosphere is equivalent to a distance of about 8 km at normal conditions [7]; therefore, absorption losses of photons during the connection with the satellites are rather small. Further research in this field seems to be aimed at applying long-range atmospheric communication lines and increasing the data transfer rate.

Acknowledgments

This work was supported by RFBR (Grant No. 14-07-00809) and by the Siberian Branch of the Russian Academy of Sciences.

5. References

[1] Shannon C E 1949 *Bell Syst. Tech. Jour* **28** 658.
[2] Bennet C H 1992 *Phys. Rev. Lett* **68** 3121
[3] Wooters W K, Zurek W H 1982 *Nature* 299 802
[4] Bennet C H Brassard G 1984 Public Key Distribution and Coin Tossing *Proc. of IEEE Int. Conf. on Comput. Sys. and Sign. Proces., Bangalore, In-dia, December 1984* pp 175-179
[5] Bennet C H, Bessette F, Brassard G et.al. 1992 *J.Cryptology* **5** 3
[6] Ursin R, Tiefenbacher F, Schmitt-Manderbach T et al. *Nature Physics*, 2007 **3**, 481
[7] Gisin N, Ribordy G, Title W et al. 2002 *Rev. of Mod. Phys.* **74** 145
[8] Kurochkin V L Ryabtsev I I and I. G. Neizvestny I G 2005 *Technical Physics* **50** 727
[9] Kurochkin V L Zverev A V Kurochkin Yu V Ryabtsev I I and I. G. Neizvestny 2011 *Russian Microelectronics* **40** 245

Influence of external electrooptical modulator biasing on gain and nonlinear distortions in analog fiber-optic links

A. Petrov[1,2], I. Ilichev[1], P. Agruzov[1], V. Lebedev[1], E. Velichko[2], A.Shamray[1]

[1]Ioffe Physical Technical Institute, RAS, Saint Petersburg, Russia
[2]Saint Petersburg State Polytechnical University, Saint Petersburg, Russia

E-mail: alexey-np@yandex.ru

Abstract. The dependences of gain and nonlinear distortions in analogue fiber-optic links on bias of an external electrooptical modulator were investigated. The increase in the gain by up to 5 dB as compared with the conventional quadrature point operation was demonstrated by shift the bias voltage applied to the external electrooptical modulator to a low transmission. Dependences of nonlinear distortions on the bias voltage of an electrooptical modulator were investigated. A minor increase in nonlinear distortions (less than 0.5 %) was observed at the conditions of a maximum gain. Proposed theoretical model is in a good agreement with the experimental data.

1. Introduction

Microwave fiber-optic links offer many advantages over coaxial cables and other metal guides. These are very low transmission losses (~ 1 dB/km), large bandwidths, small weight, and a high degree of immunity to radiation and electromagnetic interference [1]. The main drawbacks of these links are a limited dynamic range performance because of a low gain and nonlinear distortions.

It has been shown that maximal transmission coefficient (gain) and minimal nonlinear distortions in microwave fiber-optic links correspond to the quadrature bias point of the external electrooptical modulator [2]. A fairly low gain (≈ - 20 dB) in this case is mainly caused by the photodetector saturation due to a high power in the unmodulated carrier which does not convey meaningful information.

In this report we analyze the influence of bias voltage applied to the external electrooptical modulator on the main transmission characteristics of analog fiber-optic links. The gain optimization of the fiber-optic link with an external electrooptical modulator and an erbium doped fiber amplifier is performed.

2. Experimental setup and samples

In figure 1 the scheme of investigated radio-frequency (RF) fiber-optical transmission line is presented.

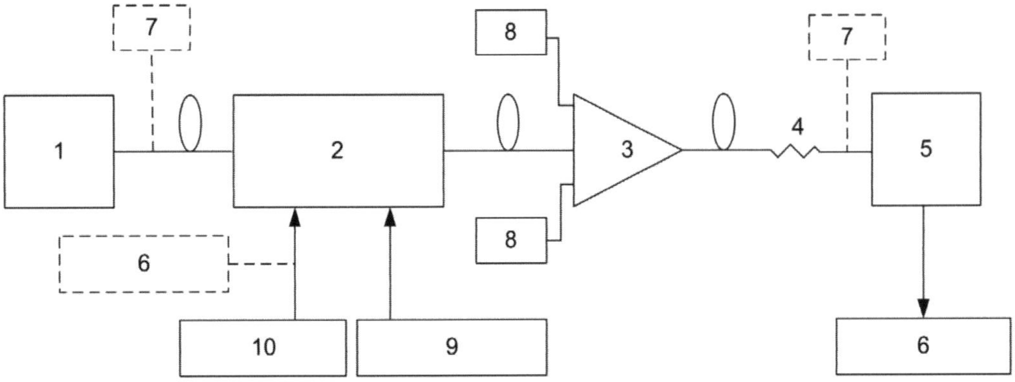

Figure 1. Experimental setup for investigation of radio-frequency fiber-optic link: 1 – laser; 2 – electrooptical modulator; 3 – erbium doped fiber amplifier; 4 – optical attenuator; 5 – wide-band photoreceiver; 6 – oscilloscope; 7 – optical power meter; 8 – laser pump diode; 9 – DC adjustable power supply; 10 – high-frequency generator.

Junction laser (1) with laser central wavelength 1551 nm and controlled output to 20 mW was used as optical carrier source. The Mach-Zehnder modulator with half-wave voltage 6 V and modulation band width to 55 GHz and optical losses 5 dB was used as external electrooptical modulator (2). The bias voltage in range ±20 V (which fixes a work point) and sine wave signal generator at radiofrequency 100 MHz were applied to the input of the modulator. After modulator the optical signal was applied to input of erbium doped optical fiber power amplifier with pump wavelength 980 nm (3). Enhanced signal was transformed to digital signal by wide-band photoreceiver (5) and registered by oscilloscope (6). Optical attenuator (4) with propagation factor ~ 10 dB was used for increasing of transmission ratio.

In the experiments we measured the relative amplitude of fundamental harmonic (for transmission ratio evaluation) and the relative amplitudes of the second and the third harmonics (for evaluation of non-linear distortion influence).

3. Results and discussion

We investigated the optimization of RF fiber-optic link by bias voltage applied to the external electrooptical modulator to the transmission minimum. We used method based on the increasing of the optical signal modulation contrast and optical carrier suppression.

The transmission rate dependence of Mach-Zehnder electrooptical modulator is detailed described in [3]:

$$\begin{aligned} I_0 &\propto \cos(\frac{\pi V_{bias}}{V_\pi}) J_0(\frac{\pi V_{rf}}{V_\pi}) \\ I_1 &\propto 2\sin(\frac{\pi V_{bias}}{V_\pi}) J_1(\frac{\pi V_{rf}}{V_\pi})\sin(\omega t) \\ I_2 &\propto -2\cos(\frac{\pi V_{bias}}{V_\pi}) J_2(\frac{\pi V_{rf}}{V_\pi})\cos(2\omega t) \\ I_3 &\propto 2\sin(\frac{\pi V_{bias}}{V_\pi}) J_3(\frac{\pi V_{rf}}{V_\pi})\sin(3\omega t) \end{aligned} , \quad (1)$$

where I_0, I_1,I_2,I_3 are the amplitudes of the zero, the first, the second and the third harmonics, J_0, J_1,J_2,J_3 are the zero, the first, the second and the third order Bessel functions of the first kind, V_π is the half-wave modulator voltage, V_{bias} is the modulator bias voltage, V_{rf} is RF modulating signal voltage, ω is the frequency of RF signal.

In case of the modulator bias voltage shift to a low transmission of the amplitude-modulated optical signal, the constant component decreases faster that the first harmonics, which leads to increase in the optical signal modulation depth.

The dependence of the gain coefficient of optical erbium doped fiber amplifier on input optical power is [4]:

$$G_{edfa} = \frac{G_0}{1 + (G_0 \langle P_{mz} \rangle / P_{\max})^{\alpha}}, \quad (2)$$

where G_0 is the amplification ratio of small signal, P_{mz} is the modulator output optical power, P_{max} is the maximal output power of amplifier in saturation regime, α is the empirical coefficient (which is close to 1).

At low modulation amplitude approximation the output modulator power may be evaluated as power of zero harmonic:

$$\langle P_{mz} \rangle = \frac{P_{laser}}{2}[1 - Jo(\pi V_{rf} / V_\pi)\cos(\pi V_{bias} / V_\pi)], \quad (3)$$

where P_{laser} is the output laser power, J_0 is the zero-order Bessel functions of the first kind.

Equation (2) describes the amplifier saturation due to increasing of the input optical power. In case of shift to a minimum transmission ratio of modulator the coefficient became equal to the amplification ratio of small signal, which increases with saturation power.

So, the transmission ratio of RF signal has the quadratic dependence on optical power and gain of optical amplifier:

$$G_{rf} \propto [G_{edfa} \cos(\pi V_{bias} / V_\pi + \frac{\pi}{V_\pi} V_{rf} \sin(\omega t))]^2 \quad (4)$$

In figure 2 the theoretical curves of transmission ratio and non-linear distortions of RF optical-fiber link, which were calculated using equations (4) and (1) taking into account the experimentally evaluated parameters of the amplifier ($G_0 = 200$, $P_{mz} = 0.2$ mW, $P_{max} = 5$ mW, $V_\pi = 6.5$ V, $V_{rf} = 1$ V, α= 1).

The increase in the link gain of up to 5 dB as compared with the conventional quadrature point operation was experimentally demonstrated. A minor increase in the nonlinear distortions (0.5 %) was observed for the relative bias voltage shift ($0.5V_\pi$) corresponding to the maximum gain.

As one can see from the presented data, the proposed theoretical model is in a good agreement with the experimental data.

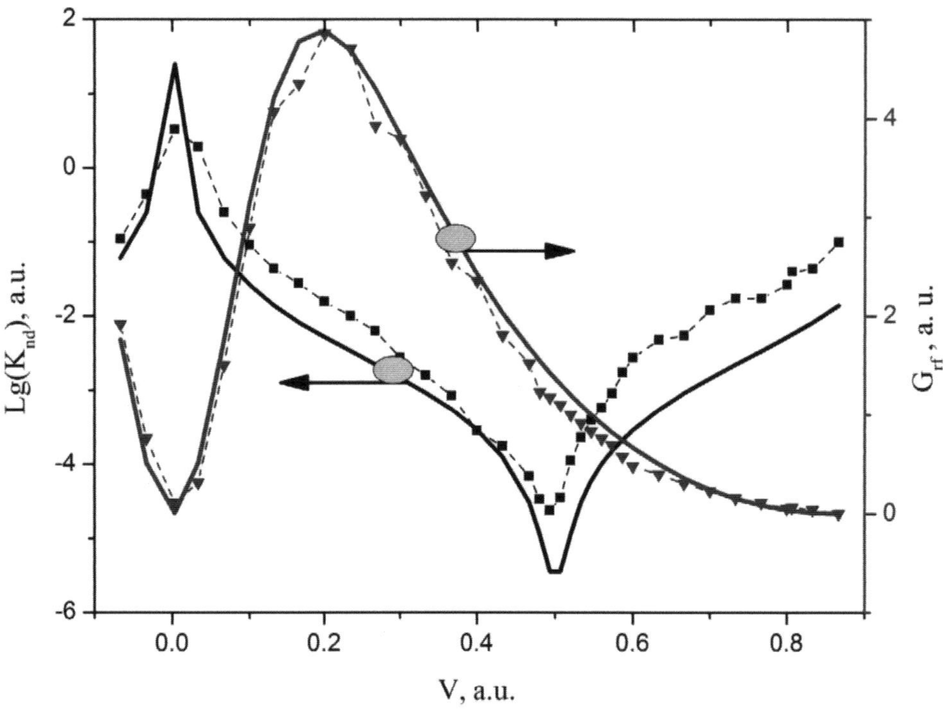

Figure 2. Nonlinear distortions (K_{nd}, squares) and gain (G_{rf}, triangles) of a radio-frequency fiber-optic link as functions of relative shift of bias voltage ($V = (V_0-V_{min})/V_\pi$) of external electrooptical modulator. Solid lines are calculated theoretical dependces and dotted lines are corresponded to experimental data.

4. Conclusion

The theoretical model describing the increase of transmission coefficient was developed. In accordance with it a shift of the modulator bias voltage to a low transmission and amplification of the amplitude-modulated optical signal by an optical amplifier operating near its saturation provide an optical carrier suppression, an increase in the optical signal modulation depth and, as a consequence, an increase in the maximum link gain limited by the photodetector saturation current. Theoretical dependences of gain and nonlinear distortion coefficient on the bias voltage of the external electrooptical modulator were calculated. They were found to be in a good agreement with the experimental data.

The results can be useful for the emerging technology of radiophotonics taking advantages of optical methods for generation, transmission, transformation, and processing of analog radiofrequency signals.

References

[1] Urick V, Rogge M, Bucholtz F and Williams K 2006 *Electron. Lett.* **42** 552
[2] Sisto M M, LaRochelle S and Rusch L A 2006 *J. Lightw. Technol.* **24** 4974
[3] Kolner B H and Dolfi D W 1987 *Appl. Opt.* **26** 3676
[4] Zhang X and Mitchell A 2000 *IEEE Photon. Technol. Lett.* **12** 28

Analysis of efficiency of solar energy conversion by tandem $Cd_xZn_{1-x}Te/Si$ solar cell

D.V. Grigoryev, K. A. Lozovoy, A. A. Pishchagin[1]

Tomsk State University, 36 Lenin Av., Tomsk, 634050, Russia

E-mail: pisch@sibmail.com

Abstract. In this work efficiency of solar energy conversion by solar cell based on tandem structure CdZnTe/Si is analyzed. It is shown that CdTe-CdZnTe-ZnTe material system is promising for creation of high-performance cascade solar cells.

Search of material systems effective for creation of high-performance monolithic multi-junction solar cells (SCs) is an urgent task of today's research. Such material systems must allow designers to intentionally select the band gap energy and prepare on silicon substrate specified semiconductor structures of required quality. The systems of A2B6 semiconductor solid solutions, in particular CdTe-CdZnTe-ZnTe material system, as well as A3B5 compounds are very promising materials for creation of high-performance solar cells [1]. For example, CdTe has a band gap of 1.5 eV that is optimum for creation of solar cells with high efficiency. This material system allows designers to change material's band gap from 1.5 to 2.28 eV. Besides that, there are some advantages of using such structures on Si substrates for creation of high-performance solar cells.

Only high-performance tandem solar cells parameters may successfully compete with those of cascade solar cells based on A3B5 materials. For creation of such high efficiency solar cells CdTe-CdZnTe-ZnTe heteroepitaxial structures of high quality are necessary. The method of molecular beam epitaxy (MBE) allows researchers to obtain heterostructures of such quality.

The aim of this work is to perform analysis of limiting efficiency of tandem solar cells based on $Ga_xIn_{1-x}P/Ge$ and $Cd_xZn_{1-x}Te/Si$ for AM0 solar spectrum and to carry out numerical modelling of such solar cells operating characteristics for AM0 and AM1.5 solar spectra. For achieving of optimum efficiency it is necessary to choose band gap energy of semiconductor solid solution ($Cd_xZn_{1-x}Te$) from the condition that photocurrents in the maximum power points of the first and the second (Si, Ge) junctions are equal. Molar percentage x, that determines materials band gap energy, was being varied during calculations. Standard expressions obtained in [2] were used.

Maximum efficiency of the solar cell is determined as

$$\eta(E_g) = \frac{V_m(E_g)J_m(E_g)}{P_0}, \tag{1}$$

where V_m, J_m are the operating voltage and current in the maximum power point, P_0 is the incident power of solar radiation. Maximum power point (V_m, J_m) is found from current-voltage characteristic

[1] To whom any correspondence should be addressed.

of solar cell by the condition $d(JV)/dV = 0$. Maximum power point is determined from the solution of the system of equations for band gap energy E_g of photoabsorption area material:

$$eV_m(E_g) = eV_{oc}(E_g) - kT\ln\left(1 - \frac{eV_m(E_g)}{kT}\right), \tag{2}$$

$$J_m(E_g) = \frac{en_{ph}(E_g)}{1 + kT/eV_m(E_g)}, \tag{3}$$

$$eV_{oc}(E_g) = E_g - kT\ln\left(\frac{A(E_g)}{en_{ph}(E_g)}\right), \tag{4}$$

$$n_{ph}(E_g) = \int_{Eg}^{\infty} \frac{dn_{ph}}{d\hbar\omega} d\hbar\omega, \tag{5}$$

where V_{oc} is the open circuit voltage, n_{ph} is the number of photons absorbed in solar cell, and $dn_{ph}/d\hbar\omega$ is the spectral density of number of photons absorbed in solar cell. Constant A depends on bad gap energy and is defined as $A = 5639E_g^2$ A/cm^2.

Results of calculations have shown that for tandem solar cell containing two CdZnTe/Si *p-n* junctions optimum value of molar percentage x for $Cd_xZn_{1-x}Te$ is equal to 0.68. At this x value band gap energy is E_g(CdZnTe) = 1.75 eV, and limiting efficiency of solar energy conversion is $\eta_{lim} = 38.9$ %. The operating voltage for these conditions is 2.03 V and operating current is 25 mA/cm^2.

Comparison of calculated solar cell characteristics with characteristics of tandem GaInP/Ge-based solar cell was conducted. For estimation of GaInP/Ge solar cell characteristics similar procedure was used. Calculations have shown that for GaInP/Ge solar cell η_{lim} is 2.4 % less and is equal to 36.5 % for GaInP band gap energy 1.41 eV. Operating voltage and current are equal to 1.3 V and 36 mA/cm^2, respectively. Additionally, limiting efficiency of solar energy conversion for CdZnTe/Si tandem solar cell is 5 % higher than for GaNPAs/Si photoconverter [3].

Conducted calculations of limiting efficiency of solar energy conversion η_{lim} have shown that CdZnTe/Si structures a very promising for creation of tandem solar cells. The same result was obtained in the work [1] where comparison of limiting efficiencies of solar energy conversion η_{lim} for CdZnTe/Si and GaInP/InGaAs/Ge structures for solar spectrum AM1.5 was carried out. The calculations of the mentioned authors have shown that limiting efficiency of CdZnTe/Si solar cell is 2.1 % higher.

Numerical modelling of CdZnTe/Si tandem solar cell operating characteristics was conducted with the assumption that bottom junction was created on the standard 300 μm thick silicon plate. The aim of numerical modelling was to determine optimum band gap of photoabsroption area of top CdZnTe cell. Numerical modelling was conducted for AM0 and AM1.5 solar spectra with the help of SCAPS program complex [4]. The structure of the solar cell under consideration is shown in the figure 1. The top junction of the tandem solar cell consists of dopped heterostructure CdZnTe/n-ZnTe. In this structure ZnTe plays a role of optical window [5]. The thickness of photoabsorption area of CdZnTe equals to 4 μm. The thickness of ZnTe layer is 0.025 μm. Due to complexity of numerical estimation of electrical current through the tunnel junction p^+-CdZnTe/n^+-Si calculations were conducted separately for the top and the bottom junctions of CdZnTe/Si tandem solar cell. During calculations of the bottom Si *p-n* junction parameters it was considered that photons with energy $E > E_g^{CdZnTe}$, where E_g^{CdZnTe} is the band gap of photoabsorption area material in the top CdZnTe/ZnTe junction, are absorbed in the top junction completely.

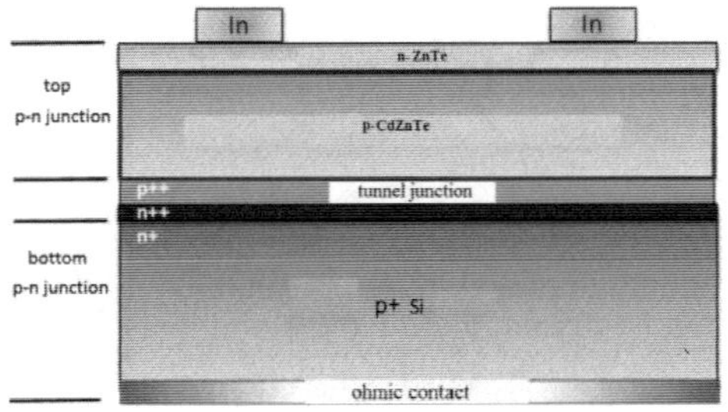

Figure 1. The structure of tandem ZnTe/CdZnTe/Si solar cell.

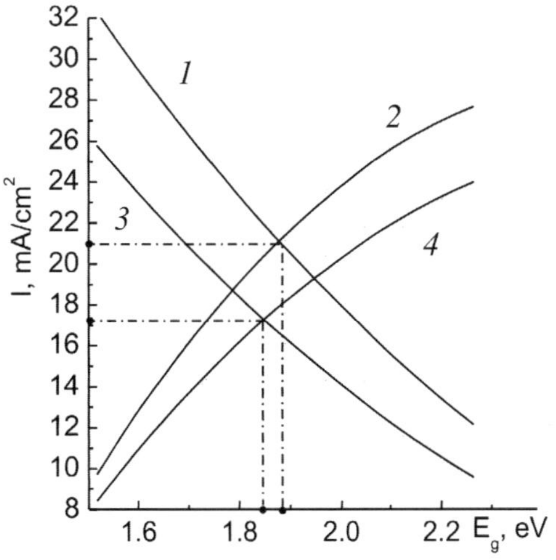

Figure 2. Dependence of tandem solar cell operating photocurrent on the band gap of the top $Cd_xZn_{1-x}Te$ junction. For the AM0 solar spectrum: 1 – $Cd_xZn_{1-x}Te$, 2 – Si. For the AM1.5 solar spectrum: 3 – $Cd_xZn_{1-x}Te$, 4 – Si.

The results of calculations of maximum power point operating current dependence on the band gap E_g^{CdZnTe} of the photoabsoption CdZnTe layer in the top *p-n* junction for AM0 and AM1.5 solar spectra are shown in the figure 2. For AM0 solar spectrum the photocurrent value for the top junction decreases from 32 to 12 mA/cm^2 with the band gap of photoabsorption area increasing from 1.5 to 2.28 eV. Meanwhile, at the same conditions value of photocurrent in the bottom Si solar cell increases from 9 to 27 mA/cm^2. Intersection of these two curves gives us optimum value of the CdZnTe band gap.

For the AM0 solar spectrum optimum value of the CdZnTe band gap is equal to 1.88 eV and photocurrent at this conditions is equal to 21 mA/cm^2. This band gap corresponds to the molar percentage $x = 0.51$ in the $Cd_xZn_{1-x}Te$.

The dependence of operating voltage in the top and the bottom *p-n* junctions on the band gap is shown in figure 3.

It may be seen that for the band gap of 1.88 eV operating voltage is equal to 1.1 V for the top $Cd_{0.51}Zn_{49}Te$ junction and 0.48 V for the bottom Si junction. Maximum output power of tandem solar cell equals to 33.1 mW/cm^2 and corresponds to the efficiency of 25.4 % for solar energy conversion. The obtained value is 13.5 % less than limiting efficiency.

For the AM1.5 solar spectrum optimum value of the CdZnTe band gap is equal to 1.85 eV and photocurrent at these conditions is equal to 17.5 mA/cm^2. This band gap corresponds to the molar percentage $x = 0.55$ in the $Cd_xZn_{1-x}Te$. Operating voltage for the band gap of 1.85 eV is equal to 1.08 V for the top $Cd_{0.55}Zn_{0.45}Te$ junction and 0.49 V for the bottom Si junction. In this case maximum output power of tandem solar cell equals to 27.4 mW/cm^2 and corresponds to the efficiency of 27.4 % for solar energy conversion.

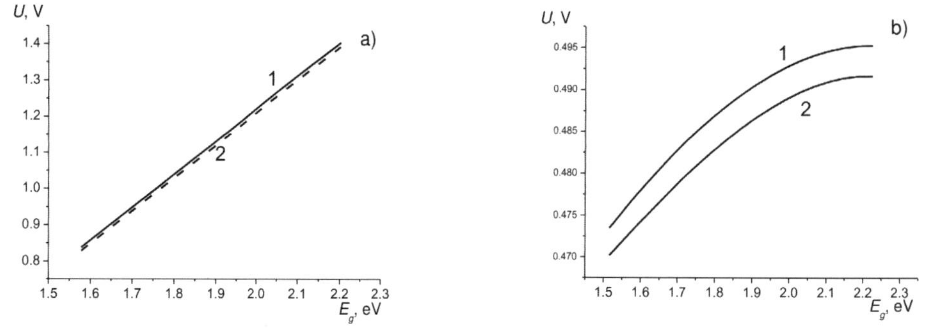

Figure 3. Operating voltage in the top and the bottom junctions of tandem solar cell as a function of band gap of the top $Cd_xZn_{1-x}Te$ junction. 1 – the AM0 solar spectrum, 2 – the AM1.5 solar spectrum. a) $Cd_xZn_{1-x}Te$ junction, b) Si junction.

Thus, the calculations for tandem solar cells based on A4B4 and A2B6 compounds have shown that limiting efficiency of solar energy conversion for the CdZnTe/Si solar cell is 2 % higher than for GaInP/Ge. It makes CdTe-CdZnTe-ZnTe material system promising for the creation of high performance cascade solar cells.

Conducted numerical modelling of operating characteristics of CdZnTe/Si tandem solar cell created on the 300 μm thick silicon substrate have shown that maximum efficiency of solar energy conversion by this tandem solar cell is equal to 25.4 % and 27.4 % for AM0 and AM1.5 solar spectra, respectively. Given conversion efficiencies correspond to the optimum band gap of the photoabsorption area in the top solar cell. These band gaps are 1.88 eV for AM0 solar spectrum and 1.85 eV for AM1.5 solar spectrum. Corresponding photoabsorption materials of the top junction are $Cd_{0.51}Zn_{0.49}Te$ and $Cd_{0.55}Zn_{0.45}Te$, respectively.

The reported study was conducted within the bounds of the National Research Tomsk State University competitive recovery program.

References

[1] Xu D, Biegala T, Carmody M, Garland J W, Grein C and Sivananthan S 2010 *J. Appl. Phys. Lett.* **96** 073508
[2] Henry C H 1980 *J. Appl. Phys.* **51** 4494
[3] Geisz J F and Friedman D J 2003 *Semicond. Sci. Technol.* **17** 769
[4] Burgelman M, Nollet P and Degrave S 2000 *Thin Solid Films* **361** 527
[5] Luque A and Hegedus S 2003 *Handbook of Photovoltaic Science and Engineering* (New York: John Wiley & Sons)

Lasing in microdisk resonators with InAs/InGaAs quantum dots transferred on a silicon substrate

E I Moiseev[1], N V Kryzhanovskaya[1], A M Nadtochiy[3], M V Maximov[1], A E Zhukov[1], I I Shostak[1], M M Kulagina[3], AV Savelyev[1,2], K A Vashanova[3], A A Lipovskii [2,1]

[1] St. Petersburg Academic University, Nanotechnology Center for Research and Education, Russian Academy of Sciences, St. Petersburg, 194021 Russia
[2] St. Petersburg Polytechnic University, St. Petersburg, 195251 Russia
[3] Ioffe Physical–Technical Institute, Russian Academy of Sciences, St. Petersburg, 194021 Russia

moiseev@spbau.com

Abstract. Semiconductor microdisk lasers with an active region based on InAs/InGaAs quantum dots before and after transferring on silicon substrate were investigated. High-temperature lasing up to 105°C in a 6-μm GaAs-based microdisk laser is demonstrated. The simple technology of transferring of microdisk laser on a silicon substrate is developed. Immersing of the transferred microdisk into glue does not lead to deterioration of surface quality and to significant increase in radiation losses. Multi-mode room temperature lasing in 6-μm microdisk laser transferred on a silicon substrate is obtained. Quasi single mode lasing is observed at 78 K.

1. Introduction

Currently there is a need to develop radiation sources for optoelectronic integrated circuits with small size, low power consumption, planar design and high temperature stability [1, 2]. The non-direct gap energy structure of Si, which is dominant in electronics, is significantly hampering the development of Si-based microlasers. Thus, the problem of integration of high performance III-V radiation sources with silicon platform remains topical. Currently developed integration methods are based on bonding procedure of InP-based microdisk on silicon substrate [3, 4]. The smallest diameter of microresonator demonstrating room temperature lasing published is 15μm. Further scaling down is limited by thermal deterioration of the laser characteristics due to weak carrier confinement in active region. At the same time ring and disk microcavities with active region based on InAs/InGaAs/GaAs quantum dots (QDs) demonstrate high thermal stability of characteristics [5, 6].

In this work we develop the simple technology of transferring of microdisk lasers with InAs/InGaAs/GaAs QDs on a silicon substrate and make it characterization in 78-380K temperature range.

2. Experiment

The structure was grown by molecular beam epitaxy with Riber 49 setup with a solid-state As source on a semiinsulating GaAs (100) substrate. For selective etching, a buffer GaAs layer was deposited and then $Al_{0.98}Ga_{0.02}As$ stop-layer was grown. The InAs/InGaAs QDs layers were inserted into a

0.2-μm thick GaAs waveguiding layer. Microdisk resonators with different diameters (3 μm, 4 μm, 6 μm and 8 μm) were formed using photolithography and etching by an Ar+ ion beam (figure 1, a). The $Al_{0.98}Ga_{0.02}As$ layer was transformed into a $(AlGa)_xO_y$ layer by selective oxidation to create optical confinement on the substrate side (for investigation of microdisks not transferred on the Si substrate). The epitaxial side of the structure was then fixed on the Si substrate by glue (figure 1, b). The GaAs substrate and buffer layers and $Al_{0.98}Ga_{0.02}As$ stop-layer were subsequently removed using an etching agents (figure 1, c). Microphotography of microdisk (D=6μm) is shown at figure 1, d.

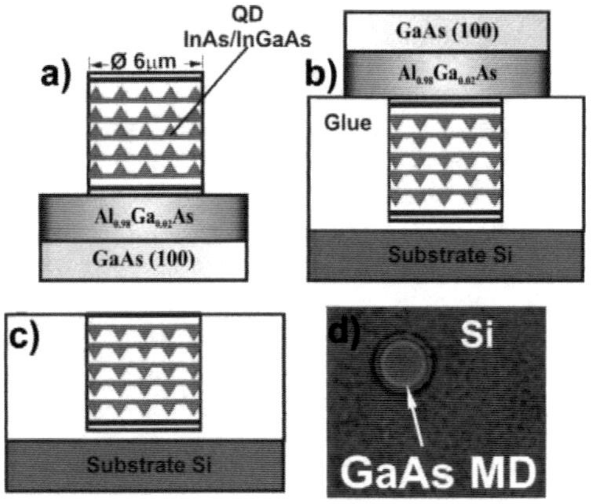

Figure 1. Schematic sequence of microdisk transfer technology: (a) photolithography and etching by an Ar+ ion beam, (b) gluing of microdisk to a Si substrate, (c) removal of a sacrificial layers; (d) microphoto of disk microresonator on silicon.

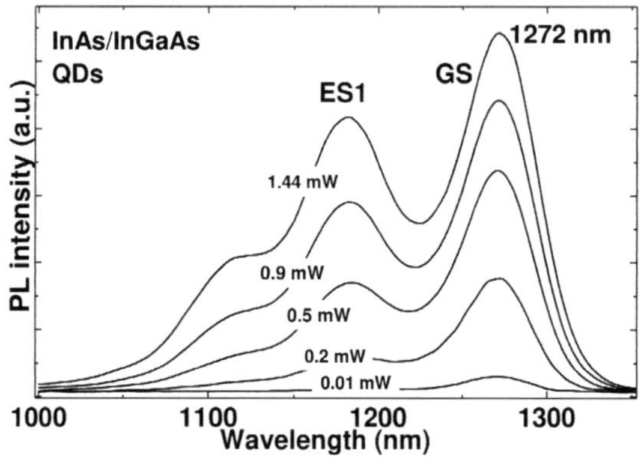

Figure 2. PL Spectra of unprocessed sample with InAs/InGaAs QDs obtained at 300 K.

The structures were investigated under optical pumping with Nd:YAG laser (λ = 532nm). The laser beam was focused using an Olymmpus LMPlan IR 100 NA 0.8 objective onto the microlaser top surface. The same objective was used to collect the PL signal, which was passed through FHR 1000 monochromator and detected by cooled InGaAs CCD array Horiba Symphony.

3. Results

The PL spectra at different excitation power of unprocessed sample with InAs/InGaAs QDs obtained at room temperature are shown at figure 2. The wavelength of PL spectra maximum is 1.27 μm and corresponds to QD ground state emission. Increase of excitation density results in linear increase of the PL intensity with slope k ~ 1 evidencing high optical quality of the QD active region. First, we investigated properties of μdisk lasers before transferring. The μPL spectra obtained at 300K from μdisk lasers with different diameters (3 μm, 4 μm, 6 μm and 8 μm) are shown at figure 3. The spectra are shifted along axis "Intensity" for clarity. Sharp lines observed at the spectra correspond to high quality whispering-gallery (WG) modes. Decrease of the resonator diameter from 8 μm to 3 μm results in increase of the resonator free spectral range (FSR) from 23 nm to 48 nm and in reducing of the WG modes quantity. One can see that in case of 3 μm microdisk resonator the number of the WG modes within the range of the InAs/InGaAs QDs gain spectra is quite low. Thus, the microcavities with diameter of less than 3 μm are candidates for single-mode radiation sources. Along with the high quality WG modes, low-quality modes exist in the disk microcavities. These modes modulate QDs spontaneous emission spectra. Ring microcavity provides suppression of these modes (figure 4). The lack of material in the center of the resonator does not deteriorate emission characteristics as compared with the disk microcavities since the high-Q WG modes are located near the resonator external rim.

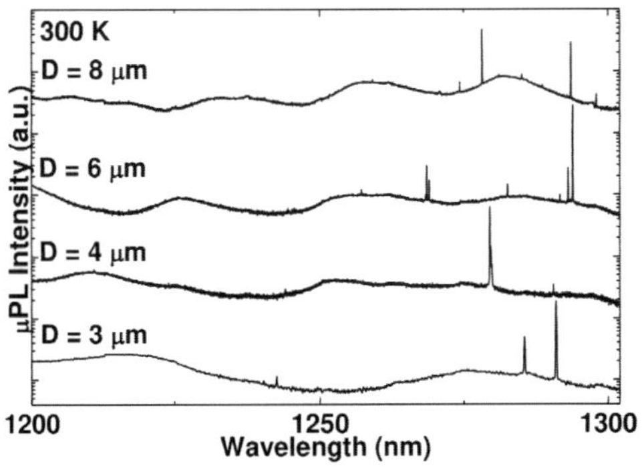

Figure 3. μPL spectra of a microdisk lasers (D = 3 μm, 4 μm, 6 μm and 8 μm) at 300 K.

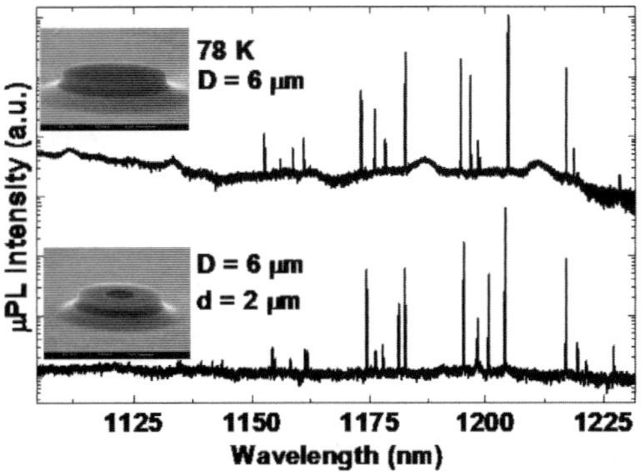

Figure 4. Emission spectrum of a ring and disk microcavity (D = 6 μm, d = 2 μm) at 78 K.

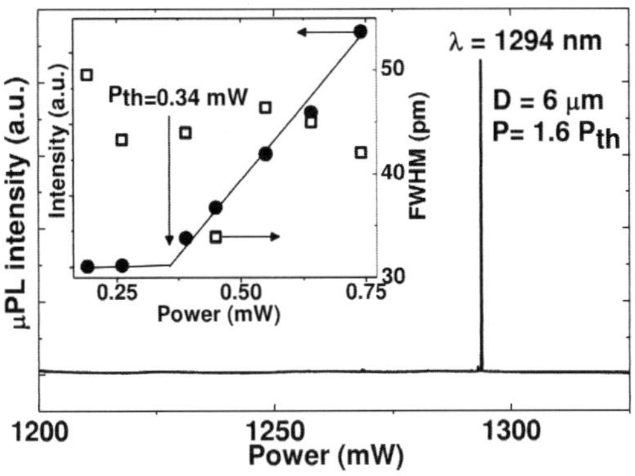

Figure 5. μPL spectrum of microdisk lasers (D = 6 μm) at 300K. Insert: Spectrally integrated emission intensity of a WG mode as a function of incident optical pump power.

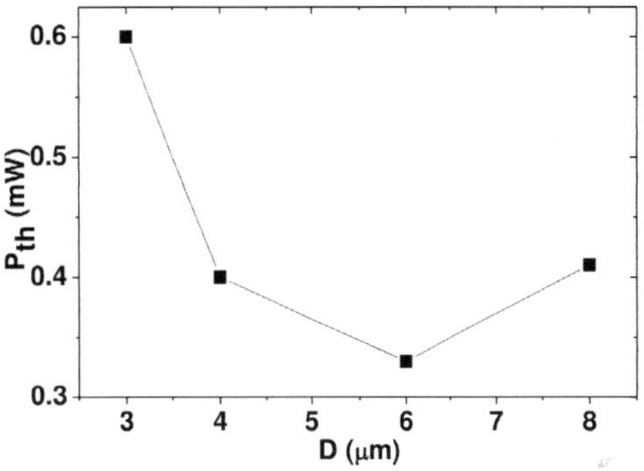

Figure 6. Dependence of the threshold power on the outer diameter

Increase of the optical pump power results in the transition to the lasing regime with rapid increase of the intensity of a WG modes of the resonators. Inset in fig. 5 shows the dependence of 1294 nm line intensity of 6 μm μdisk on the pump power (light-light curve). The threshold pump power is 0.34 mW. The FWHM of the1294 nm line near the threshold is 34 pm and it is limited by resolution of the measurement system. We can estimate the quality factor Q factor to be more than 3.4×10^4. Room-temperature threshold characteristics of the microresonators with different outer diameters were compared. The lowest threshold of 0.34-0.4 mW was found in microdisks having an outer diameter of 6 –7 μm. Both larger and smaller microdisks demonstrate higher threshold power as shown at figure 6. Lasing in the 6 μm μdisk was observed up to 105° C. Due to the temperature shift of the InAs/InGaAs QD ground state emission the lasing wavelength was observed at λ=1336 nm at 105° C.

The dependence of the threshold power on the temperature in range 20-105°C is shown at figure 7. Increase of the temperature results in increase of the threshold power (P_{th} @105°C ≈ 4 P_{th} @25°C) due to the various factors. First, temperature driven carrier escape from quantum dots increases nonradiative recombination. Second, the misalignment of the mode wavelength and the maximum of QD gain curve wavelength at certain temperature also increases threshold power.

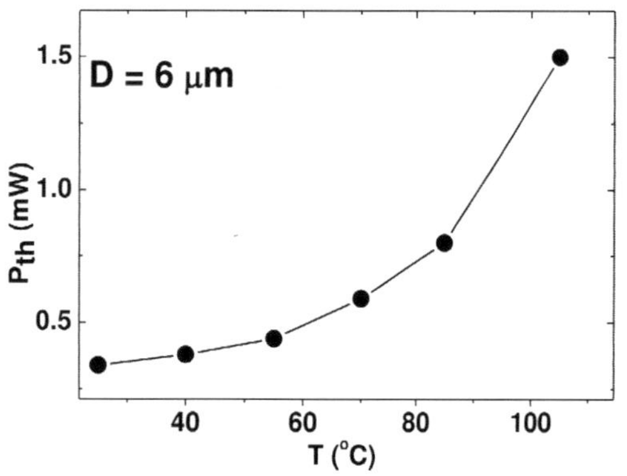

Figure 7. Dependence of the threshold power on the temperature for the disk microcavity with a diameter of 6 μm.

Then we investigated 6 μm in diameter microdisks transferred on Si substrate. μPL spectrum obtained at 300 K above threshold ($P=1.2\ P_{th}$) is shown at figure 8. The light-light curve of dominant line (λ=1296 nm) has a pronounced threshold (see inset in figure 8). The quality factor of the line $Q=\lambda/\Delta\lambda$ exceeds 3.4×10^4 at 300 K. This value is the same for initial microdisk measured before transferring. The measured threshold pump power at 300K (P_{th}^{300K}) is 0.63 mW that is higher approximately by 2 than in initial microdisks. This is allows us to conclude that developed technology of transferring and immersing of the microdisk into the glue do not lead to deterioration of microdisk surface quality or to a significant increase in radiation losses due to decrease in the contrast of the refractive index.

μPL spectra before threshold ($P=0.8\ P_{th}$) and above threshold ($P=1.3\ P_{th}$) obtained at 78K are shown at figure 9. The wavelength of dominant line is shifted to λ= 1237 nm due to the temperature driven active region band-gap shrinkage. Decrease of temperature to 78 K results in slight decrease of lasing threshold by 2 ($P_{th}^{78K}=0.5\ P_{th}^{300K}$) due to suppression of nonradiative recombination processes. The side mode suppression ratio at 78K is 14 dB. Thus, the microdisk demonstrates quasi single mode lasing. Increase of the excitation density leads to long-wavelength shift of the dominant line wavelength due to heating of the microdisk under optical pumping. We estimate the value of this heating to be 0.03 K/(W/cm^2) that is five times higher than in the initial microdisks on $(AlGa)_xO_y$ layer. We explain this heating by weak thermal conductivity of the glue and poor thermal contact with Si substrate. This problem can be solved by glue with better thermal conductivity.

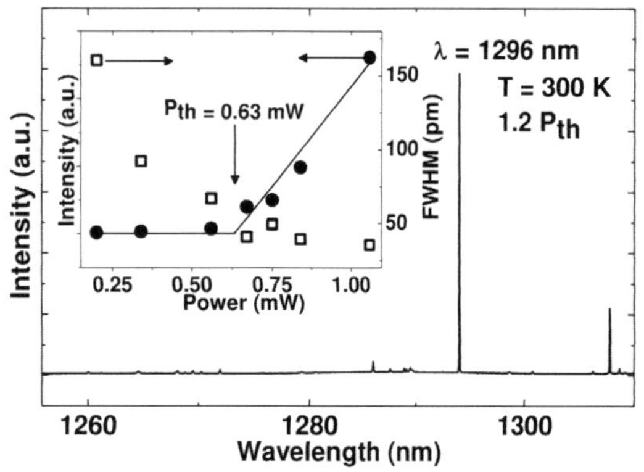

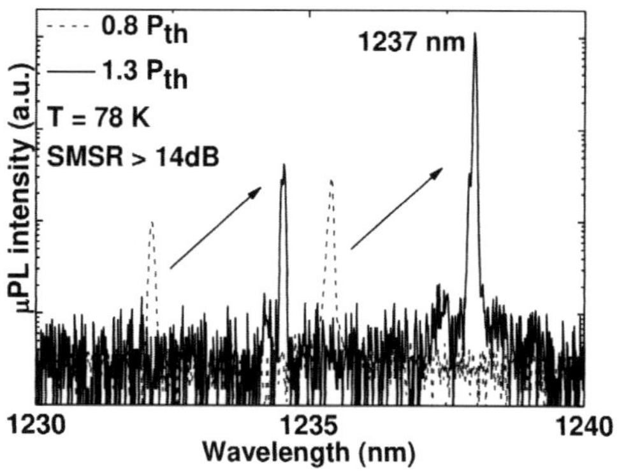

Figure 8. Emission spectrum of a microdisk laser transferred on Si substrate (D = 6 μm) at 78 K. Insert: intensity-pump curve of the microdisk.

Figure 9. Emission spectrum of a microdisk laser transferred on Si substrate at 300 K.

4. Conclusion

III-V microdisk lasers with InAs/InGaAs QDs active region before and after transferring on Si substrate are characterized in 78-380K temperature range. High-temperature lasing up to 105°C in a 6-μm GaAs-based microdisk laser is demonstrated. The method of transfer of the microdisk resonators on the silicon substrate is developed. Room temperature lasing in 6 μm microdisk laser on Si substrate is demonstrated. The quality factor Q in transferred microdisk exceeds 3.4×10^4. This value is the same for initial microdisk. This allows us to conclude that developed technology of transferring and immersing of the microdisk into the glue do not lead to deterioration of microdisk surface quality or to a significant increase in radiation losses due to decrease in the contrast of the refractive index.

5. Acknowledgments

The work is supported in different parts by «Foundation for Development of the Center for Elaboration and Commercialization of New Technologies».

References

[1] Levi A F J, Slusher R E, McCall S L, Tanbun-Ek T, Coblentz D L and Pearton S J 1992 *Electronics Letters* **28** 1010-12

[2] Baba T and Sano D 2003, *IEEE J. Sel. Top. Quantum Elecctron* **9** 1340-46

[3] Liang D, Fiorentino M, Okumura T, Chang H H, Spencer D T, Kuo Y H, Fang A W, Dai D, Beausoleil R G, and Bowers J E 2009, *Optics Express,* **17** (22), 20355–64

[4] Van Campenhout J, Rojo-Romeo P, Regreny P, Seassal C, Van Thourhout D, Verstuyft S, Di Cioccio L, Fedeli J M, Lagahe C and Baets R 2007, *Optics Express*, **15** (11) 6744–6749

[5] Munsch M, Claudon J, Malik N S, Gilbert K, Grosse P,Gerard J M, Albert F, Langer F, Schlereth T, Pieczarka M M, Hofling S, Kamp M, Forche A and Reitzenstein S 2012, *Applied Physics Letters* **100**, 031111

[6] Kryzhanovskaya N V, Zhukov A E, Nadtochiy A M, Slovinskii I A, Maximov M V, Kulagina M M, Savel'ev A V, Arakcheeva E M, Zadiranov Yu M, Troshkov S I and Lipovskii A A **46**(8), 1040-43

Noise reduction methods of single photon detector based on InGaAs/InP avalanche photodiodes

G K Krivyakin [1,2], A S Pleshkov[1,2], A V Zverev[1], I I Ryabtsev[1] V L Kurochkin[1]

1 Institute of Semiconductor Physics, Siberian Division, Russian Academy of Sciences, Novosibirsk, 630090 Russia
2 Novosibirsk State Technical University, Novosibirsk, 630090 Russia

E-mail: kurochkin@isp.nsc.ru

Abstract. This work is dedicated to the problem of noise reduction of single photon detector development based on InGaAs-InP avalanche photodiodes. Dark count probability and quantum efficiency of the detectors have been measured. We present the experimental fiber based quantum key distribution setup with phase coding of single photon states. The autocompensation two way optical scheme (plug&play) is used. The single photon source is the strongly attenuated laser pulse which goes through two paths of Mach-Zehnder interferometer where it undergoes the phase coding. Quantum channel is formed by 25 km single mode fiber. To generate the quantum key the four phases BB84 protocol is used.

1. Introduction

The telecommunication wavelength photodetectors are primarily made on the base of InGaAs photodiodes. Photon counting is the key part of quantum communication as the need arise to transmit single photons over long distances using optical fibers. InGaAs/InP single-photon avalanche diode detectors are favoured in a number of applications for single-photon detection at wavelengths around 1550 nm, e.g. quantum key distribution [1]. Gated quenching, in various realization, including the use of active quenching circuits [1,2], has been implemented in many of these applications to limit the charge flow per event to reduce the detrimental effects of the afterpulsing phenomenon. In this work the commercially available InGaAs avalanche diode ERM 547NT is used to designe as single-photon detector. We implement the noise reduction methods proposed and measure dark count probability and quantum efficiency of the detectors have been measured. Experimental results of quantum key distribution on the fiber setup obtained with these detectors are given.

2. Single photon detectors based on InGaAs avalanche diodes

To detect individual photons, avalanche photodiode (APD) operates in the Geiger mode [1,2]. For this purpose, the reverse supply voltage in them is raised above the threshold breakdown voltage of the diode. The higher the voltage above the threshold the higher the probability to detect a photon. However, voltage augmentation is usually accompanied by an increase in the noise clicks without any incoming signal (dark noise) and the probability of the noise due the so-called afterpulsing effect.

During the passage of the avalanche current after an operation of the photodiode from a photon or thermal noise pulse, the so-called traps in the semiconductor volume can be charged and then, with a delay, they begin to discharge spontaneously and lead to a new avalanche of the charge. This causes the photodetector to operate falsely. The effect of afterpulses strongly restricts the maximal click rate of photon counting.

To reduce the influence of these undesirable effects, some special solutions are implemented.. For example, APD cooling yields a significant decrease of dark noises. To decrease the probability of the appearance of afterpulses, active avalanche quenching is used. Another way is to operate in the mode with pulse supply when the voltage at the APD is sustained lower than the threshold and, to detect single photons, the voltage is raised above the threshold for a short time (for a few nanoseconds) [2,3]. After the passage of the avalanche, the voltage is decreased lower than the threshold for dead time 5–20 μs so that the traps could discharge.

The Figure 1 presents a block diagram of our quantum detector that is based on avalanche photodiode ERM547NT (VD1). Breakdown voltage of the avalanche photodiode can be adjusted with a precision 1 mV. The pulse widths of an avalanche gate (LE) and an avalanche detection strobe (QDET_IMP) can be adjusted using pulse shapers. Delay lines allow to adjust time lags from AVB_REQ pulse to LE and QDET_IMP. This time lags are required to shift an avalanche boost pulse (AVB) and LE to the time moment when the expected photon can cause the avalanche. Comparator DA3 is used to separate the avalanche pulses from the noise. The comparator discrimination threshold (Vth) can be adjusted with a precision less than 1 mV. The Figure 2 presents a timing diagram of the quantum detector. The diagram shows the se-quence of changes in the main signals of the detector. T_{le_ws} – adjusted LE pulse width. T_{avb_ws} – fixed AVB pulse width, the pulse amplitude is 3.5 V and width is 1 ns. T_{le_dls}, T_{avb_dls} – adjusted delays from AVB_REQ pulse to LE and AVB pulses, respectively.

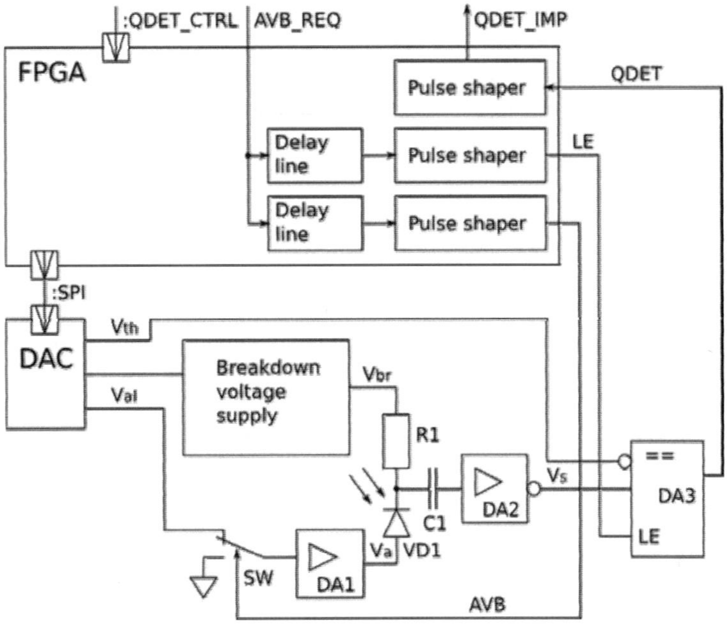

Figure 1 – The quantum detector block diagram.

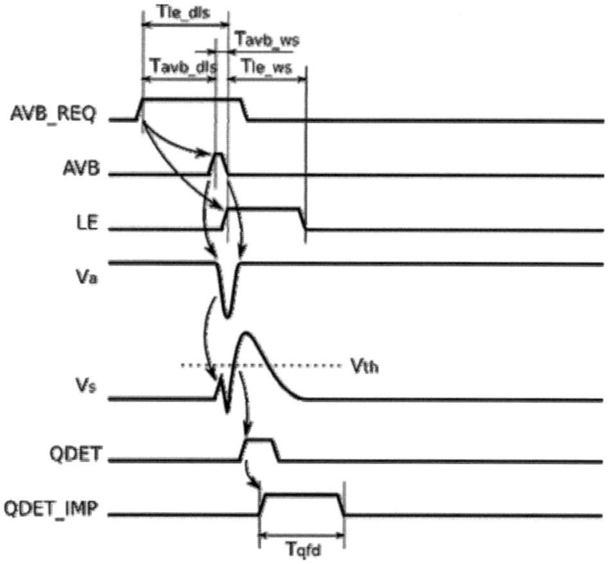

Figure 2 – Timing diagram of the quantum detector.

3. Single photon detectors: experimental results

We have measured of the dark noise and quantum efficiency for the used single-photon APD detectors ERM 547NT. Measurement of the dark noise was organized as follows. The measurements have been carried out in a dark room to protect APD from spurious photons. APD was cooled down to the -50^{0}C temperature. The repetition rate f_{PULSES} of the voltage pulses was 1 or 10 MHz. The pulses had a trapezoid form with the average duration of 2.5 ns and amplitude of 3,5V. This voltage was added to the DC bias voltage of the APD. The amplifier gave dark count pulses at the output. The average frequency of dark counts was measured by a controller FPGA. The articles often use the term the dark count probability per gate P_D was calculated from the measured f_{DURK} frequency as:

$$P_D = \frac{f_{DARK}}{f_{PULSES}} \tag{1}$$

To decrease the probability of the appearance of afterpulses after the passage of the avalanche, the voltage is decreased lower than the threshold for dead time 5–20 μs so that the traps could discharge. Figure 3 shows the plot of dark counts per second versus dead time. From this plot we obtain that optimal dead time equals to 5 μs.

Measurement of the quantum efficiency was organized as follows. The measurements have been carried out in a dark room to protect APD from spurious photons. APD was cooled down to the -50^{0}C temperature. The repetition rate f_{PULSES} of laser pulses and of the voltage pulses was set to 1 or 10 MHz. The delays of detected laser pulses and voltage pulses were synchronized. The average output power P of laser 1550 nm was measured by a calibrated Coherent power meter. The corresponding mean photon number μ emitted per laser pulse is calculated from the formula:

$$\mu = \frac{P}{\hbar\omega f_L} \tag{2}$$

where $\hbar\omega$ is the 1550 nm photon energy. The power P was set up to be 10^6 photons emitted per laser pulse on the average. After passage through a combination of fiber attenuators with the total loss of 70 dB the photon flux was attenuated to $\mu \approx 0.1$ photons per laser pulse. The photon statistics, i.e., probability P_n to find n photons in a laser pulse, is given by the Poisson distribution:

$$P_n = \frac{\mu^n}{n!} e^{-\mu} \tag{3}$$

For $\mu = 0.1$ one obtains $P_0 = 0.905$, $P_1 = 0.0905$, $\Sigma P_{n>1} = 0.0045$. This means that the probability to receive more than one photon is negligibly small in comparison with the probability of single-photon laser pulses. The output of the 70 dB attenuator was directly connected to the input of APD ERM 547NT. The average total frequency f_C of the counts was measured by a controller FPGA. The quantum efficiency η was calculated as follows:

$$\eta = \frac{1}{\mu}\left[\frac{f_C}{f_L} - P_D\right] \tag{4}$$

The measured dependences of quantum efficiency versus APD bias voltage at the 1 MHz repetition rates and -50°C temperature is presented on Figure 4.. From these dependences one finds that the APD breakdown voltage at -50°C is close to 55 V (given by the sum of the bias and gate voltages).

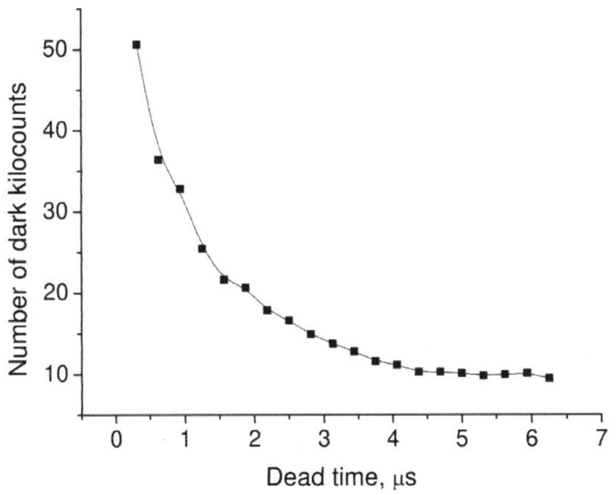

Figure 3. Measured number of dark counts per second versus dead time for quantum detector based on APD ERM 547NT.

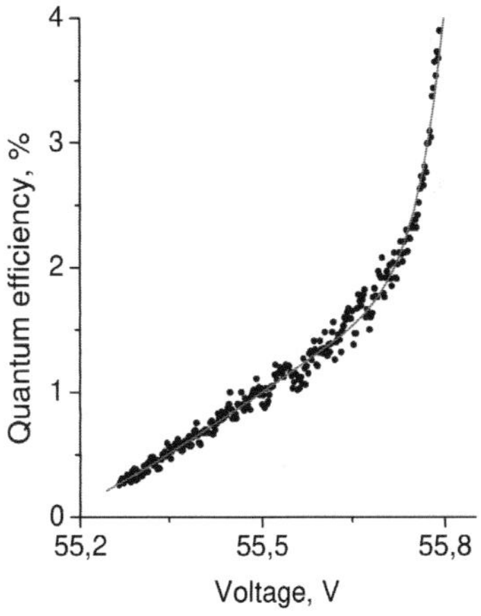

Figure 4. Quantum efficiency dependence on reverse voltage for quantum detector based on APD ERM 547NT.

4. Quantum key distribution: experimental results

From the practical standpoint, quantum cryptography is today the most mature area of quantum informatics. The essential purpose of quantum key distribution (QKD) is to create a secure communication process between two legal users, Alice and Bob, even with an eavesdropper Eve in the channel [1, 4, 5]. The first QKD protocol was proposed in 1984 [6]. This protocol is known as BB84. The first experimental demonstration was in 1992 [7].

We constructed a fiber based quantum cryptography setup to investigate the quantum key generation in the fiber channel [2,3]. This setup is controlled by the integrated processor and operates on the standard telecom wavelength. Quantum key distribution was implemented on the base of phase coding of the quantum states of single photons, emitted from a pulsed semiconductor laser, in the two alternative non-orthogonal bases. All the optical elements including laser, phase modulators and high speed attenuators are integrated to the single mode fiber. The optical scheme of the setup is designed in a two-pass autocompensating scheme [1,8]. This scheme consists of transmitter Alice and receiver Bob which are connected with the 25 km quantum channel made of the single mode fiber SMF-28. The quantum key generation process is totally guided by the standard computers (PC). These computers are setting the optolectromic components adjustments with the use of the high speed programmable array (FPGA) totally integrated in the setup. Photons were counted by single photon detector.

In this optical circuit with pulses traveling forward and back, Rayleigh backscattering can considerably increase the noise recorded by the single-photon APD detectors operating in the single-photon counting mode during quantum key distribution. Therefore, the laser does not continuously generate pulses but emits pulse trains in each transmission cycle, and the length of these trains corresponds to the length of the storage line which is inserted in Alice's optical circuit. Due to this, the single-photon pulses propagating back in the quantum channel do not overlap with the multiphoton pulses going from Bob to Alice. For a storage line 25 km long, the train contains ~2400 pulses at a laser pulse clock frequency of 10 MHz. In experiment, random numbers are generated by both sides using a pseudo-random number generator.

Alice stores the serial number of each pulse and the value of bit. Bob buffers both the serial number of the pulse and the measurement basis of the single photons detected. and sends them to a computer. Based on these data and using the public channel between the computers, Alice and Bob generate the same quantum key. The key generation process is completely operated and controlled by standard PCIs, which set the operating mode of the optoelectronic components by means of a high-integration high-speed programmable array.

We performed test quantum key distribution experiments on the our experimental setup [2,3]. Preliminary measurements were made to assess the contrast of the interferometer, which is the source of additional errors in key distribution [8]. The measurements were performed in a multiphoton mode in which there was no noise due to the self-noise of the single-photon detectors operating in the Geiger mode. The measured interferometer contrast was not worse than 98.5%, which is sufficient to minimize the errors due to imperfection of the optical circuit. On an extended fiberoptic communication line with a length of 25 km the 3.2 kbit/s bit rate was obtained on the quantum key distribution setup. The total amount of errors in the key did not exceed 8%. Considering that error rate in the quantum key distribution should be lower than 11% [1,5] result achieved on this setup can be considered as suitable for the real key distribution.

5. Conclusion

A processor-controlled fiber-optic experimental setup for quantum key distribution at telecom wavelength was designed. The self-compensated two-channel optical circuit allows operation with a communication channel 25–100 km long [1-3,8]. The setup uses specially designed single-photon detectors based on InGaAs : InP avalanche photodiodes ERM 547NT. Quantum efficiency and dark noise levels were measured. Low noise avalanche photodiodes improves the singlephoton detector parameters and allows to achieve higher bit rate of quantum key distribution. For examle the photodiodes of Princeton lightwave demonstrate outstanding parameters. Experimental studies in quantum cryptography are attracting steadily increasing interest in the world. Intensive practical studies are in progress in many scientific establishments and telecommunication companies.

Acknowledgments

This work was supported by RFBR (Grant No. 14-07-00809) and by the Siberian Branch of the Russian Academy of Sciences.

6. References

[1] Gisin N, Ribordy G, Title W et al. 2002 *Rev. of Mod. Phys.* **74** 145
[2] Kurochkin V L, Zverev A V, Kurochkin Yu V, Ryabtsev I I and I. G. Neizvestny I G 2009 *Optoelectronics, Instrumentation and Data Processin.* **45** 374
[3] Kurochkin V L Zverev A V Kurochkin Yu V Ryabtsev I I and I. G. Neizvestny 2011 *Russian Microelectronics* **40** 245
[4] Lo H-K and Zhao Y 2009 *Quantum cryptography Encyclopedia of Complexity and System Science vol 8* (New York: Springer), pp 7265-89
[5] Scarani V, Pasquinucci H, Cerf N et al. 2009 *Rev. of Mod. Phys* **81** 1301
[6] Bennet C H, Brassard G 1984 Quantum Cryptography: Public Key Distribution and Coin Tossing *Proc. of IEEE Int. Conf. on Comput. Sys. and Sign. Proces., Bangalore, India, December,* pp 175–179.
[7] Bennet C H, Bessette C H, Brassard G 1992 *J.Cryptology* **5** 3
[8] Stucki D, Gisin N, Guinnard O, et al. 2002 *New Journ. Phys.* **4** 41

Application of the time-space distributed model for the description of Q-switched regime in solid-state lasers with a saturable absorber

Maxim Belov, Leonid Burov, Liubou Krylova

Belarusian State University, Physics faculty, Nezaleznasci av. 4, Minsk 220030, Belarus

E-mail: krylova_lubou@mail.ru

Abstract. One-dimension time-space distributed model for $Nd : YAG$ laser with saturable absorber $Cr^{4+} : YAG$ is considered. The dependencies for peak power, pulse width and pulse repetition rate on pump power and resonator length are introduced.

1. Introduction

Q-switched regime in solid-state lasers has been researched sufficiently wide [1]. Nevertheless, we still face an urgent problem of laser parameters optimization and tuning for a predefined values range of output parameters. One of the ways of this tuning is varying the resonator geometry, particularly, varying the resonator length in an enow wide range, from a few millimeters up to several meters [2, 3, 4].

A major difficulty raises when one tries to exploit concentrated models that are based on resonator volume averages [5, 6]. These models either completely neglect spatial evolution of amplified radiation inside the resonator or reduce it to some effective values (e.g. per one round trip time [7]). Recently, there is a raise of spacially distributed dynamic models that allow a direct simulation of an amplified radiation pulse propagation inside the resonator [5, 6, 8, 9]. These models allow us to reveal a variety of considerable details and peculiarities of amplified pulse formation [5, 6], while concentrated models may fail.

2. Theoretical model

We present a dynamic model of formaion of amplified radiation with one spatial dimension, free of limits for the active media length, satuarable absorber length, total length, and their mutual arrangement.

Let us consider an example configuration of a laser: a $Nd : YAG$ active media of length L_g, a satuarable absorer $Cr^{4+} : YAG$ of length L_s positioned inside a resonator of total length L_{rez} closely to each other and to the input mirror, and an output mirror at a distance of L_3 from the absorber, that is $L_{rez} = L_g + L_s + L_3$. We have chosen the materials for active media and saturable absorber as their full characteristics were available in paper [10], still the model is not limit to this case. We consider transverse pumping, and base on the model [10] that is stated in terms of dimensionless values of the photon flux density (u^+ and u^-) and the population inversion of the active media (D) and the absorber (D_s).

The proposed model has the following major differences: (i) radiation amplification and absorbtion happens only in the active media and the absorber, while inside the range L_3 radiation propagates freely, (ii) a dimensionless spatial coordinate is introduced (the input mirror and the output mirror ($r_2 = 0.99$) correspond to coordinates $x = 0$ and $x = 1$ respectively, and (iii) a term describing the contribution of the amplified radiation to the lasing is accounted. Thus, we express the final equation system as follows:

$$\begin{aligned}
\frac{dD}{dt} &= \frac{\tau_c}{\alpha_L}\gamma(1-(1+(u^+-u^-))D) \\
\frac{dD_s}{dt} &= \frac{\tau_c}{\alpha_L}\gamma_1(-1-(1+\alpha(u^++u^-))D_s) \\
\frac{\alpha_L}{\tau_c}\frac{\partial u^+}{\partial t} - \frac{\partial u^+}{\partial x} &= (-1+AD+A_1D_s)u^+ + R_{lum}|D| \\
\frac{\alpha_L}{\tau_c}\frac{\partial u^-}{\partial t} + \frac{\partial u^-}{\partial x} &= (-1+AD+A_1D_s)u^- + R_{lum}|D|
\end{aligned} \tag{1}$$

The boundary conditions are:

$$u^+(0,t) = r_1u^-(0,t); u^-(1,t) = r_2u^+(1,t); u^\pm(x,0) = 0 \tag{2}$$

The constants in the system (1)-(2) are following:

$$\begin{aligned}
D &= \frac{(1+\gamma_a\tau_gI)\cdot G}{2L_gN_TI\tau_g}, D_s = \frac{G_s}{2L_sN_0\sigma_a}, u^\pm = \frac{\gamma_e\tau_gR^\pm}{1+\gamma_a\tau_gI}, \tau_c = \frac{2(L_g+L_s)n}{c} \\
G &= 2L_g\gamma_eN_2, G_s = 2L_s(\sigma_eN_{2s}-\sigma_a(N_0-N_{2s})), \alpha = \frac{\sigma^+\tau_s}{\gamma_e\tau_g}(1+\gamma_a\tau_gI) \\
\gamma &= \frac{\tau_c(1+\gamma_a\tau_gI)}{\tau_g\alpha_L}, \gamma_1 = \frac{\tau_c}{\tau_s\alpha_L}, R^\pm = u^\pm\frac{1+\gamma_a\tau_gI}{\gamma_e\tau_g}
\end{aligned} \tag{3}$$

where N_2 and N_{2s} are populations of exited levels of the active media and the saturable absorber respectevely, $R^\pm$ are values of the generation photon flux density (u^+ and u^- respectively). The value of pump power is determined as $P_p = Ih\nu_pS_p$, where ν_p is the frequency of the pump radiation, S_p is the cross-section of the pump.

Then, the lasing power is defined as $P = R^\pm h\nu_{gen}S_{gen}$, where ν_{gen} is the frequency and S_{gen} is the cross-section of the lasing. The numeric values are given in the table below.

The system (1)-(3) has been solved via explicit Euler's method for the balance equations and an explicit two-step method [11] for wave equations.

symbol	meaning	value
γ_a	absorption cross-section of pump	$8.63 \cdot 10^{-24}\ m^2$
γ_e	emission cross-section of lasing	$2.8 \cdot 10^{-23}\ m^2$
σ_a	absorption cross-section of the saturable absorber	$8.7 \cdot 10^{-23}\ m^2$
σ_e	emission cross-section of the saturable absorber	$2.8 \cdot 10^{-23}\ m^2$
τ_g	life time of the active media ions exited state	$170 \cdot 10^{-6}\ c$
τ_s	life time of the saturable absorber ions exited state	$3.4 \cdot 10^{-6}\ c$
L_g	active media length	$0.05\ cm$
L_s	saturable absorber length	look for range in the text
N_T	active media ions concentration	$2.2 \cdot 10^{26}\ m^{-3}$
N_0	saturable absorber ions concentration	$3.0 \cdot 10^{24}\ m^{-3}$
α_L	dimensionless internal losses	0.002
r_1	input mirror reflection rate	1.0
r_2	output mirror reflection rate	0.99
λ_p	pump wave length	$808\ nm$
λ_{gen}	lasing wave length	$1065\ nm$
c	speed of light in the vacuum	$3 \cdot 10^8\ m/s$
P_p	pump power	look for range in the text
S_p	pump cross-section	$1 \cdot 10^{-5}\ m^2$
S_{gen}	lasing cross-section	$1 \cdot 10^{-5}\ m^2$
n	active media and saturable absorber refraction index	1.8

3. Numerical simulation results

In figure 1 we plot a representative output power dependance P_{max} on pumping power P_p for several absorber widths.

We point out two regions. While pumping power is rather small P_{max} is almost linear, as it was observed experimentally [4]. The pumping power futher being raised, the dependence becomes nonlinear, reaches some critical value where the output power is maximal. With the increase of saturable absorber length this maximum moves towards higher pumping and output power (fig.1).

We also observe an opposite dependence of pulse duration τ on pumping power: with the increase of P_p the duration starts to decrease down its to minimal value that corresponds to output power maximum. A similar correspondence of an output power maximum and a pulse duraion minimum is found in the paper [12]. We found the length dependencies to be monotonous, moreover, if pulse duration increases almost linearly (fig.2), both the peak power and the pulse frequency decrease (fig.3). E.g. L_{rez} increase from 0.1 m. to 1.04 m. leads to a more than 20% P_{max} decrease. This result may be validated via comparison with an analytic expression for the pulse power [7], where it is stated that P_{max} has to be inversly proportional to the round trip time, and, respectively, to the resonator length. Validity of the obtained dependence is also supported by theoretical papers [7, 13] and experimental data [13, 14].

4. Conclusion

Our results show that the proposed model is applicable for passive Q-switched laser description and its parameters optimization. Despite it requieres far more computational resources, the proposed model reveals several new results, associated with the correspondance of the maximal and minimal values of photon flux density, pulse duration and with the influence of the saturable absorber length.

[1] Koechner W. 2006 *Springer* 749p.
[2] Mlynczar J., Kopczynski K., Mierczyk Z. 2008 Optica Applicata. **38** 657

[3] Feng S., Li S., Chen L., Chen W., Hu L. 2010 *Applied Optics.* **49** 3357
[4] Agnesi A., DellAcqua S., Morello C., Piccinno G., Reali G. C., Sun Z. 1997 *IEEE J. Select. Topic Quant. Electron.* **3** 45
[5] Burov L.I., Krylov G.G., Krylova L. G. 2010 *Nonlinear Phenomena in Complex Systems* **4** 368
[6] Kolpakov S.A., Barmenkov Y.O., Kiryanov A.V., Guzman-Chavez A.D., Cruz M.V., Andres J.L. 2012 *J. Opt. Soc. Am. B.* **29** 2453
[7] Degnan. J.J. 1989 *IEEE J. of Quant. Electr.* **25** 214
[8] Burov L. I., Krylova L. G. 2012 *J. Appl. Spectr.* **79** 397
[9] Azarenkov A.N., Altshuler G.S., Lebedko V.E. 1990 *Sov.J.Quant.Electron.* **20** 155
[10] Peterson P., Gavrielides A., Sharma M.P., Erneux T.1999 *IEEE J. Quant. Electron.* **35** 1247
[11] Ilina V.A., Silaev P.K. 2004. Moscow.*Numeric receipts for theoretician physicists.* **2** 118 p. (in Russian).
[12] Matia-Hernando P., Guerra J.M., Weigand R. 2013 *Laser Phys.* **23** 025003-5
[13] Spuhler G. J., Paschotta R., Fluck R., Braun B., Moser M., Zhang G., Gini E., Keller U. 1999. *J. Opt. Soc. Am. B.* **16** 376
[14] Sakai H., Kan H., Taira T. 2008 *Optics Express* **16** 19891

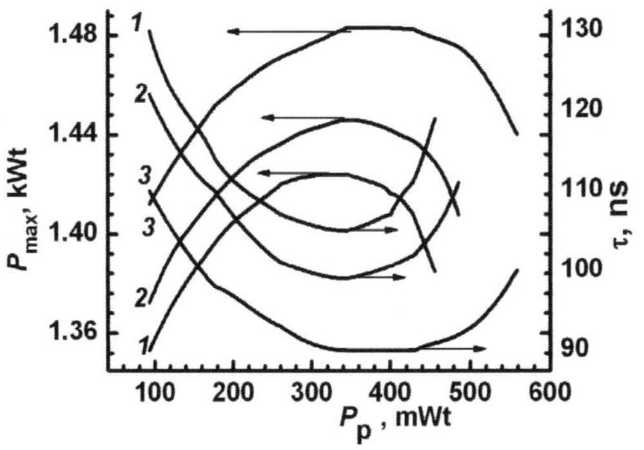

Figure 1. Generation peak power and pulse width vs. the pump power ($L_3 = 0.1$ m, $L_s = 2.4$ (1); 2.5 (2); 2.7 (3) mm).

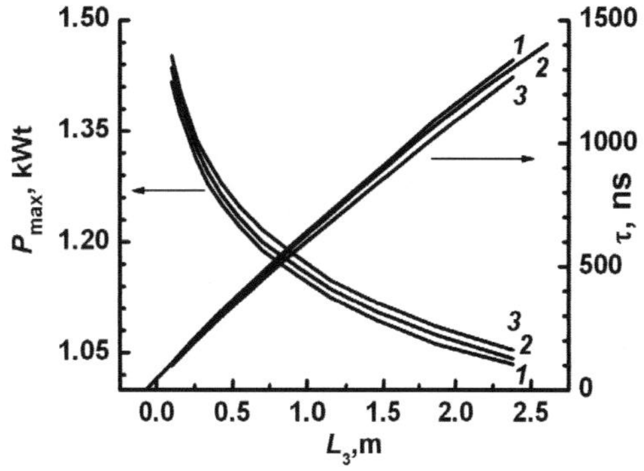

Figure 2. Generation peak power and pulse width vs. the resonator length L_3 ($P_p = 190$ mWt, $L_s = 0.25$ (1); 0.26 (2); 0.27 (3) mm).

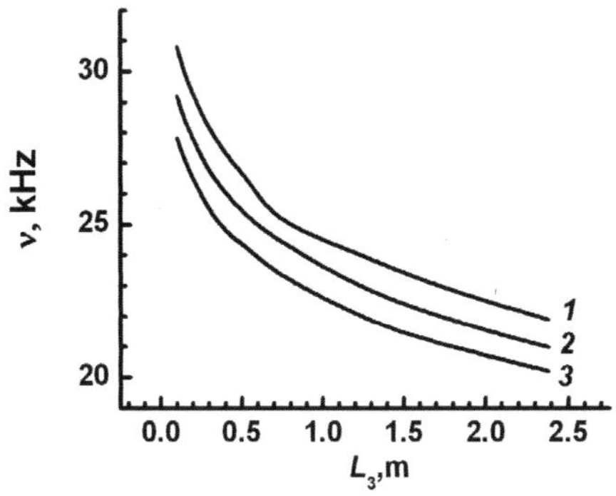

Figure 3. Pulse repetition rate vs. the resonator length L_3 ($P_p = 190$ mWt, $L_s = 0.25$ (1); 0.26 (2); 0.27 (3) mm).

THERMAL LENSING IN DIODE PUMPED Nd^{3+}:YVO_4/$LiNbO_3$ GREEN MICROCHIP LASER

M. Kerobyan[1,2], R. Kostanyan[1], V. Nersesyan[1,2], S. Soghomonyan[2]
[1]Institute for Physical Research, NAS of Armenia
[2]Spectralus CJSC, Yerevan, Armenia

E-mail: kerobyan.mk@gmail.com

Abstract. Thermal lens in Nd^{3+}:YVO_4/$LiNbO_3$ microchip green laser affects stability and working regimes of the laser. Absence of any passive or active Q-switching and mode locking make the thermal lens the only method to affect working regimes and stability of the laser cavity, which is composed of planar active and nonlinear mediums. Thus, investigation of thermal lens is crucial for the optimization of working regimes of the microchip laser.

1. Introduction

YVO_4/$LiNbO_3$ green microchip lasers are miniature, powerful and efficient source of green light [1]. Resonator of the YVO_4/$LiNbO_3$ green microchip laser is composed of two planar components. Active medium (Nd^{3+}:YVO_4) and nonlinear medium ($LiNbO_3$). Mirrors of the cavity are dielectric coatings directly applied to the surfaces of the active and nonlinear crystals. Crystals are joint by optical contacting technique.

Resonator of the microchip laser is a plane parallel. As we know, stability of optical resonators are described by $g_1 g_2$ factor, where $g_1 = 1 - L/R_1, g_2 = 1 - L/R_2$. L is the optical length of the cavity, R_1 and R_2 are radii of curvature of the mirrors. Cavity is stable when $0 \le g_1 g_2 \le 1$. In case of microchip laser without thermal lens, where mirrors are flat $R_1 = R_2 = \infty$ and $g_1 g_2 = 1$, which is example of unstable resonator. Because of quantum defect, in working regime, part of the pump power is released in the active medium as a heat. For the microchip laser, quantum defect is equal to $\eta = 1 - 808nm/1064nm = 24\%$, which means that at least 24% of pump power releases as a heat in the active medium. This heat is released in the region of the pump beam, and creates inhomogeneous distribution of the temperature. This inhomogeneous distribution of the temperature creates lensing effect in the crystal. So, we'll have resonator, which contains a positive lens.

Thermal lens affects mode size, TEM's and stability of the cavity, and in absence of any mode locking in the green microchip laser, investigation of thermal lens is important for optimization of working regimes of the green microchip laser.

2. Experimental

Experiments were conducted with 1.5mm+1.5mm YVO_4/$LiNbO_3$ microchip lasers, 808nm laser diodes from QPC, with 1W maximal power and 80um beam size. Direct measurements of focal length of thermal lens is problematic because of miniature sizes of the resonator. So focusing length was

[1] To whom any correspondence should be addressed.

measured indirectly. Beam size at the output mirror depends on the focal length of the thermal lens. Output beam divergence angle depends on the beam size at the output mirror. So, instead of measuring focal length, we measured beam divergence angle (all other parameters were kept constant, except pump power, e.g. pump beam waist, temperature of heat controller etc.). It is known from the theory, that stronger the thermal lens, the higher divergence angle. Measurement results are presented in the Fig.2. It can be seen from the Fig.2 that higher pump power leads to higher divergence angles.

To compare experimental results with the theory, we used theory of thermal lens developed by Innozenti et al [2]. Later other authors improve analytics and got more complex formulas. Nevertheless, in all calculations pump beam is considered to be a Gaussian beam. In case of green microchip laser, pump beam profile can't be considered a Gaussian. To calculate focal length of thermal lens, one needs to solve non homogeneous heat equation, where non homogeneous part has very complex form. Analytical calculations are not possible, but numerical simulations shown that the analytical formula for Gaussian pump beam suits well for this problem.

3. Theory

Using theory of equivalent resonators [3], one can represent plane-parallel resonator with positive lens with focusing length of f as a resonator with one concave and one flat mirror.

Figure 1. Schematic drawing of resonator with thermal lens and equivalent resonator. M1, M1′ and M2 are mirrors, L is the thermal lens

Radius of curvature of the first mirror is equal to the focal length of the thermal lens. Refractive indices of YVO_4 and $LiNbO_3$ are approximately the same and their difference is negligible in this case. Thus, $g_1 g_2 = 1 - nL/f$, where n is the refractive index of the material of the cavity.

Beam radius at the mirrors ω_1 and ω_2 are given by the following formulas

$$\omega_1^4 = \left(\frac{\lambda R_1}{\pi}\right)^2 \frac{R_2 - d}{R_1 - d} \frac{d}{R_1 + R_2 - d}, \omega_2^4 = \left(\frac{\lambda R_2}{\pi}\right)^2 \frac{R_1 - d}{R_2 - d} \frac{d}{R_1 + R_2 - d} \quad [1]$$

where λ is the wavelength of the circulating wave (fundamental wave), R_1 and R_2 are radii of curvatures of the mirrors, d is the length of the cavity. As long as output mirror of the cavity is flat, beam waist will be at the output mirror and equal to $\omega_0 = \omega_2$, which is given in the formula [1].

$$\omega_0 = \sqrt{\frac{\lambda f}{\pi}} \left(\frac{d}{f - d}\right)^{1/4} \quad [2]$$

This formulas are calculated for fundamental wave (1064nm). It is known from the theory of SHG that beam waist of the second harmonic wave is $\sqrt{2}$ times less than beam waist of the fundamental wave. Thus, total angular spread of output green light will be equal to

$$\Theta = \frac{\sqrt{2}\lambda}{\pi\omega_0} = \sqrt{\frac{2\lambda}{\pi}} \left(\frac{1}{df} - \frac{1}{f^2}\right) \quad [3]$$

Note that in the formula [3] $\lambda = 1064nm$ is the wavelength of the fundamental wave, not the second harmonic wave. Focal length in the formula [3] is equal to

$$f = \frac{\pi K_c \omega_p^2}{P_{ph}\gamma} \frac{1}{1-\exp(-\alpha l)} \qquad [4]$$

Where K_c is thermal conductivity of the crystal, ω_p is the beam size of the pump beam in the crystal, γ is a sum of thermo-optic coefficient and thermal expansion coefficient, α is the absorption coefficient at the pump wavelength, l is the length of the active medium, P_{ph} is the part of pump wave, which is gone as a thermal loss.

As it was mentioned above, in the experiments we have measured divergence angle of the output green light versus pump power, and in the Fig.2 measurement results are compared to the theoretically calculated value according to the formula for focal length of thermal lens in the Innozenti et al work.

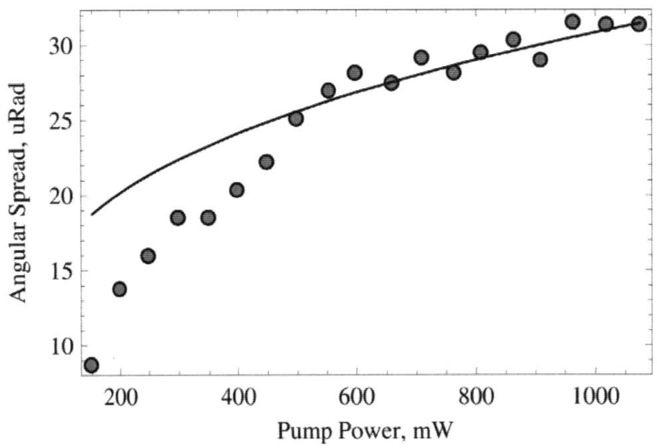

Figure 2. Angular spread of laser beam. "—" – theory, "•" – measurement.

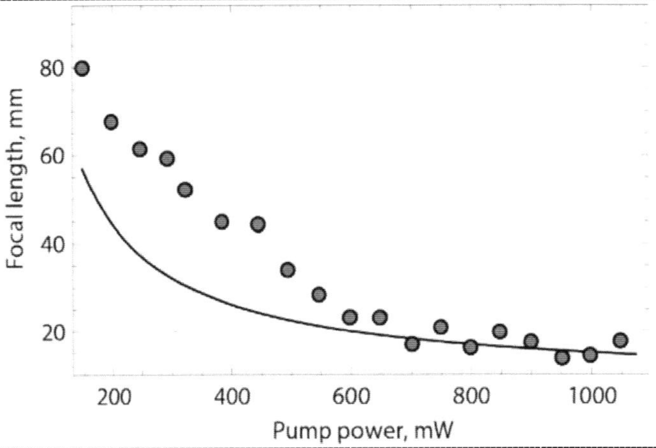

Figure 3. Focal length of thermal lens. "—" – theory, "•" – measurement.

In the Fig.3 measured and calculated focal length versus pump power is presented. Focal length is calculated from the formula [3], by expressing f trough Θ. One can see that for high values of pump power, theory and measurements are well correlated for both divergence angle and focal length of the thermal lens, unless for low pump powers. It is supposed that difference between theory and experiments in low pump powers is caused by non-Gaussian pump, which plays important role in case of low pump powers, unless in the case of high pump powers, where the shape of pump beam is not as much important. Nevertheless it is a matter for future investigation.

4. Conclusion

In this work we have investigated focal length dependence of the thermal lens on the pump power. Knowledge of focal length of thermal lens, one can calculate mode sizes on the on the mirrors, which can be used for mode matching, also it is important parameter for understanding cavity stability.

References

[1] Essaian S., Khaydarov J., Slavov S., Ter-Mikirtychev V., Gabrielyan G., Kerobyan M., Soghomonyan S., Proceedings of SPIE, 8240, 2012, pp. 82400I-1 – 82400I-8

[2] M. E. Innocenzi, H. T. Yura, C. L. Fincher, and R. A. Fields, Applied Phys. Letters, 56, 1990, pp. 1831-1833

[3] Walter Koechner, Solid-State Laser Engineering, Springer, New York, 2006

Injection currents in thin disordered organic films

V R Nikitenko, N A Sannikova, M N Strikhanov

National Research Nuclear University MEPhI (Moscow Engineering Physics Institute), 115409, Moscow, Russia

E-mail: vladronik@yandex.ru

Abstract. Analytic model of barrier-limited injection of charge carriers from metal electrodes into organic film, which was introduced by Arkhipov and co-workers, is modified, considering effects of multiple image charges and injection from both electrodes. Limits of applicability of Arkhipov's model are discussed. Variations from Arkhipov's model are important, if film thickness is comparable with Onsager length.

1. Introduction

Injection and transport of charge carriers in organics have important peculiarities, caused by hopping nature of these processes [1]. Correct description of charge injection from metal electrodes is an important part of any predictive model of operation of organic light-emitting diodes (OLEDs) and other elements of organic electronics. Although real OLEDs include numerous layers of different organic materials, single-layer structures are useful for experimental characterization of individual layers. The simple analytic model [2] describes correctly field and temperature dependence of barrier-limited injection from electrode to organic material, accounting for energetic disorder and image-force potential. However, this model considers organic material as a semi-infinite media, while thickness of an active layer of OLEDs, *L*, is typically less than 100 nm, hence it is comparable with Onsager length, $r_C = e^2/4\pi\varepsilon\varepsilon_0 kT$, where e is elementary charge, ε_0 is electric constant, ε is relative permittivity, k is Boltzmann constant and T is absolute temperature. Indeed, $r_C = 18.9$ nm and 55.7 nm, providing that $\varepsilon = 3$ and $T = 295$ K and 100 K, respectively. Consequently, one can expect deviations from results of Arkhipov's model in thin films and under low temperatures, i.e. under the condition $L \le r_C$. Indeed, effects of charge redistribution in *both* electrodes, caused by injected charge, may be important. In addition, contribution of backward current, caused by diffusion of charge carriers from opposite electrode, is not negligible at low voltages *V*, $eV < kT$. In the present work the model of ref. [2] is modified, considering these effects.

2. The model

We consider an organic layer, being placed between identical contacts (electron-only or hole-only device), providing considerable energy barriers for injection. The model of ref. [2] considers injection of charge carriers as a two-step process. First step is phonon-assisted tunnelling jump (the rate is described by Miller-Abraham's form) from the Fermi level of electrode to a localized state in organic material, in proximity to the surface. Second step is hopping diffusion in disordered system, biased by electric field, which results from Coulomb image-charge and applied voltage, V. Eventually, a carrier has to overcome a potential barrier, see the dashed line in the inset of the Figure 1. Interaction of an injected carrier with multiple image charges (except of the nearest one) results in additional term in a potential energy, which can be expressed in terms of polygamma function of zero order,

$$\psi(z) = -\gamma + \sum_{n=0}^{\infty}\left(\frac{1}{n+1} - \frac{1}{n+z}\right), \tag{1}$$

$$\delta U(y) = \left(e^2/16\pi\varepsilon\varepsilon_0 L\right)\left[2\gamma + \psi(1+y) + \psi(1-y)\right] \tag{2}$$

where $y = x/L$ is a dimensionless coordinate ($y = 0$ at injecting electrode), and $\gamma \approx 0.57721$ is Euler–Mascheroni constant. The mean potential energy of the charge carrier (excluding energy disorder), $U_{\pm}(y)$, is a superposition of charge energy in an external electric field F_0, the Coulomb interaction energy with the nearest image charge and energy of interaction with the rest of image charges $\delta U(y)$, see eq. (1):

$$U_{\pm}(y) = -\frac{e^2}{16\pi\varepsilon\varepsilon_0 L}\frac{1}{y} + \delta U(y) + \Delta \mp eF_0 y, \tag{3}$$

where field F_0 is assumed to be homogeneous, Δ is barrier height at zero field. The signs "+" and "-" correspond to the field direction from the electrode and to the electrode, respectively.

The model takes into account the currents injected from both electrodes:

$$J_{inj} = j_+ - j_-, \tag{4}$$

where j_+ and j_- are injection currents from the left (in the direction of the field, j_+) and the right (the opposite direction to the field, j_-) electrodes, respectively. A carrier can pass the layer in both directions due to diffusion, providing low applied voltage ($eV < kT$), hence it is necessary to consider both terms.

The rate of the first jumps, i. e. jumps from the Fermi level of the injecting electrode to a state with energy E, which is located at a distance x_0 from the electrode, is calculated according to the Miller-Abrahams model [3], $\nu = \nu_0 \exp(-2x_0/a_L) Bol(E)$, where $Bol(E) = \exp\left[-(E + |E|)/2kT\right]$, ν_0 is attempt-to-hopping frequency, which is determined by the properties of the organic material and the injection electrode, a_L is localization radius of wave functions. Fermi level of the injecting electrode is taken as zero point of energy.

Similar to Arkhipov's model, injection current from each electrode is an average of a rate of first jump, multiplied by the probability to overcome the potential barrier (see the solid line in the inset; this probability is calculated in the framework of one-dimentional version of Onsager's model, as in the ref. [2], but with the term δU in the energy $U_{\pm}$), over energies of hopping states and hopping distances x:

$$j_{\pm} = j_0 \frac{\int_{a/L}^{1-a/L} dy_0 \exp\left(-2\left(L/a_L\right) y_0\right) \int_{-\infty}^{+\infty} dE Bol(E) g[E - U_{\pm}(y_0)] \int_{a/L}^{y_0} dy \exp\left[U_{\pm}(y)/kT\right]}{\int_{a/L}^{1-a/L} dy \exp\left[U_{\pm}(y)/kT\right]}, \tag{5}$$

where $j_0 = eN_t\nu_0 L$, N_t - concentration of localized states, a - distance from the electrode to the nearest state hopping, $y = x/L$, $y_0 = x_0/L$, and energy distribution of localized states $g(E)$ is a Gaussian function:

$$g(E)=\frac{1}{\sqrt{2\pi}\sigma}\exp\left(-\frac{E^2}{2\sigma^2}\right), \tag{6}$$

At variance with original Arkhipov's model, the present model accounts for the finite size of a layer, see limits of integration in eq. (5), and contribution of injection from opposite electrode, see eq. (4), which occurs against applied electric force, due to diffusion. Since the field is considered to be homogeneous, the voltage is connected with field strength as follows: $V = F_0 L$.

3. Results and conclusions

Figure 1 shows I-V characteristics of thin organic layers, calculated at the following parameters: injection barrier is Δ = 0.5 eV, variance of Gaussian distribution of localized states is σ = 0.1 eV, concentration of localized states is $N_t = 10^{27}$ m^{-3}, hopping frequency factor – $\nu_0 = 1.4\cdot 10^{12} s^{-1}$, length of a first jump – $a = 0.5\cdot 10^{-9} m$, ε = 3, T = 100 K.

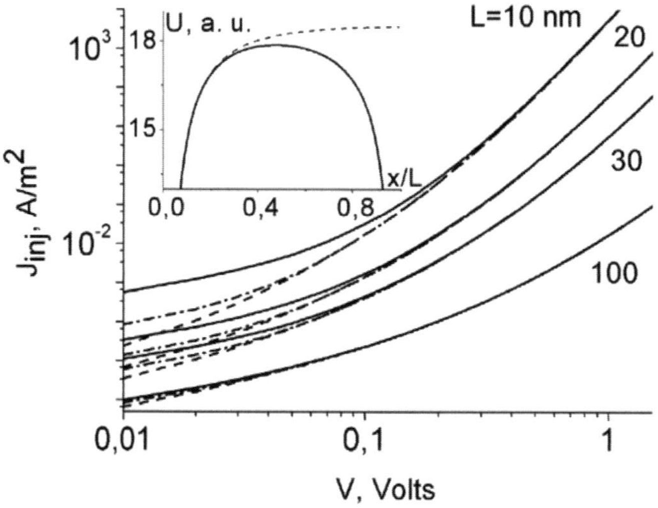

Figure 1. I-V characteristics, parametric in thickness of the layer *L*, see the figure. Solid lines shows results of the modified model, dashed lines – Arkhipov's model ($\delta U = 0$) with account of backward current and dash-dotted lines – original Arkhipov's model. Inset shows the total potential energy (solid line), and the same in Arkhipov's model ($\delta U = 0$, dashed line), the voltage is 0.01 V.

Figure 2 shows I-V characteristics for the case of different temperatures, calculated at the following values of parameters: Δ = 0.6 eV, σ = 0.1 eV and L = 20 nm.

The results shown in Figures 1 and 2 are in good agreement with the assumption that the effect of backward (diffusion) current becomes significant under the condition $eV < kT$ (see dashed and dot-dash curves, considering the criterion of difference of 2 or more times). Dashed lines in Figures 1 and 2 shows results of eqs. (4) and (5), neglecting contribution of multiple image charges ($\delta U = 0$). Original model of the ref. [2] results from eqs. (4) and (5), if, in addition, $L \to \infty$, hence $j_- \to 0$ (dash-dotted lines).

In thin films and at low voltages and temperatures effect of multiple image charges yields large increase of current at a given voltage relative to predictions of Arkhipov's model [2], due to reduction of a potential barrier, see inset in the Figure 1, in spite of negative contribution of backward current, except of high temperatures, see Figure 2. Deviation from predictions of the ref. [2] can exceed the factor 2, if $L < r_C/2$.

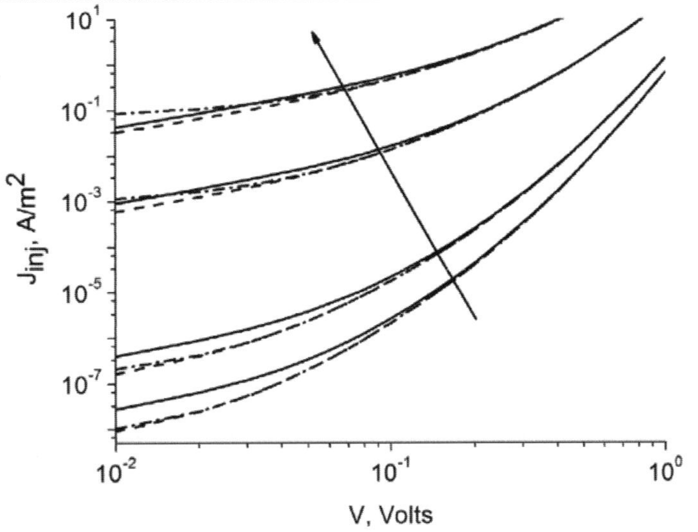

Figure 2. I-V characteristics, values of temperature, K: 80, 100, 200, 300, increases in the direction of arrow. Solid lines shows results of the modified model, dashed lines – Arkhipov's model ($\delta U = 0$) with account of backward current and dash-dotted lines – original Arkhipov's model.

The proposed analytic model, being relatively simple, is useful for testing and correct choice of parameters of numerical models in predictive modelling of OLEDs, photovoltaic elements and other devices of organic electronics.

This work was supported by Russian Ministry of Education and Science in the frames of Competitiveness Growth Program of National Research Nuclear University MEPhI, Agreement 02.A03.21.0005

References

[1] Bässler H 1993 *Phys. Stat. Sol. (b).* **175** 15
[2] Arkhipov V I, Emelianova E V, Tak Y-H, Bässler H 1998 *J. Appl. Phys.* **84** 848-856
[3] Miller A, Abrahams E 1960 *Phys. Rev.* **120** 745

Study of the characteristics of UVA LEDs grown by HVPE: active region thickness-dependent performance

E A Menkovich[1], S A Tarasov[1], I A Lamkin[1], A V Solomonov[1], S Yu Kurin[2, 3], A A Antipov[2], I S Barash[2], A D Roenkov[2], A S Usikov[4, 5], H I Helava[4], Yu N Makarov[2, 4]

[1]Saint-Petersburg Electrotechnical University "LETI", ul. Prof. Popova 5, St. Petersburg 197376, Russia
[2]Nitride Crystals Ltd., pr. Engels 27, St. Petersburg 194156, Russia
[3]Saint-Petersburg Academic University RAS, ul. Khlopina 8/3, St. Petersburg 194021, Russia
[4]Nitride Crystals Inc., 181 E Industry Court, Site B, Deer Park, NY 11729, USA
[5]Saint-Petersburg National Research University ITMO, Kronverkskiy pr. 49, St. Petersburg 197101, Russia

E-mail: menkovichea@gmail.com

Abstract. We report on results of the performance study of UVA LEDs depending on the thickness of the active region. UVA LEDs are based on GaN/AlGaN heterostructures grown on Al_2O_3 (0001) substrates by hydride vapor phase epitaxy (HVPE). It is shown that the use of thick (~ 100 nm) single layer as the active region of UVA LED is a promising concept to achieve enhanced efficiency.

1. Introduction

UV light sources are widely used in modern science, industry, medicine. In the near UV area (UVA: 320-390 nm) it is highly relevant to use UV radiation for photocatalytic purification of water and air as well as for photopolymerization processes. Nowadays mercury lamps generally serve as a source of UV radiation. Mercury lamps replacement by UVA LEDs will provide ten-fold increase of lifetime of UV devices and will lead to considerable decrease of their power consumption as well as reduce mercury pollution of the environment.

Despite the demand for UVA LEDs, they are not yet available in the world market in sufficient quantities. Currently, one of the major obstacles in the development of semiconductor UV emitters is the low luminescence efficiency in epitaxial heterostructures at high current densities, which is usually associated with self-heating of the active region that leads to a decrease in the optical emission power of the UVA LED and its lifetime [1].

In this study, we suggest a special design for an epitaxial heterostructure, which provides stable UV LED operation at high current densities.

2. Sample and experimental technique

It is known that maximum of the internal quantum efficiency (IQE) of nitride-based LED structures depends strongly on the active layer thickness [2]. In very thin active layers, thinner than 2 nm, maximum IQE is reached at relatively low current densities. In this case the electron concentration is much higher than the hole one. As a result, the carrier recombination at threading dislocations is controlled by the hole migration to the dislocation cores, lowering non-radiative recombination rate [3, 4]. As the active region is thicker (3-5 nm) electron and hole concentrations becomes nearly equal and IQE decreases [5]. Further enlargement of the active layer results in the IQE increasing accompanied by shifting of the IQE maximum to higher current densities. IQE becomes even higher than that of LEDs with 3-5 nm quantum wells (QWs) [6].

It was also shown that multiple QW structure of the active region could be ineffective for improvement of the LED IQE because of inhomogeneity of electron and hole injection in various wells and IQE−I dependence drop as a result of heating of the active region, carrier leakage and Auger recombination [2]. At a desirable current density it could be possible to choose the active layer thickness providing the maximum IQE. LEDs with a thick single active region (more than 100 nm) are promising for high-current operation to get high output optical power.

In order to check this assumption the IQE dependence of UVA LEDs on current density at different active region thicknesses was calculated and UVA LED structures with different active region thicknesses were grown by hydride vapor phase epitaxy (HVPE) and analyzed. The growth procedure included in-situ sapphire substrate Al_2O_3 (0001) treatment followed by multilayer structure growth. UVA LED structures consisted of AlN/AlGaN buffer layer, 6-8 pairs of AlGaN/AlGaN stress control layers, Si-doped n-AlGaN barrier, (Al)GaN active region (50, 80, 120 nm thick depending on epitaxial process), and Mg-doped p-AlGaN barrier. Mg-doped p-AlGaN cover layer completed the structure.

Characterization of UVA LEDs was carried out by using an automated diagnostic complex which allows performance investigation of LEDs and LED devices [1, 7-9].

3. Calculations of UVA LED efficiency

The calculations were made by using software based on finite element method. It allows two-dimensional (2D) simulation of the band diagram, electron and hole transport inside the structure, radiative and non-radiative carrier recombination rate, electric field distribution, current-voltage (I-V) and power-current (P-I) characteristics. The software implements 2D drift-diffusion model with account for such features as elastic strain in the heterostructure layers and its effect on the valence band structure, as well as existence of spontaneous electric polarization and piezoeffect in III-nitride materials. According to the model, electrons and holes obey Fermi-Dirac statistics.

A natural way to obtain the emission wavelength of 365 nm is the use of the active layer made of GaN. To ensure, firstly, a good electron carrier confinement in the active layer and, secondly, the lack of absorption of the emitted light in the contact areas the layers surrounding the active layer should be made of a wide-bandgap material − AlGaN.

We consider here a Ga-polar fully strained UVA LED structure consisting of a thick n-AlGaN contact layer ([Si] = 1×10^{19} cm^{-3}), an active region, a p-AlGaN contact layer ([Mg] = 5×10^{19} cm^{-3}). The threading dislocation density was supposed to be 10^9 cm^{-2}. We considered UVA LED structures operating at room temperature.

After the band diagram (figure 1) was computed, the simulator calculated the IQE. The calculations of the IQE maximum were made in a wide range of the active layer thickness and current density. Figure 2 shows that there are two main tendencies: 1) the current density at which the IQE maximum is achieved increases with the increase of the active layer thickness; 2) the absolute value of the IQE maximum also increases with the increase of the active layer thickness.

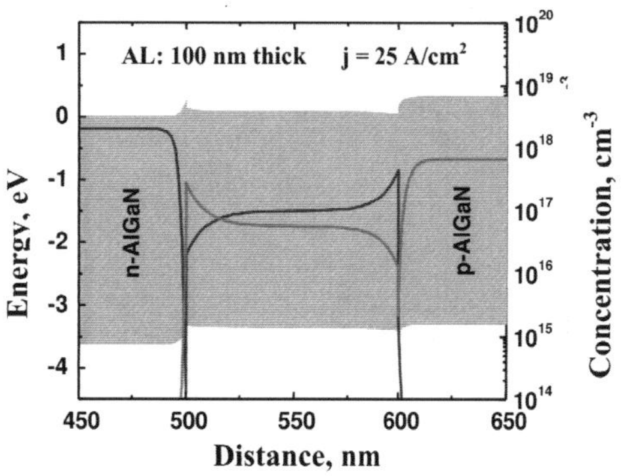

Figure 1. Band diagram of UVA LED heterostructure. Blue and red lines correspond to the concentration of electrons and holes, respectively.

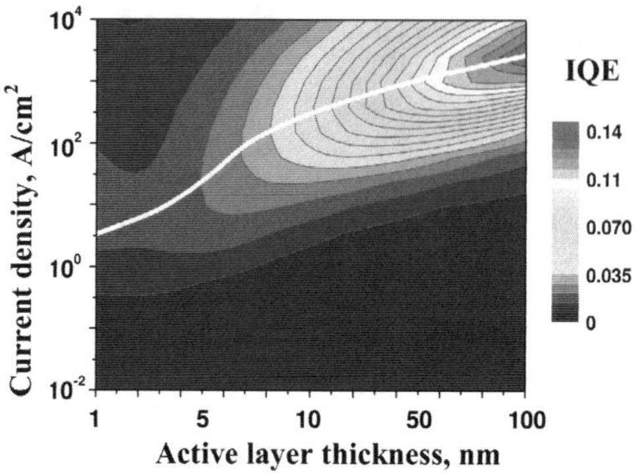

Figure 2. IQE of UVA LED structure with a single active layer as a function of its thickness and current density. White line indicates location of the IQE maximum.

It means that UVA LEDs with a thick single active region (more than 100 nm thick) are preferable to achieve enhanced efficiency.

4. Experimental results and discussions

The UVA LED structures with 50, 80, and 120 nm thick single active regions (1, 2, and 3, respectively) were investigated. Figure 3 shows the electroluminescence (EL) spectra at the operating current of 20 mA. There is a "bend" in the long-wave region of EL in type 1 structures. Its appearance is probably caused by higher defects density in the active region of the structures of this type. It leads to the existence of band tails in the density of states that distort the energy spectrum of the crystal. This affects the electroluminescence spectra and may indicate poorer structural perfection of type 1 structures that has an impact on their operational characteristics such as efficiency and optical output power.

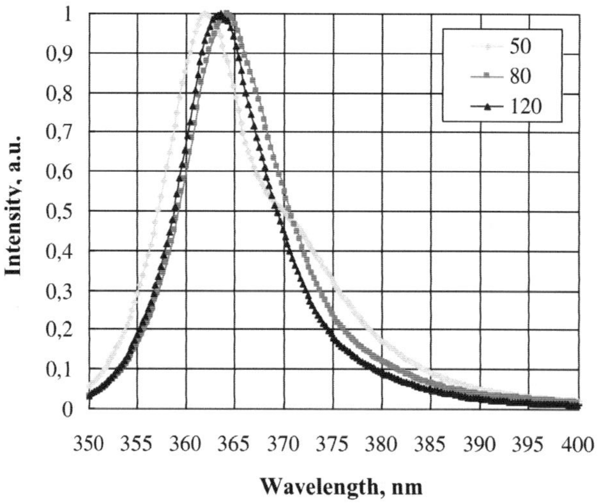

Figure 3. Dependence of luminescence intensity on emission wavelength. Figures 50, 80, 120 correspond to UV LEDs with 50, 80, 120 nm thick active regions, respectively.

Studies of the I–V characteristics showed that the voltage drop across the diode at the operating current was about 3.9 V. In general, the wavelength shift in the emission maximum was approximately 20 nm upon varying the current from 2 to 125 mA.

Figures 4, 5 show dependences of efficiency and output optical power on current. The analysis of experimental data showed that all emitters had approximately the same output optical power at the operating current. Type 3 structures revealed the highest efficiency of 1.12%. The efficiency of other structures was about 1%. The structures of type 1, as might be expected, had the lowest efficiency at high current densities. The largest efficiency decline of 0.8% was observed in the structures of this type upon varying the current from 20 to 125 mA. It was caused by significant inhomogeneity of electron and hole injection into the active region in comparison with other structures. Increasing the thickness of the active region improves uniformity of electron injection and results in higher efficiency.

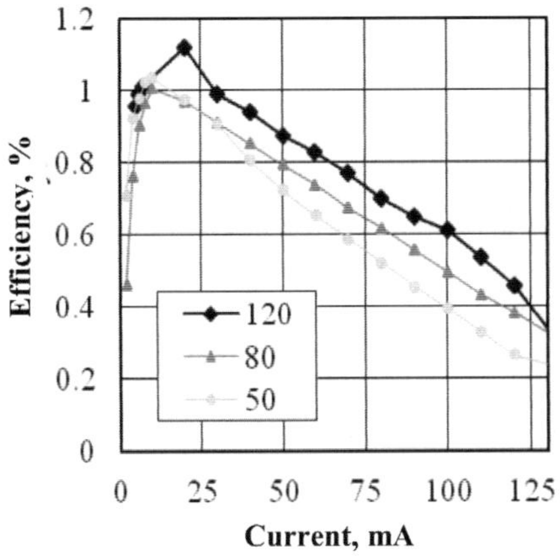

Figure 4. Dependence of the efficiency on current. Figures 50, 80, 120 correspond to UV LEDs with 50, 80, 120 nm thick active regions, respectively.

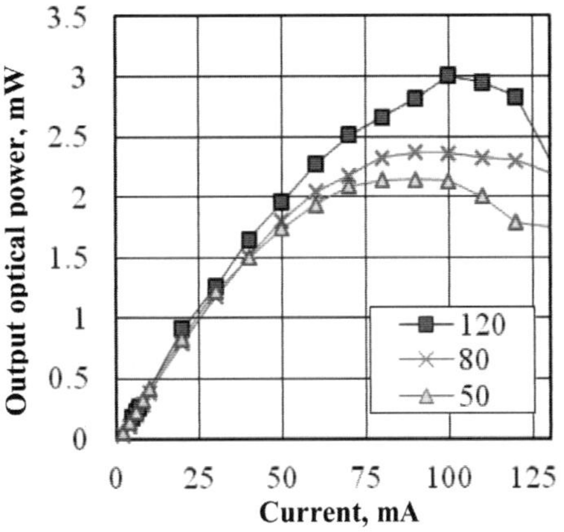

Figure 5. Dependence of the output optical power on current. Figures 50, 80, 120 correspond to UV LEDs with 50, 80, 120 nm thick active regions, respectively.

5. Conclusions

It was observed that wall-plug efficiency of UVA LEDs with 120-nm-thick active region was 1.12% at operating current of 20 mA and this value reduced to less than 1% with decreasing the thickness of the active region to 50 nm. The intensity of UV radiation demonstrated the same behavior. Thus, it is possible to conclude that UVA LEDs with thick single active layers are promising for achieving enhanced efficiency.

References

[1] F. I. Manyakhin, Izv. Vyshh. Uchebn. Zaved., Met. **2** (50), 54 (2010).
[2] S.Yu. Kurin, A.A. Antipov, A.D. Roenkov, I.S. Barash, H.I. Helava, E.A. Menkovich, S.A. Tarasov, I.A. Lamkin, N.M. Shmidt, Yu.N. Makarov, J. Phys.: Conf. Ser. **461**, 012028 (2013).
[3] K. Bulashevich, M. Ramm, and S. Karpov, Phys. Status Solidi C **6**, S804 (2009).
[4] S.Yu. Karpov and Yu.N. Makarov, Appl. Phys. Lett. **81**, 4721 (2002).
[5] D. Fuhrmann, C. Netzel, U. Rossow, A. Hangleiter, G. Ade and P. Hinze, Appl. Phys. Lett. **88**, 071105 (2006).

[6] K.A. Bulashevich and S.Yu. Karpov, Phys. Status Solidi C **5**, 2066 (2008).
[7] A.V. Solomonov, S.A. Tarasov, E.A. Menkovich, I.A. Lamkin, S.Yu. Kurin, A.A. Antipov, I.S. Barash, A.D. Roenkov, H. Helava, Yu.N. Makarov, Semiconductors **48**, 245 (2014).
[8] S. Kurin, A. Antipov, I. Barash, A. Roenkov, H. Helava, S. Tarasov, E. Menkovich, I. Lamkin, Yu. Makarov, Phys. Status Solidi C **10**, 289 (2013).
[9] E.A. Menkovich, S.A. Tarasov, I.A. Lamkin, S.Yu. Kurin, A.A. Antipov, A.D. Roenkov, I.S. Barash, H.I. Helava, Yu.N. Makarov, J. Phys.: Conf. Ser. **461**, 012027 (2013).

Comparative analysis of efficiency droop in InGaN/GaN light-emitting diodes for electrical and optical pumping conditions

V V Kopyev[1], I A Prudaev[1], I S Romanov[2]

[1] Functional Electronics Laboratory, Tomsk State University, 36 Lenin Avenue, Tomsk, 634050, Tomsk, Russia
[2] Faculty of Radiophysics, Tomsk State University, 36 Lenin Avenue, Tomsk, 634050, Tomsk, Russia

E-mail: viktor.kopev@gmail.com

Abstract. The analysis of efficiency droop in InGaN/GaN light-emitting diodes for electrical and optical pumping conditions is presented. Authors show that room temperature efficiency droop is well described by ABC-model for both pumping regimes. For low temperatures additional ballistic leakage should be considered to explain efficiency droop for the case of electrical pumping.

The quantum efficiency of light-emitting diodes (LEDs) with multiple InGaN/GaN quantum wells (MQW) is the subject of intense basic and applied research. The reason for this is an effect referred to as the efficiency droop (ED). The effect manifests itself in a decrease in the external quantum efficiency with increasing pumping current density higher than $j \approx 1–10$ A/cm^2 (at temperatures T = 300 K), and is not caused by the LED overheating. Previously an approach of study was proposed including investigation of the temperature dependences of the quantum efficiency and ED effect under different conditions of luminescence excitation [1]. This work presents recent results of study of ED effect by using electroluminescence (EL) regime and photoluminescence (PL) regime with two type of optical pumping (355 and 405 nm).

Experiments were carried out for commercial blue LED-structure grown by MOVPE on the [0001] patterned sapphire substrate. An active n-type region of the structure consisted of the $In_{0.15}Ga_{0.85}N$/GaN MQWs and barriers with the thicknesses 2.5 and 15 nm, respectively. The thickness of the p-$Al_{0.15}Ga_{0.85}N$ barrier layer was about 30 nm. According to datasheet the total thickness of LED active region was 200 nm. From this structure, planar light-emitting diodes with ohmic contacts Au/Ni and the resistance less than 10^{-2} $\Omega \cdot cm^2$ were made. The area of LED is 1 mm^2. In the experiment, we measured the dependence of the external quantum efficiency η on the current in the temperature range T = 10–300 K using a Janis cryostat, a Keithley 2636A SourceMeter, and a LeCroy104Xs oscilloscope. To prevent overheating at high current densities we used pulse regime (duration – 30 μs, frequency – 50 Hz).

Photoluminescence was excited by a pulsed YAG-laser with an average power of 35 mW (0.2-1 kHz, duration – 10 ns, 355 nm) and a semiconductor laser with a wavelength of 405 nm (LD Nichia NDV4642VFR, 0.2-90 kHz, duration – 15 ns, pulse power - 4 W). The driver for semiconductor laser were developed on a basis of avalanche S-diode [2,3]. To enhance the power density of laser radiation we used focusing by quartz lens (minimum diameter of laser spot on a sample was about 100 mkm).

Optical power was changed by neutral glass filters. The photoluminescence spectra were measured in the temperature range $T = 10–300$ K with an Ocean Optics spectrometer.

Normalized electroluminescence and photoluminescence spectra are presented in fig. 1-4. Experiments are shown that the change in the optical pumping mode (405 or 355 nm) gives no difference in photoluminescence spectra. However the shapes of luminescence spectra for electro- and photoluminescence regimes are very different. For low temperature luminescence the spectrum can be decompose to sum of Gaussian-like spectra. And there is at least one additional peak at the high energy side of photoluminescence spectrum in comparison with electroluminescence data. The appearance of additional peak leads to effective red shift of wavelength under pumping mode change from photo- to electroluminescence. The shift is equal to 10-12 nm what can be explained by scatter of quantum wells properties in different structure samples, e.g. due to InN composition inhomogeneous. Another reason of wavelength shift appearance is the peculiar conditions of carrier injection at different pumping modes. It is well known that electrical pumping is accompanied by tunneling of carriers through energy states of defects in injection barriers. Such a process leads to a lowering of the open voltage in InGaN/GaN light-emitting diodes [4]. Obviously this effect leads to a lowering of photon emission energy what is equivalent to observed wavelength shift.

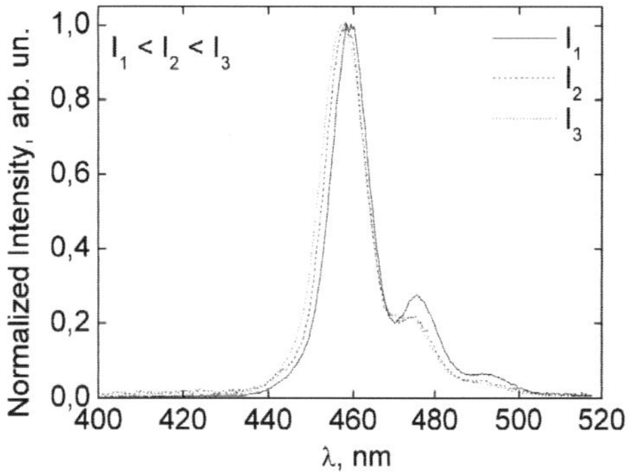

Figure 1. EL spectra for different current at T = 10 K.

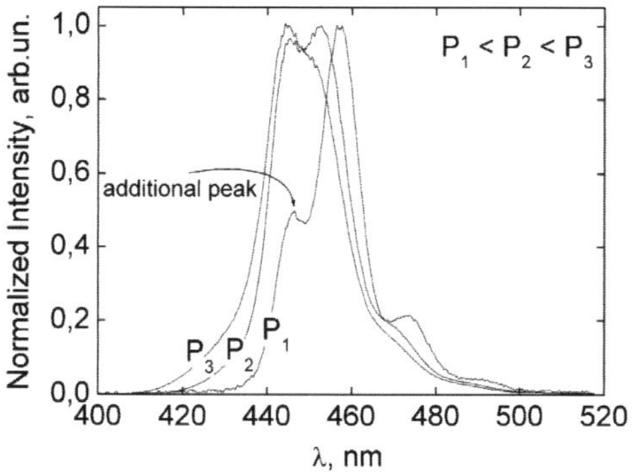

Figure 2. PL spectra for different pumping optical power at T = 10 K.

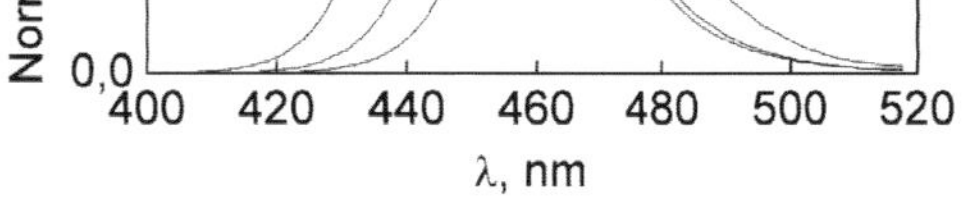

Figure 3. EL spectra for different current at T = 300 K.

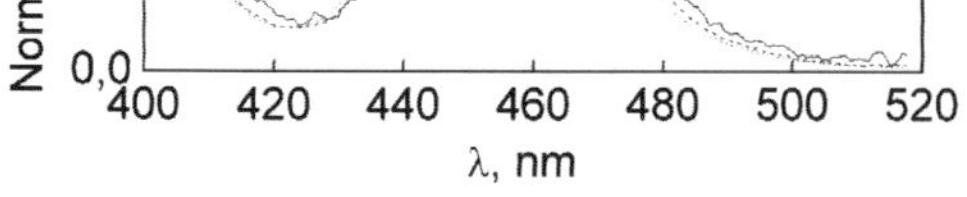

Figure 4. PL spectra for different pumping optical power at T = 300 K.

To compare EL and PL results the experimental dependences were plotted in coordinates $\eta = \eta(n_{inj})$, were η was calculated in arbitrary units from optical power P as P/n_{inj}. The value n_{inj} represents the number of injected carriers per second per unit area (injection rate). In the case of electrical pumping this value was calculated as

$$n_{inj} = \frac{I}{e \cdot S}, \quad (1)$$

where I - current, e - electron charge, and S - area of LED.

In the case of optical pumping the calculation of n_{inj} is difficult because of amount of reflected and absorbed radiation in p-layer must be considered. Moreover for accurate determination of n_{inj} the diffusion and recombination processes of nonequilibrium carriers in p-layer should be involved. Therefore to simplify the problem we used normalization which based on experimental data for wavelength shifts of luminescence spectra with pumping. It is well known that in [0001] InGaN/GaN LEDs blue shift is observed with the pumping increase, and it is explained by Stark effect [5]. We observed this effect under different pumping conditions. For all conditions luminescence spectra start to shift at the point of maximum efficiency for room temperature (or saturation of efficiency). At the electrical pumping this point corresponds to $n_{inj} = 2 \cdot 10^{18}$ $cm^{-2}c^{-1}$, and it was used for normalization of PL experimental data.

The normalization were used for room temperature because of low temperature luminescence spectra is complex, and it is unclear how to compare EL and PL data for low temperature. On the other hand it is clear blue shift for both regimes at T = 300 K as presented on the fig. 5, and it can be fitted by theoretical curve under assumption of Stark effect [6].

For determination of absolute value of quantum efficiency we also normalized experimental data. Normalization was performed by maximum efficiency which corresponds to lowest temperature. Doing so we assume that at temperature T = 10 K and low injection level an internal quantum efficiency (IQE) equals to 100% for EL and PL pumping. In an experiment this case is realized if two conditions take place. Firstly, significant leakage of carriers from multiple quantum wells to p-region should not be observed at low injection level, e.g. tunneling leakage. Secondly, at low injection level internal quantum efficiency is determined by Shockley–Read and radiative recombination process. First condition is correct in our experiment because of crystal quality of LED structures is high and tunneling leakage current is negligible in comparison with recombination current (density of threading dislocations is about $(3\text{-}7) \cdot 10^{7}$ cm^{-2}). Therefore we can use ABC-model for description of internal quantum efficiency dependence on injection level for PL-data:

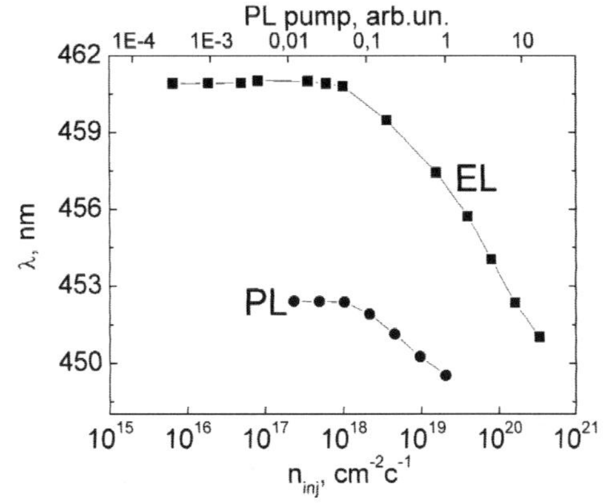

Figure 5. Dependence of normalized internal quantum efficiency on injection rate for T = 300 K; upper axis corresponds to PL data.

$$IQE = \frac{B \cdot n^2}{A \cdot n + B \cdot n^2 + C \cdot n^3}, \tag{2}$$

where A, B, and C are the temperature dependent Shockley–Read, radiative, and Auger recombination coefficients, respectively, and $n \sim n_{inj}$ (for linear recombination regime) is the concentration of non-equilibrium carriers in the MQWs.

Experimental results for quantum efficiency after normalization are presented on the fig.6 and fig.7. The highest quantum efficiencies measured at T = 300 K have similar values (~56-65%) for PL and EL regimes. Moreover the experimental curves have the same form. It follows that for EL regime, the quantum efficiency at room temperature is mainly described by ABC-model (2).

At the low temperature the experimental results for PL and EL regimes are very different. The threshold value n_{inj} for beginning of efficiency droop shifts from $2 \cdot 10^{18}$ cm^{-2}c^{-1} (for PL) to 10^{16} cm^{-2}c^{-1} (for EL). The beginning of strong blue shift at T = 10 K for EL regime is observed only at $n_{inj}=10^{20}$ cm^{-2}c^{-1}. This fact indicates that there is no sufficient concentration of carriers in quantum wells for Stark effect observation. Additional analysis of the current-voltage characteristics had shown that at low temperatures the ballistic leakage of electrons from active region is the most likely mechanism which is determined the efficiency droop in InGaN/GaN LEDs [7].

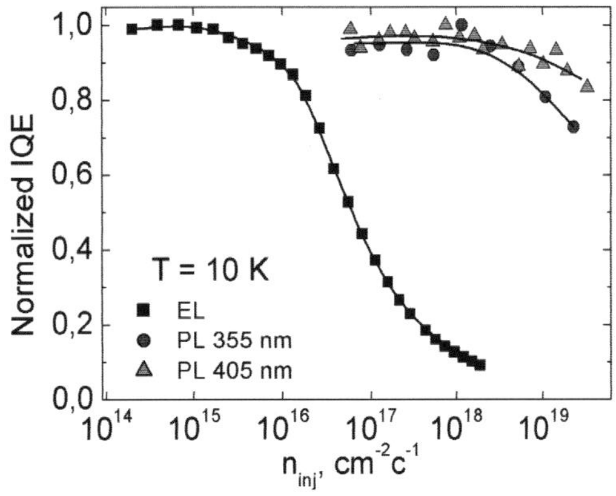

Figure 6. Dependence of normalized internal quantum efficiency on injection rate for T = 10 K.

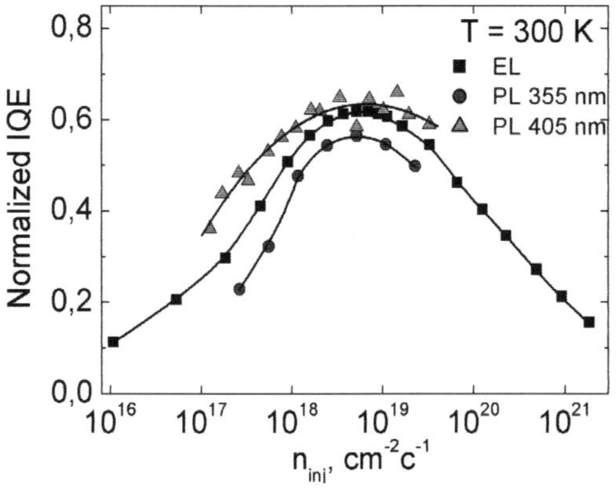

Figure 7. Dependence of normalized internal quantum efficiency on injection rate for T = 300 K.

References

[1] Prudaev I A, Romanov I S, Kopyev V V, Shirapov S B, Tolbanov O P and Khludkov S S 2013 Russian Physics Journal **56** 757
[2] Prudaev I A and Khludkov S S 2009 Russian Physics Journal **52** 163
[3] Prudaev I A, Khludkov S S, Skakunov M S and Tolbanov O P 2010 Instruments and Experimental Techniques **53** 530
[4] Bochkareva N I, Voronenkov V V, Gorbunov R I, Latyshev P E, Lelikov Yu S, Rebane Yu T, Tsyuk A I and Shreter Yu G, 2013 Semiconductors **47** 127
[5] Chichibu S F, Abare A C, Mack M P, Minsky M S, Deguchi T, Cohen D, Kozodoy P, Fleischer S B, Keller S, Speck J S 1999 Mat. Sci. Eng. **59** 298
[6] Bulashevich K A, Karpov S Yu and Suris R A 2006 Phys. Stat. Sol. (b) **243** 1625
[7] Prudaev I A, Golygin I Yu, Shirapov S B, Romanov I S, Khludkov S S and Tolbanov O P 2013 Semiconductors **47** 1382

SPbOPEN2014 IOP Publishing
Journal of Physics: Conference Series **541** (2014) 012056 doi:10.1088/1742-6596/541/1/012056

The analytical approach to optimization of active region structure of quantum dot laser

V V Korenev[1], A V Savelyev[1,3], A E Zhukov[1,2,3], A V Omelchenko[1,3], M V Maximov[1,2]
[1]Saint–Petersburg Academic University RAS, Saint–Petersburg 194021, Russia
[2]Ioffe Physico–Technical Institute RAS, Saint–Petersburg 194021, Russia
[3]Saint–Petersburg State Polytechnical University, 195251 Saint-Petersburg, Russia
E-mail: korenev@spbau.ru

Abstract. Using the analytical approach introduced in our previous papers we analyse the possibilities of optimization of size and structure of active region of semiconductor quantum dot lasers emitting via ground-state optical transitions. It is shown that there are optimal length, dispersion and number of QD layers in laser active region which allow one to obtain lasing spectrum of a given width at minimum injection current. Laser efficiency corresponding to the injection current optimized by the cavity length is practically equal to its maximum value.

1. Introduction

Efficient semiconductor lasers having broadband emission spectra, which correspond to the transparency window of standard silicon optical fibre, can be used for the wide range of practical applications, such as optical coherence tomography and ultrafast data transmission [1 – 3]. Long-wavelength InAs/InGaAs quantum dot (QD) lasers, emitting via ground state (GS) optical transitions of QDs near 1.3 μm, allow one not only to overlap this wavelength range, but also to decrease operation current of such a laser realizing lasing spectra of a given parameters [4].

At the same time, due to the possibility of usage of such semiconductor QD lasers in compact portable devices, the task of optimization of its energy consumption plays an important role. In particular, it is important to find out the ways to optimize all the key parameters of laser's active region, such as number of QD layers (Z) in it, dispersion of inhomogeneous broadening (σ) and laser cavity length (L), to obtain lasing spectrum of a given width (Ω) at minimum injection current (I_{inj}).

2. Optimization of laser active region

To elucidate the influence of the inhomogeneous broadening and of the number of QD layers in laser's active region (Z) on the injection current we have calculated the dependence of injection current optimized by the laser cavity length (I^{L}_{opt}) on the dispersion (σ) corresponding to the lasing spectrum of the given width Ω=12 meV for different values of Z – see figure 1a. The other parameters are in accord with [4 – 5]. The calculation is performed in the scopes of the model proposed in [4 – 5].

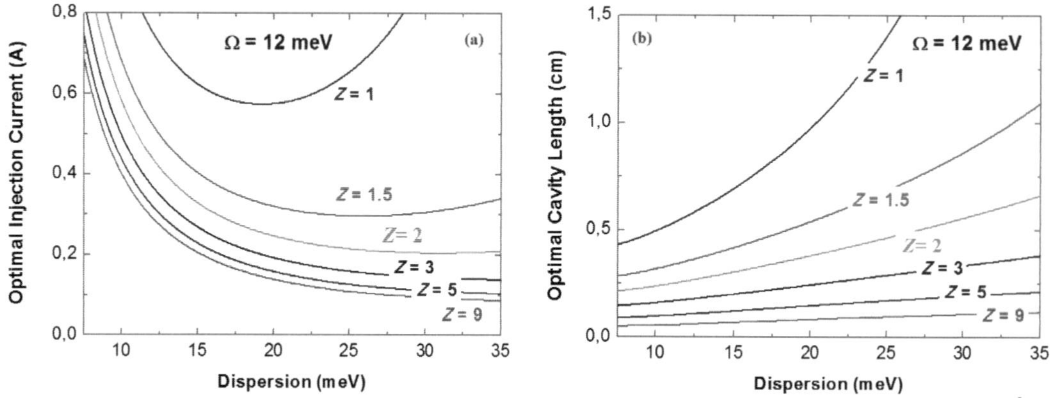

Figure 1(a, b). **(a)** The dependence of injection current optimized by the laser cavity length ($I^L{}_{opt}$) corresponding to the lasing spectrum of the given width (Ω= 12 meV) on dispersion for the different numbers of QD layers (Z). **(b)** The dependence of optimal cavity length (L_{opt}) corresponding to the $I^L{}_{opt}$ on dispersion for different values of Z.

As it can be seen from the figure 1a (see curve Z= 1), there is always an optimal value of dispersion (σ) realizing the minimum of optimal injection current ($I^L{}_{opt}$), which corresponds to the lasing spectrum of a given width. Indeed, at sufficiently small dispersion, lasing spectrum is sufficiently narrow and, therefore, the injection current (I), as well as optimal injection current ($I^L{}_{opt}$), should increase in order to attain lasing spectrum of a given width. At large dispersion, the maximum gain (G_{max}) turns out to be sufficiently small and can be even lower than the total loss (α) at a certain dispersion due to reciprocal proportionality of the maximum gain to the dispersion (G_{max} ~ $1/\sigma$). This, in the case of large dispersions, results in the increase of $I^L{}_{opt}$ with the increase of σ. Therefore, there is always an optimal value of dispersion (σ_{opt}) that allows one to realize lasing spectrum of a given width at minimum injection current – see figure 1a, e.g. curve Z= 1.

The larger is the number of QD layers in laser's active region (Z), the higher is the maximum gain (G_{max} ~ Z) and the less pronounced are the lasing spectrum narrowing, which takes place at sufficiently low dispersions, and G_{max} decrease, which is important at sufficiently high dispersions. This, in its turn, results in higher values of optimal dispersion (σ_{opt}) and lower values of injection current corresponding to the lasing spectrum of a given width. As a result, for $Z > 3$ the simple increase of dispersion leads to the decrease of optimal injection current ($I^L{}_{opt}$) and the higher is σ, the lower is $I^L{}_{opt}$ in all accessible range of dispersion – see figure 1a.

It is important to notice that the optimal values of dispersion realizing minimal injection current shown in the figure 1a correspond to the technologically accessible range of laser cavity lengths – see figure 1b. This, in its turn, makes the considered optimization very useful from the practical point of view. At the same time, in the case of extremely large dispersions exceeding 30 meV and sufficiently high output powers, the lasing via first excited optical transitions can take place and, thus, should be taken into account [6]. As it can be seen from the figure 1b at fixed Z, e.g. at Z= 5, the increase of dispersion leads to the increase of optimal cavity length corresponding to the optimal injection current ($I^L{}_{opt}$). Indeed, the higher is σ, the lower is G_{max}, which is reciprocally proportional to σ (G_{max} ~ $1/\sigma$). Therefore, the total loss (α) should decrease in order to maintain the lasing condition of the equality of the gain (G) and total loss (α): $G = \alpha$, where $G \sim G_{max}$. This, in its turn, corresponds to the laser samples with longer cavities that is in accord with the figure 1b and our further calculations – see figure 2 (Z= 5).

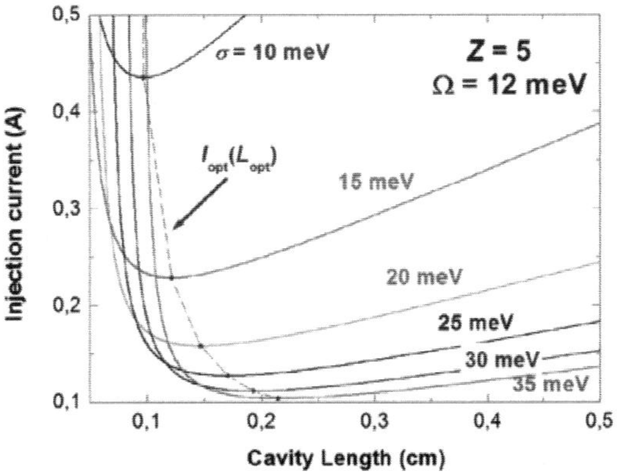

Figure 2. The dependence of the injection current (I) as a function of laser cavity length (L) for the different values of dispersion (σ) varying from 10 meV to 35 meV. Dashed line corresponds to the optimal injection current (I^L_{opt}) at optimal length (L_{opt}). Here Ω= 12 meV and Z= 5.

It is interesting to notice that at sufficiently high dispersions, when the values of G_{max} and α are practically equal, the slope of $L_{opt}(\sigma)$ dependence sharply increases above σ_{opt}. As it can be distinctly seen from the figure 1b (curve Z= 1), there is such an increase of L_{opt} for all the dispersions exceeding 19.25 meV that is equal to σ_{opt} for Z= 1 as it follows from the figure 1a. However, such an increase in the slope is not enough to increase lasing spectrum width significantly and, therefore, I^L_{opt} for Z= 1 also tends to increase – see figure 1a. The same situation is for large values of Z – see figure 2 for Z= 5. The dashed line in the latter figure corresponds to the I^L_{opt} decreasing with the increase of σ practically in all range of dispersions that is due to sufficiently high values of σ_{opt} because of large Z.

Thus, there are always optimal values of laser cavity length (L_{opt}) and dispersion (σ_{opt}) of inhomogeneous broadening allowing one to attain lasing spectrum of a given width at minimum injection current. However, at sufficiently large numbers of QD layers in laser active region (Z), the optimal value of dispersion becomes so high that the simple increase of dispersion leads to the decrease of injection current corresponding to the lasing spectrum of a given width in all accessible range of dispersions – see figure 1a. The increase of a priori given laser spectrum width (Ω) from Ω_a to Ω_b ($\Omega_b > \Omega_a$) results in the increase of optimal injection current (I^L_{opt}) as well as of optimal dispersion (σ_{opt}) because the wider is the lasing spectrum, the higher injection and dispersion are needed in order to realize it. Such an increase corresponds to the movement from left-down to the right-top corner of the figure – see figure 3.

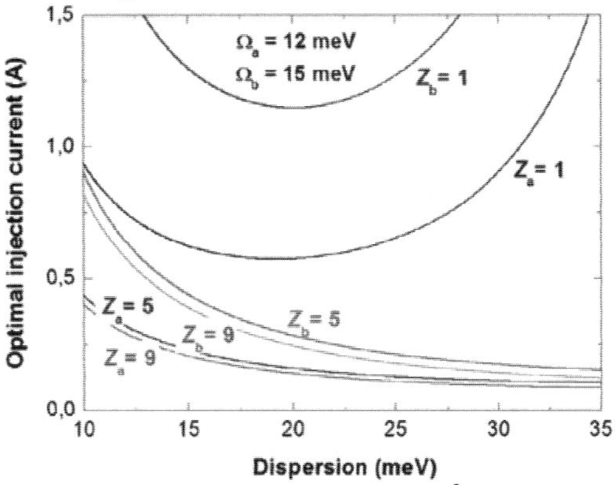

Figure 3. The dependence of optimal injection current (I^L_{opt}), which corresponds to the lasing spectrum of the given width (Ω), on dispersion (σ) for the different values of QD layers in laser active region (Z). I^L_{opt} corresponding to Ω_a= 12 meV are depicted with curves Z_a and to Ω_b= 15 meV are Z_b.

At the same time, it is very important for the practice to reveal how strong is the influence of laser cavity length (*L*) and dispersion of inhomogeneous broadening (σ) on the injection current realizing the lasing spectrum of the given width (Ω). More precisely, it is important to reveal optimization of which parameter (*L* or σ) is most crucial for the achievement of minimum injection current corresponding to the lasing spectrum of the given width (Ω) and optimization of which parameter is just fine-tune. In order to answer this question we have calculated the dependence of injection current optimized by the cavity length on dispersion $I^L_{opt}(\sigma)$ – see figure 1a, and the dependence of injection current optimized by the dispersion on cavity length $I^\sigma_{opt}(L)$ – see figure 4, for the different numbers of QD layers in laser active region (*Z*).

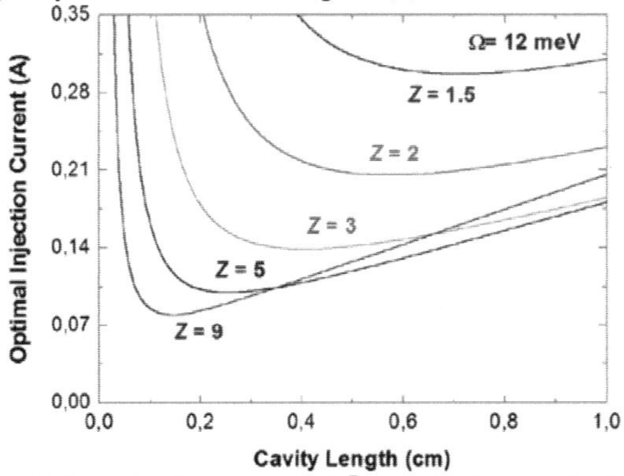

Figure 4. The dependence of the injection current (I^σ_{opt}), which is optimized by the dispersion, as a function of laser cavity length (*L*) for the different numbers of QD layers (*Z*) in laser active region.

As it can be seen from the figure 1a for *Z*= 5 the variation of dispersion from 7.0 meV to 21.5 meV results in the decrease of optimal injection current I^L_{opt} practically by 500 mA. At the same time, the variation of the cavity length from 0.1 cm to 0.5 cm in the case of *Z*= 5 results in decrease of I^σ_{opt} only by 50 mA – see figure 4. Therefore, the optimization of dispersion plays the key role and, in the case of multi-layered structures, the dispersion should be increased in order to decrease optimal injection current I^L_{opt}. However, the optimization of laser cavity length is the task of fine-tuning of injection current that is in accord with our previous results [7]. Thus, the usage of multi-layered structures (with $Z > 3$) and high dispersions (exceeding 17 – 20 meV) allow one to decrease the optimal injection current I^L_{opt} corresponding to the lasing spectrum of a given width Ω significantly.

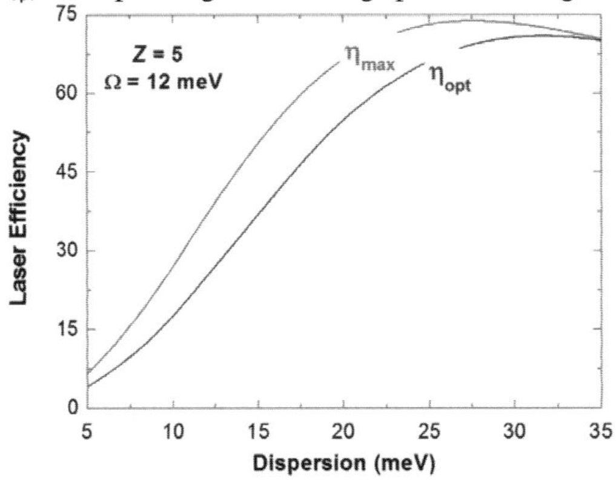

Figure 5. Laser power conversion efficiency (PCE) as a function of dispersion of inhomogeneous broadening (σ): blue curve corresponds to the PCE at the injection current optimized by the laser cavity length (I^L_{opt}), which realizes lasing spectrum of a given width (Ω); red curve corresponds to the peak value of PCE at *Z*= 5 and Ω = 12 meV.

Finally, it occurs that in the case of multi-layered structures having 3 – 5 layers of QDs, power conversion efficiency (PCE) of QD laser η_{opt} corresponding to the optimal injection current I^{L}_{opt} is practically equal to its peak value η_{max} corresponding to the specific value of laser cavity length (L_{max}), for all attainable values of dispersion, where diode opening voltage, serial diode resistivity and internal quantum efficiency are equal to 1.02 meV, 10^{-4} Ohm·cm^2 , 0.9 that is in accord with [8] – see figure 5. Thus it, is important to use multi-layered laser structures having sufficiently high dispersion in order to obtain lasing spectrum of a given width at minimum injection current and maximal dispersion which can be also further fine-tuned by the optimization of laser cavity length.

3. Conclusion

It was shown that in the case of quantum dot lasers having a small number of quantum dot layers in its active region, in particular in the case of a single plane of quantum dots, there are always optimal values of dispersion and laser cavity length allowing one to realize lasing spectrum of a given width at minimum injection current. At the same time, the usage of multi-layered structures results in the decrease of the injection current, corresponding to the lasing spectrum of the given width. The operation current can be also decreased by the increase of dispersion in all practically accessible range of dispersions.

Moreover, the optimization of dispersion plays the key role, while the optimization of laser cavity length, which has its optimal value corresponding to the minimum operation current for single- as well as for multi-layered structures, is just a fine-tune. The usage of laser structures optimized by the length allows one to decrease injection current on more than 45% on the scale ~ 0.1 A, however the optimization of dispersion takes place on the scale of ~ 1 A.

As a result, the usage of laser active regions having several layers of quantum dots and the increase of dispersion, as well as fine-tune optimization of the laser cavity length, allow one to obtain lasing spectrum of a given width at minimum operation current at power conversion efficiency practically equal to its maximum value.

Acknowledgements

This work was supported by the Program of Fundamental Research of Russian Academy of Sciences, by the Committee for Science and Higher Education of the Government of Saint–Petersburg, by the Russian Foundation for Basic Research and by the Dynasty Foundation.

References

[1] Zhukov A E and Kovsh A R 2008 *Quantum Electronics* **38** 409

[2] Zhukov A E, Kovsh A R, Nikitina E V, Ustinov V M, Alferov Zh I 2007 *Semicond.* **41** 606

[3] Djie H S, Ooi B S, Fang X, Wu Y, Fastenau J M, Liu W K, Hopkinson M 2007 *Opt. Lett.* **32** 44

[4] Korenev V V, Savelyev A V, Zhukov A E, Omelchenko A V, Maximov M V 2012 *Semicond.* **46** 645

[5] Korenev V V, Savelyev A V, Zhukov A E, Omelchenko A V, Maximov M V, Shernyakov Yu M 2013 *Proc. of SPIE* **8772** 84320W

[6] Korenev V V, Savelyev A V, Zhukov A E, Omelchenko A V, Maximov M V 2014 *Appl. Phys. Lett.* **102** 112101

[7] Korenev V V, Savelyev A V, Zhukov A E, Omelchenko A V, Maximov M V 2014 *Book of Abstracts of the 1st International School and Conference "Saint–Petersburg OPEN 2014" on Optoelectronics, Photonics, Engineering and Nanostructures* (St. Petersburg, Russia, 25–27 March 2014) vol **1** (St. Petersburg: Publishing House of St. Petersburg Academic University RAS) p 227

[8] Zhukov A E, Maximov M V 2009 *Modern Injection Lasers* (St. Petersburg: Publishing House of St. Petersburg State Polytechnical University)

Study of electric field distribution in AlGaInP light-emitting diode by Kelvin probe force microscopy

V.L. Oleynik[1], I.A. Prudaev[1], Vad. A. Novikov[2], Yu. L. Ryaboshtan[3] and P. V. Gorlachuk[3]

[1] Functional Electronics Laboratory, Tomsk State University, 36 Lenin Avenue, Tomsk, 634050, Tomsk, Russia
[2] Faculty of Physics, Tomsk State University, 36 Lenin Avenue, Tomsk, 634050, Tomsk, Russia
[3] JSC "M. F. Stelmakh Research Institute "Polyus", Vvedenskogo str., 3 bld.1., 117342, Moscow, Russia

E-mail: dozorx777@gmail.com

Abstract. This work presents results of study of different AlGaInP light-emitting diodes by using Kelvin force probe microscopy. Study of current-voltage characteristics and electric field distribution had shown that for all type of experimental samples containing multiple quantum wells the most significant series resistance of heterostructures is determined by p-isotype heterojunction and confinement p-layer.

AlGaInP light-emitting diodes (LEDs) have a complex structure comprising multiple quantum wells (MQWs), confinement layers (n:conf, p:conf), and distributed Bragg reflector (DBR). Such a complex sequence of layers may cause unacceptable effects of current limitation due to sufficient resistance of heterostructures. For optimization of LED design it is important to use measurement technique with high spatial resolution such as Kelvin force probe microscopy (KFPM). This work presents results of study of different LED structures by using KFPM. The structures differ by thickness of active region, composition of AlGaInP quantum wells, and concentration of radiation defects.

Schematically investigated LED structure is shown in fig.1. The AlGaInP heterostructures were grown on the silicon doped GaAs substrates ($n = 2\cdot 10^{18}$ cm^{-3}) with the thickness about 350 μm misoriented by 10° to the (111A) plane. A GaAs:Si layer with the electron concentration $n = 2 \cdot 10^{18}$ cm^{-3} and the thickness about 0.2 μm was formed on the GaAs substrate. For red LEDs a distributed Bragg reflector based on the silicon doped ($n = 1.5\cdot 10^{18}$ cm^{-3}) $Al_{0.45}Ga_{0.55}As/Al_{0.9}Ga_{0.1}As$ layers with the number of pairs equal 15 was used as an optical reflector. The layer thickness was chosen in accordance with the quarter-wave condition for the reflection maximum at 620–630 nm. The number of pairs was chosen so that the reflection coefficient close to 1 in the range 600–640 nm should be achieved. The composition of the layers provided non-absorbing conditions, i.e., the band gap of the reflector layers was wider than the energy transition in the GaInP QWs.

The active region of the structure was formed on the basis of the multiple quantum wells $Ga_{0.47}In_{0.53}P$, (10-20)MQW with the thickness 7 nm, and $Al_{0,27}Ga_{0,23}$ $In_{0,5}P$ barrier layers with the thickness 10 nm. The MQWs were restricted by the $Al_{0.27}Ga_{0.23}InP_{0.5}$ undoped layers, n-$Al_{0.5}In_{0.5}P$ ($n = 5\cdot10^{17}$ cm^{-3}, $d = 0.52$ μm) and p-$Al_{0.5}In_{0.5}P$ ($p = 1\cdot10^{18}$ cm^{-3}, $d = 0.52$ μm) confinement layers doped with silicon and magnesium, respectively. For yellow LED 30 MQW $(Al_{0.32}Ga_{0.68})_{0.5}In_{0.5}P/(Al_{0.54}Ga_{0.46})_{0.5}In_{0.5}P$ structure was grown.

A 9.3–9.6 μm thick window layer of the LED AlGaInP structure was made on the basis of magnesium-doped gallium phosphide (GaP) ($p = 3$–$4\cdot10^{19}$ cm^{-3}). The thickness of the window layer was chosen to improve the current spreading in the plane of the p-n-junction. Previous study had shown that electroluminescence spectra of this LED structures demonstrate high homogeneity of the solid solution in the quantum wells [1].

In this work we used atomic force microscope NTEGRA Prima (NT-MDT) to measure qualitative potential distribution in LED structures under different applied bias voltages. The measurements were conducted on a cleavage plane of LED samples. Also we measured the forward current-voltage characteristics of LEDs. The measurements were conducted in a static and pulse mode at temperature T = 300 K using Keithley 2636 source meter, and LeCroy104Xs oscilloscope.

Recently it was shown that in this structures current limitation is observed [2]. In this work additionally we studied LEDs with high series resistance which was initiated by electron beam irradiation with flux 10^{16} cm^{-2} (average electron energy is 8 MeV). In this LEDs compensation effects are observed what leads to increase of resistivity of passive regions of diodes. In experiment it is interesting to define where the most significant compensation is. Kelvin force probe microscopy can be powerful tool for such a study.

Figure 2 shows a one- and two-dimensional images of the Kelvin voltage line scans from GaAs substrate (left side) to the p-GaP (right side). The figure corresponds to equilibrium condition for LED. It is clear that resolution of KFPM is sufficiently high, and it is possible to pick out single layers of DBR with thickness of about 50 nm. Boundaries between neighbor thick layers match to sharp potential drops.

Figure 3 shows qualitative distribution of electric field in red LED with 20 quantum wells at the different applied voltages. The distribution was obtained by differentiation of experimental KFPM data. From the experimental results we can conclude that almost all voltage is applied on the MQW region at the voltages below 1.5 V. At the higher voltages electric field distribution is changed in the confinement n- and p-layers. The most significant changes of electric field distribution is observed in p:conf layer and p:conf/p:GaP heterojunction.

p-GaP
p-AlGaInP
p-AlInP
i-AlGaInP
MQW 10-30 x GaInP/AlGaInP
i-AlGaInP
n-AlInP
n-AlGaInP
DBR 15 x AlGaAs/AlGaAs
GaAs

Figure 1. AlGaInP LED structure with distributed Bragg reflector and multiple quantum wells.

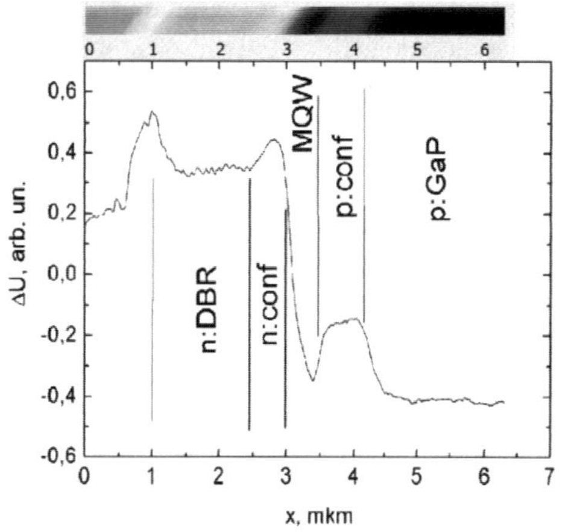

Figure 2. Potential distribution in red AlGaInP LED with 20 quantum wells under equilibrium condition.

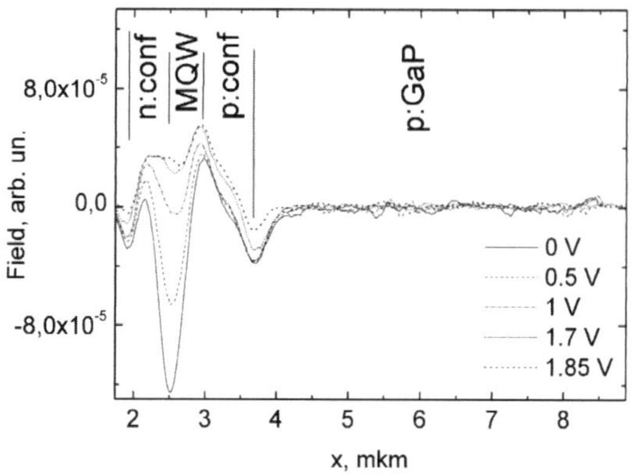

Figure 3. Electric field distribution in red AlGaInP LED with 20 quantum wells at the different applied voltages.

The study of yellow LEDs is interesting from standpoint of determination of illumination influence on KFPM measurement results. It is well known that measured potential in KFPM is affected by laser radiation reflected from cantilever in microelectromechanical system of atomic force microscope. Part of this radiation can reach the measured sample. If the band gap of semiconductor sample is less than photon energy, the absorption of radiation leads to generation of nonequilibrium carriers at the surface of semiconductor. As a result surface potential will be changed. In our case the band gap of AlGaInP solid solution in red and yellow LEDs were 1.9 and 2.1 eV respectively. The energy of laser radiation in atomic force microscope was 1.9 eV. Hence undesirable effect of potential change under optical absorption in active region of yellow LEDs should be avoided. To verify this statement we measured yellow structures under filament lamp illumination and in dark condition. Fig.4 shows measurement results. It was found that at low forward voltages potential distributions are different in the boundary between n:DBR and n:conf layers. In the rest of the structure there is good qualitative agreement of field distributions. The difference could be explained by selective absorption of laser radiation in

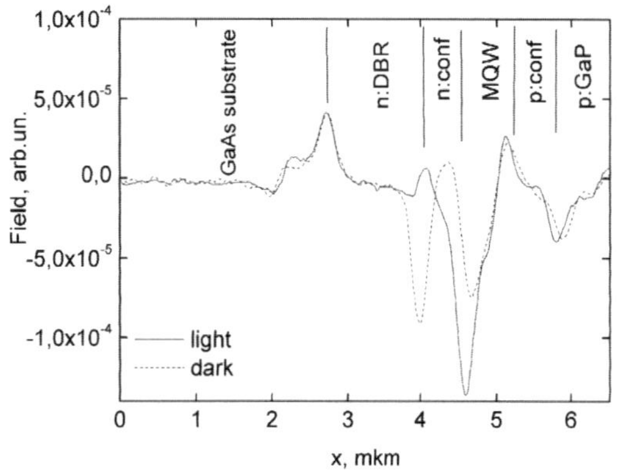

Figure 4. Electric field distribution in yellow AlGaInP LED with 30 quantum wells under lighting and dark conditions at zero applied voltage.

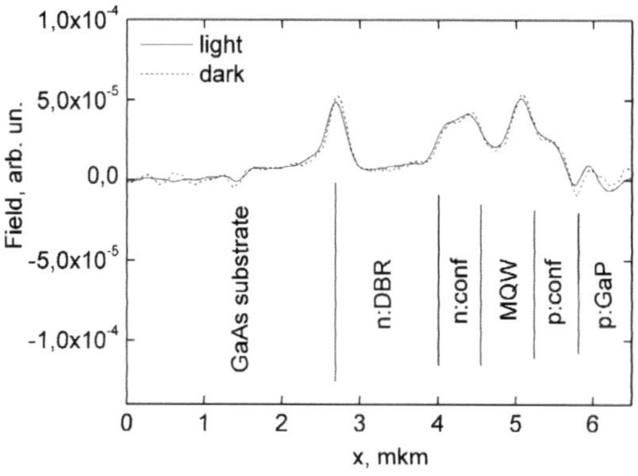

Figure 5. Electric field distribution in yellow AlGaInP LED with 30 quantum wells under lighting and dark conditions at applied voltage U = 1.9 V.

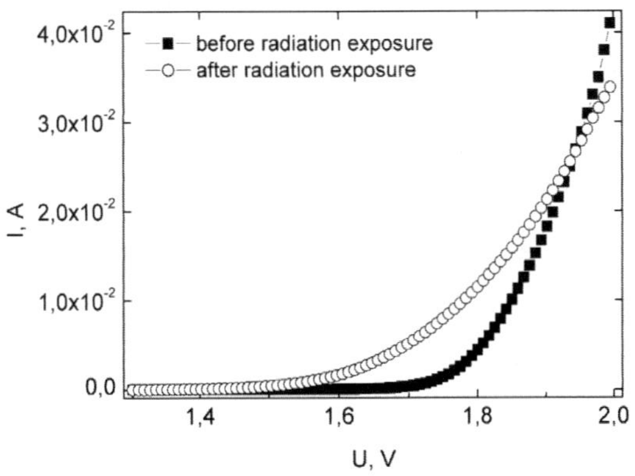

Figure 6. Current-voltage characteristics of red AlGaInP LEDs with 10 quantum wells before and after radiation exposure.

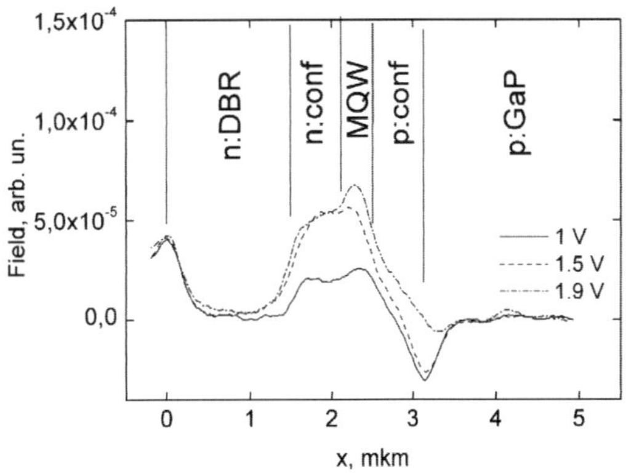

Figure 7. Electric field distribution in red AlGaInP LED with 10 quantum wells at the different applied voltages after radiation exposure.

n:DBR under dark condition. The results of measurements for high forward voltages are in good quantitative agreement (fig.5). It should be noted that for yellow LEDs at the high voltages electric field distribution is changed significantly in the confinement n- and p-layers, and in p:conf/p:GaP heterojunction.

The study of LEDs with radiation defects shown that at the voltages U < 1.9 V excess current is observed which can be attributed to nonradiative recombination through defect centers in MQW [2]. On the other hand at high voltages current limitation take place (fig.6). Decreasing of the current at voltages U > 1.9 V is associated with additional resistance (R_d) of passive regions of LED with radiation defects:

$$I = I_o \cdot \exp\left[\frac{e \cdot (U - I \cdot [R + R_d])}{n \cdot k \cdot T}\right], \tag{1}$$

where e - electron charge, k – Boltzmann constant, R – contact resistance, $I_o = I_o(T)$.

Fig. 7 shows electric field distribution in red LED with 10 quantum wells after radiation exposure. As previous cases the increasing of voltage from 1.5 to 1.9 V initiates redistribution of electric field in p:conf layer and p:conf/p:GaP heterojunction. However there is no redistribution of field in n:conf layer. This fact points to donor nature of radiation defects which compensate initial p-type conductivity of p:conf layer.

In conclusion we have shown that for different AlGaInP LEDs applied voltage drops across all confinement and MQW layers. The most significant series resistance of LEDs is determined by p:conf layer and p:conf/p:GaP heterojunction what is in agreement with previous results [3].

References

[1] Marmalyuk A A, Gorlachuk P V, Ryaboshtan Yu L, Brudnyi V N, Prudaev I A, Romanov I S and Lelekov M A 2013 Russian Physics Journal **56** 894

[2] Prudaev I A, Skakunov M S, Lelekov M A, Ryaboshtan Yu L, Gorlachuk P V, Marmalyuk A A 2013 Russian Physics Journal **56** 898

[3] Katzer Kl -D, Mertin W, Bacher G, Jager A and Streubel K 2006 Appl. Phys. Lett. **89** 103522

Influence of magnetic field on the surface waves properties in the photonic crystal / graphene structure for terahertz frequency range

A Denisultanov, S Azbite and M Khodzitsky

Department of Photonics and Optoinformatics, ITMO University, Kronverkskiy pr. 49, Saint-Petersburg 197101, Russia

E-mail: alaudi2991@mail.ru

Abstract. In this paper magneto-tunable photonic crystal with graphene layer was investigated for terahertz frequency range. The peaks of transmissivity in band-gaps of photonic crystal that caused by excitation of surface waves were obtained. The control of frequency peaks' position by magnetic field was demonstrated.

1. Introduction

Today photonic crystals (PC) are one of the most promising metamaterials. PCs allow to manipulate electromagnetic waves and therefore they can be used for development of optical devices such as filters, superlenses, polarization changers, waveguides, superprisms, etc [1-3]. Also today graphene has considerable attention among scientists due to the specific properties such as high electric and thermal conductivity etc. [4]. The photonic crystal with graphene layers has periodic narrow stop bands in frequency spectrum and that may be used as spectral-selective mirror [5]. Surface waves may be excited in the photonic crystal structure using graphene layer as the boundary medium. In this paper the influence of magnetic field on the surface waves properties in the photonic crystal bounded by graphene layer for terahertz frequency range was investigated. Peaks of transmissivity in band-gaps of photonic crystal that caused by excitation of surface waves were obtained. The possibility of frequency peaks position tuning and amplitude by external magnetic field was shown.

2. Problem statement

In this paper three one-dimensional (1D) photonic crystal structures were considered (figure 1). The first analyzed PC structure has 5 bilayer cells and each cell consists of SiO_2 and Si layer (figure 1(a)). The second structure is 5-layered SiO_2 / Si PC composite bounded by graphene layer on SiO_2 substrate (figure 1(b)). The third structure is PC - SiO_2 - graphene - PC composite (figure 1(c)). External magnetic field was applied perpendicularly to PC layers surface. The structures were excited by linearly polarized transverse electric wave (TE-wave). The wave vector k is perpendicular to the layers of each PC.

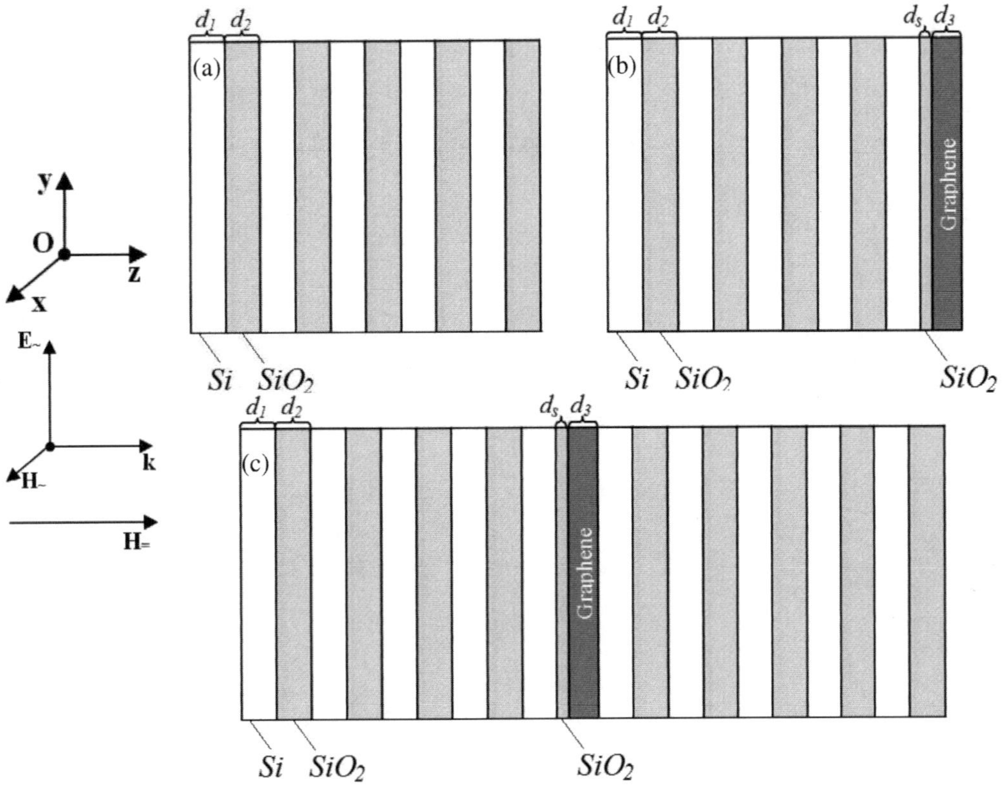

Figure 1. Schemes of photonic crystals: (a) without graphene layer; (b) with graphene layer; (c) PC-SiO_2-graphene-PC composite.

3. Optical graphene conductivity

Graphene conductivity is a sum of two contributions: (i) a Drude term σ_D describing intra-band processes, and (ii) a term σ_I taking into account inter-band transitions [6]. The inter-band transitions term can be calculated as $\sigma_I = \sigma_I' + i\sigma_I''$:

$$\sigma_I' = \sigma_0 \left(1 + \frac{1}{\pi}\arctan\frac{\hbar\omega - 2E_F}{\Gamma} - \frac{1}{\pi}\arctan\frac{\hbar\omega + 2E_F}{\Gamma}\right), \tag{1}$$

and

$$\sigma_I'' = -\sigma_0 \frac{1}{2\pi}\ln\frac{(2E_F + \hbar\omega)^2 + \hbar^2\omega^2}{(2E_F - \hbar\omega)^2 + \hbar^2\omega^2}, \tag{2}$$

where $\sigma_0 = \dfrac{\pi e^2}{2h}$, e – electron charge, E_F - Fermi energy, Γ - relaxation rate.

The Drude contribution term is following:

$$\sigma_D = \sigma_0 \frac{4E_F}{\pi}\frac{1}{\Gamma - i\hbar\omega}. \tag{3}$$

As seen from figure 2, at the terahertz frequencies the Drude term significantly exceeds the inter-band term, therefore the total conductivity may be approximately defined by the Drude term ($\sigma_g = \sigma_D$).

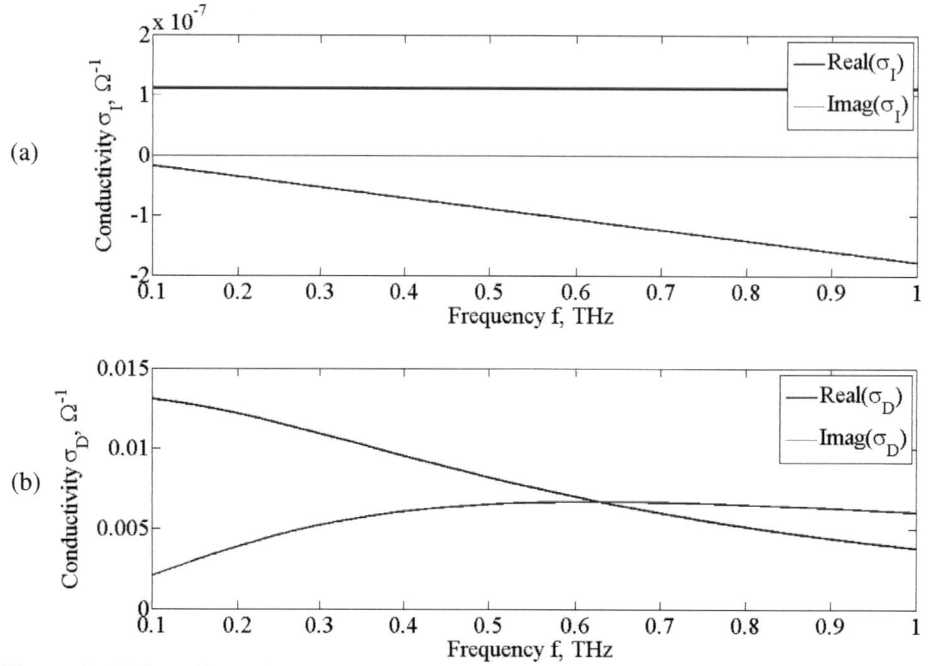

Figure 2. (Color online). Dispersion of inter-band (a) and intra-band (b) terms of conductivity.

It should be noted, that the graphene conductivity varies under external magnetic field [6]. Figure 3 shows changing of the total conductivity at various magnetic fields (0-3.6 T).

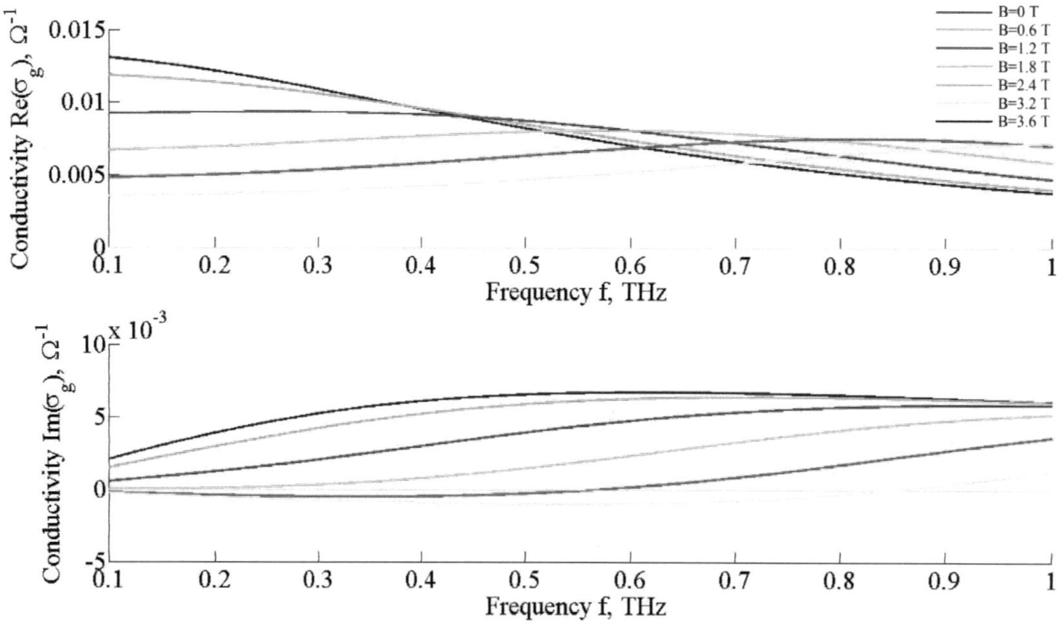

Figure 3. (Color online). Influence of external magnetic field on graphene conductivity.

3. Analytical modeling

Simulation of PC bandgap structures was performed by Matlab program using the transfer matrix method [7]. The photonic crystal structure has 5 bilayer cells, which were composed by SiO_2 and Si layers with refractive index n_{SiO_2} =2.154, n_{Si} =3.418 and thickness d_{SiO_2} =500 μm, d_{Si} =525 μm and $d_{\text{substrate}}$ =331 μm correspondingly for frequency range from 0.1 to 1 THz [8]. Therefore the transfer matrix of whole composite has the following form:

$$\hat{T} = K_{air,Si} * \Phi_{Si} * T^N * K_{Si,SiO_2} * \Phi_{sub} * (K_{SiO_2,air} + \alpha fJ), \tag{4}$$

where N is number of times. T matrix is repeated in the structure and has the following form:

$$T = \Phi_{Si} * K_{Si,SiO_2} * \Phi_{SiO_2} * K_{SiO_2,Si}. \tag{5}$$

The matrix characterizing the free propagation through some material in the structure (with $j = Si, SiO_2, \text{substrate}$) is

$$\Phi_j = \begin{bmatrix} e^{-i\omega\sqrt{\varepsilon_j}d_j/c} & 0 \\ 0 & e^{i\omega\sqrt{\varepsilon_j}d_j/c} \end{bmatrix} \tag{6}$$

The matrix characterizing the transfer through the boundary of two dielectrics is:

$$K_{i,j} = \begin{bmatrix} (\sqrt{\varepsilon_j/\varepsilon_i}+1)/2 & (1-\sqrt{\varepsilon_j/\varepsilon_i})/2 \\ (1-\sqrt{\varepsilon_j/\varepsilon_i})/2 & (\sqrt{\varepsilon_j/\varepsilon_i}+1)/2 \end{bmatrix}, \tag{7}$$

where $i, j = air, Si, SiO_2, \text{substrate}$.

The matrix $K_{SiO_2,air} + \alpha fJ$ characterizes the transfer through graphene to air, where

$$J = \begin{bmatrix} 1 & 1 \\ -1 & -1 \end{bmatrix}, \tag{8}$$

$\alpha \approx 1/137$ is fine structure constant and f is a dimensionless function, which characterizes graphene optical conductivity under magnetic field influence:

$$f = \sigma_D \frac{h}{e^2} \frac{(\Gamma/\hbar - i\omega)^2}{(\Gamma/\hbar - i\omega)^2 + wb^2}, \tag{9}$$

where E_F – Fermi energy (0.45 eV); $\Gamma = 2.6$ meV; wb is the cyclotron frequency defined as:

$$wb = \frac{e\upsilon_F^2 B}{E_F}, \tag{10}$$

where e – electron charge; υ_F – Fermi velocity in graphene ($\approx 10^6$ m/s); B is external magnetic field.

The simulated transmission coefficients of the photonic crystal and photonic crystal bounded by graphene layer of various magnetic field were shown in fig.4. As seen from fig 4, the transmission tuning of PC-graphene structure is observed at the magnetic field range from 0 to 3.6 T.

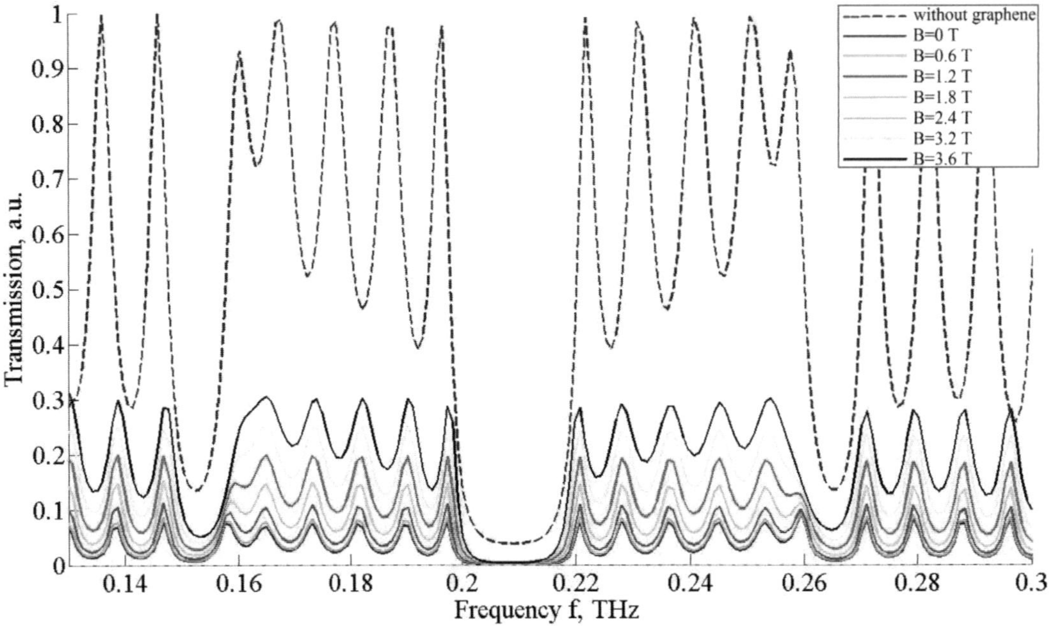

Figure 4. (Color online). The bandgap structure of photonic crystal without graphene layer and bounded by graphene.

The bandgap structure of PC (SiO_2/Si) – graphene/SiO_2 substrate - PC (SiO_2/Si) at various magnetic fields is shown in figure 5.

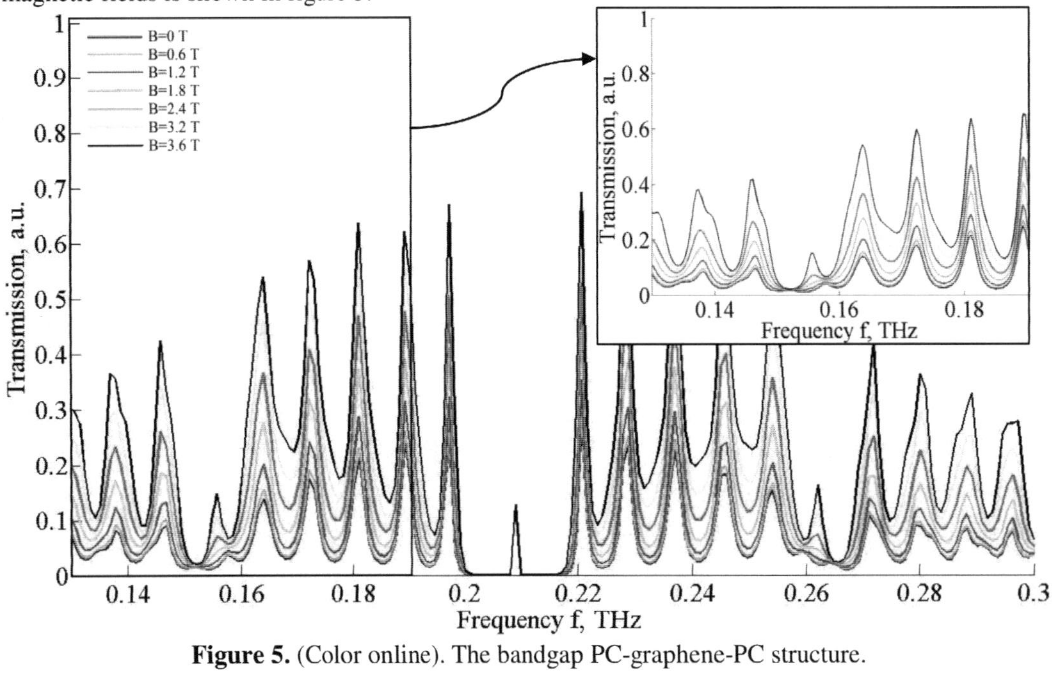

Figure 5. (Color online). The bandgap PC-graphene-PC structure.

As seen from fig. 5 peak of transmissivity in band-gap in frequency range of 0.2-0.22 THz is excited by SiO_2 substrate, it is defect mode in PC structure. The peak of transmissivity in frequency range of

0.145-0.165 THz is generated by the external magnetic field. It is caused by excitation of surface wave at the graphene boundary.

4. Conclusion

In this paper the transmission properties of photonic crystal structures on basis of graphene were investigated for terahertz frequency range. The possibility of transmission control of photonic crystal bounded by graphene layer under magnetic field influence was shown. The surface wave in the PC-graphene-PC bandgap structure was excited. The tuning of frequency position of transmitivity peak caused by surface wave excitation under magnetic field impact was demonstrated.

This work was financially supported by Government of Russian Federation, Grant 074-U01.

References

[1] Alessandro M 2012 *Photonic Crystals - Introduction, Applications and Theory* (InTech) p 344
[2] Girich A, Khodzitsky M, Nedukh S, Tarapov S 2011Experimental analysis of metamaterials' spectra to design tunable THz-GHz passive devices *NATO Science for Peace and Security Series B: Physics and Biophysics* pp. 159 – 164
[3] Tarapov S, Khodzitsky M, Chernovtsev S, Belozorov D, Merzlikin A, Dorofeenko A, Vinogradov A, Granovsky A 2010 Frequency control of the microwave Tamm state *Physics of the Solid State* **52** 7
[4] Jamie H 2013 *Graphene. Fundamentals and emergent applications* (Amsterdam: Elsevier) p 470
[5] Bludov Yu V, Peres N M R and Vasilevskiy M I 2013 Unusual reflection of electromagnetic radiation from a stack of graphene layers at oblique incidence *Journal of Optics* **15** 13
[6] Bludov Y V, Ferreira A, Peres N M R and Vasilevskiy M I 2013 A primer on surface plasmon-polaritons in graphene *J. Mod. Phys. B* **27** 1341001
[7] Bludov Yu V, Peres N M R 2013 Enhancing the absorption of graphene in the terahertz range *Europhysics Letters* **101**
[8] Grischkowsky D, Keiding S, Martin van Exter, Fattinger Ch 1990 Far-infrared time-domain spectroscopy with terahertz beams of dielectrics and semiconductors *J. Opt. Soc. Am. B.* **7** 10
[9] Rana F 2008 Graphene Terahertz Plasmon Oscillators *IEEE Trans. NanoTechnol.* **7** 91
[10] Bludov Yu V, Peres N M R and Vasilevskiy M 2010 Mechanism for graphene-based optoelectronic switches by tuning surface plasmon-polaritons in monolayer graphene *Europhysics Letters* **92** 68001
[11] Bludov Yu V, Peres N M R and Vasilevskiy 2012 Graphene-based polaritonic crystal *Phys. Rev. B* **85** 245409
[12] Jukam N and Sherwin M S 2003 Two-dimensional terahertz photonic crystals fabricated by deep reactive ion etching in Si *Applied Physics Letters* **83** 21
[13] Pitarke J M, Silkin V M, Chulkov E V and Echenique P M 2007 *Reports on Progress in PhysicsEmail alert RSS feed* **70** 1
[14] Ju L, Geng B, Horng J, Girit C, Martin M, Hao Z3, Bechtel H A, Liang X, Zettl A, Shen Y R and Wang F 2011 Graphene plasmonics for tunable terahertz metamaterials *Nature Nanotechnology* **6** 630–634
[15] Ferreira A, Viana-Gomes J, Bludov Y V, Pereira V, Peres N M R and Neto A H C 2011 Faraday effect in graphene enclosed in an optical cavity and the equation of motion method for the study of magneto-optical transport in solids *Phys. Rev. B* **84** 235410
[16] Denisultanov A Kh, Khodzitsky M K 2013 Band-gap structure for N-fold layers inside bilayer cell of photonic crystal for THz frequencies *Proceedings of SPIE* **8806** 88062I

Journal of Physics: Conference Series **541** (2014) 012059 doi:10.1088/1742-6596/541/1/012059

Investigation of device imperfection influence on measurement results in beamsplitter-based quantum random number generation schemes

A E Ivanova, S A Chivilikhin, V I Egorov and A V Gleim

St.Petersburg National Research University of Information Technologies, Mechanics and Optics, Kronverkskiy pr. 49, St.Petersburg 197101, Russian Federation

E-mail: `newiva@mail.ru`

Abstract. Quantum random number generation allows acquiring true random numbers which can be used in applications that require a high degree of randomness. This paper assesses the influence of non-ideal scheme parameters on measurement results in two quantum random number generation schemes: based on laser radiation splitting, and using vacuum fluctuations.

1. Introduction

Random number generation can be performed using algorithms, but the resulting sequences are pseudo-random by their nature and hence are not suitable for applications requiring a high degree of randomness, such as quantum cryptography [1]. For these applications it is necessary to use "true" random numbers produced by non-deterministic physical processes [2], for example, of quantum nature [3].

Existing approaches to quantum random number generation include the use of entangled photon states [4], processes of photon emission and detection [5], quantum noise of a laser [6], and separation of radiation [7]. An alternative approach is to use vacuum fluctuations of electromagnetic field [8].

Any practical random number generator operates a bit differently compared to its idealistic description. Hence, it is important to model the effects of any possible disruptions and study their impact on the output statistical features. This work is focused on investigating the case when the beamsplitter division angle is disturbed. Calculations are performed for two different methods of generation, the first based on laser radiation splitting, and the second on using vacuum fluctuations in homodyne detection. Both considered schemes include a beam splitter, and our model can be further extended to any kind of setup based on using this element.

2. Quantum random number generation based on radiation division by a beam splitter

For investigating distribution statistics obtained using a beam splitter, a probabilistic process simulation was performed. In the model, laser radiation is separated by a beam splitter and detected in two channels simultaneously (figure 1). It is shown that when radiation characterized by Poisson distribution with parameter α (describing mean photon number) passes through a beam splitter with angle θ, one of the beam splitter outputs is characterized by Poisson distribution with parameter $\alpha_1 = \alpha \cos\theta$, and the other - with $\alpha_2 = \alpha \sin\theta$.

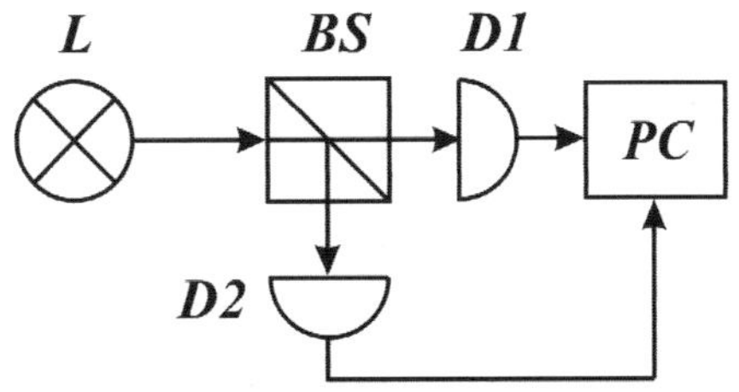

Figure 1. Scheme of random number generation based on radiation division by a beam splitter: L - laser, BS - beam splitter, D1, D2 - detectors, PC - computer.

Processing the two sequences obtained after separating radiation is performed as follows: if the photons are not detected on the first beam splitter output, and any number of them is detected on the second output, then the resulting sequence bit is considered "0", in the opposite case "1", otherwise the bit is not determined.

It was shown that for an asymmetric beam splitter a degree of final distribution uniformity depends on the beam splitter angle. By increasing deviation of the beam splitter angle (from $\theta = 45°$ value), we increase the difference in zeros and ones generation probability in the final sequence. It is necessary to calculate the beam splitter angular deviation, at which the final binary sequence is random. Statistical parameters of the binary distribution obtained from an asymmetric beam splitter are shown in figure 2. Generated sequences successfully pass tests of randomness, if the angular deviation from $\theta = 45°$ does not exceed two degrees.

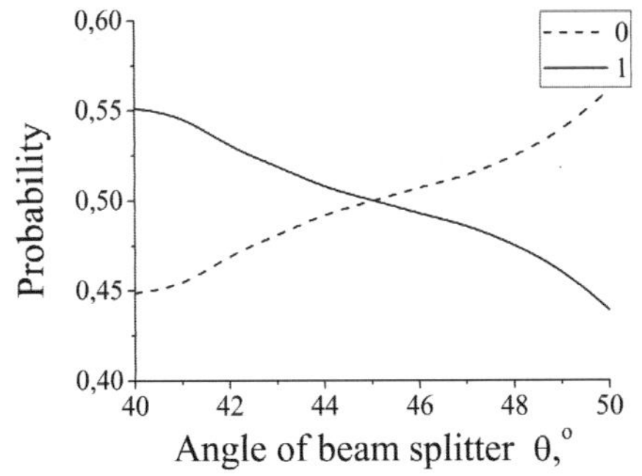

Figure 2. "Zero" and "One" occurrence probabilities in the resulting sequence, depending on the beam splitter angle.

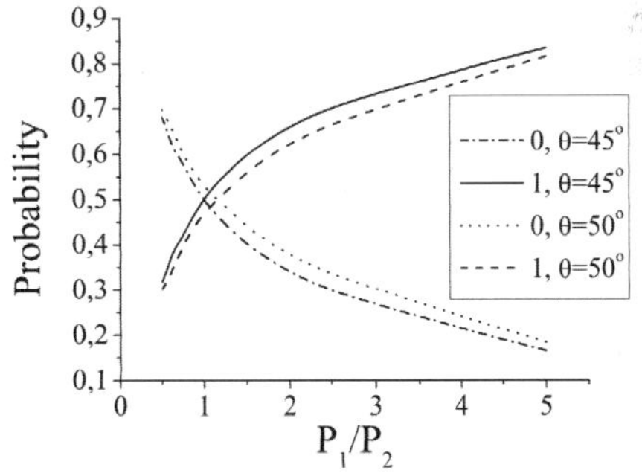

Figure 3. Probability of "0" and "1" bits occurrence in the sequence obtained using beam splitters with angles $\theta = 45°$ and $\theta = 50°$, probability of detection of the first detector $P_1 = 10\%$.

Detector parameters affect the quality of generated sequences of random numbers. When both detectors quantum efficiencies are equal, some random samples in two sequences, obtained by the beam splitter outputs zeroed. In this case, quality of the resulting sequence is not decreasing, because the changes are random and quantum efficiencies are equal for both detectors. We also considered a situation when quantum efficiencies of two detectors are different. In this case, one of the sequences produced by the beam splitter will contain more zero samples than the other. The difference in detectors parameters changes the ratio of ones and zeros in the final sequence, thus affecting the average of distribution for a given beam splitter angle, and the quality of the resulting sequence. Figure 3 illustrates the probability of "0" or "1" bits occurrence depending on detection probability ratio of the first and second detectors, P_1 and P_2. As one can see, asymmetric detectors can compensate the difference in probabilities of an asymmetric beam splitter, if their parameters are properly selected. For example, if after the beam division in one of the output sequences the amount of nonzero samples is greater than in the other, but it is detected by a device with a larger number of zero counts, an optimal balance between the probability of occurrence of zeros and ones in the final sequence of bits can be achieved.

3. Quantum random number generation based on the principles of homodyne detection

The principle of this scheme lies in extracting randomness from quantum noise that appears upon subtracting balanced detector signals received from the beam splitter outputs (figure 4). To the first splitter input a coherent state is send, and the other - a vacuum state. In this case interaction between strong laser signal and vacuum fluctuations appears in the splitter. Random numbers are obtained as a result of the received differential signal processing.

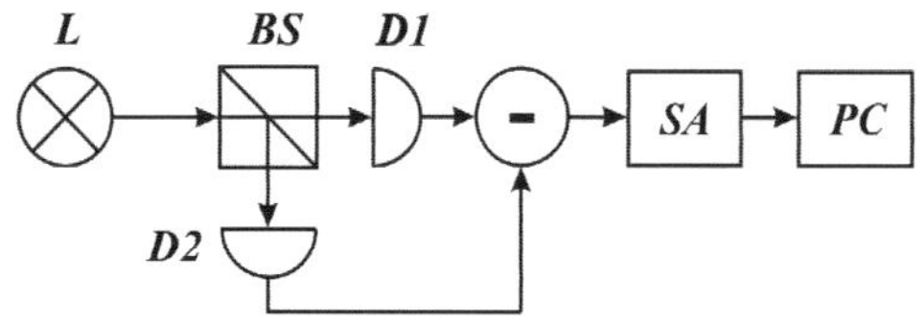

Figure 4. Scheme of random number generation based on homodyne detection principles: L - laser, BS - beam splitter, D1, D2 - detectors, SA - spectrum analyzer, PC - computer.

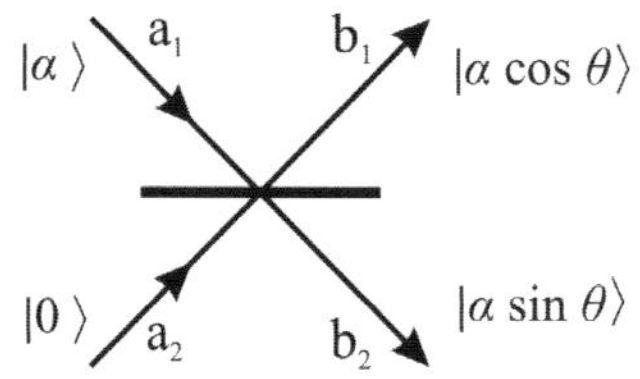

Figure 5. Scheme of a beam splitter with angle θ, where a coherent state is send to the first splitter input, and a vacuum state - to the other.

The input beam splitter signals are a_1 and a_2, the output signals are b_1 and b_2, respectively (figure 5). The output signals for the given input signals are expressed as follows

$$b_1 = a_1 \cos\theta - a_2 \sin\theta,$$

$$b_2 = a_1 \sin\theta - a_2 \cos\theta.$$

Radiation is characterized by Poisson distribution with parameter α (describing mean photon number), which in operator form is described as follows

$$|\alpha\rangle = e^{\alpha a_1^+ - \alpha^* a_1}|0\rangle,$$

where a_1^+ and a_1 are photon creation and annihilation operators at the first input of the beam splitter, $|\alpha\rangle$ - coherent state, $|0\rangle$ - vacuum state.

When a coherent state $|\alpha\rangle$ is send to the first splitter input, and a vacuum state $|0\rangle$ is send to the other, beam splitter input signal is expressed as a tensor product

$$|\alpha\rangle|0\rangle = e^{\alpha a_1^+ - \alpha^* a_1}|0\rangle_1|0\rangle_2.$$

Radiation characterized by Poisson distribution with parameter $|\alpha\rangle$ passes through a beam splitter with angle θ, one of the beam splitter outputs is characterized by Poisson distribution with parameter $|a\cos\theta\rangle$, and the other with $|a\sin\theta\rangle$. Photocurrents at the first and second detectors are defined as

$$i_1 = \gamma_1 b_1^+ b_1,$$
$$i_2 = \gamma_2 b_2^+ b_2,$$

where γ_1, γ_2 are quantum efficiencies of first and second detectors. Mean value of differential current can be expressed as follows

$$\langle \Delta i\rangle = \alpha^2(\gamma_2 \sin^2\theta - \gamma_1 \cos^2\theta).$$

In case of using a symmetric beam splitter and detectors with equal quantum efficiency mean value of the differential current was determined to be zero. We also calculated the amplitude of its deviation, which appears to be directly proportional to the intensity of laser radiation at the splitter input. For an asymmetric beam splitter and detectors with different quantum efficiencies the amplitude of differential current deviation is characterized by the following equation

$$\delta i = \alpha\sqrt{\gamma_2^2 \sin^2\theta + \gamma_1^2 \cos^2\theta},$$

where γ_1, γ_2 are quantum efficiencies of the detectors.

Relative deviation of the differential current on the deviations of detectors and beam splitter parameters dependence takes the following form

$$\frac{\Delta(\delta i)}{(\delta i)_0} = \frac{1}{2}\left(\frac{\Delta\gamma_1 + \Delta\gamma_2}{\gamma} + \frac{\Delta\gamma_1^2 + \Delta\gamma_2^2}{2\gamma^2} + \sin(2\Delta\theta)\left(\frac{\Delta\gamma_2 - \Delta\gamma_1}{\gamma} + \frac{\Delta\gamma_1^2 - \Delta\gamma_2^2}{2\gamma^2}\right)\right),$$

where $\gamma_1 = \gamma + \Delta\gamma_1$, $\gamma_2 = \gamma + \Delta\gamma_2$ are quantum efficiencies of the detectors, $\theta = \frac{\pi}{4} + \Delta\theta$ is beam splitter angle.

4. Conclusions

We investigated influence of non-ideal scheme parameters of quantum random number generation on the measurement results. Two types of generators were considered: based on beam separation and vacuum fluctuations. We show that the sequences generated by beam splitting successfully pass randomness tests, if the deviation of the beam splitter angle θ remains no more than two degrees. It was determined that in general inequality of detector parameters degrades the quality of generated sequences, but at the same time careful selection of their parameters can compensate asymmetry of the beam splitter. For quantum random number generation scheme using homodyne detection expressions describing the relationship between beam splitter input radiation and differential current were obtained. We also derived equations allowing estimation of the scheme parameters imperfection impact on the measurement results.

Acknowledgments

This work was financially supported by Government of Russian Federation, Grant 074-U01.

References

[1] Scarani V, Bechmann-Pasquinucci H, Cerf N J, Duek M, Ltkenahus N and Peev M 2009 *Rev. Mod. Phys.* **81** 1301
[2] Argyris A, Deligiannidis S, Pikasis E, Bogris A and Syvridis D 2010 *Optics Express* **18** 18763
[3] Jennewein T, Achleitner U, Weihs G, Weinfurter H and Zeilinger A 2000 *Rev. Sci. Instrum* **71** 1675
[4] Kwon O, Cho Y-W and Kim Y-H 2009 *Appl. Opt.* B **48** 1774
[5] Stipevi M and Rogina M B 2007 *Rev. Sci. Instrum.* **78** 045104
[6] Qi B, Chi Y-M, Lo H-K and Qian L 2010 *Optics Letters* **35** 312
[7] Ivanova A E, Egorov V I, Chivilikhin S A and Gleim A V 2013 *Nanosist.: fiz. him. mat.* **4** 550
[8] Shen Y, Tian L and Zou H 2010 *Phys. Rev.* **81** 063814

Optimization of the structure nanoporous TiO_2 film in a dye-sensitized solar cell

S P Malyukov and I V Kulikova and A V Sayenko and Yu V Klunnikova

Department of Electronic Apparatuses Design, Southern Federal University, Taganrog, Nekrasovsky, 44, Russia

E-mail: alex_rus.07@mail.ru

Abstract. The numerical model of dye-sensitized solar cell based on TiO_2 presented here allow to construct the theoretical I-V characteristics, determine the photovoltaic parameters and obtain important information about the main factors that influence DSSC performance. For modeling of solar cell was considered system of one-dimensional differential continuity equations, describing the charge transfer in its structure. The simulated results allow to determine the optimal TiO_2 particle diameter (radius), thickness, porosity and topological dimensions TiO_2 film.

1. Introduction

Dye-sensitized solar cells (DSSC) based on nanoporous titanium dioxide (TiO_2) are promising devices for effective, inexpensive and stable conversion of solar energy to electricity. However, a significant disadvantage of these solar cells is relatively low efficiency about 10-11 % [1].

The main component of DSSC design is nanoporous TiO_2 film, because it has a large surface area for adsorption of solar radiation-absorbing dye molecules, as well as the transport medium for generated electrons [2]. Thus, main efforts should be directed at optimizing the nanoporous structure of TiO_2 film, which will improve their photovoltaic characteristics and parameters.

2. Numerical modeling of the DSSC

The aim of this work is to develop a numerical model DSSC based on TiO_2, linking materials design parameters of the solar cell with its photoelectric behavior (I-V characteristics) and ensures the optimum TiO_2 particle diameter, thickness, porosity and topological dimensions TiO_2 film.

Figure 1 shows a design of the modeled DSSC based on TiO_2. The photoelectrode solar cell consists of a homogeneous nanoporous TiO_2 structure, with thickness d and porosity p, where a monolayer of a light-absorbing dye is adsorbed and interpenetrated by the redox mediator of iodide/tri-iodide (I^-/I_3^-) [3]. This structure is deposited on a conductive glass substrate (TCO), which defines the interface TCO/TiO_2 at x = 0. This interface is modeled as an ideal ohmic contact. The electrolyte/platinized TCO interface (counter electrode) at x = d is described by the Butler-Volmer equation electrochemical kinetic. The mobile charges considered in the present model are the electrons in the TiO_2 conduction band and the iodide and triiodide ions in the electrolyte. The charge transport in semiconductors can take place via diffusion and migration. However, due to the small dimension of the TiO_2 particles, no significant electrical potential gradient is generated, resulting only in a diffusive charge transport.

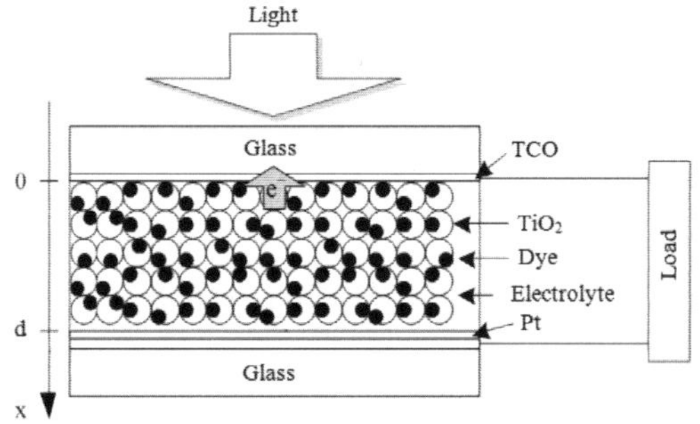

Figure 1. *Design of the modeled DSSC based on TiO$_2$*

The basis of the numerical model DSSC is a one-dimensional stationary system of differential continuity equations [4] describing the generation, transport and recombination of charge carriers in the solar cell design:

$$\begin{cases} D_{e^-}\dfrac{d^2n_{e^-}(x)}{dx^2} + G_{e^-}(x) - R_{e^-}(x) = 0; \\ D_{I^-}\dfrac{d^2n_{I^-}(x)}{dx^2} + G_{I^-}(x) - R_{I^-}(x) = 0; \\ D_{I_3^-}\dfrac{d^2n_{I_3^-}(x)}{dx^2} + G_{I_3^-}(x) - R_{I_3^-}(x) = 0; \end{cases} \quad (1)$$

where D_i is diffusion coefficient of charge carriers, n_i is density of charge carriers, x is coordinate as shown in Figure 1. The first term on the left-hand side of the equation represents the diffusion change in the concentration of charge carriers in the bulk. The second and third terms are the generation and recombination rates of charge carriers, respectively.

An integral part of the model is also the diffusion transport equation for the electron current density within the TiO$_2$ film towards the TCO [4]:

$$J_{int} = qD_{e^-}\frac{dn_{e^-}(x)}{dx}, \quad (2)$$

where J is internal current density (TCO/TiO$_2$ boundary), q is elementary charge.

In DSSC the absorption of each photon by a dye molecule is assumed to result in the electron injection into the TiO$_2$ conduction band and dye molecules regeneration according to the chemical reaction:

$$3I^- - 2e^- \rightarrow I_3^-. \quad (3)$$

Under these assumptions the generation $(G_{e^-} - R_{e^-})$ of two electrons is accompanied by generation of one triiodide ion and recombination of three iodide ions [4]:

$$(G_{e^-} - R_{e^-}) = -\frac{2}{3}(G_{I^-} - R_{I^-}) = 2\left(G_{I_3^-} - R_{I_3^-}\right). \quad (4)$$

The electron generation term in equation (8) is given by the Beer–Bouguer–Lambert law, which relates the absorption of light to the properties of the material through which the light is travelling:

$$G_{e^-} = \eta_{inj}\alpha(1 - T)\Phi e^{-\alpha x}, \quad (5)$$

where η_{inj} is electron injection efficiency from excited dye molecules to the TiO$_2$ conduction band, T is reflectance of the front surface (glass-TCO), α is absorption coefficient of the dye-sensitized nanoporous TiO$_2$ film, Φ is incident photon flux density in the range of the dye molecules absorption.

The value of the absorption coefficient of the dye-sensitized nanoporous TiO$_2$ was calculated using the expression obtained on the bases of an overlap model of TiO$_2$ nanoparticles (figure 2) [5]:

$$\alpha = 3 \cdot \varepsilon \cdot \ln 10 \cdot (1 - p) \cdot \delta_{dye} \cdot \left(\frac{1}{r} - \frac{N \cdot (2\pi r(h + Y) - 2\pi Xh)}{4\pi r^3}\right), \quad (6)$$

where ε is decimal molar absorption coefficient, d is thickness of the TiO$_2$ film, l is length of the TiO$_2$ film, w is the width of the TiO$_2$ film, r is radius of the TiO$_2$ nanoparticle, δ_{dye} is concentration of dye molecules on a flat surface, N is coordination number of particles, showing how many contacts have one particle with neighboring depending on porosity, r is radius of the TiO_2 particles, Y is overlap between two particles (typically 0.4 nm [2]), X is radius of the neck between two particles.

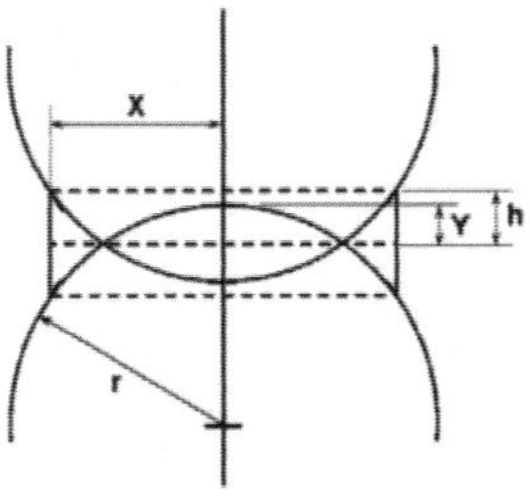

Figure 2. *Model of an overlap between two TiO_2 nanoparticles*

Incident photon flux density in the range of the dye molecules absorption was calculated according to the following expression obtained on the basis of Planck's law of blackbody radiation:

$$\Phi = \int_{\lambda_{min}}^{\lambda_{max}} C\left(\frac{R_S}{r_0}\right)^2 \left(\frac{2\pi c}{\lambda^4}\frac{F}{e^{hc/\lambda kT} - 1}\right) d\lambda, \quad (7)$$

where C is coefficient equal to 0.75, reflecting the influence of the Earth's atmosphere to the power flux density of solar radiation, R_S is radius of the Sun, T is temperature of the surface of the Sun (5780 K), r_0 is the average distance from the Earth to Sun, F is shading coefficient, λ is wavelength.

The recombination rate of electrons with triiodide ions follows a semiconductor theory:

$$R_{e^-} = \frac{n_{e^-}(x) - n_{e^-}^0}{\tau_{e^-}}, \quad (8)$$

where τ_{e^-} is electron lifetime, $n_{e^-}^0$ is dark equilibrium electron density in TiO_2.

As a result of current flow in a photoelectrochemical DSSC with a closed external circuit (short-circuit and operating conditions) an overvoltage occurs on the counter electrode, which is caused by the slowness of the charge transfer reaction in the iodide ions reduction. Overvoltage should be small because it represents a loss of voltage in the solar cell. Thus, to overcome the problems associated with the slowness of the charge transfer reaction, it is necessary to use a platinum catalyst. The electrolyte/Pt-TCO interface is an electrochemical half-cell, so it can be described by the Butler – Volmer equation linking the electrode current density ($J_{ext} = J_{int} - J_{sh}$, where J_{sh} – loss of the current density through the shunt resistor) with overvoltage η_{Pt} and concentration of charge carriers near counter electrode surface ($n_{I^-}(d)$ and $n_{I_3^-}(d)$):

$$J_{ext} = J_0\left[\frac{n_{I_3^-}(d)}{n_{I_3^-}^{oc}(d)} e^{\frac{\beta q \eta_{Pt}}{kT}} - \frac{n_{I^-}(d)}{n_{I^-}^{oc}(d)} e^{\frac{-(1-\beta) q \eta_{Pt}}{kT}}\right], \quad (9)$$

where J_0 is exchange current density at Pt electrode, β is symmetry coefficient, $n_{I_3^-}^{oc}(d)$ и $n_{I^-}^{oc}(d)$ is charge carrier concentration near the counter electrode surface.

The internal voltage of a DSSC is defined as the difference in electrochemical potentials between the boundaries x = 0 and x = d. Bearing in mind the assumptions of the present model concerning the TCO/TiO_2 interface, the respective electrochemical potential is assumed to be continuous and equals the electron Fermi level at x = 0. On the other hand, the electrochemical potential of the platinum electrode at x = d equals the electrolyte redox energy. The value of the internal voltage U_{int} can then be calculated as:

$$U_{int} = \frac{E_F(0) - E_{Redox}(d)}{q}, \quad (10)$$

where E_F is Fermi energy in TiO_2, E_{Redox} is redox energy in electrolyte.

Then value of the external voltage U_{ext} on a load can be calculated as:

$$U_{ext} = U_{int} - U_s = U_{int} - J_{ext}R_sS, \quad (11)$$

where U_S is voltage on the series resistance in the cell, S is active cell area (TiO_2 film), R_S is series resistance in the cell.

The value of the series resistance R_p was calculated according to the parameters of topological solar cell:

$$R_p = \frac{2R_\square}{w}\left(\frac{l}{2} + l_c\right), \tag{12}$$

where $R_\square$ is sheet resistance of TCO, l is length of the TiO_2 film, w is width of the TiO_2 film, l_c is path of electrons in the TCO film to an external contact.

The system of differential equations (1) DSSC model is solved numerically using the method of finite differences for the three basic modes of operation (open-circuit, short-circuit and operating modes). For example, boundary conditions at operating modes can be written as follows:

$$\begin{cases} \dfrac{dn_{e^-}(0)}{dx} = \dfrac{J_{int}}{q\,D_{e^-}}; \quad \dfrac{dn_{e^-}(d)}{dx} = 0; \\ \dfrac{dn_{I^-}(0)}{dx} = 0; \quad \displaystyle\int_0^d n_{I^-}(x)dx = n^0_{I^-}d; \\ \dfrac{dn_{I_3^-}(0)}{dx} = 0; \quad \displaystyle\int_0^d n_{I_3^-}(x)dx = n^0_{I_3^-}d. \end{cases} \tag{13}$$

The presented model allows to obtain a steady-state I-V characteristics DSSC and define its basic parameters (efficiency, short-circuit current density, open-circuit voltage).

3. Results

The I-V characteristics were obtained by the simulation at the different topological dimensions of active area (TiO_2 film) of the solar cell (figure 3). The I-V characteristics shifted to the left by increasing the active area due to the increase of the voltage drop on the series resistance (voltage loss on TCO). As the result for the parameters optimization and the efficiency increase of DSSC it is necessary topological dimensions of the active area, corresponding to the length of 0.2-0.7 cm (efficiency above 10 %) and virtually unlimited width. Length and width of the active area less than 0.2 cm causes considerable technical difficulties in its manufacture (sealing, drilling required to fill the electrolyte).

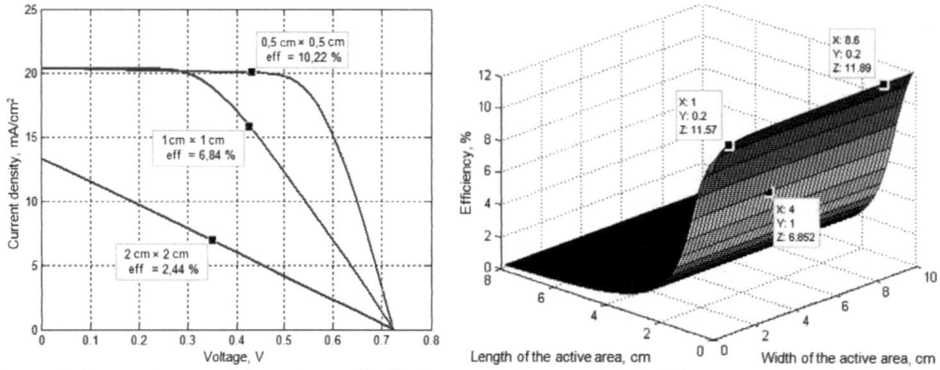

Figure 3. *Dependence of the solar cell efficiency on the topological dimensions of the active area*

Figure 4 shows the dependence of the DSSC efficiency on the TiO_2 film thickness. Amount of absorbed photons increases with increasing thickness of the TiO_2, consequently, increases the concentration of the generated electrons. However, increasing the thickness of the TiO_2 increased recombination and loss of generated electrons. Therefore, competition between the rates of generation and recombination electrons determines the optimum TiO_2 film thickness, and then begins to decrease the solar cell efficiency. Thus, the optimal TiO_2 film thickness is about 5-10 μm.

Journal of Physics: Conference Series **541** (2014) 012060 doi:10.1088/1742-6596/541/1/012060

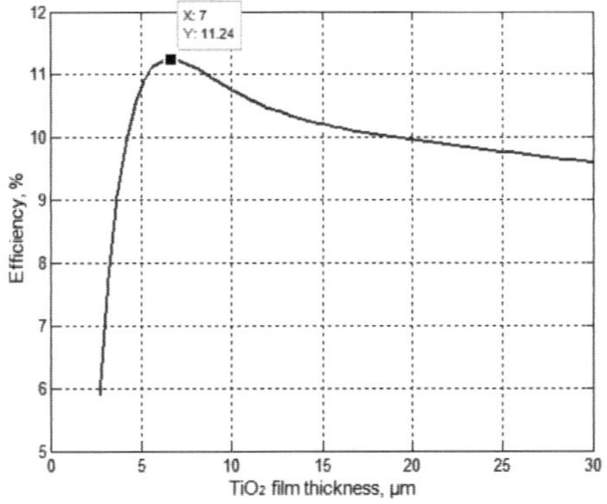

Figure 4. *Dependence of the solar cell efficiency on the TiO_2 film thickness*

Figure 5 shows the dependence of the efficiency of the solar cell on the radius of TiO_2 nanoparticles. Decrease in the radius of nanoparticles in the TiO_2 film with its constant thickness and porosity leads to an increase in short-circuit current density and efficiency of the solar cell, because it increases the surface area of the TiO_2 film, suitable for adsorption of solar radiation absorbing dye molecules. Thus, reducing the radius of the TiO_2 nanoparticles leads to an increase concentration generated electrons in the dye-sensitized TiO_2 film.

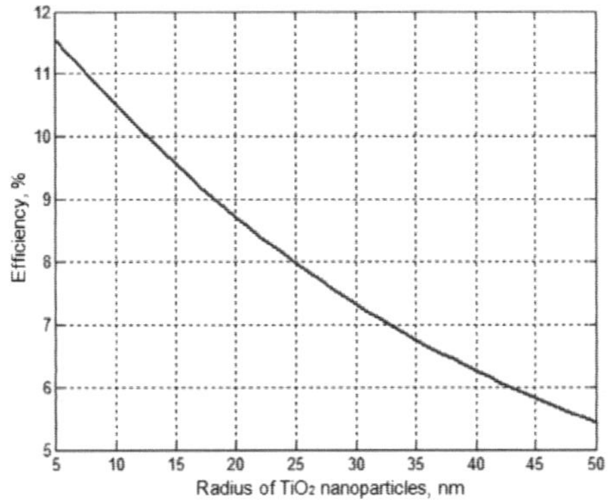

Figure 5. *Dependence of the solar cell efficiency on the radius of TiO_2 nanoparticles*

The greatest interest is the dependence of the efficiency on the TiO_2 film porosity (figure 6).

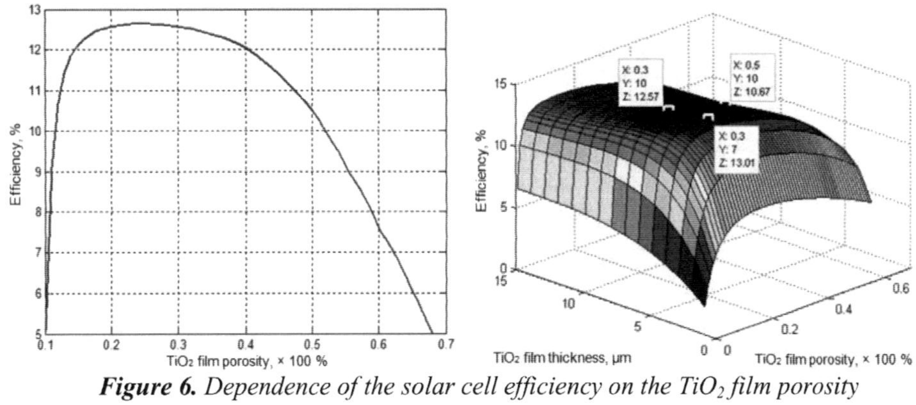

Figure 6. *Dependence of the solar cell efficiency on the TiO_2 film porosity*

Figure 5 shows that the increase in the porosity of TiO_2 electrode leads at first to a sharp increase in efficiency, achieve a maximum efficiency at 20-30 % and then decrease. The sharp increase of the efficiency in the low porosity determined a sharp increase in the solar absorption coefficient of dye-sensitized TiO_2 film due to its high internal surface area and, correspondingly, increasing the concentration of the generated electrons. With increasing porosity of TiO_2 films above 20-30 % generated electron concentration decreases. It is determined that with higher porosity the TiO_2 film consists of a smaller number nanoparticles and has a smaller surface area, which reduces light absorption coefficient and of electron diffusion in the TiO_2 film [2].

Thus, the results of modeling showed that the optimal size of topological TiO_2 film has a length 0.2-0.7 cm and unlimited width, while the optimal thickness of TiO_2 film is 5-10 μm, TiO_2 particle diameter is 10-30 nm (radius 5-15 nm) and porosity is 20-30 %. Achieving optimal parameters of nanoporous TiO_2 films improves DSSC efficiency 15-20 % in comparison with the efficiency at the standard value of porosity 50-55 % (12.57-13.01 % versus 10.67 %).

References

[1] Gratzel M 2003 Dye-sensitized solar cells *Journal of Photochemistry and Photobiology C: Photochemistry Reviews*, **4**, 145.

[2] Malyukov S P, Sayenko A V 2013 Laser sintering of a porous TiO_2 film in dye-sensitized solar cells *Journal of Russian Laser Research*, **34**, 531.

[3] Onodera M, Ogiya K 2010 Modeling of dye-sensitized solar cells based on TiO_2 electrode structure model *Japanese Journal of Applied Physics*, **49**, 73.

[4] Topic M, Campa A, Filipic M, Berginc M, Krasovec U O, Smole F 2010 Optical and electrical modeling and characterization of dye-sensitized solar cells *Current Applied Physics*, **10**, 425.

[5] Malyukov S P, Sayenko A V 2010 Simulation of light absorption TiO_2 film in dye-sensitized solar cell *Izvestiya SFedU. Engineering sciences*, **12**, 148.

Bent and branched chains of nanoresonators

A S Melikhova[1], **I Yu Popov**[1]

[1] ITMO University, Kronverkskiy pr., 49, St. Petersburg, 197101, Russia

E-mail: alina.s.melikhova@gmail.com, popov1955@gmail.com

Abstract. We study the spectral problem for bent and branched chains of weakly coupled conglobate resonators. At the joint points the δ-coupling is assumed. Our approach is based on the theory of self-adjoint extensions of symmetric operators and transfer matrix method. The structure of the spectrum is described. For the both cases it is proved that the Hamiltonian has negative eigenvalue for some values of the model parameters.

1. Introduction

One of newly created nanostructures is a chain of weakly coupled conglobate quantum resonators. An example (not unique but very interesting) of such system is so-called nano peapod (see, e.g., [1] – [5]). It is a nanotube filled by a chain of fullerene molecules. In this paper we suggest solvable model which describes spectral properties of an electron in the system of such type. Two geometries of chains are under discussion: the bent (section 2) and the branched (section 3) chains of weakly coupled resonators. Weak coupling means that neighbouring resonators interact with each other via point-like window (see, e.g., [6, 7]). Also δ-coupling in contact points of resonators is assumed. Similar problems for quantum graphs have been recently discussed (see, e.g., [8] – [10]).

We construct a model for chains of weakly coupled resonators, based on the theory of self-adjoint extensions of symmetric operators (see, [11]) and solve the spectral problem for the model Hamiltonian. Roughly speaking, this method consists of restriction of the initial operator $\hat{H}$ and it's further extension. The scheme of this procedure for simple cases you can find in [11, 12]. In our case the initial operator is the orthogonal sum of the Neumann Laplacians (we assume a system of units with $\hbar = 2m = e = 1$) for all balls that make up our system. We restrict it on the set of all smooth functions from domain of the initial operator that are equal to zero at contact points of the resonators. Its closure is the symmetric operator with infinite deficiency indices (Green's functions play the role of deficiency elements). To get the self-adjoint extension $-\tilde{\Delta}_0$ it is necessary to establish some linear relation between the coefficients of the asymptotics of functions from the domain of the adjoint operator near the contact points. Moreover, we modify the relation assuming that there is the δ-coupling condition with coupling constant α.

2. The bent chain

In this section we consider the chain with a kink of angle β (where $\beta \in \left[0; \frac{2\pi}{3}\right)$) (see, Figure 1). By analysing the direct part of such chain with paying attention to δ-coupling condition at the

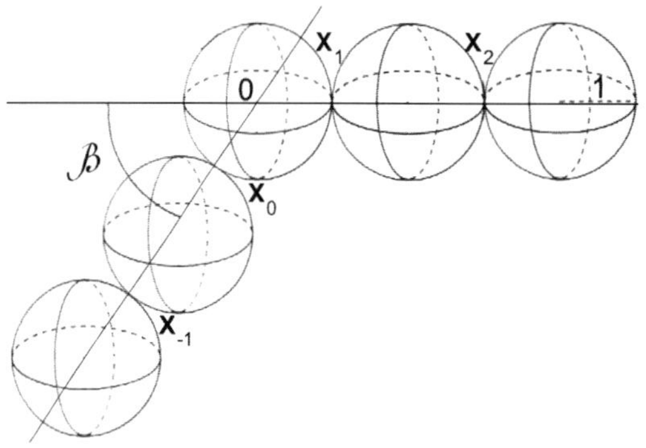

Figure 1. The bent system of weakly coupled conglobate resonators with a kink of angle β.

contact points, one can obtain the following expression for the transfer matrix:

$$\mathbf{M} = \begin{pmatrix} 0 & -1 \\ \frac{G(\mathbf{x}_{j-1},\mathbf{x}_j,\lambda)}{G(\mathbf{x}_{j+1},\mathbf{x}_j,\lambda)} & \frac{-\alpha+2\lim\limits_{\mathbf{x}\to\mathbf{x}_j}(G(\mathbf{x},\mathbf{x}_j,\lambda)-G(\mathbf{x},\mathbf{x}_j,\lambda_0))}{G(\mathbf{x}_{j+1},\mathbf{x}_j,\lambda)} \end{pmatrix}, \qquad (1)$$

where $G(\mathbf{x}_i, \mathbf{x}_j, \lambda)$ are the Green's functions; $\mathbf{x}_j$ are the contact points between resonators with numbers $(j-1)$ and j; λ_0 $(\lambda_0 < 0)$ is a parameter of the point-like window between resonators. As our chain has a kink one should investigate the place of perturbation in more details and after serial cross-linking of resonators with numbers 1, 0 and -1 one can obtain the following spectral equation:

$$G^2(\mathbf{x}_0(\beta), \mathbf{x}_1, \lambda) = G^2(\mathbf{x}_1, \mathbf{x}_2, \lambda). \qquad (2)$$

Let's analyse solutions of equation (2). First of all, as the explicit form of the transfer matrix (1) is known, we can state that solutions λ belong to the continuous spectrum if:

$$\left| \frac{-\alpha + 2\lim\limits_{\mathbf{x}\to\mathbf{x}_j}(G(\mathbf{x},\mathbf{x}_j,\lambda) - G(\mathbf{x},\mathbf{x}_j,\lambda_0))}{2G(\mathbf{x}_1,\mathbf{x}_2,\lambda)} \right| \leq 1. \qquad (3)$$

One can be referred to the discrete spectrum in the case when the above mentioned modulus is grater than 1. Secondly, it can be shown that when β equals to 0 equation (2) turns into true identity for all $\lambda \in \mathbf{R}$. So that we obtain that the model operator for straight infinite chain has empty discrete spectrum and band continuous spectrum (the bands are given by (3)).

Now we assume the existence of a bent. It means that there is a shift of point $\mathbf{x}_0$ with angle parameter β (where $\beta \in \left(0; \frac{2\pi}{3}\right)$). In this case the difference between the resolvents corresponding to the Hamiltonians of the unperturbed and the perturbed systems has finite rank, therefore, the continuous spectrum is kept. And new eigenvalues that come from this perturbation can appear only outside the bands. First, consider the negative half-axis λ. The lower bound λ_{thr} (we assume that it is small and negative) of the continuous spectrum in this case can be described as follows:

$$\lambda_{thr} = \frac{\alpha - \frac{3}{2\pi\lambda_0} + 2\sum\limits_{k=1,l=0}^{\infty} \frac{N_{lk}}{x_{lk}^2}\left(\frac{\lambda_0}{x_{lk}^2-\lambda_0} + (-1)^l\right)}{2\sum\limits_{k=1,l=0}^{\infty} \frac{N_{lk}(1-(-1)^l)}{x_{lk}^4}}, \qquad (4)$$

where x_{lk} is k-th root of the equation: $j_l'(x) = 0$, and N_{lk} are normalization factors. For small λ we can get one negative solution of (2):

$$\lambda_- = -\frac{\sum\limits_{k=1,l=0}^{\infty} \frac{N_{lk}}{x_{lk}^2}\left(P_l(-\cos\beta) - (-1)^l\right)}{\sum\limits_{k=1,l=0}^{\infty} \frac{N_{lk}}{x_{lk}^4}\left(P_l(-\cos\beta) - (-1)^l\right)},$$

and for some values of model parameters $\lambda_- < \lambda_{thr}$.

Now consider the positive half-axis λ. Due to the similar type of equation (2) and equation on the bands edges (that can be obtained from (3) by replacing the inequality sign in this equation to an equal sign) and the behaviour of the Green's function between consequent eigenvalues of the Neumann Laplacian of the ball one can conclude that there are at most two required eigenvalues in the gaps. As a result, we come to the following theorem.

Theorem 1. *The essential spectrum of the model Hamiltonian for bent chain of weakly coupled conglobate resonators consists of the continuous spectrum having band structure (the bands are given by equation (3)) and eigenvalues of infinite multiplicity which are given by the eigenvalues of the Neumann Laplacian for the ball corresponding to the eigenfunctions vanishing at the both opposite points of the ball ($\mathbf{x}_j$ and $\mathbf{x}_{j-1}$). If the bent angle β of the chain equals to 0 the discrete spectrum of the model operator is empty. Let the bent angle β of the chain belongs to $(0; \frac{2\pi}{3})$. There exist values of the model parameters β, α, λ_0 such that the model Hamiltonian has negative eigenvalue λ_- below the threshold λ_{thr}. For each gap at the positive spectral half-axis there are values of the model parameter α ensuring the existence of one or two eigenvalues in this gap.*

3. The branched chain

In this section we consider the Y-branched chain (see, Figure 2). This chain can be considered as

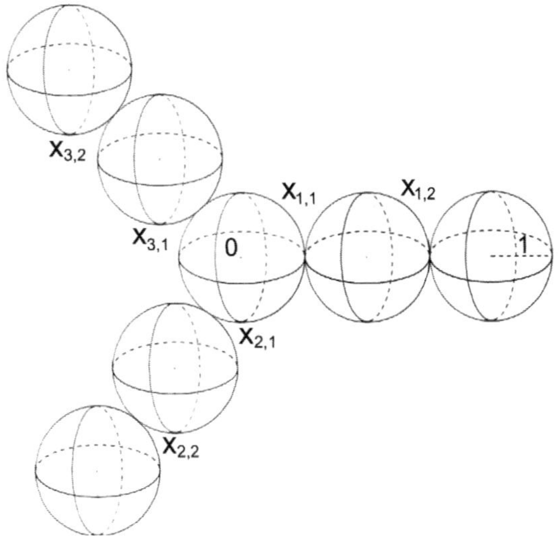

Figure 2. The symmetric Y-type branched system of weakly coupled conglobate resonators.

three semi-infinite straight chains of conglobate resonators which are connected to the "central" ball. We also assume that the system has symmetry with respect to the rotation by the angles

$\frac{2\pi}{3}, \frac{4\pi}{3}$ and that it can be laid on the plane. To identify balls of the chain we will use two indices: i — for branch's number ($i = 1, 2, 3$), j — for ball's number in one branch ($j \in \mathbf{N}$); the central ball has one index — 0.

The transfer matrix for this case has the form analogous to that in the previous case (up to balls identifying) (see (1). Cross-linking of coefficients $a^{\pm}_{i,1}$ and $a^{-}_{i,0}$ ($i = 1, 2, 3$) that refer to balls with numbers $(i, 1)$ and 0, correspondingly, leads one to the following system:

$$\begin{cases} \begin{pmatrix} a^+_{1,1} \\ a^-_{1,1} \end{pmatrix} = \begin{pmatrix} -1 \\ \frac{X}{G} \end{pmatrix} a^-_{1,0} + \begin{pmatrix} 0 \\ \frac{A}{G} \end{pmatrix} a^-_{2,0} + \begin{pmatrix} 0 \\ \frac{A}{G} \end{pmatrix} a^-_{3,0} \\ \begin{pmatrix} a^+_{2,1} \\ a^-_{2,1} \end{pmatrix} = \begin{pmatrix} 0 \\ \frac{A}{G} \end{pmatrix} a^-_{1,0} + \begin{pmatrix} -1 \\ \frac{X}{G} \end{pmatrix} a^-_{2,0} + \begin{pmatrix} 0 \\ \frac{A}{G} \end{pmatrix} a^-_{3,0} \\ \begin{pmatrix} a^+_{3,1} \\ a^-_{3,1} \end{pmatrix} = \begin{pmatrix} 0 \\ \frac{A}{G} \end{pmatrix} a^-_{1,0} + \begin{pmatrix} 0 \\ \frac{A}{G} \end{pmatrix} a^-_{2,0} + \begin{pmatrix} -1 \\ \frac{X}{G} \end{pmatrix} a^-_{3,0} \end{cases}, \tag{5}$$

where $X = -\alpha + 2 \lim_{\mathbf{x} \to \mathbf{x}_{i,1}} (G(\mathbf{x}, \mathbf{x}_{i,1}, \lambda) - G(\mathbf{x}, \mathbf{x}_{i,1}, \lambda_0))$, $G = G(\mathbf{x}_{i,1}, \mathbf{x}_{i,2}, \lambda)$, $A = G(\mathbf{x}_{2,1}, \mathbf{x}_{1,1}, \lambda)$. More detailed study of system (5) allows one to derive the equation for the Hamiltonian's eigenvalues λ:

$$(X - \mu G)^3 - 3(X - \mu G)A^2 + 2A^3 = 0, \tag{6}$$

where μ is an eigenvalue of the transfer matrix for the straight chain. The solutions of (6) belong to the continuous spectrum if:

$$\left| \frac{X}{G} \right| \leq 2, \tag{7}$$

and belong to the discrete spectrum in the following case: $|X/G| > 2$.

AS for the continuous spectrum, one can see that (7) coincides with (3), that is, the continuous spectrum is the same as in the previous case (it is natural, because local perturbation can not lead to variation of the continuous spectrum). Examine the discrete spectrum of this system. It is clear that equation (6) is a homogeneous equation with respect to $\frac{X-\mu G}{A}$ and its solutions are well known:

$$t = \frac{X - \mu G}{A} = \begin{bmatrix} -2 & (mult.\ 1), \\ 1 & (mult.\ 2). \end{bmatrix}$$

Taking into account the explicit form of μ one transforms (6) to the following equation:

$$t^2 A^2 + G^2 - tAX = 0,$$

where A, G and X are the Green's functions with positive poles x^2_{lk}. So that (due to the Green's function behaviour), one can conclude that this equation has at most one negative solution. For small negative λ one can obtain solutions of equation (6) in the following form:

$$\lambda_t = \frac{\frac{3}{8\pi\lambda_0}(2t - (t+1)\lambda_0)}{-\frac{t\alpha}{2} - \frac{3t}{4\pi\lambda_0} - \sum\limits_{k=1,l=0}^{\infty} \frac{N_{lk}}{x^2_{lk}} \left((-1)^l + \frac{t\lambda_0}{x^2_{lk} - \lambda_0} \right)}. \tag{8}$$

For certain values of parameters α, λ_0 and t derived eigenvalue λ_t can be made less than threshold λ_{thr} (as we have analogical transfer matrix in both cases of perturbation and branching, conditions (3) and (7) are equal, so the value of threshold in case of branching can be given by (4) as well). As for positive half-axis there are at most two eigenvalues in a gap. This is true due to the behaviour of the Green's functions mentioned in (6) between its poles. The result of this section is summarized in the theorem 2.

Theorem 2. *The essential spectrum of the model for Y-branched chain of weakly coupled conglobate resonators consists of the continuous spectrum having band structure (the bands are given by equation (7)) and eigenvalues of infinite multiplicity which are given by the eigenvalues of the Neumann Laplacian for the ball corresponding to the eigenfunctions vanishing at the both opposite points of the ball ($\mathbf{x}_{i,j}$ and $\mathbf{x}_{i,j-1}$). There exist values of the model parameters α, λ_0 such that the model Hamiltonian has at most one negative eigenvalue λ_t and at most two positive eigenvalues in each gap.*

Acknowledgments
The work was partially financially supported by the Government of the Russian Federation (grant 074-U01), by the Ministry of Science and Education of the Russian Federation (GOSZADANIE 2014/190, Project 14.Z50.31.0031), by grant of Russian Foundation for Basic Researches and grants of the President of Russia (state contracts 14.124.13.2045-MK and 14.124.13.1493-MK).

References
[1] Rotkin S V, Subramoney Sh 2006 *Applied Physics of Carbon Nanotubes: Fundamentals of Theory, Optics and Transport Devices* (Berlin: Springer)
[2] Yang Y, Li L and Li W 2013 *J. Phys. Chem.* C **117**, 14142–48
[3] Enyashin A N and Ivanovskii A L 2010 *Nanosystems: Phys. Chem. Math.* **1** 63–71.
[4] Krivanek O L, Dellby N, Murfitt M F, Chisholm M F, Pennycook T J, Suenaga K, Nicolosi V 2010 *Ultramicoscopy* **110** 935–45
[5] Abou-Hamad E, Kim Yo, Talyzin A V, Goze-Bac Ch, Luzzi D E, Rubio A and Wagberg T 2009 *J. Phys. Chem.* C **113** 8583–87
[6] Popov I Yu 1992 *J. Math. Phys.* **33** 3794–801
[7] Popov I Yu, Popova S L *Europhys. Lett.* **24** 373–7
[8] Duclos P, Exner P, Turek O 2008 *J. Phys. A: Math. Theor.* **41** 415206
[9] Kuchment P, Vainberg B 2006 *Commun. Math. Phys.* **268** 673
[10] Popov I Yu, Smirnov P I 2013 *Phys. Lett.* A **377** 439–42
[11] Pavlov B S 1987 *Russian Math. Surveys* **42** 127–68
[12] Albeverio S, Gesztesy F, Hoegh-Krohn R, Holden H 1988 *Solvable Models in Quantum Mechanics* (New York: Springer)

Detection of beamsplitting attack in a quantum cryptographic channel based on photon number statistics monitoring

A A Gaidash, V I Egorov, A V Gleim

St.Petersburg National Research University of Information Technologies, Mechanics and Optics, Kronverkskiy pr. 49, St.Petersburg 197101, Russian Federation

E-mail: andrewdgk@gmail.com

Abstract. Quantum cryptography in theory allows distributing secure keys between two users so that any performed eavesdropping attempt would be immediately discovered. However, in practice an eavesdropper can obtain key information from multi-photon states when attenuated laser radiation is used as a source. In order to overcome this possibility, it is generally suggested to implement special cryptographic protocols, like decoy states or SARG04. We present an alternative method based on monitoring photon number statistics after detection. This method can therefore be used with any existing protocol.

1. Introduction

Quantum cryptography systems [1] allow performing secure quantum key distribution between two or more users. The use of single photons in transmission technology provides the legitimate users (Alice and Bob) with an ability to detect eavesdropper (Eve) by monitoring quantum bit error level (QBER) on receiver side. However, in practice full security can only be guaranteed when a «true» single photon source is used [2-6], e.g. based on spontaneous parametric down-conversion [7]. Unfortunately, today these devices do not provide necessary key generation rate and stability; also they are available only in laboratory conditions. For this reason, most quantum cryptography experimental setups [1] and commercial products [8] use attenuated laser light as a source of quantum states with average probability of single photon emission per timeframe ("mean photon number") about $\mu \approx 0.1$. In this case the security condition is no longer strict due to Poisson distribution of photons of coherent light: some pulses may contain more than one photon. Probability of n-photon state emission is described as follows:

$$P(\mu, n) = \frac{\mu^n}{n!} \cdot \exp(\mu),$$

where μ – average probability of single photon emission per timeframe, n – photon number.

This fact can be easily used by Eve to successfully perform undetectable beam-splitting or photon-number splitting (PNS) attack [9] without changing QBER, and receive a part of the key which can be significant at higher μ.

In this work, we present a new method of revealing PNS attacks by monitoring the statistics of multi-photon states on receiver side. We suggest using detectors which distinguish multi-photon states and can accumulate statistics. In case of PNS attack there will be different distribution of number of multi-photon states unlike it is expected.

2. PNS attack

In our research we use an analytical model of Eve which consists of three main elements: a device which can determine multi-photon pulses without measuring [9]; a single-photon detector with high quantum efficiency (up to 100%) and a controlled beam splitter. Eve performs the PNS attack by splitting and measuring the phonons from multi-photon pulses.

For preliminary calculations if we take into account losses, there only will be higher probability of detecting multi-photon states. Emitted n-photon state has a probability to be detected:

$$\eta(n) = 1 - (1 - \eta)^n,$$

where η is total loss.

And Poisson distribution on Bob's side should be completed as:

$$P^*(\mu, n) = \frac{\mu^n}{n!} \cdot \exp(\mu) \cdot \eta(n)$$

But in this paper losses are not taken into consideration for better clarity because of theoretical priority of work. So we assume that probability of n-photon state emission equals to an appropriate probability of detection.

We estimated the impact of attack on key security by calculating the percentage of multi-photon states for different μ and the resulting fraction of key information for several widely-used two-state and four-state protocols, the most common of which are B92 and BB84. Percentage of multi-photon states $Q(\mu)$ is represented as a relation of multi-photon states detection probability and detection probability of any photon number states:

$$Q_{EVE}(\mu) = \frac{(1 - P(\mu,0) - P(\mu,1)) \cdot N}{(1 - P(\mu,0)) \cdot N} = \frac{1 - P(\mu,0) - P(\mu,1)}{1 - P(\mu,0)}$$

where $P(\mu,0)$ – detection probability of timeframe with no photons, $P(\mu,1)$ – detection probability of timeframe with one photon, N – total amount of timeframes.

In particular, in case of most common μ = 0.1 Eve can easily obtain photons from 4.9% of all timeframes. If Eve has at least one photon from current timeframe, she will know the bit in case of two-state protocol: if eavesdropper chose wrong basis, she can change it to opposite. This means that obtained photons from 4.9% of all timeframes represent equal amount 4.9% of key. In case of four-state protocols Eve has a 50% chance to chose wrong basis, so there is in two times less 2.5% of the key used respectively.

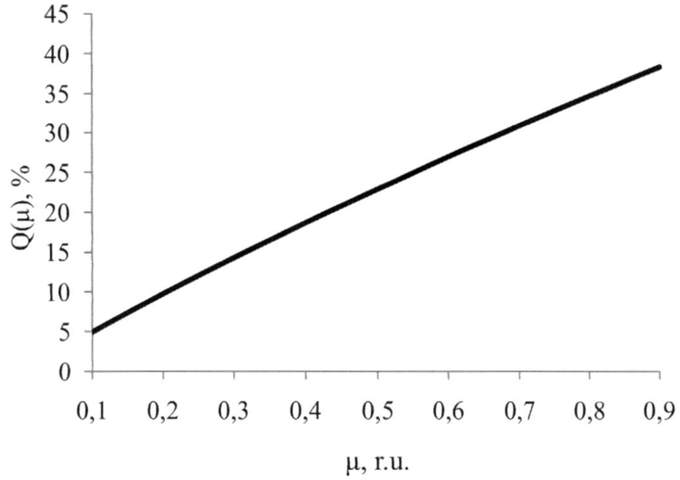

Figure 1. Percentage of multi-photon timeframes $Q(\mu)$ depends on μ. For higher μ there is a huge part of multi-photon states.

Table 1. Table represents that Eve can obtain percentage $K_{B92}(\mu)$ and $K_{BB84}(\mu)$ of the key used for two-state and four-state protocols respectively.

μ	0.1	0.2	0.3	0.4	0.5	0.6	0.7	0.8	0.9
K_{B92} (%)	4.9	9.7	14.3	18.7	22.9	27.0	30.9	34.7	38.3
K_{BB84} (%)	2.5	4.8	7.1	9.3	11.5	13.5	15.5	17.4	19.2

Today, two methods are known to overcome the PNS attack: Decoy States [10] and SARG04 [11] protocol. Although both of them were shown to be effective, they have certain drawbacks, increasing the complexity of cryptographic protocol and at the same time lowering key generation rate. Moreover, they cannot be directly applied to some setups [12].

3. Method description

It is known that some types of single photon detectors are able to distinguish multi-photon states upon detection [13, 14], thus gathering information not only about the fact of photon existence in a certain timeframe, but also about the mean number of photons arriving at the detector. In particular, superconducting nanowire single photon detectors [15] can be easily adjusted so that they would detect only the states with the number of photons exceeding the defined minimum. These detectors are also characterized by low noise (about 10 Hz) and high quantum efficiency (up to 57% at 1550 nm [14]), which makes them perfect for quantum cryptography applications.

In this work, we suggest an alternative method of monitoring the received photon number statistics by detectors which distinguish multi-photon states.

In order to uncover this attack, Alice and Bob first define input μ as well as total loss in the communication line, which allows them to calculate modified Poisson distribution at Bob's side taking losses into account. In this work we assume that input Poisson distribution does not modify in case of lossless setup. Then, using known equations for Poisson statistics, probabilities of detecting of n photons ($n > 1$) are derived. At some point of quantum key distribution process Bob adjusts the detector in order to gather statistics of multi-photon pulse arrival probability. For example, in case of a superconducting nanowire detector this can be done by changing the critical current. The received results are compared to calculated expectations and detector noise level. PNS attack is detected when certain deviations are discovered.

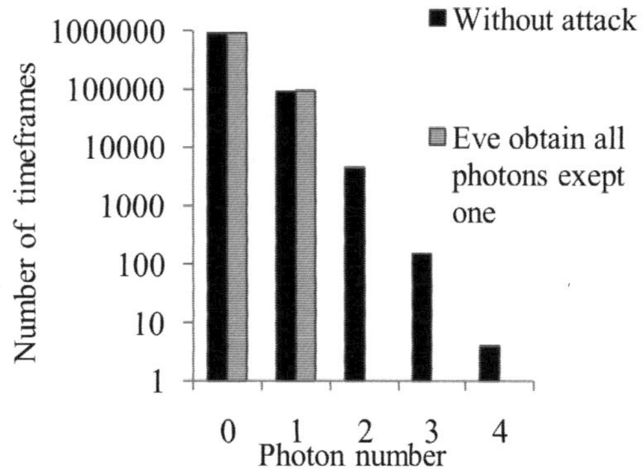

Figure 2. Number of timeframes on Bob's side as function of photons number for μ = 0.1 compared to modified distribution in case of attack, where Eve obtains all photons except one from each timeframe, which amount is 10^6

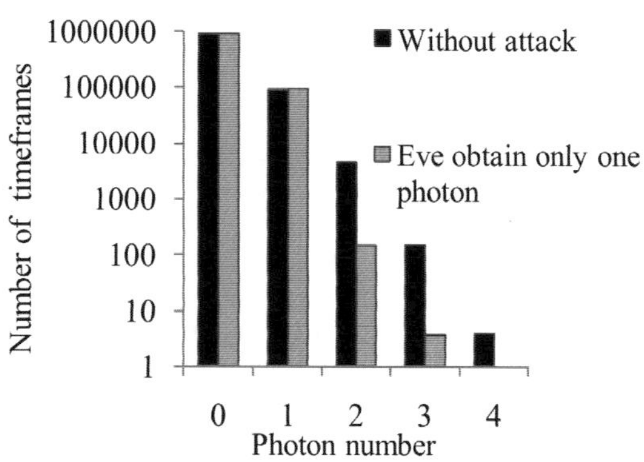

Figure 3. Number of timeframes on Bob's side as function of photons number for μ = 0.1 compared to modified distribution in case of attack, where Eve obtains only one photon from each timeframe, which amount is 10^6

So there are two extreme cases. In first one Eve obtain all photons except one, and in second case conversely obtains only one photon and leaves the rest.

In first case there will not be any timeframes with more than one photon. These changes greatly modify expected distribution and allows detect such PNS attack.

For second case the calculations show than if Eve splits only one photon from each multi-photon pulse, the number of pulses with $n > 1$ decreases more than one order of magnitude, while there appears to be a slight increase in single-photon states. These changes may be discovered when raw key rate is high enough to mask the detector noise.

If we take into account noise level on our detector about 10 Hz and necessary relation of signal to noise at least 100:1, so bitrates is to be at least 1000 bit/s. But if we want to discern the lost photons from about 5% (for $\mu = 0.1$) of all timeframes, we should rise bitrates up to 20 Kbit/s. Our estimations show that the technique can be used in systems with bitrates higher than 20 Kbit/s for two-state protocols, and 40Kbit/s for four-state protocols. Systems with those bitrates can be constructed for sure, because there are systems with bitrates up to 1Mbit/s [16].

In order to implement this method, taking into account losses, Alice and Bob must control the length of optical channel in order to make sure that Eve has not introduced a segment with lower losses.

4. Results and conclusions

As a result of research, a method of defending against PNS attacks on quantum key distribution system with coherent light source was developed. The use of detectors which distinguish multi-photon states allows revealing such attacks by analyzing statistics on receiver's side. In particular, we analysed the efficiency of PNS attack on quantum key distribution system with coherent light source and studied the ability of using these detectors to ensure security against PNS attack. The changes exceed the detector noise at bitrates higher than 20 Kbit/s and 40Kbit/s for two-state protocols and four-state protocols, respectively. Moreover, our method does not exclude possibility of combining with other known methods and can be used with any quantum cryptography setup.

Acknowledgements

This work was financially supported by Government of Russian Federation, Grant 074-U01.

References

[1] V. Scarani, H. Bechmann-Pasquinucci, N.J. Cerf *et al* 2009 *Rev. Mod. Phys.* **81**, pp 1301–1350
[2] D.C. Unitt, A.J. Bennett, P. Atkinson *et.al* 2005 *Journal of optics* **7**, pp 129
[3] Hartschuh, H. N. Pedrosa, J. Peterson, *et al* 2005 *Chem. Phys. Chem.* **6**, № 1.
[4] A. Hogele, C. Galland, M. Winger *et al* 2008 *Phys. Rev. Lett.* **100**, № 217401.
[5] T. B. Pittman, J. D. Franson, B. C. Jacobs 2007 *New J. Phys.* **9**, № 195.
[6] B. C. Jacobs, T. B. Pittman, J. D. Franson. 2006 *Phys. Rev.***74**, let. A, № 010303.
[7] A. Kalachev, D. Kalashnikov, A. Kalinkin *et al* 2007 *Laser Phys. Lett.* **4**, pp 722–725
[8] L. Lydersen, C. Wiechers, C. Wittmann *et al* 2010 *Nature Photonics* **4**, pp 686–689
[9] N. Gisin, G. Ribordy, W. Tittel *et al* 2002 *Reviews of modern physics* **74**, pp 181
[10] Hoi–Kwong Lo, Xiongfeng Ma, Kai Chen 2005 *Physical Review Letters* **94**, № 230504
[11] Valerio Scarani, Antonio Acin, Nicolas Gisin *et al* 2004 *Phys. Rev. Lett.* **92**, № 057901
[12] O. Guerreau, J.-M. Mérolla, A. Soujaeff *et al* 2003 *IEEE J.Sel.Top.Quantum Electron.* **9**, pp 1533-1540
[13] M. D. Eisaman, J. Fan, A. Migdall *et al* 2011 *Rev. Sci. Instrum.* **82**, № 071101
[14] Robert H. Hadfield 2009 *Nature photonics* **3**
[15] G. N. Gol'tsman, O. Okunev, G. Chulkova *et al* 2001 *Applied physics letters* **79**, № 6, pp 705
[16] Z. L. Yuan, A. R. Dixon, J. F. Dynes *et al* 2009 *New Journal of Physics* **11**, № 045019

An analytical method for determining the interaction energy between multiple identical spherical colloidal zinc oxide nanoparticles

A V Alfimov, E M Aryslanova, S A Chivilikhin

St. Petersburg National Research University of Information Technologies, Mechanics and Optics, Kronverkskiy, 49, St. Petersburg, 197101, Russia

E-mail: alfimov.anton@gmail.com

Abstract. A modification of the linear superposition approximation for the linearized Poisson-Boltzmann equation, describing the interaction of colloidal nanoparticles, is proposed based on the implementation of a phenomenological charge regulation model to allow for the more accurate description of the interaction at small particle separations and multi-particle interactions.

1. Introduction

The problem of analytical description of the interaction forces between colloidal nanoparticles is of a great importance for the modern science due to the wide variety of applications for nanocolloid-based materials and the problem of their toxicity [1 – 3].

The standard and well-established approach to describing the interaction between colloidal particles in electrolytes is the DLVO (Derjaguin, Landau, Verwey, and Overbeak) theory and various modifications and additions to it [1, 4]. The core of this interaction theory is the combination of short-range van der Waals attraction and long-range electrosteric repulsion due to the overlap of particles' electric double-layers (EDL). The spatial potential distribution in the EDLs of isolated and interacting particles is governed by the Poisson-Boltzmann equation, which is the Poisson equation with the volume charge density represented by the concentrations of positively and negatively charged ions obeying the Boltzmann distribution.

In the case of low surface potential φ_0 of particles: $\varphi_0 \ll kT / e$ ($\varphi_0 \ll 25$ mV at 25°C), where k - Boltzmann constant; T - temperature; e - elementary charge, the Poisson-Boltzmann equation allows for linearization and takes the form of:

$$\Delta\varphi = \kappa^2\varphi, \tag{1}$$

where $\kappa = \lambda_D^{-1}$ - inverse Debye screening length which for the simplest case of particles suspended in distilled water is given by $\kappa = (2e^2 n_0 / \varepsilon\varepsilon_0 kT)^{1/2}$, n_0 - concentration of H^+ and OH^- ions at an infinite separation from suspended particles.

The applicability of existing methods for solving the Poisson-Boltzmann problem (1) for the system under consideration is determined by the order of magnitude of the expression κa, where a is a particle radius. For the case of $\kappa a \ll 1$, which generally describes nanoparticle suspensions in weak electrolytes, the most well-studied method of Derjaguin's approximation and its modifications are not applicable. Thus the leading approach to the Poisson-Boltzmann equation for interacting nanoparticles when $\kappa a \ll 1$ is the linear superposition approximation (LSA), which suggests that the potential distribution in a system of multiple interacting particles can be presented as a sum of individual potentials of each particle involved when isolated from the others. Two main formulations of the LSA exist: the one by Bell, McCartney and Levine [5] and the one described by Ohshima [6].

Although the linearized equation (1) itself allows for the solution in a form of linear superposition of potential distributions due to each individual particle, the problem arises from the boundary conditions which are to be specified on the surfaces of every interacting particle. Thus the LSA is generally valid only at large separations between particles, when the disturbance of the surface potential of a particle by the proximity of another particle is negligible.

This paper presents a modified LSA approach (MLSA), where an effective single-particle potential distribution which reacts to the particle's surroundings through a phenomenological charge regulation model is used to compensate for the boundary conditions violation by the conventional LSA.

2. Methods

We consider a spherical nanoparticle with radius a and surface charge density σ suspended in distilled water (see figure 1). The EDL of the particle is assumed to be composed of only two main layers: the dense Stern layer of counter-ions with the thickness of δ (zone I on figure 1) and the diffuse layer of co- and counter-ions (zone II on figure 1). We make an assumption that the counter-ion concentration is distributed uniformly in the Stern layer and equal to N.

The nanoparticle and its Stern counter-ion shell forms the effective particle with radius $b=a+\delta$. We further assume that the hydrodynamic radius of the particle equals to the effective particle's radius b and, thus, the potential on the surface of the Stern layer is equal to the ς-potential.

Spherical ZnO nanoparticles were chosen as the modeling objects for the present work as it has been shown [7] that they have sufficiently low surface potential in distilled water with neutral pH.

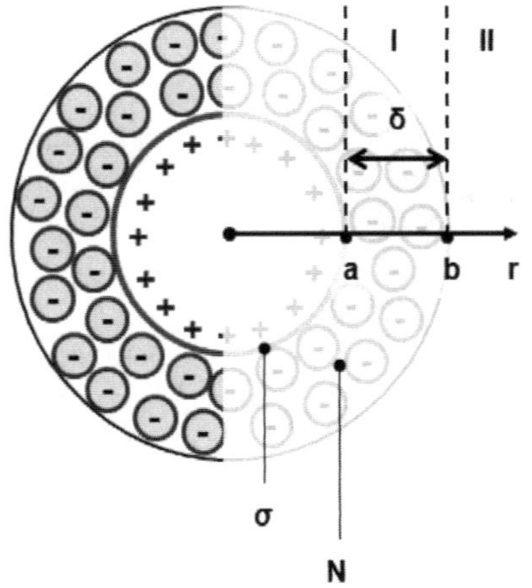

Figure 1. Schematic representation of a colloid ZnO nanoparticle and its electric double layer: I – dense Stern layer of counter-ions; II – diffuse layer of co- and counter-ions.

The Poisson-Boltzmann equations describing the discussed system are:

$$\begin{cases} \Delta\varphi_{\mathrm{I}} = eN\cdot(\varepsilon\varepsilon_0)^{-1} & a<r<b \\ \Delta\varphi_{\mathrm{II}} = \kappa^2\varphi_{\mathrm{II}} & r>b \end{cases} \tag{2}$$

with the boundary conditions:

$$\left.\frac{d}{dr}\varphi_{\mathrm{I}}\right|_{r=a} = -\frac{\sigma}{\varepsilon\varepsilon_0}, \quad \left.\frac{d}{dr}\varphi_{\mathrm{I}}\right|_{r=b} = \left.\frac{d}{dr}\varphi_{\mathrm{II}}\right|_{r=b}, \quad \varphi_{\mathrm{I}}(b)=\varphi_{\mathrm{II}}(b), \quad \varphi_{\mathrm{II}}\big|_{r\to\infty}\to 0, \tag{3}$$

where r - radial component of the spherical coordinate system connected with the centre of the particle.

The solution of the system (2) in the diffuse layer area II is given by:

$$\varphi_{II}(r) = \varphi_0 \cdot \frac{b}{r} \cdot \exp\left[-\kappa(r-b)\right], \tag{4}$$

where φ_0 - potential on the surface of the Stern layer (ς -potential):

$$\varphi_0 = \frac{1}{4\pi\varepsilon\varepsilon_0} \cdot \frac{1}{b(1+\kappa b)}\left(\sigma \cdot 4\pi a^2 - e \cdot \Omega\right), \tag{5}$$

and $\Omega = N \cdot (4/3)\pi\left(b^3 - a^3\right)$ - total quantity of counter-ions in the Stern layer.

Expression (5) needs the parameters describing charge distribution, σ (surface charge density) and Ω (ion quantity in the Stern layer), determined. To do so we introduce a phenomenological model of charge regulation which establishes the connection between the ambient concentrations of ions in the vicinity of the particle and the electrical charges on the surface and in the Stern shell of the particle.

The proposed linear model of nanoparticle's surface charge has the form of:

$$\sigma(\tilde{n}_0^+) = \sigma' + \sigma'' \cdot \tilde{n}_0^+, \tag{6}$$

where σ' is a constant part of the surface charge due to the surface features of the nanoparticle; $\sigma'' \cdot \tilde{n}_0^+$ is the part of the surface charge due to the adsorption of potential determining ions; $\tilde{n}_0^+$ is the ambient concentration of co-ions in the vicinity of the particle.

The linear model of the counter-ion total quantity in the dense shell is given by:

$$\Omega(\tilde{n}_0^-) = \omega \cdot \tilde{n}_0^-, \tag{7}$$

where $\tilde{n}_0^-$ is the ambient concentration of counter-ions in the vicinity of the particle.

Thus we obtain an expression for the effective surface potential (ς -potential) of a nanoparticle that is dependent on the ambient ion concentrations:

$$\varphi_0^{eff}\left(\tilde{n}_0^+, \tilde{n}_0^-\right) = \frac{1}{4\pi\varepsilon\varepsilon_0} \cdot \frac{1}{b(1+\kappa b)}\left[4\pi a^2(\sigma' + \sigma''\tilde{n}_0^+) - e\omega\tilde{n}_0^-\right]. \tag{8}$$

To find the three introduced charge regulation parameters, σ', σ'' and ω, one can use the experimental data on the ς -potential of the particles at different values of pH. Determining the co- and counter-ion concentrations at three different pH values and substituting them into the expression (8) with the according ς -potential values will result in a system of 3 linear algebraic equations with 3 variables. It should be noted that to find the charge regulation parameters one must also know or make an assumption on the Stern layer thickness for the three employed experimental cases.

With the connection between nanoparticle's surface potential and the ambient ion concentrations established, we now assume that under the linear superposition approximation the ion concentrations in the vicinity of a nanoparticle can be presented as a combination of the bulk concentrations of co- and counter-ions and the concentration perturbations due to the EDLs of the surrounding particles. Knowing that the ion concentrations in the EDLs of the surrounding particles are governed by the Boltzmann distribution an using the expression (8) and the $\kappa a \ll 1$ condition, we can derive an expression for the effective potential distribution of the nanoparticle under consideration:

$$\varphi^{eff}{}_i(r) = F_i\left(\left\{R_{i,j}\right\}\right) \cdot \frac{\exp(-\kappa r)}{r}, \tag{9}$$

where $\left\{R_{i,j}\right\}$ is a set of the center-to-center separations $R_{i,j}$ between the particle under consideration i and all the surrounding particles; the function F_i is determined by the effective potential distributions due to the surrounding particles:

$$F_i\left(\left\{R_{i,j}\right\}\right) = F_{ISO} - \eta \cdot \sum_{j \neq i} \varphi^{eff}{}_j\left(R_{i,j}\right), \tag{10}$$

where the F_i function value for the isolated particle F_{ISO} is given by:

$$F_{ISO} = \frac{1}{4\pi\varepsilon\varepsilon_0}\left[4\pi a^2(\sigma' + \sigma'' n_0) - e\omega n_0\right] \tag{11}$$

and

$$\eta = \frac{en_0}{4\pi\varepsilon\varepsilon_0 kT}\left[4\pi a^2 \cdot \sigma' - e\omega\right]. \tag{12}$$

Thus, combining the expressions (9) and (10), for a system of m interacting particles with the set of center-to-center separations $\{R_{i,j}\}$ we obtain a system of m linear algebraic equations with m variables:

$$F_i + \eta \cdot \sum_{j \neq i} F_j \cdot \frac{\exp(-\kappa R_{i,j})}{R_{i,j}} = F_{ISO}. \tag{13}$$

Solving the system (13) will allow for the determining of the effective potential distributions of each interacting particle using (9).

To find the potential energy of the multi-particle interaction for the nanoparticle under consideration i we use the Ohshima's LSA formulation [6]:

$$U_i\left(\{R_{i,j}\}\right) = Q^{eff}{}_i \cdot \sum_{j \neq i} \varphi^{eff}{}_j\left(R_{i,j}\right), \tag{14}$$

where $Q^{eff}{}_i = 4\pi\varepsilon\varepsilon_0 \cdot F_i$ - effective screened point-charge that creates the same potential distribution as the particle i.

3. Results and discussion

To illustrate the presented MLSA method we consider the problem of three interacting nanoparticles.

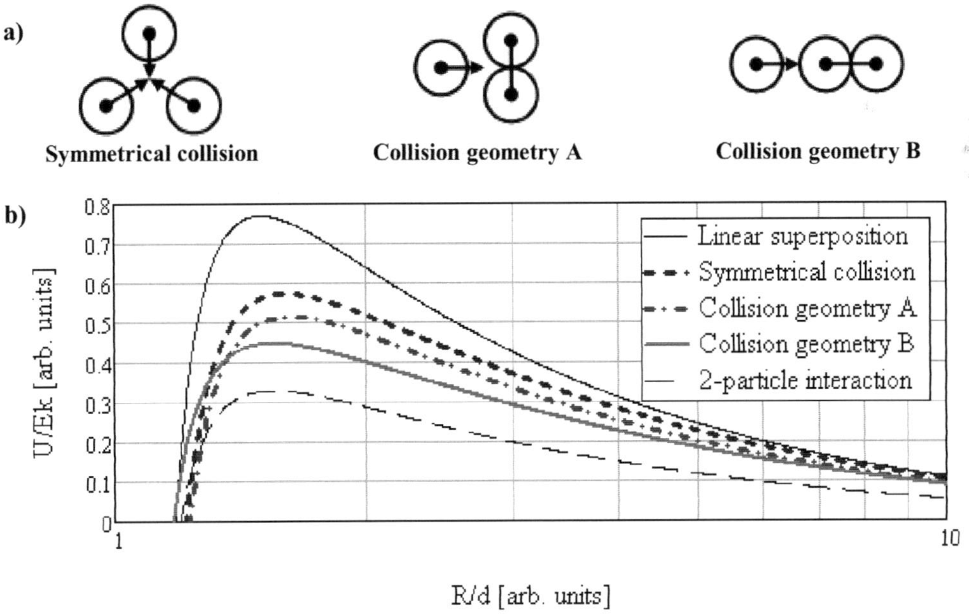

Figure 2. Three-particle interaction: a) considered interaction geometries; b) interaction potential energy U dependencies on the distance R between particles given by the LSA and the MLSA (for three nanoparticle collision geometries: symmetrical, A and B). Ek – mean square kinetic energy of nanoparticle's Brownian motion; d – particle's diameter.

Journal of Physics: Conference Series **541** (2014) 012063 doi:10.1088/1742-6596/541/1/012063

The solution to the system (13) for the case of three interacting nanoparticles has a form of:

$$F_i = F_{ISO} \cdot \frac{\left(\gamma_{i,j} + \gamma_{i,k} - \gamma_{j,k} - 1\right)\left(1 - \gamma_{j,k}\right)}{\left(\gamma_{i,j}\right)^2 + \left(\gamma_{i,k}\right)^2 + \left(\gamma_{j,k}\right)^2 - 2\gamma_{i,j} \cdot \gamma_{i,k} \cdot \gamma_{j,k} - 1} \tag{15}$$

where $\gamma_{i,j} = \eta \cdot \exp(-\kappa R_{i,j}) / R_{i,j}$.

We then use the experimental data from [7] to calculate the charge regulation parameters. To illustrate the obtained result (15) and compare it with the predictions of the conventional LSA for the case of three interacting nanoparticles we choose three specific geometries of nanoparticles collision (see figure 2,a): the symmetrical collision when the separations between each particle are equal; the geometry where separation between two particles is maintained constant and the third particle approaches them in the direction perpendicular to the line connecting "paired" particles' centers (collision geometry A on figure 2,a) and the geometry where the third particle approaches "paired particles" in the direction parallel to the line connecting their centers (collision geometry B on figure 2,a).

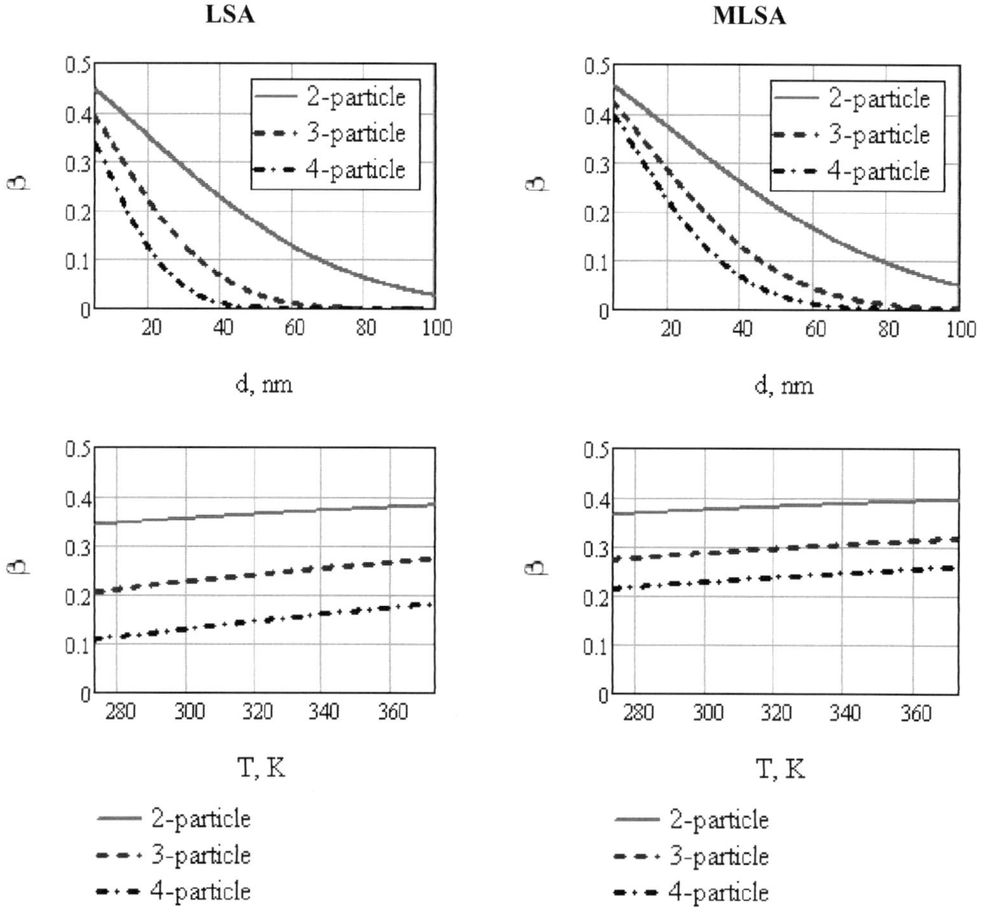

Figure 3. a, b) Attachment probability β dependencies on the particle diameter d for 2, 3 and 4 interacting nanoparticles obtained using the LSA (a) and the MLSA (b) methods. c, d) Attachment probability β dependencies on the temperature T for 2, 3 and 4 interacting nanoparticles obtained using the LSA (c) and the MLSA (d) methods. The symmetrical collision geometry was used for the multi-particle cases.

Curves showing the dependences of the nanoparticles' interaction potential on the separation between the particles obtained using the MLSA method for the three collision geometries discussed above are presented on the figure *2,a* in thick lines. The interaction potential for three nanoparticles derived using the conventional LSA is also displayed on the figure *2,a* in solid thin line; the pair interaction potential curve is also presented as a reference (thin dashed line on figure 2,a).

The potential curves presented on the figure 2 show that the increase in the potential barrier energy for the three-particle interaction in comparison with the pair interaction is sufficiently lower for the presented MLSA method than the results of the standard LSA model. This indicates that the MLSA model can overcome the problem of almost linear dependency of the barrier energy on the number of interacting particles observed in LSA. The differences between considered collision geometries also demonstrate the MLSAs ability to describe some of the orientation effects during nanocluster formation.

Assuming that the kinetic energy of nanoparticle's Brownian motion is distributed according to the Gaussian law with the mean square of $3kT/2$ ($kT/2$ per orthogonal direction), we can calculate the probability β of a nanoparticle acquiring the kinetic energy of motion towards the other particles sufficient to overcome the potential barrier. This probability has a meaning of an attachment probability. It is important to notice that due to the way the attachment probability β was introduced its maximum value equals to 0.5 as there is a 50% chance that the nanoparticle will move away from the other particles due to the random collisions with water molecules.

The dependencies of the attachment probability on the model parameters, i.e. nanoparticle size and temperature, obtained using the LSA and the MLSA for 2, 3 and 4 interacting particles are shown on figure 3. The symmetrical collision geometry was used for the multi-particle interactions. The figure shows that the MLSA generally results in smaller drop in attachment probability when considering more interacting particles when comparing it to the LSA results. Both models have good qualitative agreement with the experimental attachment probability dependancies on the temperature (see, for example, [8]).

Acknowledgments
This work was financially supported by the Government of Russian Federation, Grant 074-U01.

References

[1] Hotze E M, Phenrat T and Lowry G V 2010 *J Environ Qual* **39** 1909
[2] Zhang W, Crittenden J, Li K and Chen Y 2012 *Environ Sci Technol* **46** 7054
[3] Bian S-W, Mudunkotuwa I A, Rupasinghe T and Grassian V H 2011 *Langmuir* **27** 6059
[4] Dahirel V and Jardat M 2010 *Curr Opin Colloid In* **15** 2
[5] Bell G M, Levine S and McCartney L N 1970 *J Colloid Interf Sci* **33** 335
[6] Ohshima H 1995 *J Colloid Interf Sci* **174** 45
[7] Berg J M, Romoser A, Banerjee N, Zabda R and Sayes C M *et al* 2009 *Nanotoxicology* **3** 276
[8] Suganthi K S and Rajan K S 2012 *Int J Heat Mass Transfer* **55** 7969

Plasmon-Exciton Interaction in AuNP - Phthalocyanine Core/Shell Nanostructures.

A.Zasedatelev[1], A.Karpo[2], I. Feofanov[3], V. Krasovskii[1,2] and V. Pushkarev[4]

[1]National Research Nuclear University «MEPhI», Moscow, Russia
[2] A. M. Prokhorov General Physics Institute of the Russian Academy of Sciences, Moscow, Russia.
[3]P.N. Lebedev Physical Institute of the Russian Academy of Sciences, Moscow, Russia.
[4]Institute of Physiologically Active Compounds of the Russian Academy of Sciences, Chernogolovka, Russia.
E-mail: anton.zasedatelev@gmail.com

Abstract. We report results of theoretical modeling of photophysical properties of hybrid gold-core nanoparticles (AuNp) covered with the chlor-aluminium phthalocyanine (CAP) shell. The extinction spectra, the near field distribution and the CAP fluorescence quantum yield (QY) changing due to the vicinity to the AuNp were theoretically calculated for different values of the core/shell ratio parameter and compared to experimental data obtained for the structures under consideration.

1. Introduction

Noble metallic nanoparticles are promising building-block materials for various applications due to their tunable optical properties which depend on the size, shape and environment. Such particles support localized surface plasmon resonances, collective oscillations of surface electrons leading to the generation of a locally enhanced field. Placing chromophore molecules near to the metal surface, in the strong local field, can lead to the manifestation of such phenomena as modification of scattering and absorption spectra of the chromophore/metal structure in comparison with those of the constituents individually [1,2]; modification of the chromophore's fluorescent properties [3]; surface enhancement of Raman scattering (SERS) [4]; considerable enhancement and modification of nonlinear-optical properties and, in some cases, lasing in multilayer structures [5]. Recently, considerable interest has been focused on plasmonic nanoparticles covered with a shell of phthalocyanine (Pc) molecules. Such hybrid particles are considered as perspective structures for biology and medicine applications (e.g. photodynamic therapy of oncologic diseases), optoelectronics and photovoltaics, various gas sensors and etc. In spite of the great interest displayed by researchers to the hybrid plasmonic nanostructures [6, 7] little attention is paid to the explanation and modeling of the photophysical mechanisms occurring in such systems.

2. Theoretical model

The scheme of the model of the interaction between AuNp and CAP is presented in Fig.1. In order to explain linear optical properties of AuNp-CAP particles such as absorption, scattering and luminescence we used the joint solution approach of the following two problems. The first one is the problem of the light scattering by a multilayer spherical particle represented in Ref. 8. Applying this approach one can consider the AuNp-CAP particle as a two-layer spherical particle and calculate its near field distribution. At this step it is necessary to evaluate correct values of complex dielectric permittivity of each layer, the AuNp core and the CAP shell. The values of the AuNP dielectric permittivity were evaluated with the following analytical expression [10].

$$\varepsilon_m(\omega, r) = \varepsilon_{bulk}(\omega) + \omega_p^2 \left(\frac{1}{\omega^2 + i\omega\gamma_{\text{int}ra}^{bulk}} - \frac{1}{\omega^2 + i\omega\gamma_{\text{int}ra}^{(r)}} \right),$$

$$\gamma_{\text{int}ra}^{(r)} = \gamma_{\text{int}ra}^{bulk} + \xi \frac{\upsilon_F}{r} \tag{1}$$

where $\varepsilon_{bulk}(\omega)$ is a dielectric constant of the bulk metal, ω_p is a plasmon frequency for bulk metal, $\gamma_{\text{int}ra}^{(r)}$ is an effective damping constant (related to the frequency of electron - metal/chromophore interface scattering), $\gamma_{\text{int}ra}^{bulk}$ is a bulk metal damping constant (related to the frequency of electron scattering in the bulk metal), ξ is a parameter which is dependent on the details of the scattering process (e.g. isotropic or diffuse scattering, ξ is 1 in our case), υ_F is the velocity of the electrons at the Fermi energy, r is the radius of metal particle. The expression (1) includes the size adjusted Drude term and the interband transitions term describing the electron transitions from d – valence band to SP – conduction band in the short wavelength region. The average size of the AuNp core was equal to 18 nm, as was obtained by a scanning electron microscope. The permittivity of the CAP shell was taken equal to that of zinc Pc [11] due to nearly identical optical properties of these two compounds. CAP layer thickness was taken to be equal to 2.5.

The second step is to determine relative changes in the relaxation rates of the CAP molecule in the previously calculated near-field. In our procedure the outer shell of CAP molecules was considered as the layer of two-level quantum systems. It is known, that intensity of luminescence is proportional to rate of luminescence (γ_{em}) Thus, the equation for the luminescence intensity of CAP placed near to the AuNP was confronted to the luminescence intensity of CAP in free space determined by the following expression [9]

$$\frac{\gamma_{em}}{\gamma_{em}^0} = \frac{\gamma_{exc}[\gamma_r / \gamma]}{\gamma_{exc}^0[\gamma_r^0 / \gamma^0]} = \frac{\gamma_{exc}}{\gamma_{exc}^0} \frac{QY}{QY^0} \tag{2}$$

where $\gamma_{em}, \gamma_{ex}, \gamma_r$ are rates of luminescence, excitation, spontaneous emission, respectively, γ is the total decay rate and QY is the quantum yield of CAP molecules. The upper index "0" relates to the values of mentioned parameters in free space, whereas its absence relates to the parameters in the vicinity of the AuNp. Then the problem of changes in the rate of spontaneous decay of these systems in the self-consistent near-field was solved. This near-field was determined by solving the problem of the light scattering by the multilayer structure.

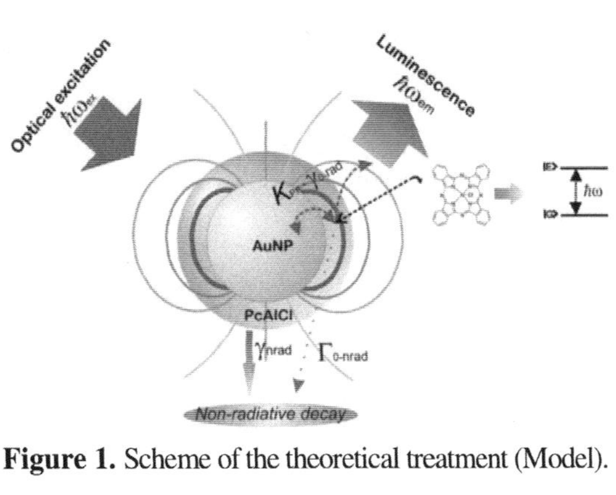

Figure 1. Scheme of the theoretical treatment (Model).

3. Results and discussion

Using the described above theoretical approaches, the extinction cross-sections for uncoated AuNP and AuNP-CAP core/shell nanostructures were determined. We also calculated the contributions of the scattering and absorption processes to the entire extinction. Below are the results of the calculated extinction spectra for AuNP with the average diameter equal to 18 nm (figure 2a), and the scattering/absorption ratio as the function of a wavelength (figure 2b). Similar curves for the particles coated with phthalocyanine molecules are shown at the figure 2 c,d.

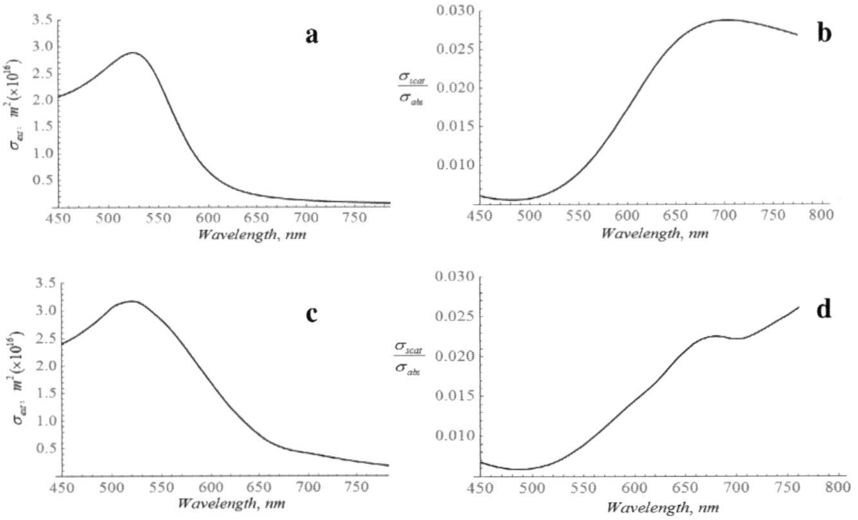

Figure 2. The extinction spectra of the AuNP – (a), the extinction spectra of the AuNP-CAP structures - (b). Scattering to absorption cross section ratio of AuNP – (c), scattering to absorption cross section ratio of AuNP-CAP structures - (d).

As can be seen from Fig.2 the extinction spectrum of the hybrid structure (Fig. 2b) is broadened with a not large increase in magnitude in comparison with the spectrum of the pure AuNp colloid. The extinction cross section also increases in a long-wavelength range of the spectrum, which is the consequence of the presence of intensive absorption band in the Pc layer due to the exciton excitation. It should be mentioned that the spectrum changing character during the formation of an outer Pc layer is not of an additive nature. This indicates to the influence of the outer layer on the AuNp field mode structure, which is dictated by free electron oscillations, i.e. LSP.
The determination of near field distributions is one of the major result of the first step calculation. The near field distributions of coated and "pure" AuNp are represented in Fig. 3.

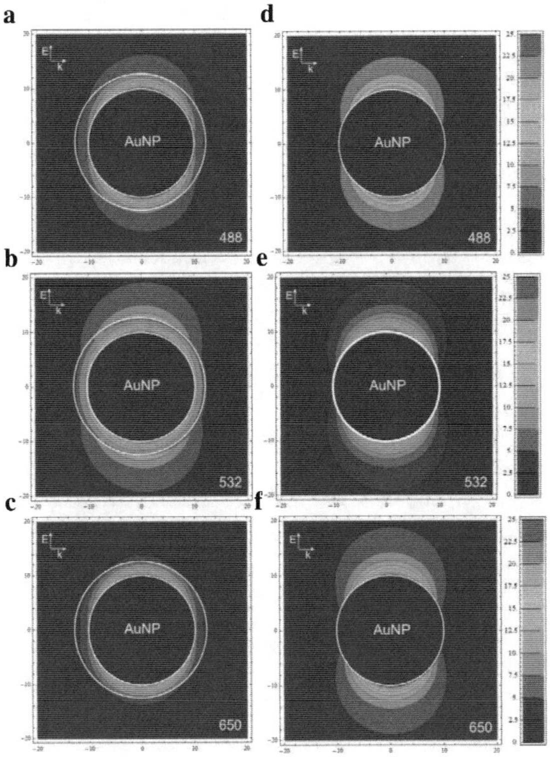

Figure 3. The intensity of near field distribution generated by coated (a,b,c) and "pure" (d,e,f) AuNp, for $\lambda_{Exc}^{(1;2;3)} = 488; 532; 650$ nm respectively.

Fig. 3a and 3b shows that the field in the core/shell structure is strongly localized in the Pc shell. The intensity enhancement reaches a 20-fold increase in comparison with the intensity of the incident field. Unlike the similar distributions of the "pure" uncoated AuNp (Fig. 3 d, e, f), a non-monotonic character of the field distribution along the radial direction is observed. In order to investigate the influence of the structure's components on optical properties of the entire system we chose the wavelength of the incident field that it could coincide with a spectral position of the exciton resonance in the Pc and the position of LSP resonance.

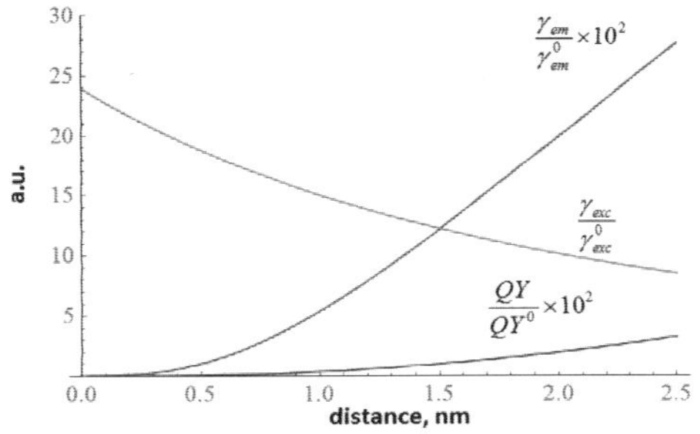

Figure 4. Relative luminescence, quantum yield and the rate of excitation as dependences on the distance from chromophore to particle.

As can be seen, the luminescence increases with the distance from the surface of the gold core. There is the strengthening coupling between the relaxation rate of the entire "particle-chromophore" system and higher order modes of the free electron oscillations when the distance decreases. It can be explained by the exciton-plasmon coupling. In our calculations we took into account the higher order modes which are up to 110.

4. Experimental section

The quenching of luminescence was also observed, which is related to the processes of energy transfer between AuNP and CAP molecules and to the subsequent Ohmic losses of energy in the AuNP. The reason of the observed quenching effect is the interaction between the molecules and nanoparticles, and is also the result of molecules aggregation on the gold particle surface. This result is consistent with the data from other studies dedicated to stabilization of AuNP with the various organic molecules. Due to the molecular aggregation and the proximity between molecules and the AuNP's surface almost full quenching of luminescence is observed [3].

Using theoretical approach described above we modeled the extinction spectrum of the experimental gold nanoparticles and AuNp-CAP structures, taking into account their real size distribution and thickness of Pc shell (2.5 nm). As illustrated in figure 5 the experimental data is well approximated by the model curve.

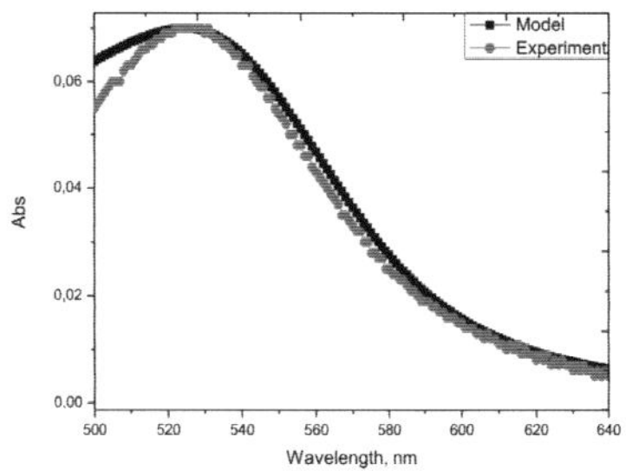

Figure 5. Model and Experimental extinction spectra of the AuNP.

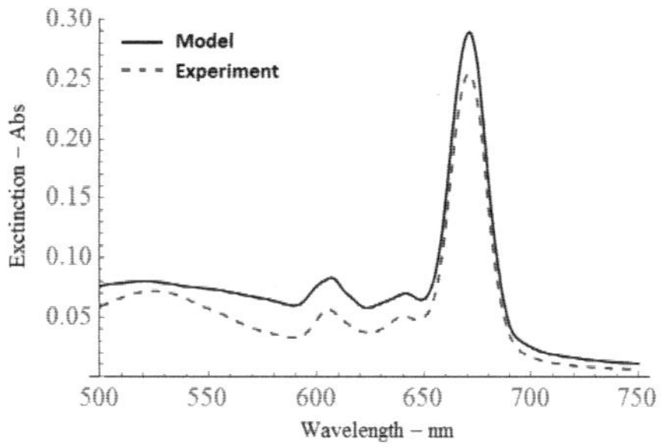

Figure 6. Model and Experimental extinction spectra of the AuNP-CAP nanostructure.

The changes observed in the extinction spectra of Core/Shell AuNP-CAP structures related to the plasmon - exciton interactions have a good agreement with the modeling results (figure 6). Figure 6 shows the experimental and model extinction spectra of the experimental obtained AuNP-CAP nanostructures. We obtained very similar results, the model spectrum includes all spectral features which were observed in the experimental dependence.

5. Conclusions

We proposed the model to describe active plasmonic spherical nanostructures, allowing modeling the linear optical properties. Using our model we calculated scattering and absorption cross sections, the near field distribution, changes in the radiative and nonradiative spontaneous decay rates, changes in the quantum yield of CAP luminescence, the total change in the luminescence. The analysis of the results let us to set mechanisms of changes in the linear optical properties which are explained throughout the plasmon-exciton interaction. The comparison experimental and theoretical results is one of the important parts of our research. We prepared the hybrid structures and determined their extinction and luminescence spectra. Good agreement between theory and experiment indicates that proposed model is able to explain linear optical properties of our AuNp-CAP structure and the same structures too.

References

[1] Tuning of the Spectroscopic Properties of Composite Nanoparticles by the Insertion of a Spacer Layer: Effect of Exciton-Plasmon Coupling, Akihito Yoshida, Langmuir 2009, 25(12), 6683–6689

[2] Strong plasmon-exciton coupling in a hybrid system of gold nanostars and J-aggregates, Melnikau et al. Nanoscale Research Letters 2013, 8:134

[3] Photoinduced Charge and Energy Transfer in Phthalocyanine-Functionalized Gold Nanoparticles, Anne Kotiaho, Riikka Lahtinen, Alexander Efimov, Hanna-Kaisa Metsberg, Essi Sariola, Heli Lehtivuori, Nikolai V. Tkachenko, and Helge Lemmetyinen, J. Phys. Chem. C 2010, 114, 162–168

[4] Surface-enhanced Raman scattering and biophysics, Katrin Kneipp1, Harald Kneipp, Irving Itzkan, Ramachandra R Dasariand Michael S Feld, J. Phys.: Condens. Matter 14 (2002) R597–R624

[5] Demonstration of a spaser-based nanolaser, MA Noginov et al., Nature 460: 1110–1112, (2009).

[6] Effect of the Interaction between Molecular Exciton and Localized Surface Plasmon on the Spectroscopic Properties of Silver Nanoparticles Coated with Cyanine Dye J-Aggregates, Akihito Yoshida and Noritsugu Kometani, J. Phys. Chem. C 2010, 114, 2867–2872

[7] Twenty-Fold Enhancement of Molecular Fluorescence by Coupling to a J-Aggregate Critically Coupled Resonator, Gleb M. Akselrod,† Brian J. Walker, William A. Tisdale, Moungi G. Bawendi, and Vladimir Bulovic, ACS NANO, VOL. 6, NO. 1, 467–471, 2012.

[8] Electromagnetic scattering for multilayered sphere: recursive algorithms, Z. S. Wu et. al., Radio Sci. 26, 1393–1401 119912.

[9] Enhancement and Quenching of Single-Molecule Fluorescence, Lukas Novotny et. al., PHYSICAL REVIEW LETTERS, 96, 113002 (2006).

[10] Optical Properties of Metal Clusters, Kreibig U., Vollmer M., Berlin: Springer, 1995.

[11] Structural and optical properties of thermally evaporated zinc phthalocyanine thin films, M.M. El-Nahaas, H.M. Zeyada, M.S. Aziz, N.A. El-Ghamaz, Optical Materials 27 (2004), 491-498

Journal of Physics: Conference Series **541** (2014) 012065 doi:10.1088/1742-6596/541/1/012065

Collimation and focusing of initially single-cycle paraxial optical beams

Danila N. Puzyrev and Arkadiy A. Drozdov

Department of Photonics and Optical Information Technology, ITMO University, Saint-Petersburg 197101, Russia

E-mail: puzurdan@gmail.com, arkadiy.drozdov@gmail.com

Abstract. The paper reports theoretical results of collimation and focusing of spatio-temporal field structure that is forming due to far-field (Fraunhofer) diffraction of paraxial wave packet emitted by single-cycle radiation source. A spatial collimation of such waves results in a peculiar shape of spatio-temporal structure with the central part moving faster than its outlying areas. During the focusing of collimated beam, a transformation from 1.5-cycle to single-cycle and again to 1.5-cycle wave occurs along a path from collimating to focusing mirrors with equal focal lengths.

1. Introduction

With the development of conversion technology of near-IR femtosecond radiation in to terahertz (THz) frequency range, it become possible to obtain single-cycle electromagnetic waves with a spectrum in the range of 0.1-10 THz [1–3]. Such radiation usually has widespread application in detection systems of drugs and explosive materials, for medical diagnostics and other applications [4–7]. Optimization of optical systems that emit, collimate and focus such waves is one of the important problem in THz science nowadays. It is also well-known that initially single-cycle pulse is transformed to 1.5-cycle pulse over the entire wave packet in far field region [8]. The properties of single-cycle THz pulses propagating through a focus of parabolic mirrors or lens have been previously studied both theoretically and experimentally [9–11]. The first noninterferometric observation of axial π phase shift for single-cycle THz pulses passing through a focus was demonstrated in the experiment [12]. The first direct observation of the 2D focused few-cycle THz field distribution by CCD based THz imaging system was shown in the work [13]. To our knowledge, detailed investigation of spatio-temporal field dynamics after collimation and subsequent focusing of initially single-cycle pulses has not been performed.

In this paper, for the case of linearly polarized radiation with Gaussian transverse spatial distribution we theoretically investigate peculiarities of collimation and focusing of spatio-temporal field structure that is forming due to Fraunhofer diffraction of paraxial wave packet emitted by single-cycle radiation source.

2. Diffraction dynamics of spatio-temporal spectrum of single-cycle optical beams

Spatio-temporal dynamics of spectrum

$$g_{x,y}(\omega, k_x, k_y, z) = \int_{-\infty}^{+\infty}\int_{-\infty}^{+\infty}\int_{-\infty}^{+\infty} E_{x,y}(t, x, y, z)\exp\left(-i\left(\omega t + k_x x + k_y y\right)\right)dtdxdy \tag{1}$$

of paraxial radiation with electric field strength $E_{x,y}(t, x, y, z)$ in an isotropic homogeneous optical medium can be described by expression [8]

$$g_{x,y}(\omega, k_x, k_y, z) = C_{x,y}(\omega, k_x, k_y)\exp\left(-ikz\left(1 - \frac{\left(k_x^2 + k_y^2\right)}{2k^2}\right)\right), \tag{2}$$

where x, y, z are coordinates of Cartesian axes (z axis coincides with wave propagation direction and x, y coincide with transverse direction), Cartesian components of wave $E_{x,y}$ and its spatio-temporal spectrum $g_{x,y}$ are denoted by x, y indexes, C_x and C_y are assumed known spectral components of radiation at $z = 0$, ω is the temporal frequency of radiation, k_x and k_y are the spatial frequencies, $k = \dfrac{\omega n(\omega)}{c}$ is the wave number, $n(\omega)$ is the refractive index of a medium, c is the speed of light in a vacuum.

In this paper we restrict the analysis of collimation and subsequent focusing of paraxial Gaussian beams, which we assume linearly polarized along axis x for simplicity. So we will assume their spectrum of the form [8]

$$C(\omega, k_x, k_y) = \pi\rho^2\exp\left(-\frac{\rho^2\left(k_x^2 + k_y^2\right)}{4}\right)G_0(\omega), \tag{3}$$

that corresponds to initial electric field distribution

$$E(t, x, y) = \exp\left(-\frac{x^2 + y^2}{\rho^2}\right)F_0(t), \tag{4}$$

where ρ is the transverse beam width, $F_0(t)$ is temporal profile of wave, $G_0(\omega)$ is its temporal spectrum at $z = 0$.

For boundary condition (3) expression (2) takes form

$$g(\omega, k_x, k_y, z) = \pi\rho^2\exp\left(-\frac{\rho^2\left(k_x^2 + k_y^2\right)}{4}\left(1 - i\frac{2cz}{\rho^2 n(\omega)\omega}\right)\right)\exp\left(-i\frac{n(\omega)\omega z}{c}\right)G_0(\omega). \tag{5}$$

Using the transform formula for spatial dependence of temporal spectrum of radiation,

$$G(\omega, x, y, z) = \frac{1}{(2\pi)^2}\int_{-\infty}^{+\infty}\int_{-\infty}^{+\infty} g(\omega, k_x, k_y, z)\exp\left(i\left(k_x x + k_y y\right)\right)dk_x dk_y \tag{6}$$

and relation (5), we can present the expression (6) for diffraction dynamics of radiation in form

$$G(\omega, x, y, z) = \frac{T(\omega, z)\omega}{T(\omega, z)\omega - i}\exp\left(-\frac{\left(x^2 + y^2\right)}{\rho^2}\frac{\left(T(\omega, z)\omega\right)^2}{1 + \left(T(\omega, z)\omega\right)^2}\left(1 + \frac{i}{T(\omega, z)\omega}\right) - ikz\right)G_0(\omega), \tag{7}$$

where $T(\omega, z) = \dfrac{\rho^2 n(\omega)}{2cz}$.

For far-field region ($z \gg \dfrac{\pi\rho^2 n(\omega)}{\{\lambda\}_{\min}}$, where $\lambda_{\min}$ is minimum wavelength from wavelength range where the majority of initial radiation energy is trapped [8]) relation (7) is simplified to form

$$G_{\mathrm{F}}(\omega, x, y, z) = T(\omega, z)\omega\left(T(\omega, z)\omega + i\right) \times \\ \times \exp\left(-\frac{(x^2+y^2)}{\rho^2}(T(\omega, z)\omega)^2 - ik\left(z + \frac{(x^2+y^2)}{2z}\right)\right) G_0(\omega). \tag{8}$$

For further analysis of paraxial Gaussian beams we restrict ourselves by the case of dispersionless media when we can present the refractive index in form $n(\omega) = n_0 = const$. Atmospheric air for THz radiation is a characteristic example of such medium [14, 15] at considered distances in this paper. We also assume the single-cycle wave of source with spectrum $G_0(\omega) = -\frac{\sqrt{\pi}}{2} E_0 \tau_p^2 i\omega \exp\left(-\left(\frac{\tau_p \omega}{2}\right)^2\right)$ [16], where τ_p is input pulse duration at $z = 0$, E_0 is correspond initial amplitude of electric field strength of wave.

Using the Fourier transform formula

$$E(t, x, y, z) = \frac{1}{2\pi}\int_{-\infty}^{+\infty} G(\omega, x, y, z)\exp(i\omega t)d\omega \tag{9}$$

and relation (8) we can find the spatio-temporal distribution of electric field in far-field region in form [see figure 1(a) below]

$$E_{\mathrm{F}}(t', x, y, z) = E_0 T(z)\frac{\tau_p^2}{\tau_{\mathrm{F}}^3}\left(2\frac{T(z)}{\tau_{\mathrm{F}}^2}\left(3 - 2\frac{t'^2}{\tau_{\mathrm{F}}^2}\right)t' + \left(1 - 2\frac{t'^2}{\tau_{\mathrm{F}}^2}\right)\right)\exp\left(-\frac{t'^2}{\tau_{\mathrm{F}}^2}\right), \tag{10}$$

where we introduce $\tau_{\mathrm{F}}^2 = \tau_p^2 + \left(\frac{\rho N_0}{c}\right)^2 \frac{(x^2+y^2)}{z^2}$ and "delay time" $t' = t - \frac{N_0}{c}\left(z + \frac{(x^2+y^2)}{2z}\right)$. Red and blue areas correspond to maximum positive and maximum negative field values of two-dimensional images. Here we take $\rho/\lambda_{\max} = 10$, where $\lambda_{\max}$ is the wavelength corresponding to the maximum of spectrum of input wave packet [16,17]. For example, for terahertz single-cycle waves this value is usually $\lambda_{\max} = 300$ μm. We also consider electric field emitted by the source (e.g., at z = 0) in the form of a single-cycle pulse with Gaussian transverse distribution [see figure 1(b) below]

$$E(t, x, y) = E_0 \exp\left(-\frac{x^2+y^2}{\rho^2}\right)\frac{t}{\tau_p}\exp\left(-\frac{t^2}{\tau_p^2}\right). \tag{11}$$

As follows from this figure and relation (10), in far field region a single-cycle beam becomes 1.5-cycle for the whole wave package. We note that this feature was also observed in [8].

3. Spatio-temporal collimation and focusing dynamics of paraxial wave packet emitted by single-cycle source of radiation

First, we set collimating mirror at distance z_0 from radiation source with focal length $f_{\mathrm{col}} = z_0$ and reflection function $R_{\mathrm{col}}(\omega, x, y) = \exp\left(ik\left(x^2+y^2\right)/2f_{\mathrm{col}}\right)$ in far field region. We neglect the effects of optical-system aberrations. Temporal spectrum of wave packet at distance z_0 from the source of radiation will be determined by following expression

$$G_{\mathrm{col}}(\omega, x, y, z_0) = G_{\mathrm{F}}(\omega, x, y, z_0) R_{\mathrm{col}}(\omega, x, y), \tag{12}$$

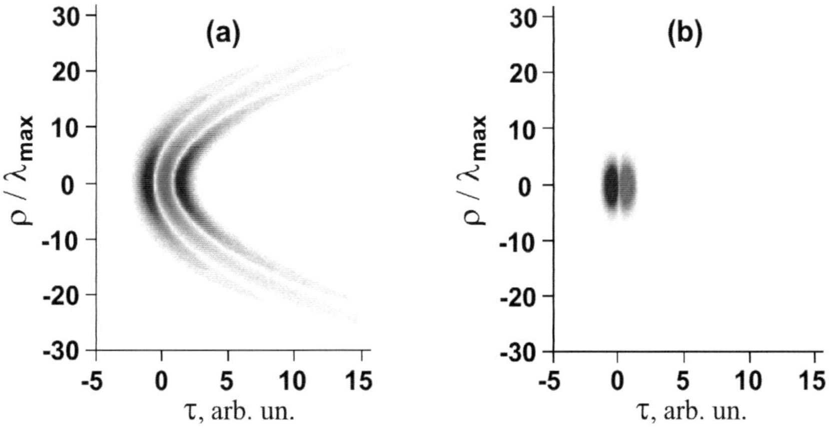

Figure 1. Spatio-temporal structure of distribution of the electric field of radiation in far-field region (10) (a) and its structure at $z = 0$ (b).

where $G_{\mathrm{F}}(\omega, x, y, z_0)$ is given by relation (8).

As follows from relation (12), the temporal spectrum of radiation immediately after collimating mirror takes form

$$G_{\mathrm{col}}(\omega, x, y, z_0) = T(\omega, z_0)\omega\,(T(\omega, z_0)\omega + i)\exp\left(-\frac{(x^2+y^2)}{\rho^2}(T(\omega, z_0)\omega)^2 - ikz_0\right)G_0(\omega). \tag{13}$$

It can be seen from (13), collimation simplifies Fourier spectrum of incident on collimating mirror radiation [see (8) and (13) for comparison]. So the spatio-temporal distribution of electric field immediately after collimating mirror will be described by relation

$$E_{\mathrm{col}}(t'', x, y, z_0) = E_0 T(z_0)\frac{\tau_p^2}{\tau_{\mathrm{F}}^3}\left(2\frac{T(z_0)}{\tau_{\mathrm{F}}^2}\left(3 - 2\frac{t''^2}{\tau_{\mathrm{F}}^2}\right)t'' + \left(1 - 2\frac{t''^2}{\tau_{\mathrm{F}}^2}\right)\right)\exp\left(-\frac{t''^2}{\tau_{\mathrm{F}}^2}\right), \tag{14}$$

where we introduce the retarded time $t'' = t - \frac{N_0}{c}z$ which is not depending on transverse coordinates now. We now consider (14) as a new boundary condition for dynamics of collimated radiation and show in figure 2 (a). It can be seen that specially shaped 1.5-cycle spatio-temporal structure with increasing duration from beam axis to periphery is formed immediately after collimating mirror.

According to (13) spatio-temporal spectrum of collimated radiation become

$$g_{\mathrm{col}}(\omega, k_x, k_y, z_0) = \pi\rho^2\frac{T(\omega, z_0)\omega + i}{T(\omega, z_0)\omega}\exp\left(-\frac{\rho^2}{4}\frac{k_x^2 + k_y^2}{(T(\omega, z_0)\omega)^2} - ikz_0\right)G_0(\omega). \tag{15}$$

The general solution for spatio-temporal spectrum of collimation dynamics by taking into account (2) can be presented in form

$$g_{\mathrm{col}}(\omega, k_x, k_y, z) = g_{\mathrm{col}}(\omega, k_x, k_y, z_0)\exp\left(-ik(z - z_0)\left(1 - \frac{(k_x^2 + k_y^2)}{2k^2}\right)\right). \tag{16}$$

Expression (16) according to (15) takes form

$$
\begin{aligned}
g_{\text{col}}(\omega, k_x, k_y, z) = \pi\rho^2 \frac{T(\omega, z_0)\omega + i}{T(\omega, z_0)\omega} \times \\
\times \exp\left(- \left(k_x^2 + k_y^2\right) \frac{\rho^2}{4T(\omega, z_0)\omega} \left(\frac{1 - iT(\omega, z_0)\omega(z - z_0)/z_0}{T(\omega, z_0)\omega} \right) - ikz \right) G_0(\omega) .
\end{aligned} \tag{17}
$$

Using the relation (17) we can describe the spatial dependence of temporal spectrum dynamics of collimated radiation at arbitrary distance after mirror by expression

$$
\begin{aligned}
G_{\text{col}}(\omega, x, y, z) = \frac{T(\omega, z_0)\omega\left(T(\omega, z_0)\omega + i\right)}{1 - iT(\omega, z_0)\omega(z - z_0)/z_0} \times \\
\times \exp\left(- \frac{x^2 + y^2}{\rho^2} \frac{\left(T(\omega, z_0)\omega\right)^2}{1 - iT(\omega, z_0)\omega(z - z_0)/z_0} - ikz \right) G_0(\omega) .
\end{aligned} \tag{18}
$$

The results of numerical simulation of spatio-temporal field evolution at different distances from collimating mirror in linear optical medium from equation (18) are shown in figure 2 (b-e). It is shown that due to diffraction the central part of this complex structure is moving faster than outlying areas.

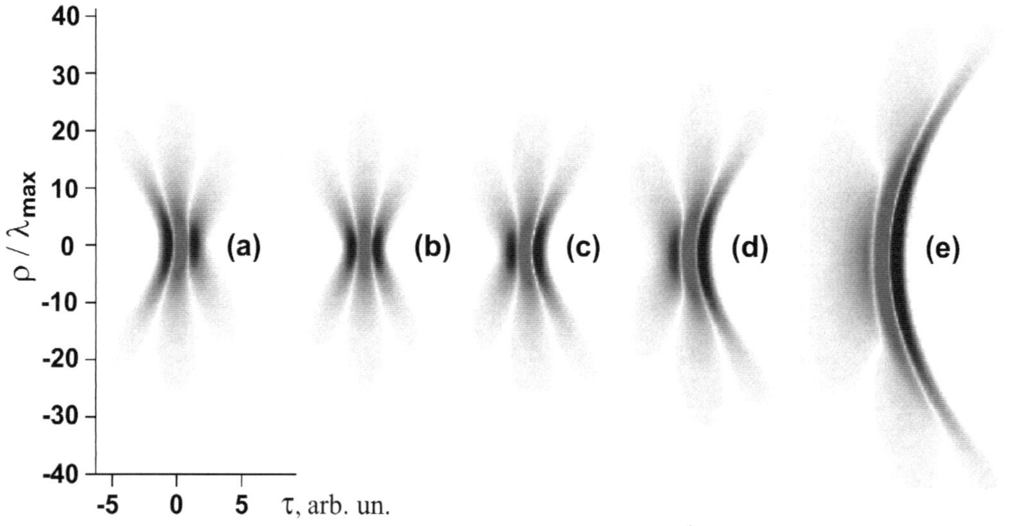

Figure 2. Spatio-temporal structure of the electric field of collimated wave packet immediately after collimating mirror (a) and during its further evolution in linear optical medium (b-e) at the distances f_{col} (b), $2f_{\text{col}}$ (c), $3f_{\text{col}}$ (d), $6f_{\text{col}}$ (e) from mirror. τ is the normalized time.

We now consider focusing features of presented in figure 2 collimated radiation by focusing mirror with focal length $f_{\text{foc}} = f_{\text{col}}$. Spatio-temporal distributions of the electric field of radiation at the focus of mirror are presented in figure 3 (a-e). Figure 3 shows that during the focusing of collimated radiation, optical wave at the focus is changing from 1.5-cycle to a single-cycle and again to 1.5-cycle pulse along the way from collimating to focusing mirrors. It can be seen from figure 3 (c), that focused electric field of radiation with an accuracy to π phase shift coincide with initial field of single-cycle wave packet when distance between collimating and focusing mirrors is twice of their focal length.

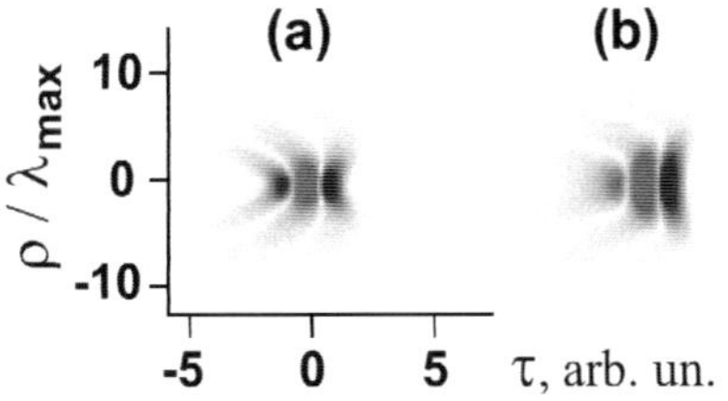

Figure 3. Spatio-temporal structure of distribution of the electric field of radiation at the focus of mirror during focusing of wave packet obtained immediately after the collimating mirror (a) and at the distances f_{col} (b), $2f_{\mathrm{col}}$ (c), $3f_{\mathrm{col}}$ (d), $6f_{\mathrm{col}}$ (e) from it.

4. Conclusion

We have theoretically investigated the features of collimation and focusing of spatio-temporal field structure that is forming due to far-field diffraction of paraxial wave packet emitted by single-cycle radiation source. We have demonstrated that spatial collimation of such waves results in a peculiar shape of the spatio-temporal structure with the central part moving faster than its outlying areas during further evolution in free space. During the focusing of collimated wave packet a transformation from 1.5-cycle to single-cycle and again to 1.5-cycle pulse occurs along a path from collimating to focusing mirrors with equal focal lengths. We have also shown that focused electric field with an accuracy to π phase shift coincide with electric field of initial single-cycle radiation when the distance between collimating and focusing mirrors is twice of their focal lengths. These features shall be taken into account for the optimization of optical systems operating with single-cycle THz pulses.

Acknowledgments

Authors would like to thank Prof. Sergei Kozlov for helpful discussions. This work was financially supported by Government of Russian Federation (Grant 074-U01).

References

[1] Bespalov V G, Gorodetski A A, Denisyuk I Y, Kozlov S A, Krylov V N, Lukomski G V, Petrov N V and Putilin S E 2008 *J. Opt. Technol.* **75** 636–642
[2] Zhang X C and Xu J 2009 *Introduction to THz wave photonics* (N.Y.: Springer Science+Business Media)
[3] Sajadi M, Wolf M and Kampfrath T 2014 *Appl. Phys. Lett.* **104** 091118
[4] Lee Y 2009 *Principles of Terahertz Science and Technology* (Corvalis: Springer Science+Business Media)
[5] Horiuchi N 2010 *Nat. Photonics* **4** 140
[6] Grachev Y V, Kuklin I A, Gerasimov I V, Smolyanskaya O A, Kozlov S A and Bespalov V G 2010 *J. Opt. Technol.* **77** 731–733
[7] Sakhnov S N, Leksutkina E V, Smolyanskaya O A, Usov A V, Parakhuda S E, Grachev Y V and Kozlov S A 2011 *Optics and Spectroscopy* **111** 257–261
[8] Ezerskaya A A, Ivanov D V, Kozlov S A and Kivshar Y S 2012 *J. Infrared Millim. Terahertz Waves* **33** 926–942
[9] Feng S, Winful H G and Hellwarth R W 1998 *Opt. Lett.* **23** 385–387
[10] Hunsche S, Feng S, Winful H G, Leitenstorfer A, Nuss M C and Ippen E P 1999 *J. Opt. Soc. Am. A* **16** 2025–2028
[11] Kužel P, Khazan M A and Kroupa J 1999 *J. Opt. Soc. Am. B* **16** 1795–1800
[12] Ruffin A B, Rudd J V, Whitaker J F, Feng S and Winful H G 1999 *Phys. Rev. Lett.* **83**(17) 3410–3413
[13] Jiang Z and Zhang X C 1999 *Opt. Express* **5** 243–248
[14] Gurtler A, Winnewisser C, Helm H and Jepsen P 2000 *J. Opt. Soc. Am. A* **17** 74–83
[15] Kulya M S, Grachev Y V, Bespalov V G and Kujanpaa V P 2013 *Nanosystems: Physics, Chemistry, Mathematics* **4** 206–213
[16] Drozdov A A, Kozlov S A, Sukhorukov A A and Kivshar Y S 2012 *Phys. Rev. A* **86** 053822
[17] Drozdov A A, Sukhorukov A A and Kozlov S A 2014 *Int. J. Mod. Phys. B* **28** 1442007

Determining influence of four-wave mixing effect on quantum key distribution

D N Vavulin, V I Egorov, A V Gleim, S A Chivilikhin

Saint Petersburg National Research University of Information Technologies, Mechanics and Optics

dima-vavulin@mail.ru

Abstract. We consider the possibility of multiplexing the classical and quantum signals in a quantum cryptography system with optical fiber used as a transmission medium. If the quantum signal is located at a frequency close to the frequency of classical signals, a set of nonlinear effects such as FWM (four-wave mixing) and Raman scattering is observed. The impact of four-wave mixing (FWM) effect on error level is described and analyzed in this work in case of large frequency diversity between classical and quantum signals. It is shown that the influence of FWM is negligible for convenient quantum key distribution.

1. Introduction

Modern quantum cryptography systems [1] that perform secret key distribution usually operate with not only a weak quantum signal in the secure channel, but also a classical (reference) signal used for clock synchronization in transmitter and receiver modules, and date transfer. However, in most experiments two different fibers are currently used for this purpose: one for the quantum and another for the classical channel. At the same time, a necessity of installing a separate optical fiber for each quantum channel may lead to significant costs and make quantum cryptography economically ineffective.

If the classical signal and the quantum signal, which is at least eight orders of magnitude less intense, are transmitted at close frequencies, a number of nonlinear effects that dramatically decrease the signal-noise ratio of the quantum channel would appear. Two effects that are most important in case of quantum communications are four-wave mixing (FWM) and Raman scattering. Several recent papers investigate the possibility of using a single fiber for quantum key distribution (QKD) and duplex information channel, with most attention paid to Raman scattering [2-10].

Quantum bit error rate (QBER), an important parameter that limits secure key generation rate and transmission distance in quantum cryptography systems, is determined by the following formula:

$$QBER = QBER_{opt} + QBER_{det} + \mathrm{QBER}_{\mathrm{Int}} + QBER_{FWM} + QBER_{SRS}.$$

The first term, $QBER_{opt}$, is due to the polarization extinction ratio resulting mainly from passive optical components within the optical system. This is a measure of the optical quality of the setup and is independent of transmission distance. The second term, $QBER_{det}$, arises from the detector dark counts. The third term, $\mathrm{QBER}_{\mathrm{Int}}$, is the intersymbol interference, where errors are caused by counts from a given clock period that leak into the adjacent period due to source jitter, detector jitter, and dispersion in the transmission medium. The third term is responsible for the FWM. The fourth term is responsible for the spontaneous Raman scattering. Typically, researchers consider only the first three

Journal of Physics: Conference Series **541** (2014) 012066 doi:10.1088/1742-6596/541/1/012066

terms, because the FWM and spontaneous Raman scattering only appear when multiplexing of classical and quantum signals. In this work we analyzed the influence of FWM for different classical signal power levels. In classical telecommunication systems FWM is the key parasitic effect in DWDM [1]. Due to this effect, the quantum key transmission at a frequency close to the frequency of the classical signal becomes greatly difficult. Therefore, in our calculations we used 1310 nm wavelength for quantum channel and 1550 nm for classical channel. In this case of distant bands not only FWM, but also Raman effects are minimized. However, quantum signal adsorption would be significantly higher, thus limiting QKD bitrate and distance.

2. Initial equations, parameters and conditions

In our calculations we used 1310 nm wavelength for quantum channel and 1550 nm for classical channel. In this case of distant bands not only FWM, but also Raman effects are minimized. However, quantum signal adsorption would be significantly higher, thus limiting QKD bitrate and distance. Initial equations for solving the stated problem were taken from [11]. These equations describe alterations of x and y components of complex amplitudes of signals passing through an optical fiber: pump field (p – classic channel, synchronization pulse), signal field (s – quantum channel) and idler wave (i – parasite wave that appears due to FWM). We used equations only for x components pump, idler and signal waves, while the y components were assumed to be zero, i.e. all three waves had the same polarization.

Since we are using only x-polarization component for the signal wavelength and pump wavelength without y-polarization component and excluding the longitudinal field components in the optical fiber (z-components), in our case the effect of FWM process is the greatest. We have shown that the effect of FWM is not critical even in this case; therefore, there is no need to examine any other wave polarization state.

Birefringence was not considered in our model. Then, we added terms that describe attenuation in the fiber, which leads us to final system:

$$\phi_{2p} = 0 \tag{1}$$

$$\phi_{2s} = 0 \tag{2}$$

$$\phi_{2i} = 0 \tag{3}$$

$$i\frac{\partial \phi_{1p}}{\partial z} + \frac{8}{9}(2\phi_{1i}\phi_{1p}\phi_{1i}^{*} + \phi_{1p}^{2}\phi_{1p}^{*} + 2\phi_{1i}\phi_{1s}\phi_{1p}^{*}e^{i\Delta kz} + 2\phi_{1p}\phi_{1s}\phi_{1s}^{*}) + i\alpha(\omega_{p})\phi_{1p} = 0 \tag{4}$$

$$i\frac{\partial \phi_{1s}}{\partial z} + \frac{8}{9}(\phi_{1p}^{2}\phi_{1i}^{*}e^{-i\Delta kz} + 2\phi_{1i}\phi_{1s}\phi_{1i}^{*} + 2\phi_{1p}\phi_{1s}\phi_{1p}^{*} + \phi_{1s}^{2}\phi_{1s}^{*}) + i\alpha(\omega_{s})\phi_{1s} = 0 \tag{5}$$

$$i\frac{\partial \phi_{1i}}{\partial z} + \frac{8}{9}(\phi_{1p}^{2}\phi_{1s}^{*}e^{-i\Delta kz} + 2\phi_{1s}\phi_{1i}\phi_{1s}^{*} + 2\phi_{1p}\phi_{1i}\phi_{1p}^{*} + \phi_{1i}^{2}\phi_{1i}^{*}) + i\alpha(\omega_{i})\phi_{1i} = 0 \tag{6}$$

Fields ϕ_{1p}, ϕ_{1s}, ϕ_{1i} here are complex amplitudes of x polarization components of pump (p), signal (s) and idler fields. Correspondingly, ϕ_{2p}, ϕ_{2s}, ϕ_{2i} are complex amplitudes of y polarization components of pump (p), signal (s) and idler fields. Amplitudes here are measured in $W^{1/2}$, so $|\phi_{1p}|^2$ is peak power of pump filed x component in Watts. In these equations we are using the following normalization: longitudinal coordinate z here is a product of physical fiber length Z, measured in km and Kerr nonlinearity coefficient γ, measured in $(W/km)^{-1}$[11].

In this work relatively small signal intensities were studied, so Kerr nonlieraity is not observed, and γ value was 1 $(W/km)^{-1}$, that is typical for optical fibers. Therefore, longitudal coordinate z is numerically equal to physical fiber length Z. Δk is a variable known from FWM theory defined below. Pump, signal and idler wave frequencies do not depend on wave polarization: $\omega_{1p} = \omega_{2p} = \omega_p$, $\omega_{1s} = \omega_{2s} = \omega_s$, $\omega_{1i} = \omega_{2i} = \omega_i$. Values of α_p (ω_p), α_s (ω_s) and α_i (ω_i) are decay coefficients for pump, signal and idler fields respectively.

Value of Δk is calculated as follows: $\Delta k = k_i + k_s - 2k_p$. Here $k_i = n_i\omega_i/c$ is propagation constant of idler wave at frequency ω_i, c – speed of light in vacuum, n_i – fiber refraction index at ω_i (similar for k_s and k_p). Wave frequencies are bounded by condition: $2\omega_p = \omega_s+\omega_i$. Here ω_p corresponds to wavelength $\lambda_p = 1.55$ µm, and ω_s – to $\lambda_s = 1.3$ µm. Thus, idler wavelength was found as:

$$\lambda_i = (\lambda_p * \lambda_s)/(2\lambda_s - \lambda_p) = 1.92 \text{ µm}$$

Indices of refraction for all wavelengths: λ_p, λ_s, λ_i were calculated using Sellmeier equation. Propagation constants k_s, k_p, k_i were calculated (with corresponding n and λ) by definition: $k = 2\pi n/\lambda$.

In our calculations we used standard decoy coefficients for telecommunication fibers (SMF-28) at target frequencies: α_s (1.3 µm) = 0.34 dB/km, α_p (1.55 µm) = 0.2 dB/km, α_i (1.92 µm) = 6.5 dB/km.

Classical (p) and quantum (s) channels are initially (z = 0) defined by their complex amplitudes, which corresponds to the following powers:

$P_{1p} = 10^{-3}$ W, $P_{2p} = 0$ W

$P_{1s} = 10^{-11}$ W, $P_{2s} = 0$ W

Idler wave is not excited: $P_{1i} = 0$ W, $P_{2i} = 0$ W

These values of complex amplitudes for classic and quantum signals were chosen keeping in mind that power of a common laser source in fiber is around 1 uW, and power of attenuated laser used as a single photon (coherent states) source is 10 pW.

It must be noted that although they describe classic (intense) light field, they can still be applied to quantum cryptography. The reason for it is that even in latest experiments dedicated to creating QKD nets [8, 9] a strongly attenuated laser radiation is used as a light source, and coherent state field with mean photon number < 1 is treated as quantum bits. For this assumption, experimental data demonstrates good compliance with theory.

The calculations were carried out for the case when the field in a quantum channel is assumed continuous. This is due to the fact that in the quantized mode we consider each photon of secret key as a separate wave which propagates together with the classical wave and generates FWM. In this case, the FWM efficiency is equal to the average QBER at long-term transfer of the key. It is responsible for the applicability of these equations to assess the impact of FWM effect.

3. Numerical calculation

Figure 1 shows the results of calculation for pumping wave (p, classical channel), signal wave (s, quantum channel) and idler wave (i) at 100km distance. For illustration the values for quantum channel and idler waves were increased in 10^8 and in 10^{17} times correspondingly.

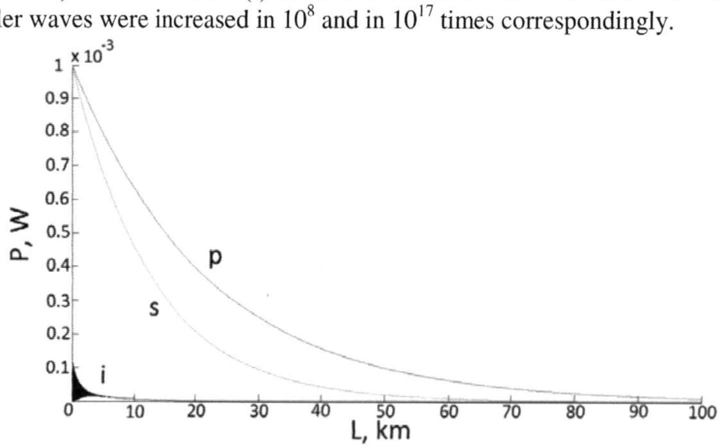

Figure 1. Calculation for pumping wave (p, classical channel), signal wave (s, quantum channel) and idler wave (i) at 100km distance.

Figure 2 illustrates the difference between intensities of signal waves (ΔPs) in presence of pumping (1 mW) and without, compared to signal wave intensity (with pumping). The initial intensity of signal wave is about 10^{-11} W, and, maximal difference between the signal intensities is around 10^{-21} W, i.e. 10 orders smaller. Consequently, the presence of classical signal has almost no affect at quantum signal.

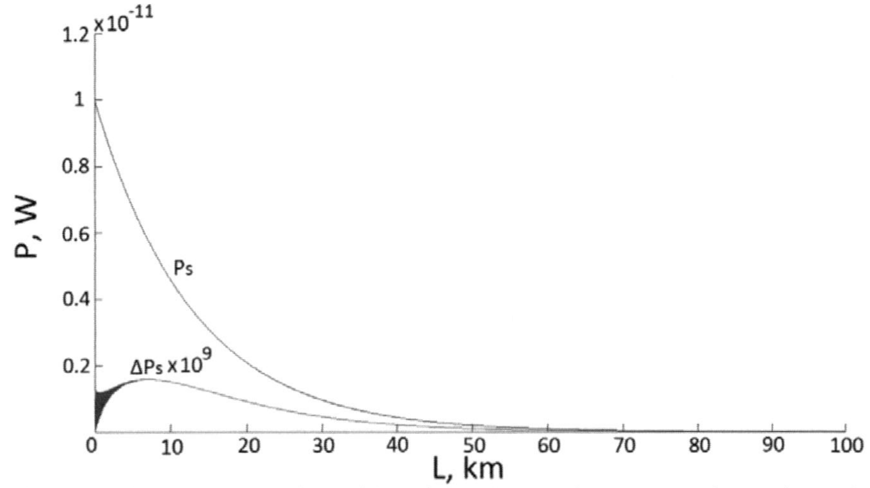

Figure 2. The difference between intensities of signal waves in presence of pumping and without it compared to signal wave intensity: 1 mW pumping

Calculation of FWM efficiency dependence on fiber length was performed for 1 mW pumping. FWM efficiency was defined as the difference between signal waves intensities with pumping and without it, divided by the intensity of signal wave in presence of pumping (figure 3). We can see that FWM efficiency is limited by some value at large fiber length.

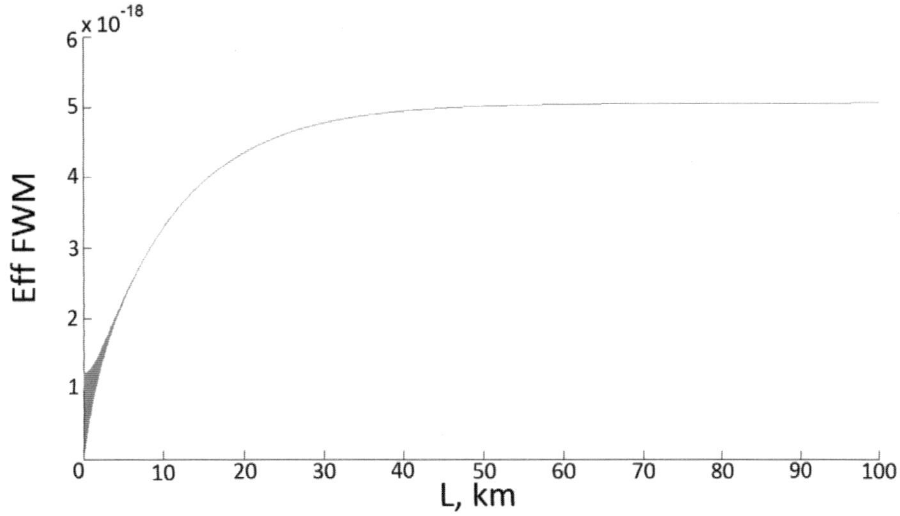

Figure 3. Difference between signal waves with pumping (1 mW) and without it, divided by signal wave intensity in presence of pumping.

4. Conclusion

It was shown that four-wave mixing effects on quantum key distribution are negligible if the classical and quantum channels are transmitted in different optical fiber transparency windows, because mean intensity of idler photons is 17-18 orders of magnitude lower than that of the quantum signal. This statement is also true in case of using DWDM in information (pump) channel. According to literature data, in this case the effects of Raman scattering can also be minimized at certain conditions. At the same time, at 1310 nm wavelength the quantum key generation rate related to losses in optical fiber would decrease about 10 times for large distances. Therefore, large frequency diversity of quantum and classical channels allows finding an optimal balance between nonlinear effect influence and raw key rate for small and medium distances (up to 100 km).

5. Acknowledgements

We would like to thank V.V. Kozlov for his participation in this work. The work was supported by the Dynasty foundation Grant for students and by the St. Petersburg government grant for students. The work was also supported by program of SPIE Scholarships.
This work was financially supported by Government of Russian Federation, Grant 074-U01.

References

[1] Scarani V, Bechmann-Pasquinucci H, Cerf N et al. 2009 The security of practical quantum key distribution. *Rev. Mod. Phys*, **81**, 1301-1350.

[2] Patel K A, Dynes J F, Choi I, Sharpe A W, Dixon A R, Yuan Z L, Penty R V, and Shields A J 2012 Coexistence of High-Bit-Rate Quantum Key Distribution and Data on Optical Fiber. *PHYSICAL REVIEW X*. **2**, 041010

[3] Lin Q, Yaman F, and Agrawal G P 2007 Photon-pair generation in optical fibers through four-wave mixing: Role of Raman scattering and pump polarization. *PHYSICAL REVIEW A* **75**, 023803

[4] Chapuran T E, Toliver P, Peters N A, Jackel J, Goodman M S, Runser R J, McNown S R, Dallmann N, Hughes R J, McCabe K P, Nordholt J E, Peterson C G, Tyagi K T, Mercer L, and Dardy H 2009 Optical networking for quantum key distribution and quantum communications, *New J. Phys.* **11**, 105001

[5] Townsend P D 1997 Simultaneous quantum cryptographic key distribution and conventional data transmission over installed fibre using wavelength-division multiplexing, *Electron. Lett.* **33**, 188–190.

[6] Choi, Young R J, and Townsend P D 2011 Quantum information to the home, *New. J. Phys.* **13**, 063039

[7] Eraerds P, Walenta N, Legr´e M, Gisin N, and Zbinden H 2010 Quantum key distribution and 1 Gbps data encryption over a single fibre, *New J. Phys.* **12**, 063027

[8] Lancho D, Martinez J, Elkouss D, Soto M, and Martin V 2010 QKD in standard optical telecommunication networks, *Lect. Notes Inst. Comput. Sci., Soc. Inf. Telecom. Eng.* **36**, 142–149

[9] Townsend P D 1998 Experimental Investigation of the Performance Limits for First Telecommunications-Window Quantum Cryptography Systems, *IEEE Photonics Technology Letters* **Vol. 10, No. 7**, 1048-1050

[10] Qi B, Zhu W, Qian L, and Lo H K 2010 Feasibility of quantum key distribution through a dense wavelength division multiplexing network, *New J. Phys.* **12**, 103042

[11] Guasoni M, Kozlov V V, Wabnitz S 2012 Theory of polarization attraction in parametric amplifiers based on telecommunication fibers *J. Opt. Soc. Am. B*, **29 №10**, 2710-2720

[12] Egorov V I, Vavulin D N, Latypov I Z, Gleim A V, Rupasov A V 2013 Analysis of a sidebands based quantum cryptography system with different detector types *NANOSYSTEMS: PHYSICS, CHEMISTRY, MATHEMATICS*, **4 (2)**, pp. 190–195

[13] Ivanova A E, Egorov V I, Chivilikhin S A, Gleim A V 2013 Investigation of quantum random number generation based on space-time division of photons *NANOSYSTEMS: PHYSICS, CHEMISTRY, MATHEMATICS*, **V. 4(4).** pp. 505-554

Simulation of beam-splitter made of metamaterials with angle spatial distribution of constitutive parameters based on transformation optics for THz frequency range

E.A. Gurvitz, A.V. Vozianova, M.K. Khodzitsky

ITMO University, dept. of Photonics and Optoinformatics, 49 Kronverskiy pr. Saint-Petersburg, Russia, 197101

E-mail: egorgurvitz@gmail.com

Abstract. New approach to design beam splitter on basis of the transformation optics using angle constitutive parameters distribution of medium was proposed. The beam splitter was numerically simulated by COMSOL Multiphysics for terahertz frequency range. The numerical simulations were carried out for ideal and reduced constitutive parameters of medium for the case of TM plane wave.

1. Introduction.

Today the metamaterials [1] are very popular structures for investigation of unusual effects in quantum physics [4], acoustics [5] and electromagnetics [6]. The most interesting properties of metamaterial are band gap structure [7], negative refractive index [2,3], near zero refractive index [9,10], inverse Dopler's effect, inverse Vavilov – Cherenkov's effect [8], special distribution of constitutive parameters [6] and etc.

Generally, composite structures can be divided on the metallic and dielectric [11-12] ones, which unit cell sizes define the medium characteristics and operating frequency range. New types of metamaterial devices are being developed to control light such as expanders, compressors, lenses with super resolution, cloaking devices and beam splitters or couplers [13-15].

Last years the researchers are interested in the field of biomedicine, spectroscopy, detection & security systems, photonic chips and micro schemes due to to development of beam splitters and couplers for the terahertz frequency range.

Usually glass plates or prisms are used for wave splitting in classical optics and photonics. In the free space optics a glass plate or a prism (Wollaston prism, Glan-Taylor prism, Rochon prism, Nicol prism and another) can be used as beam splitters by light dividing on the transmitted and reflected components in the glass plate or light separating on the ordinary and extraordinary waves in the anisotropic medium of a prism. There are numerous approaches in fiber optics, high frequency data transmission lines to split light, but generally they don't permit to control light for the sophisticated cases, for example multilayer optics schemes.

Another approach is described by transformation optics [16]. It provides strict mathematical apparatus based on the differential geometry [17] and tensor analysis [18] which allows to solve complicated tasks. Transformation optics uses Fermat's principle to control light by space-time distribution of constitutive parameters using virtual space, where light propagates with predetermined behaviour and physical space, where light control is carried out by metamaterial medium. The concept of transformation optics beam splitter/coupler was recently proposed by using emended coordinate transformations [14-15]

It is generally accepted, that the transformations between the spaces are expressed by function of radius in curvilinear system in Euclidean space. We introduce another approach based on the transformation optics and devoted to consideration of space transformation using special angle distribution for description of electromagnetic wave splitting in the metamaterial for terahertz frequency range.

The detailed description of the mathematical formalism for definition of constitutive parameters of metamaterial medium which splits electromagnetic wave by the transformation optics method is presented below.

2. Analytical results.

Coordinate transformation from physical $x^i = \{x, y, z\}$ to virtual space $x^{i'} = \{x', y', z'\}$ (1) was considered in general terms. Coordinates of Cartesian system were expressed in the terms of cylindrical polar system:

$$\begin{aligned} x &= \cos(\varphi), & x' &= \cos(\varphi'), \\ y &= \sin(\varphi), & y' &= \sin(\varphi'), \\ z &= \hat{z}, & z' &= \hat{z}. \end{aligned} \tag{1}$$

Coordinates in Cartesian system of physical space were deduced for the case of angular transformation φ(φ') preserving z and r:

$$x = x'\frac{\cos(\varphi)}{\cos(\varphi')},\ y = y'\frac{\sin(\varphi)}{\sin(\varphi')},\ z = const. \tag{2}$$

The expressions (2) used to find elements $\boldsymbol{\Lambda}^{\boldsymbol{i}}_{\boldsymbol{j'}} = \frac{\boldsymbol{dx^i}}{\boldsymbol{dx^{i'}}}$ of the transformation matrix (3) which in general defines the transformation of the coordinates from physical to virtual space.

$$\Lambda = \begin{pmatrix} \frac{d\varphi}{d\varphi'}\sin(\varphi)\sin(\varphi') + \cos(\varphi)\cos(\varphi') & \cos(\varphi)\sin(\varphi') - \frac{d\varphi}{d\varphi'}\sin(\varphi)\cos(\varphi') & 0 \\ \cos(\varphi')\sin(\varphi) - \frac{d\varphi}{d\varphi'}\sin(\varphi')\cos(\varphi) & \sin(\varphi)\sin(\varphi') + \frac{d\varphi}{d\varphi'}\cos(\varphi)\cos(\varphi') & 0 \\ 0 & 0 & 1 \end{pmatrix}. \tag{3}$$

The transformation matrix is necessary to find permittivity and permeability using well-known formulas [13] directly following from consideration of the Maxwell's equation in physical and virtual spaces:

$$\varepsilon^{i'j'} = \frac{\Lambda^{i'}_{j'}\Lambda^{j'}_{i}\varepsilon^{ij}}{\det\left(\Lambda^{i'}_{j}\right)},\quad \mu^{i'j'} = \frac{\Lambda^{i'}_{j'}\Lambda^{j'}_{i}\mu^{ij}}{\det\left(\Lambda^{i'}_{j}\right)},$$

which describe constitutive parameters and determine space transformation of light. Permittivity and permeability tensors of metamaterial medium which provide electromagnetic wave propagation and beam splitting in the medium are following:

$$\varepsilon^i_j = \mu^i_j = \begin{pmatrix} Q^2\sin^2(\varphi) + \cos^2(\varphi) & (1-Q^2)\cos(\varphi)\sin(\varphi) & 0 \\ (1-Q^2)\cos(\varphi)\sin(\varphi) & \sin^2(\varphi) + Q^2\cos^2(\varphi) & 0 \\ 0 & 0 & 1 \end{pmatrix}, \tag{4}$$

where Q is derivation ∂φ/∂φ' and φ is angle in cylindrical polar system.

Tensors of permittivity and permeability in cylindrical polar system were also derived deduced and shown below:

$$\varepsilon^i_j = \mu^i_j = diag\left[\left(\frac{2\pi}{\varphi_2-\varphi_1}\right)\ \left(\frac{\varphi_2-\varphi_1}{2\pi}\right)\ \left(\frac{2\pi}{\varphi_2-\varphi_1}\right)\right]. \tag{5}$$

Unfortunately the process of fabrication of metamaterial medium with ideal constitutive parameters is complicated task. For simplification of the fabrication process tensors of ideal constitutive parameters were reduced for the case of transverse magnetic (TM) linear polarized wave. TM wave in Cartesian coordinate system interacts with μ_x, μ_y, ε_z components of metamaterial medium. Tensors of permittivity and permeability of medium with reduced material parameters are following:

$$\mu_j^i = \begin{pmatrix} Q^2\sin^2(\varphi)+\cos^2(\varphi) & (1-Q^2)\cos(\varphi)\sin(\varphi) & 0 \\ (1-Q^2)\cos(\varphi)\sin(\varphi) & \sin^2(\varphi)+Q^2\cos^2(\varphi) & 0 \\ 0 & 0 & 0 \end{pmatrix}, \quad (6)$$

$$\varepsilon_j^i = diag[0 \quad 0 \quad 1]. \quad (6')$$

The expressions for description of the beam splitting in metamaterial medium for cylindrical polar system for reduced constitutive parameters are expressed in the following from:

$$\mu_p = \left(\frac{2\pi}{\Delta\varphi}\right)^2, \mu_\theta = 1, \varepsilon_z = 1. \quad (7)$$

3. Simulation results.

COMSOL Multiphysics software was used for the numerical simulation of wave propagation in metamaterial medium. Finite element method (FEM) using by Comsol for solving Maxwell's equations, often, requires plenty time to calculate result with sufficiently precision. Expressions (2) preserve z and for this reason simulation can be reduced from 3D into 2D without undesirable simplifications. Scheme of numerical simulation is shown on the figure 1.

For reducing of reflection on the boundary of simulated model the perfect matched layer boundaries conditions were imposed.

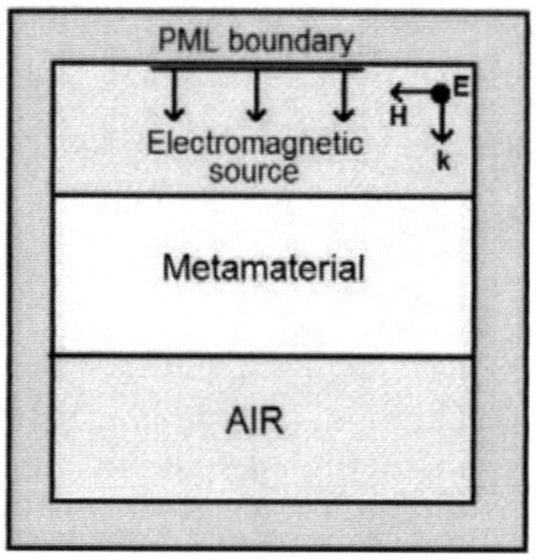

Figure 1. Numerical simulation scheme

The results of numerical simulations are presented on the figure 2 and figure 3. The splitting of the wave on the predetermined angle of 60 degrees was shown.

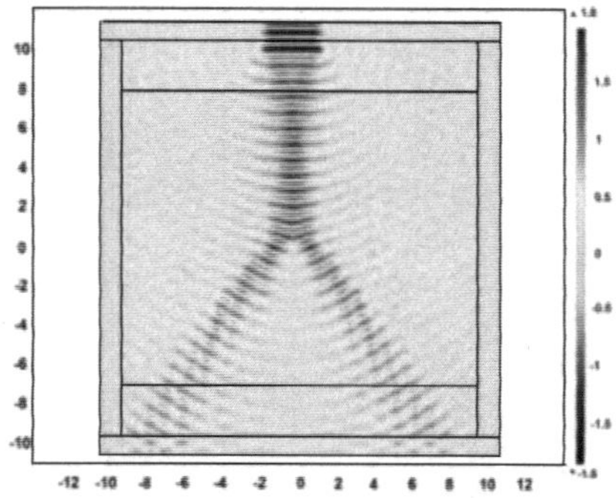

Figure 2. Wave splitting in the metamaterial medium with ideal

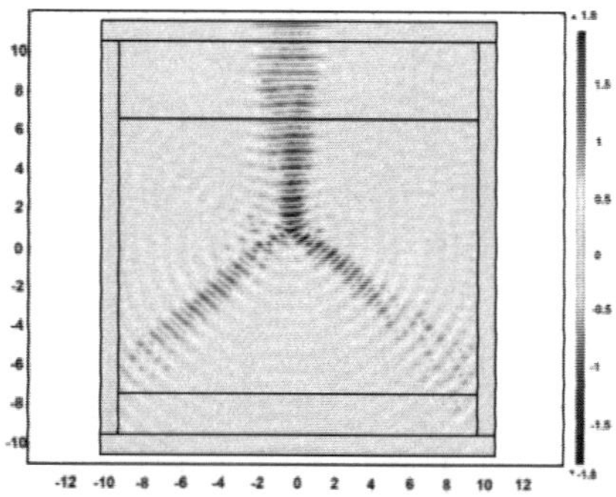

Figure 3. Wave splitting in the metamaterial medium with reduced

constitutive parameters at the frequency of 0.4 THz

constitutive parameters at the frequency of 0.4 THz

The figure 2 demonstrates the splitting of initial wave in the metamaterial medium simulated for the ideal constitutive parameters. The figure 3 shows how electromagnetic wave propagates in the metamaterial medium with reduced permittivity and permeability.

As can be seen from the numerical simulation, the medium with angle spatial distribution of constitutive parameters can be used as beam splitter.

4. Conclusion.

In this paper the model of beam splitter with angle dependence parameters based on the transformation optics mathematical apparatus was simulated for the ideal and reduced tensors of permittivity and permeability of metamaterial medium for the TM polarized electromagnetic wave. These results can be used for developments of couplers/splitters based on the metamaterials for the THz frequency range.

This work was financially supported by Government of Russian Federation, Grant 074-U01.

5. References:

[1] Filippo Capolino 2009 *Metamaterials Handbook. Theory and Phenomena of Metamaterials* (New York CRC Press)

[2] Khodzitsky M, Kharchenko A, Strashevskyi A, Tarapov S 2009 Left-handed properties of metal-ferrite composites placed into waveguide in millimetric wave range *Telecommunications and Radio Engineering*, **68** (7), pp. 561-566.

[3] Smith D, Padilla W, Vier D, Nemat-Nasser S, Schultz S, 2000 Composite Medium with Simultaneously Negative Permeability and Permittivity *Phys. Rev. Lett.* **84** 4184

[4] Rakhmanov A, Zagoskin A, Savel'ev S, Nori F 2008 "Quantum metamaterials: Electromagnetic waves in a Josephson qubit line *Phys. Rev. B* **77**, 144507

[5] Guenneau S, Movchan1 A, Pétursson1 G, Ramakrishna S 2007 *New J. Phys.* **9** 399

[6] Schurig D, Mock J, Justice B, Cummer S, Pendry J, Starr A, Smith D 2006 Metamaterial Electromagnetic Cloak at Microwave Frequencies *Science* **10** 314(5801):977-80

[7] Tarapov S, Khodzitsky M, Chernovtsev S, Belozorov D, Merzlikin A, Dorofeenko A, Vinogradov A, Granovsky A 2010 Frequency control of the microwave Tamm state *Physics of the Solid State*, **52** (7) , pp. 1427-1431.

[8] Veselago, V. 1968 The electrodynamics of substances with simultaneously negative values of [permittivity] and [permeability] *Soviet Physics Uspekhi* **10** (4): 509–514

[9] Silveirinha M, Engheta N, 2007 Design of matched zero-index metamaterials using nonmagnetic inclusions in epsilon-near-zero media *Phys. Rev. B* **75**, 075119

[10] Alù A, Silveirinha M, Salandrino A, Engheta N, 2007 Epsilon-near-zero metamaterials and electromagnetic sources: Tailoring the radiation phase pattern *Phys. Rev. B* **75**, 155410

[11] Vendik I et al 2012 Tunable Metamaterials for Controlling THz Radiation *IEEE Transactions on Terahertz Science and Technology*, Vol. **2**, № 5

[12] Zhaoa Q, Zhoua J, Zhangc F, Lippensc D 2009 Mie resonance-based dielectric metamaterials *Materialstoday,* Volume **12**, Issue 12, Pages 60–69

[13] Gurvitz E, Sedykh E, Khodzitskiy M 2012 Nonlinear cloaking at microwave frequencies *Proc. SPIE* **8455**, *Metamaterials: Fundamentals and Applications V*, 845532

[14] Rahm M, Cummer S, Schurig D, Pendry J, Smith D, (2008) Optical design of reflectionless complex media by finite embedded coordinate transformations, *Phys. Rev. Lett.* **100**, 063903

[15] Rahm M, Roberts D, Pendry J, Smith D, 2008 Transformation-optical design of adaptive beam bends and beam expanders *Optics Express*, Vol. **16**, Issue 15, pp. 11555-11567

[16] Philbin T, Leonhardt U, 2010 *Geometry and Light: The Science of Invisibility* (New York: Dover Publications)

[17] Kreyszig E, 1991 *Differential Geometry*, (New York: Dover Publications)

[18] Borisenko A, Tarapov I, 1979 *Vector and Tensor Analysis with Applications*, ed Silverman R (New York: Dover Publications)

Transmission of few-cycle pulses by Fabry-Perot interferometer

Elizaveta Buyanovskaya[1] **and Egor Kryshkovets**[2]

[1]Assistant lecturer, National Research University of Information Technologies, Mechanics and Optics, St. Petersburg 197101, Russia

[2]Laboratory assistant, National Research University of Information Technologies, Mechanics and Optics, St. Petersburg 197101, Russia

E-mail: lee.buyanovskaya@gmail.com

Abstract. We theoretically model the transmission of ultra-short optical pulses by Fabry-Perot interferometer, taking into account multiple waves re-reflection inside interferometer by examination of two superimposed waves. Based on the analytical and numerical analysis, we calculate the transformations of the optical electric field profiles and optical spectra of the pulses at the output of the interferometer.

1. Introduction

Nowadays high-intensity and extremely short optical pulses containing just few cycles of electromagnetic field are experimentally accessibly [1, 2, 3]. This opens new opportunities for applications taking advantage of strong non-destructive light-matter interactions. One potential possibility is to develop ultrafast optical transistors [4]. Optical switching could be enhanced through wave interactions in resonators [5, 6], and here we theoretically investigate the transmission of extremely short pulses through interferometer. Specifically, we analyze a Fabry-Perot interferometer (IFP) created by a nonlinear dielectric between two metal mirrors [7].

2. Theoretical model of Fabry-Perot interferometers for ultra short pulses

Theoretical investigations of ultra short pulses transmission by IFP were conducted according to technique that takes into account multiple waves re-reflection inside interferometer by examination of just two superimposed waves. We investigate ideal IFP (Fig. 1) that consists of two metal mirrors with reflection coefficient R spaced at distance L, dispersion and saturation (both mirrors and media inside interferometer) were out of the scope of current study. Here $z = z'/\lambda$, $t = t'/T_0$, where T_0 is central period of the pulse, λ is central wavelength, t, z are time and distance in seconds and meters respectively. Interferometer is considered *linear* if an intensity I is low and refraction index of the media inside interferometer N does not depend on it and *nonlinear* otherwise.

We analyze the wave transmission using short field equations accounting for the cubic nonlinearity of a dielectric material previously derived in [8] by generalizing them to account for the boundary conditions on the mirrors:

$$\begin{aligned}
&E_{+}^{(0)}(0,t) = \sqrt{1-R}\,E_{in}(t) + \sqrt{R}\,E_{-}^{(0)}(0,t)\\
&E_{refl}^{(0)}(0,t) = \sqrt{R}\,E_{in}(t) + \sqrt{1-R}\,E_{-}^{(0)}(0,t)\\
&E_{-}^{(0)}(L,t) = \sqrt{R}\,E_{in}(t)\\
&E_{out}^{(0)}(L,t) = \sqrt{1-R}\,E_{+}^{(0)}(L,t)
\end{aligned} \tag{1}$$

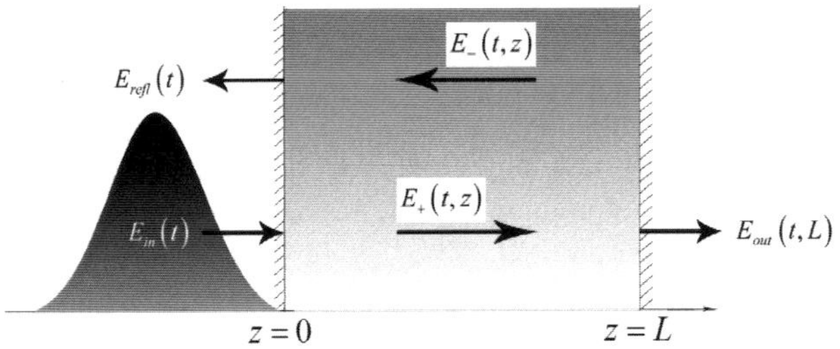

Figure 1. Schematic of ideal Fabry-Perot interferometer.

Then by applying method of successive approximations we obtain asymptotic analytical expression that describes linear propagation of the wave (the first term in RHS in (2a)) as well as nonlinear wave self-action (2b) and interaction (2c, 2d) between counter-propagating waves:

$$E_{out}(L,t) = E_{out}^{(0)}(L,t) + GE_{out}^{(1)}(L,t) = F^{-1}\left\{\frac{(1-R)F\{E_{in}(0,t)\}}{1-R\cdot exp(-i\omega\cdot 2L)}\right\} +$$
$$+ G\sqrt{\frac{1-R}{R}}\cdot\left[F^{-1}\left\{\frac{F\{s_{-}(L,t)+I_{-}(L,t)\}+\sqrt{R}F\{s_{+}(L,t)+I_{+}(L,t)\}}{R\cdot exp(-i\omega\cdot 2L)-1}\right\} + s_{-}(L,t)+I_{-}(L,t)\right] \tag{2a}$$

where

$$s_{-}(L,t) = L\left(\sqrt{\frac{R}{1-R}}\right)^3 \frac{\partial}{\partial t}\left[E_{out}^{(0)}(t-L)\right]^3,\; s_{+}(L,t) = L\left(\sqrt{\frac{1}{1-R}}\right)^3 \frac{\partial}{\partial t}\left[E_{out}^{(0)}(t-L)\right]^3 \tag{2b}$$

$$\begin{aligned}
I_{+}(L,t) = \frac{3}{2}\left(\sqrt{\frac{1}{1-R}}\right)^3 \frac{\partial}{\partial t}\Bigg[&\sqrt{R}\left(E_{out}^{(0)}(t-L)\right)^2 \int_{t-L}^{t+L} E_{out}^{(0)}(\xi-2L)\,d\xi\\
&+R\left(E_{out}^{(0)}(t-L)\right)\int_{t-L}^{t+L}\left(E_{out}^{(0)}(\xi-2L)\right)^2 d\xi\Bigg]
\end{aligned} \tag{2c}$$

$$\begin{aligned}
I_{-}(L,t) = \frac{3}{2}\left(\sqrt{\frac{1}{1-R}}\right)^3 \frac{\partial}{\partial t}\Bigg[&\left(E_{out}^{(0)}(t+L)\right)^2 \int_{t-L}^{t+L} E_{out}^{(0)}(\tau-2L)\,d\tau\\
&+E_{out}^{(0)}(t+L)\int_{t-L}^{t+L}\left(E_{out}^{(0)}(\tau-2L)\right)^2 d\tau\Bigg]
\end{aligned} \tag{2d}$$

$G = 4n_2I/N_0$ (here n_2 is coefficient of nonlinear refraction index), $F\{\}$ and $F^{-1}\{\}$ are direct and inverse Fourier transform respectively.

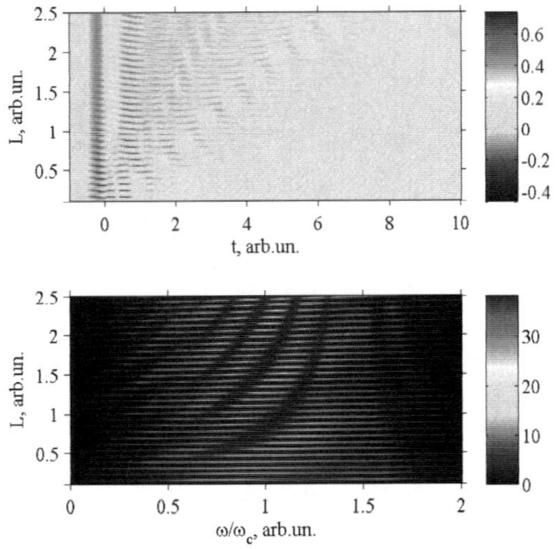

Figure 2. Temporal (above) and spectral (down) profiles of transmitted single-cycle radiation in linear case.

3. Simulation of ultra short pulses dynamics through Fabry-Perot interferometer

We perform numerical simulations for transmission of the Gaussian pulse

$$E_{in}(t) = E^{(0)}(t) = E_0 exp\left(-\frac{t^2}{\tau_0^2}\right) sin(2\pi t)$$

for various normalized pulse durations $\tau_0 = \frac{t_0}{T_0}$ (t_0 is pulse duration in fs), distances L and coefficients R for linear and nonlinear cases separately.

On the Fig. 2 typical temporal electric field profile and corresponding spectrum for different L for linear case are presented. From the figure it can be seen that in result of transmission of ultra short pulses by IFP we observe, as expected, modulated spectrum that in time domain corresponds to the sequence of subpulses. When the distance between mirrors is less than or commensurable with dimensions of single-cycle pulse in media in spectrum domain we observe bell-shaped structure more narrow than spectrum of initial pulse. In the nonlinear case (Fig. 3) the simulations predict enhanced generation of triple, quintuple and higher order harmonics.

4. Summary

In this paper we analyze the wave transmission using model that is formulated directly for the electric field without resorting to a slowly varying amplitude approximation and accounts for the cubic nonlinearity of a dielectric [9]. This allows us to analyze the dynamics of extremely short pulses, with durations down to a single field oscillation. We use method of successive approximations in order to find approximate analytical solutions, while taking into account nonlinear wave self-action and interaction between counter-propagating waves. Then based on the analytical and numerical analysis, we calculate the transformations of the optical electric field profiles and optical spectra of the pulses at the output of the interferometer.

It was shown that by varying interferometer parameters we can control transmitted spatiotemporal few-cycle structure. In result of transmission of ultra short pulses by linear IFP we observe modulated spectrum that in time domain corresponds to the sequence of subpulses that is defined by a resonator round-trip time. When the distance between mirrors is less than

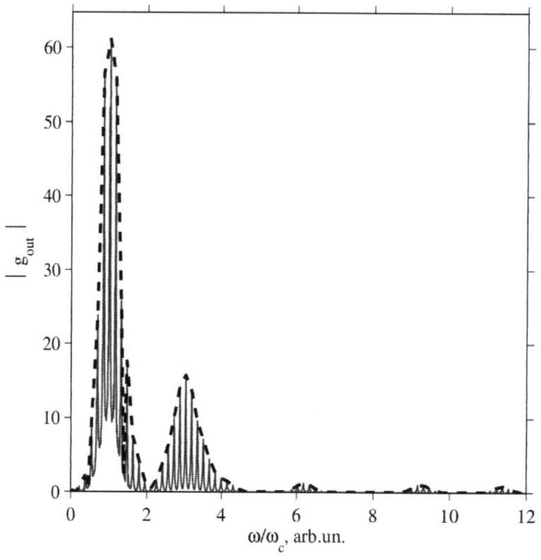

Figure 3. Spectrum of total (linear and nonlinear cases) transmitted radiation.

or commensurable with dimensions of single-cycle pulse in media in spectrum domain we observe bell-shaped structure more narrow than spectrum of initial pulse. Maximum of this structure is shifted to the short wavelength range with distance L. In time domain the single field structure with number of oscillations increasing from one and half to ten and more with coefficient R is formed. Energy of transmitted radiation is changing quasi periodically.

In the nonlinear case the simulations show enhanced generation of triple, quintuple and higher order harmonics, which is facilitated by multiple reflections and the shift of the spectrum maximum to the short wavelength region for the pulses with ultra broad spectrum [10]. The integral spectral density of transmitted radiation with decreasing distance L is changing quasi periodically like in linear case. These results indicate a possibility to create an optical transistor that works as an amplifier of few-cycle optical pulses.

Acknowledgments

Authors wish to acknowledge helpful discussions with Prof. Sergei Kozlov and financial support from Government of Russian Federation (Grant 074-U01) and from Russian Foundation for Basic Research (Grant 14-02-31367).

References

[1] Nazarkin A 2006 *Phys. Rev. Lett* **97** 163904
[2] Brabec Th and Krausz F 1997 *Phys. Rev. Lett.* **78** 3282
[3] Kozlov S and Samartsev V 2013 *Fundamentals of femtosecond optics* (Woodhead Publishing)
[4] Smith S P 1986 *Applied Optics* **25** 1550
[5] Renger J, Quidant R and Novotny L 2011 *Optics Express* **19** 1777
[6] Webb R P, Manning R J, Maxwell G D and Poustie A J 2003 *Electronics Letters* **39** 79
[7] Fabry C and Perot A 1897 *Ann. Chim. Phys.* **12** 459
[8] Buyanovskaya E and Kozlov S 2007 *J. Exp. and Theor. Phys. Lett.* **86** 297
[9] Bespalov V, Kozlov S, Shpolyansky Yu and Walmsley I A 2002 *Phys. Rev. A* **666** 013811
[10] Drozdov A, Kozlov S, Sukhorukov A and Kivshar Yu 2012 *Phys. Rev. A* **86** 053822

Journal of Physics: Conference Series **541** (2014) 012069 doi:10.1088/1742-6596/541/1/012069

Band-gap structure of photonic crystal with metasurface-teflon layers

E.A. Sedykh, A.K. Denisultanov and M.K. Khodzitsky
Department of Photonics and Optoinformatics, ITMO University,
Saint-Petersburg 197101, Russia

E-mail: man.room.harshrealm@gmail.com

Abstract: In present work the band-gap spectrum structure of photonic crystal made of metasurface/sitall/teflon layers is discussed.

Introduction

Recently a lot of attention has been dedicated towards metamaterials, especially photonic crystals [1,2,3]. Photonic crystals are composites made of alternating unit cells composed of multiple layers with various (usually dielectric) materials and can be of different configurations: 1D, 2D, 3D. Photonic crystals are useful instrument to implement such devices as frequency selective mirrors and filters. Today these devices are of high value for terahertz applications , as terahertz frequency range became extensively researched not long ago - terahertz technologies are at their infancy and effective components are to be developed. A lot of effort has recently been made towards designing filters, which can be extensively used in THz spectroscopy and tomography.

Considered structure

Present work investigates the band gap structure of photonic crystal made of alternating layers of metasurface (SRRs + sitall CT-50-1 (a counterpart of Schott Zerodur glass-ceramic) substrate) and teflon (Fig. 1). The photonic crystal consists of 10 bilayers: one having permittivity (ε_{teflon}=2.25) and thickness (h_{teflon}=300 um) of teflon and another – that of metasurface/sitall layer (experimentally obtained $\varepsilon_{sitall}(\omega)$ and h_{sitall}=550 um). SRR geometric parameters can be found in [4].

Figure 1: Considered photonic crystal structure composed of metasurface/sitall/teflon layers.

Metasurface dispersion characteristics extraction from experimental data.

Prior to numerical simulation of above mentioned photonic crystal, metasurface/sitall layer (Fig. 2) dispersion characteristics were obtained.

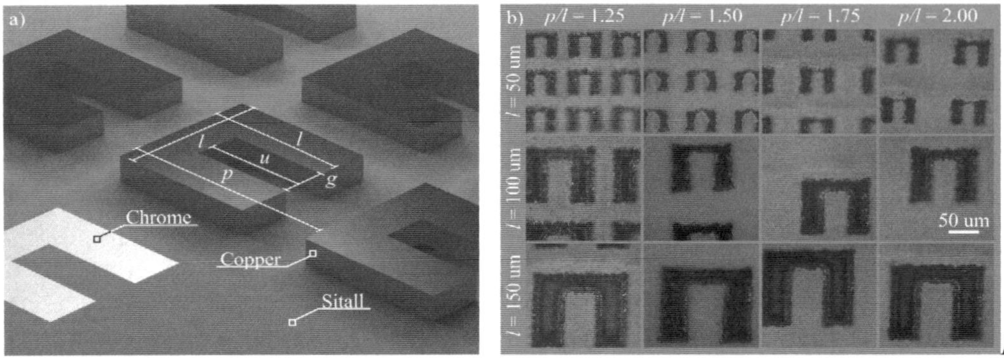

Figure 2: Metasurface unit cell dimensions (a) and photos of fabricated metasurfaces with various resonator side and relative period (b).

Time-domain spectroscopy setup (Fig. 3) was used to record transmission and reflection spectra of metasurface/sitall layers with different lengths and relative periods (different p/l ratios). This setup

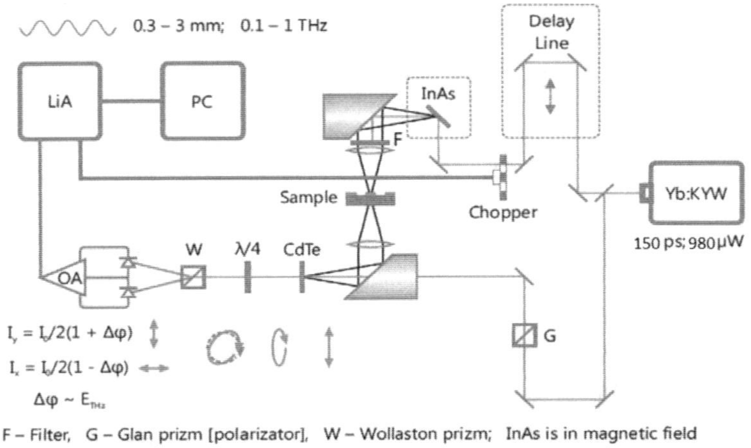

Figure 3: TDS setup used to obtain transmission spectra of the metasurface/sitall layers.

utilizes generation of THz radiation in semiconductor InAs crystal placed into strong magnetic field under femtosecond pulse excitation. Pump laser radiation is absorbed by the filter F, so only THz radiation reaches the sample, passes through it and then alters the orientation of CdTe crystal polarization plane depending on the amount of THz radiation transmitted. Thus CdTe crystal changes the state of polarization of probe pulse from linear to elliptical polarization. Quarter-wave plate is used to set the phase shift between its components. Wollaston prism is used to spatially separate orthogonal components of the probe pulse. The difference of the amplitudes of those components is registered by balanced detector. To increase signal-to-noise ratio lock-in detection (lock-in amplifier and modulator) is utilized. Delay line allows to perform time-base sweep by varying the phase lag between pump and probe pulses.

Fourier transform and Happ-Genzel apodization function were applied to those spectra. Resulting amplitude and phase spectra were normalized on the spectrum of free space. Real and imaginary parts of refractive index n(ω) and k(ω) are obtained from amplitude and phase spectra by using the following relations:

$$n_{smpl}(\nu)=1+\frac{c}{2\pi\nu d}(\phi_{smpl}(\nu)-\phi_{air}(\nu))$$
$$k_{smpl}(\nu)=1+\frac{c}{2\pi\nu d}\frac{E_{smpl}(\nu)}{E_{air(\nu)}} \quad (1)$$

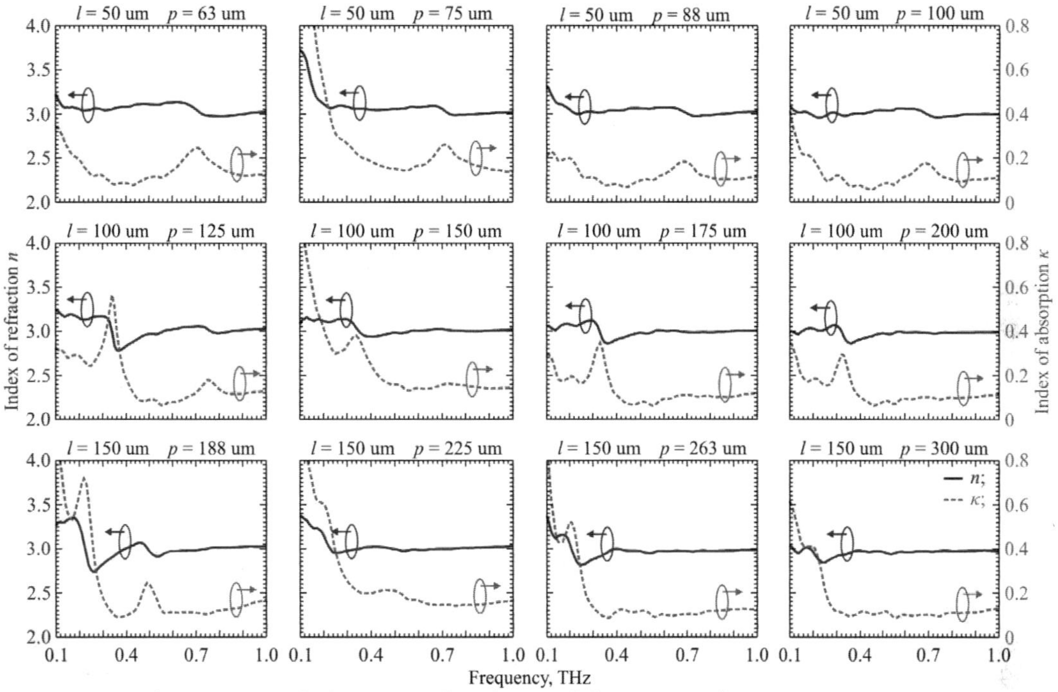

Figure 4: Refraction n and absorption k indeces of the metasurfaces.

It can be seen that metasurface/sitall n and k indeces spectra feature several regions where they change abruptly. This regions correspond to the eigenfrequencies of SRRs for lower frequencies and to plasmonic resonance for higher frequencies. Those can be most clearly seen on Fig.4 for metasurface with l=150 um and p=188 um.

Simulation and results

MATLAB program implementing transmission matrix method [5] for photonic crystal spectra acquisition was used to obtain transmission spectra (Fig. 5) of the proposed structure.

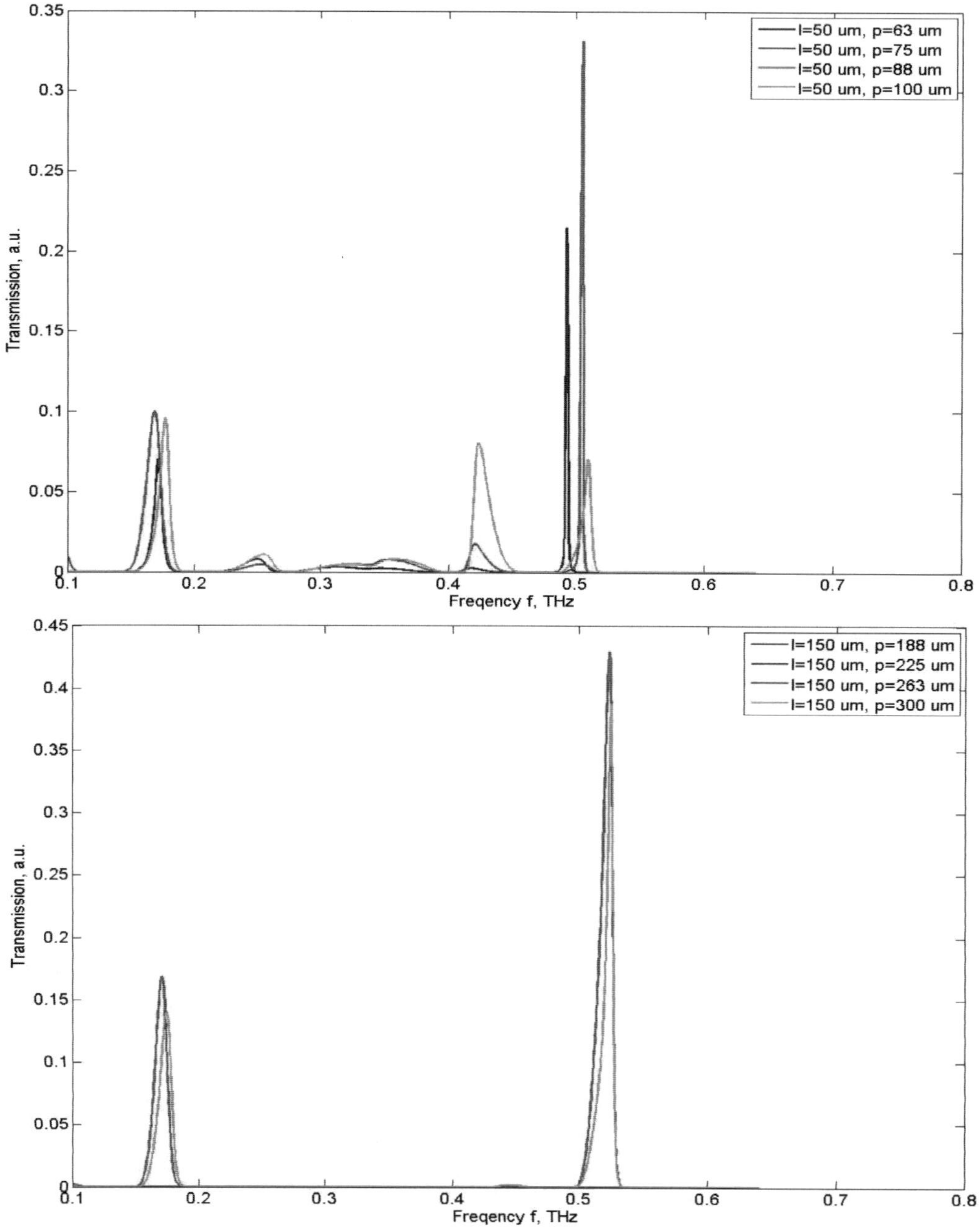

Figure 5: Transmission spectra of metasurface/sitall/teflon photonic crystals with various length and period of metasurface SRRs.

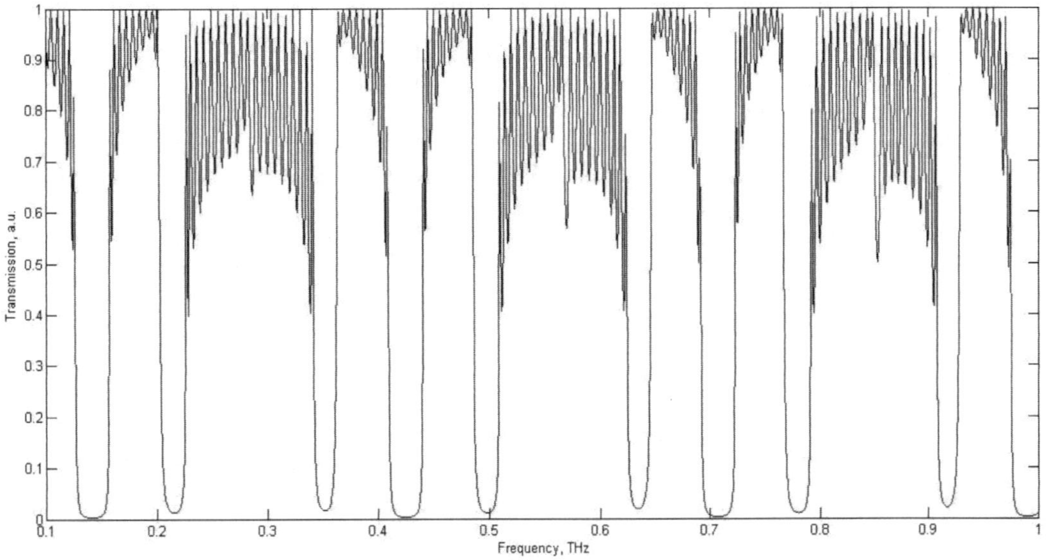

Figure 6: Transmission spectra of sitall/teflon photonic crystal without SRRs. Sitall and teflon layers have the same width as that of PC with metasurfaces.

It can be seen that resonant peaks corresponding to plasmon resonances (higher frequencies) occur near frequencies where $k_{SRR}>n_{SRR}$. Peaks corresponding to LC-resonance (lower frequencies) occur near SRR eigenfrequency. The shift of transmission peaks as compared to resonance frequencies of the metasurface is probably due to the presence of sitall and teflon layers. Compared to the transmission spectra of the photonic crystal band-gap structure of the PC with metasurfaces becomes negliable. It is also notable that such structures have Q-factor $Q=\frac{\Delta f}{f}=60$ (for metasurfaces with l=150 um) that is about 12 times greater (for both LC and plasmon resonance peaks) as that of the metasurfaces studied in [4] (see [6] for comparison).

Conclusion

Primary result of this work are numerically simulated spectra of photonic crystals with different metasurface lattice constant. This work will be used as a groundwork for development of tunable THz narrowband filters. Tunability can be realized in the variety of ways: the most trivial way is to create photonic crystal composed of areas with different relative period: photonic crystal made of such metasurfaces and layers of dielectric material can be mechanically tuned by shifting the structure along layer surface. Yet a more practical and flexible approach to realize tunable metamaterials is to use semiconductors [7] as a substrate for SRRs. It can be used to generate free carriers by the means of optical pumping or applied voltage in order to vary substrate permittivity. Tunable filters can find extensive use in THz biomedical applications – for example such filters can be use to cut THz frequencies that give negative impact on biological cells under study.

This work was partially financially supported by Government of Russian Federation, Grant 074-U01.

References

[1] M. Alessandro "Photonic Crystals - Introduction, Applications and Theory" – 2012 - p 344
[2] Tarapov S.I., Khodzitsky M., Chernovtsev S.V " Frequency control of the microwave Tamm state " - 2010 - Physics of the Solid State - 52 (7) - pp. 1427-1431.
[3] Girich A., Khodzitsky M., Nedukh S. " Experimental analysis of metamaterials' spectra to design tunable THz-GHz passive devices " - 2011 - NATO Science for Peace and Security Series B: Physics and Biophysics - pp. 159-164
[4] Y.Terekhov, M. Khodzitsky, Y. Grachev, E.Sedykh "The influence of period between U-shaped resonators on metasurface response at terahertz frequency range" – Proc. of SPIE – 2013 – Vol. 8806 – 88062Q
[5] A. Denisultanov, M. Khodzitsky "Band-gap structure for N-fold layers inside bilayer cell of photonic crystal for THz frequencies" – Proc. of SPIE – 2013 – Vol. 8806 – 88062I
[6] "Metafilms characteristics for the terahertz frequency range at the geometric parameters scaling" - Scientific and technical journal of information technologies, mechanics and optics - 2013 - issue 1 (83)
[7] R. D. Averitt, W. J. Padilla "Terahertz Metamaterial Devices" – Proc. of SPIE – 2007 – Vol. 6772 – 677209

Raman scattering in conducting metal-organic films deposited on nanoporous anodic alumina

E N Golubeva[1,3], H V Grushevskaya[1], N G Krylova[1], I V Lipnevich[1], T I Orekhovskaya[2]
[1] Belarusian State University, 4 Nezavisimosti Ave., Minsk 220030, Belarus
[2] Belarusian State University of Informatics and Radioelectronics, 6 P.Brovki Str, Minsk 220013, Belarus

Abstract. Raman scattering in dielectric nanostructures on the base of aluminum oxide was studied. A method of high dielectric nanocomposite fabrication was proposed. Metal-containing conducting Langmuir – Blodgett films were shown to shield the anodic nanoporous alumina from the action of strong electrical fields.

1. Introduction

A functional area of perspective electronic devices is a layered nanostructure exhibited quantum effects in charge and energy transfer. So, although aluminum oxide possesses an ionic conductivity, anodic aluminum oxide (AOA) film may become a p-type semiconductor at high electrode potentials in electrolytes with the oxygen-contained anions [1]. To utilize, for example, in biomedical applications, an interdigital aluminum electrode structure with AOA coating is placed in the electrolytic environment under the influence of intense electric fields of near-electrode Helmholtz layer. The described instability of AOA band structure prevents its using as an insulating layer of capacity sensors. In this paper a method for improvement of AOA operating characteristics via strong electric fields shielding by iron-containing Langmuir – Blodgett (LB) films of aromatic compounds will be proposed. In such films, energy of electrically charged quasiparticles transported along the π-conjugated double bonds chain (long-range transfer) may be spent to excite dipole-active states (exciton associated with an electromagnetic field) of iron atoms.

The polymeric LB-films belong to strongly correlated many-electron systems with linear electron spectrum (narrow-gap semiconductors). Electric charge carriers in these systems are negatively charged excitons X^- [2]. A valence band of these materials represents itself electronic states similar to p-electron orbitals with total angular momentum j = 3/2 and its projection j_z = ± 3/2, and their conduction band is s-electron orbitals with j = 1/2 and projection j_z = ± 1/2. According to current ideas [3], negatively charged excitons are formed under the action of a resonant electromagnetic field when the hole transits from the valence band to the conduction one and joins to two electrons in the conduction band. An arising three-particle excited state has zero orbital angular momentum (L = 0) and a spin S, equal to unity: S = 1 because of an absorbed photon holds unity spin and the hole adds spin 1/2. Then, one can assume that transitions from the lower Zeeman sublevel with spin "up" (s = +1/2) in the valent band for the state j = 3/2, j_z = +3/2 to upper Zeeman sublevel j = 3/2, j_z = – 3/2 with spin "down" (s = –1/2) in the valent band occur via a three-particle excited state with L = 0, S = 1.

[3] Corresponding author, E-mail: golubeva.ln.87@gmail.com

But, the total angular momentum J of the excited system having a nonequilibrium spin $S = 1$ is equal to $J = 1$. Therefore, such transitions are double forbidden. First, the law of angular momentum conservation is violated for the system, as the change of total angular momentum value from $J = 1$ to $j = 3/2$ at the decay of exciton with electromagnetic quantum emission is accompanied by the increment $\Delta = 1/2$ of the system angular momentum with respect to its initial value before excitation. Secondly, if one assumes that Δ is a spin of additional electron with spin "down" at the upper Zeeman level of the valent band, then the probability to find the electron at this level is extremely low owing to the Pauli exclusion principle. Raman spectroscopy allows investigation of forbidden transitions. However, studies, which would reveal the type of electronic states and mechanisms of charged exciton transport in LB-structures being suitable for AOA shielding, has not been performed.

The goal of this paper is to carry out Raman spectroscopy of the organometallic LB-coating deposited on AOA and to propose a mechanism of shielding coherent exciton transport for such structures under the influence of strong electromagnetic fields.

2. Materials and methods

Two-stage anodic treatment of A99 grade aluminum foil in electrolyte with sulfuric acid was carried out by the method described in [1] to fabricate nanoporous AOA with pore diameter ~ 10 nm.

Langmuir monolayers from conductive polymer oligomer molecules 3-hexadecyl-2,5-di (thiophen-2-yl)-1*H*-pyrrole (thiophene-pyrrole) with chemically bounded hydrophobic 16-link carbon chain were formed on an automated home-built Langmuir trough with controlled deposition on a substrate. Aqueous solution of ferric nitrate $Fe(NO_3)_3$ or ferric chloride $FeCl_3$ with pH adjusted by addition of hydrochloric acid HCl was used as a subphase for monolayer formation. Five or seven Langmuir monolayers in a highly ordered solid state were deposited on the AOA surface at a constant compression pressure to obtain a suspended LB-film.

Spectral studies in visible range were carried out using confocal micro-Raman spectrometer Nanofinder HE («LOTIS-TII», Tokyo, Japan) at laser excitation wavelength of 532 nm. Figures 1a and 1b demonstrate confocal microscopy images of the 5-monolayer thiophene-pyrrole LB-film deposited on nanoporous alumina and the nanoporous AOA surface, respectively.

(a) (b)

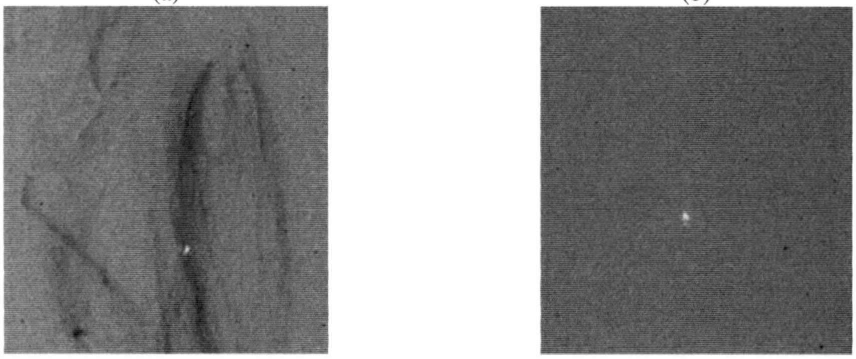

Figure 1. A confocal microscopy image of 5-monolayer thiophene-pyrrole LB-film (a) deposited on nanoporous alumina (b).

3. Model

Fe atoms in aromatic compounds of pyrrole derivatives may be in a high- or low-spin state. In the metal-center low-spin state these pyrrole rings lie in the same plane. In the high-spin state the supramolecular complex becomes dome-shaped. Therefore, we can judge about the spin state of metal center by the vibration spectra of polymer matrix. Electrochemical and thermodynamic studies of Fe-containing thiophene-pyrrole series oligomer films show that the valence of high-spin metal center is Fe^{II} [4].

It is assumed that the multiparticle excitation of the supramolecular complex consists of an electron in conduction band of the film, a hole in electron density of high-spin Fe^{II} and a negatively charged quasiparticle excitation of polymer matrix electron density that is caused by electron leaving from the valence band. Reverse electron transfer from d-electron orbitals of donor chain with π-conjugation occurs in several stages: first, there is an electron transfer to acceptor nitrogen atom of complex $[Fe(pyrrole)_3]$, and then – excitation energy transfer from the nitrogen atom of negatively charged complex $[Fe(pyrrole)_3]^-$ to the electron-hole excitation of metal center [5]. As a result, the exciton X^- consisting of an electron and an electron-hole pair is formed.

For ultrathin metal-containing polymeric 5-monolayer LB-films synthesized from thiophene-pyrrole oligomer derivative and having a thickness of monolayer conducting hydrophilic part $d \sim 0.3$ nm the estimation of the exciton radius by the formula $r_{X^-} = \frac{1}{2}\sqrt{\pi a d}$ [6] gives a value of about 1.68 nm, where a is a radius of exciton in a bulk sample (in our calculations a was chosen as large as a value of film thickness: 12 nm).

4. Raman scattering

Raman spectra of AOA excited surface states are shown in figure 2a. The surface states have Raman peak at 1058 cm^{-1}. They cause the ionic structure destruction of nanoporous AOA samples (pore diameter ~ 10 nm) with appearance of gap and are excited at high laser power P ($P \geq 3.0$ mW). These changes of quasi two-dimensional AOA band structure caused by intensive electromagnetic radiation correlate with changes of AOA electrical conductivity type from ionic to semiconducting one.

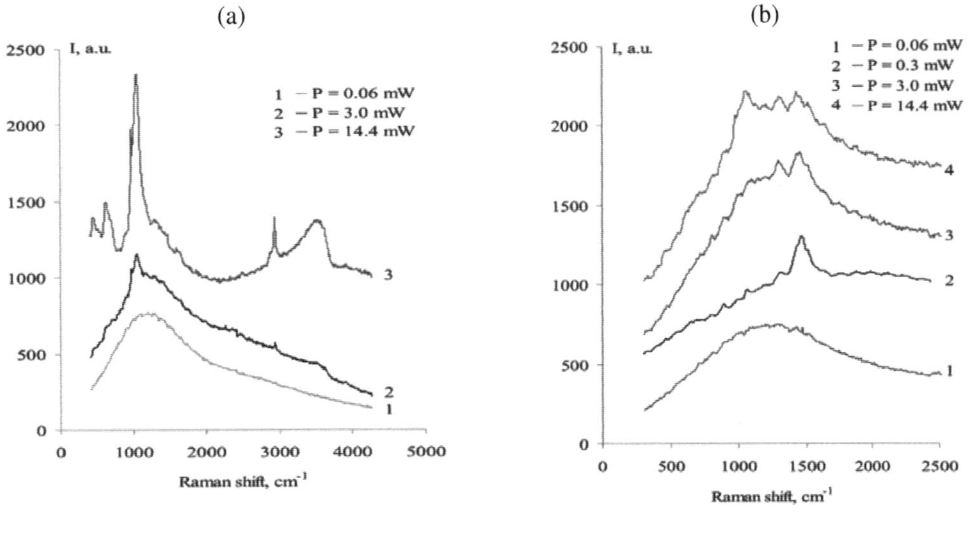

Figure 2. Raman spectra of nanoporous alumina without (a) and with LB-film coating (b).

The intensity of LB film Raman bands also depends on the power of laser radiation, as shown in figures 2b and 3. At low intensity $P \sim 0.3$ mW the bands with maxima at 893, 1063, 1310, 1432, 1472 and 1515 cm^{-1} are resolved in the Raman spectrum of LB film deposited on AOA. The Raman peaks at 1432 cm^{-1} and 1472 cm^{-1} are close to semi-circle stretch of pyrrole rings and quadrant stretch of 3-substituted pyrrole rings, respectively. The Raman peaks at 893, 1063, and 1138 cm^{-1} are close to hexadecyl-link tail carbon trans chain C-C-C out-of-phase stretches, when wide band with maximum of 1310 cm^{-1} is inter-ring C-C vibrations. All characteristic molecular vibrations, observed in the Raman spectrum of 5, 7 – monolayers LB film using 532 laser line are represented in Table 1.

Table 1. Characteristic molecular vibrations, observed in the Raman spectrum of 5, 7 – monolayers LB film using 532 laser line.

Band center for LB film (cm^{-1})	Region for vibration of similar molecular group (cm^{-1})	Assignment	Reference
556	525	ring in-plane deformation	[7]
653	586	ring in-plane deformation	[7]
686	686	thiophene-pyrrole molecule ring in-plane deformation	[4]
741		ring in-plane deformation	
811	808	ring in-plane deformation	[7]
843	844	thiophene-pyrrole molecule ring in-plane deformation	[4]
893	893	trans hexadecyl-chain vibration	[8, p.118]
1063	1063	hexadecyl-chain trans C–C–C out-of-phase stretches	[8, p.154]
1138	1138	hexadecyl-chain trans C–C–C out-of-phase stretches	[8, p.140]
1165	1148	Cl–O vibration	[4; 8, p.170]
1231	1236	CH bending	[7]
1310	1310	inter-ring C–C vibration	[9]
1342	1357	semi-circle stretch of thiophene ring	[8, p.109]
1410	1408	quadrant stretch + CH rock vibration of thiophene ring	[8, p.109]
1432	1420–1430	semi-circle stretch of pyrrole rings	[8, p.109]
1472	1480–1490	quadrant stretch + CH rock vibration of pyrrole rings	[8, p.109]
1515	1518	NH bend + C–N stretch	[8, p.151]
1581	1585–1607	C=C stretching in pyrrole rings	[9]

Increase of P from 0.3 to 3.0 mW leads to redistribution of electron density to another orbital being close but higher placed, therefore, instead of a line at 1472 cm^{-1} another one of 1460 cm^{-1} appears. Intensity of the last one is comparable with intensity of the line at 1310 cm^{-1} and 3.5 times lower than intensity of the line at 1472 cm^{-1}. Spectral changes dependency on laser beam power can be explained in the following way: quadrant stretch of 3-substituted pyrrole ring occurs at low intensities in the plane of the ring, but at high intensities out-of-plane vibrations take place. In the last case, the probability of inter-ring C–C vibrations increases, that reveals as 1310 cm^{-1} line intensity increase. Out-of-plane quadrant stretch vibrations bend the pyrrole ring plane that leads to dome-shaped form of complex. The dome-shaped complex testifies the high-spin state Fe^{II}.

5. Shielding effect

At high level of laser pumping a shielding of AOA by LB-film also takes place. For example, at P ~ 14.4 mW AOA spectral band with maximum at 1058 cm^{-1} is reduced 15 times, and the Raman peaks at 463, 632 and 2934 cm^{-1} disappear in the presence of LB film deposited on the AOA surface. The shielding is due to the presence of free charge carriers.

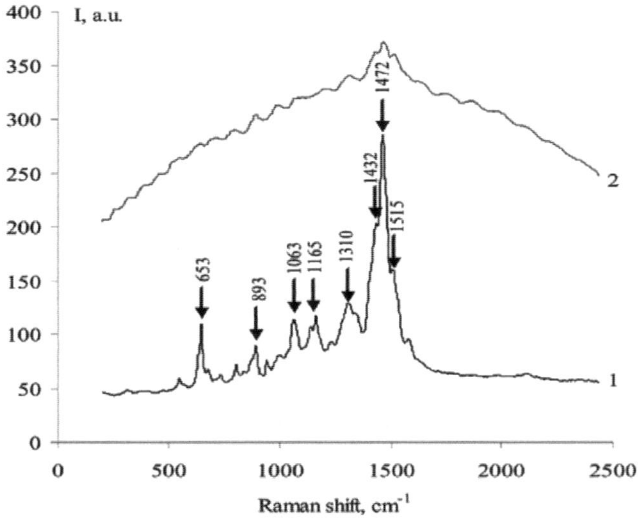

Figure 3. Raman spectra of 7-monolayer thiophene-pyrrole LB-film deposited on AOA at laser beam intensity P = 0.03 mW (curve 1), 0.6 mW (curve 2).

Electromagnetic radiation scattered by AOA diffracts on spin-polarized dipole-active states of metal centers due to the precession of the nonequilibrium spin S. The change in Raman spectrum intensity in figures 2b and 3, curves 2 has a period of 3.8 ± 0.4 nm that is close to calculated diameter of X^-. Therefore, the LB-films under consideration are 2D photon crystals whose free charge carriers are the excitons X^-.

At P ~ 14.4 mW the line at 1460 cm^{-1} shifts to 1450 cm^{-1}. This testifies a decrease of dipole polarization in the LB-structure due to the electric field shielding by the excitons.

6. Conclusion

Thus, the long-lived exciton states transfer in LB-film of metal-organic compound was established. As a result, the conducting LB-films shield the AOA layer from strong electromagnetic fields and preserve high-insulating properties of this layer.

7. References

[1] Abramov I I, Hrushevsky V V, Krylov G G, Krylova H V, Lipnevich I V and Orekhovskaya T I 2012 *Petersburg J. Electronics.* **4 (73)** 59
[2] Krylova H, Hurski L 2013 *Spin polarization in strong correlated nanosystems* (LAP LAMBERT Academic Publishing)
[3] Morton J J L, Lovett B W 2011 *Annu. Rev. Condens.Matter Phys.* **2** 189
[4] Grushevskaya H V, Lipnevich I V, Orekhovskaya T I 2013 *J. Mod. Phys.* **4** 7
[5] Hursky L I, Krylova H V, Lipnevich I V, Kozirkov Y Y, Orekhovskaya T I Method of obtaining the protecting exciton-dielectric polymer layer for narrow profile emission and absorption layers of nano-optoelectronic devices 2013 *Official Bulletin of the National Centre of Intellectual Property* **3 (92)** 14
[6] Keldysh L V 1979 *Letters to JETP(in Russian)* **29** 716
[7] Constantino C, Duff J, Aroca R 2001 *Spectrochimica Acta Part A* **57** 1249
[8] Larkin P 2011 *Infrared and raman spectroscopy: principles and spectral interpretation* (Elsevier Inc., San Diego, Oxford).
[9] Santos M J L, Brolo A G, Girotto E M 2007 *Electrochimica Acta* **52** 6141

Light bullets from Mid-IR femtosecond filament in air

E O Smetanina[1, 3], V Yu Fedorov[2], A E Dormidonov[1], V P Kandidov[1]
[1] International Laser Center, Chair of General Physics and Wave Processes, Faculty of Physics, M.V. Lomonosov MSU
Leninskie gori 1, Moscow 119991, Russian Federation
[2] Lebedev Physical Institute of the Russian Academy of Sciences, Leninskiy prospekt 53, Moscow 119991, Russia

E-mail: smetanina@physics.msu.ru

Abstract. We numerically investigated the formation of light bullets under filamentation of 3.8-μm femtosecond pulse in presence of anomalous GVD in humid air. The dispersion of humid air in infrared spectral region was characterized by model fits of the real part of refractive index from HITRAN database. The fit for the 2.8-4.2 μm region describes the areas of anomalous GVD near 3.1 μm and 4 μm in humid air. During the nonlinear propagation of femtosecond pulse in humid air, the compressed in space and time wave packet - light bullet - appears in the central time layers of the pulse and moves to the pulse tail. The duration of light bullet reaches the few-cycle value and the energy fluence in the in the light bullet cross-section is above 1 J/cm^2. Together with the light bullet formation the spectrum of the pulse broadens strongly and covers the spectral region from 3 μm to 5 μm. The interference model was used for investigation of the frequency-angular distribution of the supercontinuum spectrum components.

Introduction

The start point of the filament, the length of its plasma channels and the efficiency of the spectral conversion into supercontinuum could be controlled by the input pulse parameters according to the nonlinear medium in which the pulse propagates [1-3]. Nowadays high-power Mid-IR femtosecond sources are started to use for filament formation in the gaseous media [4-5]. The generated coherent [6] supercontinuum covers over several octaves, that is very promising for the molecular spectroscopy. The filamentation of 3.1-μm pulse under anomalous GVD was investigated numerically in [7]. In this work

[3] To whom any correspondence should be addressed.

we present the results of numerical simulation of the nonlinear propagation of the 3.8- μm femtosecond pulse in humid air.

Methods

We used two mathematical models of pulse propagation in nonlinear medium: the first model based on slow varying wave approximation [8] and the second model based on the unidirectional wave equation [9] for complex electric field. In both cases the ionization of oxygen and nitrogen was calculated according to the PPT model [10]. The order of the multiphoton ionization process for oxygen $K_{02} = 37$ and for nitrogen $K_{02} = 48$ at the considered pulse central wavelength $\lambda_0 = 3.8$ μm.
The full dispersion of humid air was taken into account according to the fits (figure 1 a) [11].

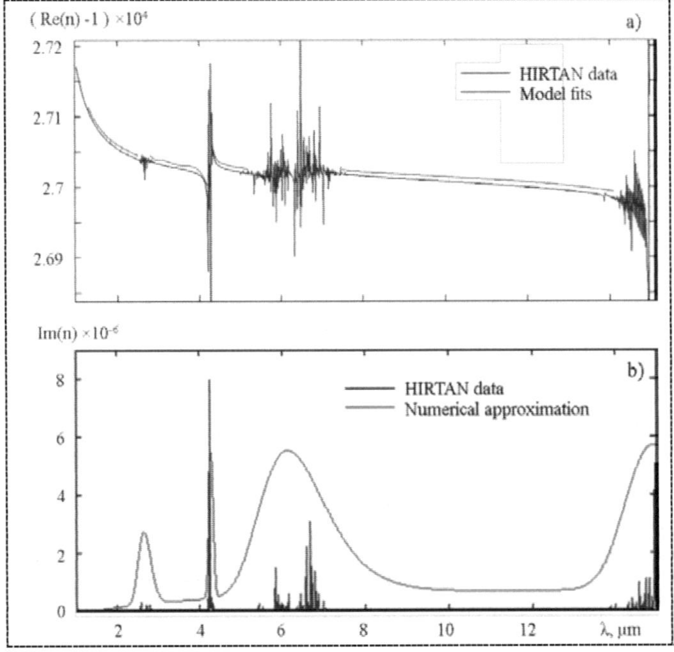

Figure 1. Refractive index of air. a) Real part of air refractive index from HIRAN database (blue line) and model fits (red line) [11]. b) Imaginary part of air refractive index from HIRAN database (blue line) [11] and numerical approximation (red line) used in this work. Temperature 290.65 K, humidity 10 % , atmospheric pressure 101330 Pa.

In our numerical simulation the resonance areas are represented as the areas with high absorption (figure 1 b). In this work we numerically investigated the formation of light bullets (LBs) under filamentation of 3.8-μm femtosecond pulse in presence of anomalous GVD in humid air.

The behaviour of the dispersion parameter $k_2 = \dfrac{\partial^2 k(\omega)}{\partial \omega^2}$ near 3.8 μm for humid air (figure 2) is close to the behaviour of k_2 near 2 μm in fused silica [12]. The anomalous DVG in fused silica plays a key role in the formation few-cycle wave packets that have the quasi-soliton properties [13].

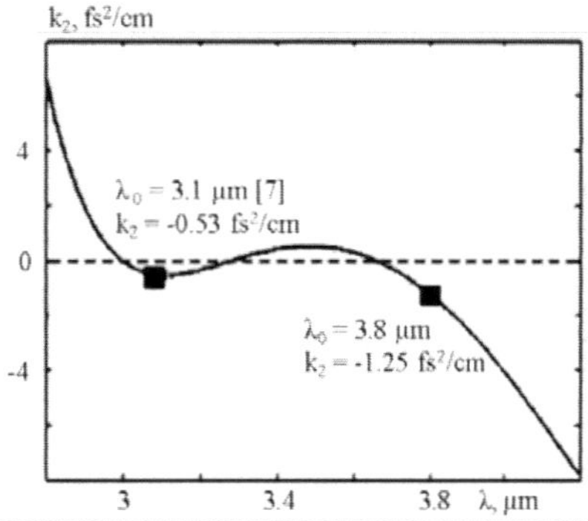

Figure 2. Dispersion parameter k_2 for humid air according to the model fits [11] in range of wavelengths: 2.8÷4.2μm.

To investigate the frequency-angular distribution of the supercontinuum spectral components during filamentation of the 3.8-μm femtosecond laser pulse we used the interference model [14]. This model is based on the assumption, that the broadened supercontinuum point source moves along the high-intensity areas in the filament. The interference of the emitted spectral components defines the modulation of the frequency-angular spectrum of the supercontinuum.

Results and discussion

The transformation of the laser pulse starts from the Kerr self-focusing of the central time layers (figure 3, z = 21.1 m). Then the self-induced laser plasma defocuses the pulse tail. The peak plasma density reaches the value $Ne_{max} = 6\times10^{14}$ 1/cm^3. In presence of the Kerr-induced positive self-phase modulation (SPM) the anomalous GVD leads to the power flows from pulse front and tail to the pulse center. Such regime of the pulse propagation under positive SPM and anomalous GVD causes to formation of the localized in space ant time wave packet - light bullet [15]. The pulse central wavelength $\lambda_0 = 3.8$ μm in our numerical simulations lays in the region of anomalous GVD in humid air [11]: $k_2 = -1.25$ fs^2/cm. The input pulse parameters: $\tau_{FWHM} = 170$ fs and $a_0 = 1$ cm. Ander the considered conditions the light bullet appears at the propagation distance about z = 21 m. During filamentation in humid air the duration and radius of LB decrease from input pulse values to the $\tau_{LB} = 30$ fs and $a_{LB} = 250$ cm (figure 3). Peak intensity reaches the value $I_{LB} = 5\times10^{13}$W/cm^2, and the energy fluence in the in the LB cross-section is above $F_{LB} = 1$ J/cm^2.

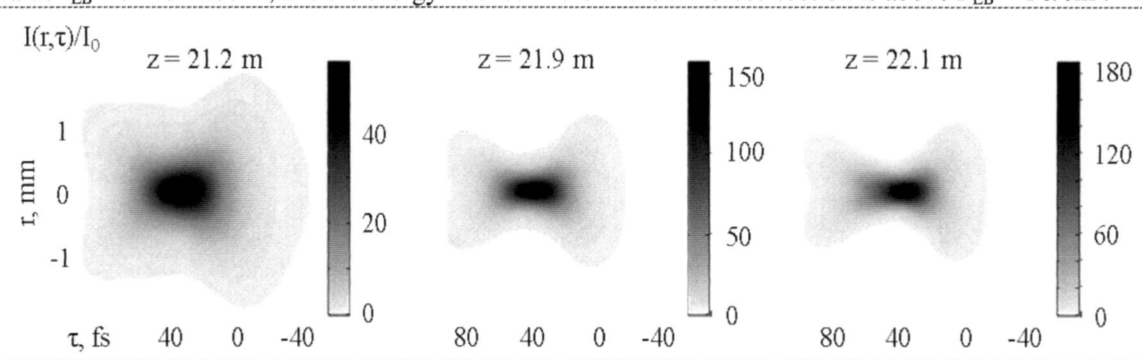

Figure 3. The spatio-temporal distribution of the intensity during the LB formation. Numerical simulation is based in the slow varying wave approximation. $I_0 = 2.7\times10^{11}$ W/cm^2. The intensity distribution is presented in the fame moving with the group velocity of the input laser pulse. Negative values of time τ corresponds to the pulse front.

The results of numerical simulation, based on both models, describe the same LB evolution during the pulse propagation. LB forms in the central time slices and than shifts to the pulse tail (figures 3,4). At the same time the spectrum of the LB broadens and the central wavelength shifts to the red side of the spectrum. At the propagation distance z = 23 m the supercontinuum spectrum covers the region from 3 μm to 5 μm. The dip in SC around 4.25 μm appears due to the absorption in the resonance region (figure 1 b) [11].

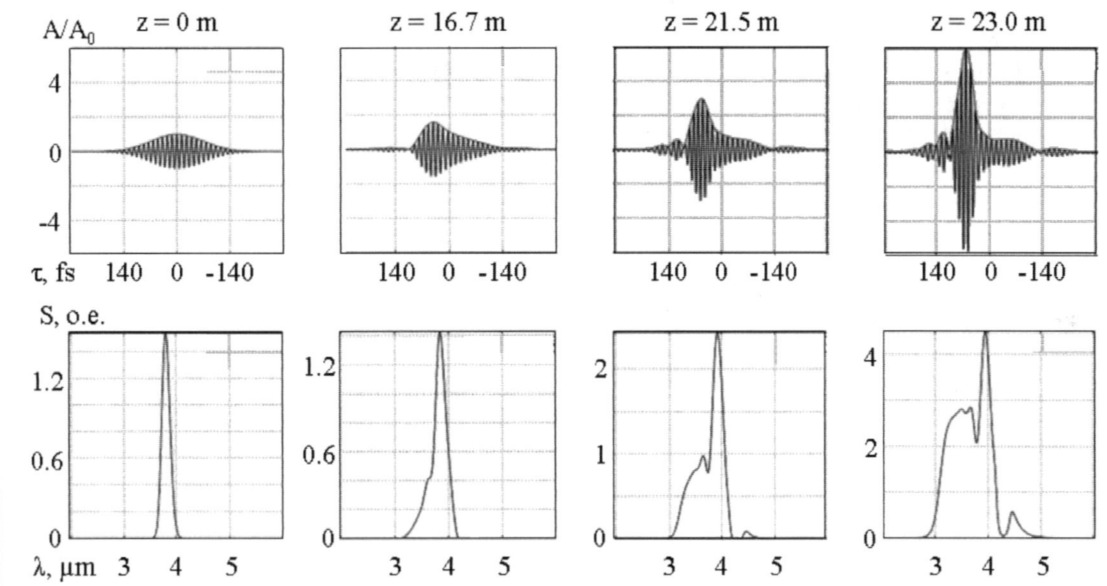

Figure 4. LB time profile (first line) and its spectrum transformation (second line). Numerical simulation is based on the unidirectional wave equation. A_0=1.4×10^7 V/cm. LB profiles are presented in the fame moving with the group velocity of the input laser pulse. Negative values of time τ corresponds to the pulse front.

Using the interference model [14] we found the frequency-angular distribution of the supercontinuum components of the 3.8-μm light bullet in humid air. We used the refractive index n(ω) from HITRAN database [11] and the numerically obtained group velocity of the light bullet υ_{LB} , that is smaller than the group velocity of the input pulse υ_g : $\upsilon_{LB} = (1-0.00005)\cdot\upsilon_g$. The length of the emitting region L in the filament also was estimated from our numerical simulation. It was assumed that the broadband source emits at all wavelengths and in all directions equally.

We found that in the conical emissions is formed in Stokes and in anti-Stokes regions of the supercontinuum spectrum $S_{interf}(\theta,\lambda)$ (figure 5) with the increasing of the emitting area length in the filament. The most strong interference maximum appears in the long-wavelength region of the supercontinuum spectrum $S_{interf}(\theta,\lambda)$. The wavelength of constructively interfering radiation increases with the increasing of the angle of its propagation. This behaviour of the long-wavelength supercontinuum components is in a good agreement with the experimental results on the registration of the conical emission of the Mid-IR radiation in the different types of filamentation processes [16,17].

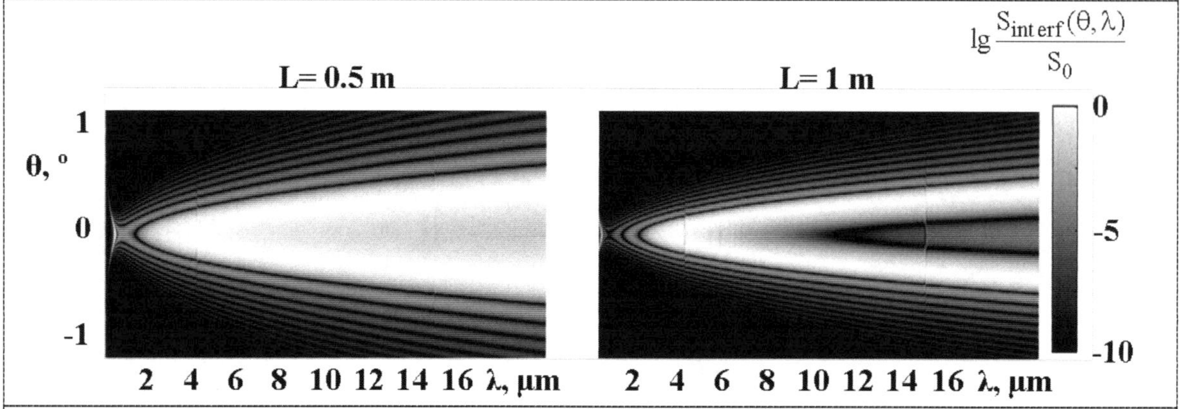

Figure 5. The transformation of the frequency-angular distribution $S_{interf}(\theta,\lambda)$ of the supercontinuum components along the emitting region in the filament. Central wavelength $\lambda_0 = 3.8$ µm. L - length of the emitting region. S_0-constant.

Acknowledgments
This work was supported by the grant of the President of the Russian Federation for state support of leading scientific schools - NSh-3796.2014.2; Grants RFBR: 13-02-90476-Ukr_f_a, 14-22-02025-Ofi_m.

4.References

[1] Couairon A and Mysyrowicz A 2007 Physics Reports **441** 47
[2] Kasparian J, Rodriguez M, Méjean G, Yu J, Salmon E, Wille H, Bourayou R, Frey S, André Y-B, Mysyrowicz A, Sauerbrey R, Wolf J-P, and Wöste L 2003 *Science* **301** 61
[3] Kandidov V P, Shlenov S A, Kosareva O G 2009 *Quantum Electronics* **39** 205
[4] Andriukaitis G, Balčiūnas T, Ališauskas S, Pugžlys A, Baltuška A, Popmintchev T, Chen M-C, Murnane M M and Kapteyn H C 2011 *Optics Letters* **36** 2755
[5] Kartashov D, Ališauskas S, Pugžlys A, Voronin A, Zheltikov A, Petrarca M, Béjot P, Kasparian J, Wolf J-P and Baltuška A 2013 *Optics Letters* **38** 3194
[6] Chin S L, Brodeur A, Petit S, Kosareva O G, and Kandidov V P 1999 *Journal of Nonlinear Optical Physics & Materials* **8** 121
[7] Shim B , Schrauth S E and Gaeta A L 2001 *Optics Express* **19** 9118
[8] Brabec T, Krausz F 1997 *Phys. Rev. Lett.* **78** 3282
[9] Kolesik M, Moloney J V 2004 *Phys. Rev. E* **70** 036604
[10] Perelomov A M, Popov V S and Terent'ev M V 1966 *Sov. Phys. JETP* **23** 924
[11] Mathar R J 2007 *J. Opt. A: Pure Appl. Opt.* **9** 470
[12] Malitson I H 1965 *JOSA* **55** 1205
[13] Smetanina E O, Kompanets V O, Dormidonov A E, Chekalin S V and Kandidov V P 2013 *Laser Physics Letters* **10** 105401
[14] Dormidonov A E, Kandidov V P *Laser Physics* 2009 **19** 1993
[15] Silberberg Y. 1990 *Optics Letters* **15** 1282
[16] Théberge F, Châteauneuf M, Ross V, Mathieu P, and Dubois J *Optics Letters* 2008 **33** 2515
[17] Nomura Y, Shirai H, Ishii K, Tsurumachi N , Voronin A A, Zheltikov A M, and Fuji T *Opt. Express* 2012 **20** 24741

Enhanced third harmonic generation using the surface states of light in periodic photonic structures

Kirill I. Zaytsev[1]*, Vladimir S. Gorelik[1,2], Gleb M. Katyba[1], and Stanislav O. Yurchenko[1]

[1]Bauman Moscow State Technical University, Moscow, Russia

[2]Lebedev Physical Institute of the Russian Academy of Sciences, Moscow, Russia

*E-mail: kirzay@gmail.com

Abstract. Third harmonic generation enhancement in periodic photonic structures was experimentally observed and theoretically interpreted. Nonlinear optical effects in opal globular photonic crystals (PC) have been studied under the femtosecond laser pumping. Strong dependence of the third harmonic generation efficiency on the ratio between the central pumping wavelength and the spectral location of the PC band gaps was found. Numerical simulations based on the finite difference time-domain technique for the solution of the Maxwell's equations were applied for investigations of the observed phenomenon origin. The simulation results have shown that the origin of the efficient nonlinear conversion is related with the surface state of electromagnetic field in PC. Interacting with the PC surface the light wavefront distorts coherently, and the effect of structure light focusing appears. Coherent wavefront distortion leads to the strong optical field localization, hence the light intensity within the certain PC regions increases. In case of the band gap pumping dramatic light redistribution appears; very sharp peaks of light intensity emerge in the region of the quartz globules, which leads to the high-efficient PC pumping.

1. Introduction and Background

An appearance of novel materials, in particular, strongly correlated optical media: periodic photonic structures, photonic crystals (PC) [1-5] and quasi-crystals [6] has attracted considerable interest in the relevant branch of optics in the recent time. The PC is a structure characterized by a spatially periodic distribution of the medium optical characteristics. The period of oscillations is comparable to the wavelength of light [7,8]. Periodicity of the structure leads to the dramatic changes of the photon behavior in a PC in comparison to the homogeneous medium: the spatial periodicity produces the zone structure of the dispersion relation, described within the Bloch's effective field theory [9-10].

PCs could be used in waveguide applications [11], including PC fibers [12], for supercontinuum generation [13], for highly efficient nonlinear light conversion, high-harmonic generation and active media pumping [14-17], and for increased Raman scattering by particles injected into the PCs [18,19]. Surface states of light in PCs are an interesting phenomenon [20-27], and it could be utilized in wide range of applications, in particular, in nonlinear optics [28]. PCs are the potential basis for the creation of new optical media and devices.

The present paper is dedicated to the investigations of the third harmonic generation efficiency in opal PC, and to the research of the origin of this phenomenon by means of the numerical simulations of light interaction with the 2D PC structure.

2. Results

2.1. Experimental observation of third harmonic generation in PC

Three samples of synthetic opal globular PC with globular diameter $a = 240.0$, 260.0 and $290.0\ nm$ were used for experimental studying of third harmonic generation efficiency. Sample pumping was implemented utilizing the femtosecond optical pulses of $Yb{:}\ KGW$ femtosecond laser, which worked either at the principal central wavelength ($\lambda_P = 1026.0\ nm$) or at the central wavelength of the second harmonic ($\lambda_P = 513{,}0\ nm$). The duration of the femtosecond pulse and pulse repetition rate were $\tau_p = 250.0\ nm$ and $f_{rep} = 200.0\ kHz$, respectively. The maximal peak power of the pulse P_{max} at the principle wavelength and at the wavelength of the second harmonic reached $0.7\ GW$ and $0.35\ GW$, respectively. During the experiment the laser radiation of the principle mode or the second laser harmonic was focused on the surface of the sample using wide-aperture optical system with ($D/f' = 1{:}1$), thus the diffraction quality of the pump spot and the peak irradiance of an object $E_{max} \approx 1{,}0\ TW/m^2$ were achieved. High pumping intensity provided by means of ultra-short optical pulses yielded maximal efficiency of nonlinear optical conversion in PC. Electromagnetic radiation scattered by the object were detected with a diffraction grating spectrometer, which spectral sensitivity corresponds to the spectral range between 190.0 and $1000.0\ nm$. Spectral resolution of described system was $\Delta\lambda = 1.0\ nm$.

Existence of three different PC structures with various locations of the main band gap λ_{BG} and two pumping wavelengths λ_P helps to study six cases of different sample pumping conditions. There are four principally different cases of the pumping wavelength λ_P position towards the samples band gap location λ_{BG}: $\lambda_P = \lambda_{BG}$, $\lambda_P \leq \lambda_{BG}$, $\lambda_P \geq \lambda_{BG}$ and $\lambda_P \gg \lambda_{BG}$. Figure 1 shows normalized intensity of the third harmonic generated in PC for all of the cases listed above.

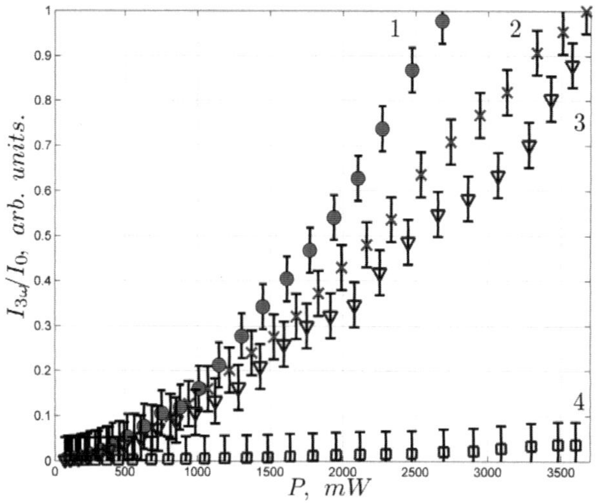

Figure 1. Dependence of normalized third harmonic intensity $I_{3\omega}/I_0$ on the pumping power P and the ratio between the central pumping wavelength λ_P and the location of the PC band gap λ_{BG}: **1** – $\lambda_P = \lambda_{BG}$; **2** – $\lambda_P \leq \lambda_{BG}$; **3** – $\lambda_P \geq \lambda_{BG}$; **4** – $\lambda_P \gg \lambda_{BG}$

Obviously, the condition of the effective third-harmonic generation is λ_P matched λ_{BG}. In this case about 10.0% of the pumping intensity is transformed to the intensity of the third harmonic. Also the efficiency of nonlinear generation in case of band gap pumping ($\lambda_P = \lambda_{BG}$) is on the order above the efficiency in case of pumping outside the band gap ($\lambda_P \gg \lambda_{BG}$).

2.2. Numerical simulation of the light interaction with the surface of PC

In order to interpret the results of the experiment (figure 1) FDTD simulation [29-31] was applied to research the structure of the surface optical modes in 2D PC. The model of the 2D PC was built of cylinders with the diameter $a = 250nm$, which is close to the diameter of the globules of the samples researched during the experiment. The matrix of cylinders is filled with air. The refractive index of the cylinder medium is $n_1 = 2.0$, and the refractive index of the air is $n_0 = 1.0$. The size of the PC model in all directions is much bigger than the incident light wavelength λ, and a PC structure periods T_x, T_y: $L_x \gg \lambda$, $L_y \gg \lambda$, $L_x \gg T_x$, $L_y \gg T_y$. The last conditions should be satisfied to prevent non-physical distortions of the modeling results, and undesirable boundary effects. The ideally plane electromagnetic wave with wavelength λ_P reached the PC surface from the left side.

During the simulation the time dependence of the electric field in the entire simulated space $E(x, y, t)$ was recorded; and the spatial distribution of the optical field intensity was computed the following way:

$$I(x,y) = \frac{1}{t' - t''} \int_{t'}^{t''} |E(x,y,t)|^2 dt, \qquad t' - t'' \gg \frac{\lambda}{c_0}, \tag{1}$$

where $c_0 = 3 \cdot 10^8 \; m/s$ is a speed of light in vacuum, t' and t'' is the time of the beginning and the end of the integration. Results of $I(x, y)$ calculations for different incident light wavelength are presented in the figure 2.

In complete accordance with the Bloch's theory of PC [1-10,32], the observed distributions $I(x, y)$ has a periodic character in case of pumping with the wavelength λ_P smaller than the spectral location of the first PC band gap λ_{BG1}: $\lambda_P < \lambda_{BG1}$ (figure 2a-g). In the opposite case ($\lambda_P > \lambda_{BG1}$) the structure of the optical field intensity $I(x, y)$ has homogenous character and does not contain any features (figure 2h-l). If the condition of the band gap pumping is satisfied ($\lambda_P = \lambda_{BG1} = 400.0 \; nm$, $\lambda_P = \lambda_{BG2} = 200.0 \; nm$) than the light intensity distribution has an average exponential decay with an increase of the sample depth x/a (figure 2c and figure 2g). One could notice dramatic lateral redistribution of optical filed which could not be predicted within the usual effective field theory and associated with an effect of coherent light wavefront distortions. The distortions of the wavefront is caused by the curvature of the globule surface, and leads to the structure focusing of light in PC and appearance of the sharp peaks of light intensity concentrated in the certain regions of PC.

The sharpest peaks of intensity corresponds to the cases of band gap pumping; the intensity of light in regions of PC globules could be 10-25 times higher than the intensity of light in the equivalent homogenous optical medium in the same pumping conditions. Note, the size of the sharp light intensity peaks is smaller than the light wavelength λ, and the reason of this phenomenon is a high-aperture focusing caused by the nature of light interaction with strongly correlated optical media. Similar effects of coherent wavefront distortion and spatial light redistribution could appear in 3D PC considering the interplay between different polarization states.

Strong field localization effect observed in case of band gap pumping could be an origin of high-efficient nonlinear light conversion and enhancement of the third harmonic generation in opal PC (figure 1). To characterize the raise of the third harmonic generation efficiency the post-processing of the simulation results was implemented.

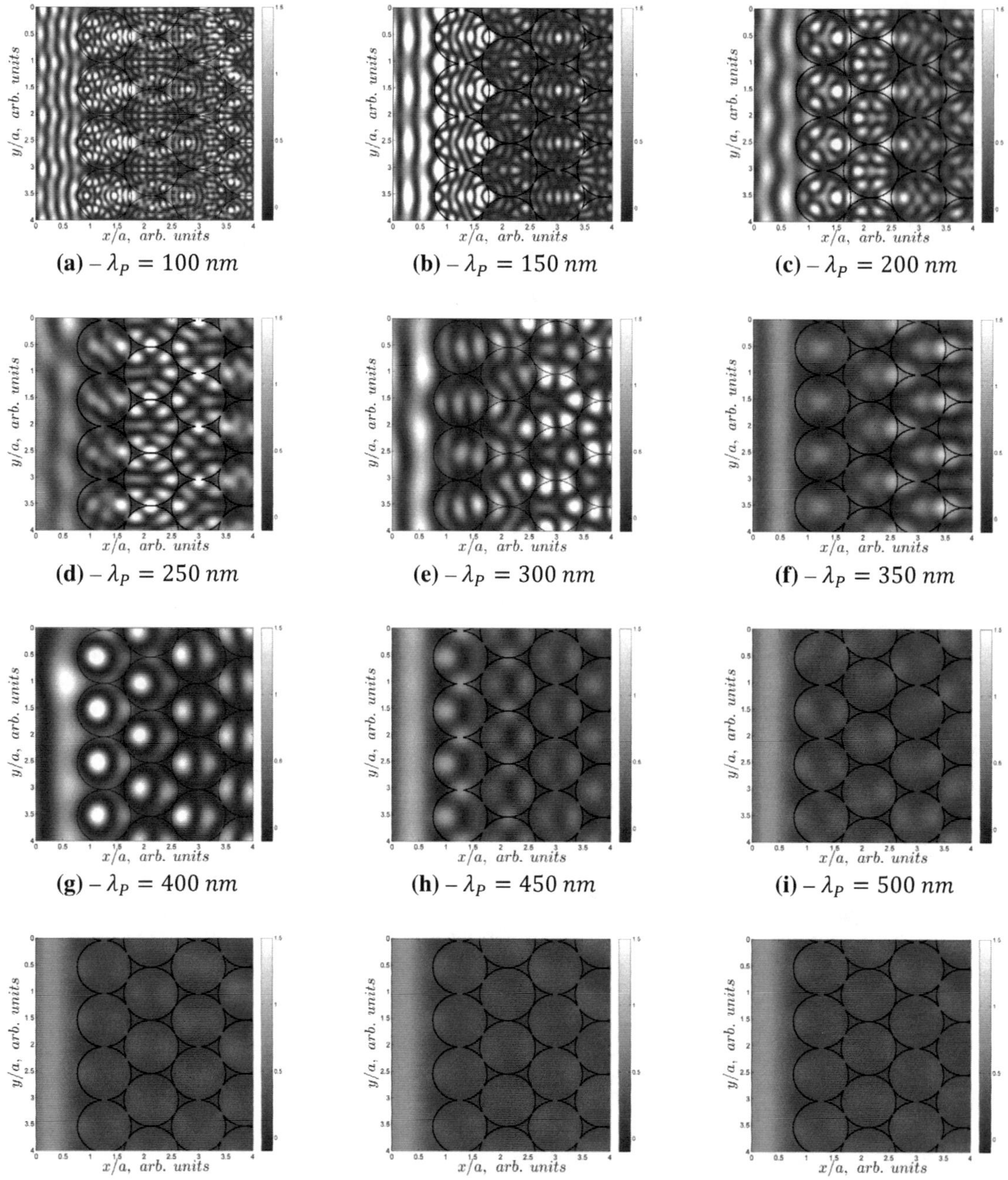

(a) – $\lambda_P = 100\ nm$ **(b)** – $\lambda_P = 150\ nm$ **(c)** – $\lambda_P = 200\ nm$

(d) – $\lambda_P = 250\ nm$ **(e)** – $\lambda_P = 300\ nm$ **(f)** – $\lambda_P = 350\ nm$

(g) – $\lambda_P = 400\ nm$ **(h)** – $\lambda_P = 450\ nm$ **(i)** – $\lambda_P = 500\ nm$

(j) – $\lambda_P = 550\ nm$ **(k)** – $\lambda_P = 600\ nm$ **(l)** – $\lambda_P = 650\ nm$

Figure 2 Spatial distribution of the electromagnetic field intensity $I(x,y)$ near the interface between the air and the $2D$ photonic crystal at different pumping wavelength λ_P: **(a)** – $\lambda_P = 100\ nm$, **(b)** – $\lambda_P = 150\ nm$, **(c)** – $\lambda_P = 200\ nm$, **(d)** – $\lambda_P = 250\ nm$, **(e)** – $\lambda_P = 300\ nm$, **(f)** – $\lambda_P = 350\ nm$, **(g)** – $\lambda_P = 400\ nm$, **(h)** – $\lambda_P = 450\ nm$, **(i)** – $\lambda_P = 500\ nm$, **(j)** – $\lambda_P = 550\ nm$, **(k)** – $\lambda_P = 600\ nm$, **(l)** – $\lambda_P = 650\ nm$

Figure 3 shows the results of calculation of the average light intensity cube in the volume of the PC globules $\langle I^3\rangle_g$ and in the volume of the PC channels filled with air $\langle I^3\rangle_{ch}$. The following equations were used for the calculation of this parameter dependencies on normalized pumping wavelength λ/a:

$$\begin{cases} \langle I^3\rangle_g = \iint\limits_{(x,y)\in V} \frac{I^3(x,y)}{I_0} dxdy, \\ \langle I^3\rangle_{ch} = \iint\limits_{(x,y)\notin V} \frac{I^3(x,y)}{I_0} dxdy, \end{cases} \tag{2}$$

where I_0 is the intensity of light in the equivalent homogenous medium, for instance, in quartz. V represents the volume plurality of the PC globules.

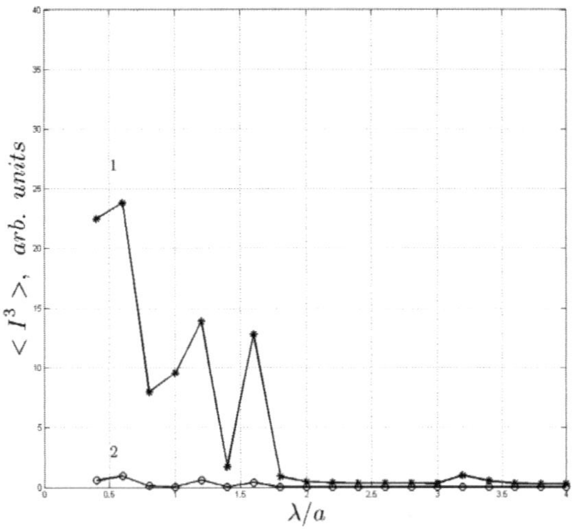

Figure 3 The dependence of the light intensity cube $\langle I^3\rangle$ **(1)** within the volume of the PC globules $\langle I^3\rangle_g$ and **(2)** within the volume of the PC channels $\langle I^3\rangle_{ch}$ on the normalized wavelength of λ/a

The intensity cube curves illustrate that the average intensity in case $\lambda_P > \lambda_{BG1}$ ($\lambda_P/a > 1.6$) is very close to the intensity in equivalent homogenous medium. In opposite case $\lambda_P < \lambda_{BG1}$ ($\lambda_P/a < 1.6$) dramatic redistribution of the light intensity appears. The intensity of quartz globule pumping raises and high efficient nonlinear conversion appears, in particular, enhanced third harmonic generation effect. One could note the peaks of $\langle I^3\rangle_g(\lambda/a)$ function in the regions of first λ_{BG1} and second λ_{BG2} band gaps, which corresponds to the normalized wavelengths $\lambda_{BG1}/a = 1.6$ and $\lambda_{BG2}/a = 0.8$, respectively.

The results of numerical simulations match the experimental results of the figure 1. The efficiency of third harmonic generation raises more than ten times in case of the first band gap pumping. The nature of the effective nonlinear conversion is associated with the effect of structure light focusing in PC, the effect of dramatic redistribution of light intensity and strong localization of optical field in certain PC regions.

There are various applications of the observed optical field localization effect including enhanced nonlinear light conversion and high-efficient pumping of active media [14-17], supercontinuum generation [13] etc. Similar effects could appear in sonic (acoustic) crystals [33], because of the close analogy between electromagnetic and sound waves. The effect of structure focusing is an important one for prediction of the destructions of the quasi-sonic [33] and quasi-photonic [6,34] crystal-like constructions under the influence of acoustic and optical waves.

Observed effects are also reveal the limitations of all effective field model approaches and Bloch's theory [1-10] to understanding periodic photonic structures, in particular, to studying strong field localizations and structure light focusing in 2D and 3D PC structures.

3. Conclusions

Third harmonic generation enhancement in PC structures was experimentally observed and theoretically interpreted. The results of the FDTD simulations of light interaction with 2D PC have shown that the origin of the efficient nonlinear conversion is associated with the surface state of electromagnetic field in PC. The structure light focusing in PC in case of band gap pumping condition is satisfied $\lambda_P = \lambda_{BGi}$ leads to dramatic increase of optical field localization and to the raise of the pumping efficiency, thus the nonlinear conversion enhancement appears.

Acknowledgments

K.I.Z. and G.M.K. are grateful for the financial support of the Russian Foundation for Basic Research (RFBR), Project No. 14-02-00256. The work of V.S.G. was supported with the grants of the Russian Foundation for Basic Research (RFBR), Project No. 13-02-90420 and Project № 13-02-00449.

References

[1] Bykov V P 1972 Spontaneous emission in a periodic structure *JETP*. **35**(2), 269 – 273
[2] Yablonovitch E 1987 Inhibited Spontaneous Emission in Solid-State Physics and Electronics *Phys. Rev. Lett.* **58**(20), 2059 – 2062
[3] John S 1987 *Phys. Rev. Lett.* **58**(23), 2486 – 2489
[4] Ho K M, Chan C T, Soukoulis C M 1990 Strong localization of photons in certain disordered dielectric superlattices *Phys. Rev. Lett.* **65**(25), 3152 – 3155
[5] Yablonovitch E, Gmitter T J, Leung K M 1991 Photonic band structure: The face-centered-cubic case employing nonspherical atoms *Phys. Rev. Lett.* **67**(17), 2295 – 2298
[6] Vardeny Z V, Nahata A, Agrawal A 2013 Optics of photonic quasicrystals *Nature Photonics*. **7**, 177 – 187
[7] Gorelik V S 2007 Optics of globular photonic crystals *Quantum Electronics*. **37**(5), 409 – 432
[8] Gorelik V S 2008 Optics of globular photonic crystals *Laser Physics*. **18**(12), 1479 – 1500
[9] Meade R D, Rappe A M, Brommer K D, Joannopoulos J D, Alerhand O L 1993 Accurate theoretical analysis of photonic band-gap materials *Phys. Rev. B*. **48**(11), 8434 – 8437
[10] Hornreich R M, Shtrikman S, Sommers C 1994 Photonic band gaps in body-centered-cubic structures *Phys. Rev. B*. **49**(16), 10914 – 10917
[11] Vlasov Y A, O'Boyle M, Hamann H F, McNab S J 2005 Active control of slow light on a chip with photonic crystal waveguides *Nature*. **438**, 65 – 69
[12] Knight J C, Birks T A, Russell P St J, Atkin D M 1996 All-silica single-mode optical fiber with photonic crystal cladding *Optics Letters*. **21**(19), 1547 –1549
[13] Herrmann J, Griebner U, Zhavoronkov N, Husakou A, Nickel D, Knight J C, Wadsworth W J, Russell P St J, Korn G 2002 Experimental Evidence for Supercontinuum Generation by Fission of Higher-Order Solitons in Photonic Fibers *Phys. Rev. Lett.* **88**(17), 173901
[14] Berger V 1998 Nonlinear Photonic Crystals *Phys. Rev. Lett.* **81**(19), 4136 – 4139
[15] Balakin A V, Bushuev V A, Koroteev N I, Mantsyzov B I, Ozheredov I A, Shkurinov A P, Boucher D, Masselin P 1999 Enhancement of second-harmonic generation with

femtosecond laser pulses near the photonic band edge for different polarizations of incident light *Optics Letters.* **24**(12), 793 – 795

[16] Balakin A V, Bushuev V A, Mantsyzov B I, Ozheredov I A, Petrov E V, Shkurinov A P, Masselin P, Mouret G 2001 Enhancement of sum frequency generation near the photonic band gap edge under the quasiphase matching conditions *Phys. Rev. E.* **63**(4), 046609

[17] Soljacic M, Joannopoulos J D 2004 Enhancement of nonlinear effects using photonic crystals *Nature Materials.* **3**(4), 211 – 219

[18] Rivoire K, Buckley S, Song Y 2012 Photoluminescence from $In_{0.5}Ga_{0.5}As$/GaP quantum dots coupled to photonic crystal cavities *Phys. Rev. B.* **85**(4), 045319

[19] Gorelik V S, Zlobina L I, Sverbil P P, Fadyushin A B, Chervyakov A V 2005 Raman Scattering in Three-Dimensional Photonic Crystals *Journal of Russian Laser Research.* **26**(3), 211 – 227

[20] Rosberg C R, Neshev D N, Krolikowski W, Mitchell A, Vicencio R, Molina M I, Kivshar Y S 2006 Observation of Surface Gap Solitons in Semi-Infinite Waveguide Arrays *Phys. Rev. Lett.* **97**, 083901

[21] Kartashov Y V, Vysloukh V A, Torner L 2006 Surface Gap Solitons *Phys. Rev. Lett.* **96**, 073901

[22] Malkova N, Ning C Z 2006 Shockley and Tamm surface states in photonic crystals *Phys. Rev. B.* **73**(11), 113113

[23] Malkova N, Ning C Z 2007 Interplay between Tamm-like and Shockley-like surface states in photonic crystals *Phys. Rev. B.* **76**(4), 045305

[24] Klos J 2007 Conditions of Tamm and Shockley state existence in chains of resonant cavities in a photonic crystal *Phys. Rev. B.* **76**(16), 165125

[25] Ishizaki K, Noda S 2009 Manipulation of photons at the surface of three-dimensional photonic crystals *Nature* **460**, 367 – 370

[26] Soboleva I V, Moskalenko V V, Fedyanin A A 2012 Giant Goos-Hänchen Effect and Fano Resonance at Photonic Crystal Surfaces *Phys. Rev. Lett.* **108**(12), 123901

[27] Afinogenov B I, Bessonov V O, Nikulin A A, Fedyanin A A 2013 Observation of hybrid state of Tamm and surface plasmon-polaritons in one-dimensional photonic crystals *Appl. Phys. Lett.* **103**(6), 061112

[28] Zaytsev K I, Gorelik V S, Khorokhorov A M, Yurchenko S O 2014 FDTD simulation of the electromagnetic field surface states in 2D photonic crystals *Journal of Physics: Conference Series.* **486**, 012003

[29] Yee K 1966 Numerical solution of initial boundary value problems involving Maxwell's equations in isotropic media *IEEE Transactions on Antennas and Propagation.* **14**(3), 302 – 307

[30] Zaytsev K I, Karasik V E, Fokina I N, Alekhnovich V I 2013 Invariant embedding technique for medium permittivity profile reconstruction using terahertz time-domain spectroscopy *Optical Engineering.* **52**(6), 068203

[31] Fokina I., Zaytsev K I, Karasik V E, Tsapenko K I 2013 Scattering of terahertz radiation in thin layers of dielectric materials *Proceedings of SPIE.* **8846**, 88460A-1

[32] Sipe J E, Bhat N A R, Chak P, Pereira S 2004 Effective field theory for the nonlinear optical properties of photonic crystals *Phys. Rev. E.* **69**, 016604

[33] Liu Z, Zhang X, Mao Y, Zhu Y Y, Yang Z, Chan C T, Sheng P 2000 Locally Resonant Sonic Materials *Science.* **289**(5485), 1734-1736

[34] Bunkin N F, Yurchenko S O, Suyazov N V, Shkirin A V 2012 Structure of the nanobubble clusters of dissolved air in liquid media *Journal of Biological Physics.* **38**, 121-152

SERS-applicable silver nanoisland film grown under protective coating

I Reduto[1], S Chervinskii[2,3], A Matikainen[3], A Baklanov[2], A Kamenskii[2] and A Lipovskii[1,2]

[1] Department of Physics and Technology of Nanostructures, St. Petersburg Academic University, 8/3 Khlopina, St. Petersburg, 194021 Russia
[2] Institute of Physics, Nanotechnology and Telecommunications, St.Petersburg State Polytechnic University, 29 Polytechnicheskaya, St. Petersburg, 195251 Russia
[3] Institute of Photonics, University of Eastern Finland, P.O.Box 111 FI-80101 Joensuu, Finland

E-mail: reduto-igor@mail.ru

Abstract. We have used recently developed out-diffusion technique of growing silver nanoisland films on glass surface to grow silver nanoislands under TiO_2 layer deposited on the glass. After covering the surface of silver ion-exchanged glasses with TiO_2 film using atomic layer deposition technique and subsequent thermal processing of the samples in hydrogen their optical absorption spectra demonstrate the absorption peak corresponding to surface plasmon resonance in grown silver nanoislands. The spectral position of the peak is shifted relatively to the peak observed in the spectra of the nanoisland film grown on the surface of ion exchanged and annealed glass samples without dielectric cover. The applicability of the silver nanoislands grown under several nm thick protective TiO_2 coating in surface-enhanced Raman scattering spectroscopy is demonstrated.

1. Introduction

Silver nanoisland films (SNF) are of interest due to their applicability in biological and chemical sensors based on surface enhanced Raman scattering (SERS) [1,2]. This enhancement is due to high local electric field of probing lightwave in the proximity of closely-spaced metal nanoparticles, and it is maximal in the spectral range corresponding to surface plasmon resonance (SPR) in the nanoislands. However a common problem in using SNF is the fragility of the films suffering from oxidation and other chemical attacks [2,3]. This results in temporal instability of the SNF that limits their main application area, SERS measurements. Luckily the nanoislands can be protected with dielectric covers deposited over formed SNF. In spite of the essential decrease in SERS efficiency due to the lower magnitude of the local electric field penetrating through the cover (spacer) the protected films are very attractive because of their stability [4,5]. In our experiments the nanoisland films were formed using metal out-diffusion from the ion-enriched glass substrate in the course of annealing in hydrogen atmosphere [6,7]. This approach allowed us to grow silver nanoislands on the glass substrates under already deposited TiO_2 cover that is to avoid any secondary processing of grown nanoisland films and to have the nanoislands protected already at the stage of their formation. The comparison of SERS signals measured using the SNF covered with TiO_2 after the nanoislands formation and the SNF grown under deposited TiO_2 layer has shown the applicability of the proposed approach. The technique developed properly suits for the manufacturing protected substrates for SERS measurements.

2. SNF formation and studies

The first stage of the SNF formation was silver-to-sodium ion exchange in soda-lime glass performed as described elsewhere [8]. In our experiments we used Menzel microscope slides [9] and 20 minutes ion exchange of this soda-lime glass samples in the melt of 5 wt.% of $AgNO_3$ and 95 wt.% of $NaNO_3$. Then the ion-exchanged samples were annealed either in hydrogen or in air, or covered with several nanometers of titania using atomic layer deposition technique (ALD) and annealed in hydrogen after the covering. In all these cases the thermal processing resulted in the reduction of ionic silver due to the interaction of the ions with partly dissociated water vapors (annealing in air) or hydrogen penetrating the glass. Due to high diffusivity hydrogen is capable of going through the TiO_2 cover as well. Because of low solubility of silver in glass matrix, solid solution of silver in the glass formed in the course of the reduction is oversaturated and tends to phase decomposition. As the glass surface, as well as glass-titania interface, is a strong sink for silver atoms, SNF is formed on the glass surface both free and covered with TiO_2 layer. The procedure to grow silver nanoislands under the cover is illustrated with Fig.1. It is worth to note that the silver-in-glass solution phase decomposition also results in the growth of silver nanoparticles inside the glass that is glass-metal nanocomposite (GMN) [10], but the formation of the GMN follows the growth of the SNF because of the strength of surface and interface sinks for neutral silver.

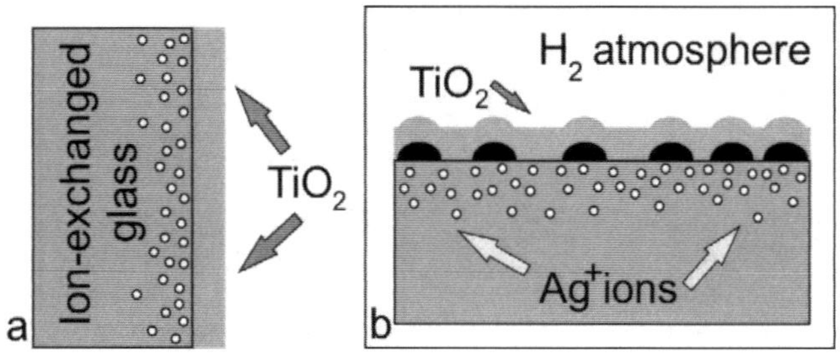

Figure 1. Schematic of metal films growth under the cover: (a) ion exchanged glass with TiO_2 cover; (b) silver nanoislands under TiO_2 cover.

Optical absorption spectra of the ion-exchanged glass samples annealed in air at temperatures from 100 to 250°C demonstrated the formation of the peak of optical absorption in the spectral range corresponding to the SPR in silver island films [11], see Fig. 2. The peak of optical absorption drops when the annealing temperature exceeds 200C. Comparison of atomic force microscope images of differently annealed samples (Fig. 3) has shown that the density of nanoislands is monotonically growing with the temperature of annealing. This allows us to conclude that the decrease in optical density of stronger annealed glass samples is because of chemical transformation of the grown nanoislands, most probably due to their oxidation.

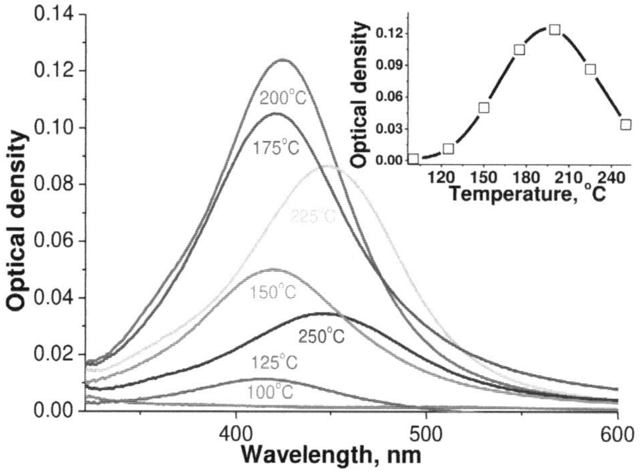

Figure 2. Optical absorption spectra of ion-exchanged glass samples annealed in air for 10 minutes at different temperatures, glass substrate spectrum is subtracted. Insert: the height of the SPR absorption peak vs annealing temperature.

Contrary, thermal processing in ion exchanged glasses in hydrogen atmosphere never resulted in the degradation of the SPR absorption. This is illustrated by Fig. 4 where we plotted the height of the SPR absorption peak for the samples differently annealed in hydrogen and an example of AFM image of the SNF grown at 78C. One can see that even low-temperature processing of the ion-exchanged glass in hydrogen results in effective formation of the nanoislands.

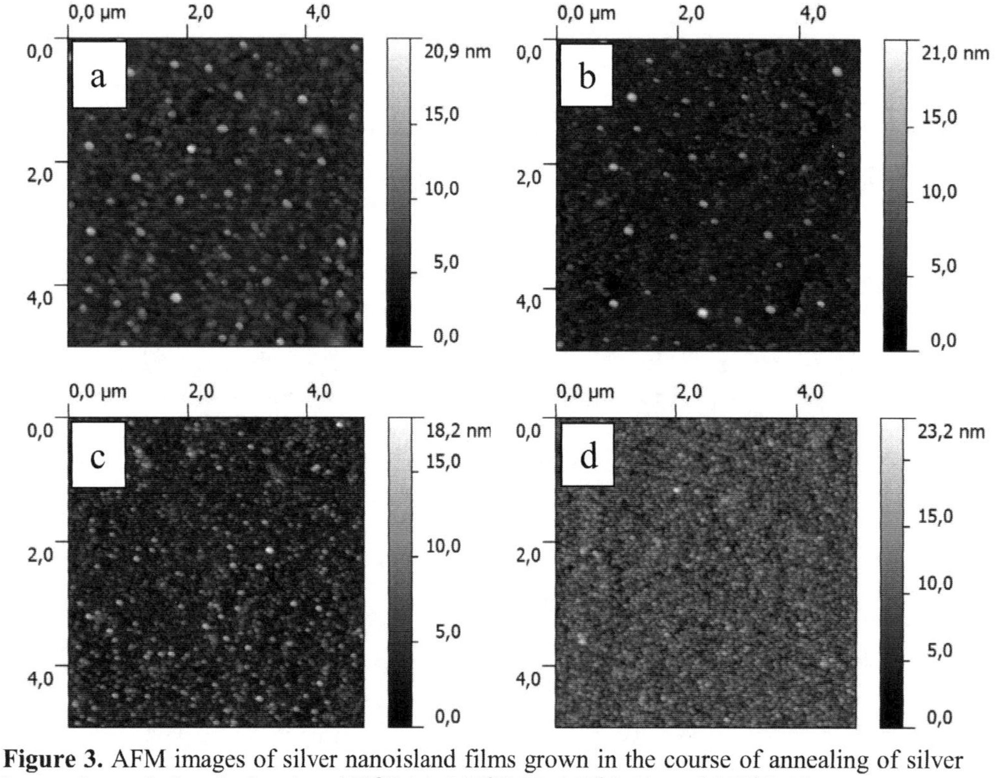

Figure 3. AFM images of silver nanoisland films grown in the course of annealing of silver ion-exchanged glass in the air at 150°C (a), 200°C (b), 225°C (c), and 250°C (d).

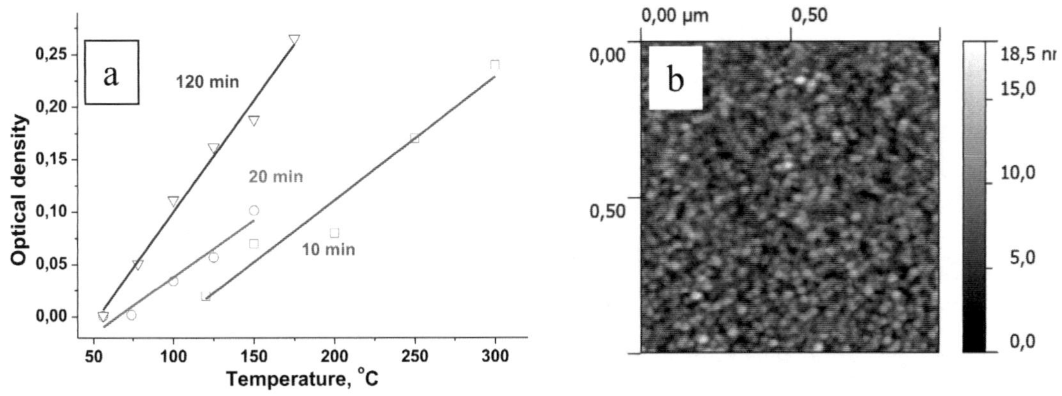

Figure 4. The height of the SPR absorption peak vs annealing temperature for the samples processed in hydrogen for 10 minutes (circles) as described in [12], 20 minutes (diamonds) and 2 hours (triangles) (a). The AFM image of the silver nanoisland films grown in the course of annealing of silver ion-exchanged glass in the hydrogen at 72°C for 120 min (b).

Results presented in Fig.2 illustrate the temperature instability of the unprotected SNF. Results on temporal instability of silver nanoparticles and the influence of their oxidation on SERS properties, which have recently been presented [13].We checked temporal stability of the SNF covered with TiO_2 film. Optical absorption spectra of the films after 6 months storage at room temperature in atmospheric air demonstrate only small change in optical absorption spectra and, respectively, in the height of SPR peak, see Fig. 5a. In Fig. 5b scanning electron microscope (SEM) image of the SNF grown under protective layer of titania formed on silver ion-exchanged glass using ALD technique is presented.

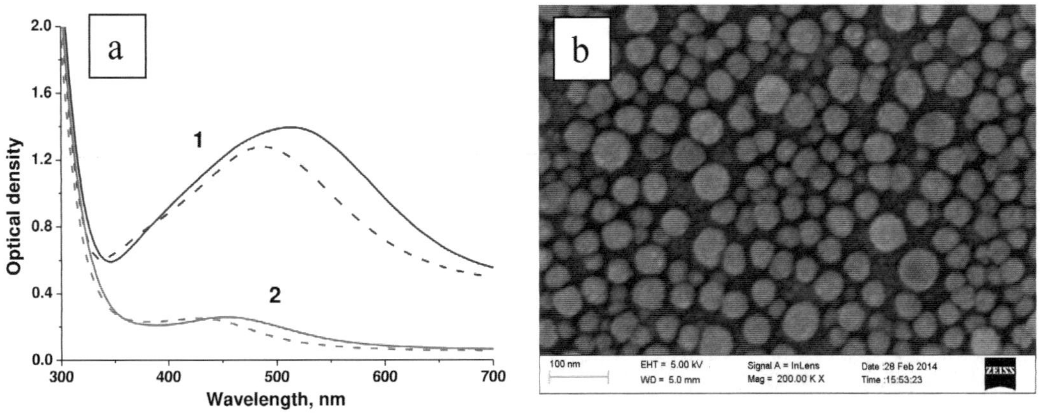

Figure 5. Optical absorption spectra of SNF under 5 nm ALD TiO_2 film, 1 – annealing at 250C for 10 min, 2 – annealing at 150 C for 10 min. Dashed lines are the spectra of the same samples measured after 6 months storage in air at room temperature (a); SEM image of SNF grown under 7.5 nm ALD TiO_2 film, the ion-exchanged glass sample was annealed in hydrogen during 30 min at 150°C and 10 min at 250°C (b).

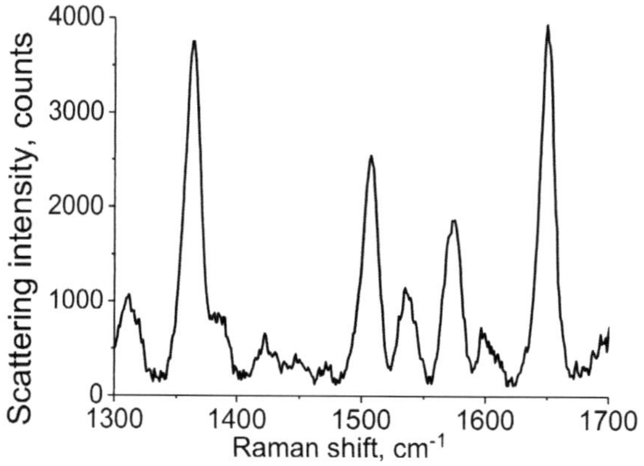

Figure 6. SERS spectrum of rhodamine 6G measured using 7.5 nm TiO_2 covered sample (annealing in hydrogen: 30min at 150°C followed by 10min at 250°C). Excitation power 50 μW at 514 nm, exposure time 10 s, and approximate illuminated spot size 5 μm.

To check applicability of the films grown under the protecting cover we have performed SERS trials. We tested the film, the SEM image of which is presented in Fig. 5 in SERS experiments with dried 5 μl droplets of 1 μM rhodamine 6G deposited on the cover surface. The experiments were carried out using an in Via Raman microscope system. We used excitation power 50 μW at 514 nm, exposure time 10 s, and approximate illuminated spot size 5 μm. SERS applicability of this SNF is illustrated with Fig. 6 where one can see clearly distinguished peaks corresponding to rhodamine 6G Raman lines. The amplitudes of the Raman peaks measured using the film grown under already deposited TiO_2 is slightly higher in comparison with the amplitudes of the Raman peaks measured using the deposition of TiO_2 film of the same thickness over already grown SNF [14].

3. Conclusions

Performed experiments have demonstrated that silver nanoisland films can be grown on the surface of silver ion-exchanged glasses using thermal processing both in atmospheric air and in hydrogen. Processing of the glasses in atmospheric air at temperatures of the annealing below 200°C provides formation of the nanoisland films demonstrating surface plasmon resonance with the amplitude of corresponding optical absorption peak growing with the increase in the temperature of the annealing. Higher temperatures of the annealing result in degradation of surface plasmon resonance in the silver nanoislands. Contrary, if the ion-exchanged glasses are annealed in hydrogen atmosphere, the annealing temperature results in the growth of the surface plasmon resonance absorption peak which saturates and does not degrade. Hydrogen annealing of silver ion-exchanged glasses with thin layer of titania allows growing of silver nanioislands beneath the deposited film providing the protection of very fragile silver nanoislands from atmosphere. Performed experiments have demonstrated that the change is surface plasmon resonance absorption of 7.5 nm TiO_2 coated silver nanoisland film after half year storage in the air is inessential. At the same time the silver nanoisland films grown under this protective layer demonstrate applicability in SERS measurements. Finally, the growth of silver nanoislands under already deposited protective cover is a prospective technique to manufacture temporary stable SERS substrates.

Acknowledgements

This study was supported the FP7 project NANOCOM, ERA.Net RUS project AN2, and Academy of Finland project #267270. The AFM studies were performed using the equipment of the Joint Research Centre «Material science and characterization in advanced technology» (Ioffe Institute, St. Petersburg, Russia).

References

[1] Otto A et al 1992 *J. Phys.: Condens. Matter* **4** pp 1143-1212
[2] Kukushkin V, Van'kov A, Kukushkin I 2013 *JEPT Letters* **98** (2) pp 64–69
[3] Dieringer J et al 2006 *Faraday Discuss* **132** pp 9-26
[4] Willets K, Van Duyne K 2007 *Rev. Phys. Chem.* **58** pp 267–297
[5] John J et al 2010 *J. Raman Spectrosc.* **41** pp 4–11
[6] Lipovskii A et al 2012 *Nanoscale Research Letters* **7** 676
[7] Chervinskii S et al 2013 *J. Appl. Phys.* **114** 224301
[8] Zhurikhina V et al 2010 *Technical Physics* **55** (10) pp 1447–1452
[9] www.menzel.de
[10] Kaganovskii Yu et al 2007 *J. Non-Cryst. Solids* **353** pp 2263–2271
[11] Kreibig U, Vollmer M 1995 Optical Properties of Metal Clusters. Berlin: Springer
[12] Chervinskii S et al 2013 *J. Appl. Phys.* **114** 224301
[13] Han Y et al 2011 *Anal. Chem.* **83** (15) pp 5873–5880
[14] Chervinskii S et al 2014 submitted to *Opt. Mat. Express*

Fabrication and optical properties of silicon nanopillars

L S Golobokova, Yu V Nastaushev, F N Dultsev, D V Gulyaev, A B Talochkin and A V Latyshev

A.V. Rzhanov Institute of Semiconductor Physics of SB RAS, 13 pr. Lavrentieva, Novosibirsk 630090, Russia

E-mail: GolobokovaLS@isp.nsc.ru

Abstract. The optical properties of ordered arrays of silicon nanopillars (Si NPs) were investigated. Electron Beam Lithography (EBL) followed by Inductively Coupled Plasma Reactive Ion Etching (ICP-RIE) was used for Si NPs fabrication. Si NPs were chemically and electrically passivated through the deposition of TiONx nanolayer. The silicon nanopillars were characterized by using scanned electron microscopy (SEM), high resolution transmission electron microscopy (HRTEM). We demonstrate that a various colors can be obtained by exploiting the resonant light scattering properties of individual Si NP. In addition the low temperature photoluminescence from Si NPs was investigated. The peak photoluminescence energy was observed at 0.83 μm and 1.14 μm. Raman scattering enhancement was found too.

1. Introduction

Silicon nanopillars are of interest due to their potential application in various fields: microelectronics [1], optoelectronics [2], photonics [3, 4], photovoltaics [5, 6], biology [7], and sensors [8, 9]. Various groups have reported strong antireflective properties in Si NPs [10, 11], and in other nanoscale structures [12]. The authors [13] have demonstrated enhanced photoluminescence (PL) and second harmonic generation (SHG) in SiNPs. SiNPs can be formed by means of different methods including vapor-liquid-solid growth, silver induced chemical etching and e-beam [14] or laser interference lithography [15] followed by dry etching. In the present paper, we studied the morphology and optical properties of Si NPs, fabricated using electron-beam lithography (EBL) followed by dry etching. The most controlled way of top-down process is using lithography to define a mask pattern followed by RIE. EBL is a well-established high resolution patterning technique. The plasma etching is capable of providing an anisotropic profile. Therefore, the diameter and the length of the Si NPs can be precisely controlled using EBL and plasma etching.

2. Experimental

The «top-down» approach was used to fabricate silicon nanopillars by means of e-beam lithography and deep reactive ion etching (RIE). Prime silicon wafer (n-type) <100> was used as a base material. The substrates were cleaned in isopropyl alcohol for 3 minute using an ultrasonic cleaner and rinsed in de-ionized water. Then, silicon substrates were cleaned with fresh solution (H_2O_2 : NH_4OH : H_2O = 1 : 1/5 : 5). They were baked at 200°C and cooled to room temperature immediately before coating. Then, e-beam resist ma-N2403 was spincoated and heated on a hotplate at 90°C for 1 minute. Various viscosities of the ma-N2403 resist were used resulting in film thicknesses of 180 and 300 nm. The resist coated wafers were exposed to pattern a dense nanopillars array. Raith 150 (Raith GmbH) was used as a lithography system for nanometer patterning. The electron dose was varied within the range of 0.01÷0.1 pC per dot. After exposure, development process was carried out using ma-D532 developer. Then the samples were rinsed in deionized water and blown to dry. Figure 1 shows a SEM image of pattern in resist as the mask of NPs before etching. The diameter of the NP pattern in the resist as a function of dose is shown in figure 2. As shown in figure 2, the resulting diameters of NP in the resist are linearly dependent on the applied point dose. The patterns for nanopillars were designed using RAITH ELPHY Quantum GDSII Editor. Silicon was etched into nanopillars using the resist as a mask in an ICP-RIE plasma etcher (Oxford Plasmalab 100). Two different fluorinated etch chemistries were utilized: a SF_6/O_2 cryogenic etch and the Pseudo-Bosch (SF_6/C_4F_8). It should be noted that the SF_6/O_2 cryogenic etching process limited the height of the Si NPs. In order to enhance the height of pillars, we utilized the Pseudo-Bosch process. The Bosch process used a two-step plasma process which alternates between etching in SF_6 flow and passivation in C_4F_8 flow. Sidewall protection is achieved by C_4F_8 gas. Polymer deposition for protecting lateral sidewalls makes for a good sidewall profile for the nanopillars. Si NPs with diameters ranging from 60 nm to 350 nm with heights from 100 to 600 nm were fabricated (pitch from 0.4 μm to 1.7 μm).

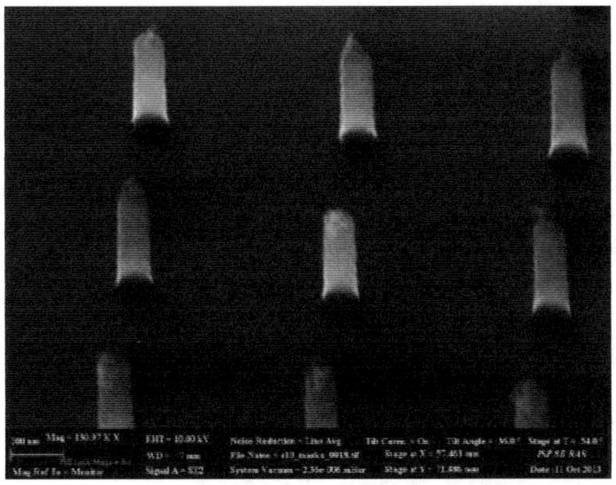

Figure 1. SEM image of the NP pattern in the resist, pitch 800 nm (tilt angle = 54°, the scale bar is 200 nm).

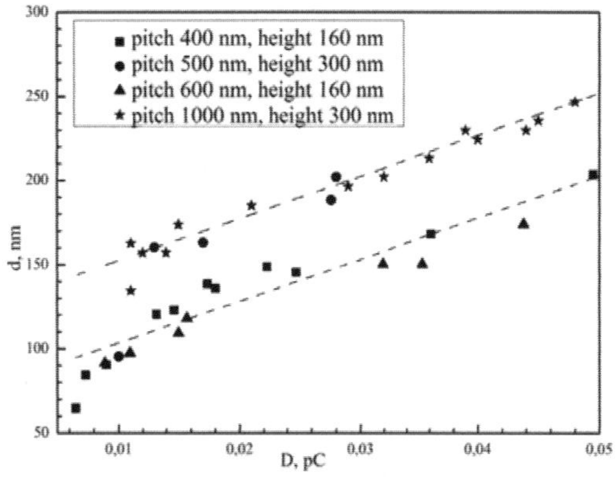

Figure 2. The diameter of NP in the resist as a function of dose value.

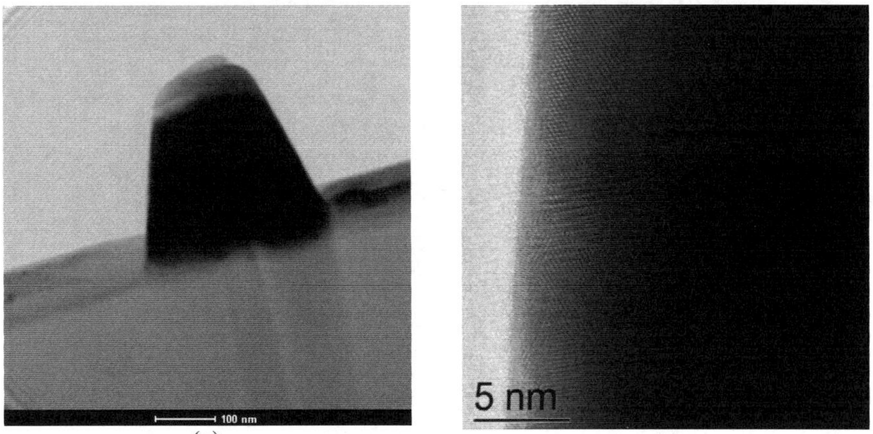

(a) (b)

Figure 3. (a) A low-magnification TEM image of a Si NP. (b) HRTEM image taken near the edge of the individual NP.

As shown in figure 4 (b, c) well aligned and patterned SiNPs were observed. Si NPs were covered with a dielectric nanolayer of TiONx, formed by plasmachemical nitridation of Ti nanolayer, in order to achieve chemical and electrical passivation and stabilization of Si NP surface. The morphologies of the samples were characterized by scanning electron microscopy (CROSS BEAM 1540XB) and high-resolution transmission electron microscopy (JEM-4000EX).

In order to analyze the structure of Si NP, HR TEM characterization was performed. The Au layer was evaporated on to the samples. The thickness of the Au layer was 300 nm. As shown in figure 3, the Si NPs are single crystalline. In figure 3 transmission electron micrograph shows that the etch sidewalls are smooth.

3. Results and discussion

In recent years, vivid colors have been demonstrated by means of SiNPs [1, 16, 17]. In this work, the sample inspection has been done with Zeiss AxioImager Z1 high-performance research microscope with darkfield (DF) and brightfield (BF) optical detection. Figure 4(a) shows a bright-field image of the vertical SiNPs arrays (50×, NA = 0.8). Figure 4 (b, c) shows SEM images of these Si NPs arrays. The pitch in all of arrays is 630nm. Figure 5(c, d) shows bright-field images of the vertical SiNPs arrays (50×, NA = 0.8). Figure 5 (a, b) shows SEM images of these Si NPs arrays (pitch 400 nm). NPs with different diameter appear distinctly different colors. The image is taken with the incident light normal to the sample. It should be stressed that generated colors are dependent on the NP diameter. The colors were created by individual SiNPs. This effect had originated not from diffractive effects of the arrays. The Si NP is a dielectric nanocavity. We guess that the SiNPs exhibit Mie resonances in the visible spectral region. Bezares F J et al [2013] investigated the behavior of the Mie resonances in cylindrical Si NPs as a function of pillar diameter (68 – 260 nm) and pitch (200 – 500 nm).

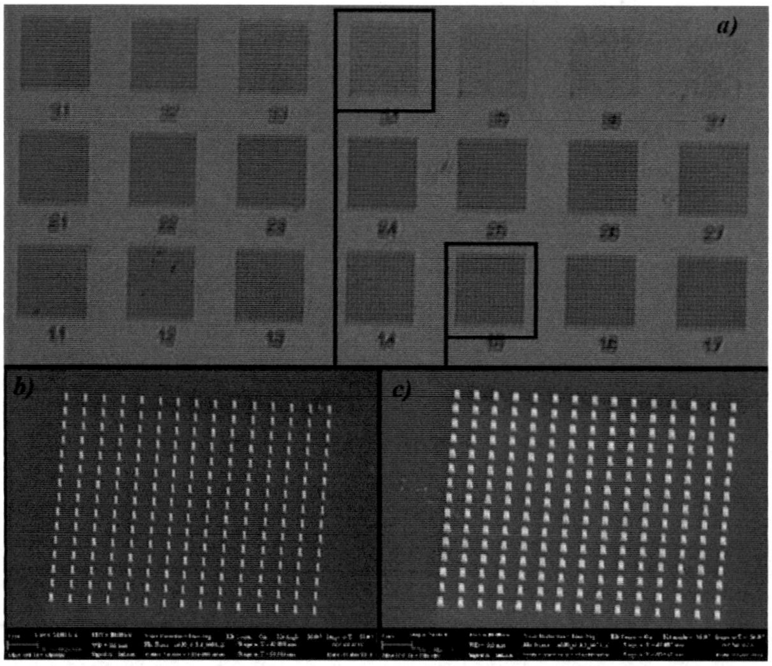

Figure 4. A bright-field image (a) and SEM images (b, c) of Si NPs array with 630 nm pitch (tilt angle = 54°, the scale bars is 1 μm).

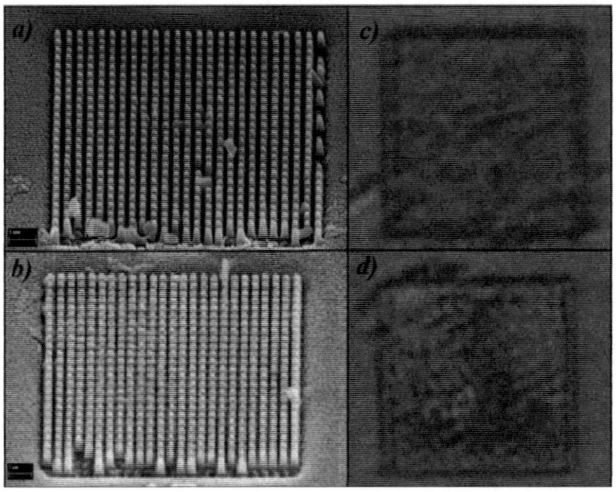

Figure 5. SEM images (a, b) and bright-field images (c, d) of Si NPs array with 400 nm pitch (tilt angle = 54°, the scale bars is 1μm). Diameters: 280 nm (a) and 330 nm (b).

Figure 7 shows a SEM image of NPs array. In contrast to above results, figure 6 shows a dark-field image of these Si NPs, the light source incident angularly to the SiNP array. The image demonstrates the color generation due to the diffractive effects. The color observed for a given array can also be tuned by changing the angle of the incident light and changing the space orientation. Tunable structural color generation from vertical silicon NPs arranged in different square lattices is realized using diffraction.

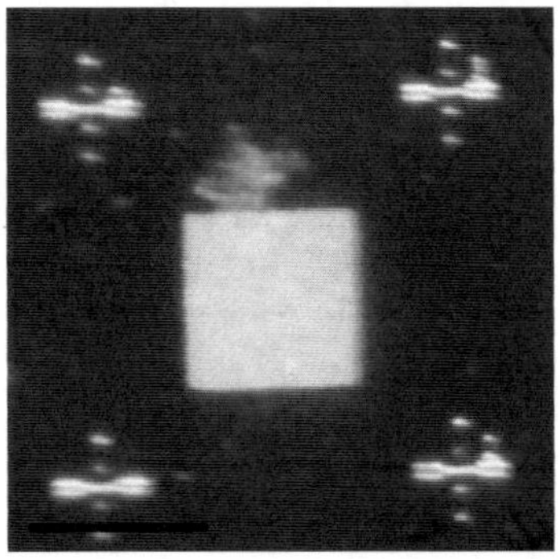

Figure 6. A dark-field optical image of Si NPs with 1 μm pitch. The scale bar is 200 μm.

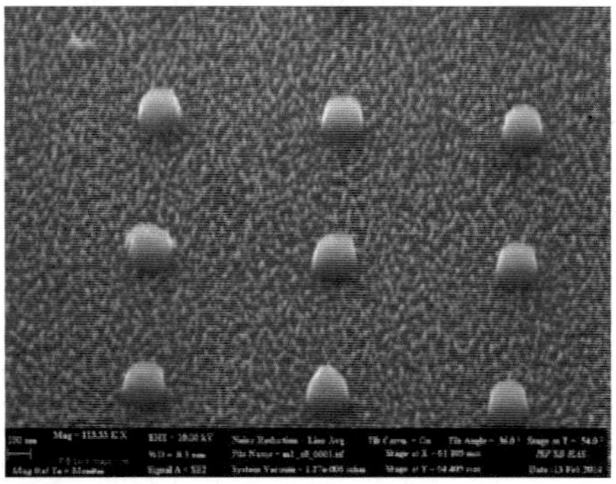

Figure 7. SEM image of Si NPs array №1 with 1 μm pitch (tilt angle = 54°). The scale bar is 100 nm.

Low temperature photoluminescence (PL) of Si NPs was investigated. Figure 8 presents the PL spectra of Si NPs. The PL spectra were measured at 4.9 K. The excitation beam from a He-Ne laser (λ = 632.8 nm) was focused onto the sample placed in a cryostat. The diameter of the focused spot was 200 μm. The peak photoluminescence energy was observed at 0.83 μm and 1.14 μm. Enhanced Raman scattering from coupled vertical Si NP arrays was observed. Enhancement at least by a factor of 40 in comparison with bulk silicon was achieved.

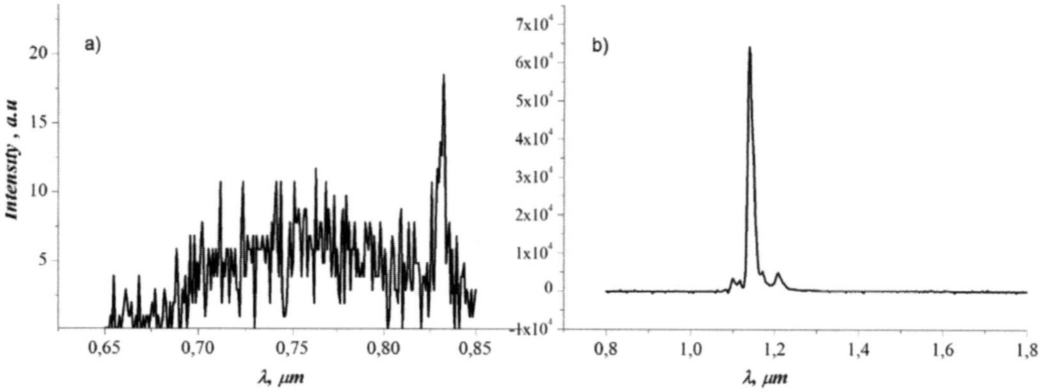

Figure 8. PL spectra of Si NPs.

4. Conclusion

In this work, ordered SiNPs arrays with controllable sizes were fabricated by means of e-beam nanolithography and dry etching. The ordered arrays of Si NPs fabricated by means etching present the advantages of being very uniform and defect free. This process allows fabricating dense NPs arrays with a wide range of dimensions. Ordered SiNPs arrays with diameters increasing from 60 to 350 nm were fabricated in square arrays. Si NPs were chemically and electrically passivated throughout the deposition of TiONx nanolayer. Using scanning electron microscopy, surface imaging and elemental analyses of structures were carried out. Vivid colors were demonstrated in Si NP. The enhancement of Raman scattering was investigated. The low temperature photoluminescence from Si NPs was investigated. The peak photoluminescence energy was observed at 0.83 μm and 1.14 μm. The enhanced optical properties can be achieved by optimizing the diameter and the pitch of Si NP. The results of the work show that Si NPs could be advantageous for nanoscale optical application.

Acknowledgements

We thank Dr. A K Gutakovsky for carrying out the TEM analysis. We thank Dr. E E Rodyakina, D A Nasimov and A S Medvedev for their assistance in sample preparation. The work was partly supported by RFBR under Project no. 13-02-01216.

References

[1] Hyunsung Park, Kenneth B Crozier 2013 *Scientific Reports* **3** 2460
[2] Green M A, Jianhua Zhao et al. 2001 *Nature* **412** 805
[3] Wei-yu Chiu, Tai-Wei Huang et al. 2007 *Optics Express* **15** 15500
[4] Butt H, Dai Q et al. 2011 *Progress in Electromagnetics Research* **113** 179
[5] Tao Song, Shuit-TongLee et al. 2012 *Nano Energy* **1** 654
[6] Zhiyong Fan, Daniel J Ruebusch et al. 2009 *Nano Res* **2** 829
[7] Murthy B R et al., *Biosensors and Bioelectronics* 2008 **24** 723
[8] Khorasaninejad M, Abedzadeh N et al. 2012 *Nano Letts.* **12** 4228
[9] Chengkuo Lee, Jayaraj Thillaigovindan et al., 2008 *Appl. Phys. Lett.* **93** 113113
[10] Xiaocheng Li, Junshuai Li 2010 *Nanoscale Res Lett* **5** 1721
[11] Y-C Lee et al. 2014 *Photonics and nanostructures – Fundamental and application* **12** 16
[12] Tsakalakos L, Balch J et al. 2007 *Appl Phys Lett* **91** 233117-1
[13] Khorasaninejad M, Swillam M A, Pillai K and Saini S S 2012 *Optics letter* **37** 4194
[14] Khorasaninejad M et al. 2012 *Journal of applied physics* **111** 044328
[15] Dominguez S et al. 2013 *Photonics and nanostructures – Fundamental and application* **11** 29
[16] Seo K, Wober M et al. 2011 *Nano Letts.* **11** 1851
[17] Walia J, Dhindsa N et al. 2013 *Small* **10** 144

Spectral properties of random LC networks with uniformly distributed entries

N A Olekhno[1], Y M Beltukov[2] and D A Parshin[1]

[1]Saint Petersburg State Polytechnical University, 195251 Saint Petersburg, Russia
[2]A. F. Ioffe Physical-Technical Institute, 194021 Saint Petersburg, Russia

E-mail: [1]`olekhnon@gmail.com`

Abstract. Disordered LC networks are an appropriate model for describing dielectric resonances in random metal-dielectric composites. The existing theoretical methods allow to study only the case when fixed values of L and C entries are randomly distributed on a lattice with probabilities p and $1 - p$ correspondingly. In this paper we generalize these approaches to arbitrary values of L and C entries and study as an example a 2d random percolative LC square lattice ($p = 0.5$) with uniformly distributed values of inverse inductances $1/L$ and capacitances C in the interval (0,1). We show that the spectral density of resonances $\rho(\omega)$ in this case has a smooth maximum at $\omega_{max} \approx 0.25$. The asymptotic behavior below the maximum is $\rho(\omega) \propto \omega$ and $\rho(\omega) \propto 1/\omega^3$ above the maximum.

1. Introduction

Metal-dielectric composites consisting of metallic nanoparticles incorporated into dielectric matrix (called cermets, such as Au-MgO, Pt-Al_20_3, Fe-Al_20_3) are attracting considerable attention last years due to their interesting optical properties both from theoretical and experimental point of view [1, 2, 3].

One can model such systems replacing a continuous medium to a large discrete random network made of capacitors C and inductances L (the latter can be in series with a weak resistor R). Such network description is rather natural and can be derived from discrete version of continuous Maxwell equations for scalar potential of electric field. This approximation is valid if all the inhomogeneity length scales such as size of the nanoparticles a and distances between them are much smaller compared to the electromagnetic lengths, namely the wavelength λ and the skin depth δ (in the case of metal). When this is not the case, the full Maxwell equations must be invoked for a correct description of the physics [4, 5].

Due to well known relation between dielectric permittivity ε and conductivity σ of the medium

$$\varepsilon = i\frac{4\pi\sigma}{\omega} \tag{1}$$

the capacitors C model dielectric bridges between metallic nanoparticles the former having conductance $G_d = i\omega C$ with capacitance

$$C \simeq \varepsilon_d a. \tag{2}$$

Here ε_d is a dielectric permittivity of dielectric. Whereas isolated metallic granules are characterized by almost purely inductive response having conductance $G_m = c^2/i\omega L$ with inductance

$$L \simeq \frac{c^2}{\omega_p^2 a}. \tag{3}$$

Here ω_p is the plasma frequency. Indeed the dielectric permittivity of dielectric ε_d is frequency independent in the investigated range, therefore according to (1) frequency dependent conductivity of dielectric $\sigma_d \propto i\omega\varepsilon_d$. However dielectric permittivity of metal $\varepsilon_m = 1 - \omega_p^2/\omega^2$. For $\omega \ll \omega_p$, $\varepsilon_m \propto -\omega_p^2/\omega^2$. Therefore frequency dependent conductivity of metall $\sigma_m \propto -i\omega_p^2/\omega$. The last approximation is valid for frequencies ω of electromagnetic radiation satisfying to condition $\omega_\tau \ll \omega \ll \omega_p$. Here ω_τ is the plasmon relaxation rate [1]. The inductance given by Eq. (3) is the so-called kinetic inductance related with kinetic energy of electrons [6]. It becomes bigger than geometric inductance $L_g \simeq a$ when size of nanoparticles $a < 30\,\text{nm}$.

Such large random dissipationless LC networks (that is networks made of random mixture of capacitances C and inductances L) have a lot of resonance clusters with any size and shape and different resonance frequencies. Due to dipole-dipole interaction between the clusters, eigenmodes in the network are not localized [7] and local electric fields show anomalous fluctuations [8]. These fluctuations are responsible for example for giant surface-enhanced Raman scattering observed in semicontinuous metal films [9].

In the papers [2, 10] considering as an example a $2d$ square lattice it was shown that the problem of resonances of a random LC network (with random positions of capacitors C and inductances L) can be reduced to some generalized eigenvalue problem. However the reduction was done only for the very particular case when all values of inductances L (and capacitances C) are equal. Then in the system there is a typical frequency $\omega_0 = 1/\sqrt{LC}$.

In the present paper we have extended these results to the general case when values of inductances L and capacitances C are also random and can fluctuate from bond to bond. We have shown that in this case frequencies of the resonances can be found from the generalized eigenvalue problem

$$\sum_j K_{ij}\varphi_j - \omega^2 \sum_j C_{ij}\varphi_j = 0. \tag{4}$$

Here $K_{ij} = 1/L_{ij}$ is the matrix of inverse inductances, C_{ij} is the matrix of capacitances between sites i and j and φ_j is a potential on site j. The diagonal elements of these matrices are equal to the minus sum of non diagonal ones

$$C_{ii} \equiv -\sum_{j\neq i} C_{ij}, \quad K_{ii} \equiv -\sum_{j\neq i} K_{ij}. \tag{5}$$

2. Kirchhoff equations for arbitrary values of L and C entries

To derive Kirchhoff equations in the general case for arbitrary values of L and C entries, we will numerate sites of our LC network using integer index i taking values from 1 to N, here N is the total number of cites (see Fig. 1). Let φ_i being the potential of site i. Also we attribute to each site i a charge q_i. It is equal to the sum of charges of capacitor plates adjoined site i (see shaded regions shown on Fig. 1 around sites 6 and 7). If no capacitor adjoins site i, then the corresponding charge $q_i = 0$. The charges q_i and the potentials φ_i are connected with each other by the system of equations

$$q_i = \sum_{j\neq i} C_{ij}(\varphi_i - \varphi_j), \tag{6}$$

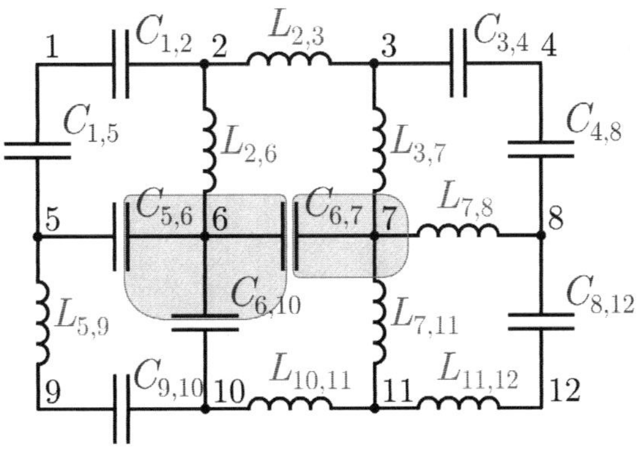

Figure 1. A piece of random LC network. Shaded are regions with charge accumulated on sites 6 and 7 respectively.

where C_{ij} is a capacity of the capacitor between sites i and j. The matrix C is by definition symmetric $C_{ij} = C_{ji}$. If there is no capacitor between sites i and j then $C_{ij} = 0$. Let us define the diagonal matrix element C_{ii} as a minus sum of the non-diagonal elements

$$C_{ii} \equiv -\sum_{j \neq i} C_{ij}, \tag{7}$$

then from Eq. (6) we get

$$q_i = -\sum_j C_{ij}\varphi_j. \tag{8}$$

The time derivative $\dot{q}_i$ by definition is equal to the sum of incoming electric currents I_{ji} to the site i from all neighboring sites j

$$\dot{q}_i = \sum_{j \neq i} I_{ji}. \tag{9}$$

Here I_{ji} is the electric current from site j to site i. The matrix I_{ji} by definition is antisymmetric $I_{ij} = -I_{ji}$. In its turn the time derivative of electric current $\dot{I}_{ij}$ is related to voltage $\varphi_i - \varphi_j$ in the corresponding inductance coil

$$L_{ij}\dot{I}_{ij} = \varphi_i - \varphi_j. \tag{10}$$

Here L_{ij} is an inductance of the coil between sites i and j. The matrix L is a symmetric matrix, $L_{ij} = L_{ji}$. If there is no coil between sites i and j, then the corresponding value of $L_{ij} = \infty$.

Now using equations (9) and (10) we can relate the second time derivative of charge $\ddot{q}_i$ with voltages $\varphi_j - \varphi_i$

$$\ddot{q}_i = \sum_{j \neq i} \frac{1}{L_{ij}}(\varphi_j - \varphi_i). \tag{11}$$

To proceed further let us introduce matrix K of inverse inductances (between sites i and j)

$$K_{ij} = \frac{1}{L_{ij}}. \tag{12}$$

The matrix K is symmetric, i.e. $K_{ij} = K_{ji}$. If sites i and j are not connected by inductance coil, then by definition $K_{ij} = 0$. Similar with the case of matrix C let us define the diagonal matrix elements of the matrix of inverse inductances K as minus sum of the non-diagonal elements (compare with Eq. (7))

$$K_{ii} \equiv -\sum_{j\neq i} K_{ij}. \tag{13}$$

Then from (11) we get

$$\ddot{q}_i = \sum_j K_{ij}\varphi_j. \tag{14}$$

Comparing Eq. (8) and Eq. (14), we obtain finally a linear system of ordinary differential equations for site potentials

$$\sum_j C_{ij}\ddot{\varphi}_j = -\sum_j K_{ij}\varphi_j. \tag{15}$$

Looking for solution of these equations in the exponential form $\varphi_i \propto \exp(i\omega t)$ we get

$$\omega^2 \sum_j C_{ij}\varphi_j = \sum_j K_{ij}\varphi_j. \tag{16}$$

It is identical to Eq. (4). Non-zero solution of this linear system of equations exist when

$$\det\left(K - \omega^2 C\right) = 0. \tag{17}$$

As a result we have reduced calculation of eigenfrequencies of our random impedance network to a generalized eigenvalue problem.

3. Resonances in the network with uniformly distributed entries

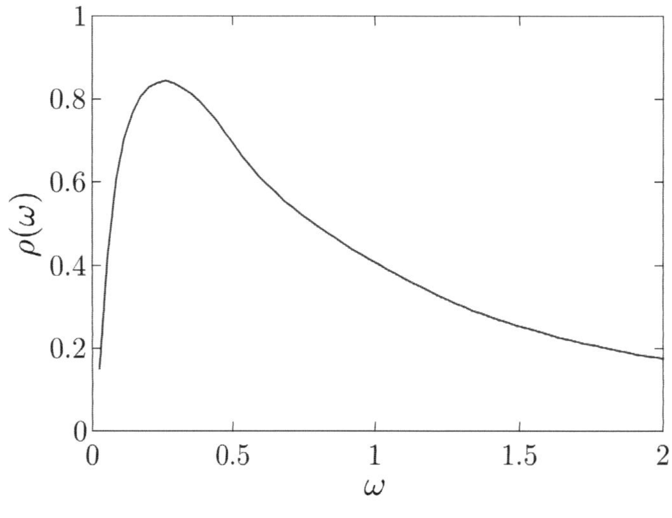

Figure 2. The spectral density of resonances $\rho(\omega)$ in the case of uniform distribution of C and $1/L$ values in the interval (0,1) with concentrations $p_L = p_C = 0.5$.

Using Eqs. (4) we as an example have considered a particular case when capacitances C and inverse inductances $1/L$ are uniformly distributed in the interval (0,1) (in appropriate units) and placed randomly with equal concentrations on a 2d square lattice. The later corresponds to percolation threshold for both types of L and C bonds. The uniform distribution of $C \propto a$ and $1/L \propto a$ values according to (2) and (3) models the realistic case when size of the metallic grains and distance between them a are distributed uniformly from zero to some maximum value. The spectral density of resonances $\rho(\omega)$ is shown on Fig. 2. It has a smooth maximum at frequency $\omega_{max} \approx 0.25$. Asymptotically at small frequencies below the maximum $\rho(\omega) \propto \omega$ and at large frequencies above the maximum the spectral density decays as $\rho(\omega) \propto 1/\omega^3$. One can show that such behavior follows directly from the chosen distributions.

Due to dual symmetry of Eqs (4) in our case the ω^2 and $1/\omega^2$ values have the same distribution function (compare with [2]). It follows from the fact that matrices K and C have the same distribution function for matrix elements (uniform in our case). To show that let us introduce the new variable

$$\lambda = \frac{\omega^2}{1+\omega^2}. \tag{18}$$

The value of $\omega = 0$ corresponds to $\lambda = 0$, whereas $\omega = \infty$ corresponds to $\lambda = 1$. The replacement of $\omega \to 1/\omega$ corresponds to replacement $\lambda \to 1 - \lambda$. The distribution of λ is shown on Fig. 3. It is a symmetric function of λ relative to the point $\lambda = 1/2$. In other words the distribution $\rho(\lambda) = \rho(1 - \lambda)$

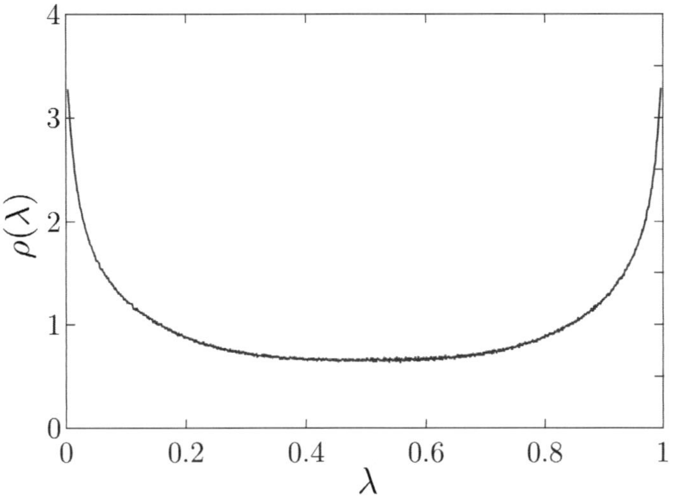

Figure 3. The spectral density of resonances $\rho(\lambda)$.

4. Conclusion

We generalized the problem of random LC networks to the case when values of capacitances C and inductances L in a lattice can fluctuate from bond to bond. As an example we have calculated the spectral density of resonances for uniform distribution of capacitances C and inverse inductances $1/L$ in the interval (0,1). Such distribution corresponds to the uniform distribution of metallic grain sizes a. One can use the developed approach for many other actual distributions.

We are very grateful to Prof. V. I. Kozub for many fruitful discussions. One of the authors (YMB) thanks Dynasty Foundation for financial support.

References

[1] Fyodorov V V 2001 *Physica E* **9** 609
[2] Jonckheere Th and Luck J M 1998 *J. Phys. A* **31** 3687
[3] Clerc J P, Giraud G, Laugier J M and Luck J M 1990 *Adv. Phys.* **39** 191
[4] Sarychev A K and Shalaev V M 2000 *Physics Reports* **335** 275
[5] Bergman D J and Stroud D 1992 *Solid State Phys* **46** 147
[6] Zhou J, Koschny Th, Kafesaki M, Economou E N, Pendry J B and Soukoulis C M 2005 *Phys. Rev. Lett.* **95** 223902
[7] Levitov L S, 1989 *Europhys. Lett.* **9** 83
[8] Brouers F, Sarychev A K, Blacher S and Lothaire O 1997 *J. Phys. A* **241** 146
[9] Brouers F, Blacher S, Lagarkov A N, Sarychev A K, Gadenne P and Shalaev V M 1997 *Phys. Rev. B* **55** 13234
[10] Clerc J P, Giraud G, Luck J M and Robin Th 1996 *J. Phys. A* **29** 4781

Plasmonic resonance of colloidal silver in nanoporous matrix

O.V.Andreeva, S.V.Saitov, N.V.Andreeva, A.I.Sidorov

St. Petersburg National Research University of Information Technologies, Mechanics and Optics, Kronverkskiy, 49, St. Petersburg, 197101, Russia

saitov.sv@yandex.ru

The object of the study in this paper - silver nanoporous silicate matrix with pore size less than 20 nm. Colloidal silver particles with volume concentration about 10^{-4} are formed within free volume of pores of silicate matrix by chemical method. Changes in the attenuation spectra of the investigated object during changing of the refractive index of free volume of pores from 1.0 to 1.5 are reviewed. Comparison of the obtained experimental data with the results of calculations was carried out.

1. Introduction

Nanoparticles of metal silver are of considerable interest in the development of plasmonic materials - artificial nanostructured materials for nanophotonics with unique functional properties that exceed the capabilities of traditional materials and systems. The basic requirements to such materials are: the absence of scattering of containing matrix and a high degree of monodispersion of metal particles. In the studied structure these requirements were achieved by using nanoporous silicate matrix with high transparency in the visible range of the spectrum (the average pore diameter is about 17 nm) and also due to the fact that metal particles are formed in the free volume of pores which are cross-cutting and occupy about 50% of the volume of the sample [1, 2].

Nanoporous silicate matrices are received from the original two-phase glass by processing in the acid and alkali. Samples represent polished plane-parallel plates or disks with the thickness of 1-3 mm. In air-dry condition samples have significant absorption in the short-wave region of the spectrum due to the scattering on porous structure and absorption by the components of the framework. During the introduction of immersion with the refractive index close to the refractive index of the frame into the free volume of pores the sample with the thickness of 1 mm has a high level of transparency in almost entire of the visible spectrum.

2. Experimental methodology

Nanoporous matrices NPM-17 with the thickness of 1 mm with free volume of pores (55±1%) were used in the work, their characteristics are shown on fig.1.

Nanoparticles of silver were formed inside the pores by photolysis of nanoparticles of $AgNO_3$. First, pores of the sample were filled with water solution of $AgNO_3$ with concentration of about one weight percent (1 w. %). Then the sample was briefly washed in distilled water to remove solution from the surface and dried to remove all water from the pores. After this was irradiation of sample by ultraviolet radiation of mercury lamp (λ = 254, 313 and 365 nm) during two hours. In the course of irradiation full photolysis of $AgNO_3$ to Ag in colloidal state took place. After formation of nanoparticles of Ag initially colorless sample has orange-red color. Range of optical density has intense peak of absorption in the spectral range from 350 to 500 nm that is caused by the plasmonic resonance of Ag nanoparticles.

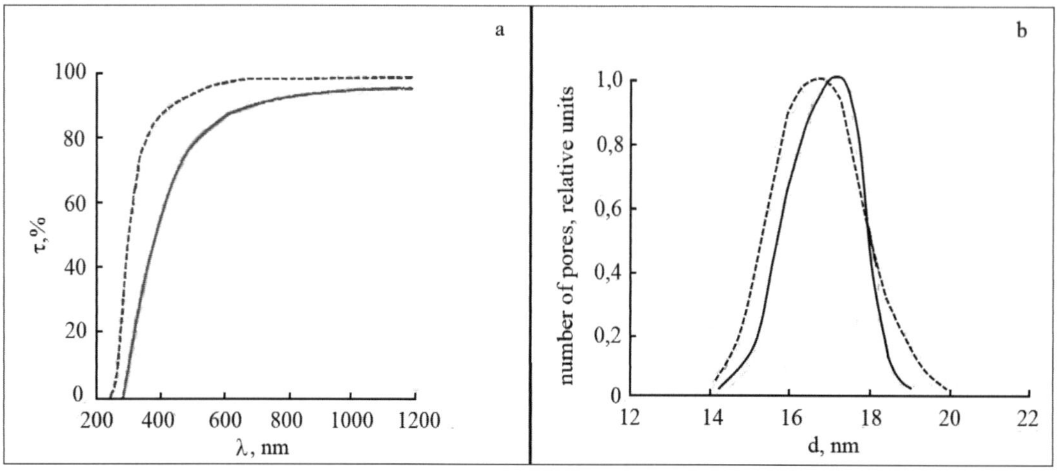

Figure 1. Characteristics of nanoporous matrices NPM-17: a - the transmission spectra of the sample with the thickness of 1 mm in air-dry condition (solid curve) and in water immersion (dashed curve); b - distribution of pores by sizes- sample with the thickness of 1 mm (solid curve), the middle part of the sample with the thickness of 0.2 mm (dashed curve).

At such method of obtaining the maximum size of silver particles is limited and can't exceed the maximum pore size of silicate matrix (i.e. the diameter of obtained particles doesn't exceed the values of 20-22 nm). Volume concentration of particles in the samples was of the order of 10^{-4}. Maximum of plasmonic resonance of such particles is located in the shortwave region of the visible spectrum and its position depends on the refractive index of the containing matrix [3, 4].

According to some researchers changing on the refractive index of the medium which contains particles from 1.5 to 2.3 leads to a significant shift of plasmonic absorption band in the long-wave region of the spectrum to the value of about 100 nm.

In this experiment the matrix containing nanoparticles of silver presents by itself a composite (heterogeneous) structure, its effective refractive index is determined by the contribution of the partial components of the refractive index of silicate glass and refractive index of the filler of free volume of pores – n_f. The increasing of the refractive index of the matrix leads to a shift of the plasmonic resonance in the long-wave region of the spectrum and its reduction - to a shift in the short-wave region of the spectrum and to increasing of the transparency of the sample in the whole visible range of the spectrum.

We know [2] that attenuation of radiation of nanoporous matrices in the shortwave spectral region due to the absorption of components of the silicate framework and to the scattering on porous structure. When filling out with water (n=1.34) free volume of pores compared to air-dry condition of matrix absorption is not changed while scattering reduces significantly and range of attenuation of matrix shifts to the short-wave region ("blue" shift), it is demonstrated on figure 2 (curves 1 and 1').

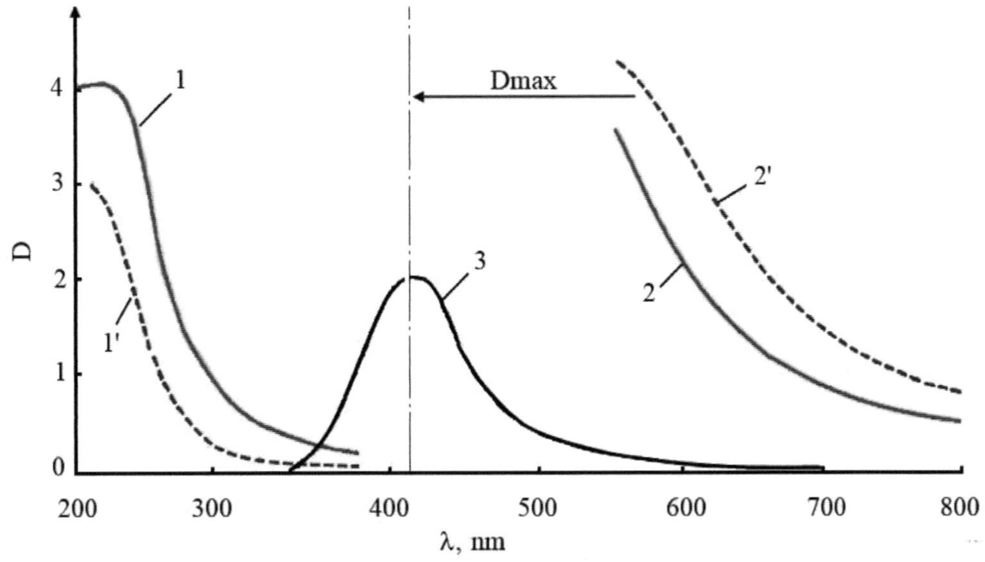

Figure 2. Spectral dependence of the optical density, D, of source nanoporous matrix (curves 1, 1') and nanoporous matrix with colloidal silver particles (curves 2, 2')in air-dry condition (curves 1, 2) and in water immersion (curves 1', 2'); spectrum of attenuation of the preparation of nanoporous matrix with particles of colloidal silver diluted with water (curve 3).

During filling out with water free volume of pores of air-dry matrix with colloidal silver particles (figure 2, curves 2 and 2') the shift of attenuation spectrum of matrix to the long-wave region of the spectrum ("red" shift) is explicitly defined.

For a given concentration of silver particles the measurements are possible in the spectral range $\lambda >$ 550 nm, where the optical density does not exceed the values of $D = 4$. In this spectral range the absorption of silicate frame and scattering on porous structure is negligible and can be ignored. Thus, the value of the attenuation spectrum caused by the absorption of silver particles, but the maximum values of optical density of the absorption spectrum in this case are impossible to determine.

For obtaining the measured values of the optical density of nanoporous medium with colloidal silver particles were prepared preparation of such medium diluted with water to the necessary concentration - on figure 2, curve 3 attenuation spectrum of such product, investigated silver particles in the water, is shown. As you can see, the maximum of the attenuation spectrum of silver particles in the water is 410 nm.

Table 1 shows the theoretically calculated values of the maximum of the spectrum of attenuation of colloidal silver particles at different values of index of refraction of the medium that surrounds the particle. The calculations were made according to the methodology described in work [5] for silver particles with the same geometric parameters. As you can see, there is good coincidence between the experimental data and theoretical calculations in the case of using water as well as oil immersion.

Table 1. Changing of wavelength of attenuation spectrum maximum of colloidal silver particles, λ(Dmax), during changing of refractive index of surrounding medium, n, theoretical calculation according to work [5].

Filler	Air (vacuum)	Water	Oil (cedar)	Immersion	
n_f	1,0	1,34	1,48	1,7	2,2
λ	370	410	430	460	520

The experimental data obtained during changing of the refractive index of free volume of pores are shown on Fig 3.

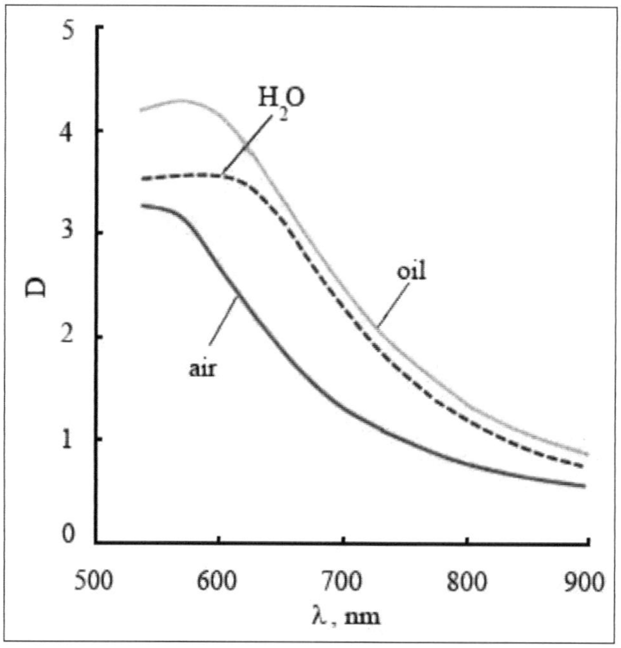

Figure 3. Attenuation spectra of nanoporous medium with colloidal silver particles during changing the filler of the free volume of pores.

As a filler are used: air, water, oil (cedar). The data presented on fig.3 demonstrate increasing of «red» shift when increasing of refractive index of the filler.

Table 2. The shift of attenuation spectra of nanoporous matrices with colloidal silver particles, (Δλ), during changing filler of free size of pores of the matrix, n.

	Conditions of measurements	Air-water	Water-oil
1	D = 2	80	20
2	D = 3	80	20
3	Calculation on Dmax	40	20

Table 2 shows the qualitative results of the definition of observed shifts in the experiments. Because the shift of the maximum can't be determined experimentally the measurements of the shift of wavelength of radiaion at changing of the refractive index of the filler at different levels of optical density (D=2 and D=3) were carried out. Experimental values of the shift of the side sections of the spectrum, of course, exceed the value of the shift of the maximum of the spectrum.

3. Conclusion

Researching of influence of the filler of free size of pores of nanoporous silicate matrix on the parameters of the attenuation spectrum of colloidal silver particles formed inside the free size of pores of this matrix is carried out. The presence of long-wavelength, «red», shift of plasmonic resonance of silver particles during increasing refractive index of free volume of pores is shown. A good coincidence between the experimental results and theoretical estimates is received.

This effect can be used for evaluation:

- effective refractive index of the medium surrounding the particle of silver;
- refractive index of the medium directly surrounding the particle of silver;
- the structural parameters of nanocomposites with particles of colloidal silver.

References

[1] Andreeva O V, Bandyuk O V 2011 Light-sensitive Media-Composites for Recording Volume Holograms Based on Porous Glass and Polymer *Holograms – Recording Materials and Applications* pp 45-70
[2] Andreeva O V, Obyknovennaya I E 2010 Nanoporistye matricy NPS-7 i NPS-17 – vozmozhnosti ispolzovaniya v opticheskom ehksperimente *Nanosistemy: fizika, himiya, matematika* pp. 37-53
[3] Klimov V V Nanoplazmonika. – 2-e izd., ispr. – M.: FIZMATLIT, 2010. – 480 p. – ISBN 978-5-9221-1205-5
[4] Sönnichsen C 2001 Plasmons in metal nanostructures *Göttingen: Cuvillier Verlag* p 134
[5] Sukhanov V I, Kuchinskii S A, Andreeva O V 1996 Diffraction efficiency of amplitude-phase holograms recorded in a silver-containing porous glass Optics and Spectroscopy pp. 779-783

Electric properties of hemispherical metal nanoparticles: influence of the dielectric substrate

S A Scherbak[1], O V Shustova[2], V V Zhurikhina[2], A A Lipovskii[1,2]

[1]Department of Physics and Technology of Nanostructures, St. Petersburg Academic University, 8/3 Khlopina, St. Petersburg, 194021 Russia
[2]Institute of Physics, Nanotechnology and Telecommunications, St.Petersburg State Polytechnic University, 29 Polytechnicheskaya, St. Petersburg, 195251 Russia

E-mail: shustova.o@gmail.com

Abstract. We calculated surface plasmon resonance position for gold and silver hemispheres placed on a dielectric substrate and covered with a thin dielectric layer. The results are applicable in metal islands film enhanced Raman spectroscopy.

1. Introduction

Surface plasmon resonance (SPR) in differently shaped metal nanoparticles is under extensive studies during last decades. This is because of both academic interest in general behavior of the SPR and its applications in optical polarizers and filters [1], nonlinear optics [2], and surface enhanced Raman scattering (SERS) [3]. A very special case of metal nanoparticles corresponds to the nanoislands forming a metal islands film (MIF) which are being widely used in SERS measurements [4]. These nanoislands are often hemispherically shaped [5], especially if they made using out-diffusion technique [6], and their SPR position determines the spectral range of SERS observations. In spite of the conventional use of MIFs in SERS the resonant properties of metal hemispheres deposited on a dielectric, conventionally glass, substrate have not been properly studied by now, but a red shift of plasmonic resonance of hemispheres deposited on a dielectric surface relatively to the SPR position in spheres deposited on the same substrates was mentioned [7].Several research papers related to the problem were concentrated on comprehensive theoretical and numerical analysis of the SPR in an isolated metal hemisphere surrounded by the air [8], spheroids deposited on a dielectric substrate [9] and a nanosphere combined of two conjoined hemispheres of different materials, one of them being plasmonic [10].Below we present an analytical study of the SPR in metal hemispheres placed on a dielectric substrate. We also use the developed theory to calculate the position of SPR in hemispherical metal nanoparticles on a soda-lime glass surface and the shift of the SPR vs the substrate permittivity.

2. Theory

Let us consider a metal hemisphere of radius a placed on a dielectric semi-infinite substrate with dielectric constant ε_{sub}(Fig.1). To examine the polarizability of the hemisphere it is common to use spherical geometry, where two separate components of polarizability exist: the axial one, $\alpha_{\|}$, when the external field is normal to the substrate and parallel to **z** axis, **E**||**z,** and the transverse component, $\alpha_{\perp}$, when the external field is parallel to the substrate surface and transverse to **z** axis, **E**⊥**z**. The latter case corresponds to the normal incidence of light.

If the object is small compared to the wavelength of light the quasistatic approximation can be used. In this case to determine the electric response of the hemisphere one has to find the potential which should satisfy the Laplace equation $\Delta\varphi=0$ throughout the space.

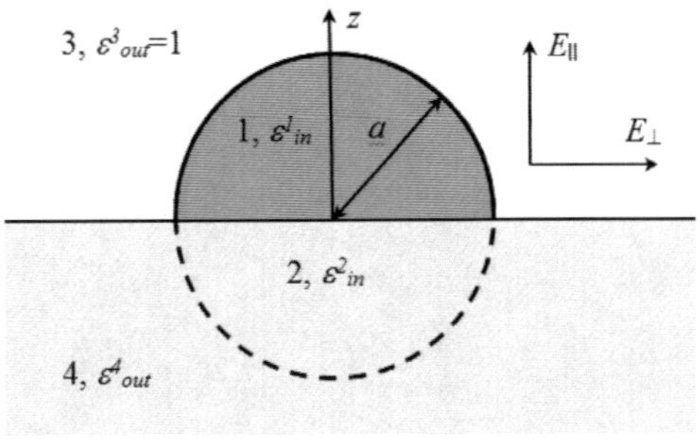

Figure 1. Scheme of the model.

Developing the approach proposed in [8] we divided the space into four regions (Fig.1): the hemisphere with dielectric permittivity ε^1_{in} (1), so-called double hemisphere in the substrate with the dielectric permittivity of the substrate ε^2_{in} (2), the upper media (3, $\varepsilon^3_{out}=1$), and the substrate (4, ε^4_{out}). To describe the problem under consideration the dielectric permittivity of the double hemisphere should be the same as the substrate permittivity: $\varepsilon^2_{in}=\varepsilon^4_{out}$. The potential corresponding to the external field parallel to the substrate (transverse case) could be written as [8]:

$$\varphi=-Er\sin\theta\cos\varphi=-ErP_1^1(\cos\theta)\cos\varphi, \tag{1}$$

where P_1^1 is the associated Legendre polynomial of the first order. To satisfy the Laplace equation in and outside the hemisphere, the potential should be written as the series of associated Legendre polynomials:

$$\varphi^1_{in}=\sum_0^\infty A^1_n r^n P^1_n(\cos\theta)\cos\varphi, \tag{2a}$$

$$\varphi^2_{in}=\sum_0^\infty A^2_n r^n P^1_n(\cos\theta)\cos\varphi, \tag{2b}$$

$$\varphi^3_{out}=\sum_0^\infty D^3_n r^{-(n+1)} P^1_n(\cos\theta)\cos\varphi-ErP^1_1(\cos\theta)\cos\varphi, \tag{3a}$$

$$\varphi^4_{out}=\sum_0^\infty D^4_n r^{-(n+1)} P^1_n(\cos\theta)\cos\varphi-ErP^1_1(\cos\theta)\cos\varphi. \tag{3b}$$

To find the unknown coefficients A_n and D_n ,one should use the boundary conditions:

$$\theta=\frac{\pi}{2}:\begin{cases}\varphi_{in}^{1}=\varphi_{in}^{2},\varphi_{out}^{3}=\varphi_{out}^{4}\\ \varepsilon_{in}^{1}\dfrac{\partial\varphi_{in}^{1}}{\partial\theta}=\varepsilon_{in}^{2}\dfrac{\partial\varphi_{in}^{2}}{\partial\theta},\varepsilon_{out}^{3}\dfrac{\partial\varphi_{out}^{3}}{\partial\theta}=\varepsilon_{out}^{4}\dfrac{\partial\varphi_{out}^{4}}{\partial\theta}\end{cases}, \quad (4)$$

$$r=a:\begin{cases}\varphi_{in}^{1}=\varphi_{out}^{3},\varphi_{in}^{2}=\varphi_{out}^{4}\\ \varepsilon_{in}^{1}\dfrac{\partial\varphi_{in}^{1}}{\partial r}=\varepsilon_{out}^{3}\dfrac{\partial\varphi_{out}^{3}}{\partial r},\varepsilon_{in}^{2}\dfrac{\partial\varphi_{in}^{2}}{\partial r}=\varepsilon_{out}^{4}\dfrac{\partial\varphi_{out}^{4}}{\partial r}\end{cases}. \quad (5)$$

Due to the special features of the Legendre polynomials ($P_n^1(0)=0$ for odd n and $-\frac{d}{d\theta}P_n^1(0)=0$ for even n) the boundary condition (4) leads to the relation between the coefficients A_n and D_n :

$$A_n^1\eta_n^1=A_n^2,\text{where}\,\eta_n^1=\begin{cases}1,n-odd\\ \dfrac{\varepsilon_{in}^2}{\varepsilon_{in}^1},n-even\end{cases} \quad (6a)$$

$$D_n^3h_n^3=D_n^4,\text{where}\,h_n^3=\begin{cases}1,n-odd\\ \dfrac{\varepsilon_{out}^4}{\varepsilon_{out}^3},n-even\end{cases}. \quad (6b)$$

Let us use the normalized radius $\rho=r/a$ and the notations $c_n=A_n^1Ea$ and $b_n=D_n^3Ea$ for the potential inside and outside the hemisphere, respectively. Then, in accordance with (6):

$$\varphi_{in}^{\begin{cases}1\\2\end{cases}}=\sum_0^\infty a^n c_n\eta_n^{\begin{cases}1\\2\end{cases}}\rho^n P_n^1(\cos\theta)\cos\varphi, \quad (7)$$

$$\varphi_{out}^{\begin{cases}3\\4\end{cases}}=\sum_0^\infty a^{-(n+1)}b_nh_n^{\begin{cases}3\\4\end{cases}}\rho^{-(n+1)}P_n^1(\cos\theta)\cos\varphi-ErP_1^1(\cos\theta)\cos\varphi, \quad (8)$$

where $\eta_n^2=1$ and $h_n^4=1$.

To match the boundary conditions at the surface of the sphere, $r=a$ (Eq.5), equations (7) and (8) should be multiplied by $P_n^1(\cos\theta)$ for the angle $\pi/2\le\theta\le\pi$ and $0\le\theta\le\pi/2$, respectively. Using the calculations made in [8] for the integration of the Legendre polynomials, one obtains:

$$\int_{-1}^{0}P_n^1(\xi)P_l^1(\xi)d\xi=(-1)^{n+l}\int_0^1P_n^1(\xi)P_l^1(\xi)d\xi, \quad (9)$$

where $\xi=\cos\theta$. Further we will use the following notation: $U_{nl}^1=\int_0^1P_n^1(\xi)P_l^1(\xi)d\xi$. Finally, the whole set of the equations could be written as:

$$\sum_n b_nU_{nl}^1[\eta_n^1h_l^3\varepsilon^1l+(-1)^{n+l}\varepsilon^2l+(n+1)(\eta_l^1h_n^3+(-1)^{n+l}\varepsilon^2)]=U_{1l}^1[(-1)^{1+l}(1-l)\varepsilon^2+\eta_l^1-l\varepsilon^1\eta_l^1], \quad (10)$$

where the calculations of U_{nl}^{1} are described in [11]. The solution of (10) is to be found numerically. This requires solving N equations and using $N\times N$ matrix.

In the far field the polarizability is defined by the dipole term of the series (3) for the potential outside the sphere:

$$\varphi_{n=1}^{3} = b_1 a^2 E a \rho^{-2} h_1^3 P_1^1(\cos\theta)\cos\varphi = b_1 a^3 E r^{-2}\sin\theta\cos\varphi . \qquad (11)$$

By comparison of (11) with the standard expression for the potential of the dipole moment **p**,

$$\varphi_{dip} = \frac{\mathbf{pr}}{r^3}, \qquad (12)$$

one obtains:

$$p_{\perp} = b_1 a^3 E = \alpha_{\perp} E . \qquad (13)$$

Sometimes it is convenient to introduce normalized transverse polarizabillity, that is

$$\alpha_{\perp n} = \frac{\alpha_{\perp}}{V} = b_1 \frac{3}{\pi} . \qquad (14)$$

The calculations above were done for the transverse case when the external field is perpendicular to the substrate. The approach can be easily extended to the case of the axial field, **E**||**z**.

3. Calculation results

To calculate the polarizability for a metal hemisphere we used experimental data on dielectric functions of Ag and Au [12]. As it was shown in [8], for positive permittivity of a hemisphere or for the permittivity 'negative enough' ($\varepsilon_{in}^{1} < -10\varepsilon_{out}^{3}$) the accuracy of the simulation using the described approach is of the order of 10^{-5}. However, the SPR in hemispherical particles is expected to be close to that in spherical particles, that is at $\varepsilon_{in}^{1} = -2\varepsilon_{out}^{3}$. Thus, the solution can be unstable in the spectral region near the SPR resonance.

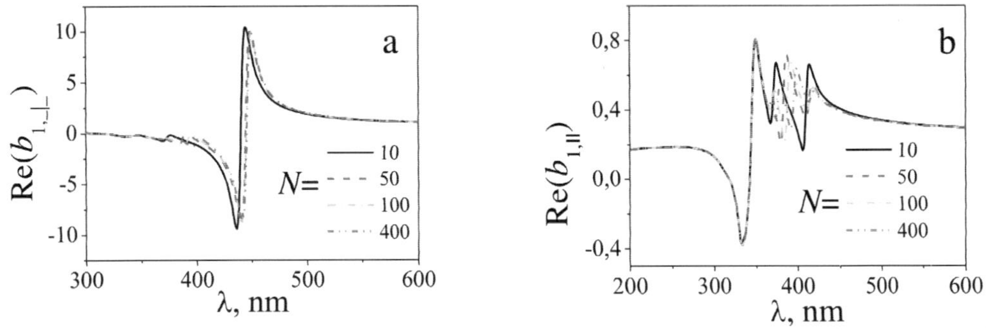

Figure 2. Real part of the coefficient b_1 for silver hemisphere on the glass in the transverse (a) and axial (b) case.

We calculated the transverse and axial polarizabilities of silver and gold hemispheres placed on a glass ($\varepsilon_{out}^{2} = 2.25$) surface for different series number N (Fig.2). As it was expected, the SPR position is very close to the instability region, however, for the transverse case (Fig.2a) the curves for different N converge. For the axial component (Fig.2b) the region of the singularity starts right from the wavelength of the SPR, so that the calculations are not correct near plasmon resonance position. For the case of gold hemisphere, the imaginary part of the permittivity is much larger than that for silver, and, as followed, the accuracy of calculations is higher (Fig.3).

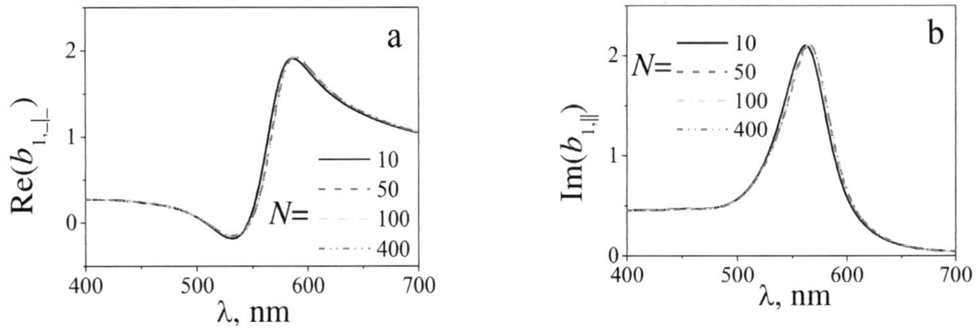

Figure 3. Gold hemisphere on the glass substrate: real part of the coefficient $b_1$in the transverse case (a) and imaginary part in the axial case(b).

The other advantage of the model presented is the possibility to use only the dipole mode to describe the response of the hemisphere at the distance greater than $2a$ from the center and to neglect the influence of the other modes (quadruple mode etc.). To demonstrate this we calculated the dipole potential and the potential from the other components in series (the full potential) in dependence of the normalized distance from the hemisphere center. The results of the calculations for the transverse case are presented in Fig.4. It is seen that at distances longer than $2a$ the difference between the dipole and full potential can be neglected. However, in the vicinity of the hemisphere the higher modes should be accounted for. Spectral dependence of the potential curves is presented in Fig.5. At the position of the SPR resonance the dipole mode dominates that makes it possible to use the dipole approximation for the calculations in this spectral region.

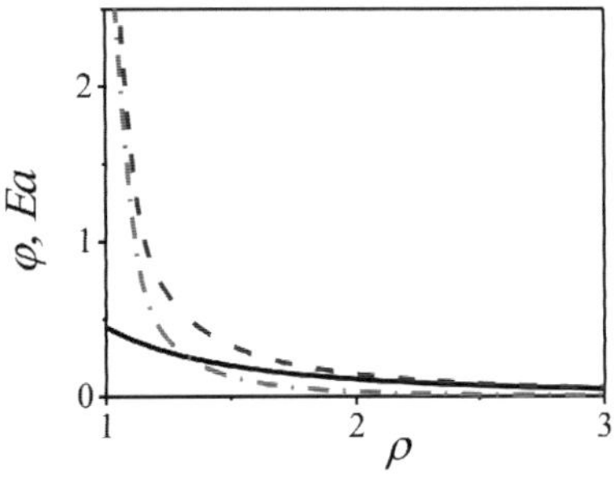

Figure 4. Dependence of the silver hemisphere potential on the distance from the center for the dipole mode (solid line), the full potential (dashed line), and the difference between them (dash dot). Wavelength of light 401 nm, glass substrate.

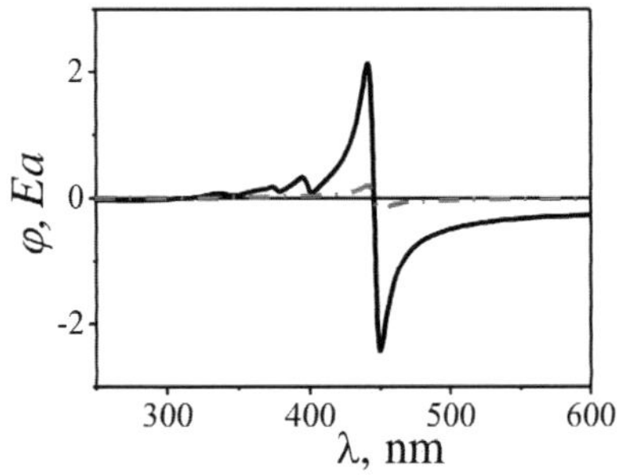

Figure 5. Spectral dependence of the silver hemisphere potential for the dipole mode (solid line) and the difference between full and dipole potentials (dash dot). Distance from the center r=$2a$, glass substrate.

The SPR position dependence on the substrate dielectric constant is presented in Fig.6 for silver and gold hemispheres. It is seen that the dependence is almost linear, and the resonance position can be tuned in a wide range by choosing substrates with different index. In the case of soda-lime glass SPR is at 445 nm for silver and at 565 nm for gold hemisphere these wavelengths are longer than for silver and gold nanospheres embedded in a glass matrix [6].

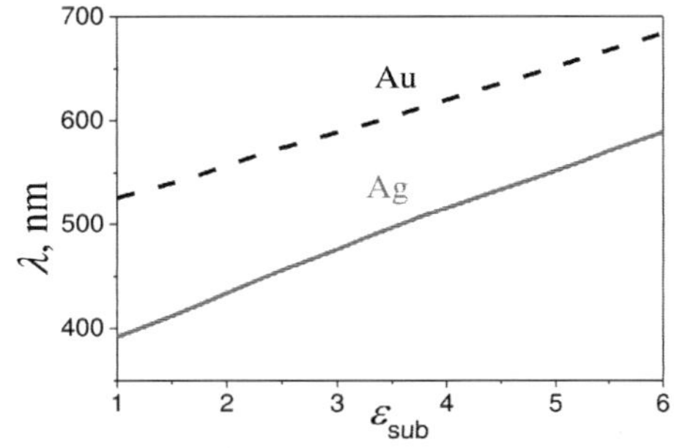

Figure 6. The SPR position of an isolated metal hemisphere on a dielectric substrate vs the substrate dielectric permittivity.

4. Conclusions

We have proposed a novel approach to analytically calculate the SPR position of a metal hemisphere on a dielectric substrate and to find its dependence on the substrate permittivity. The calculation showed that the SPR position differs from that of an isolated hemisphere and shifts towards longer wavelengths with the increase of the substrate. Dipole approximation can be successfully used to calculate the potential outside the hemisphere at distances exceeding the hemisphere diameter. These results can be used in the design of MIF substrates for SERS applications.

This study was supported the FP7 project NANOCOM, ERA.Net RUS project AN2, and Academy of Finland project #267270.

References

[1] Heger P*et al* 2004 *Proc. Conf. Advanced in Optical Thin Films(SPIE)***1** pp 21-28.
[2] Tsutsui Y, Hayakawa T, Kawamura G, Nogami M 2011 *Nanotechnology***22**p 275203
[3] AndoJ,Taka-aki Yano,Fujita K andKawata S 2013*Phys. Chem.***15**pp 13713-13722
[4] Dieringer J A*et alFaraday Discuss***132**pp 9–26
[5] Gupta G*et al* 2009*Nanotechnology***20** 025703
[6] Zhurikhina V*et al* A 2012*Nanoscale Research Letters***7**p 676.
[7] Albella P 2011 *et alNano Lett.***11**pp 3531−3537
[8] Kettunen H, Wallén H, and Sihvola A2008 *J. Appl. Phys.***103** p 094112.
[9] Fedotov V, Emel'yanov V, MacDonald K, and Zheludev N 2004 *J. Opt. A: Pure Appl. Opt.***6**pp 155–160.
[10] Alù A and Engheta N, 2009 *New J. Phys.***11**p 013026.
[11] Kettunen H, Wallén H, and Sihvola A 2007 *J. Appl. Phys.***102** 044105
[12] Johnson P and Christy R 1972 *Phys. Rev. B.***6** 12

Implementation of multiplexing in a subcarrier-wave quantum cryptography system

V V Chistyakov, A V Gleim, V I Egorov, Yu V Nazarov

St.Petersburg National Research University of Information Technologies, Mechanics and Optics, Kronverkskiy pr. 49, St.Petersburg 197101, Russian Federation

E-mail: pspuser@mail.ru

Abstract. Quantum cryptography allows distributing secure keys in a way that any eavesdropping in the channel is inevitably detected. This work is dedicated to introducing wavelength division multiplexing in a subcarrier-wave quantum cryptography system. Compared to other existing schemes, the resulting device is able to achieve higher bitrates (up to 2.26 Mbit/s at 20 km), is robust against external conditions and compatible with standard telecommunication fibres in multi-user environment.

1. Introduction

Quantum cryptography allows distributing secure keys in a way that any eavesdropping in the channel is inevitably detected [1]. Recent progress in the area has led to an increasing interest in combining this technology with telecommunication techniques, thus opening the door to a new generation of secure networks [2-5]. Subcarrier wave quantum key distribution (SCW QKD) technology [6] belongs to a group of QKD schemes most perspective for telecommunication applications. In particular, these systems are unidirectional, polarization insensitive and robust against natural conditions [7-8]. Another important advantage lies in ability to distribute several cryptographic keys using a wavelength division multiplexing technology (WDM). System with the several sidebands was developed using simultaneous phase-frequency modulation of a single central band [9]. Quality of this method is limited by losses of used modulators and complexity of the close spaced sidebands filtration system. In this work we present another principal scheme of a multichannel SCW QKD system operating at multiple bands, and experimental results that demonstrate its operation in the classical mode. This system allows distributing the keys with high speed or between more than two users, which can be useful for integrating it in telecommunication networks.

2. Theoretical analysis

In terms of SCW QKD multichannel system means that several intense carrier waves, each with its own pair of subcarriers acting as quantum channel, are transmitted through a single fiber from sender (Alice) to receiver (Bob), where they are splitted, filtered and detected. In the first part of work we estimated the impact of using WDM technique on raw key generation rate (bitrate) and quantum bit error rate (QBER) of the resulting system. It is known that bitrate is limited by detection frequency:

$$R_{raw} = \frac{1}{2} f_{bit} \mu a_{loss} \eta \ ,$$

where ½ is multiplier that means possibility of modulators states matching in B92 protocol [1], $f_{\rm bit}$ – phase-shifting frequency, μ – mean value of photon number per pulse, η – quantum efficiency of a single photon detector, $a_{\rm loss}$ means probability of photon arrival and defined through the value of total losses $A_{\rm all}$ in the channel:

$$a_{loss} = 10^{-\frac{A_{all}}{10}}$$

Total losses include losses in the channel and element's losses:

$$A_{all} = A_{channel} + A_{elem} ,$$

where $A_{\rm channel.} = L \cdot \alpha$, L – length of the channel, α – attenuation coefficient, α = 0.2 dB/km. To calculate A_{elem} we should sum up losses in the Bob's modulators, circulators, except demultiplexing system. In the single-channel SCW QKD systems this value is 10.47 dB. Losses at the sender unit shouldn't be considered, because signal in the quantum channel is attenuated to μ = 0.2 photon per pulse according to B92 protocol.

It is necessary to use SSPD as a photon counter with best combination of quantum efficiency and detection frequency. Our detector could achieve $f_{\rm bit}$ = 500 MHz and η = 16 % [10].

We considered two methods of splitting the signal at the receiver unit: by using a cascade of circulators with optical filters (serial demultiplexing system) and a regular DWDM demultiplexer (parallel demultiplexing system). The first method is preferable for less than five channels because of the lower losses. Losses in the second configuration equal to 17.44 dB.

In the DWDM systems channel spacing is 0.8 nm (100 GHz). Because of that filtration becomes much easier than WDM SCW QKD with one carrier and several subcarriers. Also because of different wavelengths attenuation coefficient becomes different for an each channel. In our calculations we compared multichannel (up to 40) systems with the regular single channel setup and serial demultiplexing system with four channels (Figure 1 a,b).

Calculation results show that in a 40 channel SCW QKD system bitrates up to 2.26 Mbit/s, 550 kbit/s, 53 kbit/s can be achieved at distances 20, 50, 100 km respectively. QBER is estimated to achieve the critical value of 11-12% at fiber length more than 200 km, therefore allowing key distribution to large distances.

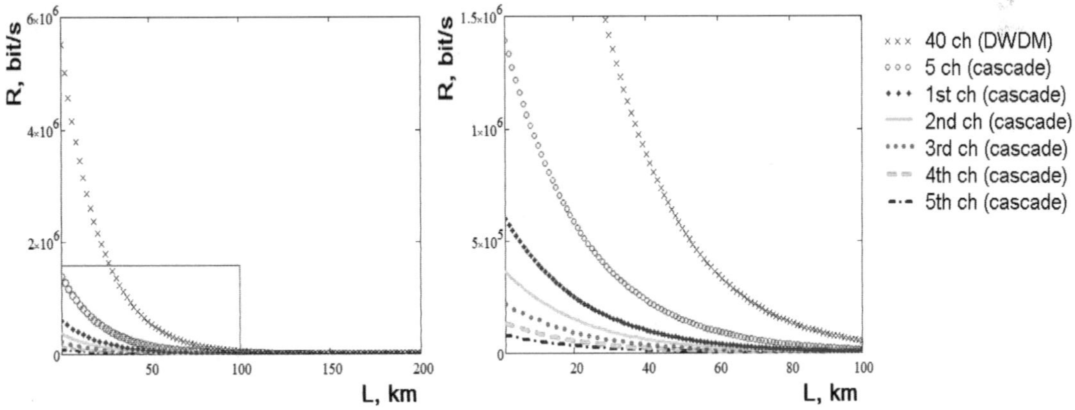

Figure 1 (a), (b). Dependence of raw key generation rate on the channel's length with the different demultiplexing configurations.

3. Experimental results

In the experimental part of work we developed a double-channel SCW QKD system prototype in two versions accordingly two stages: obtaining of the spectral characteristics (Figure 2) and oscillograms of constructive and destructive interference (Figure 3). In this setup optical carriers from two lasers (1550.12 nm and 1550.92 nm) are independently modulated at transmitter (Alice) side, combined and send to receiver (Bob). It is important that for subsequent modulation it is enough only one phase-frequency modulation system for all channels. After subsequent modulation and demultiplexing, the resulting spectrum passes through the optical filters in order to cut off the signal on subcarriers, which is then detected and analysed.

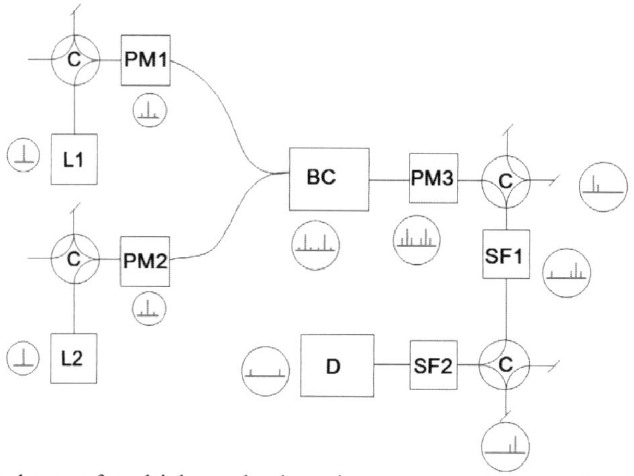

Figure 2. Principal scheme of multichannel subcarrier wave quantum cryptography system. L1, L2 – lasers, PM – phase modulator, BC – beam combiner, C – circulator, SF – spectral filter, D – detector. On the insets one can see signal spectra at different ports.

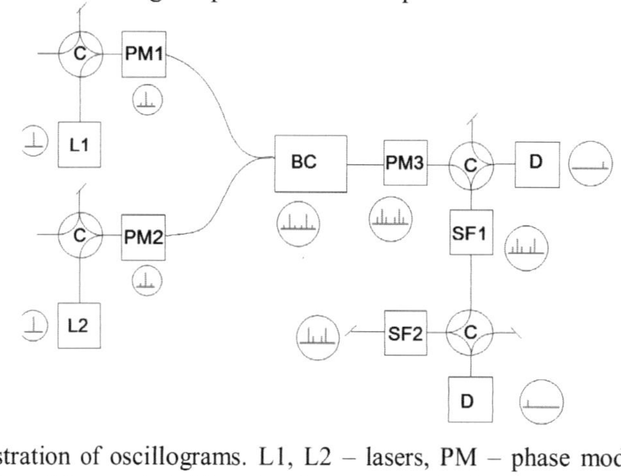

Figure 3. Demonstration of oscillograms. L1, L2 – lasers, PM – phase modulator, BC – beam combiner, C – circulator, SF – spectral filter, D – detector. On the insets one can see signal spectra at different ports.

Experiments have shown that weak signals at wavelengths as close as 0.881 nm that appear on side frequencies of the two carriers can be effectively extracted from signal spectrum in case of using π-shifted Bragg grating filters. For the first stage we constructed system in a way that sidebands were directed in the filter's transmission spectra and observed by optical spectrum analyzer (Figure 4).

For the second stage we performed simulation of quantum key distribution in the classical mode with B92 protocol [1]: it was achieved by repeatedly changing the relative phase shift introduced by the modulators in a basis [0; π] at 1 kHz rate. The dependence of the light intensity on the sidebands looks like the same frequency meander (Figure 5). Signal contrast was measured to be 20 in both channels, which corresponds to visibility 90.4% in the quantum regime.

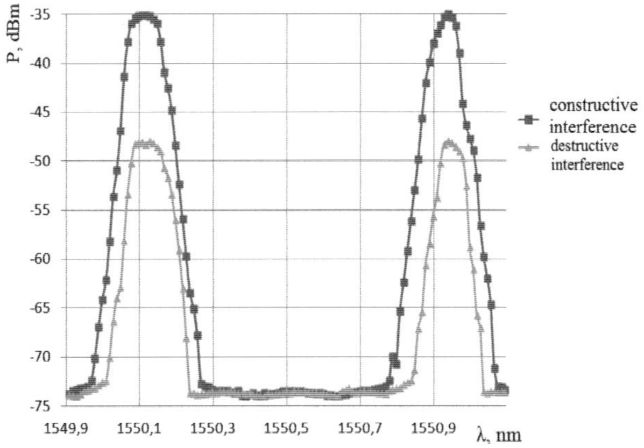

Figure 4. Spectra of constructive and destructive interference

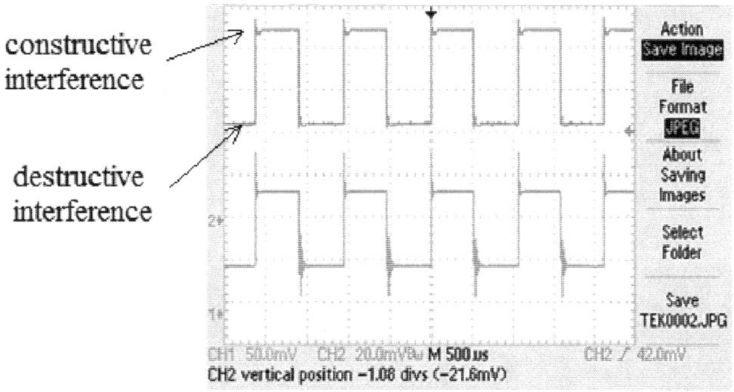

Figure 5. Oscillograms of constructive and destructive interference

4. Conclusion

We developed a multichannel SCW QKD system and performed simulation of quantum key distribution in accordance to B92 protocol. The main motivation of this work lies in the necessity of integrating QKD devices into existing telecommunication environment. Due to quantum nature of information carriers, no amplifiers can be used in the communication link, therefore, wavelength division multiplexing is useful not only for implementing multiuser setups, but also for increasing raw key generation rates. Demultiplexing configuration's selection depends on the number of channels: parallel splitting for five and higher channels or serial splitting for lower number. According to the calculations, the resulting bitrate may be as high as 2.26 Mbit/s. Combining our results with multiplexing at several subcarriers [9] would lead to even greater possibilities.

Acknowledgements

This work was financially supported by Government of Russian Federation, Grant 074-U01.

References

[1] Scarani V, Bechmann-Pasquinucci H, Cerf N J *et al* 2009 *Rev. Mod. Phys.* **81**, pp 1301-1350
[2] Townsend P D 1997 *Electron. Lett.* **33**, pp 188-190
[3] Fernandez V, Collins R J, Gordon K J, Townsend P D and Buller G S 2007 *IEEE J. Quantum. Electron.* **43**, 130
[4] Peev M, Pacher C, Alléaume R *et al* 2009 *New Journal of Physics* **11** 075001
[5] Patel K A, Dynes J F, Choi I, Sharpe A W, Dixon A R, Yuan Z L, Penty R V and Shields A J 2012 *Physical Review X* **2**, 041010
[6] Mérolla J-M, Mazurenko Yu T, Goedgebuer J-P, Porte H and Rhodes W T 1999 *Optics Letters*, Vol **24**, Issue 2, pp 104-106
[7] Guerreau O, Mérolla J-M, Soujaeff A, Patois F, Goedgebuer J-P and Malassenet F J 2003 *IEEE J. Sel. Top. Quantum Electron.* **9**, pp 1533-1540
[8] Gleim A V, Anisimov A A et al. 2014 *Bulletin of the Russian Academy of Sciences. Physics*, Vol **78**, No 3, pp 171–175
[9] Mora J, Ruiz-Alba A, Amaya W, Martínez A, García-Muñoz V, Calvo D and Capmany J 2012 *Optics Letters* Vol **37**, No 11
[10] G.N Gol'tsman, O. Okunev, et al. 2001 *App. Phys. Lett.,* **79** (6), P. 705-707

Edge modes splitting by nanostructured surface

A A Zalutskaya[1,2], A V Prokaznikov[3]

[1]Yaroslavl Demidov State University, Sovietskaya, 14, Yaroslavl, 150000 Russia
[2]MIIT, Suzdalskoe shosse, 13, Yaroslavl, 150030 Russia
[3]Yaroslavl Branch of Physico-Technological Institute RAS, Universitetskaya, 21, Yaroslavl, 150007 Russia

E-mail: an.zalutskaya@gmail.com

Abstract. In this work the regularities of edge modes splitting in electrostatic approximation were considered in the structures like paraboloid of revolution. It was demonstrated that large splitting of edge modes occur within determined interval of wave vector with splitting out a soft, zero mode.

Nano-structured systems play a sufficient role in the miniatuarizing process of devices in contemporary electronics as well as enable ones to include unusual properties of these objects into an arsenal of modern technology. Some attractive moment in nano-system usage is the circumstance that theirs properties, as a rule, are described by quantum physics laws what causes the appearance of new possibilities in realization more universal devices in addition to traditional ones.

In the present work effects that were connected with edge modes splitting by sharp formations on a nano-structured surface were studied. Analytical solution was carried out for the case of paraboloid of revolution. The problem was solved in the electrostatic approximation at the condition when wave length of incident electromagnetic radiation λ is more large than characteristic size L of scatterer, i. e. λ >> L. Laplace's equation was solved in parabolic coordinates (ξ, η, ϕ):

$$\nabla^2\Phi = \frac{1}{\xi^2+\eta^2}\left[\frac{1}{\xi}\frac{\partial}{\partial\xi}\left(\xi\frac{\partial\Phi}{\partial\xi}\right)+\frac{1}{\eta}\frac{\partial}{\partial\eta}\left(\eta\frac{\partial\Phi}{\partial\eta}\right)\right]+\frac{1}{\xi^2\eta^2}\frac{\partial^2\Phi}{\partial\phi^2}=0 \quad (1)$$

The solution is carried out by the traditional methods and looks like this:

$$\Phi_{mq}(\xi,\eta,\phi)=\begin{cases} A_{mq}J_m(\sqrt{q}\xi)I_m(\sqrt{q}\eta)e^{im\phi}, \\ \qquad \eta\le\eta_0,\ -\infty\le\xi\le\infty;\ 0\le\phi<2\pi \\ B_{mq}J_m(\sqrt{q}\xi)K_m(\sqrt{q}\eta)e^{im\phi}, \\ \qquad \eta\ge\eta_0,\ -\infty\le\xi\le\infty;\ 0\le\phi<2\pi \end{cases} \quad (2)$$

where the conditions $\eta = \eta_0, -\infty \le \xi \le \infty, 0 \le \phi \le 2\pi$ correspond to the border of paraboloid of revolution, q is a wave vector, $J_m(\sqrt{q\xi})$ is usual Bessel's function, $I_m(\sqrt{q\eta})$, $K_m(\sqrt{q\eta})$ are Bessels' functions of imaginary argument (Macdonald functions), m is a number.

The boundary conditions for the problem being solved were continuity of potential $\Phi_{mq}(\xi,\eta,\phi)$ and derivation $\varepsilon\partial\Phi_{mq}(\xi,\eta,\phi)/\partial\eta$ on the border $\eta = \eta_0$ of paraboloid of revolution, where ξ, η, ϕ are corresponding parabolic coordinates, ε is a dielectric constant of substance of paraboloid, an external medium was an air, q is a wave vector, m is a magnetic number, coordinate η describes "moving aside" of a base parabola. After demanding boundary condition being fulfilled for equation solution one can obtain in the case when some coefficients A_{mq} and B_{mq} are not equal to zero a dispersion correlation for electrostatic modes that are localized near the top of paraboloid:

$$\varepsilon(\omega) = \frac{I_m(\sqrt{q\eta_0})\, K_m'(\sqrt{q\eta_0})}{I_m'(\sqrt{q\eta_0})\, K_m(\sqrt{q\eta_0})} \tag{3}$$

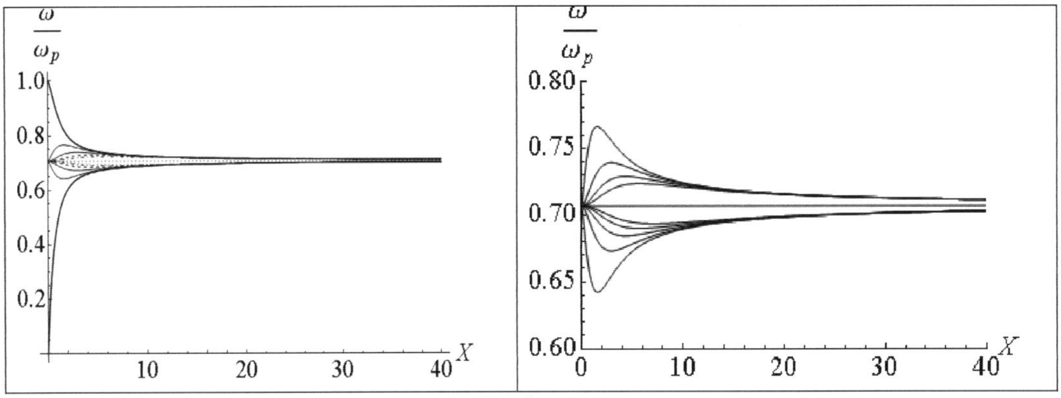

a *b*

Figure 1. Dispersion curves for dependence ω/ω_p versus $X = \sqrt{q\eta}$ with splitting of soft, zero mode for dispersion law $\varepsilon(\omega) = 1 - \omega_p^2/\omega^2$ in the case of convex paraboloid of revolution (bottom part of the graph, numbering of modes rises from bottom), and ones with splitting of dispersion curve which is approached to the bulk plasmon for concave paraboloid (upper part of the graph) – (*a*). Splitting of dispersion curves for dependence ω/ω_p versus $X = \sqrt{q\eta}$ for upper zones of convex (under $1/\sqrt{2}$) and concave (above $1/\sqrt{2}$) paraboloids for first four modes (b)

Putting dielectric constant for narrow band semiconductors (and high doped ones) in the form: $\varepsilon(\omega) = 1 - \omega_p^2/\omega^2$ one can obtain following result for dimensionless variable ω/ω_p for typical values $n = 10^{23}$ cm^{-3}, $m^* = m_0 = 9.1 \cdot 10^{-28}$ g which is presented on the Figure 1, where $X = \sqrt{q\eta}$.

For determination of dispersion curves for concave paraboloid in the form of parabolic pit on a flat interface, for example, an air and semiconductor one should change in expression for $\varepsilon(\omega)$ this value into $\varepsilon^{-1}(\omega)$, what results from a demand of continuity of derivative with respect to corresponding variable quantities on the border of two media that are changed each other by places in this case. As a result one can obtain dispersion curves that are presented on the Figure 1. Evaluation of maximal splitting region gives $\sqrt{q\eta} \le 10$. The region of maximal splitting of edge modes for an upper band for $\lambda \sim 500$ nm has an order of 800 nm and the region of modes splitting spread up to approximately 8 μm.

The character of soft mode for a convex paraboloid of revolution in approximation $\vec{q} \to 0$ ($X = \sqrt{q\eta} \to 0$) is described by following formula:

$$\omega_0(\vec{q} \approx 0) = \frac{\omega_p}{\sqrt{2}} X \left[ln(2/X) \right]^{1/2} \tag{4}$$

In the work [1] an analogous result was obtained which was interpreted as a quasi-acoustical mode which has a dispersion law of one dimensional plasmon. Thus, soft, zero mode can be identified with collective plasmon excitation and generally speaking obtained in a corresponding way be means of polarization operator.

The special structures were prepared on the base of standard silicon wafers with specific resistance equals 4.5 Ohm·cm and (100) orientation. Then vertical columns were formed by plasmo-chemical way by means of Bosch-process through chromium (Cr) mask with the thickness of 30 nm while using lift-of lithography. The result of described procedure is presented on the Figure 2.

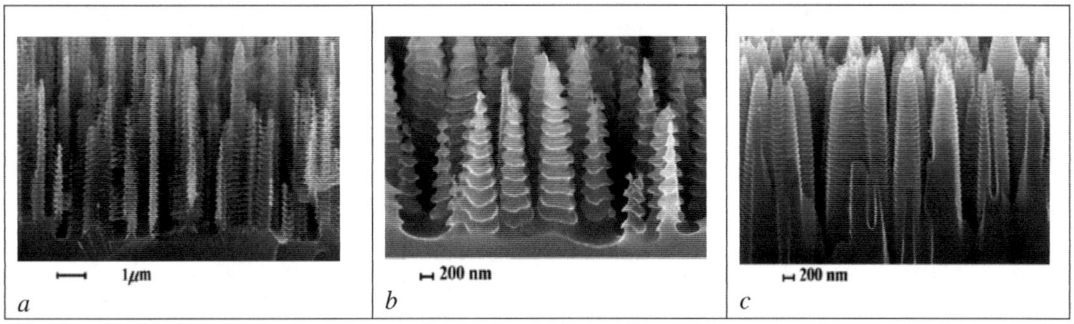

Figure 2. Yellow colored structures (*a*), green colored structures (*b*), black colored structures (*c*), that were formed by plasmochemical etching on silicon wafer.

Drude–Lorentz–Sommerfeld model in the framework of which dispersion law $\varepsilon(\omega) = 1 - \omega_p^2 / \omega^2$ was presented precisely enough describes optical properties in particular metals in infrared region. In visible region this model should be completed taking into account a response of binding electrons. Within frameworks of classical model it can be done by means of considering dielectrics or semiconductors as a sum of molecules each of them is an oscillator with the same resonant frequency ω_0. The equation of motion for an electron in Drude–Lorentz–Sommerfeld model can be presented as following:

$$\frac{\partial^2 \vec{r}}{\partial t^2} + \Gamma \frac{\partial \vec{r}}{\partial t} + \omega_0^2 \vec{r} = \frac{e\vec{E}}{m} \exp(-i\omega t) \tag{5}$$

where $\vec{r}$ is a displacement of charge along $\vec{E}$ direction relatively non-perturbed location, ω is a frequency of incident radiation, ω_0 is a frequency of a binding transition, m is the mass of a free electron, Γ is a constant which is resulted from charge carrier relaxation, moreover, for the metals on the order Γ is equal to $\Gamma = \upsilon_F/l$, υ_F is a velocity on a Fermi surface, l is a free path length of charge carriers. Let us note that for free electrons last addend in the right hand side is absent so that $\omega_0 = 0$. To analyze the character of reflection spectra during interaction of radiation with free electrons one can use the expressions for real (ε_r) and imaginary (ε_i) parts in the form $\varepsilon_r(\omega) = \varepsilon_\infty\left(1 - \frac{\omega_p^2}{\omega^2 - \tau^{-2}}\right)$; $\varepsilon_i(\omega) = \varepsilon_\infty \frac{\omega_p^2}{\omega^3 \tau_p}$,where ε_∞ is a high frequency dielectric constant, ω is a cyclic frequency of incident light, τ is a time between collisions, τ_p is a momentum relaxation time of charge carriers, ω_p is a plasmon frequency, $\Gamma = \tau^{-1}$. Taking into account formulae for ε_r and ε_i one can conclude that the result of calculations looks like that presented on the Figure 3 (we put approximately $\tau \sim \tau_p$).

For reflection coefficient at a normal fall of radiation following formula takes place $R = \frac{(n-1)^2 + \kappa^2}{(n+1)^2 + \kappa^2}$, where n is a refraction coefficient, κ is absorption coefficient (extinction coefficient), moreover:

$$n = \sqrt{\frac{1}{2}\cdot\left(\varepsilon_r + \sqrt{\varepsilon_r^2 + \varepsilon_i^2}\right)}; \quad \kappa = \sqrt{\frac{1}{2}\cdot\left(-\varepsilon_r + \sqrt{\varepsilon_r^2 + \varepsilon_i^2}\right)}. \tag{6}$$

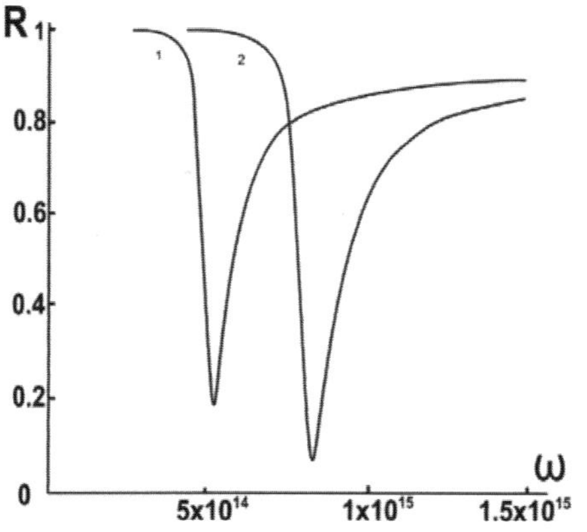

Figure 3. Dependence of refraction coefficient R against the cyclic frequency ω of incident radiation in Drude–Lorentz–Sommerfeld model for free electrons at following values of parameters: $\omega_p = 5\cdot10^{14}\ s^{-1}$, $\tau = 2\ 10^{-14}\ s$, - (1), $\omega_p = 8\cdot10^{14}\ s^{-1}$, $\tau = 2\ 10^{-14}\ s$, $\varepsilon_\infty = 11.8$ – (2).

Near the minimum of dependences for wave-guidance modes the correlation takes place for the low mode: $\sqrt{q_{\min}}\cdot\eta_0 \sim 1$, where $q_{\min} = 2\pi/\lambda_{\min}$ corresponds to the minimum of dependence ω(q). Taking the value $\eta_0^2 \sim d_0 \sim 2r_0$ (r_0 is characteristic radius of coloring structure) one can obtain for estimations the proportional dependence ($\lambda_{\min} \sim r_0$) for wave length which corresponds to the minimum on dispersion curve and for radius of curvature of scattering system. The mode with such a wave length will be captured more effectively by distorted surfaces.

References

[1] Longe P, Bose S M 1993 *Phys. Rev. B.* **48,** 18239 – 18243

Investigation of $In_{0.7}Ga_{0.3}As/In_{0.7}Al_{0.3}As$ metamorphic HEMT-heterostructures by photoluminescence spectroscopy

D. S. Romanovskiy[1], S. A. Tarasov[1], G.B. Galiev[2], S.S. Pushkarev[2]

[1]Saint-Petersburg Electrotechnical University "LETI", Prof. Popova 5, St. Petersburg 197376, Russia
[2]Institute of Ultrahigh Frequency Semiconductor Electronics, Russian Academy of Sciences, 117105 Moscow, Nagornyi proezd, 7 (5).

E-mail: DSRomanovskiy@gmail.com

Abstract. Low-temperature photoluminescence and photoreflectance have been studied in several metamorphic HEMT- (MHEMT-) heterostructures with the same active regions and different buffer layer designs grown by solid-source molecular beam epitaxy. The indium mole fraction in InAlAs/InGaAs/InAlAs single quantum well (QW) is 0.7. It was found that structures with step-graded metamorphic buffer have better quality. Also it was shown that mismatched superlattices in metamorphic buffer can influence on the half-width of photoluminescence spectra. The possible attribution of photoluminescence and photoreflectance spectral lines and their thermal behaviour are critically discussed.

Metamorphic epitaxial heterostructures with quantum well InAlAs/InGaAs/InAlAs for high electron mobility transistor (HEMT) grown on GaAs substrates have recently attracted considerable attention [1]. Such heterostructures have the following advantages in comparison with heterostructures on InP substrates: lower substrate cost, less substrate brittleness, and high flexibility for bandgap engineering. On the one hand the range of indium mole fraction 0.3–0.5 in quantum well is available only with GaAs substrates. On the other hand the range of indium mole fraction 0.5–1.0 may be achieved by using metamorphic technology either on InP substrates or on GaAs substrates, but as it was mentioned above GaAs substrates are more paying.

Metamorphic technology employs in the case of epitaxial growing of layers mismatched with substrate by the lattice parameter. It consists in epitaxial growing of thick transition layer with gradually varying composition between substrate and active region of HEMT-heterostructure concluding quantum well, doping layer and contact layer. This layer is named metamorphic buffer (MB), its thickness is about 1 micron or more. Usually the InAlAs compound is used for this purpose due to good isolated properties, but also InGaAlAs compound may be employed [2].

In lattice-mismatched structures consisting of layers with a thickness in excess of the critical thickness for plastic relaxation, misfit dislocation networks are formed. MB behaves as a virtual substrate that can be designed in order to accommodate the lattice parameter of the topmost structure to that of the underlying substrate. MB allows the reduction, by means of magnitude, of the dislocation density by confining them in definite regions of the structure, which can be kept apart from the active layers of the device. MBs may have constant or graded composition, either in a steplike or in a continuous way. Superlattices (SLs) composed of very thin alternate compressive strained and tensile strained layers are capable of blocking threading dislocations generated in metamorphic buffer thus preventing their penetration into the active region [3, 4]. It was demonstrated that threading dislocations bend near the strained superlattices and spread in the lateral direction after that [5]. In this study efficiency of step-graded and continuous MBs with internal strained superlattices has been investigated.

The heterostructures used in this study (hereinafter referred to as «samples») were grown by conventional molecular beam epitaxy on GaAs substrates. Their construction shown in figure 1 is typical for metamorphic HEMT-heterostructures with the exception of upper layer: instead of doped contact layer n-InGaAs the undoped protective layer i-InGaAs was grown for the correct measurement of electrophysical parameters. Samples possess different design of metamorphic buffer as shown on figure 2. Sample 895 was grown on GaAs substrate 2° misoriented from (1 0 0) plane towards $[0\bar{1}\bar{1}]$ direction, other samples were grown on GaAs (1 0 0) substrates.

Layer	Thickness, nm
$In_{0.7}Ga_{0.3}As$ (*protecting layer*)	7
$In_{0.7}Al_{0.3}As$ (*barrier*)	22
Si (*delta-layer*)	–
$In_{0.7}Al_{0.3}As$ (*spacer*)	6
$In_{0.75}Ga_{0.25}As$ (*quantum well*)	**16**
$In_{0.7}Al_{0.3}As$ (*barrier*)	46
$In_{0.7}Al_{0.3}As$ (*smoothing layer*)	115
$In_xAl_{1-x}As$, $x = 0.75 \rightarrow 0.70$ (*inverse step*)	50
$In_xAl_{1-x}As$ (*MB*)	$h_{МБ}$
{AlGaAs/GaAs}×5 (*SL1*)	{2.0/1.5}
GaAs (*substrate*)	–

Figure 1. General construction of heterostructures

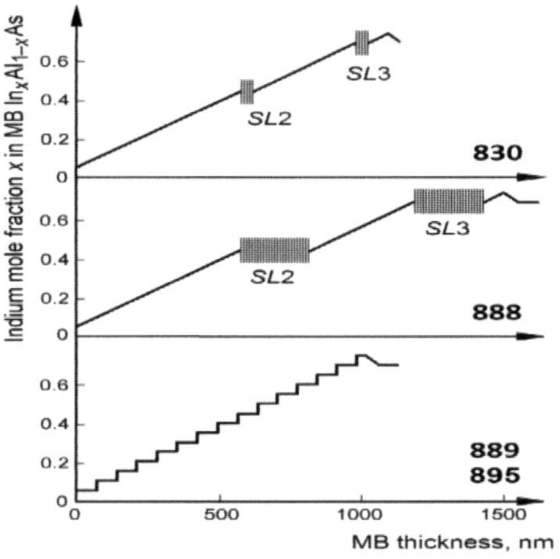

Figure 2. Design of MBs

An arrangement of the photoluminescence (PL) apparatus [6] employed in this study is shown in fig. 3. A MDR-2 monochromator scanned the sample over a range of 0.5–1.1 eV. An InGaAs detector was employed. In order to suppress the effect of second order diffraction, a long path filter was set in front of monochromator. Samples were excited using an Nd:YVO_4 laser with 5320 Å wavelength that was chopped at 218 Hz. For low-temperature measurements samples were mounted in a variable-temperature closed cycle helium cryostat (10 K < T < 300 K).

In additional to the PL research MHEMT-heterostructures have been investigated by photoreflection technique (PR). It is possible by adding to PL system a tungsten lamp with direct current supply. Care is to be taken to overlap laser's and lamp's spots on the sample and to focus reflected lamp beam into the monochromator. See more details in figure 3.

Electrophysical parameters measured from Hall effect with Van der Pau technique are summarized in table I.

Table I – Electron mobility μ_e and 2D electron concentration n_S

sample	μ_e, $cm^2/(V \cdot s)$		n_S, $10^{12} cm^{-2}$	
	300 K	77 K	300 K	77 K
830	10500	33300	1.45	1.38
888	11900	47100	1.56	1.47
889	11600	44200	1.53	1.45
895	11400	38700	2.16	2.11

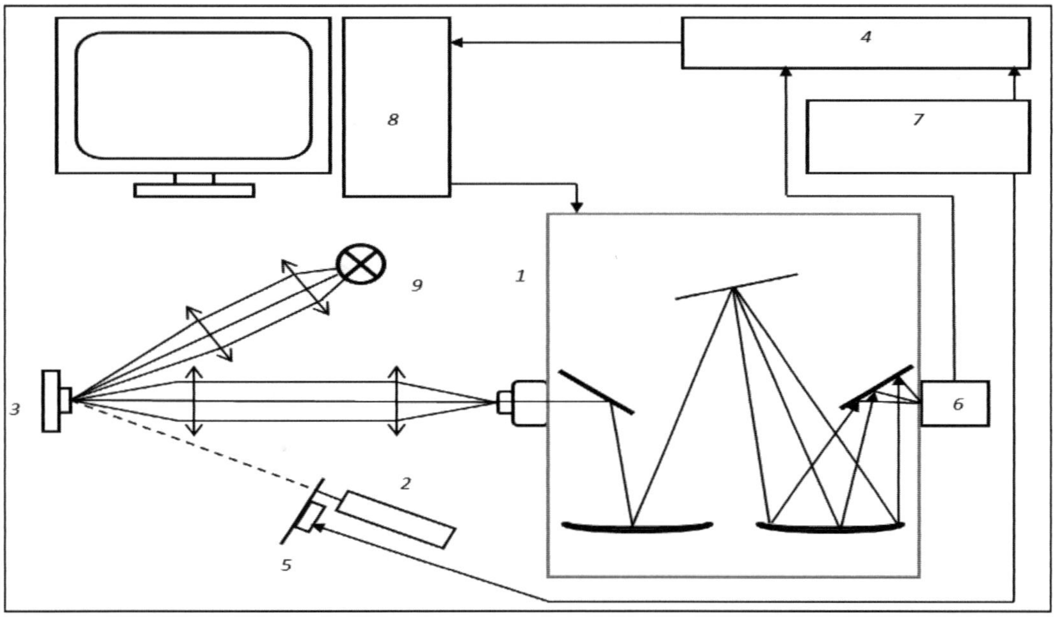

Figure 3. Simplified laboratory setup for PL and PR studies. Light scheme of PR technique is used (1 – monochromator, 2 – laser, 3 – sample in cryostat, 4 – lock-in amplifier, 5 – chopper, 6 – detector, 7 – spin controller, 8 – computer, 9 – lamp)

As we can see from spectra in figure 4 the highest intensity shows sample 889 with step-graded metamorphic buffer. There is a 5% difference in indium mole fraction between buffer layers. So we can suppose that it is the most successful construction of metamorphic buffer: it locks the most of dislocations from substrate and has minimum residual strain at the top. This supposition is agreed by Hall measurements (see table I).

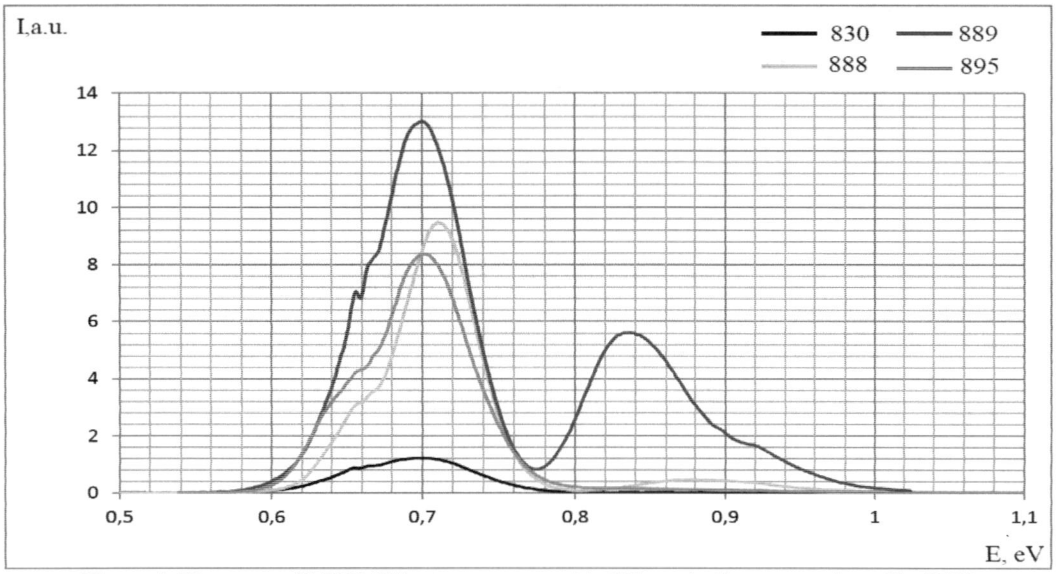

Figure 4. PL spectra of MHEMT-heterostructures with different buffer constructions (T = 77 K)

One can see that the most intensive peak of sample 888 has greater energy position in comparison with other samples for about 0.1 eV. That might mean lower indium concentration in quantum well region of sample 888. The difference in indium concentration between sample 888 and the other samples is about 1–2%, which is normal for molecular beam epitaxy method.

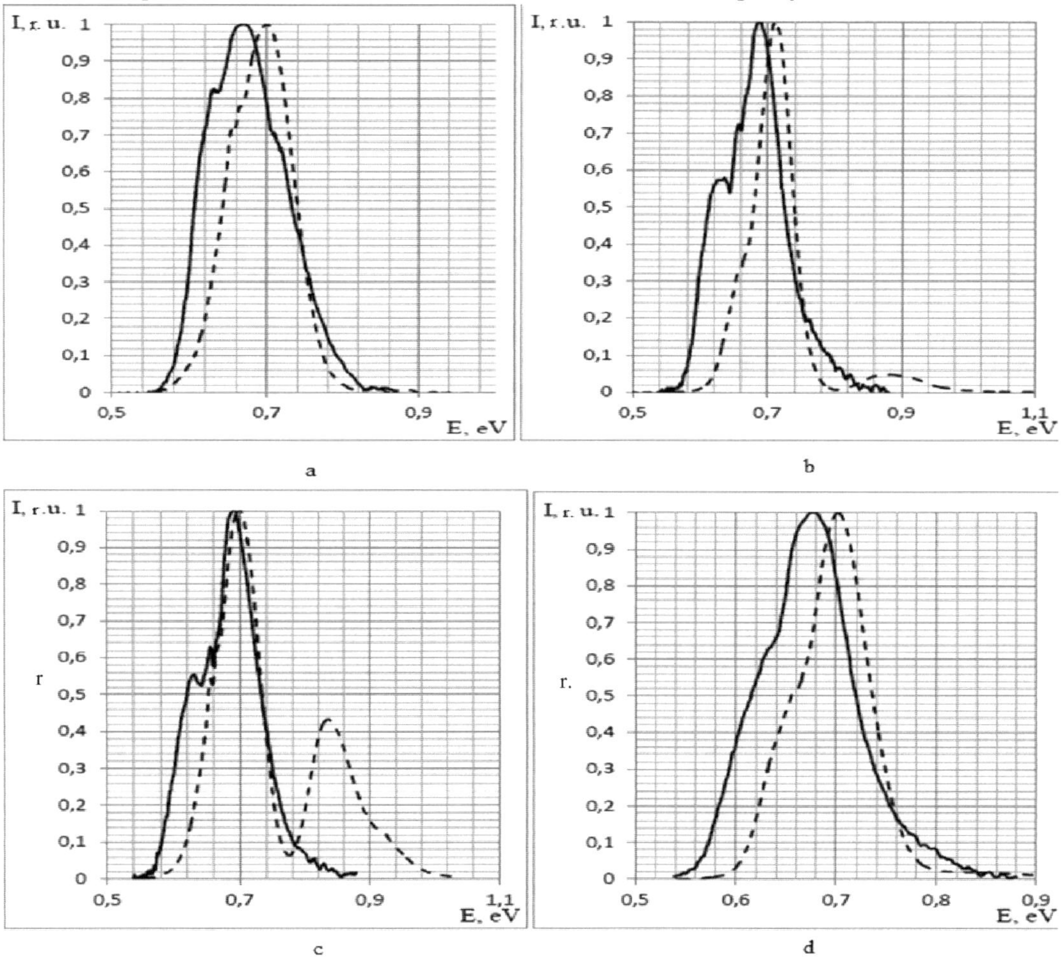

Figure 5. Relative spectra of MHEMT-heterostructures at 300 K (solid lines) and at 77 K (dashed lines) a – 830, b – 888, c – 889, d – 895

The main peaks near 0.70 eV and 0.84 eV correspond to transitions from 2-nd electron confined state to the 1-st hole confined state in QW and from conduction band to the valence band in upper region of MB respectively. These peaks become more intensive at lower temperature, whereas the feature at 0.65 eV originate hypothetically from 1-st electron confined state to the to the 1-st hole confined state transitions have an opposite behavior (see figure 5). Moreover, the lower temperature, the lower intensity it has. In our opinion this is due to a strong localization of the 1-st electron confined state carriers in QW [7]. At lower temperatures localization of electrons and holes on the 1-st electron confined state becomes stronger so the probability of the radiative recombination decreases proportionally.

Sample 889 with step-graded metamorphic buffer has the minimum half-width of spectral line. It is probably due to deficiency of great defect concentration and non-radiative centers in the barrier material near the interface [8]. So, we suggest that this sample has the best quality of interface and minimum defect states in barrier layers.

Also sample 889 has much better temperature dependence of PL intensity. Figure 6 shows absolute spectra of two samples at two different temperatures. As mentioned earlier, sample 889 has much greater intensity at both temperatures. Moreover, as compared to sample 830 the increase in intensity at lower temperature is greater too: about 10 times for sample 889 and only 2 times for sample 830. It may be due to a lot of thermal independent defects and misfits in sample 830. If so, the decrease of temperature will slightly influence on radiative transition probability. On the other hand, if sample 889 has much more perfect structure and phonon scattering is the main nonradiative transition mechanism the temperature decrease will drastically improve the probability of radiative transitions from the first level in QW, as shown in figure 6 b.

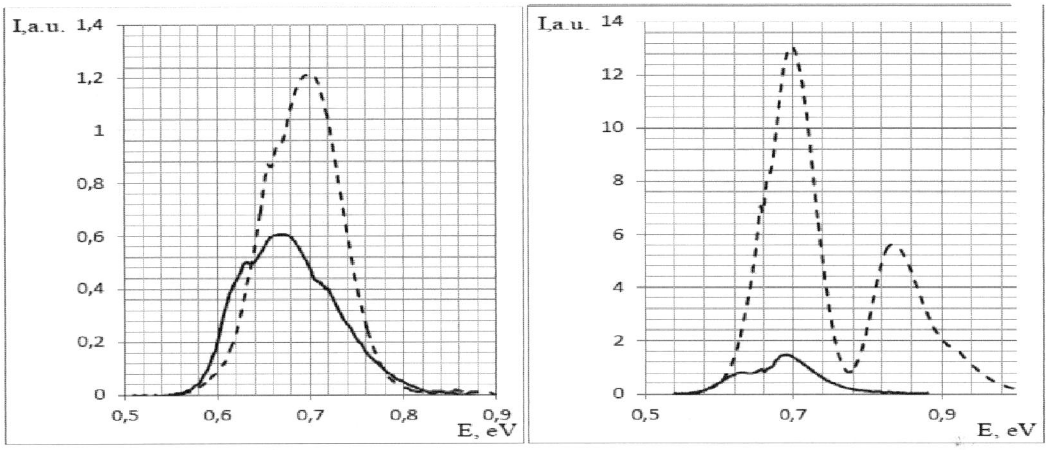

a b

Figure 6. Comparison of PL spectra intensity at 300 K (solid line) and 77 K (dashed line): a – 830, b – 889

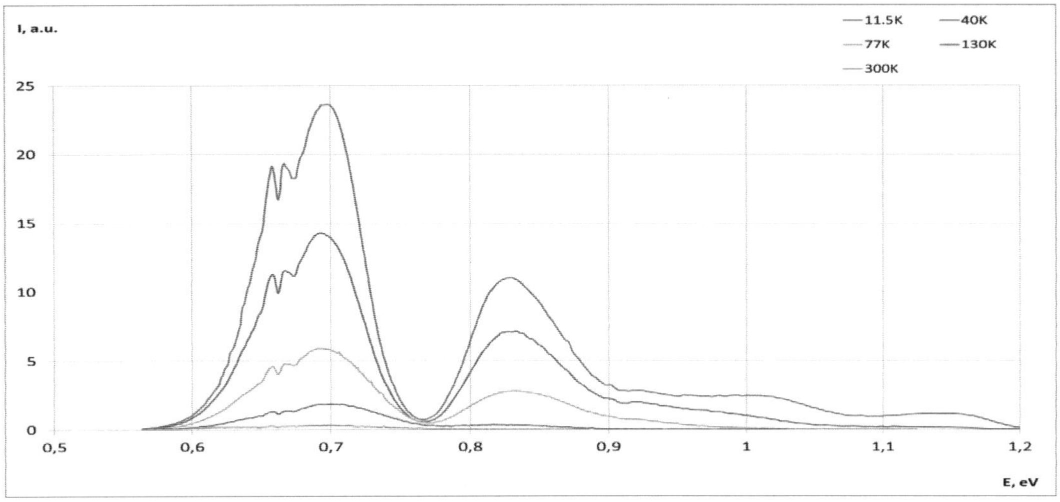

Figure 7. Typical PL spectra of MHEMT-heterostructures at different temperatures

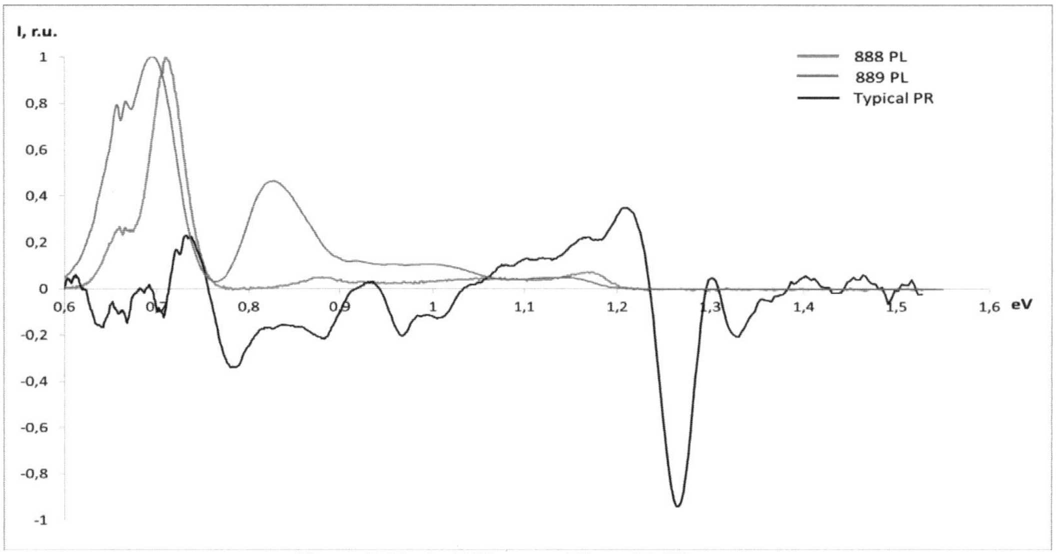

Figure 8. Typical PR spectral line with PL spectra

Figure 8 shows PR spectral line obtained at 11 K. One can see a lot of features (inflection points) that are similar to PL peaks. However there are a lot of PR features in energy region, where none of PL features observed. It is probably due to much higher sensibility of PR technique as compared to PL. Thereby PR features in non-PL energy regions may provide some additional information about subband transitions in QW or about interband transitions in MB material.

Conclusions

We have performed PL and PR investigation of metamorphic HEMT-heterostructures with In mole fraction ~ 0.7 in channel region. Transitions from conduction band to valence band in thick layers and from 1-st and 2-nd electron subband to heavy-hole subband in QW were observed in PL and PR spectra. The transition energies measured from PL spectra in good correlation with those calculated using the effective mass approximation. PL spectra show that the most successful heterostructure is sample 889 with step-graded metamorphic buffer. It has more pronounced temperature dependence of PL intensity, higher electron mobility μ_e and 2D electron concentration n_S. Also sample 889 with step-graded metamorphic buffer has the minimum half-width of spectral line. It is probably due to deficiency of great defect concentration and non-radiative centers in the barrier material near the interface. So, we suggest that this sample has the best quality of interface and minimum defect states in barrier layers.

PR spectra obtained are in good correlation with PL spectra. There are a lot of inflections at the energies similar to PL peaks. Also one can see a lot of PR features where none of PL peaks can be seen. It is normal state of things for PR. However some more evidence is required to analyze these spectra.

References

[1] New semiconductor materials. Biology systems. Characteristics and properties. [www.matprop.ru].
[2] Patent US 6818928 B2. Quaternary-ternary semiconductor devices / W.E. Hoke, P.S. Lyman ; Raytheon Company (US). – Appl. No. 10/310207 ; filling data 05.02.2002 ; publication date 16.11.2004.
[3] I.J. Fritz, P.L. Gourley, L.R. Dawson, J.E. Schirber. Electrical and optical studies of dislocation filtering in InGaAs/GaAs strained layer superlattices // Appl. Phys. Lett. – 1988. – Vol. 53. – P. 1098–1100.
[4] Patent US 7009224 B2. Metamorphic long wavelength high-speed photodiode / R.H. Johnson, J.K. Guenter, J.R. Biard ; Finisar Corporation. – Appl. No. 10/413186; filling data 14.04.2003 ; publication date 07.03.2006.
[5] G.B. Galiev, I.S. Vasil'evskii, S.S. Pushkarev, E.A. Klimov, R.M. Imamov, P. A.Buffat, B. Dwir, E. I. Suvorova. Metamorphic InAlAs/InGaAs/InAlAs/GaAs HEMT heterostructures containing strained superlattices and inverse steps in the metamorphic buffer // J. Cryst. Growth. – 2013. – Vol. 366. – P. 55–60.
[6] S. A. Tarasov, I. E. Gracheva, K. G. Gareev, O. E. Gordyushenkov, I. A. Lamkin, E. A. Men'kovich, V. A. Moshnikov, A. V. Presnyakova. Atomic Force Microscopy and Photoluminescence Analysis of Porous Metal Oxide Materials // Semiconductors, 2012, Vol. 46, No. 13, pp. 1584–1588.
[7] PL features of heterostructures $In_{0.53}Ga_{0.47}As/In_{0.52}Al_{0.48}As$ with double-side doping/A.F. Tsatsulnicov, V.M. Ustinov, A.U. Egorov et al.//Fizika I Tekhnika Poluprovodnikov. – 1996. – Vol. 30. – issue 10.– P. 1814–1821.
[8] The effect of potential fluctuations on the optical properties of InGaAs/InAlAs superlattices [Текст] / L.C. Pocas, E.M. Lopes, J.L. Duarte et al.//Journal of Applied Physics. – 2005. – Vol. 97. – P. 103518.

Intersubband light absorption in tunnel-coupled GaAs/AlGaAs quantum wells for electrooptic studies

D A Rybalko[1], M Ya Vinnichenko[1], L E Vorobjev[1], D A Firsov[1], R M Balagula[1], V Yu Panevin[1], M M Kulagina[2], A P Vasil'iev[2]

[1] St Petersburg State Polytechnical University, St Petersburg 195251, Russia
[2] Ioffe Physical-Technical Institute, St Petersburg 194021, Russia

e-mail: dimarybalko91@mail.ru

Abstract. Structures with doped multiple tunnel-coupled quantum wells GaAs/AlGaAs were grown. They were designed to investigate the intersubband absorption and modification of absorption spectra under transverse electric field. The samples were formed in mesa configuration. Preliminary results for the temperature modification of the intersubband absorption spectra were obtained.

1. Introduction

The optical properties of nanostructures with quantum wells (QW) change under the electric field influence. Study of optical properties under electric field associated with the intersubband electron transitions in tunnel-coupled quantum wells attracts attention due to the need of development of optoelectronic mid-IR devices (detectors and sources of radiation). Longitudinal electric field applied in the plane of the QW structure can heat the electrons, therefore electron distribution between the conduction band subbands may change. Changes in optical properties can be expected in this case [1]. Transverse electric field applied along the growth axis of the QW structure can influence QW optical properties more significantly. It changes the QW potential profile, which can cause not only the redistribution of electrons between the subbands in the conduction band, but also a spectral shift of the intersubband absorption peaks and changes of the spectral peak width [2]. The changes of the real part of dielectric permittivity of QW layers under the electric field can be used, for example, for the fast frequency modulation of the quantum cascade laser emission.

This work presents the results of design development of samples with tunnel-coupled quantum wells, intended to investigate the intersubband absorption under transverse electric field. Preliminary experimental results show changes of spectra with temperature which can be connected with the redistribution of electrons between subbands.

2. Sample and experiment

Structure was MBE grown on semi-insulating GaAs substrate and contained 100 pairs of double tunnel-coupled quantum wells GaAs/$Al_{0.25}Ga_{0.75}As$. The narrow and wide quantum walls have widths of 4.5 and 6.6 nm, respectively; the barrier is 3 nm thick (see schematic representation of the conduction band at figure 1). According to the theoretical calculations, the distance between the first two subbands e1 and e2 is about 30 meV. It is known that intersubband selection rules for electron transitions are not applicable in case of tunnel-coupled quantum wells. The energy of the e1-e3 transition is about 125 meV, energy of e2-e3 one is about 95 meV. The wide quantum well was doped with Si in the central part of quantum well (~2.5 nm) up to the surface electron concentration of $4 \cdot 10^{11}$ cm^{-2}. The active area containing quantum wells was located between the doped GaAs contact layers with electron concentration of 10^{18} cm^{-3}. The top and bottom contact layers have a thickness of 1.5 and 0.5 μm, respectively. There is $Al_{0.9}Ga_{0.1}As$ stop-layer right above the doped bottom layer that stops the selective etching of the structure.

Let us note that the application of the transverse electric field leads to the bending of valence and conduction bands (see fig. 1). Thus one quantum well shifts in relation to the other quantum well. In the equilibrium the transitions from the ground state of the wide quantum well e1 to the states with higher energy e2, e3 will primarily take place. If we apply electric field of enough magnitude, then the position of the ground state of narrow quantum well e2 may match the position of the ground state of wide quantum well e1. If we continue to increase the electric field, then ground state of narrow quantum well can become lower than the ground state of wide quantum well. In this case transitions from the ground state of the narrow quantum well e2 to the upper subband e3 will be more pronounced.

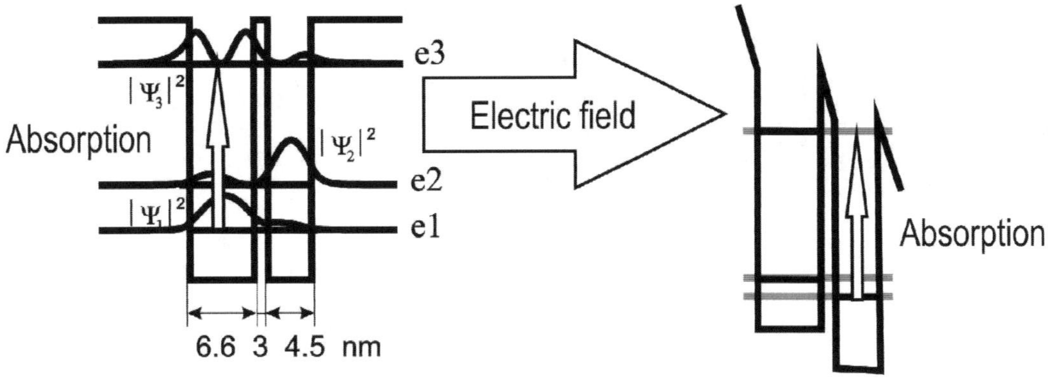

Figure 1. Influence of the transversal electric field on the intersubband absorption in tunnel-coupled quantum wells.

Postgrowth processing of structures included standard contact photolithography (direct and explosive), ion beam etching and chemical etching. Mesa structure with 7x10 mm^2 size was etched. The image from the light mask to the structure was transferred using the photoresist AZ 4562 mask with beam of Ar+ ions from the ion beam etching equipment MIM TLA2 (Technics). The ion etching was 2.8 μm deep. After that structure was etched to the stop-layer using the selective chemical etching. A mixture of sodium citrate ($Na_3S_6H_5O_7$) and 33% hydrogen peroxide (H_2O_2) at a ratio of 5:1 was used as an etchant. The stop layer was removed by the mixture of HCl:H_2O = 10:1. The Cr(400Å)-Au(4000Å) conductive pads were formed using two-layer resist (LOR10B and AZ 1518) mask by thermal vacuum deposition on the VUP-5M equipment. Multipass sample geometry was created in order to study the absorption of *s*- and *p*-polarized emission separately.

Intersubband absorption spectra were studied using vacuum Fourier spectrometer Bruker Vertex 80v for two polarizations of light (*s*- amd *p*-). The globar was used as a source of infrared light emission. The absorption signal was measured with the pyroelectric detector. The sample was placed in the closed-cycle cryostat Janis PTCM-4-7 that allowed us to obtain results in temperature range from 4 to 320 K.

3. Results and discussions

Scanning electron microscope (SEM) image of the cross-section of the structure is shown in figure 2. The active region that contains 100 double tunnel-coupled quantum wells and etched top and bottom contact layers are clearly visible. The walls of the mesa structure are not vertical, since the proposed measurement technique does not require it.

Figure 2. SEM image of the cross-sectional mesa structure.

The intersubband absorption spectra for different lattice temperatures are shown in figure 3. In accordance with the intersubband transitions selection rules, the absorption of radiation is possible only for the *p*-polarized light (which has the polarization component perpendicular to the plane of the structure). The shortwave absorption peak (see figure 3) corresponds to the electron transitions from the ground subband e1 to the third subband e3. Its position (e1-e3 = 130 meV for temperature of liquid helium) is in a good agreement with the theoretical calculation. Position of the longwave absorption peak has a deviation from the calculation results. We attribute it to some minor discrepancy in the calculation parameters. Both subbands (e1 and e3) are localized mainly in the wide quantum well. Change of lattice temperature partially simulates the effect of the electric field. Small distance between the first two subbands e1 and e2 results in the increase of the population of the second subband with the temperature.

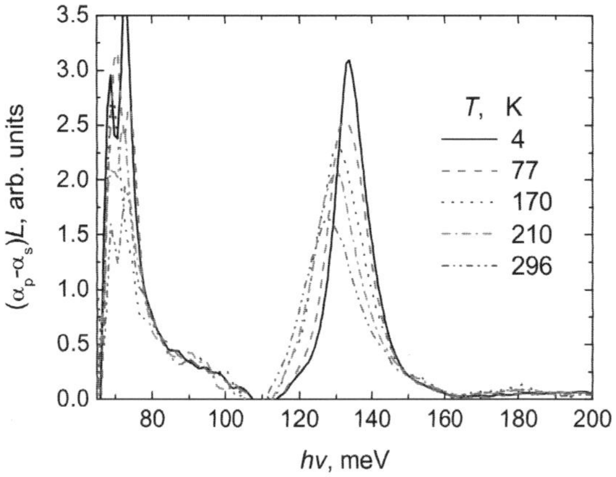

Figure 3. Intersubband absorption spectra (*L* - optical length), measured at different temperatures.

Analogous investigations have been carried out for a similar structure, but with wider quantum wells with another solid solution composition: GaAs/$Al_{0.38}Ga_{0.62}As$. The intersubband absorption spectra for different lattice temperatures for this model structure are shown in figure 4. One can observe the appearance of the intersubband peak near 130 meV and its increase with temperature. Its nature also can be explained by the redistribution of electrons from the ground state of quantum well to the second subband due to the small energy gap between them. So we attribute this peak at 130 meV to the transition e2-e3. At low temperatures (4-77 K, fig. 4), electrons are mainly located in the ground state and only transitions e1-e3 are possible.

The observed temperature modification of the absorption spectra indicates the decrease in the electron density on the first subband with the temperature that corresponds to the filling of the second subband.

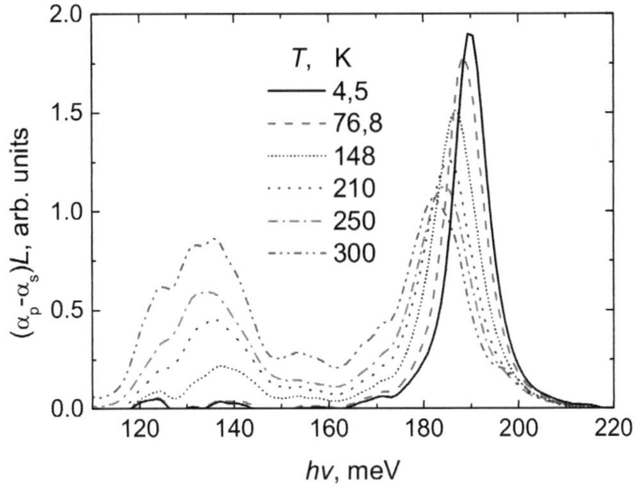

Figure 4. Intersubband absorption spectra (*L* - optical length), measured at different temperatures for similar structure.

Acknowledgments
This work has been supported by the Russian Foundation for Basic Research (Grants № 14-02-31489, 14-02-00336).

References

[1] V.L.Zerova, L.E.Vorob'ev, D.A.Firsov, E.Towe 2007 *Semiconductors* **41** 596.
[2] E. Dupont, D. Delacourt, V. Berger, N. Vodjdani, M. Papuchon 1993 *Appl. Phys. Lett.* **62** 1907.

Far- and near-infrared photoluminescence from n-GaAs/AlGaAs multiple quantum wells

L E Vorobjev[1], D A Firsov[1], V Yu Panevin[1], A N Sofronov[1], R M Balagula[1], I S Makhov[1] and A P Vasil'ev[2]
[1]St. Petersburg State Polytechnical University, Politechnicheskaya str., 29, St. Petersburg, 195251, Russia
[2]Ioffe Physical Technical Institute, Politechnicheskaya str., 26, St. Petersburg, 194021, Russia

Email: makhoviv@gmail.com

The results of experimental investigations of impurity-assisted photoluminescence at low temperatures from n-GaAs/AlGaAs multiple quantum wells both in near- and far-infrared (terahertz) spectral ranges are presented. The optical electron transitions from impurity ground state to heavy hole subband in near-infrared spectral range are revealed. The depopulation of the donor ground state due to these transitions allowed us to observe photoluminescence in terahertz spectral range related to electron transitions from the first electron subband to donor ground state as well as to intracenter optical transitions. Experimental results in far- and near-infrared spectral ranges are in good agreement with the results on THz photoconductivity and energy spectrum calculation.

1. Introduction.

Binding energy of shallow impurities in semiconductors is about units or tens of meV [1]. Thus, the possible optical transitions between band and impurity can be observed in terahertz (THz) spectral range. Many papers are devoted to investigations of different principles of terahertz radiation emitting with assistant of impurity states. Amongst them - terahertz radiation emitting via resonant and localized impurity states under carrier heating in strong lateral electric field in uniaxially stressed bulk p-Ge [2], in GaAs/GaAsN:Be microstructures with internal strain [3], in n-doped GaAs/AlGaAs QWs [4]. In condition of powerful impurity-band optical pumping the terahertz radiation was achieved on band-impurity transitions in Si:Bi [5]. Recently, the interest to terahertz radiation emitting under interband optical pumping of bulk semiconductors is rising. The terahertz photoluminescence (PL), caused by electron radiative transitions from acceptor levels to valence band in bulk Ge [6] as well as by transitions from conduction band to ionized donor states in bulk GaAs [6,7] and in GaN epilayers [8] was recently observed under optical interband pumping. The depopulation of impurity ground states was achieved due to the impurity-assisted electron–hole recombination which was observed in the interband PL spectra in near infrared spectral range. The similar mechanism of terahertz radiation emitting via impurity state in QWs is not studied yet. At the same time, the binding energy of impurity states in QWs is increasing in comparison with bulk material allowing to easily tune the spectrum of emitting radiation with changing the QW parameters. That is why the present work is devoted to investigation of impurity related terahertz radiation in n-GaAs/AlGaAs QW structure under condition of interband optical pumping. We present spectra of both intraband terahertz (far-infrared) and interband near-infrared emission where the impurity-assisted electron transitions in quantum wells are clearly observed.

2. Samples and experimental setups.

In this work we present results for sample with doped QWs in comparison with its substrate which has no any epitaxial layers. Sample was MBE grown on GaAs semi-insulating substrate on a 0.2 μm GaAs buffer layer and consisted of 50 layers of 30 nm GaAs quantum wells separated with 7 nm $Al_{0.3}Ga_{0.7}As$ barriers. Quantum wells were doped in the 4 nm layer with surface donor (Si) concentration of $3\cdot10^{10}$ cm^{-2}. Doped region was shifted on 6 nm from QW center. Samples had a 20 nm GaAs cap layer doped up to $5\cdot10^{17}$ cm^{-3} with a silicon.

For optical study, the sample was mounted into the low vibration closed cycle Janis PTCM-4-7 cryostat based on the pulse tube thermodynamic cycle. Sample temperature can be varied from ~4 K to 320 K. QW sample was studied separately from substrate. Interband optical excitation of the sample was attained with solid state CW laser (λ = 532 nm, P = 8 mW). For directing the laser beam to the sample inside of the cryostat with an incidence angle of approximately 45° to growth axis, we used mirrors, lens and fused silica cryostat window. Far-IR emission was studied through the polymethylpentene (TPX) cryostat window, near-IR emission was studied through the fused silica cryostat window.

Far-IR photoluminescence spectra were studied with Fourier transform technique using vacuum Fourier transform infrared (FTIR) spectrometer Bruker Vertex 80v operating in step-scan mode. FTIR spectrometer had a polyethylene (PE) entrance window and a kit of exchangeable terahertz beamsplitters. There was a distance about 2 cm between cryostat TPX window and spectrometer PE window. Here we put ~100 μm black PE filter to block the pumping light. The intensity of the far-IR emission was measured with Si bolometer system (made by Infrared Laboratories, Inc.) cooled with liquid helium. Bolometer cryostat had a vacuum optical coupling with a FTIR spectrometer. Bolometer had a PE window and two internal interchangeable filters namely 0.5 mm thin PE filter and 1 mm thick crystalline quartz filter.

We have used two optical configurations of the far-infrared experimental setup with different spectral throughoutputs. The first configuration with the combination of 0.5 mm polyethylene bolometer filter with 6 μm multilayer Mylar beamsplitter was used to investigate the spectral range 4-40 meV. The second one is the combination of 25 μm Mylar beamsplitter with crystalline quartz bolometer filter. Second configuration considerably increased setup transmission in 2-14 meV spectral range but blocked higher frequencies.

Near-IR PL spectra were studied with grating monochromator Horiba Jobin Yvon FHR-640 operating in spectrograph mode with holographic grating with 1200 groves per mm and liquid nitrogen cooled CCD camera.

3. Experimental results and discussion.

The results of PL study in the far-IR spectral range for the sample, as well as for substrate are presented in figure 1 (please, see legend on the plot to recognize the curves). It should be noted that the real emission spectra can slightly differ from spectra presented in our paper because of spectral dependency of photodetector sensitivity, windows and beamsplitter transmission using in the first optical configuration of the setup. Anyway, comparing PL spectra for sample and substrate, one can distinguish wide emission band 14-30 meV inherent to both of them. This fact probably means, that in this spectral range we have deal with some impurity related emission from the bulk layers of the sample such as semi-insulated (probably compensated) substrate which as well could contain unintentional carbon impurity accompanying Czochralski grown process of bulk GaAs [9]. Really, the thickness of QW layers is not too high, so the exciting laser radiation can reach the substrate and produce impurity involved optical transitions in according to the mechanism described for bulk semiconductors in Ref.'s [6-8]. Slight difference of the QW sample and the substrate spectra in 13-30 meV range could be explained by the presence of additional impurities in QW sample "matrix" (buffer, cap and probably barrier layers of the structure) which as well could contain unintentional impurities due to MBE growth process [11]. In the near-IR spectra of the sample and substrate shown, correspondingly, in figures 2 and 3, the bulk band-impurity interband radiative recombination is marked with label "Matrix". Matrix luminescence in sample and substrate looks similar and lies, approximately, on 15-30 meV (at the level of the half of maximum) lower then GaAs interband transition energy (marked with arrow Eg_{GaAs} in figure 3). That interband recombination depopulates impurity levels and allows nonequilibrium carrier capture from band to impurity, which could leads to terahertz photon emission. Interband transitions at the photon energy corresponding to energy gap of the bulk GaAs (1.519 eV [1]) is not clearly observed, due to exciton formation effects. Exciton binding energy for bulk GaAs is 4.2 meV [11] and emission line marked in figure 3 with arrow *Xe-h* could be connected with heavy free exciton. Also, the emission line marked with arrow *A-X* lying on 2.9 meV lower in figure 3 could be connected with acceptor (carbon) bound exciton [12].

Returning to far-IR spectra presented in figure 1 one can see that the weaker emission band near 4-13 meV is a feature of QW spectrum only and is not observed on substrate spectrum. Using second optical configuration of experimental setup with improved throughoutput in 2-14 meV spectral range, we have studied the QW emission spectral feature more elaborately at two temperatures (see figure 4). It gives us broad spectra which overlap with photoconductivity spectra of the same QW structure, presented in Ref. [4] and cover both of the expected optical transitions of electrons: from ground QW subband to impurity ground state (*e1-1s*) and intracenter transition $2p_{x,y}$-*1s* (their energies are calculated in Ref. [4] and marked with arrows in figure 4). One should not wonder that in our spectra different impurity related transitions are broadening and cannot be resolved clearly. It is well known, that narrow impurity lines could be observed in pure samples only, with doping level significantly less then in our QW structure (see, for example Ref.[13]).

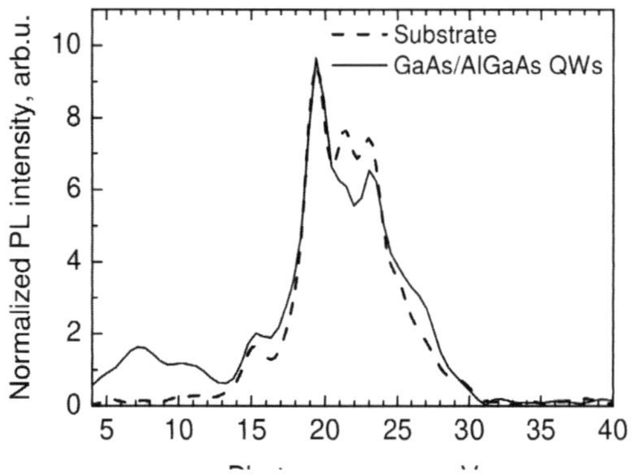

Figure 1. THz PL spectra for the diferent samples (see legend on the plot) in the photon energy range of 4-40 meV, measured in the first optical configuration. T ≈ 5 K.

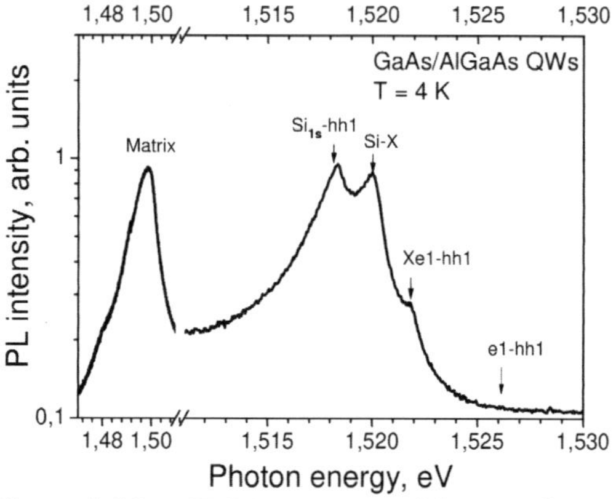

Figure 2. Near-IR PL spectrum of the sample.

Results of PL investigations of the sample in the near-IR spectral range are shown in the figure 2. Accordingly to calculation of electron (*e*) and heavy hole (*hh*) energy states in our QW, we expected the interband transition *e1-hh1* at the photon energy of 1.526 eV, but due to exciton formation effects this transition is not clearly observed in our spectra. For such a wide QW as our one exciton binding energy is almost equal to bulk GaAs aforesaid value and emission line marked in figure 2 with arrow *Xe1-hh1* could be connected with heavy free exciton for the first QW subband. Also, the emission line marked with arrow *Si-X* in figure 2 could be connected with donor bound exciton. The emission line marked with arrow Si_{1s}-*hh1* in figure 2 could be connected with electron-hole radiation recombination between ground donor state and first heavy hole subband with the photon energy about 1.5183 eV. That recombination process forerun mentioned above *e1-1s* and $2p_{x,y}$-*1s* optical transitions observed in far-IR PL spectra. The energy separation between *e1-hh1* and Si_{1s}-*hh1* transitions in figure 2 is about 8 meV. This value corresponds with binding energy of silicon in our QW calculated in [4].

The observed in figure 4 temperature quenching of the terahertz PL can be explained with ionized donor capture probability decrease or with the increase of the carrier ejection from the neutral donor with temperature increase. Earlier, the similar temperature behavior of terahertz luminescence was observed in Ref. [5] in bulk doped semiconductor.

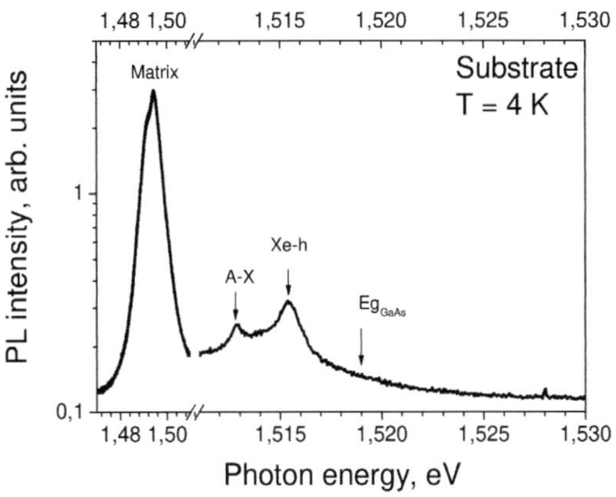

Figure 3. Near-IR PL spectrum of the substrate.

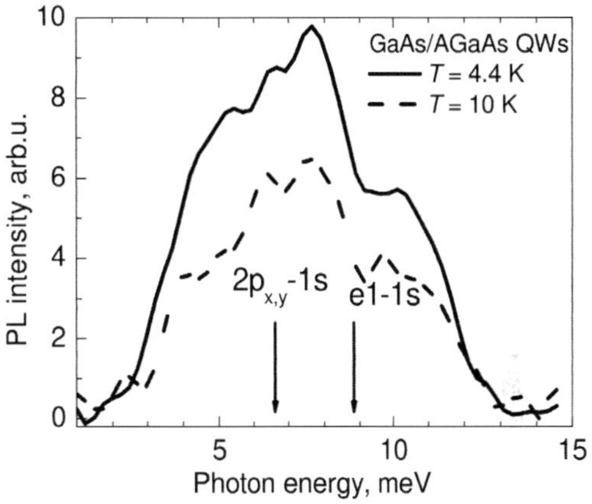

Figure 4. THz PL spectrum of the QWs sample at the different temperatures (see legend on the plot), measured in the second optical configuration suitable for 2-14 meV range.

4. Summary.

In the present work, the far- and near-infrared emission under optical interband excitation related to impurity states in silicon doped GaAs/AlGaAs quantum well structure was observed and investigated. In the near-infrared photoluminescence spectra of the QW the emission line associated with electron transitions from ground donor state to ground heavy hole subband is detected. Optical transitions from the ground electron subband *e1* as well as from excited impurity state $2p_{x,y}$ to the getting free ground impurity state *1s* becomes possible. Corresponding PL lines have been observed in far-IR spectra and their spectral positions are in good agreement with electron energy state calculations in QW including impurity states as well as with photoconductivity spectra for the same structure (see Ref. [4]).

References

[1] Levinshtein M E , Rumyantsev S L, Shur M 1996 *Handbook Series on Semiconductor Parameters* **1,2**
[2] Altukhov I V, Kagan M S, Korolev K A et al 1994 *JETP Lett.* **59**, 455
[3] Shalygin V A, Vorobjev L E, D A Firsov et al 2007 *Appl. Phys. Lett.* **90** 161128
[4] Firsov D A, Shalygin V A, Panevin V Yu et al 2010 *Semiconductors* **44**(11) 1394
[5] Pavlov S G, Hübers H-W, Rümmeli M H et al 2002 *Appl. Phys. Lett.* **80**(25) 4717
[6] Andrianov A V, Zakhar'in A O, Livanov Yu L, Kipa M S 2010 *Pis'ma v ZhETF* **91**(2) 102
[7] Zakhar'in A O, Andrianov A V, Egorov A Yu, Zinov'ev N N 2010 *Appl. Phys. Lett.* **96** 211118
[8] Zakhar'in A O, Bobylev A V, Andrianov A V 2012 *Semiconductors* **46**(9) 1135
[9] Wakefield B, Leigh P A, Lyons M H et al 1984 *Appl. Phys. Lett.* **45** 66
[10] Dingle D, Weisbuch C, Stormer H L et al 1982 *Appl. Phys. Lett.* **40** 507

Journal of Physics: Conference Series **541** (2014) 012082 doi:10.1088/1742-6596/541/1/012082

[11] Nam S B, Reynolds S B, Litton C W et al 1976 *Phys. Rev. B* **13**(2) 761
[12] Bludau W, Wagner E 1976 *Phys. Rev. B* **13**(12) 5410
[13] Lewis R A, Cheng T S, Henini M et al 1996 *Phys. Rev. B* **53**(19) 12829

Journal of Physics: Conference Series **541** (2014) 012083 doi:10.1088/1742-6596/541/1/012083

Effect of growth temperature of GaN:Mg layer on internal quantum efficiency of LED structures with InGaN/GaN quantum wells

I S Romanov[1], I A Prudaev[1], V V Kopyev[1], A A Marmalyuk[2], V A Kureshov[2], D R Sabitov[2], A V Mazalov[2]

[1]Tomsk State University, Lenin Avenue 36, 634050 Tomsk, Russia
[2]Sigm Plus Co., Vvedenskogo Str. 3, 117342 Moscow, Russia

rff.romis@gmail.com

Abstract. The results of studies of quantum efficiencies for photoluminescence and electroluminescence regimes of blue light-emitting diode structures with InGaN/GaN quantum wells are presented. Experimental samples differed in growth temperature of p-GaN emitter layer. It is shown that increasing of growth temperature of p-GaN layer in temperature range 940–1060 °C leads to decreasing of internal quantum efficiency due to diffusion of magnesium atoms from p-GaN into quantum wells. At the electroluminescence regime quantum efficiency of samples with low temperature p-GaN is limited by insufficient crystal quality or low solubility of magnesium in emitter layer.

1. Introduction

Magnesium is the most commonly used acceptor impurity to form p-conductivity layers in growth process of nitride semiconductors by metal organic chemical vapor deposition (MOCVD). In gallium nitride the Mg initiates level with sufficiently large ionization energy (about 150 meV [1]) what is greater than average thermal energy of carriers at room temperature. As a result a small portion of acceptors are ionized, contributing holes to valence band. Furthermore during epitaxial growth the chemical compensation of Mg acceptors by hydrogen atoms occurs. As a rule to decompose chemical bonds of Mg-H complexes thermal annealing of light-emitting diode (LED) structures in inert gas is used.

In LED structures p-GaN layer usually is grown immediately after active region which consist of InGaN/GaN multiple quantum wells (MQWs). Therefore growth parameters of p-GaN layer effect on crystal quality of active region and characteristics of whole LED structure. In the present work blue LED structures with InGaN/GaN MQW with different p-GaN layers are investigated. The purpose of this paper is to establish general patterns of temperature growth influence on efficiency of InGaN/GaN LED structures.

2. Experimental details

The InGaN/GaN MQW LED samples were grown on (0001) sapphire substrate by a MOCVD method. As precursors of III-group elements the following metal organic compounds were used: trimethylindium $In(CH_3)_3$, trimethylgallium $Ga(CH_3)_3$, triethylgallium $Ga(C_2H_5)_3$. As nitrogen precursor an ammonia (NH_3) was used. For acceptor dopant (magnesium) bis(cyclopentadienyl)magnesium was used. For donor dopant (silicon) silicomethane was used. The LED structures consisted of a low temperature (LT) nucleation layer, a 2 µm unintentionally doped GaN layer, a 2 µm Si-doped GaN layer, five InGaN/GaN QWs (3/10 nm), and a 140 nm Mg-doped GaN. Figure 1 shown schematic cross-section of the epitaxial layers for investigated InGaN/GaN MQW LED structures. All experimental samples were grown under the same conditions except growth temperature of p-GaN emitter layer, which varied for different samples in the range T = 940–1060 °C.

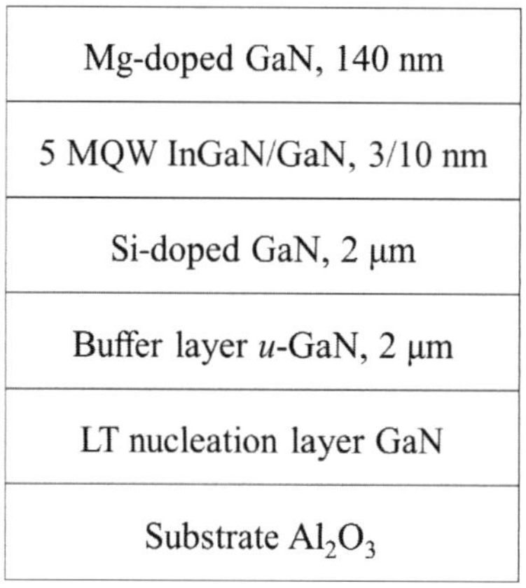

Figure 1. Schematic cross-section of the epitaxial layers for investigated InGaN/GaN MQW LED structures.

In experiment we measured internal quantum efficiency for photoluminescence and external quantum efficiency (EQE) for electroluminescence regimes. The internal quantum efficiency (IQE) were measured by photoluminescence (PL) method at different excitation power and temperatures [2]. The dependence of IQE on excitation power were measured in temperature range 10–300 K. Pulse YAG laser with wavelength 355 nm were used as a pumping source. Detailed description of the measurement condition are presented in [3,4]. EQE was measured at current density 1 A/cm^2. The samples for EQE measurements were made with indium contacts fused in LED structures in inert gas at temperature T = 350 °C

3. Experimental results

Figure 2 shows dependence of normalized PL intensity on excitation power for samples with different growth temperature of p-GaN layer: 940, 990 and 1060 °C. The obtained dependences qualitatively the same for all samples and can be explained by the ABC-model of recombination:

$$IQE = \frac{B \cdot n^2}{A \cdot n + B \cdot n^2 + C \cdot n^3}, \tag{1}$$

where A, B, and C are the temperature dependent Shockley-Read, radiative, and Auger recombination coefficients, respectively, and n is the concentration of non-equilibrium carriers in the MQW (n is proportional to excitation power for linear recombination regime).

At a low excitation level, the quantum efficiency drastically decreases with increasing T due to the temperature dependence of the nonradiative recombination rate. At a high excitation level, the Auger recombination rate starts to play a significant role leading to a decrease in IQE with increasing excitation power. A characteristic feature of the dependences IQE on temperature is the shift of the maximum to higher excitation intensities with increasing T.

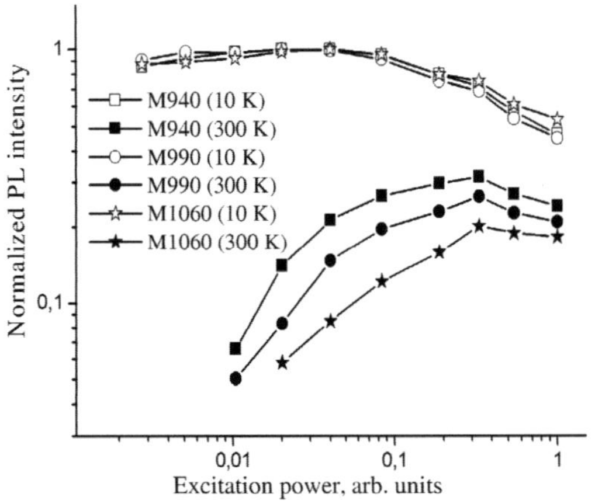

Figure 2. Dependence of normalized PL intensity on excitation power for investigated samples at measurement temperature 10 and 300 K.

Increasing of growth temperature of p-GaN layer from 940 to 1060 °C leads to reduction of IQE of LED structures (figure 3, left axes). Decreasing of IQE for sample with growth temperature of p-GaN layer 1060 °C was about 50 % relative to the sample with growth temperature of p-GaN layer 940 °C. Probably this is due to Mg diffusion into the active region with increasing of growth temperature of p-GaN layer. An additional argument for this assumption is broadening of PL spectra (figure 3, right axes) with increasing of growth temperature of p-GaN layer. In [5] it is shown that Mg diffusion leads to interdiffusion of In-Ga atoms in InGaN layers, which should leads to an increase of full width at half maximum (FWHM) of PL spectra.

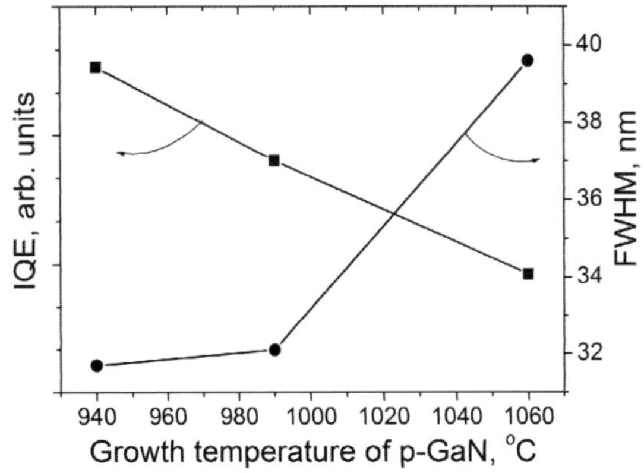

Figure 3. Dependence of IQE and PL FWHM on growth temperature of p-GaN layer.

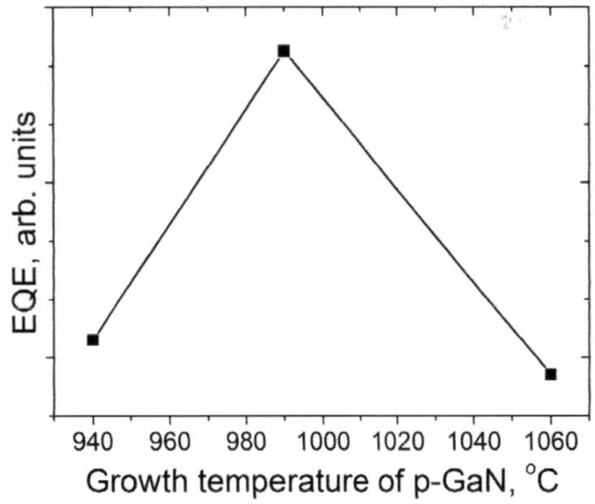

Figure 4. Dependence of EQE on growth temperature of p-GaN layer.

Figure 4 shows dependence of EQE on growth temperature of p-GaN layer for investigated InGaN/GaN MQW LED structures. The maximum EQE is observed at the temperature 990 °C. Lower value of EQE at the growth temperature 940 °C of p-GaN can be explained by worse crystal quality in comparison to samples grown at temperature 990 °C. On the other hand it can be explained by low injection efficiency. The external quantum efficiency of electroluminescence can be written as the product

$$\eta = \gamma \cdot \eta_i \cdot \eta_{ext}, \quad (2)$$

where γ is the injection coefficient, η_i is the internal quantum efficiency, and η_{ext} is the extraction efficiency. Injection coefficient is a complex parameter which is depend on structure design, and in particular it is determined by doping level of p-GaN emitter. It is well known that solubility of impurities in solids decreases at the lower temperatures. Probably in our case temperature lowering leads to reduction of γ due to Mg solubility decrease.

Also it should be noted that for growth temperature of p-GaN T = 1060 °C the lower EQE is observed in comparison to samples grown at temperature 990 °C. This fact can be explained by faster diffusion of Mg from p-GaN into quantum wells at higher temperatures. Obviously Mg atoms initiate the centers of nonradiative recombination in quantum wells what leads to lowering of EQE.

4. Conclusion

Thus, according to the results of this study, increasing of growth temperature of p-GaN layer from 940 to 1060 °C leads to reduction of IQE of blue LED structures with InGaN/GaN MQW. This fact can be explained by Mg diffusion into active region of LED structures. The investigation of electroluminescence have shown that dependence of EQE on growth temperature is described by curve with a maximum. At the low growth temperature EQE is limited by low crystal quality or low Mg solubility which cause insufficiently injection coefficient of holes from p-GaN into MQWs. At the higher temperatures EQE is limited by diffusion of Mg from p-GaN into quantum wells of active region.

Acknowledgements

This study was supported by the Russian Foundation for Basic Research, project No. 14-02-31180.

References

[1] Yuan C, Salagaj T, Gurary A, Zawadzki P, Chern C S, Kroll W, Stall R A, Li Y, Shurman M, Hwang C Y, Mayo W E, Lu Y, Pearton S J, Krishnankutty S and Kolbas R M 1995 *J. Electrochem. Soc.* **142** 738–742

[2] Watanabe S, Yamada N, Nagashima M, Ueki Y, Sasaki C, Yamada Y, Taguchi T, Tadatomo K, Okagawa H and Kudo H 2003 *Appl. Phys. Lett.* **83** 4906–08

[3] Prudaev I A, Romanov I S, Kop'ev V V, Shirapov S B, Tolbanov O P and Khludkov S S 2013 *Russ. Phys. J.* **56** 757–759

[4] Romanov I S, Prudaev I A, Marmalyuk A A, Kureshov V A, Sabitov D R and Mazalov A V 2013 *Russ. Phys. J.* **56** 760–762

[5] McCluckey M D, Romano L T, Krusor B S, Johnson N M, Suski T and Jun J 1998 *Appl. Phys. Lett.* **73** 1281–83

Elongated quantum dots of Ge on Si formation modelling

K. A. Lozovoy, A. P. Kokhanenko, A. V. Voitsekhovskiy
Tomsk State University, 36 Lenin Av., Tomsk, 634050, Russia

E-mail: lozovoymailbox@gmail.com

Abstract. Quantum dots (QDs) of Ge on Si grown using the method of molecular beam epitaxy (MBE) are examined in this paper. A comparative analysis of growth kinetics of elongated QDs with different length to width ratio was carried out. Calculations of pyramidal and wedge-like clusters formation energy were made. The increase in islands' surface energy, elastic strain relaxation, and the decrease in the atoms' attraction to substrate were taken into account. By using a well-known model based on the generalization of classical nucleation theory it was shown that elongated islands emerge after pyramidal clusters but begin to dominate in the QDs array on the later stages of growth. Calculations of QDs surface density and size distribution function for wedge-like clusters with different length to width ratio were performed. The existence of a special geometry of islands was discovered. Surface density and the average size of islands reach points of extremum for this geometry. Theoretical conclusions correlate with the experimental results.

1. Introduction

Heterostructures with QDs of Ge on Si have great importance for the development of semiconductor technology. Growth of Ge on Si attracts much attention from researchers in connection with the many possible uses of Si/Ge-based semiconductor materials in optoelectronics. Materials containing nanodimensional inclusions of Ge embedded in Si matrices make it possible to create light-emitting devices and photodetectors that may successfully compete with traditional optoelectronic devices based on A^{III}-B^{V} compounds.

Molecular-beam epitaxy is one of the most promising Si/Ge heterostructures production methods. There are two basic types of Ge QDs on Si(001) surface. They are so called hut- and dome-clusters. The morphology of Ge islands during the growth process can be controlled by changing the temperature of substrate, Ge deposition rate, and the total thickness of Ge layer.

Investigations of Ge QDs arrays on Si(001) have shown that the ensemble of hut-clusters is not homogenous [1, 2]. There are numerous morphologically different types of hut-clusters. Two basic kinds of hut-clusters are pyramidal (hut-) and elongated (wedge-) clusters, which are pyramids with a square or rectangular base, respectively. Both types of clusters have the same width to height ratio that is approximately equal to 10. The width of wedge-like cluster is usually less than the lateral size of the hut-cluster. Since these two types of clusters have a different morphology, structural transitions between them are restricted. It was also revealed that wedge-like clusters dominate in QDs arrays grown at low temperatures, and their fraction increases with the growth of deposited Ge thickness [1, 2]. But the role of different types of clusters in quantum dots arrays formation kinetics is still poorly researched.

The aim of this work is to develop the kinetic model of differently shaped self-organized Ge/Si(001) QDs parameters calculation and to compare growth kinetics of pyramidal and wedge-like clusters with various length to width ratios.

2. Theoretical model, results and discussion

For calculations we used the kinetic model of Ge QDs on Si growth developed in [3, 4]. This model, based on a generalization of classical nucleation theory, allows us to define temperature dependencies of QDs surface density N and size distribution function for different growth rates V. For the estimation of QDs growth kinetics, first of all, thermodynamic parameters of Ge/Si system, such as free energy and equilibrium thickness of wetting layer, were defined.

The free energy of pyramidal islands with a rectangular base and constant contact angle φ is a function of the number of atoms i in the island and is represented by

$$\Delta F(i) = Ai^{2/3} - B\zeta i, \tag{1}$$

where $\zeta = h/h_{eq} - 1$ is the wetting layer superstress, h is the Ge wetting layer thickness, h_{eq} is its equilibrium thickness. In this equation, free energy is expressed in $k_B T$ units, where T is the substrate temperature and k_B is the Boltzmann constant.

With respect to different geometrical shapes of QDs parameters A and B will be defined as

$$A = \frac{r[\gamma(\varphi)/\cos\varphi - \gamma(0)]\alpha l_0^2}{k_B T}, \tag{2}$$

$$B = \frac{[1 - Z(\varphi)]\lambda \varepsilon_0^2 l_0^2 d_0}{k_B T} \ln\left\{\frac{\Psi_0}{d_0[1 - Z(\varphi)]\lambda\varepsilon_0^2}\right\}, \tag{3}$$

where r is the island's length to width ratio, $\gamma(0)$ and $\gamma(\varphi)$ are the specific surface energies of the base and the lateral faces of pyramid, φ is the angle between the lateral face and the base of the island, l_0 is the mean distance between atoms on the surface, $Z(\varphi)$ is the coefficient of elastic relaxation, λ is the material's modulus of elasticity, ε_0 is the lattices' mismatch, d_0 is the height of one monolayer (ML), Ψ_0 is the wetting energy density on the substrate surface, and α is the geometrical factor depending on the island's shape:

$$\alpha = \left(\frac{12 d_0 ctg\varphi}{(3r-1)l_0}\right)^{1/3}. \tag{4}$$

Parameter α defines the relationship between the lateral width of the island L and the number of atoms in it:

$$i = \left(\frac{L}{\alpha l_0}\right)^3. \tag{5}$$

The value of elastic relaxation coefficient Z may be calculated within the Ratsh–Zangwill model. This model assumes that the effective parameter of lattice mismatch decreases from underlying to top layers because top layers contain additional relaxed atoms near the island edge [5, 6].

Then the islands' nucleation rate I, QDs surface density N, rate of atoms' arrival to the island, and QDs size distribution function were calculated.

Using the Zeldovich formula [7], for the nucleation rate elongated islands we will obtain the following dependence on the wetting layer superstress

$$I(\zeta) = \frac{a}{\tau l_0^2}\zeta(\zeta+1)e^{-\Delta F(i_c)}, \tag{6}$$

where i_c is the critical size at which the free energy is at its maximum:

$$i_c = \left(\frac{2A}{3B\zeta}\right)^3, \tag{7}$$

τ is the characteristic time of atoms incorporation processes:

$$\tau = \frac{3l_0^2 \nu}{4(r+1)\alpha BD}, \tag{8}$$

D is the coefficient of diffusion of atoms from the wetting layer to the islands, ν is the cut-off parameter of the elastic strain field [5].

Parameter a is defined as

$$a = \frac{3B}{4\sqrt{\pi A}}. \tag{9}$$

Calculations of parameters of pyramidal and wedge-like clusters with different length to width ratios r were carried out. For calculations the following material values were selected [3, 5]: $\lambda = 1.27\cdot 10^{12}$ dyn/cm^2, $\varepsilon_0 = 0.042$, $d_0 = 0.145$ nm, $l_0 = 0.395$ nm, $\Psi_0 = 450$ erg/cm^2, $\gamma(0) \approx \gamma(\varphi) = 800$ erg/cm^2, $\varphi = 20°$, $h_{eq} = 3$ ML, $\nu = 10$, $D = 6\cdot 10^{-13}$ cm^2/s.

From a comparison of the nucleation rates for islands with square and rectangular bases it is shown that elongated islands with higher r emerge later and their nucleation progresses more intensively (see figure 1). It may explain, in our opinion, the increase of wedge-like clusters fraction in QDs arrays during the later stages of growth [1, 2, 8].

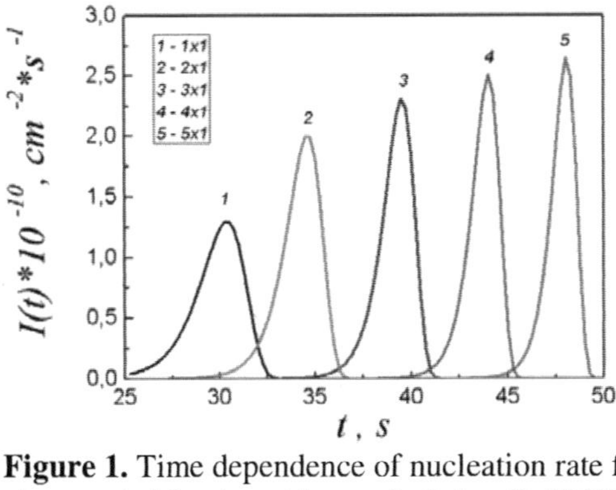

Figure 1. Time dependence of nucleation rate for elongated islands with $r = 1..5$ in Ge/Si(001) system at substrate temperature $T = 470$ °C and deposition rate $V = 0.07$ ML/s.

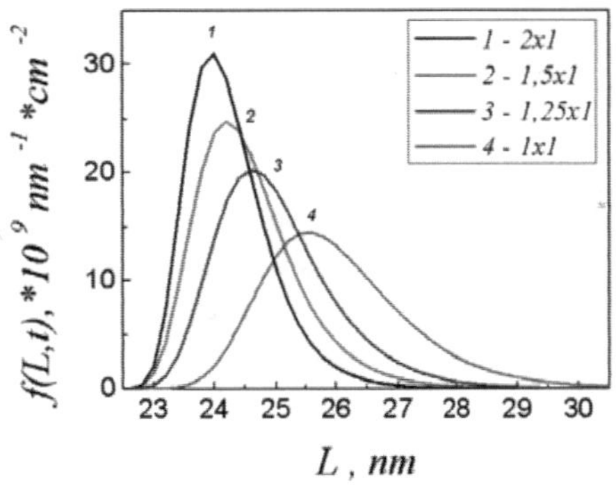

Figure 2. Size distribution function for wedge-clusters of Ge on Si(001) with $r = 1..2$ at substrate temperature $T = 470$ °C and deposition rate $V = 0.07$ ML/s.

Modelling of QDs growth dynamics allows us to estimate surface density and size distribution function for the differently shaped islands. Results of numerical experiments show that for the same growth conditions the average size of elongated islands with a small length to width ratio r is smaller than for pyramidal clusters (see figures 2 and 3). Their size variation is also smaller and surface density is higher, which makes them more attractive for possible device applications.

It must be noted, however, that surface density and average size of the elongated islands change non-monotonically with the ratio increasing (see figure 4).

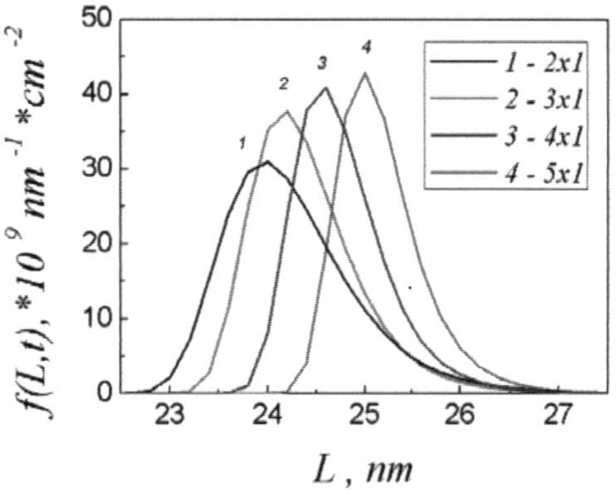

Figure 3. Size distribution function for wedge-clusters of Ge on Si(001) with $r = 2..5$ at substrate temperature $T = 470$ °C and deposition rate $V = 0.07$ ML/s.

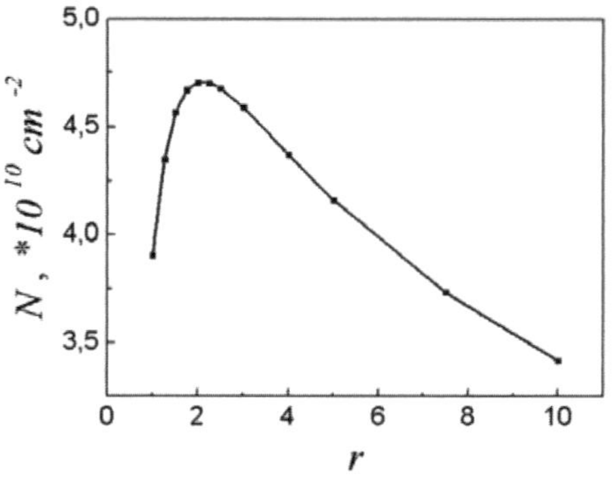

Figure 4. Surface density of wedge-like clusters of Ge on Si(001) surface as a function of r at substrate temperature $T = 470$ °C and deposition rate $V = 0.07$ ML/s.

There is a singular geometry of QDs with $r \approx 2$ for which the average size of the dot reaches its minimum and surface density inversely has its maximum value (for the given growth conditions). With the increase of the length to width ratio from 2 to approximately 7, the surface density decreases gradually and the average size increases, both reaching at $r \approx 6$–8 values that correspond to those of pyramidal clusters with square base. With the further increasing of r the tendency of average size growth and surface density reducing remains constant.

3. Conclusions

Thus, an attempt to develop a kinetic model of the initial stages of growth of different shaped QDs of Ge on Si(001) was made. This model allows us to define the quantum dots size distribution function and their surface density for a given growth temperature, Ge deposition rate and other growth conditions. These are the characteristics which indirectly influence the parameters of nanoheterostructures-based semiconductor optoelectronic devices. Based on theoretical estimations it has been shown that there is an optimal type of QDs with the length to width ratio of 2:1. For this cluster geometry in the equal growth conditions maximum surface density is obtained and the average dot size has its minimum value.

References

[1] Arapkina L V and Yuryev V A 2010 *Physics-Uspekhi* **53** 279–90
[2] Aqua J, Berbezier I and Favre L 2013 *Phys. Rep.* **522** 59–189
[3] Dubrovskii V G 2006 *Semiconductors* **40** 1123–30
[4] Lozovoy K A, Voytsekhovskiy A V, Kokhanenko A P and Satdarov V G 2014 *Surf. Sci.* **619** 1–4
[5] Osipov A V, Schmitt F, Kukushkin S A and Hess P 2002 *Appl. Surf. Sci.* **188** 156–62
[6] Ratsch C and Zangwill A 1993 *Surf. Sci.* **293** 123–31
[7] Kukushkin S A and Osipov A V 1998 *Physics-Uspekhi* **41** 983–1014
[8] Kaniewska M, Engström O, Karmous A, Oehme M, Petersson G and Kasper E 2013 *Solid State Electron.* **83** 99–106

Journal of Physics: Conference Series **541** (2014) 012085 doi:10.1088/1742-6596/541/1/012085

Determination of InSb/AlInSb quantum well energy spectrum

M S Mironova, O S Komkov, D D Firsov and G F Glinskii

Micro- and Nanoelectronics department, St. Petersburg Electrotechnical University "LETI", 5 Professora Popova Street, St. Petersburg 197376, Russia

E-mail: mironova.m.s@gmail.com

Abstract. The InSb/AlInSb single quantum well energy spectrum is determined both theoretically and experimentally (from the photoluminescence spectra). The 8-band Kane model is used to take into account the dispersion law nonparabolicity. The mechanical strain is considered by the deformation potentials theory. The photoluminescence measurements were carried out in the 12-160K temperature range and at various excitation powers. All the observed features of the photoluminescence spectra are well described by the theoretical calculations.

1. Introduction

The InSb/AlInSb quantum-well nanostructures for microwave high electron mobility transistors (HEMT) have higher electron mobility than that of conventional HEMT made of other III-V materials. This is achieved due to the highest electron mobility in InSb compared to any other III-V compound. Recent active studies of InSb/AlInSb heterostructures have been conducted by QinetiQ research group [1] and Intel [2]. In [2] the first prototype of a 305 GHz transistor with an n-InSb channel was reported. Formation of a two-dimensional InSb channel with record-breaking electron mobility was also mentioned in [3]. Additionally, InSb/AlInSb quantum-well nanostructures are used to make infrared light emitting diodes [4].

2. Theory

The InSb/AlInSb quantum well energy spectrum was calculated using the 8-band Kane model [5] for different quantum well widths and temperatures. Mechanical strain due to pseudomorphic growth was considered in the effective Hamiltonian by the deformation potentials theory [6]. The bulk InSb energy dispersion law in [010] and [110] directions for both strained and unstrained cases is shown on fig. 1. The splitting of the heavy-hole and light-hole bands is clearly seen, as well as the energy shift of conduction band minimum. Energy gap for InSb pseudomorphically grown on $Al_{0,11}In_{0,89}Sb$ increases from 0.234 eV to 0.245 eV (at 20K). The temperature dependence was taken into account using the Varshni law. The necessary parameters for calculations were taken from [7-9]. The effective Schrödinger equation for quantum well was solved in the momentum space considering wave vectors only within the Brillouin zone so as to eliminate spurious solutions [10]. We used periodic boundary conditions assuming a single quantum well as a superlattice with a large barrier thickness [11]. The band parameters space dependence was contained within the effective kp-Hamiltonian operator corresponding to BenDaniel-Duke boundary conditions at interfaces.

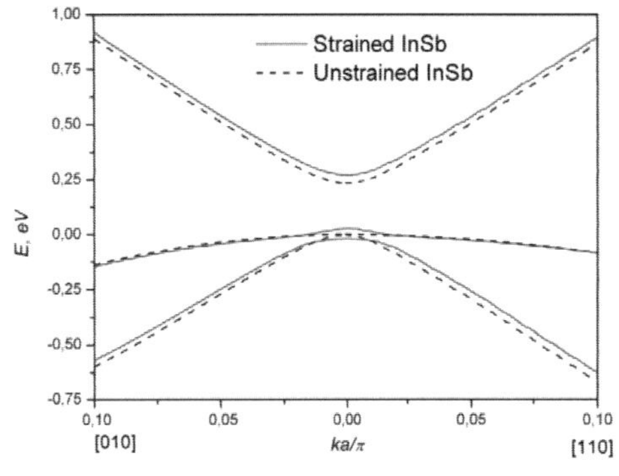

Figure 1. The bulk InSb band structure near the Γ-point with (solid lines) and without (dashed lines) mechanical strains calculated using the Kane model. The mechanical strains correspond to InSb/$Al_{0,11}In_{0,89}Sb$ quantum well (T = 20 K).

3. Experiment

Structures containing InSb/AlInSb single quantum wells were grown by molecular beam epitaxy (MBE) on GaAs semi-insulating substrates with AlSb buffer layers (fig. 2). The photoluminescence (PL) experimental setup was based upon a VERTEX 80 Fourier-transform infrared spectrometer (FTIR) operating in step-scan mode, coupled with a lock-in amplifier. This allowed us to detect the modulated PL signal, while excluding constant blackbody radiation from the spectra.

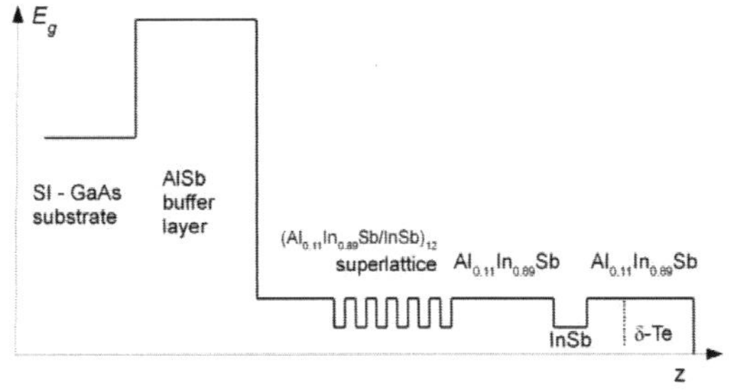

Figure 2. A sketch of the studied samples band gap profile along the growth direction.

The PL measurements were carried out in the 12-160K temperature range; excitation was performed with a chopped beam of an 809 nm laser diode of variable power. An example of the obtained PL spectra for various excitation densities is shown on Fig. 3. Excitation density values are shown for a constant laser beam (without the mechanical chopping).

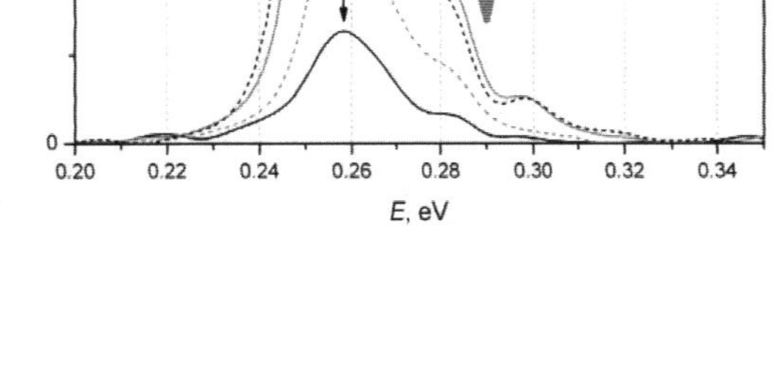

Figure 3. PL spectra of an InSb/$Al_{0.11}In_{0.89}Sb$ single quantum well at various laser excitation densities. Arrows mark the ground-state transition energies E_m .

4. Analysis

In order to eliminate the heating effect on the band gap energy we extrapolated the excitation power dependence of the PL peak energy E_m to zero excitation power (fig. 4). This procedure had been made for the spectra measured at minimum temperature (12 K). Then we calculated the ground-state transition energy ΔE for different quantum well widths so as to find out the sample quantum well width (fig. 5).

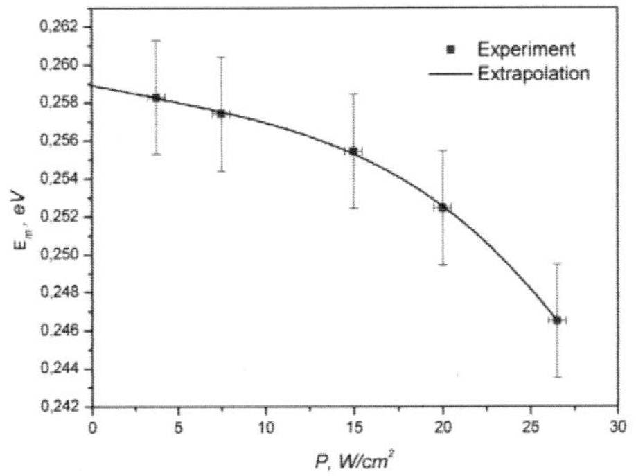

Figure 4. The excitation power dependence of the PL peak energy for InSb/$Al_{0.11}$InSb quantum well

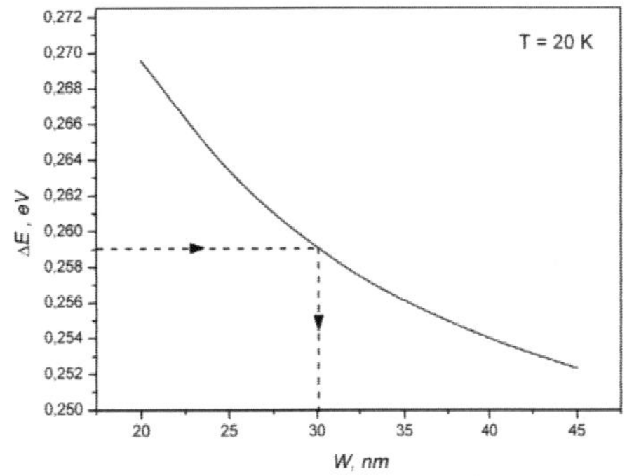

Figure 5. The theoretical dependence of the ground-state transition energy on the quantum well width for InSb/$Al_{0.11}In_{0.89}Sb$ quantum well. Dashed lines illustrate the sample quantum well width determination.

Since the real sample temperature is supposed to be higher than the temperature inside the cryostat and there is no big difference in the PL spectra at 12 K and 25 K (fig. 8), the calculations were performed for 20 K temperature. The sample quantum well width is found to be about 30 nm, which agrees with the TEM measurements. The accuracy depends on spectral resolution and in our case it is not better than 4 nm.

In contrast to a simple quantum well with parabolic dispersion law, the sample 2D density of states is not a step-like function. This is due to the strong nonparabolicity of the dispersion law, which is caused by kp-interaction and mechanical strains (fig. 1). This fact may influence the excitation power dependence of the PL peak energy. However, we assume that the main effect on the PL peak position is due to the energy band gap temperature dependence. Thus, to estimate the sample temperature for different excitation powers we calculated the quantum well energy spectrum temperature dependence taking into account the Varshni law (fig. 6). An example of the quantum well energy spectrum is shown on fig. 7.

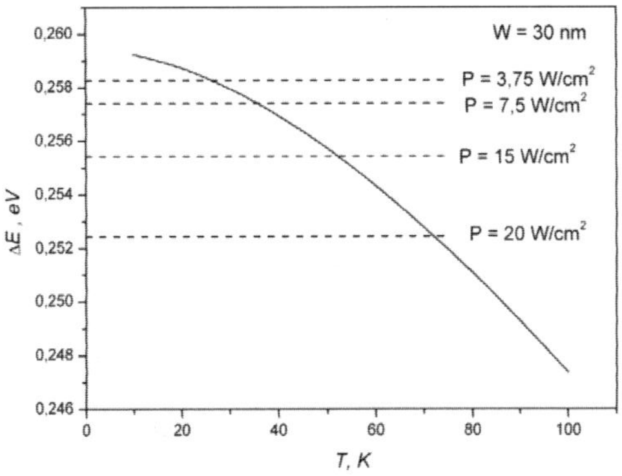

Figure 6. The theoretical temperature dependence of the ground-state transition energy for InSb/$Al_{0.11}In_{0.89}Sb$ quantum well (W=30 nm). Dashed lines mark the PL peak position for various excitation powers.

We compared the theoretical results with the PL spectra at different temperatures (fig. 8) taking into account that the real sample temperature is higher than the temperature inside the cryostat. Thus, to mark transitions at 12 K, calculations for 55 K were made, as follows from fig. 6.

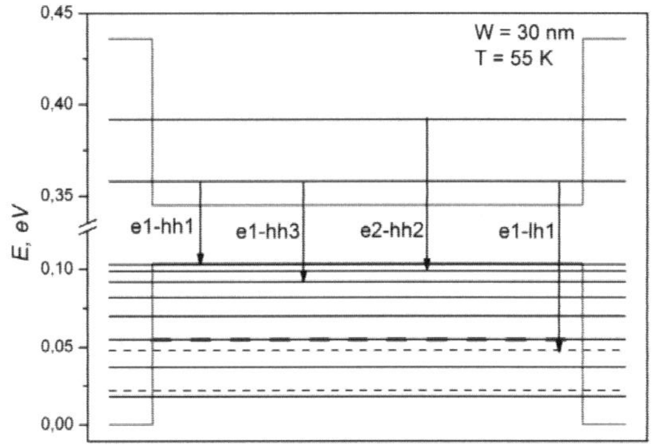

Figure 7. The energy spectrum of a InSb/$Al_{0.11}In_{0.89}Sb$ quantum well calculated using the 8-band Kane model (k_x=k_y=0). (Solid lines represent electron and heavy-hole energy levels. Dashed lines represent light-hole energy levels. The light-hole band is shown by the thick dashed line.)

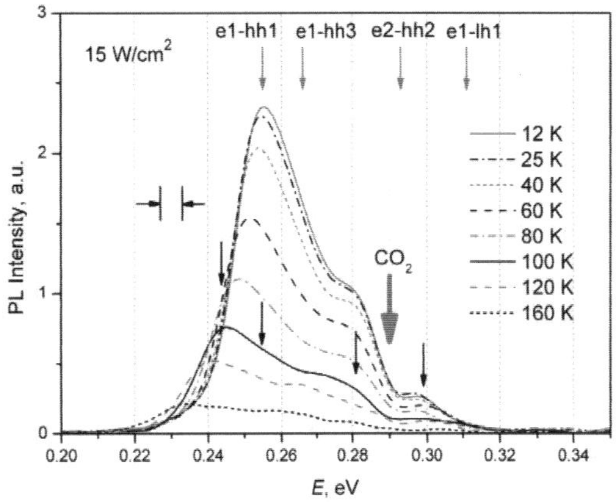

Figure 8. PL spectra of an InSb/$Al_{0.11}In_{0.89}Sb$ single quantum well at various temperatures. Arrows mark transition energies shown on fig. 8 for the 30 nm-thick quantum well at 55K (grey) and 120 K (black).

The theoretical calculations are in a good agreement with the transition energies observed on the PL spectra. However, within the performed measurements it is not possible to experimentally establish the transition energies within the CO_2 absorption region. Unlike the parabolic approximation, the Kane model allowed us to explain all the observed features on the PL spectra.

Acknowledgments
The authors thank the MBE group under the direction of S.V. Ivanov at the Ioffe Physical Technical Institute for providing the QW structures for this study.

References

[1] Orr J M S, Gilbertson A M, Fearn M, Croad O W, Storey C J, Buckle, Emeny M T, Buckle P D and Ashley T 2008 *Phys. Rev.* B **77** 165334
[2] Ashley T, Buckle L, Datta S, Emeny M T, Hayes D G, Hilton K P, Jefferies R, Martin T, Phillips T J, Wallis D J, Wilding P J and Chau R 2007 *Electron. Lett.* **43**(14) 777
[3] Debnath M C, Mishima T D, Santos M B, Hossain K and Holland O W 2012 *J. Appl. Phys.* **111** 073525
[4] Nash G R and Mirza B I 2013 *Appl. Phys. Lett.* **102** 011127
[5] Kane E O 1957 *J. Phys. Chem. Solids* **1** 249
[6] Bir G L, Pikus G E 1974 *Symmetry and strain induced effects in semiconductors* (New York: Wiley)

[7] Dai N, Khodaparast G A, Brown F, Doezema R E, Chung S J and Santos M B 2000 *Appl. Phys. Lett.* **76** 3905
[8] Vurgaftman I, Meyer J R and Ram-Mohan L R 2001 *J. Appl. Phys* **89** 5815
[9] Komkov O S, Semenov A N, Firsov D D, Meltser B Ya, Solov'ev V A, Popova T V, Pikhtin A N and Ivanov S V 2011 *Semiconductors* **45** 1425
[10] Glinskii G F and Mironova M S 2013 *J. Phys.: Conf. Ser.* **461** 012040
[11] Glinskii G F, Lakisov V A, Dolmatov A G and Kravchenko K O 2000 *Nanotechnology* **11** 233

Cyclometalated complexes of platinum metals – the new luminescent activators of OLED

M V Nikolaeva, E A Katlenok, M S Khakhalina, M V Puzyk and K P Balashev

Department of Chemistry, Herzen State Pedagogical University, St. Petersburg, 191186, Russia

E-mail: nikolaeva.mr@gmail.com

Abstract. Systematic investigations of the spectral and luminescent properties of the mixed-ligand cyclometalated Pd(II), Pt(II), Rh(III) and Ir(III) complexes are carried out. Based on these investigations some complexes may be recommended as activators of OLED.

1. Introduction

Aromatic heterocyclic molecules are characterized by intense luminescence, photostability and electrochemical reversible electron transfer processes. For this reason they are widely used in Organometallic Light Emitting Devices (OLED) [1–5]. It is possible to modify them in two ways: by introducing different substituents or by cyclometalating their aryl substituted derivatives [6–14].

The main problem of producing metal complex luminescent activators of OLED is nonradiative process behavior of the low-energy metal-centered electronically excited (d-d*)-type states. Therefore, many researchers started to pay attention to cyclometalated complexes of platinum metals to solve this problem. Energy in their electronically excited (d-d*) states increases due to metal–carbon bond formation.

There are several papers which show the wide application of cyclometalated complexes of platinum metals as OLED activators, biomaterial luminescent labels, optical sensors, materials with liquid-crystalline and photoreactive properties and photocatalytic system components for solar energy conversion to chemical or electrical one [1–5]. The aim of the present paper is to study the influence of ligand and metal structure, composition and nature on the spectral and luminescent characteristics of the cyclometalated square-planar Pt(II) and Pd(II) and octahedral Rh(III) and Ir(III) complexes. They contain the deprotonated form of the cyclometalating (C^N) ligand (2-phenylpyridine (Hppy), 7,8-benzo(h)quinolone (Hbzq), 2-tolylpyridine (Htlp), 4-phenylpyrimidine (Hppm), 1,7-phenanthroline (1,7-Hphen), 2-phenylbenzothiazole (Hbt), 2-phenylbenzoxazole (Hpbo) and 1-phenylpyrazole (Hppz)) and the noncyclometalating ligand (ethylenediamine (En), ammonium (NH_3), acetate (OAc^-), trifluoroacetate (OTF^-), diethyldithiocarbamate (dtc^-), 2-mercaptopyridinate (pyt^-), 2-hydroxypyridinate (pyo^-), acetylacetonate ($acac^-$), 2-mercaptobenzothiazolate (mbt^-) and 2-mercaptobenzoxazolate (mbo^-) ion).

The synthesis of these complexes was based on the substitution reactions of the chloride ligands with the corresponding cyclometalating and other ligands. The electronic absorption spectra were measured on an SF-2000 spectrophotometer (LOMO, St. Petersburg, Russia), and the luminescence was studied using a Fluorat-02-Panorama spectrofluorimeter (Lumeks, St. Petersburg, Russia).

2. Electronic absorption spectra of ligands and mononuclear complexes

Photo- and electrostimulated processes with complexes are classified within the model of localized molecular orbitals according to the preferred localization of involved orbitals on the metal or ligand [2]. The investigation results of the electronic absorption spectra of complexes and free heterocyclic ligands [6–14] demonstrate:

- a bathochromic shift of the long-wavelength (π-π*) optical transitions in a row (in nm) Hppz (255), Hptpy (301sh.), Hbt (312sh.), Hbzq (346) free heterocyclic ligands (sh. – shoulder);
- together with a bathochromic shift (< 1000 cm^{-1}) of the intraligand (IL) $^1(\pi$–$\pi^*)$ optical transitions complexes are characterized by a bathochromic shift of vibrationally structured long-wavelength absorption caused by both cyclometalated ligand and noncyclometalating ligand nature as well as metal nature (Pd(II) < Rh(III) < Pt(II) < Ir(III));
- in accordance with the increase of the spin-orbital coupling efficiency of Ir(III) compared to Rh(III) [3–5] together with the spin-allowed there also exists the spin-forbidden optical transitions for octahedral Ir(III) complexes in the long-wavelength region;
- substitution of σ-donor ethylenediamine or ammonium with π-donor chelated ligands results in a small ($\Delta\nu$ ~ 1000 cm^{-1}) bathochromic shift of the long-wavelength absorption bands of Ir(III) complexes;
- the absorption of complexes at the longest wavelengths is assigned to the metal-to-ligand charge transfer (MLCT) optical transition between the mainly metal-centered HOMO and the LUMO mainly localized on the imine component of the cyclometalated ligand.

3. Electrochemical properties of mononuclear complexes

Reduction and oxidation voltammograms of cyclometalated complexes are characterized by one-electron processes of electron transfer on the cyclometalated ligand unoccupied π*-orbitals and electron removal from metal-centered d-orbitals of complexes. In accordance with the presence of one or two cyclometalated ligands in the inner sphere of square-planar and octahedral complexes their reduction voltammograms are marked by one and two waves:

1) $[M(C^\wedge N)(L^\wedge L)]^Z + e^- = [M(C^\wedge N^-)(L^\wedge L)]^{Z-1}$ (M = Pd(II), Pt(II));
2) $[M(C^\wedge N)_2(L^\wedge L)]^Z + e^- = [M(C^\wedge N)(C^\wedge N^-)(L^\wedge L)]^{Z-1}$,
$[M(C^\wedge N)(C^\wedge N^-)(L^\wedge L)]^{Z-1} + e^- = [M(C^\wedge N^-)_2(L^\wedge L)]^{Z-2}$ (M = Rh(III), Ir(III)).

Metal-centered nature of one-electron oxidation processes of complexes:

1) $[M^{II}(C^\wedge N)(L^\wedge L)]^Z - e^- = [M^{III}(C^\wedge N)(L^\wedge L)]^{Z+1}$ (M = Pd(II), Pt(II));
2) $[M^{III}(C^\wedge N)_2(L^\wedge L)]^Z - e^- = [M^{IV}(C^\wedge N)_2(L^\wedge L)]^{Z+1}$ (M = Rh(III), Ir(III)),

is supported by cathode bias of wave potential according to metal nature (for octahedral complexes $[Ir(bzq)_2pyt]$ 0.39 V and $[Rh(bzq)_2pyt]$ 0.75 V, for square-planar complexes $[PtEn(tlp)]PF_6$ 0.89 V and $[PdEn(tlp)]PF_6$ 1.07 V).

Thus the electrochemical data point to metal-centered nature of the HOMO and ligand-centered nature of the LOMO of complexes that is consistent with the assignment of their long-wavelength band in absorption spectra to the metal-to-cyclometalated-ligand charge transfer optical transition.

4. Spectral and luminescent properties of ligands and mononuclear complexes

Heterocyclic ligands' metalation results in their fluorescence quenching in the ultraviolet region due to metal spin-orbital coupling. Complexes' photoexcitation in glassy solutions leads to their phosphorescence in the visible region (luminescence maximum is from 418 to 530 nm for [PtEn(ppz)]Cl and [Pt(bt)(acac)] respectively) (table 1). Phosphorescence excitation spectra of complexes agree with their absorption spectra.

Like the bathochromic shift of free ligands' fluorescence, complexes' phosphorescence has also a bathochromic shift in the visible region during a change of ligand nature. For square-planar and octahedral complexes with ammine ligands substitution of Pd(II) and Rh(III) with Pt(II) and Ir(III) results in a bathochromic shift of spectra ($\Delta\nu$ = (900±100) cm^{-1}) while luminescence spectra of Pd(II) and Rh(III) as also Pt(II) and Ir(III) complexes are similar [14].

Table 1. Luminescent characteristics of mono- (square-planar and octahedral) and binuclear Pd(II), Rh(III), Pt(II) and Ir(III) complexes (at 77 K, in $CH_3OCH_2CH_2OH$:C_2H_5OH (1:1) solution).

Square-planar complexes	λ_{max}, nm	τ, μs	Octahedral complexes	λ_{max}, nm	τ, μs
$[PtEn(ppz)]^+$	416	40	$[RhEn(tlp)_2]^+$	462	114
$[Pd(NH_3)_2ppm]^+$	455	150	$[RhEn(pbo)_2]^+$	482	90
$[Pd(NH_3)_2ppy]^+$	455	150	$[RhEn(bzq)_2]^+$	491	1720
$[Pd(NH_3)_2tlp]^+$	467	420	$[IrEn(tlp)_2]^+$	485	6
$[PdEn(tlp)]^+$	469	410	$[IrEn(pbo)_2]^+$	508	5
$[PtEn(tlp)]^+$	489	20	$[IrEn(bzq)_2]^+$	524	25
$[PtEn(bzq)]^+$	484	307	Binuclear complexes	λ_{max}, nm	τ, μs
$[PdEn(bzq)]^+$	476	8600	$[Pt(tlp)(\mu\text{-mbo})]_2$	481/574	9/2
$[Pd(NH_3)_2(bzq)]^+$	476	7200	$[Pt(tlp)(\mu\text{-mbt})]_2$	482/588	18/8
$[Pd(NH_3)_2(1,7\text{-phen})]^+$	453	3600	$[Pt(bt)(\mu\text{-pyt})]_2$	625	6
$[PdEn(pbo)]^+$	486	130	$[Pt(bt)(\mu\text{-mbo})]_2$	531/631	not exp
$[PtEn(pbo)]^+$	506	16	$[Pt(bt)(\mu\text{-mbt})]_2$	678	5
$[PdEn(bt)]^+$	510	150	$[Pd(bzq)(\mu\text{-mbt})]_2$	673	274
$[PtEn(bt)]^+$	528	10	$[Pd(bt)(\mu\text{-mbt})]_2$	683	306
[Pt(bt)(acac)]	530	9	$[Pd(tlp)(\mu\text{-OAc})]_2$	476/740	392

As against Pt(II) and Ir(III) complexes which luminesce in glassy (77 K) and liquid (293 K) solutions, Pd(II) and Rh(III) complex luminescence is quenched in liquid solutions. The temperature quenching the luminescence of Pd(II) and Rh(III) complexes in liquid solutions (293 K) is assigned to the thermally activated population of the relative low-energy metal-centered excited (d-d*) states undergoing rapid nonradiative deactivation [9–12].

The energy increasing of metal-centered excited states of Pt(II) and Ir(III) complexes determines their luminescence in the visible region in glassy and liquid solutions. In agreement with an increased 5d-orbitals' splitting of octahedral complexes (in comparison with square-planar complexes) [2–5, 11–14], the temperature quenching the luminescence of Ir(III) complexes is less than Pt(II).

In agreement with the luminescence quenching of the complexes as a result of the thermally activated population excited (d-d*) states for complexes with 1-phenylpyrazole, the temperature quenching the luminescence leads to its absence for $[PdEn(ppz)]ClO_4$ and $[RhEn(ppz)_2]ClO_4$ not only in liquid but in glassy solutions. Relative energy increase of excited (d-d*) states for [PtEn(ppz)]Cl determines the low-temperature luminescence and its quenching in liquid solutions. Further energy increase of excited (d-d*)-type states of Ir(III) complexes based on 1-phenylpyrazole determines their luminescence [8] both in glassy and in liquid solutions.

The relatively long luminescence decay time of the complexes (from 3 to 8600 μs) and its sensibility to oxygen presence in solution indicate on the spin-forbidden character of radiative process as a result of phosphorescence [2–4, 6, 11].

Monoexponential phosphorescence decay kinetics and the phosphorescence excitation spectrum agreement with the absorption spectrum of complexes show the occurrence of radiative decay of photoexcitation energy from the lowest energy electronically excited triplet state of complexes. Increasing the efficiency of singlet-triplet intersystem crossing of the complexes' excited states (in comparison with free luminophore) points to the effective interaction of luminophore with platinum metals characterized by significant spin-orbital coupling.

Independently of the metal and the cyclometalating ligand nature, the low-temperature phosphorescence spectra of complexes are characterized by a main progression with frequency ν = (1460±50) cm^{-1} corresponding to the C=C/C=N stretching vibrations of the cyclometalated ligand. As against the bathochromic shift of long-wavelength MLCT bands in the excitation spectra of Rh(III) complexes in comparison with Pd(II) ($\Delta\nu$ 1400 cm^{-1}) and Pt(II) complexes in comparison with Ir(III) ($\Delta\nu$ 2000 cm^{-1}), the phosphorescence spectra are virtually unchanged (figures 1 and 2). The Huang–

Rhys parameter (S) estimation from intensity ratio of the vibrational components in the low-temperature phosphorescence spectrum [14] and the singlet-triplet splitting energy (E_{ST}) for the intraligand (π-π^*) excited state of complexes show their practical independence from the metal nature (S = (0.8±0.1), E_{ST} ~ 11600 cm^{-1}). The obtained estimations allowed us to classify complexes' phosphorescence to the intraligand optical transition. At the same time the bathochromic shift of phosphorescence spectra of Ir(III) and Pt(II) complexes in comparison with Rh(III) and Pd(II) complexes at (900±100) cm^{-1} as well as tenfold cutting phosphorescence decay time indicate on the mixed IL/MLCT character of complexes' excited state responsible for phosphorescence.

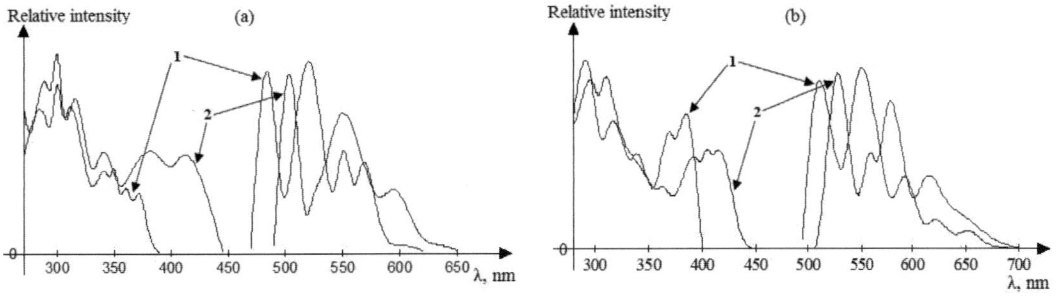

Figure 1. Phosphorescence (curves to the left) and excitation of emission (curves to the right) spectra at 77 K: (a) $[PdEn(pbo)]PF_6$ (1) and $[PtEn(pbo)]PF_6$ (2), (b)$[RhEn(bt)_2]PF_6$ (1) and $[IrEn(bt)_2]PF_6$ (2).

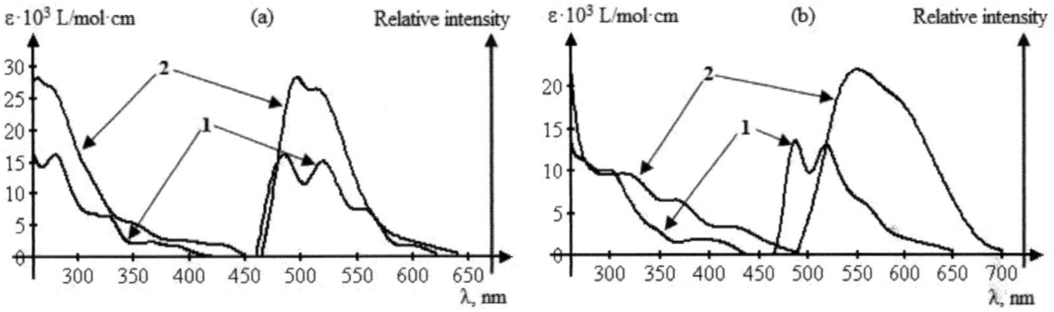

Figure 2. Electronic absorption (curves to the left) and luminescence (293 K) (curves to the right) spectra: (a) $[PtEn(tlp)]PF_6$ (1) and $[IrEn(tlp)_2]PF_6$ (2), (b) $[PtEn(bzq)]PF_6$ (1) and $[IrEn(bzq)_2]PF_6$ (2).

Thus spectral and luminescent properties of cyclometalated complexes of platinum metals with σ-donor ethylenediamine or ammonium ligands are defined by phosphorescence from the energy lowest mainly intraligand triplet state. At the same time π-donor chelated ligands enhance mixing degree of intraligand electronically excited state with charge transfer state due to interaction with d_π-metal orbitals. It results in the bathochromic shift of phosphorescence spectrum.

5. Optical and electrochemical properties of binuclear cyclometalated Pd (II) and Pt (II) complexes

In comparison with mononuclear complexes, the presence of metal–metal bond in binuclear cyclometalated complexes leads to a nature change of the HOMO due to formation of dσ*-molecular orbital. This determines, together with the IL (π-π^*) and the MLCT (d-π^*) spin-allowed optical transitions, additional absorption in the the long-wavelength region (405–512 nm) related to the metal–metal-to-cyclometalated-ligand charge transfer (MMLCT) (dσ*-π^*) optical transition [10].

In agreement with the energy increase of Pt(II) 5d-orbitals (in comparison with Pd(II) 4d-orbitals) dσ* MO also have higher energy that leads to a bathochromic shift of their optical MMLCT-type bands in contrast with Pd(II) complexes.

Change of nature and energy of the HOMO during the formation metal–metal chemical bond in binuclear $[Pd(C^{\wedge}N)(\mu\text{-}OAc)]_2$ and $[M(C^{\wedge}N)(\mu\text{-}N^{\wedge}L)]_2$ complexes is reflected in cathode bias of electrooxidation wave potentials of binuclear complexes in comparison with mononuclear complexes.

Oxidation voltammograms of binuclear Pt(II) complexes are characterized by two one-electron waves corresponding to the sequent oxidation process behaviour of two palladium centres with the formation binuclear Pd(III) complex. Electrooxidation of binuclear Pt(II) complexes proceed as a result of two-electron process of binuclear Pt(III) complexes' formation:

$[Pt^{II}(C^{\wedge}N)(\mu\text{-}N^{\wedge}L)]_2 - 2e^- = [Pt^{III}(C^{\wedge}N)(\mu\text{-}N^{\wedge}L)]_2^{2+}$.

6. Luminescent properties of binuclear complexes with metal–metal chemical bond

Photoexcitation of glassy solutions (77 K) of binuclear Pd(II) complexes in the absorption bands' region results in intraligand vibrationally (π-π*) phosphorescence similar to the phosphorescence of $[Pd(C^{\wedge}N)En]PF_6$ complexes. At the same time together with intraligand phosphorescence in the red region there is a weak structureless band (figure 3) classified to MMLCT-type phosphorescence [10]. The phosphorescence excitation spectrum in the blue and red region is the same and agree with the absorption spectrum of complex. Phosphorescence decay kinetics is nonexponential that agree with radiative process behaviour from two thermally non-equilibrium electronically excited IL and MLCT states (table 1).

Photoexcitation of binuclear Pt(II) complexes both in glassy (77 K) and in liquid (293 K) solutions leads to complexes' phosphorescence only in the red region (figure 4). The phosphorescence excitation spectrum of complexes agree with the absorption spectrum. Exponential phosphorescence decay kinetics confirms radiative process behaviour of photoexcitation energy degradation from one electronically MMLCT excited state.

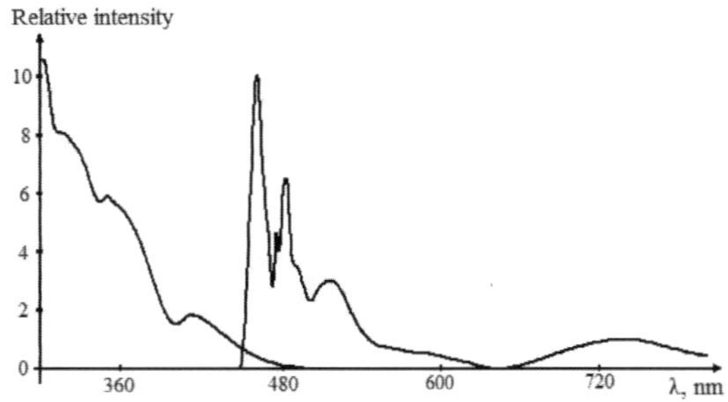

Figure 3. Absorption (293 K) (curves to the left) and low-temperature (77 K) luminescence (curves to the right) spectra of $[Pd(tlp)(\mu\text{-}OAc)]_2$.

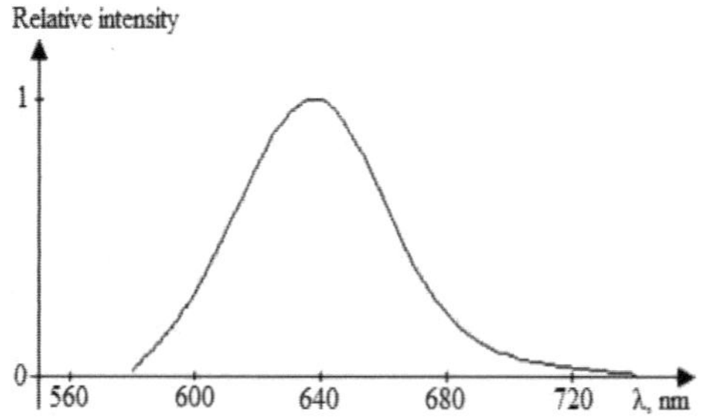

Figure 4. Low-temperature (77 K) phosphorescence spectrum of $[Pt(bzq)(\mu\text{-}pyt)]_2$ in $CH_3OCH_2CH_2OH:C_2H_5OH$ (1:1) solution.

7. Conclusion

Thus mixed-ligand square-planar $[M(C^{\wedge}N)(L\text{–}L)]X$ (M = Pt(II), Pd(II); X = ClO_4^-, PF_6^-, Cl^-) and octahedral $[M(C^{\wedge}N)_2(L\text{–}L)]X$ (M = Rh(III), Ir(III)) complexes with heterocyclic cyclometalated $((C^{\wedge}N)^-$ = ppz, pbo, bt, tlp, bzq, ppy, ppm, 1,7-phen) and noncyclometalated ((L–L) = En, NH_3, $acac^-$, dtc^-, OAc^-, OTF^-, pyt^-, mbt^-, pyo^-, mbo^-) ligands are synthesized and classified by the composition, structure and optical properties.

Obtained results of this study demonstrate that mixed-ligand cyclometalated mono- and binuclear complexes of platinum metals form a family of complexes with long-lived electronically excited states. Localization of electronically excited states on the metal complex fragments in complex composition, together with the control and regulation ability of their optical parameter as a result of change of heterocyclic ligands' nature and metal coordination centres in a wide range, determines their application perspectiveness as activators of organic light-emitting systems.

References

[1] Krasovitskiy B M and Bolotin B M 1984 *Organic Luminofores* (Moscow: Khimiya) [in Russian]
[2] Balashev K P, Puzyk M V, Kotlyar V S and Kulikova M V 1997 *Coord. Chem. Rev.* **159** 109
[3] Lamansky S, Djurovich P, Murphy D, Abdel-Razzaq F, Lee H-E, Adachi C, Burrows P E, Forrest S R and Thomson M E 2001 *J. Am. Chem. Soc.* **123** (18) 4304
[4] Brooks J, Babayan Y, Lamansky S and Thompson M E 2002 *Inorg. Chem.* **41** (12) 3055
[5] Chen T-S 2008 *J. Organomet. Chem.* **693** (19) 3117
[6] Khakhalina M S, Rodionova O A and Puzyk M V 2009 *Opt. Spectrosc.* **106** (4) 529
[7] Katlenok E A, Puzyk M V and Balashev K P 2012 *Opt. Spectrosc.* **113** (2) 131
[8] Katlenok E A, Ivanova E V, Puzyk M V and Balashev K P 2012 *Opt. Spectrosc.* **113** (3) 279
[9] Katlenok E A and Balashev K P 2012 *Russ. J. Gen. Chem.* **82** (9) 1583
[10] Katlenok E A and Balashev K P *Russ. J. Gen. Chem.* 2013 **83** (1) 128
[11] Nikolaeva M V and Puzyk M V 2013 *Opt. Spectrosc.* **114** (2) 247
[12] Nikolaeva M V and Puzyk M V 2013 *Opt. Spectrosc.* **115** (6) 839
[13] Nikolaeva M V 2013 *Izvestia: Herzen University Journal of Humanities & Sciences* **163** 49 [in Russian]
[14] Katlenok E A and Balashev K P 2013 *Opt. Spectrosc.* **114** (5) 756

Photoinduced absorption and photoconductivity of Ge/Si quantum dots in mid-infrared range under interband excitation

R M Balagula[1], A N Sofronov[1], V Yu Panevin[1], D A Firsov[1], L E Vorobjev[1], A A Tonkikh[2,3] and P Werner[2]

[1] St.Petersburg State Polytechnical University, Polytechnicheskaya str., 29, St. Petersburg, Russia.
[2] Max Planck Institute of Microstructure Physics, Weinberg 2, 06120 Halle (Saale), Germany
[3] Institute for Physics of Microstructures RAS, 603950, Nizhny Novgorod

E-mail: rmbal@spbstu.ru

Abstract. Photoinduced absorption and lateral photoconductivity of Ge/Si quantum dots with different doping levels are studied under interband optical excitation. The obtained spectra of absorption and photoconductivity are in good agreement. Observed photoconductivity is attributed to hole escape from ground and excited states while interlevel transitions do not impact the photoconductivity signal. Temperature dependence of photoinduced photoconductivity was measured. The decrease of photoresponse with the temperature increase is attributed to the hole recapture into the quantum dots.

1. Introduction

Ge/Si quantum dots (QD) can be used to develop near-IR and mid-IR radiation detectors, and a possibility of integration into existing Si technology is their advantage. A lot of research groups in the world are engaged in research of optical and photoelectrical phenomena in these structures, and their results tend to be useful and applicable for device development [1, 2]. At the same time details of these phenomena in Ge/Si quantum dots are not clear enough now. Studies of optical and photoelectrical properties of Ge/Si QD structures under the external photoexcitation of charge carriers may improve this situation since it is possible to observe the contribution of excited states occupied with nonequilibrium photoexcited carriers. In this paper, we present the results of both photoinduced mid-infrared photoconductivity and photoinduced absorption studies in Ge/Si quantum dots with different levels of boron doping. Photoexcitation with interband light allowed us to vary the number of holes per QD.

2. Samples and methods

Samples containing self-assembled Ge/Si quantum dots were studied. They were grown by molecular beam epitaxy on both-sides-polished Si (100) substrates with a resistivity of 1 kΩ·cm. Growth temperature was 600°C. The structures consisted of a 100 nm silicon buffer layer, on which ten layers of Ge QDs separated by 15 nm Si interlayers were grown. Seven Ge monolayers (MLs) were deposited at a rate of 0.14 ML/s in order to form QDs. A 100 nm Si cover layer was deposited to complete the structure fabrication process. The average content of germanium in the QD material was about 60–65%. A surfactant (Sb) was used during the growth process in order to increase the density of the QD array. The structure was locally doped with boron. Doping was performed out of the Ge dot layer at 5 nm distance from it. We had at our disposal undoped structures and structures with boron doping levels of $4 \cdot 10^{11}$ and $8 \cdot 10^{11}$ cm^{-2}. The form and dimensions of QDs were found using transmission electron microscopy. Average height of QDs was 2.7 nm, and average size of QD base was 14 nm. Results of atomic force microscopy of specially grown samples with one QD level and no surface silicon layer showed that QD density was $2 \cdot 10^{11}$ cm^{-2}.

The undoped structure was used in the photoinduced absorption studies. Doped structures were used in the photoconductivity measurements.

Samples were mechanically processed in order to create multipass geometry. Sample sides were polished at 45° angle, and radiation lighted normally on these sides. This allowed light to pass the structure multiple times due to internal reflection in the sample. This also made polarization dependent studies possible.

The electrical contacts made of thin gold stripes were formed on the sample surface to measure photoconductivity.

Fourier transform IR vacuum spectrometer Bruker Vertex 80v with globar as a source of broadband mid-infrared radiation was used for spectral measurements. Samples were mounted into the liquid nitrogen-cooled cryostat with ZnSe windows. In both photoconductivity and absorption measurements the non-equilibrium charge carriers were generated by solid-state YAG:Nd laser with frequency doubling. Liquid nitrogen-cooled MCT detector was used in the absorption measurements. ZnSe metal grid polarizer was used to study photoresponse for two different light polarizations. Photoconductivity signal was amplified with low-noise current preamplifier SR570, which was also used to bias a sample and compensate the change of DC signal with excitation level. Amplified signal was then registered by lock-in amplifier SR830. In the photoconductivity studies we had to modulate globar radiation instead of exciting laser radiation (as in the photoinduced absorption measurements) because in the latter case giant interband photoconductivity signal would make it impossible to detect intraband photoconductivity.

3. Experimental results

Photoinduced absorption spectra were measured for two light polarizations and different pumping levels at 77 K. Spectrum for the highest excitation level is shown in figure 1.

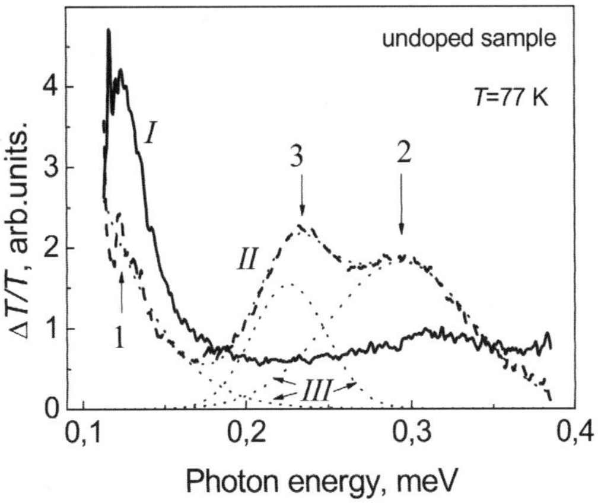

Figure 1. Photoinduced absorption spectra under interband excitation for two light polarizations at 77 K. Solid line – s-polarized light, dashed line – p-polarized light. Dotted lines – Gaussian fit. Arrows show observed peak positions.

The absorption spectrum for s-polarized light (curve *I*) demonstrates one low-energy peak 1, which we attribute to the hole transitions from ground to excited localized state. The absorption spectrum for p-polarized light (curve *II*) demonstrates two peaks corresponding to transitions of holes from ground state (peak 2) and from excited state (peak 3) to continuum. Curves *III* show Gaussian fit of the absorption coefficient of the observed peaks.

Inhomogeneous broadening of the absorption peaks can be attributed to the variation of QD geometry.

The peak positions correlate well with previously obtained data for the equilibrium light absorption in the doped structures [3], and with the theoretical estimates of energy spectrum [4]. But in the described experiments nonequilibrium charge carrier concentration pumped by the external photoexcitation exceeded the values of equilibrium hole concentration existing due to the original doping. Thus one can see the absorption peak corresponding to the previously unoccupied excited states.

Photoinduced photoconductivity spectra were measured in Ge/Si quantum dots at liquid nitrogen temperature (figure 2). Again, external photoexcitation allowed us to create nonequilibrium population of QD states. Both doped samples demonstrate a peak in photoconductivity spectra for z-polarized light. Peak position correlates well with absorption measurements considering that absorption of z-polarized light leads to escape of holes from QDs. This fact is also proved by previously obtained data for equilibrium photoconductivity and absorption in Ge/Si QDs with higher doping level [5].

The nature of the observed absorption peaks is confirmed by the photoconductivity specifics. As such only the peak at the energy of about 300 meV is observed at the photoconductivity spectra, because the absorption of radiation at this energy results in the appearance of free charge carriers in the Si matrix. The intersubband absorption (at approximately 100 meV) doesn't have an influence on the photoconductivity spectra because the involved holes are still localized inside the QD.

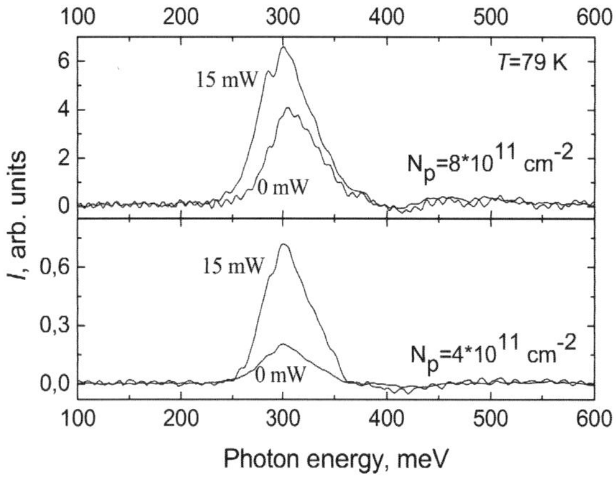

Figure 2. Photoconductivity spectra for z-polarized light for two samples under different excitation levels (shown in the plot) at 79 K.

For the sample with lower doping level photoconductivity signal increases with excitation level and peak position remains the same, while for the sample with higher doping level the signal increase is not so fast and the observed peak broadens to the lower energy range. This broadening with an excitation can be attributed to the contribution of hole transitions from previously unoccupied QD excited states.

The obtained temperature dependence of photoconductivity signal is shown in figure 3. Measurements of a photocurrent at different temperatures show that photoconductivity signal diminishes with the temperature increase. Temperature suppression of photoconductivity can be related to the increase of recapture rate of free holes to the quantum dots at the presence of a potential barrier for holes at the heterointerface. The band bending can occur in silicon around the QD as a result of redistribution of mechanical strain [6] and due to the influence of selective doping.

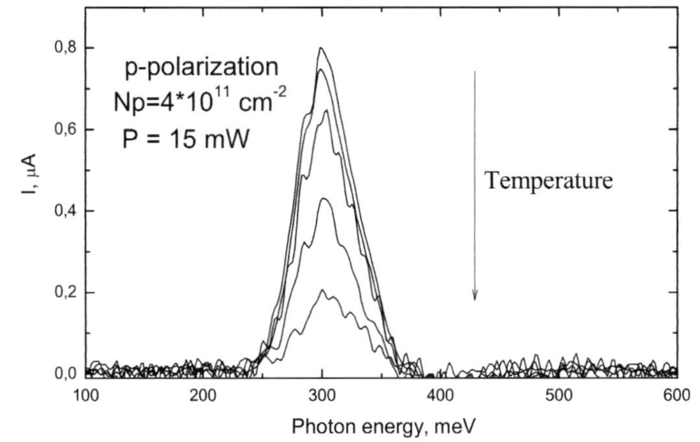

Figure 3. Temperature dependence of photocurrent spectra for z-polarized light for the sample with doping level of $4{\cdot}10^{11}$ cm^{-2} (excitation level is shown on the plot). T = 77, 85, 95, 105, 115 K.

The presence of a barrier for holes at the Ge/Si interface can also qualitatively explain a slight blue shift of photoconductivity compared to the absorption peak (2) in figure 1. Light with low photon energy transfers holes from QD ground state to the states under the QD at the top of silicon valence band. To contribute to the photocurrent holes must overpass the barrier while nothing prevents them from falling back into the quantum dot (transition I on figure 4). Light with higher photon energy transfers holes to the higher states above the barrier so they can easily contribute to the photoconductivity (transition II on figure 4).

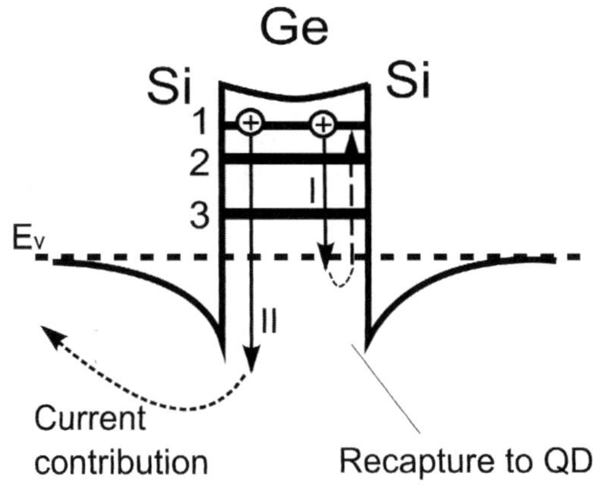

Figure 4. Schematics of excitation processes in the QD structure.

4. Conclusion

In conclusion, in this work the lateral photoconductivity and absorption spectra were studied under the conditions of interband optical excitation. Absorption spectra reveal contribution of hole transitions from ground and excited states under their nonequilibrium occupancy. Spectrum of the photocurrent is in good agreement with the absorption spectrum. The absence of the low energy peak related to hole interlevel transitions corroborates the fact that only holes escaping from quantum dots may contribute to the photoconductivity.

Acknowledgments
This work has been supported by the Russian Foundation for Basic Research (Project №13-02-12203).

References

[1] Yakimov A I, Timofeev V A, Bloshkin A A, Kirienko V V, Nikiforov A I, Dvurechenskii A V 2012 *J. Appl. Phys.* **112** 034511
[2] Schilling J, Talalaev V, Tonkikh A, Fuhrmann B, Heyroth F, Otto M 2013 *Appl. Phys. Lett.* **103** 161106
[3] Vorobjev L E, Firsov D A, Shalygin V A, Panevin V Yu, Sofronov A N, Yakimov A I, Dvurechenskii A V, Tonkikh A A, Werner P 2012 *Semiconductors* **46** 1529
[4] Anikeeva M S, Vinnichenko M Ya, Firsov D A, Vorobjev L E, Tonkikh A A 2012 *St. Petersburg State Polytechnical University Journal. Physics and Mathematics* **158** 9
[5] Panevin V Yu, Sofronov A N, Vorobjev L E, Firsov D A, Shalygin V A., Vinnichenko M Ya, Balagula R M, Tonkikh A A, Werner P, Fuhrman B, Schmidt G 2013 *Semiconductors* **47** 1574
[6] Yakimov A I, Dvurechenskii A V, Nikiforov A I, Bloshkin A A, Nenashev A V, Volodin V A 2001 *Phys. Rev.* B **73** 115333

Electronic energy transfer from the semiconductor quantum wire excitons to an organic media

Y A Strokova and M G Kucherenko

Orenburg State University, Orenburg, 460018, Russia

E-mail: clibph@yandex.ru, strokova.yuliya@yandex.ru

Abstract. We consider an electronic energy transfer from a semiconductor quantum wire to an organic material and discuss the dependence of the energy transfer time on the configuration parameters

1. Introduction

The electronic energy transfer from the Wannier-Mott excitons confined in a semiconductor quantum well and quantum dot to the organic material has been examined in a number of papers [1-3]. It has been already demonstrated that the nanoobject shape and surface curvature significantly influence on its spectral and transfer properties. In our research we explore the electronic energy transfer from the semiconductor quantum wire (QWR) excitons to the surrounding organic molecule matrix, which is presented as a solid homogenous media and characterized by the field energy resonant absorption in the exciton frequency region. In addition to the previous work [4] we consider two cases of the QWR polarization and the influence of the configuration parameters.

2. Electric field potential

We consider the QWR of a circular section. It consists of the core of radius R_c surrounded by the semiconductor barrier of thickness $\Delta = R - R_c$. The QWR core and barrier have the same dielectric constant ε_{in}, which is taken to be real, while the dielectric constant ε_{out} of the surrounding organic media is taken to be complex. We define Z-axis as a longitudinal axis of the wire.

The polarization vector $\mathbf{P}(r,z)$ of the QWR is created by the one-dimensional Wannier-Mott excitons strong confined in the radial direction

$$\mathbf{P}(r,z) = \frac{\mathbf{d}_{vc}}{\sqrt{a_B L R_c^{\,2}}} J_0^2(k_r r)\exp(ik_z z), \qquad (1)$$

where $\mathbf{d}_{vc}$ is the vector matrix element of the interband transition dipole moment, a_B is the exciton Borh radius, L is a quantum wire length, $J_0(x)$ is the Bessel function, k_r and k_z are the wave vectors of the radial and longitudinal motions.

The electric field potential $\Phi(r,\varphi,z)$ complies with the Poisson equation within the quantum wire

$$\varepsilon_{in}\, \nabla^2 \Phi_{in}(r,\varphi,z) = -4\pi\, \rho(r,\varphi,z), \qquad (2)$$

and to the Laplace equation in the absorbing media

$$\nabla^2\Phi_{out}(r,\varphi,z)=0. \tag{3}$$

We consider two cases of the quantum wire polarization, the first case when $\mathbf{d}_{vc}$ is parallel to Z-axis and the second one when they are perpendicular. We will refer these two cases Z and X polarization respectively. In this way the volume charge density $\rho(r,\varphi,z)$ in the Poisson equation is given by

$$\rho_M(r,\varphi,z)=-\mathrm{div}\,\mathbf{P}_M(r,z), \tag{4}$$

here and hereinafter $M=Z,X$ indicate the case of polarization.

The electric field potential depends on coordinate z in the following way

$$\Phi(r,\varphi,z)=u(r,\varphi)\exp(ik_z z). \tag{5}$$

Using (5), we have the inhomogeneous and homogeneous Helmholtz equations for $u_{in}(r)$ and $u_{out}(r)$ correspondently

$$\left[\nabla^2_{r,\varphi}-k_z^2\right]u(r,\varphi)=\begin{cases}-4\pi\rho_M(r,\varphi)/\varepsilon_{in}, & r<R\\ 0, & r>R,\end{cases} \tag{6}$$

where

$$\rho_Z(r)=-\frac{\mathbf{d}_{vc}}{\sqrt{a_B L R_c^2}}ik_z J_0^2(k_r r), \tag{7}$$

$$\rho_X(r,\varphi)=-\frac{2\mathbf{d}_{vc}}{\sqrt{a_B L R_c^2}}k_r J_0(k_r r)J_1(k_r r)\cos\varphi=\rho_X(r)\cos\varphi, \tag{8}$$

with the boundary condition

$$\left.\frac{\partial u_{in}}{\partial r}\right|_{r=R-0}=\frac{\varepsilon_{in}}{\varepsilon_{out}}\left.\frac{\partial u_{out}}{\partial r}\right|_{r=R+0}=f(r,\varphi), \tag{9}$$

where function $f(r,\varphi)$ is unknown and determined from the condition of equality of u_{in} and u_{out} on the QWR boundary

$$\left.u_{in}\right|_{r=R-0}=\left.u_{out}\right|_{r=R+0}. \tag{10}$$

Note, that the polarization vector in the barrier with $R_c<r<R$ is zero.

For solving the equations (6) we use the Green's functions. The solution of the inhomogeneous Helmholtz equation can be determined as the sum of the circulation and area integrals

$$u(r,\varphi)=\oint_L G_2^{r<R}(r,\varphi,R,\varphi')f(R,\varphi')Rd\varphi'+\int_S G_2^{r<R}(r,\varphi,r',\varphi')F(r',\varphi')r'dr'd\varphi', \tag{11}$$

where $F(r,\varphi)=4\pi\rho_M(r,\varphi)/\varepsilon_{in}$, $G_2^{r<R}(r,\varphi,r',\varphi')$ is the Green's function of the interior Neumann problem in a circle for the Helmholtz operator and has the following form

$$G_2^{r<R}(r,\varphi,r',\varphi')=\frac{1}{2\pi}\left(K_0(k_z R_{MM_0})-\frac{K_0'(k_z R)}{I_0'(k_z R)}I_0(k_z r)I_0(k_z r')-\right.$$
$$\left.-2\sum_{n=1}^{\infty}\frac{K_n'(k_z R)}{I_n'(k_z R)}I_n(k_z r)I_n(k_z r')\cos(\varphi-\varphi')\right) \quad (12)$$

where $R_{MM_0}=\sqrt{r^2+r'^2-2rr'\cos(\varphi-\varphi')}$, $K_n(k_z\xi)$ and $I_n(k_z\xi)$ are the modified Bessel function of the second kind.

The solution of the homogeneous Helmholtz equation is only the contour integral from equation (11) and the Green's function of the exterior Neumann problem outside the circle for the Helmholtz operator has the similar form as (12) with the replacement $K_n(k_z\xi)\leftrightarrow I_n(k_z\xi)$ with $\xi=R,r,r'$.

The received expression for the electric field potential in the surrounding media is

$$\Phi_{out}^{(M)}(r,\varphi,z)=f_M(R,\varphi)\frac{K_n(k_z r)}{k_z K_n'(k_z R)}\frac{\varepsilon_{in}}{\varepsilon_{out}}\exp(ik_z z), \quad (13)$$

where

$$f_M(R,\varphi)=u_F^{(M)}(R,\varphi)\;k_z\left[\frac{K_n(k_z R)}{K_n'(k_z R)}\frac{\varepsilon_{in}}{\varepsilon_{out}}-\frac{I_n(k_z R)}{I_n'(k_z R)}\right]^{-1}, \quad (14)$$

$$u_F^{(M)}(R,\varphi)=\frac{4\pi}{\varepsilon_{in}}\left[\frac{I_n(k_z R)K_n'(k_z R)}{I_n'(k_z R)}-K_n(k_z R)\right]g_M(\varphi)\int_0^{R_c} I_n(k_z r)\,\rho_M(r,z)\,r dr \quad (15)$$

where
$$g_M(\varphi)=\begin{cases}1, & M=Z\\ \cos\varphi, & M=X,\end{cases}$$

$n=0$ and $n=1$ in the case of Z and X polarization correspondently. The dependence on angle φ appears only in the case of X polarization.

3. Energy transfer time

The total power W transferred from QWR to the organic material and dissipated in it is given by the volume integral over the external area [5]

$$W=\frac{\omega L}{8\pi}\mathrm{Im}\,\varepsilon_{out}(\omega)\int_{r>R}\int_0^{2\pi}\left|\vec{\nabla}\Phi_{out}^{(M)}\right|^2 r\,dr\,d\varphi=\frac{\hbar\omega}{\tau}, \quad (16)$$

where τ is the characteristic time of the energy transfer from an exciton to the organic media.

We plot τ_Z and τ_X as functions of the quantum wire barrier thickness Δ and core radius R_c in figures 1,2 and as a function of k_z in figure 3.

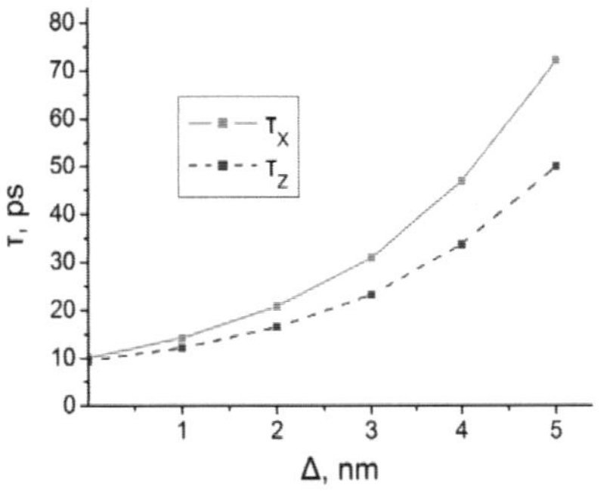

Figure 1. The electronic energy transfer time τ (ps) versus the quantum wire barrier thickness Δ (nm). $R_c = 5\,\text{nm}$, $\mathbf{d}^{vc} \approx 0.1\,ea_B$, $k_z = 3 \cdot 10^{-6}\ \text{cm}^{-1}$, $\varepsilon_{in} \approx 5$, $\varepsilon_{out} \approx 4 + 3i$.

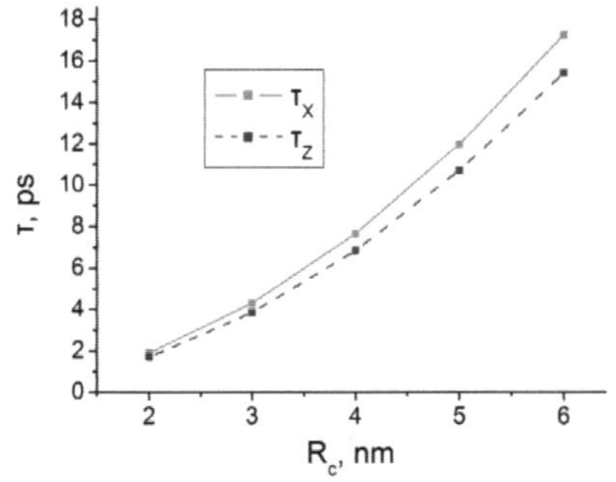

Figure 2. The electronic energy transfer time τ (ps) versus the quantum wire core radius R_c (nm). $\Delta = 0.5$ nm. All other parameters are the same as in Figure 1.

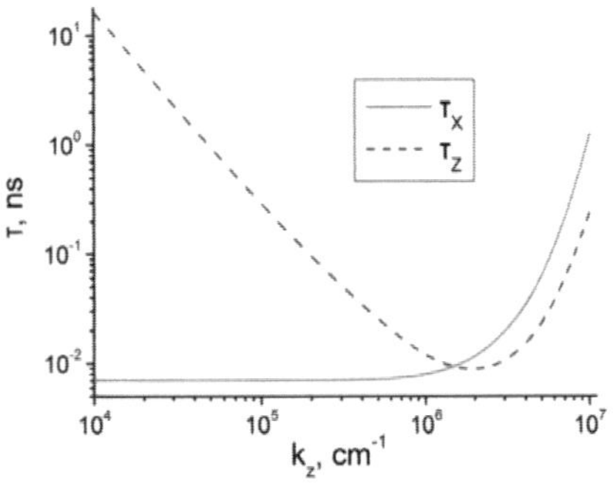

Figure 3. The electronic energy transfer time τ (ps) versus the wave vector longitudinal motion k_z (cm^{-1}). $R_c = 5\,\text{nm}$, $\Delta = 0.5\,\text{nm}$ Other parameters are the same as in Figure 1.

4. Conclusion

It is seen from the plots, that the characteristic energy transfer time increases with the core radius R_c and in the case of fixed value of R_c it can be regulated by the quantum wire barrier thickness Δ. The QWR barrier influence is strong, because the barrier isolates the excitons from the surrounding media separating them into the distance Δ.

In addition to the QWR barrier influence, the characteristic energy transfer time τ dependence on the wave vector k_z has been examined. This value is a spatial frequency (inverse value of the de Broglie wavelength) of the one-dimensional exciton free longitudinal motion. The volume charge density (2) $\rho(z) \sim \cos(k_z z)$ is modulated along the coordinate z exactly this frequency with the charge alternating in sign. For this reason the electric field potential (13) attenuates on the radial direction outside the QWR, and the attenuation rate is proportional to k_z. In other words, the volume charge density $\rho(z)$ frequent oscillations along the longitudinal axis cause an efficient decrease of electric field potential and strength at removal from the QWR boundaries.

It has been ascertained that the characteristic energy transfer time τ dependence on the wave vector k_z significantly differs for the Z and X polarizations. This difference is particularly large and over three orders for small values of k_z. In the case of X polarization the time dependence on k_z is monotonous and in the case of Z polarization it has an extremum spatial frequency ($\sim 10^6\ \text{cm}^{-1}$) with which the most efficient energy transfer matches. It should be noted that in the paper [2] dedicated to the energy transfer from semiconductor quantum wire well to the surrounding organic media this fact was not marked. For the considered cylindrical geometry in the case of X polarization the rate of energy transfer to a near-surface organic layer is determined as it follows from (16) by an asymptotic expression $W \sim \left|K_1'(k_z R)\right|^2$ and doesn't include another factors strongly depending on k_z. By this fact the monotonous dependence $W(k_z)$ is explained. In the case of Z polarization the specified dependence is determined by more difficult expression $W \sim k_z^2 \left|K_0'(k_z R)\right|^2$, which explains the function $W(k_z)$ extremum. The ascertained energy transfer time behavior has the following physical explanation. The interaction $V(r)$ between surrounding organic molecules characterized by the electronic dipole moment $\mathbf{p}_M$ and the QWR exciton electric field $\mathbf{E}(r > R)$ is determined by the scalar product of the vectors $V(r) = -\mathbf{p}_M \mathbf{E}(r)$. The electric field $\mathbf{E}(r > R)$ attenuates slower on the radial direction with the spatial frequency k_z decreasing. As a result the interaction $V(r)$ is considerable at longer distances from the QWR surface. The energy transfer rate for $k_z \to 0$ is monotonous increases. In the considered case of X polarization the weak field attenuation is due to increase of a length $L_z = 1/k_z$ of the region in which the volume charge density $\rho_X(z) \sim \cos(k_z z)\cos\varphi$ has the constant sign. With the fixed angle φ the charge of one sign is not compensated by the charge of an opposite sign over relatively long distance $L_z = 1/k_z$, thus the electric field $\mathbf{E}(r > R)$ extension beyond the QWR boundaries is significant $E \sim k_r / r \exp(ik_z z)$ for $k_z R << 1$. In the case of Z polarization also as in the case of X polarization the modulation period increases with $k_z \to 0$, but the modulation amplitude monotonically decreases to zero $\rho_Z(z) \sim k_z \cos(k_z z)$. Thus in the case of Z polarization it is observed the electric field strength decrease with $k_z \to 0$ according to the approximate asymptotic expression $E \sim k_z / r \exp(ik_z z)$ (for $k_z R << 1$). The energy transfer rate decreases.

Thus in our research the parametric dependencies of the energy transfer rate from one-dimensional QWR excitons to the surrounding organic media with the known complex dielectric constant are obtained. The variable parameters include the barrier thickness Δ, the QWR core radius R_c, the wave number k_z of the exciton longitudinal free motion. It is revealed the significant difference in the energy transfer rate with $k_z \to 0$ between two cases of polarization.

The work has been supported by the Ministry of education and science of Russia (government job, basic part) and by RFBR grants (project № 14-02-97000).

References

[1] Agranovich V M, La Rocca G C and Bassani F 1997 *JETP Lett.* **66** 748-51
[2] Basko D, La Rocca G C, Bassani F and Agranovich V.M. 1999 *Eur. Phys. J.* B **8** 353-62
[3] Agranovich V M and Basko D M 1999 *JETP Lett.* **69** 250-254
[4] Strokova Y A and Kucherenko M G 2013 *15th AllRussian Youth Conf. Proc.* Saint-Petersburg, 25-29 November 2013 (Saint-Petersburg: publishing house of Polytechnical University) p 66
[5] Landau L D and Lifshits E M 2003 *Electrodynamics of Continuous Media* vol 8 (Moscow: publishing house FizMatLit)

Size distributions and scaling relations for heterogeneous nucleation and growth of atomic chains

Yury Berdnikov [1], **Vladimir G Dubrovskii** [1,2,3]

[1] St.Petersburg Academic University, Khlopina 8/3, 194021 St.Petersburg, Russia
[2] Ioffe Physical Technical Institute RAS, 26 Polytekhnicheskaya ,194021 St.Petersburg, Russia
[3] ITMO University, Kronverkskiy pr. 49, 197101 St. Petersburg, Russia

E-mail: yury.berdnikov@gmail.com

Abstract. An analytical model for one-dimensional islands growth was developed being inspired by experiments on growth of metallic atomic chains. On the basis of reported observations and simulations we considered the case of heterogeneous nucleation with linear on size rates for growth and decay. In this case the exact solution of the system of rate equation leads to Polya distribution for lengths of chains and corresponding scaling behaviour. These analytically computed dependences demonstrated good matching with experimental results.

1. Introduction

The group-III and IV metals [1, 2, 3, 7], several transition metals (like Mn, Pb)[8, 12] and some mixed dimers (like $In - Sn$) [13], being deposited in low coverage regime, are known to assemble into linear atomic chains . The substrate structure plays the key role in this case. The formation of metal chains was observed on $Si(100) - 2 \times 1$ surface composed by silicon dimers arranged into rows. The deposited metal atoms arrange into the chains in a direction perpendicular to these substrate rows. Resulting surfaces could be observed by atomic resolved scanning tunnelling microscopy(STM). The STM technique gives the direct access to analysis of chains density, structure and such statistical characteristics as length distributions and their scaling [4, 14, 15]. The latter one could be also extracted from previously reported numerical simulations [7, 16, 5] and calculations based on analytical model which is presented in the following sections. A schematic image of metallic atomic chains on $Si(100) - 2 \times 1$ substrate is shown on Fig.1. Where red full spheres corresponds to substrate atoms of Si dimers and semi-full spheres denotes chain atoms.

2. Analytical model

On the Si surface so called C-type defects play the important role. According to *Hossain et al.* and *Okano and Oshiyama* [19, 20] a defect of type C is interpreted as a dissociated water molecule with the H and OH group bonded to Si atoms of two adjacent surface dimers. These defects are reported to be chemically active and play the role of nucleation centres [3, 6].

Thus the theoretical model should imply the heterogeneous nucleation involving growth and decay of atomic clusters via the scheme $A_sB + A_1 \Leftrightarrow A_{s+1}B$ with $s = 0, 1, 2, 3....$ Here, B denotes the nucleation center, A_1 is the free metallic atom and A_sB is the cluster consisting of s monomers and one nucleation center. By definition, $A_0B \equiv B$ is the free seed. Since C-type defects are not appear or disappear during the metal deposition the total number of all clusters plus the number of free nucleation centres conserves in time and equals the total number of nucleation centres, so that the size distribution $f_s(t)$ can be normalized to one:

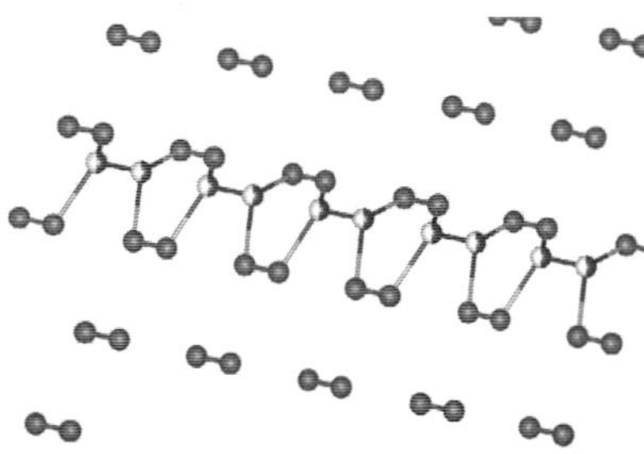

Figure 1. Schematic image of metal atomic chain (semi-full spheres) grown at room temperature on $Si(100)-2\times1$ substrates (full spheres).

$$\sum_{s\geq 0} f_s(t) = 1 \tag{1}$$

Various cases of scenario of atom attachment to chain were discussed and simulated by *Albao et al.* [7]. According to their approach a randomly deposited atom diffuse along the chain after reaching it, so the probability of atom attachment is proportional to chain length. For our model it means that (as in [21]) the growth (k_s^+) and decay (k_s^-) rate for whatever material depends linearly on the number of monomers in the cluster:

$$\begin{aligned} k_0^+ &= k^+b \\ k_s^+ &= k^+(a-1+s) \\ k_s^- &= k^-s;\ s\geq 1 \end{aligned} \tag{2}$$

with some positive dimensionless a, b , and the constants k^+, k^- having the dimensions inverse time. The process could be considered in terms of the dimensionless time $\tau = k^-t$. Within these notations the set of corresponding discrete rate equations for the size distribution $f_s(\tau)$ (according to [21]) could be expressed as:

$$\begin{aligned} \tfrac{\partial f_0}{\partial \tau} &= -b(\zeta+1)f_0 + f_1 \\ \tfrac{\partial f_1}{\partial \tau} &= b(\zeta+1)f_0 - f_1 - a(\zeta+1)f_1 + 2f_2 \\ \tfrac{\partial f_s}{\partial \tau} &= (a-2+s)(\zeta+1)f_s - 1 - (a-1+s)(\zeta+1)f_s + (s+1)f_s;\ s\geq 2 \end{aligned} \tag{3}$$

Where $\zeta \equiv (\frac{k^+}{k^-})f_1$ is a function of time with a clear sense of the effective supersaturation of a free monomer vapor with the concentration f_1.

In view of (2) the growth rate is:

$$\frac{ds}{d\tau} = (a-1)(\zeta+1) + \zeta s \tag{4}$$

which distinguish the cases of $a > 1$ and $a < 1$. The latter case, with a positive supersaturation, leads to the growth instability such that smaller clusters decay and larger clusters grow, as in nucleation theory, with a supersaturation-dependent critical size $s_c = (1-a)(\zeta+1)/\zeta$. Since such critical size and such behaviour were not observed in experiments, further we assume $a > 1$ which corresponds to stable unlimited or limited growth [17, 18].

Time evolution of mean clusters size, $< s >= \sum_{s\geq 1} f_s s$, follows the equation:

$$\frac{d<s>}{d\tau} = \zeta <s> + (1-a)(\zeta+1)(qf_0 - 1) \tag{5}$$

with parameter $q \equiv \frac{b+1-a}{1-a}$

The material balance in neglect of desorption leads to:

$$\Theta = < s > + (\frac{k^-}{k^+})\zeta \tag{6}$$

Where Θ is the normalized coverage.

For further analysis we introduce the generating function for discrete size distribution function f_s:

$$f(x,\tau) = \sum_{s=0}^{\infty} f_s(\tau) x^s \tag{7}$$

with properties $f(0,\tau)$ and $f(1,\tau) = 1$.

Differentiating Eq. (7) with respect to τ and using (3) we arrive to differential equation for generating fuction $f(x,\tau)$:

$$\frac{\partial f}{\partial \tau} = (x-1)\left[(\zeta+1)\frac{\partial f}{\partial x} - (1-a)(\zeta+1)f + (b+1-a)f_0\right] \tag{8}$$
$$f(x,\tau=0) = 1$$

Further it is more convenient to use the new function $F(x,\tau) = f(x,\tau) - qf_0(\tau)$ In these terms the exact solution of (8) has a form:

$$F(x,\tau) = (1+\alpha(\tau)(1-x))^{1-a}\left[1 - q - q\int_0^{\tau} d\tau'(1+\frac{e^{-z(\tau')}\alpha(\tau')}{\Psi})^{1-a}\frac{\partial f_0}{\partial \tau'}\right] \tag{9}$$

Here we denote the functions $\Psi(x,\tau)$ as:

$$\Psi(x,\tau) = e^{-z(\tau)}(\frac{1}{1-x} - \alpha(\tau)) \tag{10}$$

with

$$z(\tau) = \int_0^{\tau} d\tau' \zeta(\tau') \tag{11}$$
$$\alpha(\tau) = e^{z(\tau)} \int_0^{\tau} d\tau'[\zeta(\tau')+1]e^{-z(\tau')} \tag{12}$$

We now consider the asymptotic solution to Eq.(8) which corresponds to the approximation $1 + \frac{e^{-z(\tau)}\alpha(\tau)}{\Psi} \simeq 1$ under the integral. This reduces the integral to $1 - qf_0$. This kind of approximation is equivalent to applying the Laplace method for calculating the integral, so the validity requires $\frac{\partial f_0}{\partial \tau'}$ to be a steep function near $\tau = 0$. In this case the solution has the form:

$$F(x,\tau) = (1+\alpha(\tau)(1-x))^{1-a}[1 - qf_0(\tau)] \tag{13}$$

For chains size distribution it means:

$$f_0(\tau) = \frac{(1+\alpha(\tau))^{1-a}}{1+q(1+\alpha(\tau))^{1-a} - q} \tag{14}$$
$$f_s(\tau) = \frac{<s>(\tau)}{s} f_{s-1}[\alpha(\tau)],\ s \geq 1 \tag{15}$$

Putting the free parameter $b > 0$ to $b = a - 1$ (which requires the stable case $a > 1$ [21, 18]) we arrive to Polya distribution for lengths of chains:

$$f_s(\tau) = f_s^{(b)}[\alpha] = \frac{\Gamma(b+s)}{\Gamma(b)s!}\frac{\alpha^s}{(1+\alpha)^{b+s}},\ s \geq 0, \tag{16}$$

$$< s >= b\alpha \tag{17}$$

Here it is worth for further applications to notice that the Polya distribution is a kind of universal function which could be the purely decreasing with s (for $b < 1$) or have a maximum (for $b > 1$). Both such distributions were observed in experiments.

3. Scaling

The derived form of size distribution could be rewritten in a scaling form. For this let us use the Stirling formula and the large s and α asymptote $\frac{\alpha^s}{(1+\alpha)^s} = e^{-s/\alpha}$. In this way we arrive at the continuum form of the Polya distribution:

$$f^{(b)}(s,\alpha) = \frac{s^{b-1}}{\Gamma(b)\alpha^b}e^{-s/\alpha} \tag{18}$$

This form of size distribution preserves the previously fixed normalisation and correct mean size:

$$\int_0^\infty dsf^{(c)}(s,\alpha) = 1;\ \int_0^\infty ds\ sf^{(c)}(s,\alpha) = c\alpha \tag{19}$$

Now the eq. (16) could be written in the desired scaling form:

$$f^{(b)}(s,< s >) = \frac{1}{< s >}\phi_b\left(\frac{s}{< s >}\right)\ ;\ \phi_b(x) = \frac{b^b}{\Gamma(b)}x^{b-1}e^{-bx} \tag{20}$$

with b-dependent scaling function $\phi_b(x)$.

The obtained scaling $f^{(b)}(s,< s >) = \frac{1}{<s>}\phi_b\left(\frac{s}{<s>}\right)$ at small enough ζ when $< s > \simeq \Theta$ satisfy well-known scaling relation $f^{(b)}(s,< s >) = \frac{\Theta}{<s>^2}\phi_b\left(\frac{s}{<s>}\right)$ suggested by *Vicsek and Family* [9] and discussed in quite recent review of *Einax et al.* [10].

It is important to mention that the scaling given by Eq.(20) is the intrinsic property of the growth model with the size-linear rate constants and does not require any other assumptions. It works in case heterogeneous nucleation but could an object of further investigation whether this condition is necessary or the model, in some extent, could be applied in case of homogeneous nucleation.

4. Comparison with experiment

For the systems of metal atomic chains described in first section the experimentally obtained data could be fitted with analytical solution derived in previous sections.

The most of theoretical models predict the monomodal size distributions with single well-defined maximum around the mean size while the purely decreasing distributions also were observed in experiments [7]. The monomodal distributions are more usual for experiments. They were observed in many systems, for example, for one-dimensional Mn islands on $Si(100) - 2 \times 1$ surface reconstruction [11] or for Si homoepitaxy at low temperatures [22].

The systems of Mn atomic chains grown on $Si(100) - 2 \times 1$ substrates was found to be stable at room temperature [11]. The time-evolution of length size distributions of Mn chains is illustrated by Fig. 2. Due to the ambiguity of single Mn atom identification the size s enters in arbitrary units which correspond to distance between dimers of Si. The distributions are fitted with our model, Eq. (16), for three different deposition times: 2s, 7s and 12s. The value of the parameter

b remains around 4 while α (which is proportional to mean size) is changing with time. The mean sizes in all three cases was calculated according to Eq. (17) and summarized in Table 1. The values obtained from the model fit was found to be inconsiderably smaller than the values extracted from experimental measurements which could be either due to the ambiguity of size units or due to lack of statistics (the data only about 100 - 300 chains were collected for each deposition time). Nevertheless, our simple model seems to fit the data quite well.

Let us next discuss the scaling of experimentally observed distributions. The Fig. 3 (a) shows the scaling form of dependences on Fig. 2 corresponding to 7s and 12s of deposition. The points of $f(s) < s >$ extracted from experimental data (ref. [11]) is plotted versus scaled size $\frac{s}{<s>}$ and fitted in accord with Eq.(20). For two cases of different times of deposition but the same growth conditions the dependences in a scaled form collapsed into one curve corresponding to $b = 4$. Large dispersion around the maximum could originates from lack of statistics, slight ambiguities in chain length measurement or impact of surface terraces.

Table 1

Dep. time	α	Model $<s>$	Measured $<s>$
2s	0.614	2.5	3.27
7s	1.24	4.96	5.3
12s	1.61	6.44	6.86

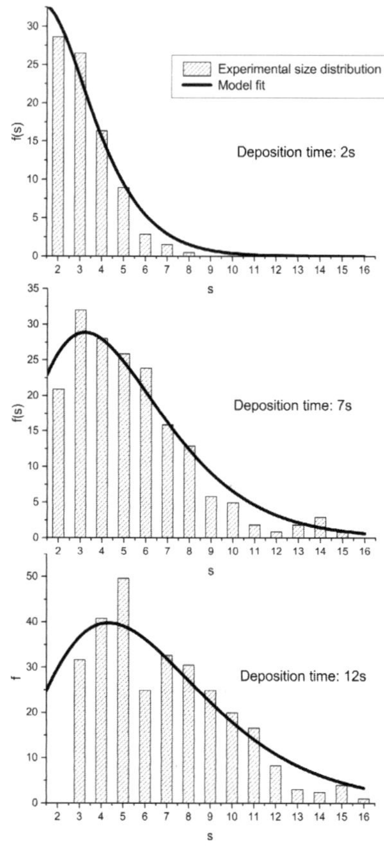

Figure 2. Time evolution of size distribution of Mn atomic chains, the experimental reported in [8]

The process of Mn lines described previously could be compared with others for similar systems of other deposited materials on the same type of substrates. Obviously in these processes the value of b parameter could be different. The Fig. 3 (b) shows the experimental points and corresponding fits of scaled size distributions for Mn (discussed above), In (ref. [3]) and two samples of Ga (ref. [7]) corresponding to coverages of 0.095 ML and 0.130 ML. All these processes were carried out under different growth conditions and, thus, not surprisingly, demonstrate different values of b. Scaled distributions has no peak when $b < 1$. When $b > 1$ and decreases the maximum shifts towards small sizes. In practice it could shifts to non-physical range when s smaller than 2. So here the importance of unique property of Polya distribution (which could have or not to have the peak in positive range) is revealed.

In conclusion, the model for heterogeneously nucleated clusters with size-linear grow and decay rates has been presented in exactly solvable case. The obtained solution in asymptotic case leads to Polya form for size distributions with shapes controlled by dimerization parameter b. This obtained form is rather universal and provides the description for purely decreasing distributions when $b < 1$ as well as for monomodal ones with well defined peak when $b > 1$.

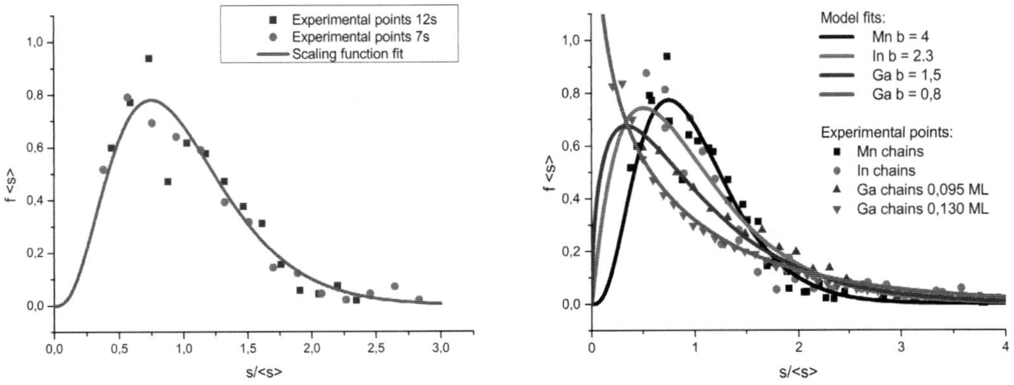

Figure 3. Fits of experimental data for different systems of metallic rows on $Si(100) - 2 \times 1$ by presented model (Eq.(20)): (a) - Scaled size distributions of Mn chains reported in [11], (b) - Comparison of scaled length distribution in different processes for In [3], Ga [7] and Mn [11] chains.

This property became useful in comparison of theoretical dependences and the distributions measured in experiments. The theory demonstrated agreement with experimentally observed time evolution of Mn chains length distribution and scaling for the systems of In, Ga and Mn lines on $Si(100) - 2 \times 1$ substrates.

This work was supported by FP7 projects NANOEMBRACE (Grant Agreement 316751) and grants of the Russian Foundation for Basic Research.

References

[1] M. M. R. Evans and J. Nogami 1999 *Phys. Rev.* B **59 (11)** 7644
[2] H. Itoh, J. Itoh, A. Schmid, and T. Ichinokawa 1993 *Phys. Rev.* B **48 (19)** 14663
[3] J. Javorsky, M. Setvin, I. Ostadal, P. Sobotik, M. Kortla 2009 *Phys. Rev.* B **79** 165424
[4] H. Brune, 1998 *Surf. Sci. Rep.* **31**, 121 .
[5] M. A. Albao et al. (2013) *Surface Science* **617** 7380
[6] P. Kocan, L. Jurczyszyn, P. Sobotik, and I. Ostadal, 2008 *Phys. Rev.* B **77**, 113301
[7] M. A. Albao, et al., 2005 *Phys. Rev.* B **72** 035426
[8] K. R. Simov, C. A. Nolph, and P. Reinke 2012 *J. Phys. Chem. C* **116** 16701678
[9] T. Vicsek and F. Family, 1984 *Phys. Rev. Lett.* **52**, 1669
[10] M. Einax, W. Dieterich, and P. Maass 2013 *Rev. Mod. Phys.* **85**, 921
[11] H. Liu, P. Reinke 2008 *Surface Science* **602** 986992
[12] L. Jure, L. Magaud, J.-M. Gomez-Rodriguez, P. Mallet and J.-Y. Veuillen 2000 *Phys. Rev.* B **61 (24)**, 16903
[13] P. Sobotk, M. Setvn, P. Zimmermann, P. Kocan, and I. Ostadal 2013 *Phys. Rev.* B **88** 205406
[14] M.C. Bartelt and J.W. Evans, 1992 *Phys. Rev. B* **46**, 12675
[15] D.D. Vvedensky, 2000 *Phys. Rev. B* **62**, 15435 .
[16] J.W. Evans, P.A. Thiel, and M.C. Bartelt, 2006 *Surf. Sci. Rep.* **61**, 1 .
[17] V.G. Dubrovskii and N.V. Sibirev, 2014, *Phys. Rev. B* **87**, 195426
[18] V.G. Dubrovskii and M.V. Nazarenko. 2010, *J. Chem. Phys.*, **132**, 114507
[19] M. Z. Hossain, Y. Yamashita, K. Mukai, and J. Yoshinobu 2003 *Phys. Rev.* B **67** 153307
[20] S. Okano and A. Oshiyama 2004 *Surf. Sci.* **554**, 272
[21] V.G. Dubrovskii 2014 *Nucleation Theory and Growth of Nanostructures* Springer-Verlag, Berlin Heidelberg
[22] Y.W. Mo, J. Kleiner, M.B. Webb, and M.G. Lagally, 1991 *Phys. Rev. Lett.* **66**, 1998

Peculiarities of optical and ODMR Spectra of Nitrogen-Vacancy Color Centers in Diamond Crystals

A K Dmitriev and A K Vershovskii

Ioffe Physical Technical Institute, Russian Academy of Sciences, St. Petersburg, 194021 Russia

E-mail: alexdmk777@gmail.com.

Abstract. The optical and ODMR spectra of nitrogen vacancy (NV) centers in diamond were investigated as applied to the development of micro-scale diamond magnetometer sensor. It is shown that the NV^0 center (unlike the NV^- center) has an absorption line in the UV range. The inversion of the zero phonon line of the NV^- center was observed and explained. ODMR signals in high magnetic fields (as compared with inner fields in the crystal) as well as in weak fields also were investigated. We propose the solution to the problem of the magnetometer sensitivity reduction at low magnetic field, that is of interest for quantum magnetometry applications.

1. Introduction

NV centers in diamond crystals are of enormous interest for modern quantum optics, informatics, cryptography, and magnetometry. First of all, it concerns the negatively charged NV^- centers by virtue of the possibility of excitation and observation of ODMR (optically detected magnetic resonance) in them [1]. In this work we took the first step to the application of these centers to the task of precision measurement of sub-millimeter-scale magnetic fields by investigating their photoluminescence and ODMR spectra.

2. Experimental investigation of the luminescence spectra of NV centers in diamond

Photoluminescence spectra were recorded for two specimens of diamond, which were provided by the Lebedev Physical Institute of the Russian Academy of Sciences. Specimens were prepared at that institute as follows: synthetic diamonds (from Element Six; type SDB 1085 60/70; crystal volume, ~0.01 mm3) were irradiated by electron beams with intensities of 5×10^{18} electrons/cm^2 (specimen 1) and $\sim 10^{19}$ electrons/cm^2 (specimen 2) and, after that, were annealed for 2 h in an argon atmosphere at a temperature of 800° C. As a pump source, we used the second harmonic of a tunable MaiTai Ti-Sa femtosecond laser radiation (345-520 nm), which was generated by an optical-frequency doubler.

The photoluminescence spectra of specimens 1 and 2 have qualitative similarity. First, their integral photoluminescence intensities do not tend to zero with decreasing pumping wavelength λ_P (as could be expected if one assumes that each of the NV centers has one absorption band in the visible range of the spectrum), but, rather, at $\lambda_P < 390$ nm, they begin to grow. At $\lambda_P > 490$ nm, a "classical" pattern is observed (Fig. 1): against the background of a broad photoluminescence band, narrow peaks of zero-phonon lines are seen, which indicates that the specimens contain electrically neutral NV^0 and negatively charged NV^- centers (the wavelength of the zero-phonon line is 637 nm for NV^- centers and 575 nm for NV^0 centers). However, as λ_p is decreased, the amplitude of the zero-phonon line of NV^- centers decreases and then the phonon line peak changes into a dip, the amplitude of which increases as λ_p is further decreased.

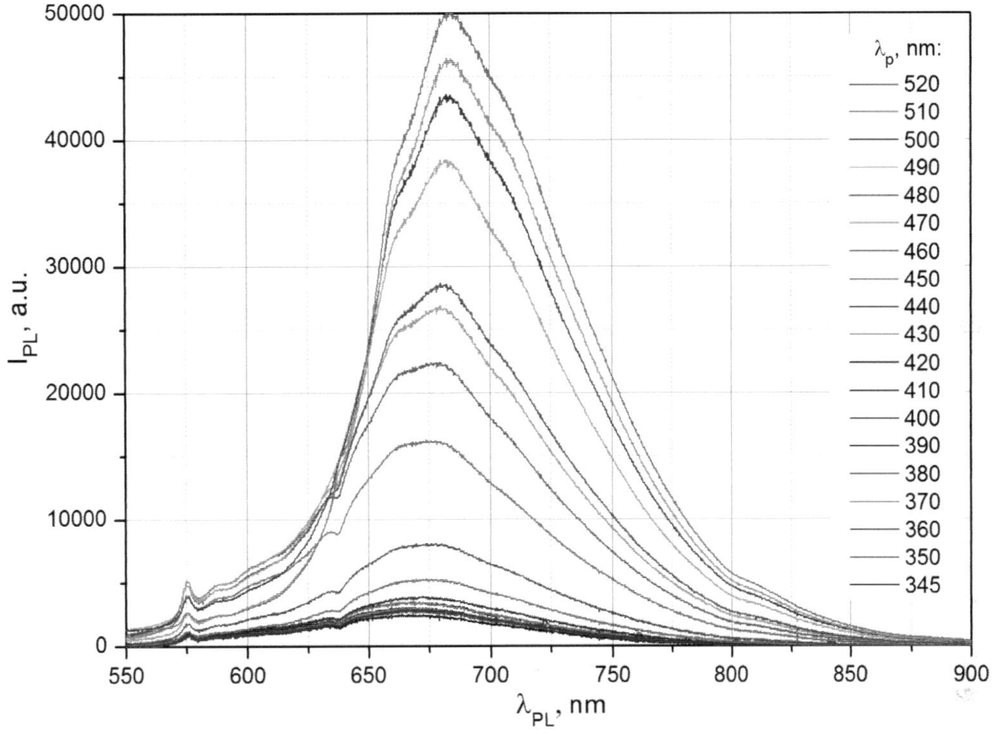

Figure 1. Photoluminescence (Pl) spectra of a diamond crystal with a high content of NV centers (specimen 1). The wavelength of the zero-phonon line is 637 nm for NV^- centers and 575 nm for NV^0 centers. The pumping wavelength λp varies from 520 nm to 345 nm.

We have observed zero-phonon line atop of rather wide ODMR pedestal. The dependence of thus obtained zero-phonon line amplitudes on the pumping wavelength allowed us to conclude that NV^0 centers have an additional absorption line in the UV range [2], which as far as we know, was not observed before. The existence of UV-absorption line was evidenced both by an increase of the NV^0 zero-phonon line amplitude as well as by an increase in the photoluminescence integral intensity, as the pumping wavelength was decreased from 390 to 345 nm. In contrast to NV^0 centers, NV^- centers do not reveal absorption lines in the near UV range. Moreover, as the pumping wavelength is decreased to 495 nm and below, an inversion of the zero phonon line in the luminescence at a wavelength of 637 nm is observed (Fig. 2).

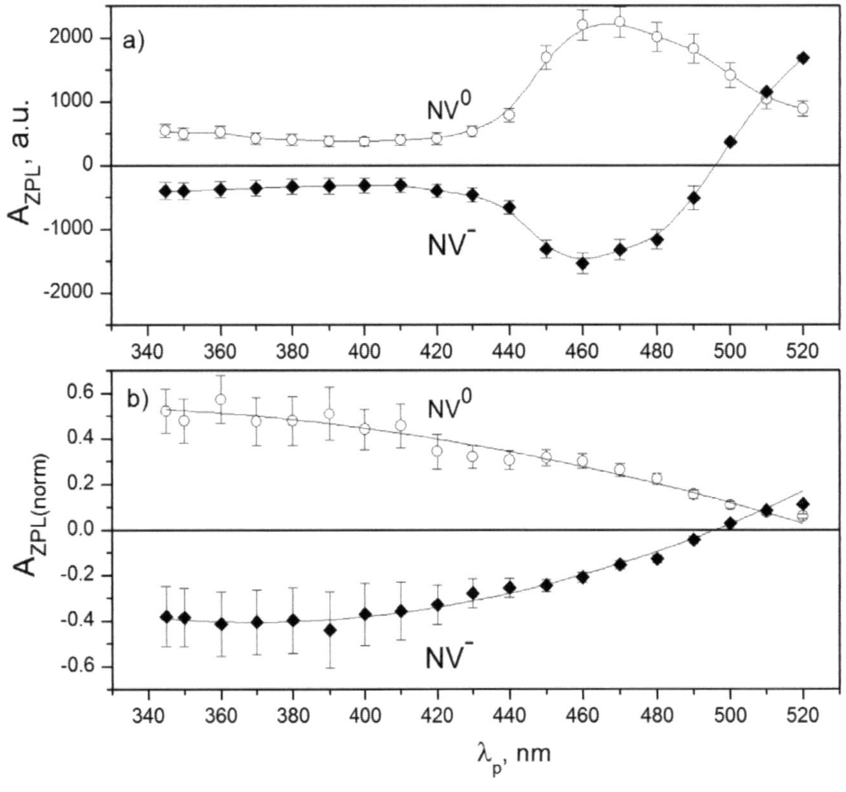

Figure 2. Dependences of the amplitudes of the zero-phonon line (ZPL) of NV^0 and NV^- centers (specimen 1) on the pumping wavelength λp: (a) without normalization and (b) normalization to the average amplitude of the photoluminescence spectrum of the specimen in the spectral interval 550 ÷ 900 nm. Each point corresponds to the line on Fig.1.

This effect can be explained as follows: since typical spectra of NV^- centers are red shifted by 62 nm with respect to the spectra of NV^0 centers [3], the luminescence band of NV^0 centers overlaps both the zero-phonon line of NV^- centers and, partially, the broad phonon absorption band of NV^- centers. Therefore, under the conditions when NV^0 centers absorb the UV pumping radiation and reemit it in the broad band in the range $\lambda \geq 575$ nm, NV^- centers efficiently absorb the light reemitted by NV^0 centers. In particular, this absorption leads to the formation of a narrow dip at the wavelength of the zero-phonon line of NV^- centers (637 nm) in the spectrum of the total photoluminescence of centers of the two types.

3. Observation of ODMR signals in NV^- centers in diamond

ODMR signals were investigated both in high field (as compared with inner fields in the crystal, i.e. above 200 *mkT*), as well as in weak magnetic fields. ODMR resonances were excited with slow

100% amplitude modulation of the microwave field; it allowed us to avoid broadening of the ODMR signal by the modulation frequency.

Frequency splitting of magneto-dependent levels of the ground state of NV^- centers against the magnetic field was measured in the fields (0 ÷ 1) mT and then approximated by formula [3]:

$$2\pi \cdot f_\pm = [E^2 + (g_s\mu_B B_z)^2]^{1/2} = [E^2 + (\gamma \cdot B_z)^2]^{1/2}, \quad (1)$$

where E – parameter of transverse zero-field splitting, $\gamma = g_s\mu_B$ – gyromagnetic ratio. For the observable sample $E = 2\pi \cdot (3.2 \pm 0.1) MHz$, and gyromagnetic ratio $\gamma = 2\pi \cdot (1.6 \pm 0.1) 10^{10}\ Hz/T$ (Fig.3). Present value is an effective gyromagnetic ratio averaged over all orientations of the crystal. Therefore, according to (1), in typical terrestrial field $B = 5 \cdot 10^{-5} T$ the maximal achievable sensitivity of a magnetometer based on this sample falls not more than four times compared to the sensitivity achievable in strong fields.

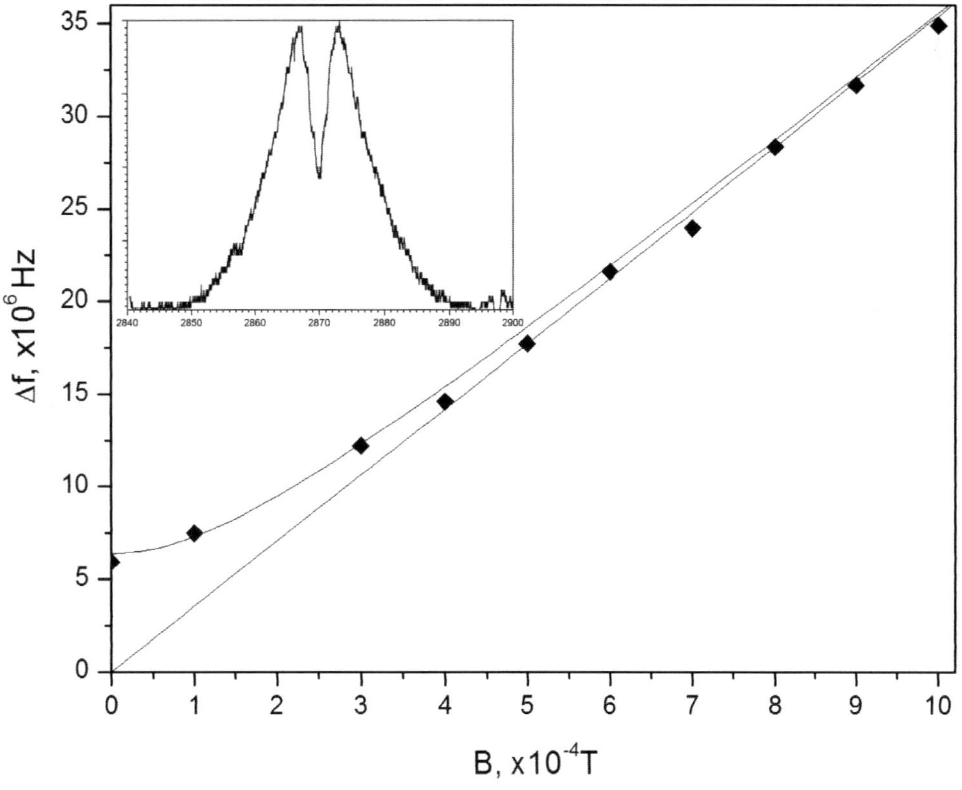

Figure 3. Frequency splitting of magneto-dependent levels of the ground state of NV^- centers against the magnetic field. Inset: zero-field ODMR signal.

Our results prove the feasibility of the sub-millimeter range precision magnetometry of weak magnetic fields ($B \geq 20\ \mu T$) using ODMR in NV^- centers. For this purpose a differential balanced scheme of interrogation of two magnetic microwave transitions ($m_F = 0 \leftrightarrow m_F = -1$ and $m_F = 0 \leftrightarrow$

$m_F = +1$) can be used, which would largely eliminate the temperature dependence of the magnetometer readings. At the same time, the sum of these magnetic transitions frequencies does not carry an information about the magnetic field. Therefore, it can be used for the temperature measurement instead, allowing the final adjustment of the magnetometer readings.

4. Acknowledgments

Authors are grateful to O.S. Vasyutinskii (Laser Center of Ioffe Physical Technical Institute) for support, and to V.L. Velichanskii and S.A. Zibrov (the Lebedev Physical Institute) for providing specimens and fruitful discussions. This work was supported by the Russian Foundation for Basic Research, project no. 13 02 00589.

References

[1] J. M. Taylor, P. Cappellaro, L. Childress, L. Jiang, D. Budker, P. R. Hemmer, A. Yacoby, R. Walsworth, and M. D. Lukin, *Nat. Phys* **4**, 810 (2008).

[2] V. A. Astapenko, A. S. Baturin, V. L. Velichanskii, S. A. Zibrov, E. V. Korostylev, and L. P. Sukhanov, *Nanostrukt. Mat. Fiz. Model.* **8** (1), 17 (2013).

[3] V. M. Acosta, D. Budker, P. R. Hemmer, J. R. Maze, R. L. Walsworth, *Optical Magnetometry*, Ed. by D. Budker and D. F. Kimball (Cambridge Univ. Press, Cambridge, 2013).

Investigation of lead-free thin films based on barium titanate for electrocaloric devices

E V Abrashova[1], A K Fominykh[2]

[1] Postgraduate , St. Petersburg State Electrotechnical University "LETI", Pr. Popova 5, St. Petersburg, 197376 Russia

[2] Student, St. Petersburg State Electrotechnical University "LETI", Pr. Popova 5, St. Petersburg, 197376 Russia

[1] E-mail: katabr@mail.ru

[2] E-mail: anatolyfom@gmail.com

Abstract. Lead-free thin films were synthesized by sol-gel for possible use in solid-state coolers. Surface morphology of the layers was obtained by atomic force microscopy (AFM). Electrophysical properties were investigated by impedance spectroscopy.

1. Introduction

Due to the rapidly deteriorating global environment problem of eliminating the use of refrigerants is relevant world over. Devices based on solid-state cooling elements are alternative solution of using compression and evaporative coolers. Using such elements is especially important for miniature devices. Solid-state coolers, which are based on thermoelectric effects, in particular, the electrocaloric are the most perspective for diverting large heat flux from small surfaces of modern devices. Currently, elements which are controlled by an electric field are developed and investigated actively [1].

Electrocaloric effect is manifested in increasing the temperature of the substance when creating therein an electric field E and reduction of the temperature when switching off the field under adiabatic conditions. This effect is strongly pronounced in ferroelectric materials due to the possibility of domain reorientation. Electrical domains have a random direction and the polarization vector of the total sample P is equal to zero when no electric field. If you turn on a constant electric field, the dielectric is polarized, and the dipole moments of molecules or domains are reoriented. Main line of domains appears, entropy decreases and the temperature of the sample increases. When you turn off the external field, the dipoles get random orientation again. Due to this internal energy is reduced and the temperature decrease under adiabatic conditions [2]. It is obvious that high electric field determines maximum effect. These conditions are achieved in thin films easier.

Found that the maximum electrocaloric effect in ferroelectrics observed near their phase transition. Ferroelectric-paraelectric transition does not occur abruptly, but in a certain temperature range to not single crystals. In this case percolation cluster is formed for the ferroelectric phase. Ferroelectric effect is an especially strong upon reaching the percolation threshold [3]. Thus, a material having a phase transition temperature close to the operating temperature does not require special heating of the active substance.

Currently ferroelectrics with a perovskite structure with a lead content [4-6], known for its toxic effects, are the most commonly used electrocaloric materials. The authors [5] reported about achieving of the giant electrocaloric effect $\Delta T = 12$ K on the films $PbZrO_{0,95}Ti_{0,05}O_3$ thickness of 350 nm near the phase transition temperature of 242 ° C, as well as in films $0.9*PbMg_{1/3}Nb_{2/3}O_3-0.1*PbTiO_3$ temperature change is $\Delta T = 5$ K at a voltage of 25 V at 260 nm thick films near the phase transition temperature of 60 ° C [6]. Therefore, the problem of finding an alternative lead-free materials having the electrocaloric properties is very actual.

Among the promising materials for considered purposes can be distinguished, for example, NH_4HSO_4 [7]. Another promising material is known ferroelectric material - barium titanate. Currently working on the development of optimal design of cooling elements on the basis of electrocaloric effect are conducted actively [8,9].

2. Purpose

The main purpose of this work was obtaining thin lead-free electrocaloric films. For this aim, the composition of compounds based on barium titanate $(Ba_x Ca_{1-x})TiO_3$ и $Ba(Sn_y Ti_{1-y})O_3$ in different ratios, known for its ferroelectric properties was selected. Analysis of T-X- diagrams of ternary systems determined the choice of material [10]. Sol-gel method to obtain the investigated structures was chosen because it is relatively economical, affordable for a wide selection of components, precision calculation of the composition, as well as it allows for obtaining enough homogeneous thin films. This method is widely used to produce gas-sensitive layers [11,12], functional transparent, protective and adsorption coatings [13,14], etc.

3. Experiment

As a precursor to obtain the desired material were used barium acetate, calcium acetate, tin acetate and titanium isopropoxide. The dipole structure was formed by using acetyl acetone molecules in combination with titanium isopropoxide. Ethylene glycol and 2-methoxyethanol were used to obtain a specific viscosity and stability of the solution. Sol matured at room temperature. Sol solution with a different maturation time (from 2 to 300 hours) has been applied by the dispenser on glass substrates and annealed after centrifugation at 600 °C from 30 minutes to 1 hour.

Surface morphology was investigated by atomic force microscopy (AFM) in tapping mode (NTEGRA, NT-MDT, Zelenograd, Russia). Electrophysical properties were obtained using impedancemeter "Z-500 P" (ELINS) in the frequency range from 1 Hz to 500 kHz.

4. Results

Homogeneous by visual inspection, transparent, colorless thin films were obtained in the experiment. Analysis of the morphology of the surface with AFM showed that the film surface formed objects with elliptical shape. Their average diameter is about of 120-150 nm (Figure 1). The sample had a fractal structure with a fractal dimension of 2.50, calculated by triangulation with program Gwyddion. Supposedly, such a structure was formed on the basis polymerization centers of titanium compounds, in the initial stage of forming the sol solution. Perhaps this process step leads to the production dipole structure of the sample, which will determine the electrical properties of thin films.

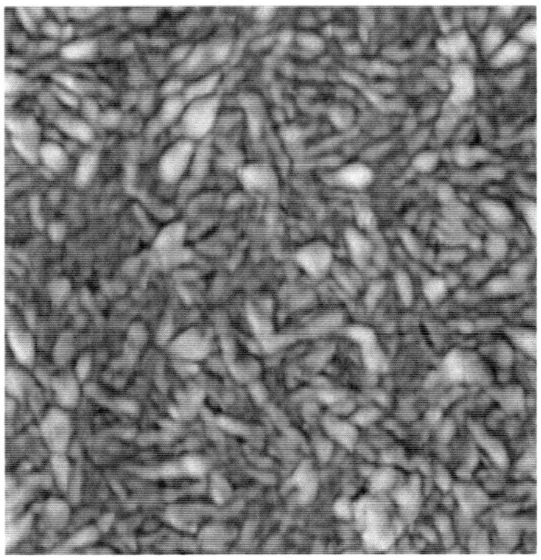

Fig. 1. The image of surface of the structure obtained by using the AFM, 2 × 2 mkm.

Study electrical properties of the films of the composition $(Ba_xCa_{1-x})TiO_3$ and $Ba(Sn_yTi_{1-y})O_3$ confirmed the existence of dipolar relaxation processes (Fig. 2). They are manifested on the graph of the complex part of impedance of the structure Im from its real part Re, at high frequencies (section II), which can be approximated by a circle segment, unlike section I, can be approximated by a straight line.

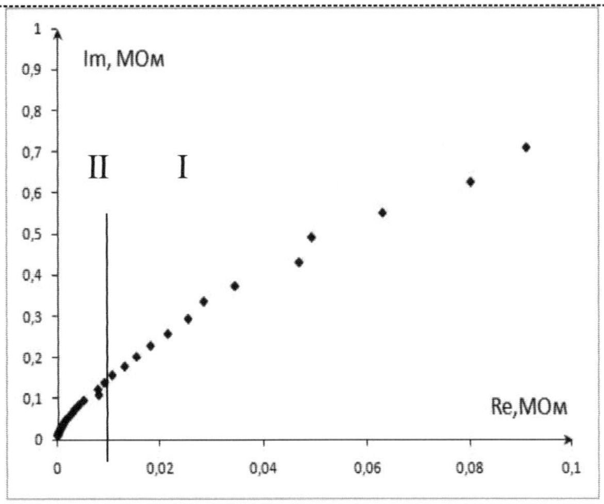

Fig. 2. Dependence of the complex part of impedance of the structure Im from its real part Re at T = 20 ° C.

The changing in the character of these curves was revealed at different temperatures. At temperatures down to 220 ° C character of the curve determined by the influence constant phase element (CPE), while at temperatures above 250 ° C distributed capacitance affects on the dependence stronger. Moreover phase transition point of the compound in question lies in the same temperature range.

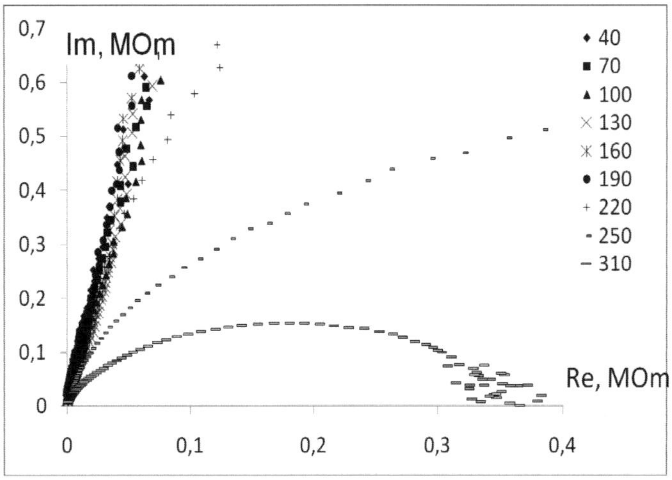

Fig. 3. Dependence of the complex part of impedance of the structure Im from its real part Re at different temperatures.

For samples with different relative content of Ba and Ca was revealed (Fig. 4) that at higher relative content of Ba (composition number 2) the nature of the curves changes at lower temperatures. Probably because that barium compounds are more active than calcium compounds. Thus, addition of calcium compounds improves the thermal stability properties of the films, but based on its physical characteristics, increases the Curie temperature and reduces the dielectric constant of the composition. Knowing that the influence of CPE is defined by ions and the diffusion coefficient, it is logical to assume that the calcium compounds bind free ions in the material.

5. Conclusion.

The study revealed that the morphology of investigated films had a fractal structure consisting of objects in the form of an ellipse having a linear dimension of about 120-150nm.

Availability dipole relaxation processes required to use this material in electrocaloric devices was discovered during the investigation of the electrophysical properties of the material by impedance spectroscopy.

It was found that the introduction of calcium compounds stabilizes the electrical properties of the investigated compositions with increasing temperature

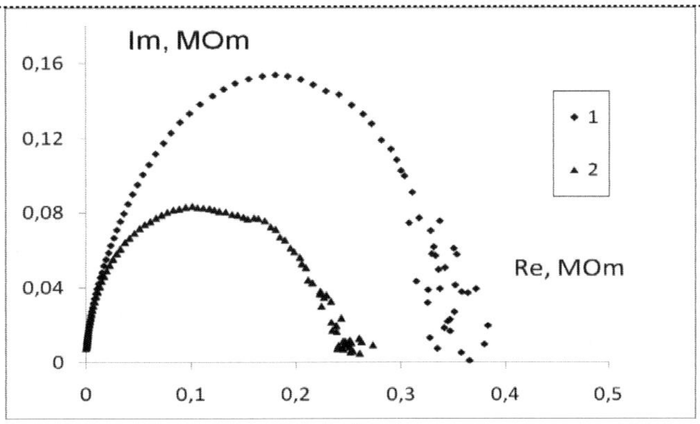

a)

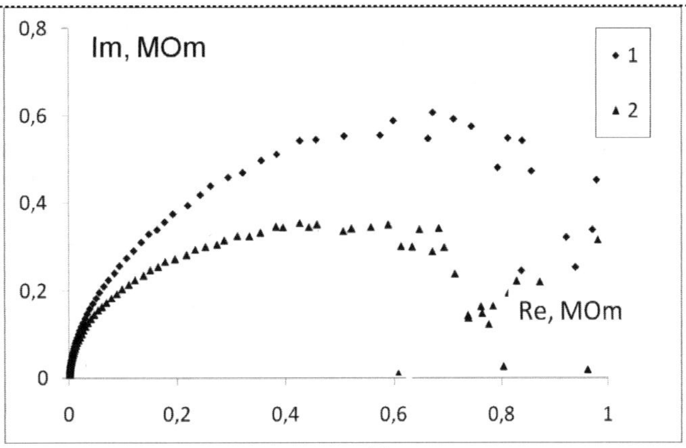

b)

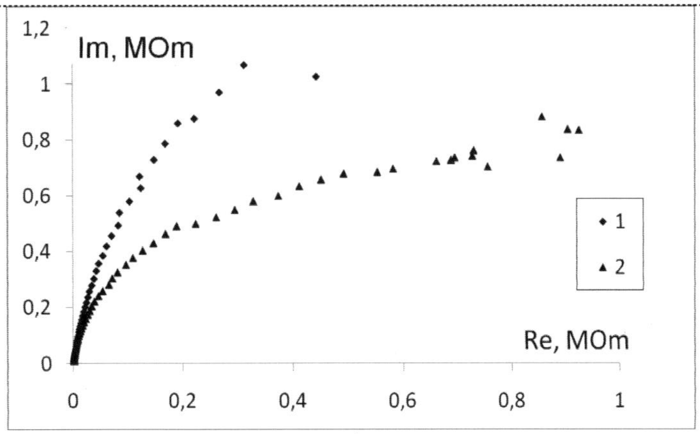

c)

Fig. 4. Dependence of the complex part of impedance of the structure Im from its real part Re samples composition $(Ba_x\ Ca_{1-x})TiO_3$ and $Ba(Sn_y\ Ti_{1-y})O_3$ with different relative proportions Ba and Ca (1 : x = 0,2 ; 2 : x = 0,3) at different temperatures: a) T = 310°C, b) T = 250°C, c) T = 220 °C

References

[1] A. B. Ustinov, P. I. Kolkov, A. A. Nikitin, B. A. Kalinikos, Y. K. Fetisov, G. Srinivasan // Ferrite–Ferroelectric Phase Shifters Controlled by Electric and Magnetic Fields / Technical Physics, 2011, Vol. 56, No. 6, pp. 821–825

[2] L.P. Bulat// Solid-state cooling systems / Thermoelectricity, №3, 2007, p. 15-22

[3] A.I.Maximov, V.A.Moshnikov, Yu.M.Tairov, O.A.Shilova/ Fundamentals of sol-gel technology of nanocomposites 2-nd ed. / St Petersburg./ Elmore, 2008. 255 p. in rus

[4] Biaolin Peng, Huiqing Fan, Qi Zhang //A Giant Electrocaloric Effect in Nanoscale Antiferroelectric and Ferroelectric Phases Coexisting in a Relaxor $Pb_{0,8}Ba_{0,2}ZrO_3$ Thin Films at Room temperature/ Adv. Funct. Mater./2013/p.1-6

[5] Mischenko A. S., Zhang Q., Scott J. F., Whatmore R. W., Mathur N. D. // Giant Electrocaloric Effect in Thin-Film $PbZrO_{0,95}Ti_{0,05}O_3$ / Science 3, March , V. 311, №. 5765, 2006 , P. 1270-1271.

[6] Mischenko A. S., Zhang Q., Scott J. F., Whatmore R. W., Mathur N. D. Giant electrocaloric effect in the thin film relaxor ferroelectric 0.9PbMg1/3Nb2/3O3–0.1PbTiO3 near room temperature // Appl. Phys. Lett., 89, 2006, P.242912.

[7] I.N.Flerov, E.A.Mihaleva// Electrocaloric effect and anomalous conductivity conductivity of ferroelectric NH_4HSO_4/ Solid State Physics, v.50, N 3, 2008, p. 461-466

[8] A.V. Es'kov, S.F. Karmanenko, O.V. Pahomov , A.S.Starkov // Modeling solid cooler with the electrocaloric elements / Solid State Physics, 2009, v 51, N 8, p. 1483-1486

[9] A.A. Semenov, O.V. Pahomov, P.Yu. Belyavskiy, A.V. Es'kov, S.F. Karmanenko, A.A. Nikitin // Investigation of electrocaloric response dynamics in ferroelectrics using ferromagnetic resonator / Journal of Technical Physics, 2012, v 82, N 1, p. 59-62

[10] S.S.Gorelik, M.Ya.Dashevskaya // Materials science of semicondactors and dielectrics /Moscow, 2003, 480p.

[11] Irina E. Gracheva, Grazyna Olchowik, Kamil G. Gareev, Vyatcheslav A. Moshnikov, Vladimir V. Kuznetsova, Jan M. Olchowikc // Investigations of nanocomposite magnetic materials based on the oxides of iron, nickel, cobalt and silicon dioxide / Journal of Physics and Chemistry of Solids, 74 , 2013, p 656–663

[12] V.V.Tomaev, L.N.Gar'Kin, V.P.Miroshkin, V.A. Moshnikov // Investigation into the gas sensitivity in nanostructured films based on tin dioxide by impedance spectroscopy / Glass Physics and Chemistry. 2005. T. 31. № 2. C. 246-251

[13] I E.Gracheva, V.A. Moshnikov, E.V.Maraeva, et al // Nanostructured materials obtained under conditions of hierarchical self-assembly and modified by derivative forms of fullerenes / Journal of non-crystalline solids Volume: 358 Issue: 2 Pages: 433-439 DOI: 10.1016

[14] E.V.Abrashova, I.E.Gracheva, V.A.Moshnikov // Functional nanomaterials rased on metal oxides with hiererhical structure / Journal of Physics: Conference Series. 2013. T. 461. C. 012019

SPbOPEN2014 IOP Publishing
Journal of Physics: Conference Series **541** (2014) 012092 doi:10.1088/1742-6596/541/1/012092

Chain of point-like potentials in $\mathbb{R}^3$ and infiniteness of the number of bound states

A.A.Boitsev, I.Yu.Popov and O.V.Sokolov

St. Petersburg National Research University of Information Technologies, Mechanics and Optics, 49 Kronverkskiy, St. Petersburg, 197101, Russia

E-mail: `boitsevanton@gmail.com, popov1955@gmail.com`

Abstract. Infinite chain of point-like potentials having the Hamiltonian with infinite number of eigenvalues below the continuous spectrum is constructed. The background of the model is the theory of self-adjoint extensions of symmetric operators in the Hilbert space. The analogous example of the Hamiltonian is obtained for the system of three-dimensional waveguides coupled through point-like windows.

1. Introduction

The problem of existence of infinite number of eigenvalues of the Schrödinger operator. It appears in different physical situations. Particularly, in three-particle problem it is known as the Efimov effect [1], [2], [3]. Strict mathematical justification for the existence of infinitely many bound states below the continuous spectrum for short-range perturbations of a much larger class of spinorbit Hamiltonians, which includes, in particular, the above Rashba Hamiltonian was obtained in [4]. The authors uses min-max principle in the spirit of [5]. In the present paper we construct an operator having infinitely many different eigenvalues below the continuous spectrum. The result can be interested from the point of view of functional model due to close relation between the properties of the characteristic functions and the operator spectrum [6] More precisely, we construct a chain of zero-range potentials with specific intensities which gives us the operator in question. This type of potential has useful advantage- it allows us to construct the spectral equation in an explicit form, correspondingly, the spectral analysis of the operator is simpler. Point-like interaction is introduced by a conventional way [8], [7]. The construction is based on the theory of self-adjoint extensions of symmetric operators. It is widely used for the description of various quantum systems.

Let us describe briefly the model operator construction. Consider a chain of n delta-potentials in $\mathbb{R}^3$. The positions of the potentials are x_m. First, we restrict the Laplace operator onto the set of smooth functions vanishing at the points x_m. The obtained operator is symmetric non-self-adjoint. To construct a self-adjoint extension, it is more convenient to deal with the corresponding restriction of the adjoint operator instead of the extension of the symmetric operator. There are several ways of extensions descriptions, e.g., boundary triplets method ([9], von Neumann formulas ([11]), Krein resolvent formula ([10]). We will use here a variant of the second approach which allows one in the case of semi-boundedness of the Hamiltonian to present an element from the domain of the adjoint operator in the following form :

$$\psi(x) = \psi_0(x) + \sum_{m=1}^{n} \tilde{a}_m \cdot \frac{e^{-ik_0|x-x_m|}}{4\pi\,|x-x_m|}, \tag{1}$$

where ψ_0 belongs to the domain of the Friedrichs extension of the initial symmetric operator, $\tilde{a}_m$ is some constant, $k_0 = \sqrt{\lambda_0}$, λ_0 is a regular value of the spectral parameter (we choose real negative value of λ_0, $\Im k_0 > 0$). Let $\psi_0(x_m) = b_m, m = 1, 2...n$. We will not describe all possible self-adjoint extensions, but will consider a particular extension determined by the following correlation between $n-$vectors:

$$b_i = \alpha_i \cdot \tilde{a}_i, \quad i = 1, ..., n. \tag{2}$$

Here the matrix parameterizing the extension is chosen in the diagonal form, $\alpha_j, j = 1, ...n$, are some real parameters. The physical meaning of the parameters are the potential intensities.

2. Main result

The main result of the paper is the following theorem.

Theorem. The Laplace operator in $L_2(\mathbb{R}^3)$ perturbed by infinite chain of zero-range potentials posed at the points $(0,0,0), ...x_m, ...,$

$$x_1 = (0,0,0) \qquad x_m = x_{m-1} + (m^m \cdot 2^{2m^2}, 0, 0), m = 2, 3, ...,$$

of intensities, correspondingly, α_m,

$$\alpha_m = \frac{1}{2^m} + \frac{1}{4\pi} - \frac{ik_0}{4\pi},$$

has infinite number of eigenvalues below the continuous spectrum.

Note that the analogous consideration can be performed for another system. Consider two half spaces of $\mathbb{R}^3$ and let us deal with the Laplace operator in $L_2(\mathbb{R}^3_-) \oplus L_2(\mathbb{R}^3_+)$ with the Neumann boundary condition at the plane. Let the set of points $(0,0,0), ...(x_m, 0, 0), ...,$ belongs to the plane. Let us restrict the operator on the set of smooth functions vanishing near the chosen points. It is symmetric operator with infinite deficiency indices. Elements from the domain of the adjoint operators have the form:

$$\begin{pmatrix} \psi_+(x) \\ \psi_-(x) \end{pmatrix} = \begin{pmatrix} \psi_{0,+}(x) \\ \psi_{0,-}(x) \end{pmatrix} + \begin{pmatrix} \sum_{m=1}^{n} \tilde{a}_{m+} \cdot \frac{e^{-ik_0|x-x_m|}}{2\pi|x-x_m|} \\ \sum_{m=1}^{n} \tilde{a}_{m-} \cdot \frac{e^{-ik_0|x-x_m|}}{2\pi|x-x_m|} \end{pmatrix}.$$

To construct a selfadjoint extension it is necessary to determine a linear relation between the coefficients $a_{m\pm}$ and $b_{m\pm}$ (analogously to the procedure for point-like potential described above). This procedure gives us a model of point-like windows in the plane (see [12], [13]). The type of the chosen self-adjoint extension is related with the type of physical coupling through the windows. It is easy to see that in this case one can obtain the analogous theorem. Moreover, one can obtain similar theorem for the case when the half spaces are replaced by straight waveguides coupled through a system of point-like windows [14]. In this situation it is necessary to replace

$$\frac{e^{-ik_0|x-x_m|}}{2\pi\,|x-x_m|}$$

by the Green functions for the waveguides. The theorem is obtained due to the coincidence of the singularities of the Green functions for the waveguides and for the free space. We will not formulate the theorem for this case because the changes in respect to the above written theorem is rather evident.

3. Main steps of the proof

Our goal is to construct such infinite chain that gives us infinite number of eigenvalues. To ensure it we consider a limiting procedure ($n \to \infty$) and make a special choice of potentials positions x_m and potentials intensities α_m. More precisely, we construct the chain step by step. At each stage we add one center to the chain and look for the behavior of the eigenvalues. The addition of the zero-range potential leads to the appearance of new eigenvalue and to the shift of the eigenvalues obtained at the previous stage. We choose the position and the intensity of the added zero-range potential in a proper way to avoid gluing of the eigenvalues. The resulting Hamiltonian with infinite chain of zero-range potentials is obtained as a strong resolvent limit of the constructed sequence of operators.

Let us choose the following positions of the centers:

$$x_1 = (0,0,0) \qquad x_m = x_{m-1} + (m^m \cdot 2^{2m^2}, 0, 0), m = 2, 3, ...,$$

In accordance with the description of the operator domain given above we seek the eigenfunction corresponding to the eigenvalue λ in the form:

$$\sum_{m=1}^{n} a_m \cdot \frac{e^{-ik|x-x_m|}}{4\pi\,|x-x_m|}, \tag{3}$$

$k = \sqrt{\lambda}$. Let us determine parameters $\tilde{a}_m, b_m$ from the definition (1). Due to the fact that the singularities of $\frac{e^{-ik|x-x_m|}}{4\pi\,|x-x_m|}$ and $\frac{e^{-ik_0|x-x_m|}}{4\pi\,|x-x_m|}$ are identical, one gets $\tilde{a}_m = a_m$. To get the coefficient b_m one should subtract the singular term $\frac{e^{-ik_0|x-x_m|}}{4\pi\,|x-x_m|}$.

Let us introduce the notations

$$t = \frac{ik}{4\pi} \qquad t \in \mathbb{R},\ t > 0, \qquad \Delta e_{i,j} = \frac{e^{-4\pi t|x_i - x_j|}}{4\pi\,|x_i - x_j|} \tag{4}$$

Due to (4) the relations for b_i are as follows

$$b_i = \left(t + \frac{1}{4\pi}\right) a_i + \sum_{j=1, j\neq i}^{n} \Delta e_{i,j} \cdot a_j. \tag{5}$$

In accordance with our choice of the extension (2) one has

$$b_i = \alpha_i \cdot a_i, \quad i = 1, ..., n \tag{6}$$

Let us choose the extension parameters (potential intensities α_j) by a special way

$$\alpha_j = \frac{1}{2^j} + \frac{1}{4\pi} - \frac{ik_0}{4\pi}$$

Subtracting (6) from (5), one obtains that the system of equations has non-trivial solutions if and only if

$$\widetilde{f}_n(t) = \begin{vmatrix} t-\beta_1 & \Delta e_{1,2} & \Delta e_{1,3} & \dots & \Delta e_{1,n} \\ \Delta e_{1,2} & t-\beta_2 & \Delta e_{2,3} & \dots & \Delta e_{2,n} \\ \Delta e_{1,3} & \Delta e_{2,3} & t-\beta_3 & \dots & \Delta e_{3,n} \\ \dots & & & & \\ \Delta e_{1,n} & \Delta e_{2,n} & \Delta e_{3,n} & \dots & t-\beta_n \end{vmatrix} = 0 \tag{7}$$

This equation (7) doesn't have more than n roots. Our goal is to show that the equations has exactly n roots.

Below we assume that $t \in (0;1)$.

Expanding the determinant $\widetilde{f}_n(t)$ along the last column, one gets

$$\widetilde{f}_n(t) = \widetilde{f}_{n-1}(t) \cdot (t - \beta_n) + \epsilon_n(t). \tag{8}$$

Making algebraic transformations and using the Hadamard inequality, we come to the following estimation

$$|\epsilon_n(t)| < \frac{1}{2^{2n^2}}. \tag{9}$$

Let

$$d(n) = \frac{1}{2^{2n^2}} \tag{10}$$

Remark. If $\widetilde{f}_n(t)$ has n roots in the case when $\epsilon_n(t) = d(n)$ then it has n roots when $|\epsilon_n(t)| < d(n)$.

Consider the case when $\epsilon_n(t) = d(n)$. One can see that

$$\widetilde{f}_1(t) = t - \frac{1}{2},$$

$$\widetilde{f}_n(t) = \widetilde{f}_{n-1}(t) \cdot \left(t - \frac{1}{2^n}\right) + d(n),$$

Hence,

$$\widetilde{f}_n(t) = (\dots(((t - \tfrac{1}{2})(t - \tfrac{1}{2^2}) + d(2))(t - \tfrac{1}{2^3}) + d(3)) \\ (t - \tfrac{1}{2^4}) + \dots)(t - \tfrac{1}{2^n}) + d(n).$$

Simple transformation gives us

$$\widetilde{f}_n(t) = \prod_{k=1}^{n}\left(t - \frac{1}{2^k}\right) + \sum_{i=2}^{n-1} d(i) \cdot \prod_{k=i+1}^{n}\left(t - \frac{1}{2^k}\right) + d(n). \tag{11}$$

Let us introduce the notation

$$f_n(t) = \prod_{k=1}^{n}\left(t - \frac{1}{2^k}\right), \tag{12}$$

$$D_n(t) = \sum_{i=2}^{n-1} d(i) \cdot \prod_{k=i+1}^{n}\left(t - \frac{1}{2^k}\right) + d(n), \tag{13}$$

correspondingly,

$$\widetilde{f}_n(t) = f_n(t) + D_n(t). \tag{14}$$

The function $f_n(t)$ has n zeroes in $(0;1)$. To show that $\widetilde{f}_n(t)$ also has n zeroes in this interval, we prove that for any interval of the form $\left(\frac{1}{2^m}, \frac{1}{2^{m-1}}\right)$ there exists $t \in \left(\frac{1}{2^m}, \frac{1}{2^{m-1}}\right)$ such that $f_n(t) \cdot \widetilde{f}_n(t) > 0$. Here $m = 2, 3, \dots n$.

Consider the intervals where $f_n(t) > 0$. They are as follows $\left(\frac{1}{2^{2m+1}}; \frac{1}{2^{2m}}\right)$, $m = 1, \dots [n/2] - 1$. It can be shown that $\widetilde{f}_n(t) > 0$ for these intervals.

Consider the intervals where $f_n(t) < 0$. They are $\left(\frac{1}{2^{2m}}; \frac{1}{2^{2m-1}}\right)$, $m = 1..\lfloor n/2 \rfloor$. Fix m, so we have $t \in \left(\frac{1}{2^{2m}}; \frac{1}{2^{2m-1}}\right)$. Let $t_0 = \frac{1}{2^{2m}} + \frac{1}{2^{2m+1}}$ be the middle of the interval. It can be shown that $\widetilde{f}_n(t_0) < 0$.

As a conclusion, we have proved that any interval of the form $\left(\frac{1}{2^m}, \frac{1}{2^{m-1}}\right)$ contains a point such that $f_n(t) \cdot \widetilde{f}_n(t) > 0$, where $m = 2, 3, ...n$. Then, from the fact that a continuous on $[a, b]$ function $f(x)$ with the condition $f(a) \cdot f(b) < 0$ has at least one root on (a, b) we obtain the fact that $\widetilde{f}_n(t)$ has exactly n roots.

4. Final steps

As a result, we come to the conclusion, that the constructed Hamiltonians (with n zero-range potentials) has n different eigenvalues below the continuous spectrum for any fixed n, and the limiting procedure ($n \to \infty$) does not lead to the gluing of the eigenvalues (due to the fact that they are localized in non-intersecting intervals). The required Hamiltonian is obtained as a strong resolvent limit of this operator sequence. Namely, let

$$X = \{x_j \in \mathbb{R}^3, j \in \mathbb{N}\}, X_n = \{x_j \in \mathbb{R}^3, j \in \mathbb{N}, j \le n\},$$

$$\alpha : X \to \mathbb{R}.$$

Let $-\Delta_{X,\alpha}$, $-\Delta_{X_n,\alpha}$, be the Laplace operators with potentials posed at the points from X (correspondingly, X_n) of intensities (extension parameters α_j, see (2)) determined by α (see above). We denote the corresponding resolvent by $R(\lambda)$, $R_n(\lambda)$.

Lemma 2. The resolvent $R(\lambda)$ is a strong limit of the sequence of operators $R_n(\lambda)$:

$$R(\lambda) = s - \lim_{n \to \infty} R_n(\lambda), \qquad \lambda \in \mathbb{C} \setminus \mathbb{R}.$$

Proof. The statement of the lemma follows from the theorem 3.1.1.1 [8] due to the fact that in our case (for our choice of X and α)

$$\inf_{j \neq j', \quad j,j' \in \mathbb{N}} |x_j - x_{j'}| = d > 0$$

and α is bounded.

Hence, the Hamiltonian with the constructed infinite chain has infinite number of different eigenvalues.

Thus, the proof of the theorem is completed.

Acknowledgements

This work was partially financially supported by the Government of Russian Federation (grant 074-U01), by Ministry of Science and Education of the Russian Federation (GOSZADANIE 2014/190, Project 14.Z50.31.0031), by grant of Russian Foundation for Basic Researches and grants of the President of Russia (state contracts 14.124.13.2045-MK and 14.124.13.1493-MK).

References

[1] Efimov V.N. Bound states of three resonances of interacting particles. Nuclear Phys. **12** (5) (1970) 1080-1091.
[2] Yafaev D.R. On the theory of the discrete spectrum of the three-particle Schrodinger operator. Mat. Sborhik.**94** (4) (1974) 567-593.
[3] Tamura H. The Efimov effect of three-body Schrodinger operators. J. Funct. Anal. **95** (2) (1991) 433-459.

[4] Bruning J. , Geyler V., Pankrashkin K. On the number of bound states for weak perturbations of spinorbit Hamiltonians. J. Phys. A: Math. Theor. **40** F113 - F117 (2007).

[5] Yang K., de Llano M. Simple variational proof that any two-dimensional potential well supports at least one bound state. Am. J. Phys. **57**, 85-86 (1989).

[6] Sz-Nagy B., Foias C., Bercovici H., Kerchy L. Harmonic Analysis of Operators on Hilbert Space. Springer, Berlin, 2010.

[7] Pavlov B.S. Extensions theory and explicitly solvable models. Uspekhi Mat.Nauk **42** (6), 99-131 (1987).

[8] Albeverio S., Gesztesy F., Hoegh-Krohn R. and Holden H. with an appendix by Exner P., *Solvable Models in Quantum Mechanics: Second Edition.* (AMS Chelsea Publishing, Providence, R.I. 2005) .

[9] Berndt J., Malamud M. M., Neidhardt H. Scattering matrices and Weyl functions. *Proc. London Math. Soc.* **97** (3), P. 568–598 (2008).

[10] Alonso A., Simon B. The Birman-Krein-Vishik theory of selfadjoint extensions of semi-boundedoperators. J.Oper. Theory. **4**, 251-270 (1980).

[11] Birman M. S., Solomyak M. Z. Spectral Theory of Self-Adjoint Operators in Hilbert Space (D. Reidel Publishing Company, Dordrecht-Boston-Lancaster, 1987).

[12] Popov I.Yu. Extensions theory and localization of resonances for domains of trap type. Math. USSR-Sbornik **71** (1), 209-234 (1992)

[13] Popov I.Yu. The resonator with narrow slit and the model based on the operator extensions theory. J. Math. Phys. **33** (11), 3794-3801 (1992)

[14] Popov I.Yu. Asymptotics of bound states and bands for laterally coupled waveguides and layers. J. Math. Phys. **43** (1), 215-234 (2002).

Optical phenomena associated with surface plasmon polaritons in GaN-based microstructures

D S Mostovoy, G A Melentyev, V A Shalygin, M D Moldavskaya and L E Vorobjev

St. Petersburg State Polytechnical University, 29 Polytechnicheskaya Street, St. Petersburg 195251, Russia

E-mail: denmostovoy@mail.ru

Abstract. Surface plasmon polaritons in *n*-GaN are investigated. Experiments on optical reflection demonstrate the possibility of surface plasmon polariton excitation on the surface of the grating structure using terahertz radiation. Study of terahertz radiation emission under lateral electric field shows that the scattering of hot surface plasmon polaritons on the grating makes a significant contribution to the radiation intensity. In the certain spectral range this emission mechanism prevails.

1. Introduction

One of the most promising materials for terahertz (THz) emitters is gallium nitride. Emission of THz radiation has been observed in various micro- and nanostructures based on GaN. THz radiation emission due to intracenter optical transitions was studied in lightly doped *n*-GaN epitaxial layers [1]. Thermal radiation of the hot electrons in the THz range was investigated in single heterojunction AlGaN/GaN [2]. THz emission due to plasmon excitation was observed in GaN/AlGaN HEMT structures with submicron gate length [3]. It is promising to obtain THz radiation emission by means of surface plasmon polaritons (SPP). In [4] the emission of THz radiation was detected in degenerate *n*-InN epitaxial layers with the random grating formed by topographical defects. It was shown that the observed radiation is due SPP scattering on the epilayer inhomogeneities. A similar mechanism is anticipated for intentionally formed grating.

In the present paper, the optical phenomena in heavily doped *n*-GaN epitaxial layers with a regular grating formed on their surface has been studied. The main goal of the study is to reveal the effects associated with the scattering of the SPP on the regular grating.

2. Experimental details

GaN epitaxial layers with a different concentration of Si donors ($5{\cdot}10^{18}$ and $3.6{\cdot}10^{19}$ cm^{-3}) were investigated. As is well known, at such high doping level an impurity band is formed in GaN:Si epitaxial layers [5]. That is why free electrons are not frozen under samples cooling from room to helium temperature and their concentration is constant and equal to the donor concentration. The epilayer thickness d was in the range from 6 to 10 μm. The diffraction grating on the surface of epitaxial layers was fabricated by means of photolithography and had a period of 86 μm. The grating depth H varied from 2 to 4.5 μm. The reference samples with a flat semiconductor/air interface were also fabricated and investigated.

The reflection spectra were studied at room temperature using Fourier spectrometer and pyroelectric detector. Globar was used as radiation source, the incidence angle θ was about 11°, and the plane of incidence was perpendicular to the grooves of the grating. The spectra were recorded for *p*- and *s*-polarized radiation.

Under applying lateral electric field, the THz radiation emission from the sample has been observed. Pulsed voltage was applied to the samples. The pulse duration and repetition frequency were about 2.5 μs and 87 Hz, respectively. Spectral studies of the emitted radiation were performed at a temperature of 12 K using Fourier spectrometer. A silicon bolometer was used as a detector. The THz radiation was collected in the direction perpendicular to the sample surface ($\theta = 0°$).

The field dependence of the integral intensity of the THz luminescence was measured at $T = 4.2$ K. Measurements were carried out also in a pulse regime (with pulse duration of 2 μs and repetition frequency less 1 Hz). By means of a Ge:Ga detector, the radiation was detected at $\theta = 0°$ in the spectral range from 2.2 to 7 THz.

3. Results and discussion

It should be emphasized that we investigate co-called "non-radiative" mode of surface plasmon polaritons. At any frequency, the wave vector of such mode is larger than the wave vector of light in vacuum $k_0 = \omega/c$. The dispersion law for SPP at the plane interface between two semi-infinite mediums can de described by following expression [6]:

$$k_{SPP}(\omega) = \mathrm{Re}\left(k_0 \sqrt{\frac{\varepsilon_1(\omega)\varepsilon_2(\omega)}{\varepsilon_1(\omega)+\varepsilon_2(\omega)}} \right), \tag{1}$$

where ω is angular frequency, $\varepsilon_1(\omega)$ and $\varepsilon_2(\omega)$ are permittivities of two semi-infinite mediums. In our case we assume that $\varepsilon_1(\omega)$ is permittivity of heavily doped *n*-GaN and consider the spectral range where $\mathrm{Re}(\varepsilon_1(\omega))$ is negative. The permittivity of vacuum $\varepsilon_2 = 1$. In the samples with grating, SPP with a certain frequency can be excited with the help of THz radiation with the same frequency if the following phase-matching condition for SPP and photons is satisfied [7]:

$$k_{SPP} = k_0 \sin\theta + m\frac{2\pi}{a},\ m = \pm 1, \pm 2, \pm 3, \dots , \tag{2}$$

where θ is angle of radiation incidence, a is grating period.

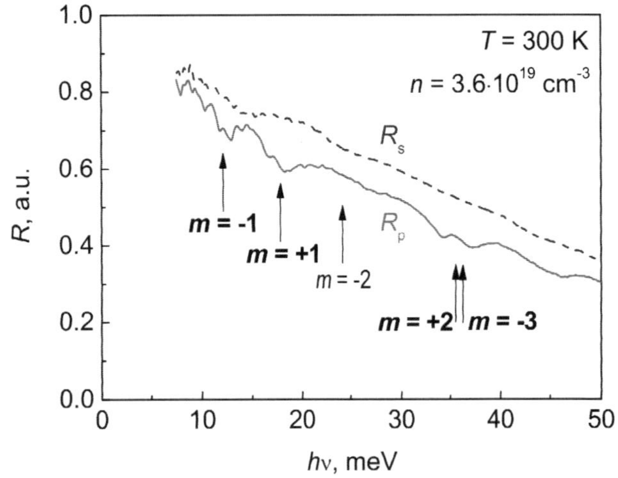

Figure 1. The reflectance spectra for *p*- and *s*-polarized radiation (solid and dashed lines, respectively). The epilayer thickness *d* is 6 μm, grating depth *H* is 4.5 μm, concentration of free electrons is $3.6 \cdot 10^{19}$ cm^{-3}, incidence angle θ is 11°. The arrows denote expected dip positions in accordance with equation 2.

Journal of Physics: Conference Series **541** (2014) 012093 doi:10.1088/1742-6596/541/1/012093

Experimental reflection spectra for two linear polarizations are presented in figure 1. One can see a few dips on the reflection spectrum for *p*-polarization. The reflection spectrum for *s*-polarization demonstrates no dips. The observed polarization dependence of the reflectance indicates that the dips can be associated with the excitation of the SPP, as the latter are electromagnetic waves of *TM*-type. According to the simulation, the spectral positions of the dips correspond to the phase-matching condition for the SPP and photons (2).The features of the reflection spectrum for *p*-polarization proves SPP initiation under THz irradiation of the samples.

THz emission was detected in the samples under lateral electric field. We associate the microscopic origin of the emission with a deviation of the electron and SPP ensembles from equilibrium. We measured experimentally a spectral radiation density and compared its spectrum for the sample with grating, *I*(*H*), with one for the reference sample without grating, *I*(0). The both samples had the same free electron concentration ($5 \cdot 10^{18}$ cm^{-3}). As it was found earlier, the THz radiation emission in the reference sample is related to the blackbody-like radiation emission from hot electrons, similar to [2]. It has been found that the ratio *I*(*H*)/*I*(0) (figure 2) exceeds unity in the spectral range from 8 to 41 meV and reaches about 3 around the photon energy of 11 meV. We conclude that the increase of the spectral radiation density in the sample with grating results from hot SPP scattering on the grating.

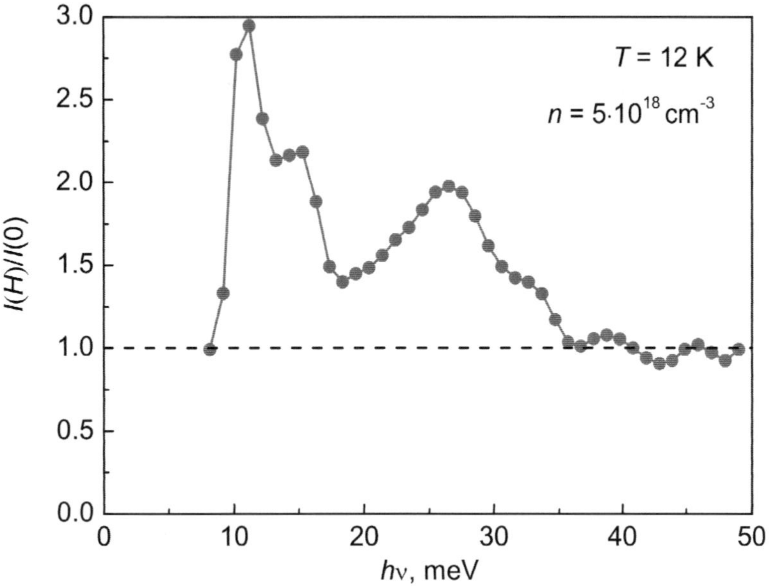

Figure 2. The spectral radiation density ratio for sample with grating *I*(*H*) and reference sample without grating *I*(0). The epilayer thickness *d* is 10 μm, grating depth *H* is 2 μm, concentration of free electrons is $5 \cdot 10^{18}$ cm^{-3}, electric power is 1.4 kW, $\theta = 0$.

Dependence of integral intensity of the THz radiation on electric power (figure 3) demonstrates twofold increase of the THz emission from the samples with regular grating as compared with the samples without grating in the spectral range of Ge:Ga detector (2.2 – 7 THz). It also proves the significant contribution of the SPP scattering to the THz radiation emission.

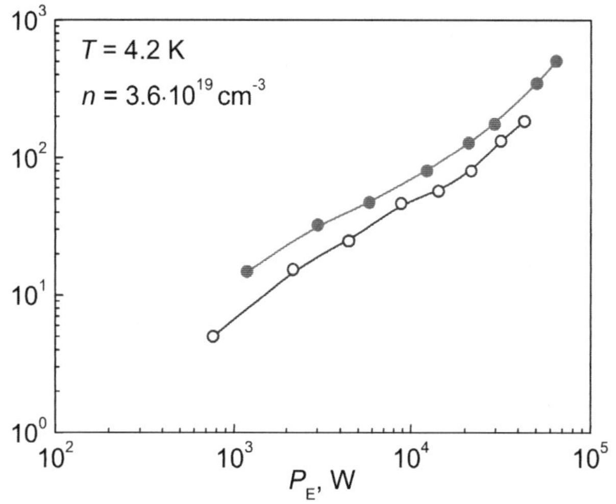

Figure 3. The dependence of integral intensity of the THz radiation on electric power. The filled circles correspond to the sample with grating, open circles correspond to the sample without grating. The epilayer thickness d is 6 μm, grating depth H is 4.5 μm, concentration of free electrons is $3.6{\cdot}10^{19}$ cm^{-3}, $\theta = 0$.

4. Conclusion

Spectra of reflection, spectra of THz radiation emission and integral intensity of the THz electroluminescence were investigated in heavily doped GaN:Si epitaxial layers. Samples with grating on the surface of the epitaxial layers were studied. Reference samples with a flat surface were studied as well.

Experiments on optical reflection demonstrate the possibility of surface plasmon polariton excitation on the surface of the grating structure using terahertz radiation. Spectral study of terahertz radiation emission under lateral electric field shows that the scattering of hot surface plasmon polaritons on the grating makes a significant contribution to the radiation intensity. Investigation of integral intensity of the THz electroluminescence demonstrates twofold increase of THz emission in the samples with grating as compared with the samples without grating (in the spectral range from 2.2 to 7 THz). This proves feasibility of THz emitter based on hot surface plasmon polaritons.

Acknowledgments

The authors are grateful to H. Nykänen, L. Riuttanen, O. Svensk and S. Suihkonen (Aalto University) for the microstructure growth. This work was supported by the Russian Foundation for Basic Research and Russian Federation President grant for young scientists.

References

[1] Shalygin V A *et al* 2009 *J. Appl. Phys.* **106** 123523
[2] Shalygin V A, Vorobjev L E, Firsov D A, Sofronov A N, Melentyev G A, Lundin W V, Nikolaev A E, Sakharov A V and Tsatsulnikov A F 2011 *J. Appl. Phys.* **109** 073108
[3] El Fatimy A *et al* 2010 *J. Appl. Phys.* **107** 024504
[4] Shubina T V, Gippius N A, Shalygin V A, Andrianov A V and Ivanov S V 2011 *Phys. Rev.* B **83** 165312
[5] Agekyan V F, Vorob'ev L E, Melentyev G A, Nykänen H, Serov A Yu, Suihkonen S, Filosofov N G and Shalygin V A 2013 *Phys. Solid State* **55** 296
[6] Teng Y-Y and Stern E A 1967 *Phys. Rev. Lett.* **19** 511
[7] Beaglehole D 1969 *Phys. Rev. Lett.* **22** 708

Algorithm of molecular computing on the base of membranes

D S Nikiforov and I Yu Popov

St. Petersburg National Research University of Information Technologies, Mechanics and Optics, Kronverkskiy, 49, St. Petersburg, 197101, Russia

E-mail: dmitri_nikiforov@mail.ru

Abstract. Parallelism of membrane computing can be used for efficient solving NP-complete computational problems. A problem of such type known as questionnaire problem is considered. One should choose the sequence of yes-no questions which allows to reveal the system configuration. Algorithm based on membrane computing is suggested.

1. Introduction

Membrane computing is rapidly developing area of Natural Computing [1, 2] introduced by Paun in [3]. It is inspired by the structure and the vital functions of a living cell. The perspectives in this field are related with the progress of bio- and nanotechnologies which allows one to construct structures and materials needed for these procedures [4]. This model is based on a set of membranes bounding regions where some chemical reactions are proceed. Membranes can contain other membranes and chemical reagents. We consider the cell-like membrane structure where there is a top-level membrane called skin membrane. It is surrounded by environment. Membrane is called elementary if it doesn't contain other membranes. In mathematical definition reagents are called objects and reactions are called rules. At each step, rules are applied to objects in maximal parallel manner. It means that the multiset of rules is applied to multiset of objects and there are no more rules that can be applied to the remaining objects. The computation halts when no rule can be applied to any object in any membrane. The multiset of objects caught in the membrane marked as output region is considered as a result of computation. Membrane computing gives one a possibility to trade time for space and solve computationally hard problems in feasible time [5]. Algorithms for SAT [6,7], Subset Sum [8], Knapsak [9], Bin Packing [10] and Partition [11] were suggested. It is known that any NP-complete problem can be reduced to another one but it can be non-trivial so construction of specific algorithms for various problems is interesting and important. Algorithm for questionnaire problem is suggested in this paper.

2. Questionnaire problem

To make the procedure clear we give a simple example (the toy task) of the questionnaire problem, which is related with known logical task. There are two cities: A and C. Inhabitants of A always tell the truth, inhabitants of C in turn tell the truth and a lie. Observer E is in one of these cities. He meets one person and wants to know in what city is he, and from what city is this person. How many questions and in what sequence should be asked if the person answers only yes or no?

For our example, let the initial set of questions contain the following ones:

(1) Am I in A?

(2) Am I in C?

(3) Are you from A?

There are 6 possible configurations of system because there are two types of C inhabitant's behavior: he can tell the truth answering the first question or he can tell a lie. We denote these options respectively C+ and C−. Shannon entropy of the system is $log_2 6$. It means that two questions are not sufficient to obtain the answer and there exists an appropriate sequence of three correct questions because $2 < log_2 6 < 3$. There are three ones in our set (we assume that we have correct set of questions) so the only thing to do is to find the right order of them. Table 1 shows the answers to the questions asked in the following order 1, 2, 3: + means yes, − means no.

Table 1. Answers for valid sequence of questions {1, 2, 3}.

	E in A	E in C
Person from A	+ − +	− + +
Person from C+	+ + −	− − −
Person from C−	− − +	+ + +

One can see that plus-minus sequences are unique for each cell, so all possible configurations are distinguishable. In contrast, table 2 shows the answers to the sequence of questions that leaves some configurations undistinguishable.

Table 2. Answers for "bad" sequence of questions {2, 3, 1}.

	E in A	E in C
Person from A	− + +	+ + −
Person from C+	− + +	+ + −
Person from C−	+ − −	− − +

In our model all possible sequences of questions are processed in the same time. Each of them corresponds to membrane substructure of the following type:

$$[_0[_{1_n}[_{2_n}[_{3_n}]_{3_n}[_{3_y}]_{3_y}]_{2_n}[_{2_y}[_{3_n}]_{3_n}[_{3_y}]_{3_y}]_{2_y}]_{1_n}[_{1_y}[_{2_n}[_{3_n}]_{3_n}[_{3_y}]_{3_y}]_{2_n}[_{2_y}[_{3_n}]_{3_n}[_{3_y}]_{3_y}]_{2_y}]_{1_y}]_0$$

Common notation for membrane structures is used: all stuff placed between opening and closing brackets with same label is considered as content of membrane with this label. This example refers to the sequence {1, 2, 3}. 1_n, 1_y, 2_n, 2_y, 3_n, 3_y are the labels of membranes corresponding to negative (with suffix n) and positive (y) answers for the first, the second and the third questions from the list. Tree representation of this structure is shown in figure 1.

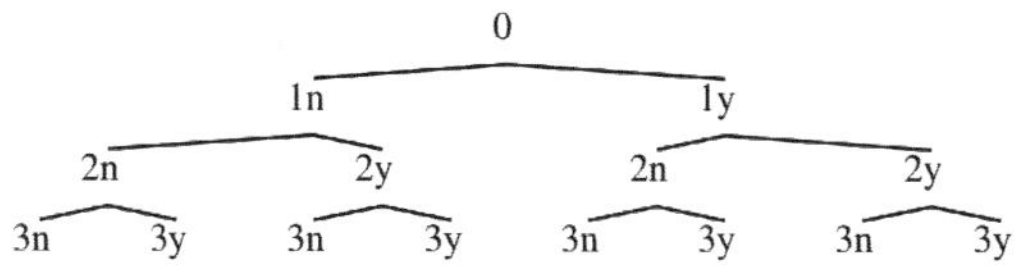

Figure 1. Hierarchy of membranes for sequence {1, 2, 3}.

0-membrane is a container. In our example the skin membrane, denoted as 00, contains $3! = 6$ 0-membranes for every possible order of questions. In the initial configuration each 0-membrane contains objects a_1,a_2, ..., a_6 corresponding to table cells listed from left to right and from top to bottom (see table 3) each as a single copy. a_1 corresponds to situation when E is in A and person from A is asked and so on. 0-membrane also contain four copies of object t corresponding to the turn of the truthful answer for C+ and the turn of the false answer for C−. Object f coming later is the opposite to object t. There are objects $finish$ located in every elementary membrane in a single copy.

Table 3. Each a-object corresponds to one cell.

	E in A	E in C
Person from A	a_1	a_2
Person from C+	a_3	a_4
Person from C−	a_5	a_6

We use *in* and *out* communication rules in our model. *In* communication rules introduce an object into the specified membrane and possibly object can be modified. *Out* rules send object out of the membrane and possibly object can be modified. Following *in* rules send a-objects to 1_y or 1_n membranes according to person's answer on the first question:

$$a_1\ [_{1_y}]_{1_y} \rightarrow [_{1_y} a_1\ ]_{1_y}$$

$$a_2\ [_{1_n}]_{1_n} \rightarrow [_{1_n} a_2\ ]_{1_n}$$

$$a_3 t\ [_{1_y}]_{1_y} \rightarrow [_{1_y} a_3 f\ ]_{1_y}$$

$$a_3 f\ [_{1_n}]_{1_n} \rightarrow [_{1_n} a_3 t\ ]_{1_n}$$

$$a_4 t\ [_{1_n}]_{1_n} \rightarrow [_{1_n} a_4 f\ ]_{1_n}$$

$$a_4 f\ [_{1_y}]_{1_y} \rightarrow [_{1_y} a_4 t\ ]_{1_y}$$

$$a_5 t\ [_{1_n}]_{1_n} \rightarrow [_{1_n} a_5 f\ ]_{1_n}$$

$$a_5 f\ [_{1_y}]_{1_y} \rightarrow [_{1_y} a_5 t\ ]_{1_y}$$

$$a_6 t\ [_{1_y}]_{1_y} \rightarrow [_{1_y} a_6 f\ ]_{1_y}$$

$$a_6 f\ [_{1_n}]_{1_n} \rightarrow [_{1_n} a_6 t\ ]_{1_n}$$

There are two rules for two cells corresponding to person from A. Object a_1 goes to 1_y membrane because positive answer on the first question is always given if E is in A. In the opposite a_2 goes to 1_n because person from A answers "no" on the first question if E is in C. The same logic applies to objects a_3, ..., a_6 but their routes depend on the turn of truthful or false answer given by C+ or C−. The presence of t or f objects is taken into consideration to represent this fact. Analogous rules involve 2_n, 2_y, 3_n and 3_y membranes.

Introduced rules send a-objects from 0-membrane deeper into membrane structure. They go from root to leaves if tree representation is considered. Consider the configuration after three steps made. All a-objects came to elementary membranes. If the order of questions was appropriate each elementary membrane contains at most one a-object otherwise some of them can contain several a-objects. Then the next *out* rules come into play:

$$[_i a_n finish\ ]_i \rightarrow [_i]_i q_i$$

$$[_j q_i\ ]_j \rightarrow [_j]_j q_{ji}$$

$$[_k q_{ji}\ ]_k \rightarrow [_k]_k q_{kji}$$

where a_n is arbitrary a-object; i, j, k are in (1, 2, 3); q-objects are special objects tracking the order of questions (q_1, q_2, q_3 ,q_{12}, q_{13}, q_{23}, q_{21}, q_{31}, q_{32}, q_{123}, q_{231}, q_{312}, q_{321}, q_{213}, q_{132}). Three steps later q-objects with three indexes will arrive to 0-membranes. There will be 6 q-objects in 0-membranes corresponding to valid sequence of questions but there will be less than 6 q-objects in 0-membranes corresponding to "bad" sequence of questions. All q-objects can be popped to 00-membrane if it is considered as an output region and one can count their multiplicities.

Consider the questionnaire problem in general. There are n cities of type A (A_1, A_2, ..., A_n) and m cities of type C (C_1, C_2, ..., C_m). The number of possible system configurations is $(n+m)(n+2m)$. The lower bound for necessary and sufficient minimal number of questions is $log_2(n+m)(n+2m)$. The initial set of questions contains not less than mentioned number of questions. It is possible that no fitting sequence can be found in this set in general. In this case when computation halts there will be insufficient number of q-objects in 00-membrane. The structure, initial configuration and rules are constructed in the same way as in example. The difference lies in a greater number of a and q objects needed and deeper membrane structure.

Further generalization leads to functional rules. We used two objects t and f one by one to alternate a-objects' behavior according to the turn of truthful or false answer: $t \rightarrow f \rightarrow t$ and so on. Rules with arbitrary alternation can be suggested, e.g. involving $t_1 \rightarrow \ldots \rightarrow t_n \rightarrow f_1 \rightarrow \ldots \rightarrow f_m$ and so on. Such rules can handle arbitrary determinate answering strategy. Another point of interest is usage of nondeterminism in rules' definition and application.

The work was partially financially supported by the Government of the Russian Federation (grant 074-U01), by the Ministry of Science and Education of the Russian Federation (GOSZADANIE 2014/190, Project 14.Z50.31.0031), by grant of Russian Foundation for Basic Researches and grants of the President of Russia (state contracts 14.124.13.2045-MK and 14.124.13.1493-MK).

References

[1] Ciobanu G, Perez-Jimenez M J and Paun G 2008 *Applications of Membrane Computing* (Berlin: Springer)
[2] Popov I Yu, Vorobyova A V and Blinova I V 2014 *Int. J. Bioinform. Res. Appl.* **10** 145-56
[3] Paun G 2000 *J. Comput. System Sci.* **61** 108-43
[4] Rambidi N G 2007 *Nanotechnologies and Molecular Computers* (Moscow: Fizmatlit)
[5] Gutierrez-Naranjo M A, Perez-Jimenez M J and Romero-Campero F J 2007 *Theoret. Comput. Sci.* **371** 54-61
[6] Perez-Jimenez M J, Romero-Jimenez A and Sancho-Caparrini F 2003 *Proc. 5th Workshop on Descriptional Complexity of Formal Systems (Budapest)* 284-94
[7] Paun G 1999 *Centre for Discrete Mathematics and Theoretical Computer Science Research Report* **102**
[8] Perez-Jimenez M J and Riscos-Nunez A 2005 *New Generat. Comput.* **23** 367-84
[9] Perez-Jimenez M J and Riscos-Nunez A 2004 *Lecture Notes in Comput. Sci.* **2933** 250-68
[10] Perez-Jimenez M J and Romero-Campero F J 2004 *Proc. 2nd Brainstorming Week on Membrane Computing (Seville)* 414-30
[11] Gutierrez-Naranjo M A, Perez-Jimenez M J and Riscos-Nunez A 2005 *Soft Comput.* **9** 673-8

A change in the electro-physical parameters of narrow-gap CdHgTe solid solutions acted upon by a diffusion discharge in the air, argon and nitrogen at atmospheric pressure

D.V. Grigoryev, E. A. Stavkov[1]

National Research Tomsk State University, 36 Lenin Av., Tomsk, 634050, Russia

E-mail: eastavkov@sibmail.com

Abstract. The effect of a nanosecond volume discharge forming in an inhomogeneous electrical field in various gas environments at atmospheric pressure on the CdHgTe(MCT) epitaxial films of the p-type conduction is studied. The measurement of the electrophysical parameters of the MCT specimens upon irradiation shows that a layer exhibiting the n-type conduction is formed in the near-surface region of the epitaxial films. Analysis of the preliminary results reveals that the foregoing nanosecond volume discharge in various gas environments at atmospheric pressure is promising for modification of electro-physical MCT properties.

At present ternary semiconductor MCT compounds are one of the major materials for producing IR photodetectors for the 3–5 and 8–14 μm wavelength ranges [1]. The main tendencies in the development of modern technologies are aimed at production of highly effective multi-element photodetector devices based on the epitaxial MCT material grown by molecular-beam epitaxy (MBE). Along with examination of initial properties of epitaxial MBE MCT films, a controlled change of the material parameters is important for producing semiconductor structures with specified properties.

Electric discharges of different types as well as electron beams are now widely used for the modification of near-surface layers of various materials [2]. It was reported that a high-voltage diffuse discharge at a low gas pressure and a low power could be used for the inactivation of microbiological cultures [3]. As is known [4], a volume discharge can be generated using inhomogeneous electric field in gases at atmospheric pressure. For this purpose, high-voltage (≥100 kV) pulses of nanosecond duration are applied to a gas-filled interelectrode gap. A specific feature of such discharges in air at atmospheric pressure is the accompanying X-ray emission and the formation of runaway electron beams [5].

In recent years, it was established that a volume discharge could be generated in air at atmospheric pressure in an inhomogeneous electric field for both polarities of the high-voltage pulses applied to an electrode with a small radius of curvature [6]. Another feature of volume discharges in inhomogeneous electric fields, which are initiated by supershort avalanche electron beams (SAEBs), is the possibility of reaching high levels of specific power (up to 800 MW/cm^3 [7]) deposited in the discharge. In such regimes, runaway electron beams emitted from the discharge plasma are

[1] To whom any correspondence should be addressed.

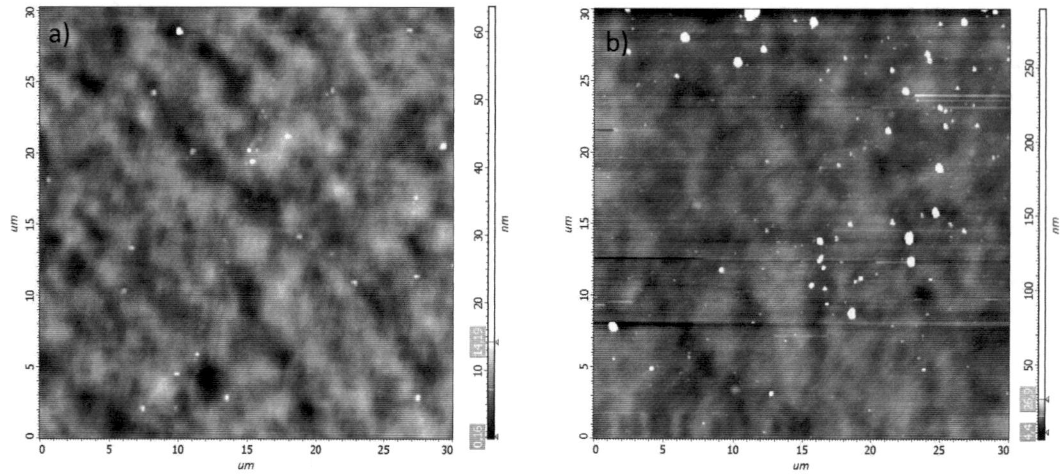

Figure 1. Surface topography CdHgTe samples before (a) and after (b) volume-discharge action.

characterized by the beam current amplitudes reaching tens and hundreds of amperes (behind the foil), while the current pulse duration (defined as the pulse full width at half maximum, FWHM) does not exceed 100 ps [8]. Such discharges in inert gases are also accompanied by high-intensity vacuum ultraviolet (VUV) emission. Thus, the generation of SAEB-induced volume discharge in gases at high pressures leads to a complex action of the high-density nanosecond discharge plasma (with a specific deposited power of several hundred megawatt per cubic centimeter), SAEBs (with a broad energy spectrum), shock wave and radiation in various spectral ranges (including UV and VUV emitted from the discharge plasma) on the anode.

The aim of this work is to study the effect of a nanosecond volume discharge in various gas environments at atmospheric pressure on the electro-physical properties of epitaxial film CdHgTe grown by molecular beam epitaxy.

Investigation of the discharge influence for semiconductors was carried out on narrow-band material $Cd_xHg_{1-x}Te$. To perform the investigations, epitaxial $Cd_xHg_{1-x}Te$ (x=0.22) films were grown by molecular beam epitaxy at the Semiconductor Physics Institute of the Siberian Branch of the Russian Academy of Sciences in Novosibirsk. The material was grown on a GaAs substrate backing in the (013) orientation with buffer layers of ZnTe and CdTe. After growth, the films had n-type conductivity with carrier concentration 10^{14} cm^{-3} and mobility ~ 10^5 $cm^2V^{-1}s^{-1}$. For conversion to p-type conductivity, the films were annealed in a neutral atmosphere of hydrogen or helium. After annealing, the samples had p-type conductivity ($p = 1.1 \div 2.5\times10^{16}$ cm^{-3}, $\mu_p = 300 \div 500$ $cm^2V^{-1}s^{-1}$).

The experiments were performed using a discharge chamber. A distance from a sample holder to a tubular electrode was 20 mm. The interelectrode voltage was supplied from a pulser of the RADAN-220 type, which generated voltage pulses with an amplitude of ~230 kV (in the open-circuit regime), a current pulse duration (defined as the pulse full width at half maximum) of ~ 2 ns (on a matched load), and a leading front width of ~ 0.5 ns. The results of discharge current measurements showed that the current pulse amplitude was ~3 kA and the total duration of the discharge current pulse was ~30 ns (the first half-period of the discharge current pulse had a duration of ~8 ns). The $Cd_xHg_{1-x}Te$ specimens were irradiated in the pulsed-periodic mode at the pulse repetition rate 1 Hz. The action was realized through 100–1200 pulses for a series of specimens irradiated in air and 600, 1200 pulse irradiation in an atmosphere of argon and nitrogen. Measurements of the electro-physical parameters of the samples before and after irradiation were made at the temperature of liquid nitrogen using the Hall Electromotive Force (EMF) method in the Van-der-Pau configuration. The surface irradiated MCT sample has been studied by atomic force microscopic (AFM) system Solver HV.

Analysis of the surface structure of the initial and irradiated epitaxial films (fig. 1) showed, that

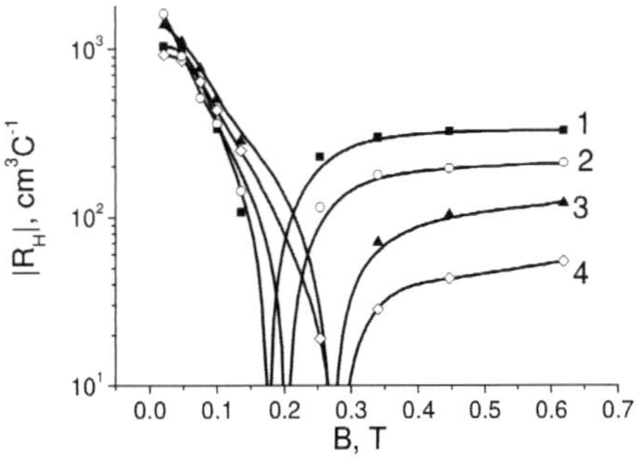

Figure 2. Hall-coefficient field dependence of CdHgTe epitaxial films after volume-discharge in the air at atmospheric pressure action. The number of the pulses: 0 – (1), 100 – (2), 200 – (3), 400 – (4).

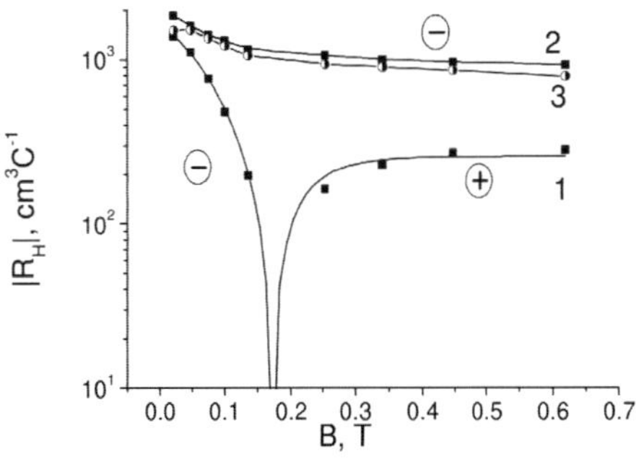

Figure 3. Field dependence of the Hall coefficient of epitaxial films of MCT before irradiation (1), after exposure to 600 pulses of a volume discharge in air (2) and 3 months after irradiation (3).

the quality of the specimen surface upon volume-discharge action is not affected, that is, no changes in the surface structure of the irradiated specimens occur. The surface roughness slightly increases from 1.46 to 2.2 nm.

The measurement of electrophysical parameters of the MCT epitaxial-film samples after exposure discharge pulses in the air atmosphere showed that for all the samples an increase in conductivity. The specimens irradiated by 100 – 400 pulses, however, exhibit a decrease in the Hall coefficient. In so doing, the field dependence of the Hall coefficient is characterized by a shift of the inversion point of the Hall coefficient sign to the region of higher magnetic fields from 0.17 T to 0.28 T (fig. 2). An increase in the number of volume-discharge pulses up to 600 results in the inversion of the Hall-coefficient sign in the range more than 0.2 T (fig. 3). A still further increase in the number of pulses exposure leads to a decrease in the value of the Hall coefficient. Also the lack of relaxation of electrical parameters of irradiated specimens within 3 months is noted (fig. 3, curve 3).

It has been suggested that, on or near the surface of the film formed by a layer of highly conductive n-type, whose parameters are such that the measured field dependence of the Hall coefficient corresponds to the n-type conductivity.

For specimens irradiated in argon and nitrogen, there is a change in the field dependence of the Hall coefficient as in the case of irradiation in air. For the specimens irradiated at 1200 pulses under a nitrogen atmosphere after exposure in the field of 0.2 T (fig. 4a). The observed inversion of the sign of the Hall coefficient, which is not on the original field dependence, for a specimens irradiated with 600 pulses in an argon atmosphere, after exposure to the field dependence of the Hall coefficient noticeable change in the sign of a 0.2 T (fig. 4b). The field dependence of the specimens corresponds to the field dependence of the material n-type conductivity. It has been suggested that, on or near the surface of the film formed by a layer of highly conductive n-type similar to the case of irradiation in air, whose parameters are such that the measured field dependence of the Hall coefficient corresponds to the n-type conductivity. It also marked relaxation of the electrical parameters of epitaxial films irradiated to baseline values within 6 months (fig. 4, curves 3).

Based on the Petriz model [9] the theoretical calculation of the Hall coefficient field dependence was carried out. Theoretical calculation and experimental data are in good agreement. The calculation showed that the integrated conductivity of the resulting highly conductive n-type layer is for air 8.6×10^{-3} ohm^{-1} at 600 pulses and 9.6×10^{-3} ohm^{-1} at 1200 pulses,, for nitrogen 10^{-2} ohm^{-1} at 600 pulses and $10^{-4}ohm^{-1}$ at 1200 pulses, for argon 5×10^{-3} ohm^{-1} at 600 pulses and 5×10^{-5} ohm^{-1} at 1200 pulses.

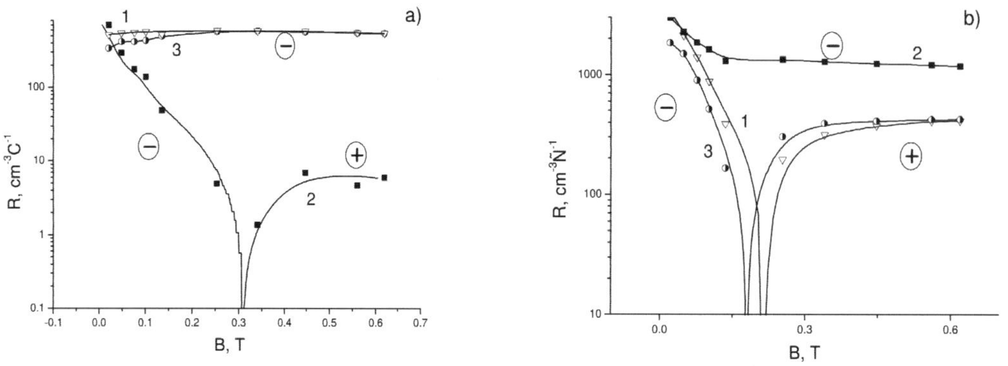

Figure 4. Field dependence of the Hall coefficient of epitaxial films of MCT to (1), after (2) and after 3 months (3) after exposure to a volume discharge in the atmosphere: a) -nitrogen (1200 pulses), b) - argon (600 pulses).

Thus, our experimental data show that the action of pulses of nanosecond volume discharge in air, argon and nitrogen at atmospheric pressure leads to changes in the electrophysical properties of CdHgTe epitaxial films. These changes are due to formation of a near-surface high-conductivity layer of the *n*-type conduction. The electrophysical parameters of this layer are such that the measured field dependence of the Hall coefficient of p-type CdHgTe epifilms corresponds to the n-type material conductivity. This result suggests it is possible to use such discharge in various gases for the controlled change of the properties of CdHgTe narrow-band solid solutions.

The authors thank M. A. Shulepov for irradiation of sample.

The reported study was conducted within the bounds of the National Research Tomsk State University competitive recovery program.

References

[1] Rogalskii A 1999 Infrared detectors (Amsterdam : Gordon and Breach)

[2] A. N. Didenko, Yu. P. Sharkeev, E, V. Kozlov, et al., *Long-Range Effects in Ion-Implanted Metallic Materials* (NTL, Tomsk, 2004)

[3] S. N. Buranov, S. V. Voevodin, I. A. Voevodina, et al., Teplofiz. Vys. Temp. **43**, 367 (2005)

[4] Yu. L. Stankevich and V. G. Kalinin, Dokl. Akad. Nauk SSSR **177** (1) 72 (1967) [Sov. Phys. Dokl. **12**, 1148 (1967)]

[5] L. V. Tarasova, L. N. Khudyakova, T. V. Loiko, and V. A. Tsukerman, Zh. Tekh. Fiz. **44**, 564 (1974) [Sov. Phys. Tech. Phys. **19**, 351 (1974)]

[6] I. D. Kostyrya, V. M. Orlovskii, V. F. Tarasenko, et al., Pis'ma Zh. Tekh. Fiz. **31** (11), 19 (2005) [Tech. Phys. Lett. **31**, 451 (2005)]

[7] S. B. Alekseev, V. P. Gubanov, I. D. Kostyrya, et al. Kvantovaya Elektron. (Moscow) **34**, 1007 (2004)

[8] V. F. Tarasenko, V. G. Shpak, S. A. Shunailov, and I. D. Kostyrya, Laser Part. Beams **23**, 545 (2005)

[9] R.L. Petritz, Physical Review, **110**(6), 1254, (1958)

Journal of Physics: Conference Series **541** (2014) 012096 doi:10.1088/1742-6596/541/1/012096

Dielectric spectroscopy of polymer nanocomposites based on tetrazol and KNO_3

R A Castro, E N Lushin

Department of Physical Electronics, Herzen University, Moika Embankment, Saint-Petersburg 191186, RUS

E-mail: lushin.ev@gmail.com

Abstract. For tetrazole polymers by dielectric spectroscopy the existence of three relaxation processes in the temperature range T=273–423 K is revealed, the values of relaxation and structural parameters are determined: activation energy E_A and glass transition temperature T_g.

1. Introduction

Compounds based on tetrazole complex differ with complex of unique properties: low sensitivity to shock and friction, high thermal stability, low toxicity and great flatulence. These qualities determine the prospects of the use of polymers and copolymers based on vinyltetrazoles as components of condensed energy-systems for aerospace industry. Currently, there is information about the structure and physico-mechanical properties of this class of materials, but to date there had been no research of the electrical and dielectric properties and their relation to the structural features.

2. Experimental details

The detail analysis of influence of concentration of the modifier on relaxational properties of polymer composites based on tetrazole was aim of this work. The subject of research was poly-N-methylallil-5-vinyl tetrazole (MPVT-A) having the following chemical structure:

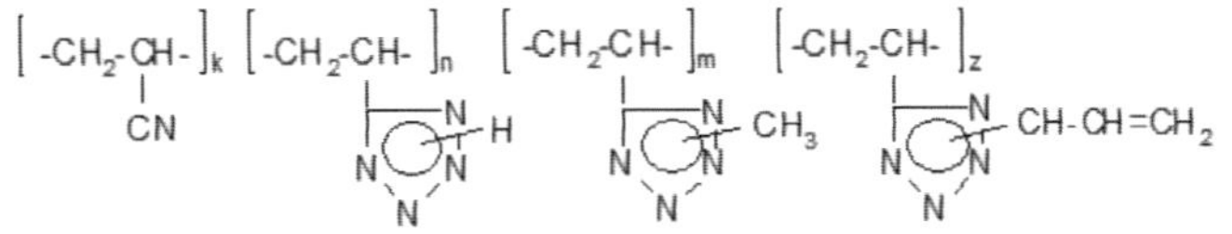

where k = 5%, n = 1%, m = 90%, z = 4%.

Dimethyl phthalate was used as plastifier.

As modifiers were used KNO_3 (55%) and KNO_3 (70%), which accounted for 65% of the weight of the polymeric binder. The hardening of the samples was carried out by low-temperature hardening agent to di-N-oxide-1.3-dinitro-2,4,6-trietilenbenzol (TON-2).

The DSC (differential scanning calorimetry) curves of samples belonging to these three samples were recorded to detect the glass transition temperature T_g (Fig. 1).

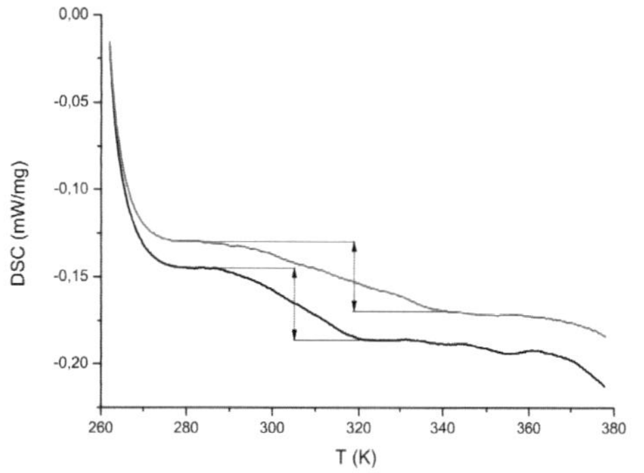

Figure 1. DSC analysis for nanocomposite: tetrazol + KNO_3. (Studies were performed at the Center of Thermal Analysis and Calorimetry of St. Petersburg State University)

Measurements of dielectric spectra were performed on installation of "Concept 41" of the NOVOCONTROL Technologies GmbH Co, which consists of a frequency analyzer impedance measuring cell, temperature control system, automatic data collection system and Dewar vessel with system of evaporation and submission of gaseous nitrogen.

3. Result discussion

There have been revealed the existence of three relaxation processes in the temperature range T=273–423 K (α, β and γ-processes), there have been determined the values of relaxation and structural parameters: the activation energy E_A (Fig. 2) and the glass transition temperature T_g (by the method, proposed in [1]) (Fig. 3) (See Table 1).

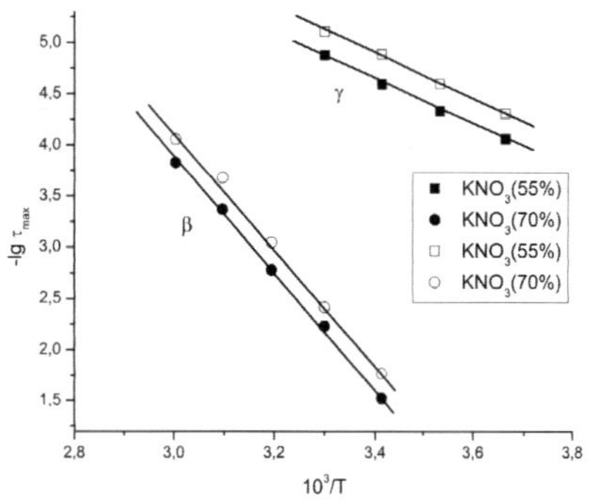

Figure 2. The temperature dependence of the relaxation times corresponding to the maxima of the dielectric loss.

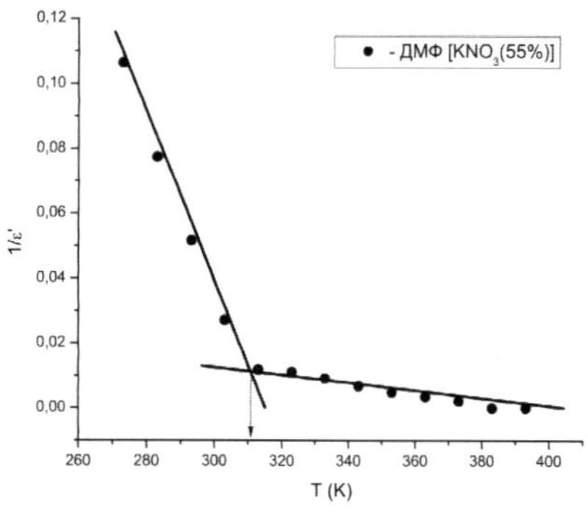

Figure 3. Determination of glass transition temperature of the nanocomposite: tetrazol + KNO_3

As can be seen from the table, the concentration change of KNO_3 changes the value of the activation energy of β-process, whereas for γ-process the value of this parameter remains unchanged. With the increase of filler concentration, a slight decrease in the glass transition temperature occurs (associated with the transition to a glassy state). Since the position of the glass transition temperature is associated with the occurrence of macromolecular chains mobility, in such a case, the findings indicate that the formation of the adsorptive bonds between polymer

Table 1. Glass transition temperature and activation energy of the nanocomposite: tetrazol + KNO_3.

Concentration of KNO_3 (%)	T_g, K	Activation energy β, eV	Activation energy γ, eV
70	306	1.22	0.39
55	312	1.06	0.39

molecules and the surface, as well as conformational changes at the phase boundary of polymer - filler, result in changes of the relaxation polymer properties [2]. Raising of the glass transition temperature indicates a noticeable restriction of chains mobility, equivalent to their flexibility decrease due to the additional bonds formation, or changes in the molecular conformation. This restriction of mobility (and hence, the raised glass transition temperature) should be the greater, the greater the number of polymer molecules is involved in the interaction with the surface.

A common feature to all the amorphous semiconductors is that the ac conductivity $\sigma_{ac}(\omega)$ increases with frequency according to

$$\sigma_{ac}(\omega) = \sigma_i(\omega) - \sigma_{dc} = A\omega^s, \qquad (1)$$

where ω is the angular frequency, s is the frequency exponent and A is a constant independent of frequency [3]. Fig. 4 and Fig. 5 shows the frequency dependence of ac conductivity $\sigma_{ac}(\omega)$ for samples with different KNO_3 concentration. It is clear from the figure that $\sigma_{ac}(\omega)$ increases linearly with frequency. The same behavior of the frequency dependence of $\sigma_{ac}(\omega)$ was obtained for all the investigated films. Values of the frequency exponent s were obtained from the slopes of these lines of the figure (s= 0,75-0,85).

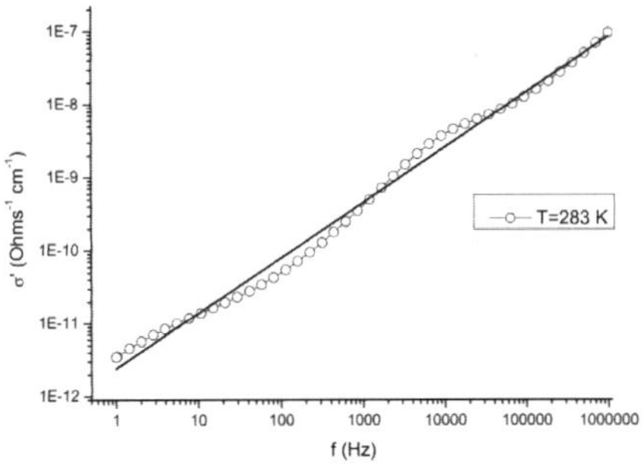

Figure 4. Frequency dependence of conductivity $\sigma_{ac}(\omega)$ for sample KNO_3(55%).

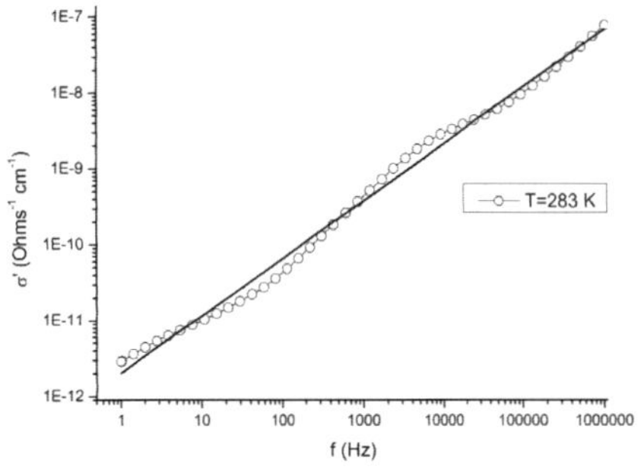

Figure 5. Frequency dependence of conductivity $\sigma_{ac}(\omega)$ for sample KNO_3(70%).

According to correlated barrier hopping (CBH) model, values of the frequency exponent s is ranged from 0.7 to 1 at room temperature and is found to decrease with increasing temperature. This is in good agreement with the obtained results, so the frequency dependence of $\sigma_{ac}(\omega)$ can be explained in terms of CBH model. The expression for s derived on the basis of this model is given by Elliott [4,5] as

$$s = 1 - (6\ kT/E_g), \qquad (2)$$

where k is the Boltzmann constant, T is the temperature in Kelvin and E_g is the optical band gap of the material.

According to the Austin-Mott formula [6], based on CBH model, ac conductivity $\sigma_{ac}(\omega)$ can be explained in terms of the hopping of electrons between pairs of localized states at the Fermi level. $\sigma_{ac}(\omega)$ is related to the density of states $N(E_F)$ at the Fermi level by

$$\sigma_{ac}(\omega) = (\pi/3)[N(E_F)]^2 kTe^2\alpha^{-5}[ln(\nu_{ph}/\omega)]^4, \quad (3)$$

where α is the exponential decay parameter of localized states wave functions, and ν_{ph} is the phonon frequency. By assuming $\nu_{ph} = 10^{12}$Hz and $\alpha^{-1} = 10$Å[7], the density of states is calculated and it is found that $N(E_F)$ has values of the order 10^{22}eV^{-1} cm^{-3}, which increases with frequency and temperature.

The temperature dependence $\sigma_{ac}(\omega)$ at different frequencies for used composites is shown in Fig. 6.

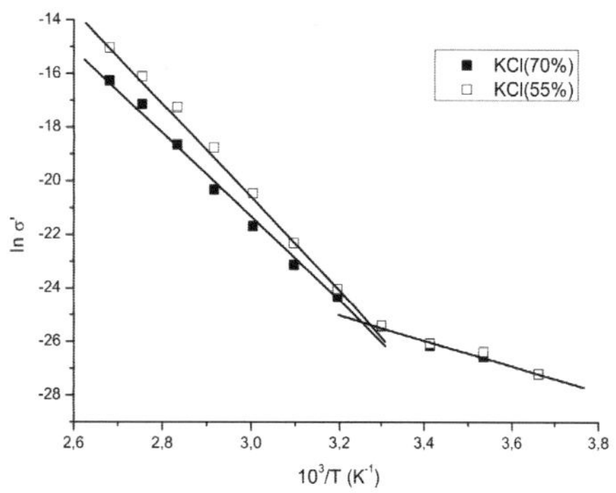

Figure 6. The temperature dependence conductivity.

It is clear from the figure that $\sigma_{ac}(\omega)$ decreases linearly with the reciprocal of absolute temperature in two temperature regions. This suggested that the ac conductivity is a thermally activated process from different localized states in the gap or its tails. The activation energy of conduction (E_σ) is calculated at different frequencies from the slopes of lines of Fig. 5 using the well-known equation $\sigma = \sigma_0\ exp(E_\sigma/kT)$. For 55% concentration the values (E_σ): ($E_{\sigma 1}$)= 1,54 eV, ($E_{\sigma 2}$)= 0,41 eV, for 70% concentration the values (E_σ): ($E_{\sigma 1}$)= 1,41 eV, ($E_{\sigma 2}$)= 0,40 eV.

References

[1] Lushin E, Castro R St. Petersburg State Polytechnical University Journal. Physics and Mathematics. 2013;90–93:182.
[2] Kalinin M, Petrekov P, Lushin E, Tereshchenko I, Sapozhnikova T BTI. Bijsk, Altai State Technical University Publishing House, 2010; 59-61:160.
[3] Durand B, Taillades G, Pradel A, Ribes M, Badot JC, Belhadj–Taher. N. J Non-Cryst Solids 1994;172-174:1306.
[4] Elliott SR. Philos Mag B 1978;36:129.
[5] Elliott SR. Philos Mag 1978;37:135.
[6] Austin LG, Mott NF. Adv Phys 1969;18:41.
[7] Agarwal SC, Guha S, Narasimhan KL.J Non–Cryst Solids 1975;18:429.

Prediction of phase equilibria in the W–Si system in the generalized lattice model

G A Panov and M A Zakharov

Department of Solid State Physics and Microelectronics, Novgorod State University, 173003, Novgorod the Great, Russia

E-mail: mef.gena@rambler.ru

Abstract. The present work is devoted to the phase diagrams calculation of the W–Si system within the framework of the generalized lattice model taking account of volume effects. The theoretically calculated phase diagram is compared with the corresponding experimental diagrams.

1. Introduction

Investigation of the thermodynamic properties and phase equilibria in the W–Si system is of great importance for microelectronics. For example, the compound WSi_2 used in the creation of Schottky barriers, heterostructures, rectifying and ohmic contacts in VLSI. At high temperature heat treatment of contact in W–Si the effect of gettering of defects and impurities was observed [1, 2]. In this connection an important problem is to study the composition of the coexisting phases in the binary system W–Si in a wide range of temperatures and concentrations. The aim of this work is a theoretical calculation of the phase diagram of the W–Si system within the generalized lattice model [3].

2. Model

The phase diagram of the binary system W–Si refers to the phase diagram of the eutectic type with several intermediate phases of constant and variable composition. The phase diagram of the eutectic type with intermediate phase of variable composition is schematically presented in Figure 1.

It is easy to see that the left and right parts of the initial phase diagram are similar to a conventional phase diagram of the eutectic type. The fundamental difference between both parts of the phase diagram in Figure 1 and the phase diagram of the eutectic type is that the mole fractions x of the B component (reckoned along the concentration axis) vary in the range $0 \leq x \leq n/(m+n)$ for the left subsystem A–A_mB_n and in the range $n/(m+n) \leq x \leq 1$ for the right subsystem A_mB_n–B. As a consequence, the lengths of the segments that correspond in the concentration axis to the left and right parts of the diagram are not equal to unity, whereas the length of the concentration axis in any phase diagram should be equal to unity. These differences do not permit one to directly use the well-developed methods for calculating phase diagrams in the case of simulation of the components of phase diagrams for systems with intermediate phases. The method of calculation of phase diagrams of such type has been proposed in paper

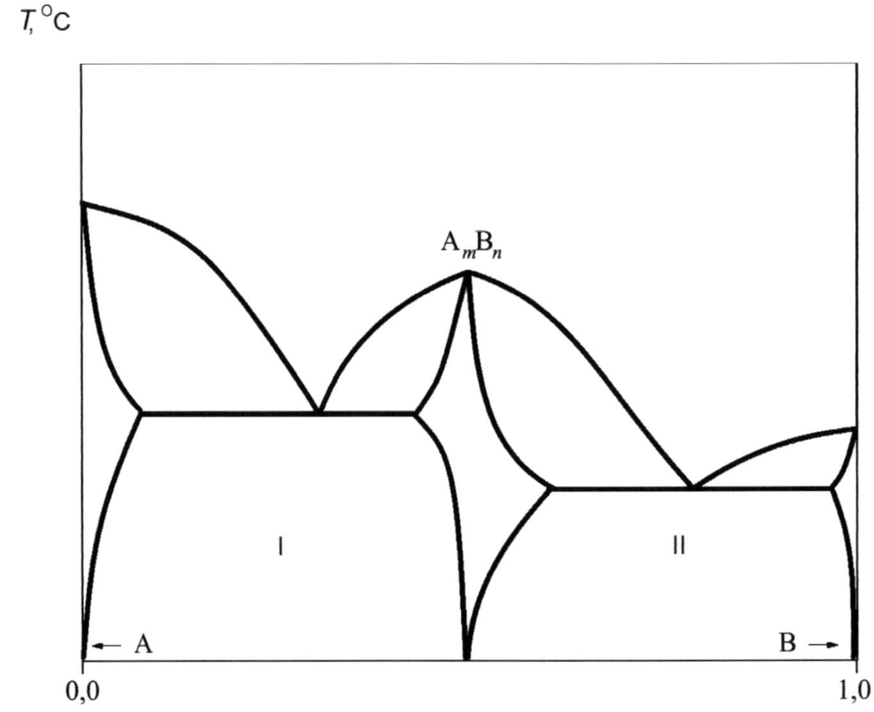

Figure 1. Phase diagram of the eutectic type with intermediate phase variable composition.

[4]. The method is based on the nonlinear transformations of the concentration axes and makes it possible to correctly reduce the calculation of the phase diagrams with several intermediate phases to a sequence of phase diagrams of individual subsystems. These cases were studied in detail in papers [4,5]. For this reason, we do not write out the explicit expressions for the nonlinear transformations of the concentration axes, which can be found in the papers mentioned above.

According to the basic principles of the generalized lattice model, the chemical potentials of a homogeneous binary solution (per mole of a material) can be represented in the following form [6]

$$\begin{aligned}\mu_1 &= \mu_{10} + RT\ln x + W\lambda\left(\frac{1-x}{x+\lambda(1-x)}\right)^2,\\ \mu_2 &= \mu_{20} + RT\ln(1-x) + W\left(\frac{x}{x+\lambda(1-x)}\right)^2,\end{aligned} \tag{1}$$

where μ_{i0} is the standard chemical potential of the ith component, R is the universal gas constant, T is the temperature of the system, x is the mole fraction of the first component, W is an analog of the energy of mixing in the generalized lattice model, $\lambda = \omega_2/\omega_1$, and ω_i is the characteristic atomic volume of the ith component.

Further let us consider the subsystem of an A–A_mB_n binary solution of eutectic type (see Figure 1). We introduce three new parameters W, G and U, where W is an analog of the energy of mixing in the luiqid phase, G is an analog of the energy of mixing in α solid solution based on atomic component A and U is an analog of the energy of mixing in β intermediate phase based on molecular component A_mB_n. Then relationships (1) determine the chemical potentials of components in luiqid phase. The chemical potentials of components in solid solutions are considered in a similar way.

We can obtain all parameters of general lattice model for binary solutions of the eutectic type taking into account the chemical equilibrium in the three-phase system, i.e., equating the chemical potentials of the components in the solid and liquid phases.

Equations of the chemical equilibrium in the three-phase system form a closed system, whose solution allows us to find the concentration dependences of the liquidus branches, solidus branches and solvus branches and, hence, to construct the phase diagrams of binary system A–A_mB_n in the framework of the generalized lattice model.

It should be noted that the chemical equilibrium in the subsystem of an A_mB_n–B binary solution is considered in a similar way.

3. Phase diagram calculation

The table presents the calculated parameters of the generalized lattice model for the W–Si binary system.

Table 1. Parameters of the generalized lattice model.

A–B	q_1	q_2	λ	$\widetilde{W}$, K	$\widetilde{U}$, K	$\widetilde{G}$, K
Si_2W–Si_3W_5	25.0	5.0	0.109	3411	1654	-
Si_3W_5–W	2.0	25.0	2.437	10491	8303	-

Here $\widetilde{W} = W/R$, $\widetilde{U} = U/R$, and $q_i = \Delta H_i/RT_i$, where ΔH_i are the latent heats of the liquid–solid transition for the pure components, T_i stands for the melting temperatures of the pure components. The parameters q_i were determined with the use of the reference data on the latent heats of the liquid–solid transition.

In addition it should be noted that explicit account of "volume" effects in the W–Si system with intermediate phases of constant composition (Si_2W) and variable composition (Si_3W_5) was carried by introducing "specific" the atomic volumes of W and Si within the generalized lattice model.

The corresponding phase diagram calculated with these parameters are depicted in Figure 2. It should be noted that the results obtained from the theoretical calculations are in agreement with the experimental data available in the literature [7–9].

Thus, comparison with experimental data shows that the theoretical approach can be used to adequately description and prediction of the phase equilibria in the W-Si binary system.

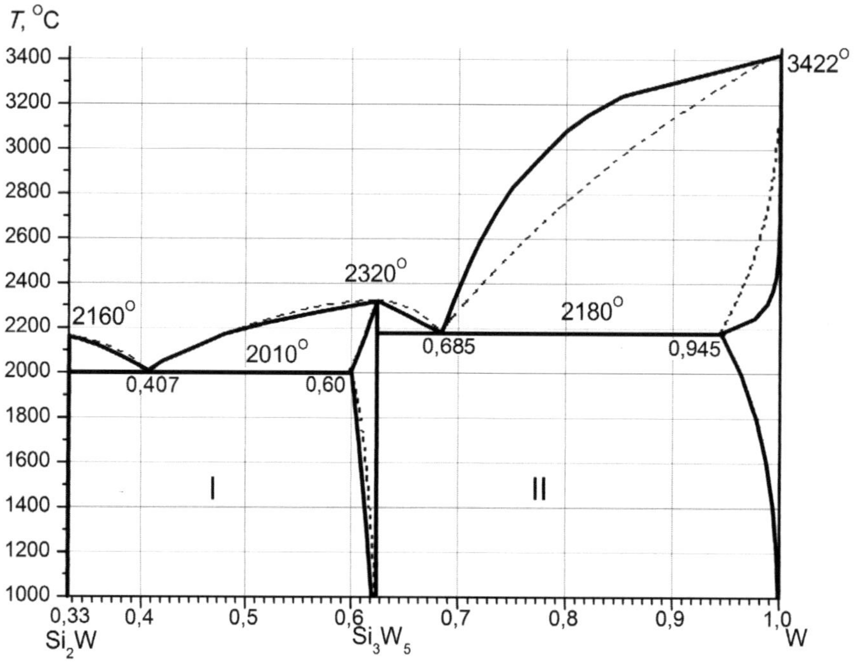

Figure 2. Phase diagram of the W–Si system. Dashed lines are the experimental data taken from [7-9]. Solid lines indicate the results obtained from theoretical calculations.

References

[1] Gorelenok A T, Krukov V L and Furmanov G P I A 1994 *Technical Physics Letters* **20** 21
[2] Voronkova G M, Zuev V V and Kiruhin A D 1997 *Neorg. Mater.* **33** 1873
[3] Zakharov A Yu, Zakharov M A and Loginova O V 2004 *Int. J. Quant. Chem.* **100** 435
[4] Zakharov M A 2007 *Phys. Solid State* **49** 2312
[5] Panov G A and Zakharov M A 2013 *J. of Phys.: Conf. Ser.* **461** 012003
[6] Korzun E L and Terekhov S V 1987 *Zh. Fiz. Khim.* **61** 1186
[7] Kieffer R, Benesovsky F and Gallistl E 1952 *Z. Metallkd.* **43** 284
[8] Maksimov V A and Shamrai P I 1969 *Neorg. Mater.* **5** 1136
[9] Kocherzhinskii Yu A, Kulik O G, Shishkin E A and Yupko L W 1973 *Dokl. Akad. Nauk SSSR* **212** 642

Diffuse reflectance measurements using lensless CMOS imaging chip

I. Schelkanova*1, A. Pandya*1, D. Shah*3, L. Lilge*2,3, A. Douplik*1,4,5

[1] Physics Department, Ryerson University, Toronto, Canada
2 Princess Margaret Cancer Centre and Department of Medical Biophysics, University of Toronto, Toronto, Canada
3 Techna Institute for the Advancement of Technology for Health, UHN Microfabrication Centre, Toronto, Canada
4 Keenan Research Centre of the LKS Knowledge Institute St. Michael Hospital, Toronto, Canada
5 Erlangen Graduate School in Advanced Optical Technologies, Friedrich Alexander University Erlangen-Nuremberg, Erlangen, Germany

e-mail: irina.schelkanova@ryerson.ca

Abstract. To assess superficial epithelial microcirculation, a diagnostic tool should be able to detect the heterogeneity of microvasculature, and to monitor qualitative derangement of perfusion in a diseased condition. Employing a lensless CMOS imaging chip with an RGB Bayer filter, experiments were conducted with a microfluidic platform to obtain diffuse reflectance maps. Haemoglobin (Hb) solution (160 g/l) was injected in the periodic channels (grooves) of the microfluidic phantom which were covered with ~250 μm thick layer of intralipid to obtain a diffusive environment. Image processing was performed on data acquired on the surface of the phantom to evaluate the diffuse reflectance from the subsurface periodic pattern. Thickness of the microfluidic grooves, the wavelength dependent contrast between Hb and the background, and effective periodicity of the grooves were evaluated. Results demonstrate that a lens-less CMOS camera is capable of capturing images of subsurface structures with large field of view.

1. Introduction

Functional capillary density (FCD) and heterogeneity of microvascular pattern (MVP) are widely used clinical parameters associated with a number of medical conditions such as inflammation, benign lesions, early mucosal cancers, pathophysiology of sepsis, and quality of the tissue perfusion [1]–[4]. It is due to the complex relationships between microvascular evolution and an early detection of cancer, assessment of its grade and prediction of treatment outcome still remain a formidable challenge of modern medicine [4].

The normal mucosa is composed of an avascular epithelium and a subjacent vascularized dermis. Dense arborescent vessels nourish the epithelial layer via a network of thin vessels, which rise to the surface of the mucosa and form intrapapillary capillary loops (IPCLs). Individual IPCLs are periodically distributed in the dermal papilla space with an average interval of about 100-180μm, which corresponds to the maximum distance that oxygen diffuses from a vessel [5], [6]. In mucosal layers, this capillary network is located, on average, 350 μm beneath the surface [7], [8].

With the onset of mucosal cancer, the progressive neovascularization leads to an increase in microvessel density. This process results in gradual transformation of the organized microcirculation network, inducing structural changes of the local capillary structure, and disturbing the spatial pattern of the initially periodically distributed haemoglobin [6].

Diffuse reflectance spectroscopy is widely used to evaluate tissue averaged haemoglobin changes in a volume sampled by the light traveling from the source to the detector. The spectroscopic technique offers a potential to determine optical properties of a sample, and/or quantify relative concentrations of the chromophores of interest integrated over the interrogated volume. A spatially resolved diffuse reflectance spectroscopy (SRDR spectroscopy) based on a fibre bundle probe can potentially provide a sufficient resolution to capture the structural distribution of the capillary loops but it lacks proper visualization of the tissue [9].

Current article is focused on the results of the development of a novel miniature, spatially resolved diffuse reflectance technology based on a lensless CMOS imaging chip to monitor spatial parameters of subsurface microfluidic pattern. These experiments are designed to test the ability of the method to evaluate the width of the haemoglobin grooves in the microfluidic phantom, to estimate the difference in haemoglobin contrast as reordered by the RGB channels of the conventional CMOS chip, and to determine the periodicity of the embedded structure.

2. Method

2.1 Equipment

A Motic AE31 Inverted Microscope, on-board RGB CMOS chip (Motic, China) was used. The CMOS matrix had physical dimensions of 6.8cm x 5.3cm (the active dimensions are less but this information was not provided by the manufacture), and a maximum resolution of 3 mega pixels (MP) (2048 x 1536) which corresponds to a pixel pitch of ≈ 3.2μm, neglecting packaging area losses at the chip boundary. The CMOS read-out offered variable resolutions ranging from 0.2 MP, 0.8MP, 3MP, corresponding to an effective pixel pitch of 12.5μm, 6.3μm, and 3.2μm respectively. Spectral resolution of the CMOS was determined by the spectral width of the Red-Green-Blue channels (≈ 100 nm). The analog to digital conversion had an 8-bit resolution. Software Motic Images Plus 2.0 for PC software package was used for image acquisition. The visualization of images and the image processing was performed by an algorithm created in MATLAB (MathWorks, USA).

Illumination of the sample was achieved with two 0.400 mm core optical fibers (Silica fibers, NA 0.22, Thorlabs Inc., USA) coupled to a broadband white light source, HPLS-30-02 (Thorlabs Inc, US). Placing the fibres at the extreme edges of the CCD (as shown in Fig. 1) results in a non-uniform illumination of the sample. For analysis, a region of interest was chosen manually which portrayed the lowest variation in illumination and a background compensation was performed as described in the "Image processing" section. Although non-uniform illumination can be compensated with background subtraction techniques, a diffusive illumination setup is planned to be designed for future experiments to provide uniform illumination.

Microvasculature-simulating microfluidic phantoms with narrow periodic grooves filled with haemoglobin extracted from red blood cells were used in the experiments. The microfluidic structure was produced in collaboration with the Microfabrication Laboratory at University Health Network, Techna Institute (Toronto, Canada). The device was designed to reproduce capillary spatial pattern with close to realistic physical dimensions: on average, 34μm wide groove, separated by 82μm spacing, the groove's height - 30μm.

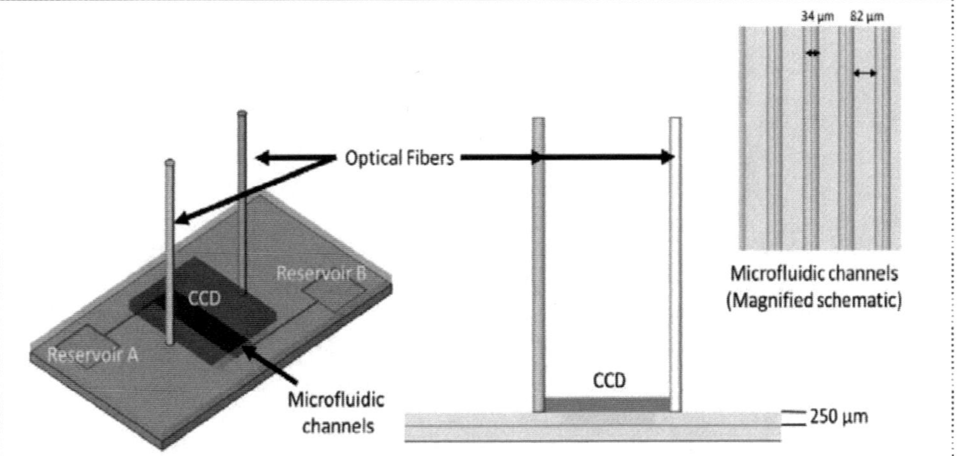

Figure 1. Design of the experimental setup for the diffuse reflectance measurements: an axonometric view and a side view. The magnified schematic of the microfluidic channels shows an average value of the groove's width, 34µm, a separation of approximately 82 µm. In Z-direction (depth), the groove is 30µm. The total area covered by the microfluidic channels was 5mm x 5.75mm.

The phantom was manufactured from a clear PDMS (silicone) material, n=1.56 in the visible spectrum without addition of scattering particles. The microfluidic channels were covered by a layer of intralipid to mimic the biotissue light diffusive properties. Original solution of intralipid (20%) (I141, Sigma-Aldrich Inc, USA) was diluted in deionized water to obtain 1% and 2% intralipid concentrations, which were used within 10 minutes of the preparation time. Based on the scattering properties of the intralipid reported in the literature, [10],[11],[12] 1% and 2% concentrations were selected to match the scattering properties (µs') of the epithelium [13].

The thickness of the intralipid layer on the top of the grooves was ≈ 250µm. The solution of the haemoglobin (160 g/l, Sigma Aldrich, USA) was injected in the reservoir A (2mm x 2mm) by means of an 25 gauge needle and a syringe (Fig.1). Another needle was placed in reservoir B (2mm x 2mm) to facilitate the flow of the haemoglobin from reservoir A to reservoir B, and to obtain a uniform distribution of haemoglobin in the microfluidic channels.

2.2 Image Processing

Denoising of the images was performed by a 2D Wiener filter implemented in MATLAB, with a pixel neighbourhood of the dimensions 5 x 5 pixels. The Wiener filter is designed as an adaptive filter which performs little smoothing where the variance is large, and more smoothing where the variance is small. This helps to preserve edges and the high frequency components of the image. Images were rotated for the alignment of channels vertically to facilitate the calculation of microfluidic channel widths. Line scanning was performed on the images and due to the nature of illumination in our experiments, a low pass filter was implemented in the form of a moving average with a window size of 30 samples per scan line to estimate the background compensation required. It was estimated that the spatial frequency of the channels was much higher than the frequency pertaining to the variance of the background illumination across the illumination field. Hence, illumination variance compensation was achieved by the subtraction of a low-passed output from the raw image data line by line. The resultant data was mapped as a binary image [-1, 1] and the zero crossings in each line were obtained to calculate the channel widths. The zoomed section of the Fig.2 represents a sample of processed output.

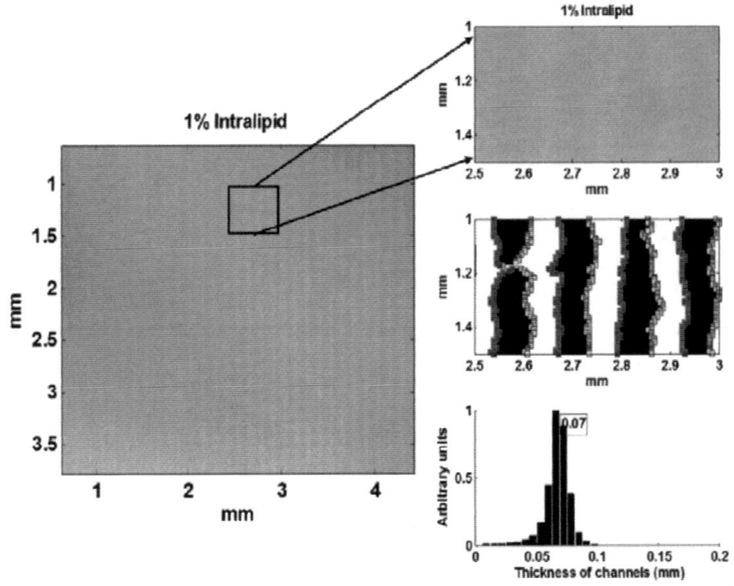

Figure 2. A portion of the original image of the microfluidic channels covered by 1% intralipid layer (top right) is selected for further image processing: creation of a binary mask, and defining the boundaries of the channels (middle right). The extracted values of the thicknesses of the channels are presented in the form of a histogram (bottom right), where the peak values of the distribution is provided in the box.

3. Results

In measurements of the diffuse reflectance signal with on-surface CMOS device, two optical fibres irradiate the surface of a sample while the RGB CMOS camera captures an image. Some part of unpolarised light returning from the subsurface layers of the microfluidic phantom, passes through a set of red, green, and blue filters integrated in the CMOS chip array. The corresponding image is stored digitally as three matrixes representing the red (600–700 nm), green (500–600 nm), and blue (400–500 nm) colour planes, respectively [14].

Evaluation of the thickness of the grooves in three different experimental conditions are shown in Fig.3. Left column contains two images (original and binarized) and a histogram of the condition when no diffusive media (intralipid) was added on top of the capillary layer, whereas the middle, and the right column show the results for a low and high scattering layers over the grooves of the microfluidic. The intralipid layer on the top of the microfluidic phantom produces an intensity gradient at the boundaries of the haemoglobin channels. As can be seen in the histograms (bottom plots), the thickness of the channels' images acquired through the intralipid layer demonstrates broadening of their width due to the longer paths of light (Fig.3, central, and right bottom charts).

Results of the comparative analysis of the RGB channel intensities for the grooves and background are displayed in Fig.4. At first, the intensity values in the grooves and the background were normalized to the 8-bit integer maximum (255) to obtain values in the range [0, 1]. The distribution of the normalized intensities (X-axis) in the grooves (solid lines) and background (dashed lines) are colour-coded, and plotted (Fig 4). The distributions were normalized (Y-axis) by the respective maximum frequency to account for variations in the field of view as well as for the better representation of the peak separations. Percent difference representing the peak shifts in intensities for each colour channel was calculated to observe the contrast for each spectral band.

Journal of Physics: Conference Series **541** (2014) 012098 doi:10.1088/1742-6596/541/1/012098

Periodicity of the haemoglobin grooves in three experimental conditions (no intralipid, 1% and 2% intralipid on top of the microfluidic channels) was evaluated by obtaining the prominent spatial frequency contained within the line scans performed on the field of view depicted in Fig. 3. Fast Fourier Transform (FFT), corresponding to a sampling interval of 0.0063mm (effective pixel pitch), was performed line-by-line from top to bottom, and a mean spectrum was obtained to represent the most prominent frequency as shown in Fig. 5. The peak in the spectrum was observed at an 8 mm^{-1} mark, corresponding to a spatial separation of 125μm. This value closely resembles the actual pitch of the haemoglobin carrying grooves (34μm + 82μm = 116μm).

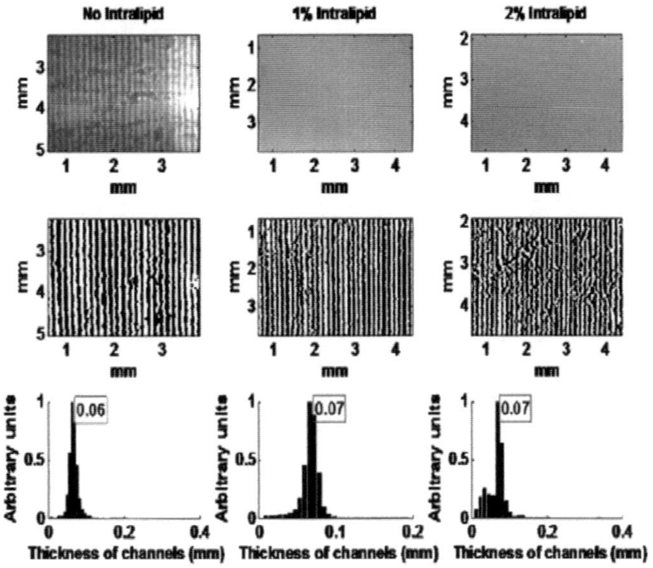

Figure 3. Top panel: Original images of the microfluidic phantoms: no intralipid (left), intralipid 1%, intralipid 2% (right). Binarized images of the region of interest for three experimental conditions are in the centre row of the figure. Histograms (bottom panel) of the recovered widths of the grooves, corresponding to the binary images above.

4. Discussion

A novel method of the application of a lensless CMOS imaging chip for the on-surface measurements of the diffuse reflectance signal has a potential for further development into a clinical technique, which can provide both, structural and metabolic information about functional heterogeneity of microvascular pattern (MVP) and capillary density (FCD). Since application of the lensless CMOS chip for the on-surface measurements of the diffuse reflectance data is a novel approach, the most important research question to be answered is how the camera can capture subsurface images of embedded structures. The main objective of these experiments is to evaluate the image quality of the periodic channels under various scattering conditions, using intralipid as a diffusive media topping the surface of the haemoglobin grooves.

The experimental results have confirmed that the lensless CMOS imaging set-up can be used to acquire the diffuse reflectance signal from the subsurface of a microfluidic phantom with the embedded periodic structure. The collected signal contains both components of the reflected light, i.e., surface reflectance and subsurface reflectance [14]. To separate these two components a set of polarizers will be included in the future system, so the surface and subsurface information could be de-coupled by the polarization gating techniques.

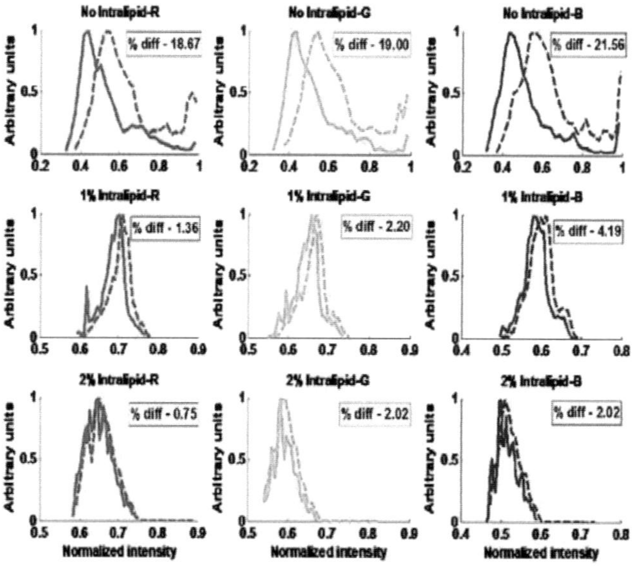

Figure 4. Distribution of intensity values for Red, Green, Blue .ipg imaging channels of the target (grooves, solid lines) and the background (dashed lines) for three experimental conditions: no intralipid (left), 1% intralipid (center), 2% intralipid (right) are plotted. Percent difference between the target and background intensities are displayed on the graph.

Noise reduction in the images was achieved by using an adaptive 2D Wiener filter which was chosen for its ability to preserve the dimensions of the original spatial pattern. The noise in the images was assumed to be Gaussian, and performance of the image processing was manually verified. This approach allowed us to recover the spatial frequency of the grooves which was found sufficiently close to the original dimensions.

Although the 100nm bandwidth of the RGB channels is too broad to be considered for spectroscopic applications, the absorption contrast of the Red, Green and Blue channels have been qualitatively compared. In tissue spectroscopy, wavelengths dependent absorption is linked to the quantitation of the relative concentrations of tissue chromophores. For hemoglobin, there are several absorption peaks present in the green and blue regions of the spectrum. Hence, it was expected that, despite the breadth of the spectral bands, we should observe an enhanced contrast for the Green-Blue channels of the RGB .jpg image as compared to the Red channel values. Indeed, the largest peak difference (contrast) between the grooves and background intensities was found for the Blue and Green spectral bands, while the red component demonstrated relatively insignificant shift of the curve (Fig.4). Virtually no significant difference between the Blue and Green contrasts was observed when the scattering properties of the phantoms were increased. That is the condition when the narrow band imaging would have greater advantage. In future modification of the system, a monochrome CMOS imaging chip would provide more clinically valuable information as it enables the experiment to be fine-tuned for the absorption by a specific chromophore using a selected spectral band. Switching multiple spectral bands opens up advantages of the hyperspectral imaging.

The periodicity of the pattern was evaluated by observing the FFT spectrum. All the images were rotated before processing to align the microfluidic channels parallel to the direction of line scans (Y-axis in Fig. 1). The microfluidic phantom was placed on a flat surface, and the channels were parallel to the CMOS chip plane. The difference in the measured spatial frequency and the actual spatial frequency was ≈ 7.2%. We believe this difference is attributed to the uneven orientation of the layers to the

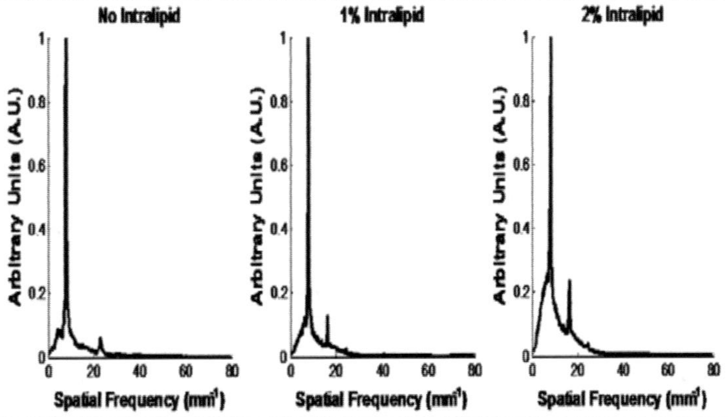

Figure 4. Periodicity of the microfluidic channels obtained by taking a Fast Fourier Transform (FFT) of the filtered scan lines.

direction of line scans, non-uniform illumination, and the scattering top layer (in case of the intralipid layers). Since our set-up contained a linear pattern of grooves, 1D Fourier transform was sufficient to provide an estimate of the periodicity of the pattern. Also, the variability in the thickness estimation can be used to qualitatively evaluate the periodicity of the pattern. This shows that we can design application-based computational techniques to evaluate the periodicity of the microvascular patterns. In future, correlation matrix based method can be implemented to assess the periodicity, and the quantitative analysis (e.g. covariance measures on thickness estimation) can be performed to obtain a measure of periodicity as a clinical parameter.

There are several advantages of the usage of the CMOS imaging chip for the diffuse reflectance measurements of the subsurface space. The larger coverage area facilitates collection of the optical signal from a large field of view, whereas the small sampling interval increases the method's spatial resolution. Its pixel size enables "continuity" of the signal, and satisfies sampling criteria for robust 2D resolution of the microvasculature in epithelial tissue. The interrogated depth and blood contrast can be improved by exploiting the Optical Clearing technique [15].

Future modifications of the device will target improvements of the design, and reproducibility of the measurements. The uniform illumination of the sample is an important factor in prevention of the saturations spots, which cause losses of imaging area [16]. Overall, the technique has potential for subsurface imaging.

5. Conclusion

The main application of the technique considered in this paper is a targeted imaging of the spatial parameters of subsurface microfluidic pattern. The experiments with a lensless, on-surface CMOS chip imaging on the intralipid microfluidic tissue phantoms have demonstrated a potential of such 2D spatially resolved method. It offers the advantage of the real time, non-invasive imaging of the structural architecture of subsurface patterns.

6. Acknowledgments

This work was supported in part by the Ryerson University Health Research Grant and the NSERC Personal Discovery Grant, Canada. The research was also partially supported by the Ontario Graduate Scholarship, Ryerson University, Toronto, Canada and by the Erlangen Graduate School in Advanced Optical Technologies (SAOT), German National Science Foundation (DFG), Germany, in the framework of the excellence initiative.

7. References

[1] K. Yao, G. K. Anagnostopoulos, and K. Ragunath, "Magnifying endoscopy for diagnosing and delineating early gastric cancer," Endoscopy, vol. 41, no. 5, pp. 462–467, 2009.

[2] C. Schmidt, C. Lautenschläger, B. Petzold, Y. Sakr, G. Marx, and A. Stallmach, "Confocal laser endomicroscopy reliably detects sepsis-related and treatment-associated changes in intestinal mucosal microcirculation," Br. J. Anaesth. , Jun. 2013.

[3] B. Nico, V. Benagiano, D. Mangieri, N. Maruotti, A. Vacca, and D. Ribatti, "Evaluation of microvascular density in tumors: Pro and contra," Histol. Histopathol., vol. 23, no. 5, pp. 601–607, 2008.

[4] E. Sabo, A. Boltenko, Y. Sova, A. Stein, S. Kleinhaus, and M. B. Resnick, "Microscopic analysis and significance of vascular architectural complexity in renal cell carcinoma," Clin. Cancer Res., vol. 7, no. 3, pp. 533–537, 2001.

[5] I. M. Braverman and A. Yen, "Ultrastructure of the capillary loops in the dermal papillae of psoriasis," J Investig Dermatol, vol. 68, no. 1, pp. 53–60, 1977.

[6] Y. Kumagai, M. Toi, and H. Inoue, "Dynamism of tumour vasculature in the early phase of cancer progression: outcomes from oesophageal cancer research," Lancet Oncol., vol. 3, no. 10, pp. 604–610, 2002.

[7] I. M. Braverman and A. Yen, "Ultrastructure of the human dermal microcirculation: II. The capillary loops of the dermal papillae," J. Invest. Dermatol., vol. 68, no. 1, pp. 44–52, 1977.

[8] R. A. O. Bennett, R. N. Pittman, and S. M. Sullivan, "Capillary spatial pattern and muscle fiber geometry in three hamster striated muscles," Am. J. Physiol. - Hear. Circ. Physiol., vol. 260, no. 2 29–2, pp. H579–H585, 1991.

[9] G. Saiko, A. Pandya, I. Schelkanova, M. Sturmer, R. J. Beckert, and A. Douplik, "Optical Detection of a Capillary Grid Spatial Pattern in Epithelium by Spatially Resolved Diffuse Reflectance Probe: Monte Carlo Verification," Selected Topics in Quantum Electronics, IEEE Journal of, vol. 20, no. 2. pp. 1–9, 2014.

[10] S. T. Flock, S. L. Jacques, B. C. Wilson, W. M. Star, and M. J. C. Van Gemert, "Optical properties of intralipid: A phantom medium for light propagation studies," Lasers Surg. Med., vol. 12, no. 5, pp. 510–519, 1992.

[11] H. J. van Staveren, C. J. M. Moes, J. van Marie, S. A. Prahl, and M. J. C. van Gemert, "Light scattering in Intralipid-10% in the wavelength range of 400-1100 nm," Appl. Opt., vol. 30, no. 31, pp. 4507–4514, 1991.

[12] P. D. Ninni, F. Martelli, and G. Zaccanti, "Intralipid: Towards a diffusive reference standard for optical Phantoms," Phys. Med. Biol., vol. 56, no. 2, pp. N21–N28, 2011.

[13] V. V Tuchin, Handbook of optical biomedical diagnostics. SPIE Press, 2002.

[14] M. J. Leahy, Microcirculation Imaging. Wiley, 2012.

[15] V. V Tuchin, Optical Clearing of Tissues And Blood. SPIE Press, 2006.

[16] P. Sun, R. Q. Yang, F. H. Xie, J. Q. Ding, F. Q. Zhang, and X. P. Cao, "A method for determining optical properties of human tissues by measuring diffuse reflectance with CCD," in Progress in Biomedical Optics and Imaging - Proceedings of SPIE, 2010, vol. 7845.

The discrete spectrum of the multiparticle Hamiltonian in the framework of the Hartree-Fock approximation

I F Melikhov[1], **I Yu Popov**[1]
[1] ITMO University, Kronverkskiy pr., 49, St. Petersburg, 197101, Russia
E-mail: ivan.melikhov@gmail.com, popov1955@gmail.com

Abstract. The paper deals with the problem of quantum particles storage in nanolayered structures. The system of a few interacting electrons placed into deformed waveguide is studied. Different interaction models — δ-type and Coulomb type — and different 2D geometries of the waveguide are under investigation. The ability to store or localize particles ("bound state") corresponds to the existence of the discrete spectrum of the system Hamiltonian. The dimension of the problem growth with the number of particles. To avoid these difficulties we use Hartree-Fock method which gives the system of one particle equations. The equations are solved with the help of FEM. As the result we have the dependence of the minimal deformation coefficient which keeps bound states on the number of particles and its total spin. Comparison of the storage efficiencies is made.

1. Introduction

The problem of the existence of the system's bound state is very important in many physical questions. One of its illustration is quantum particle storage problem. For example, it may be applied in a hydrogen accumulator based on nanolayered structure. One layer of such structure can be considered as a waveguide. In the terms of math, the ability to store a particle means the existence of the discrete spectrum of the corresponding Hamiltonian. It is known that single particle placed into straight waveguide has no bound states, so it can't be stored or localized. Introducing a small deformation of the waveguide causes the eigenvalues appearance [1]-[7]. For many practical applications it's necessary to store more than one particle. That leads to the multi-particle system [8], [9]. Straightforward solving of Schrödinger equation for such system is heavy task even for numerical methods. We use Hartree-Fock approximation to simplify the equations and solve it by FEM.

2. Governing equations

Consider a number of interacting electrons placed into a nanolayer or a waveguide. If we ignore the spin-orbit interaction then the Hamiltonian of the multi-particle system is written as follows

$$\hat{H} = \sum_k \left(-\frac{\hbar^2}{2m}\nabla_k^2 + U_k \right) + \frac{1}{2}\sum_{\substack{j,k \\ j\neq k}} V_{jk} = \sum_k \hat{H}_k + \frac{1}{2}\sum_{\substack{j,k \\ j\neq k}} \hat{H}_{jk}, \quad (1)$$

where m is particle's mass, U_k is the external field potential, V_{jk} is the potential of the particles interaction. The multi-particle Schrödinger equation for (1) needs to be simplified. An approach based on the Hartree method is described in [10]. The drawback of this method is symmetry of the system's wavefunction with respect to the interchange of particles spatial coordinates. Thus, it is not fully correct model for fermions.

In this article we offer an approach based on the Hartree-Fock method which allows one to avoid the described drawback. The idea of the method is to look for the system's wavefunction in the form of Slater determinant:

$$\psi(x_1, x_2, \ldots, x_n) = \frac{1}{\sqrt{n!}} \begin{vmatrix} \psi_1(x_1) & \psi_2(x_1) & \ldots & \psi_n(x_1) \\ \psi_1(x_2) & \psi_2(x_2) & \ldots & \psi_n(x_2) \\ \ldots & \ldots & \ldots & \ldots \\ \psi_1(x_n) & \psi_2(x_n) & \ldots & \psi_n(x_n) \end{vmatrix}, \quad (2)$$

where ψ_i is the single particle function, n is the number of particles, x_i represents the set of four (or three) coordinates: three (or two) spatial ones and a spin one.

Substituting (2) into (1) and using the variational principle, after simplification one can obtain the *Hartree-Fock equations*:

$$\hat{F}^{\uparrow}\psi_k^{\uparrow} = \left[\hat{H}_k + \sum_{j=1}^{n^{\uparrow}} \left(\hat{J}_j^{\uparrow} - \hat{K}_j^{\uparrow} \right) + \sum_{j=1}^{n^{\downarrow}} \hat{J}_j^{\downarrow} \right] \psi_k^{\uparrow} = E_k^{\uparrow}\psi_k^{\uparrow}, \qquad k = \overline{1, n^{\uparrow}} \quad (3)$$

$$\hat{F}^{\downarrow}\psi_k^{\downarrow} = \left[\hat{H}_k + \sum_{j=1}^{n^{\uparrow}} \hat{J}_j^{\uparrow} + \sum_{j=1}^{n^{\downarrow}} \left(\hat{J}_j^{\downarrow} - \hat{K}_j^{\downarrow} \right) \right] \psi_k^{\downarrow} = E_k^{\downarrow}\psi_k^{\downarrow}, \qquad k = \overline{1, n^{\downarrow}} \quad (4)$$

$$\hat{J}_j^{\uparrow\downarrow}\psi_k(r_k) = \left[\int \overline{\psi_j^{\uparrow\downarrow}(r_j)} V_{jk} \psi_j^{\uparrow\downarrow}(r_j)\, dr_j \right] \psi_k(r_k), \quad \hat{K}_j^{\uparrow\downarrow}\psi_k(r_k) = \left[\int \overline{\psi_j^{\uparrow\downarrow}(r_j)} V_{jk} \psi_k(r_j)\, dr_j \right] \psi_j^{\uparrow\downarrow}(r_k),$$

where the superscripts $\uparrow$ and $\downarrow$ denote to particles with different spins, ψ_i is a single particle function, r_i is the spatial coordinate of a particle. Now the $n^{\uparrow}$ and $n^{\downarrow}$ lowest eigenfunctions of the operators $\hat{F}^{\uparrow}$ and $\hat{F}^{\downarrow}$, respectively, are to be found. Equations (3)–(4) are solved numerically in the way similar to the one described in [10].

If all eigenvalues $E_1^{\uparrow\downarrow}, \ldots, E_n^{\uparrow\downarrow}$ are less than the threshold, the bottom of the continuum spectrum of single particle, then the bound state exists. The threshold can be found by $E_{tr} = \pi^2\hbar^2/(2mL^2)$, where L is the waveguide width at infinity.

3. Results

We studied two types of particles interaction, δ-type and Coulomb, which are described by the following potentials:

$$V_{ij}^1 = U_\delta \delta(r_i - r_j), \qquad V_{ij}^2 = \frac{U_c}{|r_i - r_j|},$$

where U_δ and U_c are interaction intensity constants. Physically, they can be changed by the effective mass and charge of the particles.

All the problems are solved in 2D. In the case of δ-interaction three geometries were studied: i) waveguide with locally deformed boundary (Fig. 1), ii) two waveguides coupled through a window (Fig. 2), iii) bent waveguide (Fig. 3). For all these cases the existence of single particle bound states has been proved [1] – [7]. The Coulomb interaction was studied for the coupled waveguides.

Comparing three types of the system geometry in the case of δ-interaction we can conclude that the bent waveguide is the less efficient one. It can store only two particles with opposite

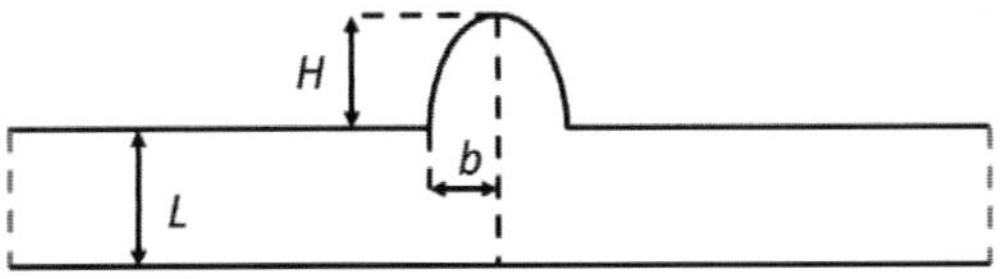

Figure 1. Deformed waveguide.

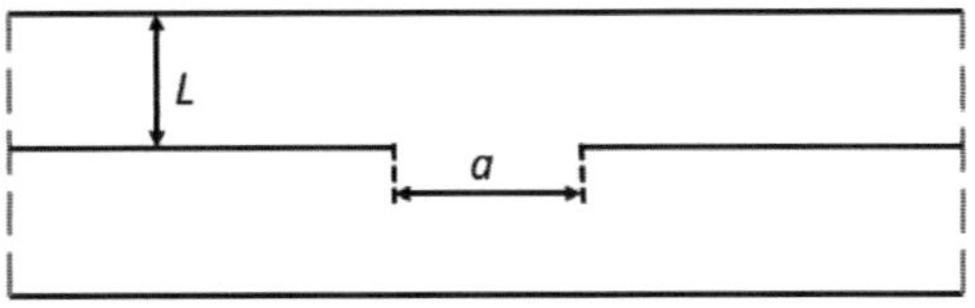

Figure 2. Window-coupled waveguides.

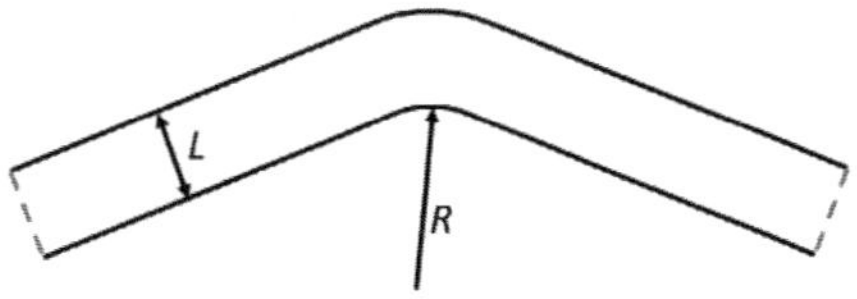

Figure 3. Bent waveguide.

spins. Placing more particles or changing one particle's spin leads to destroying of the bound state.

The window coupled layers appear the most efficient in the sense of particle storage. Increasing of the interaction intensity leads to destroying of the bound state. From the other hand, increasing of the window width leads to the rise of the discrete spectrum threshold. The correlation between the intensity and the deformation in the case of two particles in a window-coupled layers is shown on Fig. 4. To compare two types of potentials the relationship between the constants U_δ and U_c must be established:

$$U_\delta = \alpha U_c, \qquad \alpha = \frac{\iint \delta(|r_1 - r_2|)\, dr_1\, dr_2}{\iint \frac{1}{|r_1 - r_2|}\, dr_1\, dr_2}.$$

The domains below the curves corresponds to the existence of the bound states. One can see that for particles with collinear spins there is a threshold deformation which allows bound states. The system of two particles with collinear spins for δ-interaction represents two single particles and doesn't depend on the interaction intensity.

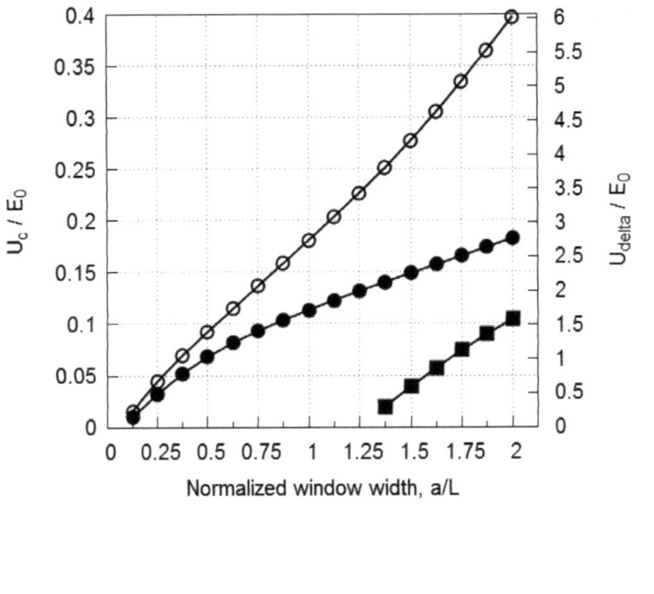

Figure 4. Domain on the dimensionless parameters plane $(a/L, U_c/E_0)$ corresponding to the existence of eigenvalues of two-particle Hamiltonian for coupled layers (below the curve), $E_0 = \hbar^2/(2mL^2)$. ● Coulomb interaction, zero total spin, ○ δ-interaction, zero total spin, ■ Coulomb interaction, unit total spin.

Increasing of particles number or total system's spin leads to larger deformation required to store the system. It is interesting to find what (minimal) deformation is sufficient for storage of a given system. On Fig. 5 the minimal window width keeping bound states is shown. The result on the plot is computed for $U_c = 0.01$ and corresponding $U_\delta = 0.066$. The lower points corresponds to the lower total spin. One can see that for small interaction intensity the δ-type and Coulomb potentials gives one similar results. It can be seen that the required deformation depends on the number of the particles with the same spin. Thus, it is more efficient to store even number of particles with zero total spin than one particle less.

Fig. 6 shows the distribution of the electron density for the system of two particles with opposite spins in the layers with the window width $a/L = 1/2$.

The results of this work can be used to estimate the optimal parameters of a storage nanosystem.

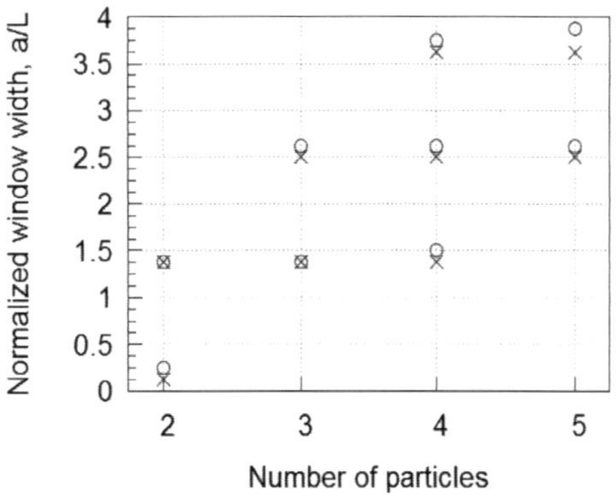

Figure 5. Minimal window width which keeps bound states for the case of layers coupled through a window, lower points correspond to smaller total spin.
○ Coulomb interaction, × δ-interaction.

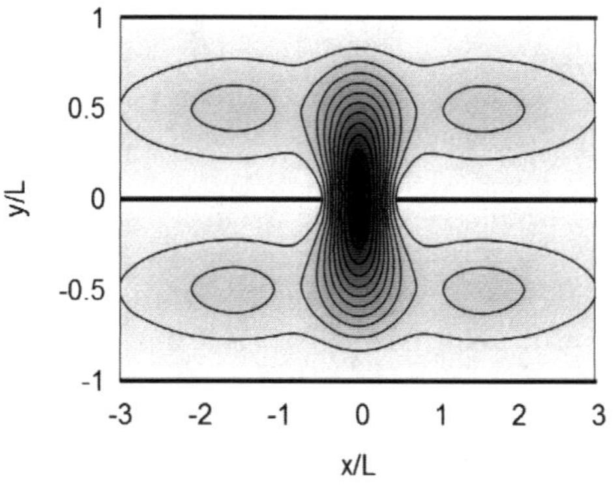

Figure 6. Electron density distribution of the two-particle state for layers coupled through a window.

Acknowledgments

The work was partially financially supported by the Government of the Russian Federation (grant 074-U01), by the Ministry of Science and Education of the Russian Federation (GOSZADANIE 2014/190, Project 14.Z50.31.0031), by grant of Russian Foundation for Basic Researches and grants of the President of Russia (state contracts 14.124.13.2045-MK and 14.124.13.1493-MK).

References

[1] Duclos P, Exner P and Stovichek P 1995 *Annales de l'I.H.P.* A **62**, 81–101
[2] Duclos P, Exner P and Krejcirik D 2000 *Ukrainian J. Phys.*, **45**, 595–601
[3] Exner P and Vugalter S A 1996 *Ann. Inst. H. Poincare* **65** 109–23.
[4] Exner P and Vugalter S A 1997 *Lett. Math. Phys.* **39**, 59–68
[5] Exner P, Seba P, Tater M and Vanek D 1996 *J. Math. Phys.* **37** 4867–87
[6] Popov I Yu 1999 *Rep. on Math. Phys.* **43** 427–37.
[7] Popov I Yu 2000 *Phys. Lett.* A **269** 148-153
[8] Exner P and Vugalter S A 1999 *J. Math. Phys.* **40** 4630–38
[9] Linde H 2006 *J. Phys.* A **39** 5105–14
[10] Popov S I, Gavrilov M I and Popov I Yu 2012 *Phys. Scr.* **86** 035003

Assessment of surface roughness of substrates subjected to plasma-chemical etching

M S Glyanko[1], A V Volkov[1,2] and S A Fomchenkov[1]

[1]Department of Nanoengineering, Samara State Aerospace University National Research University, 34 Moskovskoye shosse, Samara 443086, Russia
[2]Image Processing Systems Institution, 151 Molodogvardejskaya st., Samara 443001, Russia

E-mail: snake7732@yandex.ru

Abstract. The model based on fractional Brownian surface for substrate surface roughness assessment subjected to plasma-chemical etching is proposed.

1. Introduction

Surface roughness is an important factor to take into account in variety problems of science and technology, for instance, aerodynamic calculation of bodies moving in the rarefied gas or study of flows in channels and vessels. Roughness relates to such properties of a surface as microhardness, friction and aero- and hydrodynamic resistance.

Substrate surface roughness control is an important technological operation in the process of diffractive optical elements (DOE) microrelief formation. By a diffractive optical element we mean transmissive or reflective plate with a thin phase microrelief calculated within the theory of diffraction. A Diffraction grating and a zone plate are classical examples of DOE. Roughness item, even smaller than the smallest element of diffractive microrelief, can cause changes in DOE optical parameters. Thus, developing models and methods for the estimation of surface roughness of optical substrates is an actual problem.

There is a variety of models used to account for the roughness in applied and theoretical problems that can be divided into three main groups:

- regular model;
- probabilistic model;
- fractal models.

In the regular models roughness is considered as a set of elements of identical size and shape. But due to its physical features real roughness has the property of irregularity, that makes its mathematical description considerably complicated [1]. So, probabilistic models of roughness were developed, for instance, Greenwood-Williamson model that is widely used in contact mechanics to study friction between two surfaces. In this model, surface roughness is modeled by hemispheres with the same radii of curvature. The distribution of roughness height relative to the plane of contact follows the Gaussian distribution function. However, the statistical approach has its own drawbacks due to the great complexity of calculations with the need to calculate continual integrals, which are approximated by integrals of high multiplicity. At present fractal models of surface roughness are used increasingly. The advantage of this group of models is the relative simplicity of calculations and the possibility of accounting for nanoscale roughness [2].

2. Real surface roughness modelling using fractals

Various factors influence on the process of optical substrates surface roughness formation. These are surface treatment methods such as- mechanical, chemical, plasma and plasma-chemical etching and others. Generally, the methods are independent of each other and have approximately the same scales. This fact allows us to apply to the process of surface roughness formation probability theory Central limit theorem according to which the resulting distribution is close to normal. Therefore, in this work we proposed a model for the simulation of surface roughness based on the fractional Brownian surface.

By definition, a Gaussian process $Z(x, y)$ is called a two-dimensional fractal Brownian motion with parameter H, where 0 < H < 1, if meet the following conditions:

- the function $Z(x, y)$ is almost always continuous;
- the random variable $Z(x+\Delta x, y+\Delta y) - Z(x, y)$ has a Gaussian distribution with zero mean and variance $\sigma^2(\Delta x^2 + \Delta y^2)^{2H}$, where $t_2 > t_1$ and σ is a positive constant.

The most accurate simulation of the Brownian surface is achieved by the Fourier transform of spectral density. In figure 1 and 2 you can see the simulation results of the algorithm for the Hurst parameter values H = 0.2 and H = 0.5.

Figure 1. Fractional Brownian surface for Hurst parameter value H = 0.2.

Figure 2. Fractional Brownian surface for Hurst parameter value H = 0.5.

Considering the problem of the difference between the real surface and a perfect fractal surface we use a structural function (1) [3]:

Journal of Physics: Conference Series **541** (2014) 012100 doi:10.1088/1742-6596/541/1/012100

$$S_I = \frac{I}{(K-I)^2} \sum_{i=1}^{K-I} \sum_{j=1}^{K-I} \left(\left| X_{i+I,j} - X_{i,j} \right| + \left| X_{i,j+I} - X_{i,j} \right| \right) \tag{1}$$

The graph of S_I against I for a perfect fractal while built in the double logarithmic scale will represent a straight line. Hurst parameter may be determined by the tangent of the angle of line slope. The wider the scope of I changes where the graph is well approximated by a straight line, the stronger fractal properties of the surface exhibits.

3. Experiments

In our study we used optical glass substrate with copper coating subjected to the procedure of plasma-chemical etching (figures 3 and 4). The elevation map of surface irregularities was obtained using three-dimensional surface structure analyzer NewView Zygo 7300, scan size – 140 um × 110 um.

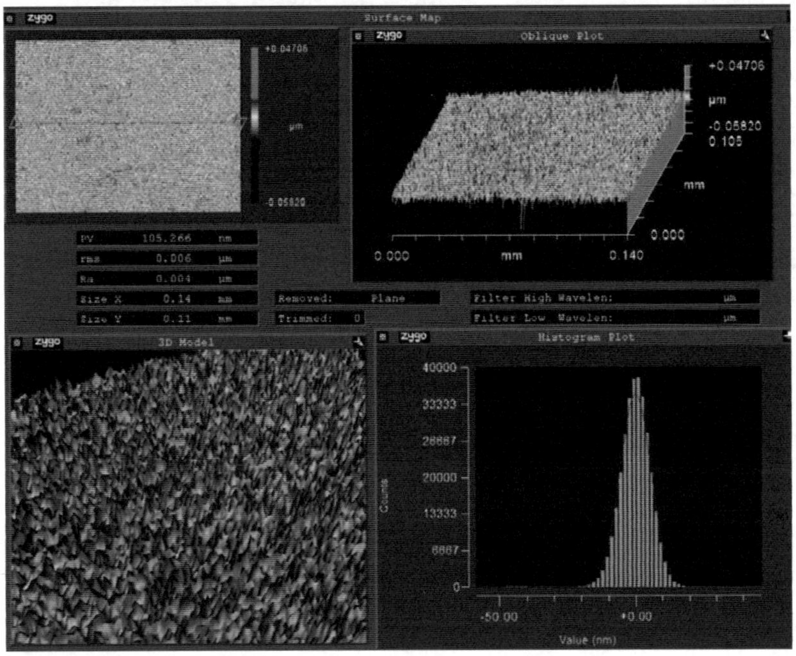

Figure 3. The substrate surface subjected to plasma-chemical etching. The etching process durability is 3 minutes.

4. Conclusion

The effectiveness analysis of proposed model for roughness estimation of the surfaces subjected to the operation of plasma-chemical etching proves that it is in accordance with real data obtained from the chosen type substrate and treatment method. According to the research results plasma-chemical etching operation forms the surface roughness whose height distribution can be considered Gaussian with a high confidence. It allows us to use the model for estimating of the roughness parameters of substrates subjected to the operation of plasma-chemical etching. It also can be used to determine optimal values of plasma-chemical etching parameters that lead to increase the percentage of work diffractive optical elements.

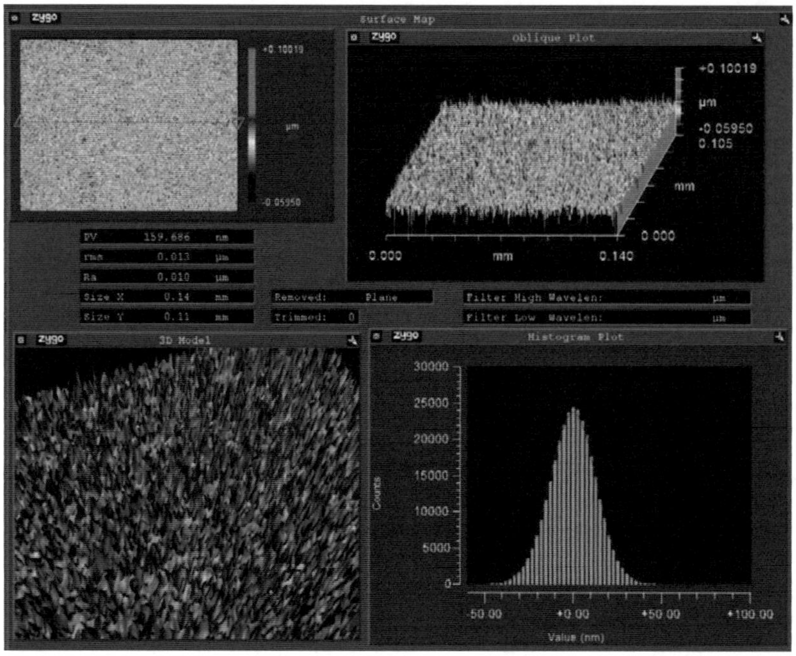

Figure 4. The substrate surface subjected to plasma-chemical etching. The etching process durability is 5 minutes.

References

[1] Husu A, Vittenberg Yu and Palmov V 1975 *Surface Rougness (Theoretical-Probabilistic Approach)* (Moscow: Science) p 24

[2] Gerasimova O, Borisov S and Protsenko S. 2004 *Matem. Mod.* **16** 40-43

[3] Korolenko P, Maganova M. and Mesniankine A 2004 *New Methods for the Analysis of the Stochastic Processes and Structures in Optics. The fractal and multifractal methods and wavelet transformations* (Moscow: Moscow State University, Research Institute of Nuclear Physics) p 36

[4] Volkov V, Kazansky N, Kolpakov V. 2001 *Comp. optics.* **21** 121-125

Quantum state tomography of the microwave field

G P Miroshnichenko, O M Korneeva

Saint-Petersburg National Research University of Information Technologies, Mechanics and Optics, 49 Kronverksky Avenue, 197101 Saint Petersburg, Russia

gpmirosh@gmail.com, korneevaom@yandex.ru

Abstract. We propose a new method for reconstructing the density matrix of a single-mode field of photons in the cavity using the discrete photodetection, which allows studying the quantum field in the microwave range. For reconstruction it is necessary to know the phase quadrature moments of all orders. A method of obtaining the first-order moment is discussed here.

1. Introduction

Recently the interest in quantum computation and quantum information has been increased, therefore much attention is given to quantum tomography, which allows reconstructing the density matrix of quantum field. In quantum mechanics it is impossible to develop a procedure that would allow to fully recover system state by several measurements, due to significant restrictions from the Heisenberg uncertainty principle and no-cloning theorem (except in the case when the quantum state coincides with the eigenstate of the measured value). Any measurement of a quantum system changes it, and the no-cloning theorem forbids the creation of an exact copy of the system without a priori knowledge of its condition. For recovering state of field the quantum tomography is used. The theory of quantum tomography of photon states in optical range is well developed in [1], [2], [3]. This theory is based on optical homodyne method. In the microwave range there are no photon counters. Therefore, the state of the microwave field in the cavity is measured by classic radio-frequency methods [4]. In [5] a quantum endoscopy method, which allows to obtain the wave function of a pure quantum state of the microwave field in the resonator is suggested.

In this paper we proposed a new method of quantum tomography of the photon field in the cavity using the discrete photodetection [6], an analogue of optical homodyne tomography [1], which in contrast to previous methods allows to recover the quantum field in the microwave range.

2. Procedure of tomography

Quantum tomography reconstructs the unknown density matrix ρ_f of field in the cavity [1]. For this purpose it is necessary to determine the k-order moments set $\{M_k(\phi)\}$ of phase quadrature operator $K(\phi)$ of field

$$M_k(\phi) = Tr_f\left[\left(K(\phi)\right)^k \rho_f\right] \tag{1}$$

Here ϕ – operator's phase, Tr_f – a trace on the photon states,

$$K(\phi)=\frac{a^{+}e^{-i\phi}+ae^{i\phi}}{\sqrt{2}}, \tag{2}$$

where a^{+},a – the photon creation and annihilation operators. Characteristic function for phase quadrature operator [7], [8], [9] has the following form

$$\chi(\eta,\phi)=Tr_f\left[\mathrm{e}^{i\eta K(\phi)}\rho_f\right]=\sum_k\frac{(i\eta)^k}{k!}Tr_f\left[\left(K(\phi)\right)^k\rho_f\right]=\sum_k\frac{(i\eta)^k}{k!}M_k(\phi).$$

The probability $w(x,\phi)$ of finding the quadrature in eigenstate $|x\rangle$ at the measure is

$$w(x,\phi)=\frac{1}{2\pi}\int_{-\infty}^{+\infty}\chi(\eta,\phi)\mathrm{e}^{-i\eta x}d\eta=Tr_f\left[\frac{1}{2\pi}\int_{-\infty}^{+\infty}\mathrm{e}^{-i\eta x}\,\mathrm{e}^{-i\eta K(\phi)}\rho_f d\eta\right]=\langle x|\rho_f|x\rangle$$

This procedure reconstructs the diagonal elements of the density matrix ρ_f in eigenbasis of the phase quadrature . Wigner function

$$W(u,v,s)=\frac{1}{4\pi^2}\int_{-\infty}^{+\infty}\int_{-\infty}^{+\infty}\int_{0}^{\pi}w(x,\phi)\exp\left(s\eta^2/8+i\eta\left(x-u\cos\phi-v\sin\phi\right)\right)|\eta|\,dx\,d\eta\,d\phi$$

$$u=K(0),\quad v=K\left(\frac{\pi}{2}\right),$$

is related to the density matrix ρ_f, which can be restored by using the inverse Radon transform [7].

3. Process of detection

During the process of discrete detection [6], investigated cavity probed by single atoms-probes that passed through the cavity one at a time at a given speed. The probe has two energy states: the ground (unexcited) state $|0\rangle$ and the excited state $|1\rangle$. Interaction time of investigated field and the probe is stabilized using a velocity selector. It will vary according to initial velocity of the atom. After its escape from the cavity, the probe contains information about the quantum field mode inside the cavity. Further, laser pulse has an effect on atom within time μ. Interaction time of atom and the laser is chosen so that the atom would come in a superposition of states [2] with a given phase ϕ

$$\left(|0\rangle,|1\rangle\right)\rightarrow\left(\frac{|0\rangle+ie^{-i\phi}|1\rangle}{\sqrt{2}},\frac{|0\rangle+ie^{i\phi}|1\rangle}{\sqrt{2}}\right) \tag{3}$$

Exposure to the laser pulse is equivalent to the influences of beam splitter in optical tomography with penetration/reflection probability is 50/50 percent [2]. After all interaction, the atom gets to the detectors (ionization chamber) which record its energy state. We will conduct a lot of these experiments on the detection of an atom by the known initial state, whereupon it is possible to determine the escape frequency of atom in the ground and excited states. This information will be used for calculation of quantum field quadrature in the cavity. The interaction between atom and facility for discrete photodetection is shown in figure1.

Cavity for microwave fields has a high finesse. Therefore in this work we will not take into account the field relaxation in the cavity, we assume the efficiency of the detector being unitary. Atom is prepared in the excited state $|1\rangle$ beforehand, the transition frequency between the energy states of the atom is ω_a. The eigenfrequency of the field in the cavity is $\omega=\omega_a$. The atom-cavity interaction time is t. Then the Hamiltonian of an atom-field system has the form

$$H_{af}=\omega_a|1\rangle\langle 1|+\omega a^{+}a+\gamma\left(a^{+}|0\rangle_a\langle 1|+a|1\rangle_a\langle 0|\right),$$

where γ – the constant of atom-field interaction.

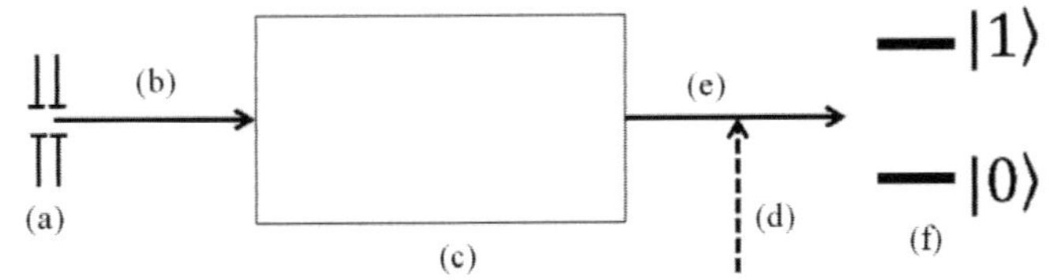

Figure 1. Facility for discrete photodetection
(a) – velocity selector, (b) – atom in excited state, (c) – cavity, (d) – laser beam, (e) – atom after the interaction with the cavity, (f) – selective detector on energy state.

After atom passed through the cavity, the classical laser pulse with a frequency ω_l ($\omega = \omega_l$) over time μ affects it. The interaction Hamiltonian of the atom and the laser is as follows

$$H_{al} = \omega_\alpha |1\rangle\langle 1| + \delta\left(e^{i\phi}|0\rangle\langle 1| + |1\rangle\langle 0| e^{-i\phi}\right)\cos(\omega_l \mu)$$

δ – the interaction constant between the atom-probe with a laser, ϕ – the laser impulse phase. The laser impulse converts the atomic basis in superposition (3). After two interactions, full evolution operator of the atom-field density matrix has the form (we consider the non-dimensional time $\tau = t\gamma$)

$$R(\tau,\phi) = \frac{1}{\sqrt{2}}\begin{bmatrix} \cos\left(\tau\sqrt{a^+a}\right) - e^{i\phi}\dfrac{\sin\left(\tau\sqrt{aa^+}\right)}{\sqrt{aa^+}}a & -ia^+\dfrac{\sin\left(\tau\sqrt{aa^+}\right)}{\sqrt{aa^+}} - ie^{i\phi}\cos\left(\tau\sqrt{aa^+}\right) \\ -ie^{-i\phi}\cos\left(\tau\sqrt{a^+a}\right) - i\dfrac{\sin\left(\tau\sqrt{aa^+}\right)}{\sqrt{aa^+}}a & -ie^{-i\phi}\,a^+\dfrac{\sin\left(\tau\sqrt{aa^+}\right)}{\sqrt{aa^+}} + \cos\left(\tau\sqrt{aa^+}\right) \end{bmatrix},$$

The equation for the probability of the atoms detection in the state $|\beta\rangle$, $\beta = 0,1$, when the atom was conditionally prepared in the state $|\alpha\rangle$, $\alpha = 1$, are

$$P_{\beta\alpha}(\tau,\phi) = Tr_f\left[\langle\beta|R(\tau,\phi)|\alpha\rangle\rho_f\langle\alpha|R^+(\tau,\phi)|\beta\rangle\right] \tag{4}$$

When

$$\tau\sqrt{aa^+} \ll 1 \tag{5}$$

the probability of the atoms detection (4) transforms to the form

$$P_{01}(\tau,\phi) \approx Tr_f\left[\left(\frac{1}{2} + \frac{ae^{i\phi} + a^+e^{-i\phi}}{2}\cdot\tau\right)\rho_f\right],\ P_{11}(\tau,\phi) \approx Tr_f\left[\left(\frac{1}{2} - \frac{ae^{i\phi} + a^+e^{-i\phi}}{2}\cdot\tau\right)\rho_f\right] \tag{6}$$

From (6) by the fixed phase ϕ

$$\frac{\Delta DP(\tau,\phi)}{\sqrt{2}\Delta\tau} = tg\theta \approx Tr_f\left[K(\phi)\rho_f\right] = M(\phi) \tag{7}$$

where θ – the slope angle of dependence $DP(\tau,\phi) = P_{01}(\tau,\phi) - P_{11}(\tau,\phi)$ from τ the interaction time of the atom with the cavity. We obtained connection of the first moment of quadrature with the detection probability an atom in the ground and excited states by the fixed phase ϕ. For the full procedure of tomography we must know the k-th order moments set.

4. Simulation of experiment

Simulation of experiment was conducted to verify the connection by the equation (7) of the first moment of the phase quadrature with the probability of atom detection in each state. According to the formula (1) by the condition (5) and the known ρ_f density matrix of the field in the cavity, we evaluated the set $M_{teor}(\phi)$ of theoretical quadrature moments for the different phases ϕ. The density matrix is generated by a Poisson distribution

$$\rho_f = e^{-\xi^2} \sum_{i,j} \frac{\xi^i \xi^{*j}}{\sqrt{i!j!}} |i\rangle\langle j| \tag{8}$$

The density matrix is the infinite size, therefore, the coherent state amplitude ξ is specially selected, so that the elements of the density matrix ρ_f is quickly decreases with increasing the number of photons *i*, *j*. The escape probability of atom P_{01} in the ground and P_{11} excited states is determined by the formula (4).

The first experimental moment $M_{ex}(\phi)$ can be found by the equation (7). For this we calculate $P0(\tau,\phi)$ the escape frequency of atom in the ground state and $P1(\tau,\phi)$ – in excited state by the fixed phase ϕ. We assume that these frequencies are probabilities $P0(\tau,\phi) \approx P_{01}(\tau,\phi)$, $P1(\tau,\phi) \approx P_{11}(\tau,\phi)$.

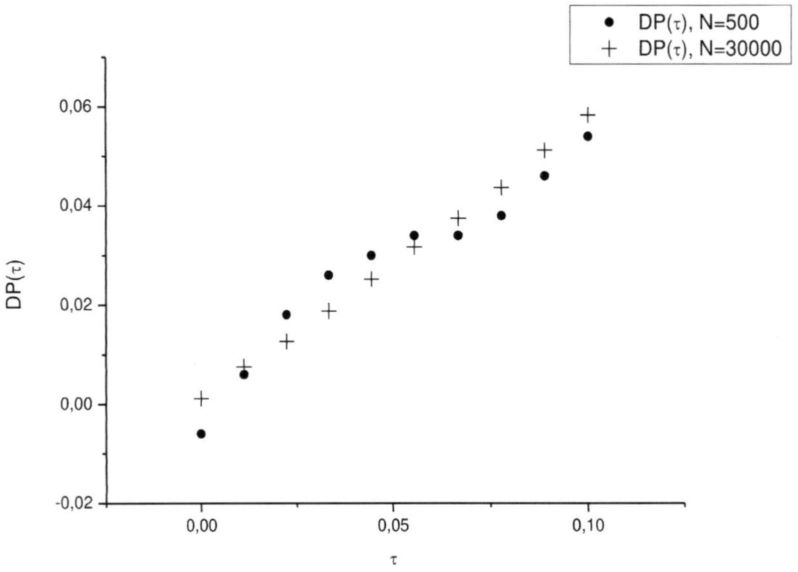

Figure 2. Dependence of $DP(\tau)$ from τ by the fixed phase $\phi = 0.064$ and number of experiments N = 500 и N = 30000

We can derive the set of the theoretical and experimental moments of the phase quadrature of field in the cavity by means of varying the phase ϕ from 0 to π. On figure 3 it exhibit that the experimental moment has been calculated by the assumption from this paper near to the theoretical moment.

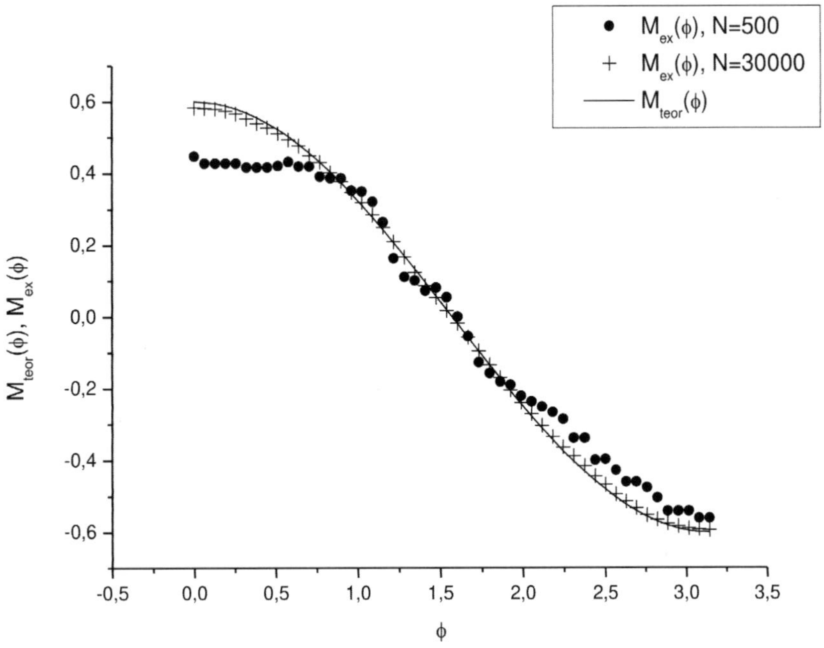

Figure 3. Dependence of theoretical $M_{teor}(\phi)$ and experimental first quadrature moment $M_{ex}(\phi)$ on phase ϕ. For receipt of every quadrature moment at fixed phase was considered 10 time points (time interaction between the atom and the cavity). The number of phase points is 50. The number of experiments to determine the frequency of atom detection in each state is 500 and 30000

5. Conclusion

In this paper we research the technique that allows us to study the microwave field by the method which is equivalent to the optical homodyne tomography. The relation between the first phase quadrature moments of the quantum field with the frequency of detection atom in the ground and excited state was received. The check that the proposed method can recover the first moment of the phase quadrature of the field was performed. For a complete tomography procedure it is necessary to know all the moments of the quadrature. They can be defined similarly.

Acknowledgments

This work was financially supported by the Government of Russian Federation (grant 074-U01) and by the Ministry of Science and Education of the Russian Federation (project 14.Z50.31.0031)

References

[1] Lvovsky A I, Raymer M G 2009 *Rev. Mod. Phys.* **81** 299–332
[2] D'Ariano G M, Paris M, Sacchi M 2003 *Advances in Imaging and Electron Physics* **128** 205–308
[3] D'Ariano G M, Leonhardt U, Paul H 1995 *Phys. Rev. A* **52** 1801
[4] Dicke R H 1946 *Rev. Sci. Instrum.* **17** 268
[5] Bardroff P J, Mayr E, Schleich W P 1995 *Phys. Rev. A* **51** 4963
[6] Miroshnichenko G P 2007 *JETP* **104** 724–735

[7] Vogel K, Risken H 1989 *Phys. Rev. A* **40** 2847
[8] Smithey D T, Beck M, Raymer M G, Faridani A 1993 *Phys. Rev. Lett.* **70** 1244
[9] Smithey D T, Beck M, Cooper J, Raymer M G, Faridani A 1993 *Phys. Scr.* **48** 35

Analysis of applicability of the transport level concept in organic materials, providing correlated disorder

O V Porvatkina, V R Nikitenko and M N Strikhanov
National Research Nuclear University MEPhI (Moscow Engineering Physics Institute), 115409, Moscow, Russia

E-mail: OVPorvatkina@mephi.ru

Abstract. Investigations of charge transport in disordered organic materials are actively developing area of science today. We investigated hopping carrier transport in disordered organic materials according to the model of correlated disorder (namely, the model of dipole glass). Numerical Monte-Carlo simulation of a random walk of charge after it's jump from deep localized state was performed. The applicability of transport level concept was held.

1. Introduction

Study of disordered organic materials and their fundamental properties, in particular charge transport, is actively developing area today. This is due to wide application of such materials, for example, in organic light emitting diodes (OLEDs). Transport of charge carriers in disordered organic materials occurs by phonon-assisted tunneling of charge carriers among localized states (hopping transport). At present, two models are widely used in theoretic description of hopping transport. The first is Gaussian disorder model (GDM) [1], which assumes non-correlated Gaussian distribution of localized states. The second is the correlated disorder model (CDM) [2]. In accord with this model, neighbor states forms clusters of states with proximal energies. The transport level concept is a useful approximation, which reduces description of hopping transport to multiple-trapping formalism. This concept was successfully applied in GDM [3], but it is doubtful in CDM [4]. However, in the work [4] only first jump from initial state has been considered, while a random walk inside a region around initial state should be analyzed [5]. In this work an adequate Monte-Carlo (MC) simulation is performed and applicability of transport level concept is reasoned.

2. Simulations

The material has been modeled, as usual [1,2,5], by the sites of simple cubic lattice (lattice constant is a). For each jump, the waiting time is calculated and final state is determined, as it was described previously [1]. Three concentric cubic regions have been considered around initial state – external, border and internal, see Figure1. We consider random walk inside the internal region, until the distance from the initial site does not exceed a certain value r_{max}. Sites of the border region (thickness of 2-3a, i.e. the maximal permitted hopping distance) are used in the procedure of evaluation of the probability to escape the internal region. The value of r_{max} should be large enough to provide absence of return to initial state, in the case of carrier's escape from this region. Randomly oriented dipole with momentum $\mathbf{p}_j$ was generated at any site j. Energies of sites of internal and border regions were calculated as energies of electrostatic interaction of charge e with ensemble of randomly oriented dipoles:

$$E_i = \sum_{j \neq i} \frac{e(\mathbf{r}_j - \mathbf{r}_i)\mathbf{p}_j}{\kappa |\mathbf{r}_i - \mathbf{r}_j|^3}, \qquad (1)$$

where $\kappa = 4\pi\varepsilon\varepsilon_0$, ε is relative dielectric constant. This approach yields long-range correlation of energies [2]. Discrete set of initial energies, E_i^{init}, has been considered. If the energy of the central site, E_i, falls in the interval from E_i^{init} - $kT/2$ to E_i^{init} + $kT/2$, kT is the thermal energy, the initial energy taken to be E_i^{init}. If the energy E_i does not falls into the mentioned interval,

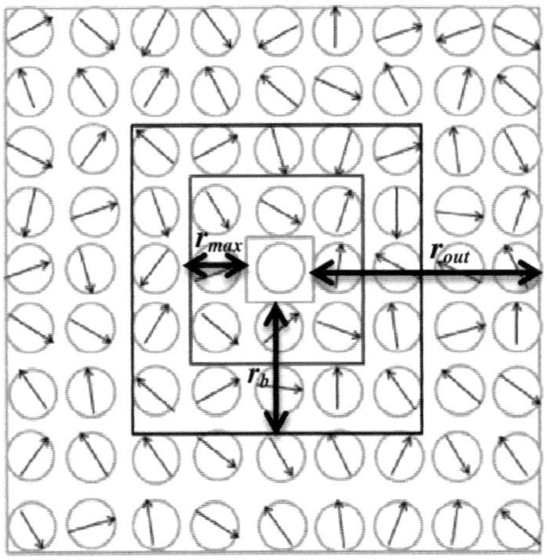

Figure 1. Schematic of the MC-model. Square frames shows (from outside to inside): boundaries of external, border and internal regions. Circles - lattice sites, arrows - randomly oriented dipoles. The central site in a frame is the initial position of a carrier.

then the distribution of dipoles at the lattice should be given again. The result is approximately Gaussian, although spatially correlated, distribution with the variance σ, which is related to the value of dipole momentum *p*, as it was derived previously [2]. The minimal size of external region, which provides practical independence of energies inside internal region from this size, has been found as r_{out} = $11a$, while the values of r_{max} = $3a$ and r_b = $5a$, (additionally, r_{max} = $5a$, r_b = $7a$ and r_{out} = $15a$) have been employed in calculations, see Figure 1. The main modelled quantity is a release time, i.e. time of carrier's escape outside the internal region. In addition, an average energy of the escaping state (i.e. an energy of the state, after the jump to which the carrier is gone beyond the internal region) was calculated.

Miller–Abrahams hopping rates from site i to site j are presumed [1-4]:

$$\omega_{i\to j} = \omega_0 \exp\left(-2\gamma r_{ij} - \frac{E_j - E_i + |E_j - E_i|}{2kT} \right), \qquad (2)$$

where ω_0 is the attempt-to-escape frequency, $1/g$ is the decay length of the localized wave function, E_i is the energy of the carrier at the site i, r_{ij} is the distance between sites i and j. The values T = 295 K, $2\gamma a = 10$ and $\varepsilon = 3$ have been used in calculations. Algorithm of MC-modeling of a random walk has been described in previous works [1,2].

3. Results and discussion

There are two ways to introduce a transport level [3] – real (physical) or formal (effective) transport level. The *real* transport level is a characteristic energy of final states after release of the carrier from deep initial states, this level is virtually independent on energies of initial states. One has to mention that the "escaping" jump is not a fastest jump typically, because of probability of return to initial state [1,5]. *Effective* transport level is just a formal quantity in expression for release rate, irrespective to distribution of energies of real "escaping" states. Anyway, the following dependence of mean release time on the energy of the initial state E_i should be valid, similar to multiple trapping model, if transport level concept is applicable:

$$< t_{esc} > (E_i) = t_0 \exp\left[(E_{tr} - E_i)/kT\right], \tag{3}$$

where $t_0 = \omega_0^{-1} \exp(2\gamma a)$ and E_{tr} are constants.

Figure 2 shows, in accord with the ref. [4], that mean energy of escaping states, which provides carrier's escape from the internal region, depends on E_i strongly. Indeed, variation of the average escaping energy exceeds a thermal energy *kT*. Energies of escaping states strongly correlate with initial energies, as it expected in a system with correlated disorder. Moreover, the former states lies below in energy, than the latter ones, hence escape occurs via downward jumps predominantly. Therefore, *real* transport level does not exist.

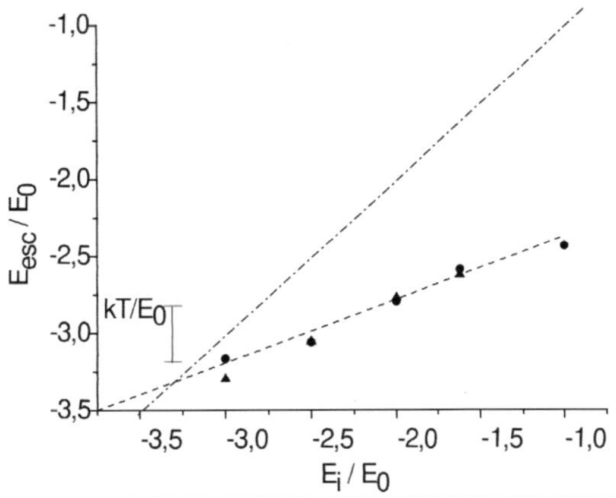

Figure 2 Dependence of average energy of escaping states on initial energy. Circles and triangles are simulation results for $r_{\max} = 3a$ and $5a$, respectively. The average energy of the final states and initial energy are normalized by characteristic energy of Gaussian distribution, $E_0 = \sqrt{2}\sigma = 0.075\,\text{eV}$. Dash-dotted line shows the equality $E_{esc} = E_i$.

However, Figure 3 shows correctness of Eq. (3) at various dimensions of the region of charge carrier's walks. Values of E_{tr} at different magnitudes of r_{max} are in mutual agreement within the minimum energy scale for this problem, i.e. thermal energy kT, hence E_{tr} is virtually independent on the size of the region. Validity of Eq. (3) indicates the applicability of transport level concept to description of charge transport in organics with correlated disorder, providing that transport level is an effective quantity. The reason is that the rate-limiting step of carrier's release from a small cluster of rather deep states is a thermally activated jump to a state, which is not nearest neighbor of initial state, hence it's energy is correlated weekly with energy of initial state, reminiscent to Gaussian disorder model, see inset in the Figure 3.

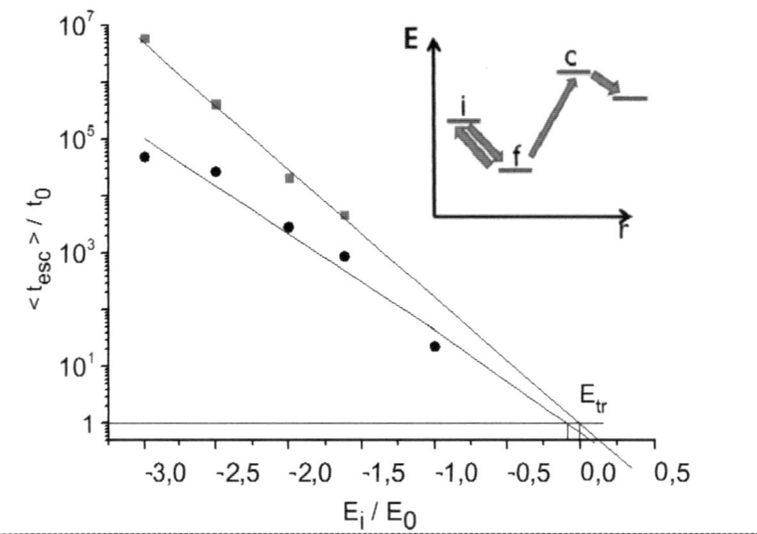

Figure 3. Dependence of mean release time on initial energy. Circles and squares are simulation results for $r_{max} = 3a$ and $5a$, respectively. Release time and initial energy are normalized by the time t_0 and characteristic energy of Gaussian distribution, $E_0 = \sqrt{2}\sigma = 0.075\,\text{eV}$, respectively. Inset is the schematic of typical escape of a carrier from a deep state i.

The work was supported by Russian Ministry of Education and Science in the frames of Competitiveness Growth Program of National Research Nuclear University MEPhI, Agreement 02.A03.21.0005.

References

[1] Bässler H Charge transport in disordered organic photoconductors 1993 *Phys. Status Solidi B* **175** 15-56

[2] Novikov S V, Dunlap D H, Kenkre V M, Parris P E, Vannikov A V 1998 Essential role of correlations in governing charge transport in disordered organic materials *Phys. Rev. Lett.* **81** 4472-75

[3] Nikitenko V R, von Seggern H, Bässler H 2007 Non-equilibrium transport of charge carriers in disordered organic materials *J.Phys.: Condens. Matter* **19** 136210-15

[4] Sergey V Novikov, George G Malliaras 2006 Transport energy in disordered organic material *Phys. Stat. Sol. (b)* **243** 387-90

[5] Nikitenko V R, Strikhanov M N 2014 Transport level in disordered organics: An analytic model and Monte-Carlo simulations *J. Appl. Phys.* **115** 073704

Dependence of the oscillations amplitude on the thickness of magnetostrictive-piezoelectric bilayer structure in the theory of magnetoelectric effect

T A Galichyan[1,2] and D A Filippov[1]
[1] Yaroslav-the-Wise Novgorod State University,
173003 Velikiy Novgorod, Russia

E-mail: tigrangalichyan@yahoo.com

Abstract. Dependence of the amplitude of mechanical oscillations on the thickness of magnetostrictive-piezoelectric bilayer structure in the form of rectangular plate is presented for the theory of linear magnetoelectric effect. The case of longitudinal orientation of the electric and magnetic fields was considered for structures in the form of a rectangular plate. The frequency dependence of the magnetoelectric effect is obtained using motion equation and electrostatic equations for both phases taking into account the boundary conditions on the interface. The theoretical dependencies of the displacement and stress distributions over the thickness of the sample in the magnetostrictive and piezoelectric phases are presented. These dependencies have a nonlinear character, and their accounting leads to a noticeable contribution to the magnitude of the effect.

Keywords: magnetoelectric effect, layered structure, magnetostriction, piezoelectricity.

1. Introduction

Magnetostrictive-piezoelectric structures have attracted much attention because in these structures there are effects which can be absent in magnetostrictive and piezoelectric components separately. They appear due to mechanical interaction between the magnetostrictive and piezoelectric subsystems. Magnetoelectric (ME) effect is one of such effects, which is caused by the elastic interaction of magnetostrictive and piezoelectric subsystems in bilayer structures. The action of the magnetic field in the magnetostrictive phase induces mechanical deformations which pass via mechanical coupling into the piezoelectric phase and leads to a change of the polarization of the sample.

The theoretical consideration of the ME effect in magnetostrictive-piezoelectric composites has two main methods at the present time. The method of effective parameters is one of these methods. Using this method, theory of the ME effect in bulk and multilayer composites was presented in works [1-6]. In these papers, an expression for ME voltage coefficient was defined by the method of effective parameters and the frequency dependence was analyzed. One of the disadvantages of the method of effective parameters is its usage limitation. The method is applicable when the characteristic size of the structural units of the composite is much smaller than the length of the acoustic oscillations, so that

[2] **Corresponding author:** Tigran Galichyan, postgraduate, research fields: condensed matter physics.

the composite can be considered as a homogeneous medium. Thus, the method of effective parameters is applicable for bulk composites. Bilayer structures attract much attention because of the value of ME effect, which is greater than in bulk composites [7]. But the method of effective parameters can not be used for these structures, because the thickness of the layers is comparable with the acoustic wavelengths. The second method is based on the simultaneous solution of the equations of motion and the constitutive relations for magnetostrictive and piezoelectric phases taking into account the boundary conditions on the interface. It was previously presented in [8-13]. The problem of this method is the consideration of the boundary condition between magnetostrictive and piezoelectric layers. These boundary conditions are presented formally by an interface coupling coefficient in [8-11] or by an assumption of ideal coupling [12,13]. The ME effect was investigated in laminated composite structure in work [14] where strain and ME voltage coefficient distributions were presented in piezoelectric, but in this work, the amplitude change of the oscillations over the thickness of the sample (in direction perpendicular to the interface) was not assumed. Recently the wave propagation and the ME effect in bilayer magnetostrictive-piezoelectric structure has been presented in case of a perfect bonding taking into account the changes of the amplitude in oscillations over the thickness of the sample in works [15-17]. But in these works, the spatial distribution of displacement and stress on the thickness of the sample were not analyzed and the influence of inhomogeneous distribution on the value of the effect was insufficiently analyzed.

In this paper, the dependencies of displacement and stress on the thickness of the sample are presented and their influence on the ME voltage coefficient is analyzed. The expressions for the ME voltage coefficient in the cases of longitudinal and transverse orientations of electric and magnetic fields were obtained using these dependencies and the open circuit condition. It is shown that the consideration of the change of oscillations amplitude over the thickness leads to a noticeable contribution to the magnitude of the effect.

2. Model and basic equations

As a model, we consider a bilayer structure in the form of a rectangular plate of length L and width W, consisting of mechanically interacting magnetostrictive and piezoelectric layers of thickness ${}^{m}t$ and ${}^{p}t$, the values which are not assumed to be small (figure 1). Thin metal contacts are applied on the top and the bottom of the plate.

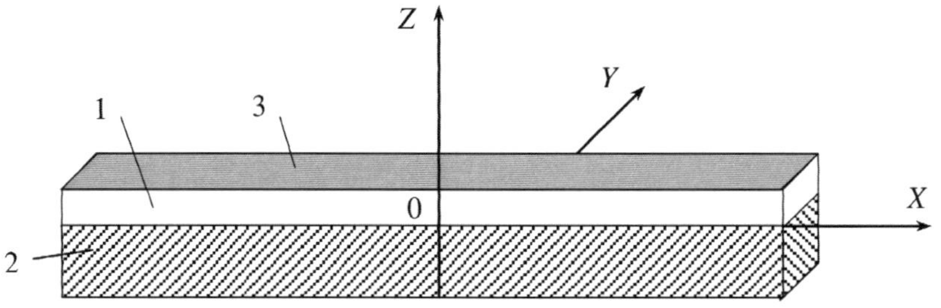

Figure 1. Schematic view of the sample. 1-Magnetostrictive phase with ${}^{m}t$ thickness, 2-Piezoelectric phase with ${}^{p}t$ thickness, 3-Electrodes.

Let us consider that the origin of the coordinates coincides with the boundary of the partition of magnetostrictive and piezoelectric layers. Preliminarily, the piezoelectric layer is polarized perpendicularly to the contact (Z axis). In the case of longitudinal orientation of the electric and magnetic fields (longitudinal effect) the magnetic fields (bias $\boldsymbol{H_0}$ and alternating $\boldsymbol{H}$ with the frequency ω) coincide in direction with the polarization vector $\boldsymbol{P}$. We restrict our consideration to the planar vibrations propagating along the X axis. The alternating magnetic field causes elastic oscillations in the magnetostrictive component, transferring through the interface of the partition to the piezoelectric component through the shear stresses, which brings to interconnected oscillations of magnetostrictive and piezoelectric subsystems. As there is a nonuniformity along Z axis and the width of the plate is small (Y axis), under the first approximation we can assume that the displacements along plane Y are homogeneous and the nonzero components of stress are only ${}^{\alpha}T_{xx}$ and ${}^{\alpha}T_{xz}$. The value of the stresses will depend on the thickness of the sample, perpendicularly to the interface between magnetostrictive and piezoelectric phases. Therefore the equations of motion for the x projection of the ${}^{\alpha}u_x$ displacement vector for the magnetostrictive and piezoelectric phases are of the form

$$ {}^{\alpha}\rho\frac{\partial^2\, {}^{\alpha}u_x}{\partial t^2} = \frac{\partial^{\alpha}T_{xx}}{\partial x} + \frac{\partial^{\alpha}T_{xz}}{\partial z}, \qquad (1) $$

where upper-case index "α" is correspondingly "m" for ferrite and "p" for piezoelectric, ${}^{\alpha}\rho$ is the density of ferrite and piezoelectric layers, ${}^{\alpha}T_{xx}$ and ${}^{\alpha}T_{xz}$ are the components of the stress tensor. The constitutive equations for the magnetostrictive and piezoelectric phases are

$$ {}^{p}S_{xx} = \frac{1}{{}^{p}Y}{}^{p}T_{xx} + {}^{p}d_{xx,z}\,{}^{p}E_z, \qquad (2) $$

$$ {}^{p}S_{xz} = \frac{1}{{}^{p}G}{}^{p}T_{xz}, \qquad (3) $$

$$ {}^{p}D_z = {}^{p}\varepsilon_{zz}\,{}^{p}E_z + {}^{p}d_{xx,z}\,{}^{p}T_{xx}, \qquad (4) $$

$$ {}^{m}S_{xx} = \frac{1}{{}^{m}Y}{}^{m}T_{xx} + {}^{m}q_{xx,z}\,{}^{m}H_z, \qquad (5) $$

$$ {}^{m}S_{xz} = \frac{1}{{}^{m}G}{}^{m}T_{xz}, \qquad (6) $$

where ${}^{\alpha}S_{xx}$ and ${}^{\alpha}S_{xz}$ are the strain tensor components; ${}^{p}E_z$ and ${}^{m}H_z$ are the vector components of the electric and magnetic fields; ${}^{p}D_z$ is the vector component of the electric displacement; ${}^{\alpha}Y$ and ${}^{\alpha}G$ are the Young's and shear moduli; ${}^{p}d_{xx,z}$ and ${}^{m}q_{xx,z}$ are the piezoelectric and piezomagnetic coefficients; ${}^{p}\varepsilon_{zz}$ is the tensor component of permittivity.

Using the solution of equation for the displacement vector of the medium and the condition of mechanical equilibrium at the free surfaces of the plate, i.e. points $x = \mp L/2$, we obtain the expression for displacement of the magnetostrictive and piezoelectric media in final forms

$$ {}^{m}u_x = \left[\exp(-2\,{}^{m}\kappa)\exp({}^{m}\chi z) + \exp(-{}^{m}\chi z)\right]B\sin(kx), \qquad (7) $$

$$ {}^{p}u_x = \left[\left(\cos({}^{p}\chi z) - \mathrm{tg}({}^{p}\kappa)\sin({}^{p}\chi z)\right)\left(1 + \exp(-2\,{}^{m}\kappa)\right)\right]B\sin(kx), \qquad (8) $$

where $B = \dfrac{{}^{m}Y\,{}^{m}t\,{}^{m}q_{xx,z}\langle {}^{m}H_z\rangle + {}^{p}Y\,{}^{p}t\,{}^{p}d_{xx,z}\langle {}^{p}E_z\rangle}{k\cos(\kappa)\left(1+\exp(-2\,{}^{m}\kappa)\right)\left({}^{m}Y\,{}^{m}t\dfrac{\mathrm{th}({}^{m}\kappa)}{{}^{m}\kappa} + {}^{p}Y\,{}^{p}t\dfrac{\mathrm{tg}({}^{p}\kappa)}{{}^{p}\kappa}\right)}$, $\kappa = kL/2$ and ${}^{\alpha}\kappa = {}^{\alpha}\chi\,{}^{\alpha}t$ are non-dimensional parameters; ${}^{m}\chi^2 = -2(1+\nu)\left[\dfrac{\omega^2}{{}^{m}V_L^2} - k^2\right]$, ${}^{p}\chi^2 = 2(1+\nu)\left[\dfrac{\omega^2}{{}^{p}V_L^2} - k^2\right]$, $\dfrac{1}{{}^{\alpha}V_L^2} = \dfrac{{}^{\alpha}\rho}{{}^{\alpha}Y}$; ${}^{\alpha}V_L$ are the velocities of longitudinal waves respectively in the magnetostrictive and piezoelectric medium and ν is the Poisson's ratio.

Expressing the stress tensor components ${}^{\alpha}T_{xx}$ and ${}^{\alpha}T_{xz}$ by the strain tensor components in Eqs. (7), (8), (10) and (11), we obtain the final expressions

$$ {}^{m}T_{xx} = {}^{m}Y\left[kB\cos(kx)\left(\exp(-2^{m}\kappa)\exp({}^{m}\chi z) + \exp(-{}^{m}\chi z)\right) - {}^{m}q_{xx,z}\,{}^{m}H_{z}\right], \quad (9) $$

$$ {}^{m}T_{xz} = {}^{m}G\,{}^{m}\chi\left[\exp(-2^{m}\kappa)\exp({}^{m}\chi z) - \exp(-{}^{m}\chi z)\right]B\sin kx, \quad (10) $$

$$ {}^{p}T_{xx} = {}^{p}Y\left[kB\cos(kx)\left(1 + \exp(-2^{m}\kappa)\right)\left(\cos({}^{p}\chi z) - \mathrm{tg}({}^{p}\kappa)\sin({}^{p}\chi z)\right) - {}^{p}d_{xx,z}\,{}^{p}E_{z}\right], \quad (11) $$

$$ {}^{p}T_{xz} = -{}^{p}G\,{}^{p}\chi\left(1 + \exp(-2^{m}\kappa)\right)\left[\sin({}^{p}\chi z) - \mathrm{tg}({}^{p}\kappa)\cos({}^{p}\chi z)\right]B\sin kx. \quad (12) $$

As can be seen in Eqs. (7) and (8), the solution is represented as plane waves whose amplitude is changed over the thickness of the sample. These dependencies have nonlinear character and depend on the frequency of oscillations in general. It is easy to show that in the case of low frequencies, when the dimensionless parameters ${}^{m}\kappa$ and ${}^{p}\kappa$ are less than one, the amplitude ceases to depend on the thickness of the sample. Thus, the results obtained earlier in [12], where the change of oscillations amplitude was not taken into account, take place only at low frequencies and thin layers. Using Eqs. (7) and (8), the distributions of relative displacements ${}^{\alpha}u_{x} = \dfrac{{}^{\alpha}u_{x}(z)}{{}^{\alpha}u_{x}(0)}$ on the thickness of magnetostrictive and piezoelectric phases are presented in figure 2 and the distribution of relative displacement in the piezoelectric phase is presented for three different cases of frequency (figure 3) in bilayer nickel–lead zirconate titanate (Ni-PZT) structure.

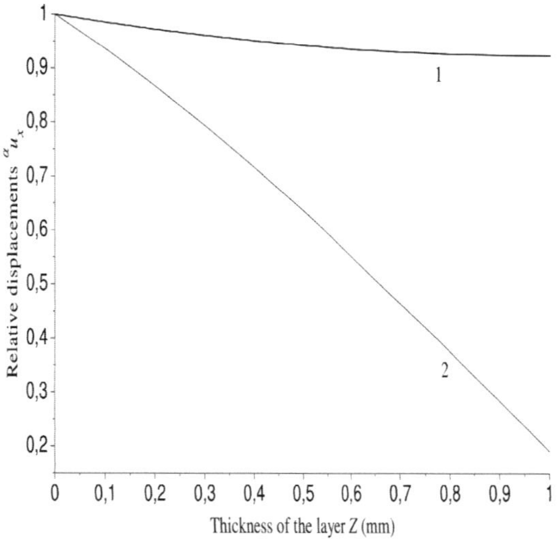

Figure 2. Relative displacement distributions on the thickness of magnetostrictive (line 1) and piezoelectric layers (line 2). Frequency of the ac magnetic field is f = 300 kHz.

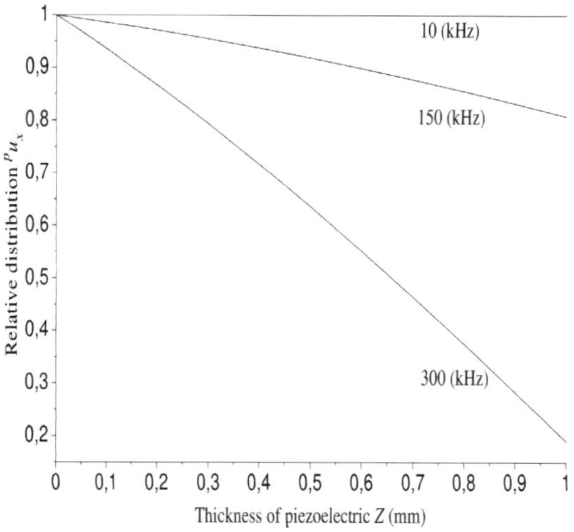

Figure 3. Relative displacement distribution on the thickness of piezoelectric layer in cases of different frequencies. Frequencies of the ac magnetic field are f = 10 kHz, f = 150 kHz, f = 300 kHz.

The following parameters of the structure were used in the calculations; for nickel – ${}^{m}Y = 204\,\mathrm{GPa}$, ${}^{m}\rho = 8900\,\mathrm{kg/m^3}$, ${}^{m}q_{xx,z} = 1156\cdot10^{-12}\,\mathrm{m/A}$; and for the PZT: ${}^{p}Y = 65\,\mathrm{GPa}$, ${}^{p}\rho = 7600\,\mathrm{kg/m^3}$, ${}^{p}d_{xx,z} = -175\cdot10^{-12}\,\mathrm{m/V}$, ${}^{p}\varepsilon_{zz}/\varepsilon_0 = 1750$, ac magnetic field H = 100 Oe. Using Eqs. (9)-(12), the distributions of relative stresses $\left({}^{\alpha}T_{xx} = \dfrac{{}^{\alpha}T_{xx}(z)}{{}^{\alpha}T_{xx}(0)} \text{ and } {}^{\alpha}T_{xz} = \dfrac{{}^{\alpha}T_{xz}(z)}{{}^{\alpha}T_{xz}(0)}\right)$ on the thickness of magnetostrictive and piezoelectric phases are presented for the same structure in figures 4 and 5.

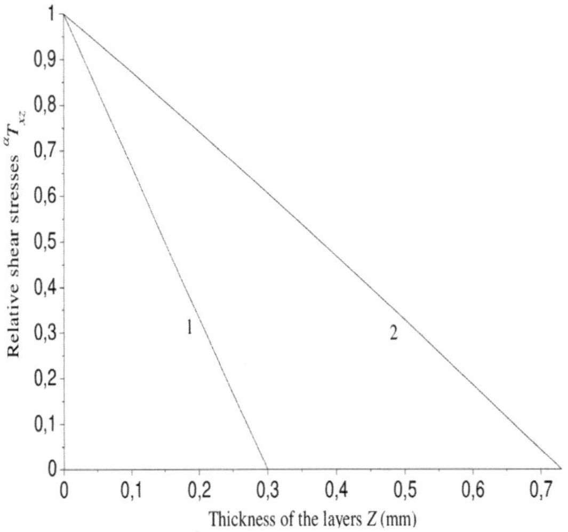

Figure 4. Relative shear stress distributions on the thickness of magnetostrictive (line 1) and piezoelectric layers (line 2). Frequency of the ac magnetic field is f = 300 kHz.

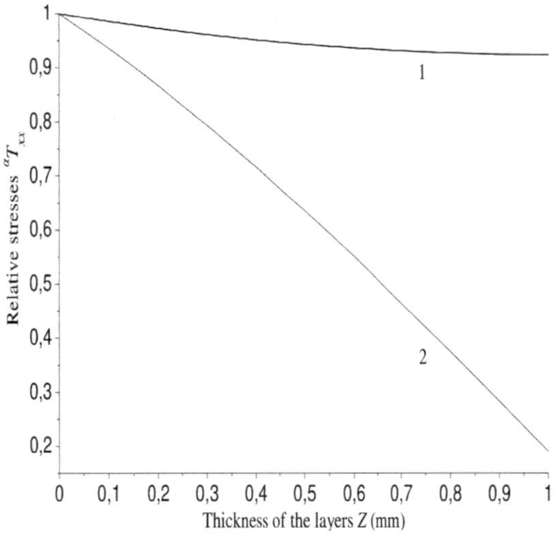

Figure 5. Relative stress distributions on the thickness of magnetostrictive (line1) and piezoelectric (line 2) layers. Frequency of the ac magnetic field is f = 300 kHz.

Magnetoelectric voltage coefficient is defined as the ratio of the average value of the electric field to the average value of the external magnetic field, which induced it, i.e.:

$$\langle \alpha_E \rangle = \langle E \rangle / \langle H \rangle , \tag{13}$$

where $\langle E \rangle = U/({}^{m}t + {}^{p}t)$ is the average value of the electric field of the structure, U is the potential difference between the electrodes.

In order to define a theoretical expression for the ME voltage coefficient, we use the method developed earlier [15,16]. Substituting the expression in Eq. (11) into Eq. (4), using the condition of open circuit and the fact that the potential difference between the electrodes is defined as $U = \langle {}^{p}E_z \rangle {}^{p}t$, we obtain the expression for the ME voltage coefficient in the case of longitudinal orientation of the fields in the final form

$$\alpha_{E,L} = \frac{{}^{p}Y\,{}^{p}d_{xx,z}\,{}^{m}q_{xx,z}}{{}^{p}\varepsilon_{zz}\Delta_L} \frac{{}^{m}Y\,{}^{m}t}{{}^{m}Y\,{}^{m}t\dfrac{\mathrm{th}({}^{m}\kappa)}{{}^{m}\kappa} + {}^{p}Y\,{}^{p}t\dfrac{\mathrm{tg}({}^{p}\kappa)}{{}^{p}\kappa}} \frac{\mathrm{tg}(\kappa)}{\kappa} \frac{\mathrm{tg}({}^{p}\kappa)}{{}^{p}\kappa} \frac{{}^{p}t}{{}^{m}t + {}^{p}t}, \tag{14}$$

where $\Delta_L = 1 - K_p^{\,2}\left(1 - \dfrac{{}^{p}Y\,{}^{p}t}{{}^{m}Y\,{}^{m}t\dfrac{\mathrm{th}({}^{m}\kappa)}{{}^{m}\kappa} + {}^{p}Y\,{}^{p}t\dfrac{\mathrm{tg}({}^{p}\kappa)}{{}^{p}\kappa}} \dfrac{\mathrm{tg}(\kappa)}{\kappa}\dfrac{\mathrm{tg}({}^{p}\kappa)}{{}^{p}\kappa}\right)$, with $K_p^2 = \dfrac{{}^{p}Y({}^{p}d_{xx,z})^2}{{}^{p}\varepsilon_{zz}}$ the squared coefficient of electromechanical coupling.

3. Results and discussion

The theoretical investigation was conducted for bilayer nickel–lead zirconate titanate (Ni-PZT) structure in the form of a rectangular plate. As can be seen from figure 2, the amplitude of oscillations in magnetostrictive phase is practically unchanged over the thickness of its layer. This is due to the alternating magnetic field which excites oscillations of the magnetostrictive phase simultaneously throughout the thickness of the magnetostrictive phase. Oscillations in the piezoelectric phase are excited due to the shear deformations which are transferred via the ferrite-piezoelectric interface.

As can be seen from figure 3, the distributions strongly depend on the frequency of oscillations. The dependencies have a nonlinear character at frequencies f = 150 kHz and f = 300 kHz. It is shown that in the case of low frequency (f = 10 kHz) the amplitude ceases to depend on the thickness of the layer and it has a linear character. Thus, the results obtained in [12] are incomplete and they can take place only at low frequencies and thin layers. Figure 4 shows that the shear stresses have the maximum values on the interface (Z = 0) and are equal to zero at the top and the bottom surfaces of the sample in accordance with the theory.

Calculation of the ME voltage coefficient, with the use of Eq. (14) obtained in this work, leads to a value not equal to the one obtained earlier in [12], where the inhomogeneity of the oscillations amplitude over the thickness was not taken into account. Thus, the value of ME voltage coefficient is $\alpha_{E,L}$ = 513 mV/cm·Oe, whereas the use of the expression obtained in [12] leads to overestimated value $\alpha_{E,L}$ = 860 mV/cm·Oe. Thereby, the consideration of inhomogeneity of oscillations amplitude at frequency of 300 kHz causes a change of the ME voltage coefficient. This value differs more than 40% from the result calculated with an assumption that the oscillations amplitude is the same over the thickness of the sample. Both models yield the same values for the ME voltage coefficient at low frequencies.

4. Conclusion

The inhomogeneity of the structure, associated with the presence of the ferrite-piezoelectric interface, leads to the change of displacement and stress amplitudes over the thickness of the sample. This inhomogeneity can be neglected, and we can assume that the amplitude is not changed over the thickness of the sample only at low frequencies. However, this change makes a significant contribution in the value of the ME voltage coefficient at the region of frequencies near 100 kHz.

References

[1] Harshe G, Dougherty J P and Newnham R E 1993 *Int. J. Appl. Electromagn. Mater.* **4** 145-59
[2] Harshe G, Dougherty J P and Newnham R E 1993 *Int. J. Appl. Electromagn. Mater.* **4** 161-71
[3] Bichurin M I, Petrov V M and Srinivasan G 2002 *J. Appl. Phys.* **92** 7681-83
[4] Osaretin I A and Rojas R G 2010 *Phys. Rev.* B **82** 174415
[5] Bichurin M I, Filippov D A, Petrov V M, Laletsin V M, Paddubnaya N N and Srinivasan G 2003 *Phys. Rev.* B **68** 132408
[6] Filippov D A, Bichurin M I, Petrov V M, Laletin V M and Srinivasan G 2004 *Phys. Solid State* **46** 1674-80
[7] Filippov D A, Laletin V M and Srinivasan G 2012 *Tech. Phys.* **57** 44-47
[8] Filippov D A 2004 *Tech. Phys. Lett.* **30** 983-86
[9] Filippov D A 2005 *Phys. Solid State* **47** 1118-21
[10] Chang C M and Carman G P 2008 *J. Intell. Mater. Syst. Struct.* **19** 1271-80
[11] Filippov D A, Srinivasan G and Gupta A 2008 *J. Phys.: Condens. Matter* **20** 425206
[12] Bichurin M I, Petrov V M, Averkin S V and Filippov A V 2010 *Phys. Solid State* **52** 2116-22
[13] Wang Y, Hasanyan D, Li M, Gao J, Li J, Viehland D and Luo H 2012 *J. Appl. Phys.* **111** 124513
[14] Hasanyan D, Wang Y, Gao J, Li M, Shen Y, Li J and Viehland D 2012 *J. Appl. Phys.* **112** 064109
[15] Filippov D A, Laletin V M and Galichyan T A 2013 *Phys. Solid State* **55** 1840-45
[16] Filippov D A, Galichyan T A and Laletin V M 2013 *Appl. Phys.* A doi:10.1007/s00339-013-7957-z
[17] Galichyan T A and Firsova T O 2013 *J. Phys.: Conf. Ser.* **461** 012016

Relaxation of conductivity in amorphous multi-component chalcogenides of copper and silver

V E Zaikova, N V Melnikova, K V Kurochka and O L Kheifets

Ural Federal University, Institute of Natural Science, 51 Lenin ave, Ekaterinburg, 620000, Russia

e-mail: vasilisazaykova@gmail.com

Abstract. Amorphous materials in the system $Cu_{1-x}Ag_xGeAsSe_3$ ($0.6 \leq x \leq 1.0$) show polarization phenomena. The time dependences of the conductivity and the ionic component of the conductivity in the studied materials were investigated.

1. Introduction

Multicomponent copper and silver chalcogenides have been known as promising materials with the ionic conductivity by silver and copper ions. The electric properties of glassy materials $Cu_{1-x}Ag_xGeAsSe_3$ had been studied in the broad ranges of pressure, temperature and electric field frequency [1-3]. Studied materials are mixing (electron - ionic) semiconductors with a predominant share of the ionic component of the conductivity. This paper is devoted to studies of the relaxation processes in $Cu_{1-x}Ag_xGeAsSe_3$ with x varying from 0.6 to 1.0.

2. Results and Discussion

The glassy materials $Cu_{1-x}Ag_xGeAsSe_3$ were obtained by quenching from the melt [2]. Glass surfaces were studied with scanning electron microscopes JEOL–JSM6390LA, Auriga CrossBeam Zeiss and microanalysis of samples was carried out using an energy-dispersive X-ray microanalyser on a fresh fracture of samples. These studies confirmed that all bulk glasses were homogeneous.

Studied materials have the ionic conductivity therefore it assumed that there the ion-relaxation polarization appears. For verifying this assumption, the relaxation time of carriers was estimated from analyzes of the frequency dependences of real and imaginary part of impedance (figure 1). The relaxation time of carriers is ~ $6\ 10^{-5}$ s. This time is typical for ion-relaxation polarization in amorphous materials, which is caused by the excess transfer of weakly bound ions under the effect of an electric field to distances on the scale of sizes that describe the structure.

The relaxation processes, connected with the migration polarization, in amorphous $Cu_{1-x}Ag_xGeAsSe_3$ were investigated by means of DC measurements in the cells with blocking and semi-reversible electrodes relative to ion component of the conductivity. Studied materials have a mixed electronic and ionic conductivity, the ionic component of conductivity varying from 72 to 99 percent depending on the concentration of silver and copper. At the first moment, when a constant voltage is applied to samples, the samples show the total (electronic and ionic) conductivity, and then the conductivity decreases until the constant level (electronic conductivity) (figure 2a, 2b). It is due to a gradual reduction of the ionic component of the conductivity in connection with the polarization phenomena, which is due to formation at the boundary sample - electrode double electric layer.

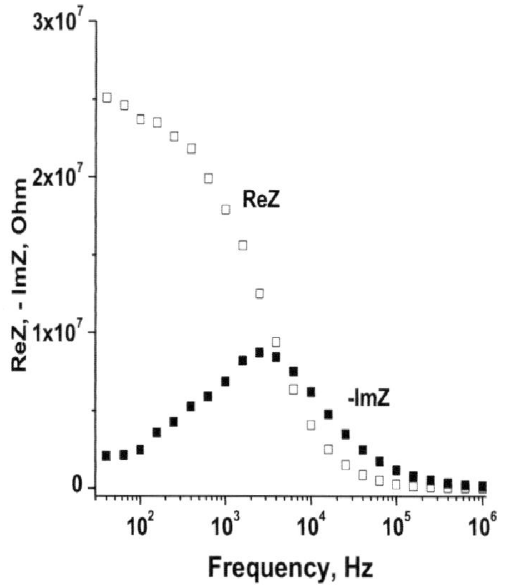

Figure 1. The frequency dependence of real and imaginary part of impedance of $Cu_{1-x}Ag_xGeAsSe_3$ with x=0.6. The measurements were obtained using Solartron 1260A at room temperature under atmospheric pressure.

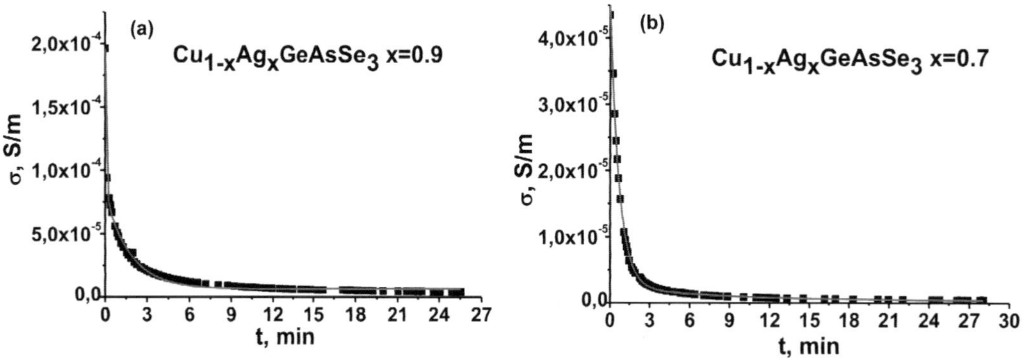

Figure 2. (a) – The time dependence of conductivity $Cu_{1-x}Ag_xGeAsSe_3$ with x=0.9. (b) – The time dependence of conductivity $Cu_{1-x}Ag_xGeAsSe_3$ with x=0.7. The measurements were obtained using Solartron ModuLad at room temperature under atmospheric pressure using semi-reversible copper electrodes with the graphite filter.

The time dependence of conductivity of the amorphous $Cu_{1-x}Ag_xGeAsSe_3$ is best approximated by the exponential function as in the equation (1).

$$\sigma_i = a + b \cdot e^{\frac{t}{t_1}} + c \cdot e^{\frac{t}{t_2}} \tag{1}$$

It can be assumed, that the times t_1 and t_2 characterize the relaxation processes in the amorphous $Cu_{1-x}Ag_xGeAsSe_3$. One time (t_1) is always bigger than the other (t_2). One can assume, that the first, smaller time (t_1) characterizes the relaxation process at the boundary sample - electrode; the other, bigger time (t_2) characterizes the relaxation process related with the diffusion flow of ions, which is probably determined including by the atomic structure of the amorphous $Cu_{1-x}Ag_xGeAsSe_3$ [4]. Also, at single exponential approximation, an average relaxation time has been obtained. This average time (t) is closely to the time t_1. It means that the main contribution to the conductivity relaxation is introduced by the processes at the boundary sample – electrode.

These times t_1 and t_2 vary depending on the electrode. When the conductivity time dependences were measured using blocking graphite electrodes these times generally appeared bigger then when measurement were carried out using semi-reversible copper electrodes with the graphite filter on cathode (table 1). It can be assumed, that this fact confirms appearance of ionic conductivity by copper ions and also it confirms the assumption that the main contribution to the conductivity relaxation is introduced by the processes at the boundary sample – electrode.

Table 1. Comparison of the relaxation time in measurements using copper electrodes with the graphite filter on cathode and using graphite electrodes.

	Copper electrodes with the graphite filter on cathode			Graphite electrodes		
x	t_1, sec	t_2, sec	t, sec	t_1, sec	t_2, sec	t, sec
0,95	11,60465	201,7216	13,6002	39,5662	597,188	44,5444
0,9	4,59127	108,7045	42,00854	36,6004	411,6132	169,8272
0,85	33,26861	311,9522	166,58	161,8194	1350,912	496,0305

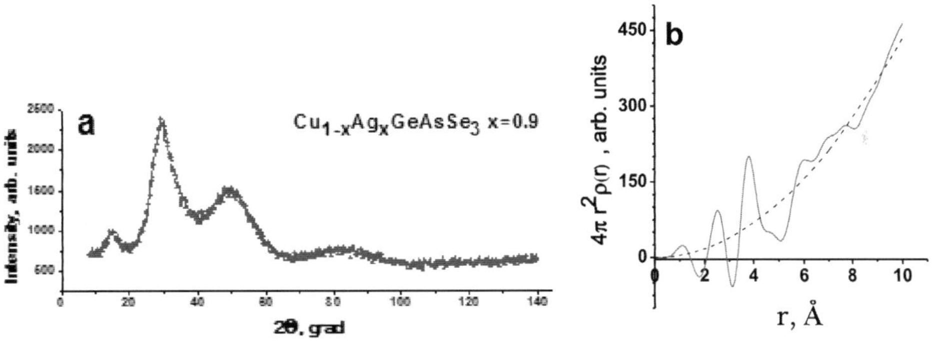

Figure 3. (a) - Powder diffraction patterns of compounds $Cu_{1-x}Ag_xGeAsSe_3$.with x = 0.9. (b) – The experimental radial distribution function of amorphous atoms $Cu_{1-x}Ag_xGeAsSe_3$ with x = 0.9.

For verifying the assumption that the relaxation processes in amorphous $Cu_{1-x}Ag_xGeAsSe_3$ is connected with their structure, the detail structure analyzes of studied materials is necessary. X-ray qualification of materials (figure 3a) has been performed with diffractometer Shimadzu XRD 7000 (monochromatic radiation CuKα). The experimental radial distribution function RDF of $Cu_{1-x}Ag_xGeAsSe_3$.with x = 0.9 (figure 3b) was obtained from the X-ray analysis data. The initial analysis of experimental RDF was carried out in ideas of the fragmentary model [5, 6]. This analysis showed that studied materials have layered structure. This layers is formed by randomly alternating tetrahedrally coordinated atoms of germanium and silver via selenium and arsenic atoms [4]. It should be mentioned that the part of the silver atoms are not included in described structure units, and it can easily move between layers. This conception of studied materials structure is verified by the analysis of the structure using the Raman scattering methods. The Raman scattering was obtained using Probe NanoLaboratory NTEGRA Spectra (NT-MDT, Russia), which comprises high resolution confocal scanning laser microscope, Raman spectrometer, and scanning probe microscope (SPM). The Raman scattering of stone chip of amorphous $AgGeAsSe_3$ ($Cu_{1-x}Ag_xGeAsSe_3$, x=1) is presented on figure 4. Since investigated materials are multicomponent and it is very difficult to analyze all possible bonds, only the most basic reflexes will be considered. There are two intensity maxima at 140 sm^{-1} and 193 sm^{-1}, which may correspond to vibrations of Ge-As and Ge-Se (tetrahedra with four-coordinated germanium atom in the centre) bonds, respectively [7]. Also there are two small maxima at 210 sm^{-1} and 240 sm^{-1}, which may correspond to vibration of As-Se ($AsSe_{3/2}$ pyramids) and Se-Se bonds, respectively [7].

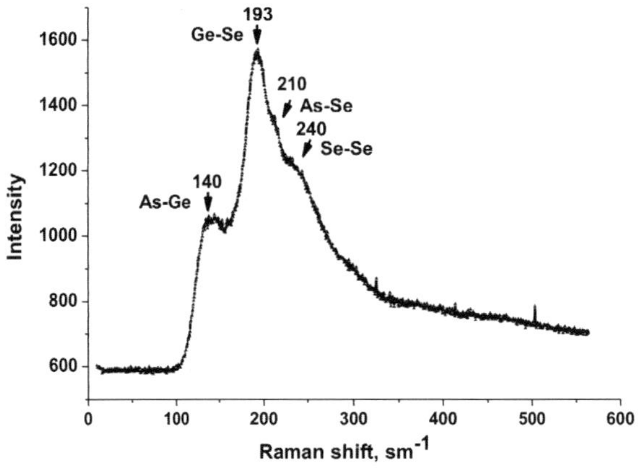

Figure 4. Raman spectra of $AgGeAsS_3$. The He-Ne laser with λ = 633 nm and power 30 mW was used as a pumping source.

3. Conclusion

The time dependences of conductivity were analyzed. It can be approximated by exponential functions, and exponential indices characterize relaxation processes in studied materials. Relaxation processes are determined by atomic structure and processes at the boundary sample – electrode. And the main contribution to the conductivity relaxation is introduced by the processes at the boundary sample – electrode.

Acknowledgments
The study was supported by the Russian Foundation for Basic Research, projects 13-02-00633.

References

[1] Kheifets O L, Melnikova N V, Filippov A L, Shakirov E F, Babushkin A N and Nugaeva L L 2012 *PHYS SOLID STATE+* **54** p 1562
[2] Melnikova N, Kheifets O, Babushkin A and Sukhanova G 2011 EPJ *Web of Conferences* **15** 03004
[3] Kheifets O L, Melnikova N V, Pinigina K S and Filippov A L 2014 *B. Russ. Acad. Sci. Phys.* **78** p 76
[4] Zinchenko E N, Aleinikova K B and Melnikova N V 2012 *Book of Abstr. of VIII Int.Conf. on Amorphous and Microcrystalline Semiconductors* (St.-Petersburg: Polytechnical University) p 215
[5] Aleinikova K B et. al. 2005 *IND LAB - USSR+*. **71** p 27
[6] Aleinikova K B et. Al 2007 *CRYSTALLOGR REP+* **52** p 338
[7] Wang R P, Smith A, Prasad A, Choi D Y, and Luther-Davies B 2009 *J APPL PHYS* 106, 043520

Conditional evolution of vacuum state in dynamical Casimir effect

Sheremetyev V.O., Trifanova E.S., Trifanov A.I.

Saint Petersburg National Research University of Information Technologies, Mechanics and Optics, Kronverksky, 49, Saint Petersburg, 197101, Russia

E-mail: alextrifanov@gmail.com

Abstract. We analytically investigate the process of photons creation in the empty cavity with trembling ideal conductive boundary. Quantum state of the intracavity field is monitored by the two-level atom passing through the system and undergoing the subsequent state measurement in the ionizing chamber. The resulted field evolution conditioned by atomic state detection result is of the most interest.

1. Introduction

The dynamical Casimir effect (DCE) is the process of photons creation out of vacuum due to the motion of the cavity mirrors. The most intensive studies in this area were made about two decades ago [1, 2, 3] and accompanied with observation of Casimir force in laboratory [4]. The main difficulty in experimental verification of DCE was the very high frequency variation of moving boundaries [5], which prevented photons creation using mechanically driven cavity mirrors. In the consequence of this fact different ideas based on imitation of boundary moving were proposed [6], [7]. Recently, Wilson and coworkers reported about the successful experiment on photon creation using DCE effect in superconductive circuit [8]. That work initiated a discussion about the most effective measurement protocol of created quanta detection and modification of intracavity field statistics due to the interaction with measurement apparatus [9]. In [10] the approximate analytical solution for the problem of created photon detection was done and the main values, which characterized the field statistics were obtained for the case of interaction with two-level atom.

Here we estimate analytically the conditional state evolution of created intracavity field in the process of indirect photodetection with two-level atom-pointer in use. The measurement protocol was organized as follows: two-level atom-pointer prepared in its ground state passing through the cavity with trembling boundary and interacts with generated Casimir photons. After interaction it undergoes the state detection in ionizing chamber. The unitary atom-field evolution is under investigation. For the simplest case of pure squeesing we obtain the Jaynes-Cummings Hamiltonian in the renormalized vacuum representation. Under this approximation we get the field evolution operators, conditioned by the atomic state detection result. Thereby we can predict the field state and all its statistical properties. Also it is possible to investigate the quantum trajectory for the field state using atomic state detection statistics, repeating the interaction many times.

2. The model

In [3] it was shown that in general there are two factors, which contribute to the process of photon creation: squeezing and relativistic acceleration due to the fast boundary motion. Here we will investigate the conditional dynamics of intracavity quantum field state, assuming that acceleration and mode interaction effects are small compared to squeezing and consider the last one as an only source of Casimir photons.

Following [10] we start our description of the system dynamics using Hamiltonian in its simplest form:

$$H_0 = \omega(t) a^\dagger a + \frac{\Omega}{2}\sigma_z + i\xi(t)\left(a^{\dagger 2} - a^2\right) + g\left(a^\dagger \sigma_- + a\sigma_+\right), \tag{1}$$

where $\omega(t)$ is instantaneous frequency of intracavity field, Ω is the frequency of atomic transition, a is the field annihilation operator, $\sigma_+ = |e\rangle\langle g| = \sigma_-^\dagger$ is the atomic operator and $\xi(t)$ is squeezing parameter [1]:

$$\xi(t) = \frac{1}{4\omega(t)}\frac{d\omega}{dt}. \tag{2}$$

Let us choose the time dependence of cavity field frequency in the harmonic low $\omega(t) = \omega_0(1 + \varepsilon \sin \eta t)$, with frequency modulation η and modulation depth ε. We'll describe the regime corresponding to the exponentially growth of cavity photons $\varepsilon\omega_0 t \gg 1$ [8], which gives

$$\xi(t) \approx \frac{\varepsilon\eta}{4}\cos(\eta t). \tag{3}$$

Then we write Hamiltonian (1) in the form, corresponding to the following state transformation [10]

$$|\varphi\rangle = V(t)|\psi\rangle, \quad V(t) = \exp\left[-it\eta/2\left(a^\dagger a + \sigma_z/2\right)\right], \tag{4}$$

which gives us $H = V^\dagger(t) H_0 V(t) - iV^\dagger(t)\dot{V}(t)$ or

$$H = x a^\dagger a + \frac{\Delta + x}{2}\sigma_z + i\frac{\varepsilon\nu}{4}\left(a^{\dagger 2} - a^2\right) + g\left(a^\dagger\sigma_- + a\sigma_+\right). \tag{5}$$

Here $\nu = \eta/2$, $x = \omega_0 - \nu$, $\Delta = \Omega - \omega_0$ and the rotating wave approximation (RWA) was applied.

Unitary transformation of (5) with squeezing operator $S = \exp\left[\frac{1}{2}\left(\beta^* a^2 - \beta a^{\dagger 2}\right)\right]$, where

$$\beta = re^{i\theta}, \quad r = \frac{1}{2}\operatorname{arctanh}\frac{\varepsilon\nu}{2x}, \quad \theta = \pi/2, \tag{6}$$

gives the following form of Hamiltonian $H_B = S^\dagger H S$ with renormalized vacuum state:

$$H_B = \mu\left(a^\dagger a + \frac{1}{2}\right) + \frac{\alpha}{2}\sigma_z + g\left[\left(a\cosh(r) - ia^\dagger \sinh(r)\right)\sigma_+ + \left(a^\dagger\cosh(r) + ia\sinh(r)\right)\sigma_-\right], \tag{7}$$

where $\alpha = \Delta + x$ and

$$\mu = \frac{\sqrt{4x^2 - \varepsilon^2\nu^2}}{2}.$$

From (6) it is easy to obtain that only for $|\varepsilon\nu/2x| < 1$ this unitary transformation is allowable.

3. In the bounds of Jaynes-Cummings model

Being in the region of parameters which allows the system Hamiltonian in the form (7) we can obtain its simplest form assuming that $|\mu - \alpha| \ll \mu, \alpha$. Writing (7) in interaction picture and applying RWA we reach the following expression for H_B which has the form of Jaynes-Cummings Hamiltonian:

$$H_{J-C} = G\left(e^{i\delta t} a\sigma_{+} + e^{-i\delta t} a^{\dagger}\sigma -\right), \tag{8}$$

where $\delta = \mu - \alpha$ and $G = g\cosh(r)$.

From expression (8) it is easy to obtain the operator of conditional field state evolution. Expanding evolution operator $U(t)$, which correspond to the solution of Schrodinger equation with Hamiltonian (8) in the following composition

$$U(t) = \sum_{p,q\in\{g,e\}} U_{pq}(t)\,|q\rangle\langle p|, \tag{9}$$

it is easy to obtain the system of operator-valued differential equation [11]:

$$i\frac{d}{dt}U_{pq}(t) = \sum_{r\in\{g,e\}} \langle p|\,H_{J-C}\,|r\rangle\,U_{rq}(t)\,. \tag{10}$$

This system has a simple analytical solution. Namely, for the atom prepared in its ground state $|g\rangle$ and detected in ground $|g\rangle$ or excited state $|e\rangle$ we obtain

$$U_{gg}(t) = \frac{e^{i\delta t/2}}{\sqrt{\gamma}}\left(\sqrt{\gamma}\cos\frac{\sqrt{\gamma}t}{2} - i\delta\sin\frac{\sqrt{\gamma}t}{2}\right), \quad \gamma = \delta^2 + 4G^2a^{\dagger}a, \tag{11}$$

$$U_{ge}(t) = -i\frac{Ge^{-i\delta t/2}}{\sqrt{\delta^2/4 + G^2aa^{\dagger}}}\sin\left(t\sqrt{\delta^2/4 + G^2aa^{\dagger}}\right)a. \tag{12}$$

Operators U_{gr}, $r \in \{g, e\}$ have the following physical interpretation: they map initial state $|\varphi(0)\rangle$ of intracavity field onto the final state $|\varphi(t)\rangle$, conditioned by detection result r. This final state however is not normalized and full transformation is

$$|\varphi_r(t)\rangle = \frac{U_{gr}(t)\,|\varphi(0)\rangle}{P_r(t)}, \quad P_r(t) = \langle\varphi_r(t)\mid\varphi_r(t)\rangle\,. \tag{13}$$

Here $P_r(t)$ is the probability to detect the atom in state $|r\rangle$ after interaction lasted time t.

4. Analytical and numerical results

In this section we represent the results followed from the numerical analysis of the evolution generated by Hamiltonian (5). In particular the basic statistical characteristics of the monitored intracavity field will be investigated under the assumption that atom was found in its excited state. The system parameters was chosen in a such way to satisfy the condition which gives approximate analytical formula (8). Namely, the following values of parameters are used (in units 10^9 Hz):

$$x = 1, \quad \varepsilon\nu = 1.7, \quad \Delta = 0.36, \quad g = 0.1. \tag{14}$$

The probability $P_e(t)$ of event, corresponded to the atomic state detection result $|e\rangle$ is given by Eq. (13) and is shown on Fig. 1(a) as a function of dimensionless time gt. It increases monotonically and reaches the value 0.015 on the concerned interval. The mean cavity photon number $\langle n(t)\rangle = \langle\varphi_e(t)|\,a^{\dagger}a\,|\varphi_e(t)\rangle$ us a function of time is shown in Fig.1 (b). This quantity, conditioned by detection result overcome the value 6 at $gt \approx 2$. For this instant of time the

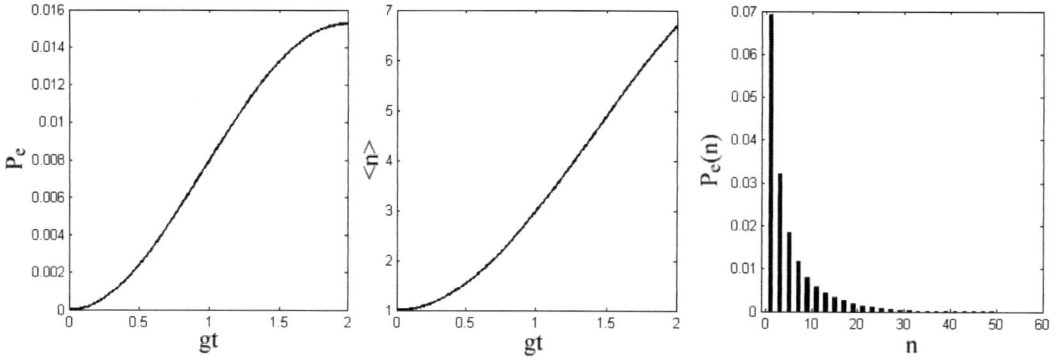

Figure 1. Excited atomic state detection: probability (a) and mean photon number (b) as a functions of dimensionless time gt; (c) photon number probability distribution for $gt \approx 2$. All parameters were taken from (14); $r = 0.63$ and $\mu = 0.53$ in units 10^9 Hz.

mean photon number probability distribution $P_e(n)$ is presented in Fig. 1(c). This distribution contain the nonvanishing probability only for odd numbers of photons, which may be explained easily. The matter is that in the absence of atom the intracavity field is in squeezed vacuum state (which has the distribution with only even photons). When atom is found in its excited state, the photon number in cavity decreases by one.

In Fig. 2(a–c) we show the Mandel factor (a) $Q = \left[\langle(\Delta n)^2\rangle - \langle n\rangle\right] / \langle n\rangle$, second order correlation function $g^{(2)} = \langle a^\dagger a^\dagger aa\rangle / \langle n\rangle^2$, and field quadratures $\langle(\Delta X)^2\rangle$ and $\langle(\Delta P)^2\rangle$, where $X = (a + a^\dagger)/2$ and $X = (a - a^\dagger)/2i$ as a functions of time. This quantities describe the nonclassicallity of the generated intracavity field interacted with atom during certain time interval. From here it is easy to see that statistical properties of the field changes dramatically [9] due to the interaction with atom. Indeed, in the beginning of this process Mandel factor $Q < 0$ and $\langle(\Delta P)^2\rangle$ decreases to the values, corresponded to squeezed state. Then Q increases and at the point $gt \approx 0.5$ ($g^{(2)} \approx 1$) become equal to unity, which corresponds to the Poissonian field statistics. For $gt > 0.5$ statistics of the field become super-Poissonian.

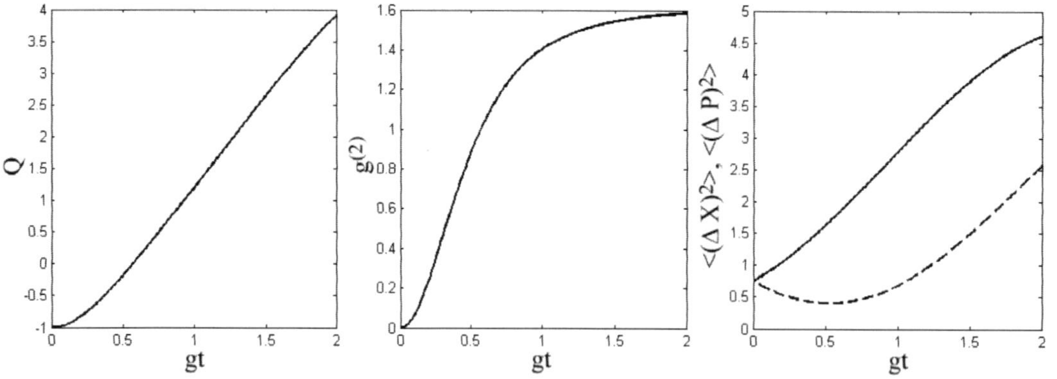

Figure 2. Excited atomic state detection: Mandel factor Q (a), second order correlation function $g^{(2)}$ (b), field quadratures $\langle(\Delta X)^2\rangle$ and $\langle(\Delta P)^2\rangle$ (c) as a functions of dimensionless time gt. All parameters were taken from (14); $r = 0.63$ and $\mu = 0.53$ in units 10^9 Hz.

5. Conclusion

We investigated the conditional state evolution of the field, generated in dynamical Casimir effect under the indirect photodetection protocol. Using two-level atom as a microscopic part of measurement apparatus allows us to obtain the approximate analytical expressions for conditional evolution operators and to estimate basic statistical characteristics of monitored intracavity field. For the following it would be interesting to investigate the case of nonunitary interaction between atom and field due to the presence of environment.

Acknowledgments

The work was partially financially supported by the Government of the Russian Federation (grant 074-U01), by the Ministry of Science and Education of the Russian Federation (GOSZADANIE 2014/190, Project 14.Z50.31.0031), by grant of Russian Foundation for Basic Researches and grants of the President of Russia (state contracts 14.124.13.2045-MK and 14.124.13.1493-MK).

References

[1] Law C K 1994 *Phys. Rev. A* **49** 433-537
[2] Dodonov V V and Klimov A B 1996 *Phys. Rev. A* **53** 2664-2682
[3] Schützhold R Plunien G and Soff G 1998 *Phys. Rev. A* **57** 2311-2318
[4] Lamoreaux S K 1997 *Phys. Rev. Lett.* **78** 5-8
[5] Dodonov V V 2010 *Physica Scripta* **82** 038105
[6] Yablonovitch E 1989 *Phys. Rev. Lett.* **62** 1742-1745
[7] Braggio C et all 2005 *Europhys. Lett.* **70** (6) 754760
[8] Wilson C M et all 2011 *Nature* **479** 376-379
[9] Dodonov V V 2011 *Physics Letters A* **375** 4261-4267
[10] Dodonov V V 2012 *Phys. Rev. A* **85** 2311-2318
[11] Trifanov A I and Miroshnichenko G P 2013 *Nanosystems: Phys., Chem., Math.* **4** (5) 635-647